# Stratigraphy: A Modern Synthesis

Andrew D. Miall

# Stratigraphy: A Modern Synthesis

 Springer

Andrew D. Miall
Department of Geology
University of Toronto
Toronto, ON
Canada

ISBN 978-3-319-24302-3          ISBN 978-3-319-24304-7    (eBook)
DOI 10.1007/978-3-319-24304-7

Library of Congress Control Number: 2015952024

Springer Cham Heidelberg New York Dordrecht London

Printed on acid-free paper

Springer International Publishing AG Switzerland is part of Springer Science+Business Media
(www.springer.com)

*"For Meredith, Henry and Owen"*

# Preface

The stratigraphic record is the major repository of information about the geological history of Earth, a record stretching back for nearly 4 billion years. Stratigraphic studies fill out our planet's plate-tectonic history with the details of paleogeography, past climates, and the record of evolution, and stratigraphy is at the heart of the effort to find and exploit fossil-fuel resources.

The exploration of this history has been underway since James Hutton first established the basic idea of uniformitarianism toward the end of the eighteenth century, and William Smith developed the stratigraphic basis for geological mapping a few decades later. Modern stratigraphic methods are now able to provide insights into past geological events and processes on time scales with unprecedented accuracy and precision, and have added much to our understanding of global tectonic and climatic processes. But it has taken 200 years and a modern revolution to bring all the necessary developments together to create the modern, dynamic science that this book sets out to describe.

It has been a slow revolution, but stratigraphy now consists of a suite of integrated concepts and methods, several of which have considerable predictive and interpretive power. It is argued in Chap. 1 of this book that the new, integrated, dynamic science that stratigraphy has become is now inseparable from what were its component parts, including sedimentology, chronostratigraphy, and the broader aspects of basin analysis. In this chapter, the evolution of this modern science is traced from its nineteenth-century beginnings, including the contributions that such special fields as facies analysis, fluid hydraulics, plate tectonics, and the reflection-seismic surveying method have made to its evolution.

The following are just some of the major features of the stratigraphy of the early twenty-first century: Sequence stratigraphy has become the standard methodology for documentation, mapping and interpretation, replacing the old descriptive practices of lithostratigraphy; reflection-seismic methods, including the use of 3-D seismic and the application of seismic geomorphology, have become steadily more advanced tools for subsurface exploration and development; the Geological Time Scale is being standardized with the universal adoption of the system of Global Stratigraphic Sections and Points (GSSPs) and has become much more precise, with the incorporation of several new methods for evaluating deep time.

The basic field and subsurface observations on which stratigraphy is based are described in Chap. 2. Facies analysis methods are detailed in Chap. 3, and the recognition of depositional environments by facies methods is described in Chap. 4. Chapter 5 provides a succinct summary of sequence models for siliciclastic and carbonate sediments, and Chap. 6 describes modern mapping methods for use in surface and subsurface studies, including seismic methods. The synthesis of all this material is detailed in Chap. 7, which includes a discussion of the current attempts to standardize sequence-stratigraphic terminology and the Geological Time Scale.

Chapter 8, the concluding chapter of the book, focuses on the new understanding we are acquiring about the processes by which the stratigraphic record preserves elapsed geologic time. Refinements in chronostratigraphic methods are revealing the importance of breaks in the sedimentary record and the ubiquity of missing time, and are revealing an important disconnect between sedimentation rates and preservational processes operating at the present day versus those we interpret from the rock record. This calls for a significant modification in the way that we apply the traditional principles of uniformitarianism to our reconstructions of geologic history.

The new synthesis that is the subject of this book is offered for advanced undergraduate and graduate training and for use by professionals, particularly those engaged in mapping and subsurface exploration and development.

Toronto                                                                                      Andrew D. Miall
April 2015

# Revision History

My first book, *Principles of Sedimentary Basin Analysis*, has gone through three editions, published successively in 1984, 1990, and 1999. A year or so ago, I realized that it might be time for a new edition. In reviewing the changes that have taken place in the whole broad field of sedimentary geology since that last edition, it became clear that stratigraphy is the area that has undergone the most significant changes in the last decades, and that is what I decided would most usefully be treated at length in this book.

Stratigraphy has undergone a revolution that has brought together multiple developments dealing with different themes and concepts in sedimentary geology and basin analysis. Chapter 1 includes a new section in which I trace the evolution of these many themes, and attempt to show how they have come together during the last few decades (since about 1990).

The text of Chaps. 2 and 4 from *Principles* has been updated and becomes Chaps. 2–4 in the present book. Chapter 3 of *Principles*, which dealt with dating and correlation, and the formal methods for the definition and naming of units, has been substantially rewritten and incorporates much of the material I wrote for "Sophisticated Stratigraphy," a review prepared at the invitation of the Geological Society of America (Miall 2013). It has been moved further along in the present book, appearing as Chap. 7, the point being that stratigraphy should now be seen as a science that synthesizes sedimentary geology, and which therefore requires that the subject is best addressed once the work of sedimentological description and interpretation is underway.

Chapters 5 and 6 of the present book are those that have undergone the most complete rewriting, to reflect the major changes in the science since the previous edition. Sequence stratigraphy (Chap. 5) has become the standard method for formal description and paleogeographic interpretation, and mapping methods (Chap. 6) are now dominated, at least in the petroleum industry, by the techniques of the reflection-seismic method, including 3-D seismic and the interpretive methods of seismic geomorphology.

The book culminates with Chap. 8, which is intended primarily as a review of current research into the nature of deep time as preserved in the sedimentary record. It is partly based on two research publications (Miall 2014b; Miall 2015) that focus on modern data dealing with sedimentation and accommodation rates, and the implications of these data for stratigraphic interpretation. The chapter concludes with a review of the current advanced research into cyclostratigraphy and astrochronology.

## References

Miall, A. D., 2013, Sophisticated stratigraphy, in Bickford, M. E., ed., The web of geological sciences: Advances, impacts and interactions: Geological Society of America Special Paper 500, p. 169-190.

Miall, A. D., 2014b, The emptiness of the stratigraphic record: A preliminary evaluation of missing time in the Mesaverde Group, Book Cliffs, Utah: Journal of Sedimentary Research, v. 84, p. 457-469.

Miall, A. D., 2015, Updating uniformitarianism: stratigraphy as just a set of "frozen accidents", in Smith, D. G., Bailey, R., J., Burgess, P., and Fraser, A., eds., Strata and time: Geological Society, London, Special Publication 404, p. 11-36.

# Acknowledgments

Colleagues who assisted with the earlier editions of *Principles* by critically reading parts or all of the manuscript include Tony Tankard, Andy Baillie, Guy Plint, and Ray Ingersoll. I am eternally grateful for their wise advice, and if they choose to do so they will find significant portions of the book that describe basic methods, such as the foundations of facies analysis methods, largely unchanged in this book.

Reviewers for the journal articles from which material in Chaps. 7 and 8 was drawn included Felix Gradstein, Ashton Embry, Bruce Wilkinson, Tony Hallam, Alan Smith, Brian Pratt, Gerald Bryant, John Holbrook, Chris Paola, Pete Sadler, Robin Bailey, Dave Smith, John Howell, and Torbörn Törnqvist. The historical section in Chap. 1 was reviewed by Bill Fisher, Ron Steel, Bob Dalrymple, and Martin Gibling. Chapters 5–7 were critically read by David Morrow. I extend my thanks to all these individuals. Any remaining errors or omissions remain my responsibility.

Once again I must acknowledge the enthusiastic support and encouragement of my wife, Charlene Miall. Her insights into the scientific method and the sociology of science have been particularly invaluable. My children, Chris (partner Natalie) and Sarah (partner Brad) have consistently been supportive.

# Contents

# The Scope of Modern Stratigraphy

## Contents

## 1.1  The Importance of Stratigraphy

It could be argued that in some respects **Stratigraphy** is the most important component of the science of Geology. Here's why:

McLaren (1978) provided nine reasons why the study of Stratigraphy with, at its center, an accurate geological time scale, is important:

> [Stratigraphy supplies unique and essential information regarding:] (1) rates of tectonic processes; (2) rates of sedimentation and accurate basin history; (3) correlation of geophysical and geological events; (4) correlation of tectonic and eustatic events; (5) are epeirogenic movements worldwide [?]... (6) have there been simultaneous extinctions of unrelated animal and plant groups [?]; (7) what happened at era boundaries [?]; (8) have there been catastrophes in earth history which have left a simultaneous record over a wide region or worldwide [?]; and (9) are there different kinds of boundaries in the geologic succession [?] (That is, "natural" boundaries marked by a worldwide simultaneous event versus "quiet" boundaries, man-made by definition).[question marks added]

Doyle and Bennett (1998, p. 1) stated that "Stratigraphy is the key to understand the Earth, its materials, structure and past life. It encompasses everything that has happened in the history of the planet" In this statement is the recognition that the stratigraphic history of layered sedimentary rocks preserved on the continents, and on the ocean floors constitutes the documented record of Earth history. No other branch of geology can provide this information.

Berggren et al. (1995, p. v) explained that the "essence of Stratigraphy and its handmaiden Geochronology" is to "understand the dynamic relationship which certainly exists between the evolution of ocean-continental geometries and concomitant changes in the climate and ocean circulation system and the evolution of life itself" by situating "the progression of events in this intricately related system in a precise temporal framework."

Torrens (2002, p. 251) pointed out a unique and essential component of stratigraphy: "The science of geology is all about time. ... So stratigraphy must first and foremost concern questions of time. It is the only area of geology that

© Springer International Publishing Switzerland 2016
A.D. Miall, *Stratigraphy: A Modern Synthesis*, DOI 10.1007/978-3-319-24304-7_1

is truly unique, other branches of geology are too often borrowed bits of physics, chemistry or biology."

The world's first stratigrapher was William Smith, a canal surveyor, who produced the first regional geological map in 1815, covering England, Wales and part of Scotland. The construction and refinement of geological maps, and the documentation of the subsurface for the purposes of petroleum and mineral exploration, have constituted two of the primary activities of practicing stratigraphers worldwide for the past 200 years. Dating and correlating the rocks have formed an integral part of this work, and questions about the nature of the time signal preserved in stratigraphic successions, and the developments of methods to investigate it have constituted a large part of this activity. William Smith's principal theoretical contribution—which is what made geological mapping possible—was the recognition of the reliability of the fossil record: the same assemblages of fossils always occur in the same order, and thus were born both the method of relative age-dating and the first reliable method for correlation on the basis of time.

As I demonstrated in my review paper for the Geological Society of America ("Sophisticated Stratigraphy": Miall 2013), modern stratigraphic methods are now providing extraordinary insights into the history of our Earth. Modern methods of age-dating and modern analytical methods have revolutionized the business of **historical geology**. In the drive to develop and apply ever more precise laboratory methods to geological samples, whether this be with the aim of age dating or the reconstruction of past climates, it is all too easy to lose sight of exactly where samples come from and what was their field context. How typical and how representative are samples, relative to the variability of their field setting? How were they situated with respect to breaks in sedimentation, the record of rare events, or disturbance induced by bioturbation or syn-depositional tectonism? Specialists in quantitative methods, particularly geophysicists, and those using numerical and statistical methods, including models and simulations, may be particularly susceptible to these hazards. One of the key elements of the stratigraphic data base is therefore **field context**. What is the stratigraphic and sedimentologic setting of the rocks that we are using to make these sophisticated interpretations? This is one of the unique characteristics of the science of stratigraphy, of particular relevance to the reconstruction of events in past time.

Until the 1960s, stratigraphy was largely a descriptive science, concerned primarily with the documentation of the lithologic and biostratigraphic successions of sedimentary basins as a basis for locating and exploiting fossil fuel and mineral deposits. Textbooks on petroleum geology, such as Levorsen (1954), contain some of the most advanced and detailed treatments of stratigraphy as actually practiced in the field. In many respects, William Smith's focusing on

basic mapping survived as a central focus through several of the revolutions that were took place in the earth sciences, beginning in the 1960s. However, over the last fifty years a profound change in approach has taken place, initially under the rubric of **Sedimentology**, which took sedimentary geologists away from description and classification into a focus on processes (Seibold and Seibold 2002, provided a detailed history from a European perspective; see also Middleton 2005). Only in recent years have **Stratigraphy**, **Sedimentology** and **Basin Analysis** come together to provide a dynamic, unified approach to the study of sedimentary basins. It would now be accurate to state that "Stratigraphy IS Sedimentology and Sedimentology IS Stratigraphy."

The evolution of modern methods can be understood as a series of separate developments that partially overlapped in time and which have gradually coalesced to create, in my review written for the Geological Society of America (Miall 2013), what I called "Sophisticated Stratigraphy." Middleton (2005, p. 628) suggested that:

> Only after 1950 was it common to find specialists who studied sedimentary rocks, but declined to be called stratigraphers, and since 1977 an increasing number of specialists refuse to make a hard distinction between sedimentology and at least some aspects of paleontology and stratigraphy, which they include together as "sedimentary geology."

Some of the developments in the study of sedimentary rocks were initiated many years ago, but it has only been since sequence stratigraphy matured as a standard descriptive and mapping method during the 1990s that it has become apparent that it has drawn on, exploited, and pulled together these earlier developments that commonly tended to be considered and written about in isolation. The historical evolution of these concepts is summarized in the next section.

## 1.2   The Evolution of "Sophisticated Stratigraphy"

The roots of modern, dynamic stratigraphy go back to the recognition of the concept of **facies** in the early nineteenth century, but it is argued here that the modern era began with the increased understanding of **fluid hydraulics** and **cyclic sedimentation** and the evolution of the **facies model** concept in the 1960s. The evolution is broken down below into fourteen steps. A critical fifteenth strand of development concerns the developments of concepts about geologic time and the increasing accuracy and precision with which geologist can now reconstruction the ages of events in the distant geological past. This topic is addressed in Sect. 7.8. These strands of development did not take place in isolation; however, they represent separate concepts or areas of specialization, which took some time to come together into the unified, integrated science that is now practiced. This summary is intended only to touch on the main highlights. A full historical and analytical account remains to be written. Many of the publications listed have become classics, with many hundreds of citations.

### 1.2.1 Beginnings (Nineteenth Century)

Middleton (2005) divided the history of sedimentology into six periods or stages. The first stage ended about 1830 with the publication of Lyell's (1830) masterwork, that led to the general acceptance of **uniformitarianism**, or **actualism**, as the basis for geology. What follows in this section falls into his second period. The subsequent discussion does not adhere to his subdivision into "periods" because I focus on specific themes which overlapped in time.

Two key early developments were the recognition of the concept of **facies** (Gressly 1838), and the establishment of **Walther's Law** (Walther 1893–1894). Teichert (1958), Middleton (1973) and Woodford (1973) reviewed the history and use of the concepts in light of contemporary ideas. Note the dates of these papers (1958, 1973), in light of the stages of development summarized below, because they help to explain the chronological evolution of modern stratigraphic thought and theory. Walther's law is discussed further in Sect. 3.4.1.

Developments in biostratigraphy were enormously important, in establishing some of the basic ideas about stratal succession, relative ages, and correlation. The evolution of the concepts of zone and stage are discussed in detail elsewhere (Hancock 1977; Miall 2004), topics that are not repeated here. Stratigraphic paleontology was a central theme of stratigraphy until relatively recent times. In fact, the first professional society in the field of sedimentary geology, the *Society of Economic Paleontologists and Mineralogists*, founded in Tulsa in 1931, emphasized this fact in the title of the society. Paleontology and mineralogy were important elements of petroleum geology and basin analysis until the seismic revolution of the 1970s, mainly because of their use in the identification and correlation of rock units in petroleum-bearing basins.

### 1.2.2 Cyclic Sedimentation (1932–1968)

Implicit in the early work on facies and on Walther's Law is the concept of recurrence of certain environments and their deposits. The idea of cyclicity became explicit with the study of the Carboniferous deposits of the US Midcontinent in the early 1930s, which consist of repetitions of a coal-bearing clastic-carbonate succession. These came to be called cyclothems. Wanless and Weller (1932, p. 1003) are credited with the original definition of this term:

> The word "cyclothem" is therefore proposed to designate a series of beds deposited during a single sedimentary cycle of the type that prevailed during the Pennsylvanian period

Shepard and Wanless (1935) and Wanless and Shepard (1936) subsequently attributed the cyclicity to cycles of sea-level change, an explanation that has never been challenged.

The beginnings of an understanding of the significance of the lithofacies signatures of common environmental settings is implicit in the paper by Nanz (1954), where coarsening- and fining-upward trends extracted from some modern sedimentary environments in Texas are presented. There is no discussion of repetitiveness or cyclicity in this paper, but the work was clearly foundational for the very important papers by Nanz's Shell colleagues that followed less than a decade later (see Sect. 1.2.6).

Duff and Walton (1962) demonstrated that the cyclothem concept had become very popular by the early 1960s. For example, Allen (1962, 1964), who is credited as one of the two originators of the meandering-river point-bar model for fluvial deposits, used the term cyclothem for cycles in the Old Red Sandstone in his first papers on these deposits. Duff and Walton (1962) addressed the widespread use (and misuse) of the term cyclothem, and discussed such related concepts as modal cycle, ideal cycle, idealized cycle, and theoretical cycle, the differences between cyclicity, rhythmicity and repetition, and the possible value of statistical methods for refining cyclic concepts. They speculated about the possibility of repeated delta-lobe migration as a cause of cyclicity, in contrast to the prevailing interpretation of the cycles as the product of sea-level change.

With Carboniferous coal-bearing deposits as the focus, two edited compilations dealing with cyclic sedimentation made essential contributions to the birth of modern sedimentology at about this time. Merriam (1964), based in Kansas, provided a focus on the US Midcontinent deposits, while Duff et al. (1967) dealt at length with European examples. The Kansas publication included a study of cyclic mechanisms by Beerbower (1964) that introduced the concepts of **autocyclic** and **allocyclic** processes. Autocyclic processes refers to the processes that lead to the natural redistribution of energy and sediment within a depositional system (e.g., meander migration, shoreline progradation)—the preference is now to use the term **autogenic** because they are not always truly cyclic—whereas **allocyclic** (**allogenic**) processes are those generated outside the sedimentary system by changes in discharge, load, and slope. Beerbower (1964) was dealing specifically with alluvial deposits in this paper, but his two terms have subsequently found universal application for other environments and their deposits. The term **allogenic** is now used to refer to processes external to a sedimentary basin, including eustasy, tectonism and climate change.

Another important contribution at this time was that by Visher (1965). The purpose of his paper was to build on the ideas contained in Walther's Law to highlight the importance of the vertical profile in environmental interpretation. He provided detailed descriptions of the profiles for six clastic environments, regressive marine, fluvial (channel or valley-fill), lacustrine, deltaic, transgressive marine, and bathyal-abyssal, drawing on both modern settings and ancient examples. This was, therefore, one of the first comprehensive attempts to apply the principles of actualism (uniformitarianism) to sedimentological interpretations. Interestingly (and this highlights one of the arguments of this chapter that some ideas develop as separate lines of research, which take time to come together), Visher's paper makes no reference to what are now the classic papers on Bouma's turbidite model (Bouma 1962), or Allen's (1964, 1965) work on alluvial deposits, which include his block diagram of a fluvial point bar. However, Beerbower's (1964) description of autocyclicity and Duff and Walton's (1962) speculation about deltaic processes (neither of which are referenced by Visher) indicate the beginnings of what shortly became a flood of new work providing the basis for the facies model revolution. Early applications of these ideas to the interpretation of the subsurface are exemplified by Berg's (1968) study of an interpreted point-bar complex constituting a reservoir unit in Wyoming.

### 1.2.3  Basin Analysis and the Big Picture (1948– 1977)

Driven in large measure by the needs of the petroleum industry to understand subsurface stratigraphic successions, geologists devised a number of ways to explore the broader origins of a basin fill and understand its paleogeographic evolution. Until the plate-tectonics revolution of the 1970s, basins were interpreted in terms of the **geosyncline theory**, which reached its full expression in this period with the definition of a range of classes based on structural and stratigraphic attributes (Kay 1951), many of which, as the plate-tectonics paradigm subsequently revealed, had little to do with the actual dynamics of continental crust.

Whereas the facies model revolution of the 1970s (Sect. 1.2.7) dealt with sedimentology on the relatively small scale of individual depositional systems (rivers, deltas, submarine fans, reefs, etc.), paleogeographic reconstruction for industry meant attempting to understand entire basins. Provenance studies based on detrital petrography were central to this work, hence the title of the first specialized

journal in this field, the *Journal of Sedimentary Petrology*, founded in 1931. Isopachs revealed broad subsidence patterns, and (for outcrop work) regional paleocurrent studies confirmed regional transport patterns, even in the absence of the understanding of the hydraulics of sedimentary structures that came later with the development of the flow regime concept (Sect. 1.2.5). Krumbein (1948) pioneered the generation of lithofacies maps based on such indices as a clastic-carbonate ratios. Dapples et al. (1948) demonstrated how these maps could be used to deduce tectonic controls in a basin. The subject of stratigraphy meant classical lithostratigraphy. The books and reviews by Pettijohn (1949, 1962; Potter and Pettijohn 1963) and Krumbein and Sloss (1951) and Levorsen's (1954) textbook on petroleum geology exemplify this approach.

However, some interesting new ideas that we would now classify under the headings of basin architecture, accommodation and sequence stacking patterns began to emerge, although little of this work was widely used at the time, it being only from the perspective of modern sequence methods that we can look back and see how a few individuals were ahead of their time. Rich (1951) described what we would now term the continental shelf, the continental slope and the deep basin as the **undaform, clinoform,** and **fondoform**, respectively, and provided descriptions of the processes and resulting sedimentary facies to be expected in each setting. The only one of his terms to survive is **clinoform**, although now it is used as a general term for deposits exhibiting a significant depositional dip, rather than as a term for a depositional environment. Van Siclen (1958) examined the late Paleozoic cyclothems where they tip over the southern continental margin which, at that time, lay within what is now central Texas. His work includes a diagram of the stratigraphic response of a continental margin to sea-level change and variations in sediment supply that is very similar to present-day sequence models. Oliver and Cowper (1963, 1965) may have been the first to specifically identify "clino" beds in the subsurface using Rich's concepts in a stratigraphic reconstruction based on petrophysical log correlation. This work is discussed in Sect. 6.2.3. Curray (1964) was among the first to recognize the importance of the relationships between sea-level and sediment supply. He noted that fluvial and strandplain aggradation and shoreface retreat predominate under conditions of rising sea level and low sediment supply, whereas river entrenchment and deltaic progradation predominate under conditions of falling sea level and high sediment supply. Curtis (1970) carried these ideas further, illustrating the effects of variations in the balance between subsidence and sediment supply as controls on the stacking patterns of deltas, concepts that are now encapsulated by the terms **progradation, aggradation** and **retrogradation**. Frazier (1974) subdivided the Mississippi deltaic successions into **transgressive, progradational,** and **aggradational** phases, and discussed autogenic (delta switching) and gla-cioeustatic sedimentary controls.

Perhaps it is because Texas specializes in bigness; this may be the explanation why some critical concepts concerning large-scale sedimentological environments were first developed there. The location of petroleum research laboratories, such as that of Shell Oil in Texas (referred to below) may also have been very influential. I refer to the concept of the **depositional system**, the concept that takes sedimentological analysis beyond the shoreface or the river meander or the reef talus slope to an analysis that encompasses entire systems. Fisk's (1944) work on the lower Mississippi valley and delta is an early example of this approach, but it was the later work of William L. Fisher that better exemplifies this next step and was more influential. The work he and his colleagues carried out on the deltas and other depositional systems of the Texas coast (Fisher et al. 1969, 1972) established a whole different scale of operation. Application of current subsurface stratigraphic methods to part of the Eocene section of the Gulf Coast (Fisher and McGowen 1967) demonstrated that existing rivers and deltas along a huge swath of the Gulf Coast had occupied essentially the same map locations for about 40 million years. The depositional systems approach provided the foundation for the **systems tracts** that became a critical part of sequence stratigraphy twenty years later. Lastly, in a paper that appears in the famous memoir that introducedseismic stratigraphy to the geological community (Payton 1977), Brown and Fisher (1977) summarized the ideas of this important group of stratigraphers at the Bureau of Economic Geology (at the University of Texas) and helped to bridge the intellectual next step from large-scale sedimentology to sequence stratigraphy.

### 1.2.4   The Meaning of "Facies" (1949–1973)

The concept of **facies** and the importance of **Walther's Law** were well understood and used in continental Europe during the nineteenth century, according to Teichert (1958), but did not become widely used in the English-speaking world until the 1930s.

On November 11th 1948 a conference was organized by the Geological Society of America in New York to discuss "Sedimentary facies in geologic history." This was a landmark event, the outcome of which was a *Geological Society of America* memoir (Longwell 1949) that marked the beginnings of several important developments. The memoir begins with a lengthy paper by Moore (1949) which set the scene by describing and illustrating, with the use of a block diagram, the various facies present within a modern carbonate reef complex in Java, from which he derived this definition:

> Sedimentary facies are areally segregated parts of different nature belonging to any genetically related body of sedimentary deposits.

The paper includes numerous examples of complex stratigraphic relationships from the Phanerozoic record of the United States, illustrating the inter-tonguing of facies of a wide range of environments. Moore's paper also includes an interpretation of the cyclicity exhibited by the cyclothems of the Mid-Continent, accompanied by a diagram showing how different facies develop as a result of repeated transgression and regression. Other papers by E.D. McKee, E.M. Spieker, and others, provide many other examples of complex stratigraphy, indicating that by this time there was a sophisticated understanding of the diachronous nature of facies in the stratigraphic record, and its control by sea-level change. The concluding contribution in this memoir is a lengthy paper by Sloss et al. (1949) in which the concept of the **sequence** is first described (see Sect. 1.2.9).

A decade later, Teichert (1958, p. 2719), working from Gressly's original discussion, explained the derivation of the term Facies:

> Facies is a Latin word meaning face, figure, appearance, aspect, look, condition. It signifies not so much a concrete thing, as an abstract idea. The word was introduced into geological literature by Nicolaus Steno (1669, pp. 68–75) for the entire aspect of a part of the earth's surface during a certain interval of geologic time.

In his abstract Teichert (1958, p. 2718) provided this succinct definition:

> [Facies means] the sum of lithologic and paleontologic characteristics of a sedimentary rock from which its origin and the environment of its formation may be inferred.

Teichert (1958) asserted that the concept of **facies associations** and the importance of **vertical facies successions** were well understood by nineteenth-century European geologists.

Interest in the work of the founders of modern sedimentology was renewed in the 1960s, with the new developments in the study of modern sediments, structures and environments. Woodford (1973, p. 3737) translated Gressly's (1838) "second law" as follows:

> Facies of the same petrographic and geologic nature assume, in different formations, very similar paleontologic characteristics and even succeed each other generally across a more or less numerous series of formations lying one upon the other.

Middleton (1973, p. 981) provided a translation of Walther's methodology from the original German. Walther referred to it as **"ontology"** (**actualism**, or **uniformitarianism**, in modern useage) as follows:

> It consists in trying to investigate the events of the past through modern phenomena. From being (existence), we explain becoming (genesis).

Middleton's (1973, p. 982) translation of Walther's original statement of his Law is as follows:

> The various deposits of the same facies-area and similarly the sum of the rocks of different facies-areas are formed beside each other in space, though in a cross-section we see them lying on top of each other. ... it is a basic statement of far-reaching significance that only those facies and facies-areas can be superimposed primarily which can be observed beside each other at the present time.

Middleton (1973, p. 980) suggested that "Walther must be named with Sorby, Gilbert, Grabau, and a few others, as one of the founders of the modern sciences of sedimentology and paleoecology," although he pointed out that whereas Walther's work was cited and acknowledged in much of the pioneer work in the early 20th century, in the first modern treatment of the subject of facies (Longwell 1949) there was no explicit mention of Walther or his law. He had a much greater influence in Russia, where facies studies were termed **"comparative lithology"**.

## 1.2.5  Fluid Hydraulics and Sedimentary Structures (1953–1976)

A key step in the development of modern sedimentology was the emergence of the idea that sedimentary structures represent measureable and repeatable physical processes, and that they therefore provide information on depositional environments and processes. Early work on the subject included the observations by Sorby (1859, 1908) and Gilbert (1884, 1899) on sedimentary structures, and Gilbert's experimental work (Gilbert 1914). Sorby (1852) was the first to recognize the utility of crossbedding for determining current directions (Sect. 6.7.1). However, as Allen (1993) pointed out, it was not until the appearance of the synthesis by Potter and Pettijohn (1963) that the richness and significance of the preserved record caught the general attention of sedimentary geologists.

A necessary first step towards a modern study of sedimentary structures is accurate description and classification. McKee and Weir (1953) made an important contribution in this direction, with their description of the scales of structures, their internal architecture and bounding surfaces. It is in this paper that the familiar terms **planar-** and **trough-cross-stratification** first appear. A decade later, a comprehensive classification by Allen (1963a) introduced a set of Greek letters for different types of crossbedding, a system that was widely used for some time. Several illustrated atlases of sedimentary structures also appeared during this period (Pettijohn and Potter 1964; Conybeare and Crook 1968), indicating that sedimentary geologists were coming to grips with the full range of preserved and observable structures.

By the 1950s, sedimentary geologists had become more widely aware of the directional information contained in sedimentary structures, and some pioneering studies of what came to be known as **paleocurrent analysis** were being performed. For example, Reiche (1938) analysed eolian crossbedding, Stokes (1945) studied primary current lineation in fluvial sandstones, and several authors were dealing with grain and clast orientation (e.g., Krumbein 1939). Pettijohn (1962) provided an overview of the subject, with many examples of the different techniques for analysis and data display that were then in use. Curray (1956) published what became the standard work on the statistical treatment of paleocurrent data.

Meanwhile, several pioneers were attempting to make sedimentary structures in the laboratory, in part as a means to understand the sedimentary record. There was also an interest in understanding fluid hydraulics from an engineering perspective, to aid in the construction of marine facilities, such as bridges and breakwaters. Kuenen and Migliorini (1950), in a classic paper, brought together flume experiments and observations of the ancient record to demonstrate that graded bedding could be generated by turbidity currents. As with many such contributions, it had been preceded by observations and suggestions by many other authors, but this was the paper that brought these observations together into the comprehensive synthesis that made it the benchmark contribution that it became. The term **turbidite** was subsequently coined by Kuenen (1957). McKee (1957), following his many years observing cross-stratification in outcrop, particularly in fluvial and eolian deposits in the Colorado Plateau area, experimented with the formation of cross-stratification by traction currents in a flume.

The critical theoretical breakthrough at this time was the series of flume experiments carried out by the US Geological Survey to study sediment transport and the generation of bedforms. This resulted in the definition of the **flow regime concept**, and the recognition of **lower-** and **upper flow regimes** based on flow characteristics (particularly the structure of turbulence), sediment load and resulting bedforms (Simon and Richardson 1961). At this time, Allen (1963b) reviewed the observational work of Sorby and made one of the first attempts to interpret sedimentary structures in terms of flow regimes. However, the most important next step was a symposium organized by Middleton (1965), which brought together current field and experimental studies in a set of papers that firmly established flow regime concepts as a basic tool for understanding the formation of hydraulic sedimentary structures formed by traction currents as preserved in the rock record.

Middleton (1966, 1967) extended the work of Kuenen with further experiments on turbidity currents and the origins of graded bedding, work that was ultimately to lead to a significant new classification of sediment gravity flows, of which it was now apparent that turbidity currents were only one type (Middleton and Hampton 1976). Reference is made in the first of these papers to field observations of turbidites by Walker (1965), a reference which marks the beginning of a significant professional collaboration, to which I return later.

Walker's (1967, 1973) field experience with turbidites led to a proposal for the calculation of an index of the proximal-to-distal changes that occur down flow within a turbidite. This marked an attempt at an increasingly quantitative approach to the study of sedimentary structures, although this index was not to survive an increasing knowledge of the complexities of the submarine fan environment within which most turbidites are deposited.

The important new developments in this field were well summarized in a short-course, organized by the Society for Sedimentary Geologists, the manual for which provides an excellent review of the state of knowledge at this time (Harms et al. 1975). This review contains the first description and definition of hummocky cross-stratification (HCS), and the recognition of this structure as a key indicator of combined-flow (unimodal and oscillatory) storm sedimentation. Additional details regarding the history of development of ideas about flow-regime bedforms, hydrodynamic sedimentary structures, and sediment gravity flows, are provided in Sects. 3.5.4 and 3.5.5.

## 1.2.6 Early Studies of Modern Environments (1954–1972)

As noted above, references to modern depositional settings appear in much of the early stratigraphic literature, but in the 1950s studies of "the modern" became more focused. Much of this was due to the recognition by some oil companies of the value of understanding the primary origins of petroleum-bearing rocks. A leader in this field was the research team at Shell Development Company.

Some of the earliest of these studies of modern environment were carried out in carbonate environments, including the work of Illing (1954), and Newell and Rigby (1957) on the Great Bahamas Bank, and Ginsburg and Lowenstam (1958) on the Florida platform. This, and other work on ancient carbonates (referred to below) led to two approaches to the classification of carbonate rocks (Folk 1962; Dunham 1962) that are still used today. In fact, these two papers (which appeared in the same SEPM Special Publication) are amongst the most important of the "classic" papers mentioned in this chapter, because of their long survival. Later studies of the Bahamas and Florida by Purdy (1963) and Ball (1967) contributed much to the subsequent growth of facies models for carbonate platforms and reefs.

The other outstanding set of classic works consists of the research on the Texas coastal plain by Bernard, Leblanc and their colleagues at Shell, building on the preliminary work of Nanz (1954). The first facies model for barrier islands emerged from the work of these individuals on Galveston Island (Bernard et al. 1959, 1962) and the point-bar model for meandering rivers is also attributed to this group, based on their studies of the Brazos River (Bernard et al. 1959; Bernard et al. 1962).

The Mississippi River and Delta is one of the largest of modern fluvial-delta systems, and its location in the centre of one of the most important, well-populated, industrial and tourist regions of the United States, in a petroleum province that generates a quarter of the US domestic supply, has led to intensive environmental and geological studies. The stratigraphic significance and complexity of the deposits of this system were first brought to geologists' attention by the detailed work of Fisk (1944). From the point of view of the growth of sedimentology the studies of Frazier (1967) were more significant, providing architectural block diagrams that illustrated the growth of distributaries in a river-dominated delta. Later studies by Fisher et al. (1969, 1972) broadened the scope of delta studies to other regions of the Texas coast and to other deltas worldwide, providing an essential basis for the subsequent development of formal delta facies models. Shepard et al. (1960) edited a collection of broader studies of the Gulf Coast.

Exploration methods for the continental shelf and deep oceans were primitive, until the introduction of side-scan sonar methods and improvements in navigation. The GLORIA sonar system was developed in 1970, but did not receive widespread use for geological purposes until it was adopted by the US Geological Survey in 1984 at the commencement of a program to map the newly established US Exclusive Economic Zone. The Deep Sea Drilling Project (DSDP) began in 1968. Extensive use of seismic stratigraphic techniques had to await the developments taking place in Shell, Exxon and BP, as noted below (in particular, the work of Vail et al. 1977). Sedimentological studies of the continental shelves and slopes and the deep basin were being carried out at this time, but the main breakthroughs in sedimentological analysis came from studies of the ancient sedimentary record, and are referred to below.

## 1.2.7 Facies Model Concept (1959–2010)

By the late 1950s a key idea was emerging that environments could be categorised into a limited number of depositional configurations, which are amenable to basic descriptive summaries. The first explicit use of the term **"facies model"** was in a conference report by Potter (1959, p. 1292). He opened the report with the following words:

A discussion concerning sedimentary rocks was held at the Illinois State Geological Survey on 4-5 Nov. 1958, for the purpose of pooling the knowledge and experience of the group concerning three topics: the existence and number of sedimentary associations; the possibility of establishing a model for each association that would emphasize the areal distribution of lithologic units within it; and the exploration of the spatial and sequential relations between the associations.

Later, on the same page, this definition is provided:

A facies model was defined as the distribution pattern or arrangement of lithologic units within any given association. In the early stages of geological exploration, the function of the model is to improve prediction of the distribution of lithologic types.

Note that the essential basis for a facies model is the recognition of a distinctive facies association. Much work to identify these associations now ensued.

A mention should be made here of the term **process-response model**. This term has sometimes been used with essentially the same meaning as facies model. Whitten's (1964, p. 455) discussion of this term quoted from Krumbein and Sloss (1963, p. 501), who:

suggested that in the search for "... generalizing principles it is a useful philosophical device to recognize *models*- actual or conceptual frameworks to which observations are referred as an aid in identification and as a basis for prediction."

The journal *Sedimentology* was founded by the *International Association of Sedimentologists* (IAS) in 1962. The editor was Aart Brouwer from the University of Leyden in the Netherlands, representing what had become a strong Dutch school of sedimentological studies. All the early work on tidal flat sedimentation emerged from this school (e.g., Van Straaten 1954). The then President of the IAS, the American marine geologist Francis Shepard said this, in the Preface on p. 1 of v. 1 of the new journal:

As this is written, there appear to be several primary purposes in sedimentological studies. One is to relate more completely the present day sediments to ancient sedimentary rocks. Although much has been done in this field recently, there are numerous types of sedimentary rocks for which no equivalent has yet been found in the sediments of today and some correlations need careful reexamination to see if they are correctly interpreted. Another need is for more careful study of sedimentary structures that are often obscured both in old and recent sediments. These structures can be very useful in interpreting paleoclimates and conditions of deposition of ancient sediments. A third important field to investigate is the geochemistry of sediments. Some of the early indications from the chemical nature of sediments have proven misleading and are in need of further study to explain apparent anomalies. Fourth, the rates of sedimentation can be given much more study with all of the new radioactive counting methods.

In an introductory assessment of sedimentary studies immediately following the Preface, editor Brouwer (1962,

pp. 2–3) reviewed the early history and origins of the separate discipline now called *Sedimentology*:

Essential parts are derived from sedimentary petrography, others from stratigraphy and still others have a purely palaeontological source. Perhaps stratigraphy takes a more or less central position, and many definitions recently given of stratigraphy (Hedberg 1948; Weller 1960; and others) seem to include nearly all of sedimentology, at least of ancient rocks. This is quite understandable, as sedimentary rocks are the stratigrapher's natural environment. Three modem textbooks, whose scope is mainly sedimentological, have "stratigraphy" in their title (Krumbein and Sloss 1951; Dunbar and Rodgers 1957; Weller 1960).

The reference to sedimentary petrography should be noted here. The first journal to deal specifically with sedimentological topics, the Journal of Sedimentary Petrology, was founded in 1931, and initially dealt exclusively with petrographic studies, including studies of detrital composition and provenance, and diagenesis. The scope of the journal gradually widened, and the name was changed to the *Journal of Sedimentary Research* in 1994. According to Middleton (2005) the term **Sedimentology** was coined by A.C. Trowbridge in 1925 and first used in print by Waddell in (1933), but did not come into common usage until the 1950s.

Now began a focused program to identify specific lithofacies and lithofacies associations by direct comparison between modern sediments and the preserved record. The comparison went both ways, determined in large measure by the initial interests of the researcher. One of the first of these studies was that by Beales (1958, p. 1846) who proposed the term Bahamite for "the granular limestone that closely resembles the present deposits of the interior of the Bahamas Banks described by Illing (1954)." Although this new term did not become part of the sedimentological lexicon, the methods pioneered by Beales and his colleagues were about to become part of the mainstream.

Two classic studies appeared in the early 1960s, Bouma's (1962) turbidite model and Allen's (1964) point-bar model for meandering river deposits. Both are concerned primarily with interpretation of the rock record, but make extensive reference to deposits and structures forming at the present day.

There appeared a flood of new work during the 1960s and 1970s making use of the new facies model concepts. Potter (1967) reviewed sandstone environments. He stated (Potter 1967, p. 360):

The facies-model concept with its emphasis on the existence of relatively few recurring models represents cause-and-effect "deterministic geology"-an approach that attempts to relate distribution and orientation of sand bodies in a basin to measurable, causal factors.

However, much of Potter's discussion dealt with grain size and other petrographic issues, and discussions about the shape and orientation of sand bodies (of importance for stratigraphic-trap prospecting) rather than facies modeling, as this term has come to be understood.

An edited compilation that appeared in the middle of this period (Rigby and Hamblin 1972) provides another good snapshot of the state of sedimentology at this time. It opens with a brief review of the topic of "environmental indicators" by H.R. Gould and this is followed by a classification of sedimentary environments by E.J. Crosby, and by eleven chapters providing details of seven depositional environments (three chapters on alluvial sediments and one discussing the use of trace fossils). There were also several important new textbooks published during this period (e.g., Blatt et al. 1972; Reineck and Singh 1973; Wilson 1975; Freidman and Sanders 1978; Reading 1978). That by Blatt et al. (1972) contains the first summary of depositional environments specifically focused on the concept of the facies model (and using that term in the chapter heading).

The critical contribution at this time was the development by Walker (1976) of a formal, theoretical description of the concept of the facies model and its value as a summary and a predictor. Central to this work was a new concept that environments could be characterised by a discrete and limited number of specific facies states. Drawing on Middleton's (1973) restatement of Walther's Law, Walker emphasized the importance of the vertical succession of facies, and introduced the **facies relationship diagram**, a semi-quantitative expression of the range of vertical transitions revealed by careful vertical measurement of a stratigraphic succession. Reference was made to a detailed study of de Raaf et al. (1965), which was the first to employ the concept of facies states and the use of a facies relationship diagram. Another study of vertical facies relationships at this time was that by Miall (1973) using the basic concepts of Markov Chain Analysis.

Walker's (1976, Fig. 4) diagram summarizing the construction of a facies model as a process of "distilling away the local variability" to arrive at the "pure essence of environmental summary" has been much reproduced (Fig. 3.9 is the current version, from James and Dalrymple 2010).

Walker's (1976) paper appeared first in a new journal, *Geoscience Canada* (founded and edited by his colleague at McMaster University, Gerard Middleton), and was intended as the introductory paper in a series of invited articles written mainly by Canadian sedimentologists dealing with specific environments and facies models. The series was later published as a stand-alone volume (Walker 1979) which became a best-seller and subsequently, under changing editorships, went into four editions (Walker and James 1984, 1992; James and Dalrymple 2010). Its success was due in large measure to the concise nature of the descriptions, the elegant diagrams, and the emphasis on the nature of the vertical profile, making this a very practical approach for

undergraduate teaching and for work with well logs and cores. A close competitor was the edited volume compiled by Reading (1978), a book written at a more advanced, graduate to professional level by him and some of his graduate students at the University of Oxford. This book went into two later editions (1986, 1996).

Amongst the other widely used facies models that appeared in the *Geoscience Canada* series (and subsequently in Walker 1979) was a treatment of continental shelf sedimentation highlighting the rock record of hummocky cross-stratification, and a simple and elegant model for submarine fans based almost entirely on ancient fan deposits in California and Italy. In this book, carbonate facies models were compiled and co-authored by Noel P. James, who became a co-editor of later editions. **Ichnology**, the study of trace fossils, evolved into an enormously valuable subsurface facies-analysis tool, allowing detailed analysis of sedimentary environments in drill core, as well as throwing much useful light on the significance of stratal surfaces, with the preservation of evidence of non-deposition and early lithification (Frey and Pemberton 1984; McEachern et al. 2010).

By the mid-1970s the stage was set for Sedimentology to flourish. The Walker (1979) *Facies Models* volume, and Reading's (1978) textbook were enormously influential. However, through the 1980s it remained largely isolated from the "big-picture" concepts that were emerging from the plate-tectonics revolution, and developments in seismic stratigraphy. These I discuss below. Textbooks that appeared during this period (e.g., Miall 1984; Matthews 1984a, b; Boggs 1987) deal with all these topics essentially in isolation, as separate chapters with little cross-referencing. As I argue below, it took the maturing of sequence stratigraphy to bring these topics together into what we may now term sophisticated stratigraphy.

### 1.2.8  The Impact of the Plate-Tectonics Revolution on Basin Studies (1959–1988)

The plate-tectonics revolution explained where and why basins form, provided a quantitative basis for their subsidence and uplift behavior, and elucidated the relationships between sedimentation and tectonics. As far as sedimentary geology is concerned, the revolution was not complete until the mid-1970s, when the re-classification of basins in terms of their plate-tectonic setting reached maturity. However, some important preliminary studies pointed the way.

Bally (1989, pp. 397–398) noted the work of Drake et al. (1959) "who first tried to reconcile modern geophysical-oceanographic observations with the geosynclinal concept" and that of Dietz (1963) and Dietz and

Holden (1974) who were the first to equate Kay's "mio-geosyncline" with the plate-tectonic concept of an Atlantic-type passive continental margin. Mitchell and Reading (1969) made one of the first attempts to reinterpret the old tectono-stratigraphic concepts of **flysch** and **molasse** in terms of the new plate tectonics.

But it was John Bird and John Dewey, in two papers published in 1970, who completely revolutionized our understanding of the origins of sedimentary basins (and much of the rest of geology) with reference to the geology of the Appalachian orogen, in particular, that portion of it exposed throughout the island of Newfoundland. Dickinson (1971) made reference to all of this work in his own first pass at relating sedimentary basins to plate tectonics.

These breakthroughs of the 1970s initiated a worldwide explosion of studies of basins and tectonic belts exploring the new plate-tectonic concepts. Through the 1970s, a series of books and papers was published containing the results (Dickinson 1974a; Dott and Shaver 1974; Burk and Drake 1974; Strangway 1980; Miall 1980b, 1984). One of the more important of these contributions was a paper by Dickinson (1974b) which constituted the first comprehensive attempt to classify sedimentary basins of all types in terms of their plate-tectonic setting. This paper was particularly notable for the extensive treatment of arc-related basins, and was followed up by a more detailed paper on this subject (Dickinson and Seely 1979) that remained the standard work on the subject for many years. This latter work was based in part on the recognition of a series of arc-related sedimentary basins within the Cordillera (Dickinson 1976), especially the Great Valley basin of California, which has long served as a type example of a forearc basin (e.g., Ingersoll 1978a, b, 1979).

Miall (1984, p. 367) argued that, by the application of judicious simplification and by skillful synthesis we can systematize the descriptions of depositional systems (their facies assemblages and architecture), structural geology, petrology, and plate-tectonic setting into a series of **basin models**, for the purpose of interpreting modern and ancient sedimentary basins. Dickinson (1980, 1981) used the term **petrotectonic assemblages** with the same meaning. These basin models are then a powerful tool for interpreting regional plate-tectonic history.

Another important era in the field of basin analysis was initiated by the development of quantitative, geophysically-based models of crustal subsidence, commencing in the late 1970s. The importance of these models to the development of stratigraphy was that they provided the basis for the development of quantitative models of basin subsidence and accommodation generation that greatly improved our understanding of large-scale basin architectures. The main breakthrough in the development of a modern extensional-margin basin model was made by McKenzie (1978), based in part on his studies of the

subsidence of the Aegean Sea. This classic paper introduced the concept of crustal stretching and thinning during the initial sea-floor spreading event, and showed quantitatively how this could account for the subsidence history of Atlantic-type margins. Many of the important early tests of this model were carried out on the Atlantic margin of the United States. Stratigraphic data were obtained from ten Continental Offshore Stratigraphic Test (COST) wells drilled on the continental shelf off New England between 1976 and 1982, and led to the development of formal backstripping procedures (Watts and Ryan 1976; Steckler and Watts 1978; Watts 1989) and to simple computer graphic models of subsiding margins (Watts 1981) that were very useful in illustrating the development of the basic architecture of Atlantic-type margins. Dewey (1982) emphasized their simple two-stage development: the early phase of rifting, typically capped by a regional unconformity, followed by a thermal relaxation phase which generates a distinctive pattern of long-term onlap of the basement.

An important modification of the McKenzie model was to recognize the importance of simple shear during continental extension, as expressed by through-going extensional crustal detachment faults (Wernicke 1985). This style of crustal extension was first recognized in the Basin and Range Province of Nevada, and was suggested by preliminary seismic data from the facing continental margins of Iberia and the Grand Banks of Newfoundland (Tankard and Welsink 1987).

The North Sea basin is the best studied rift basin, and has provided many insights regarding subsidence styles and structural geology (White and McKenzie 1988)

Turning to the other major class of sedimentary basins, those formed by flexural loading of the crust, it was Barrell (1917) who was the first to realize that "the thick nonmarine strata of the Gangetic plains accumulated in space made available by subsidence of the Indian crust beneath the mass of thrust plates of the Himalayan Range" (Jordan 1995, p. 334). Price (1973) revived the concept of regional isostatic subsidence beneath the supracrustal load of a fold-thrust belt that generates the marginal moat we now term a foreland basin (a term introduced by Dickinson 1974b), based on his work in the Southern Canadian Cordillera. Beaumont (1981) and Jordan (1981) were the first to propose quantitative flexural models for foreland basins, constraining the models with detailed knowledge of the structure and stratigraphy of the studied basins. It is clear that the crust must have mechanical strength for a wide foredeep, such as the Alberta Basin or the Himalayan foreland basin, to be created. The classic architecture of a foreland basin is defined by the isopachs of the sediment fill, which is that of an asymmetric lozenge, with a depocenter adjacent to the location of the crustal load, tapering along strike and also thinning gradually away from the orogen towards the craton. Two major developments contributed to our current understanding of these basins. Firstly, exploration drilling and reflection-seismic data led to an understanding of the structure and dynamics of the fold-thrust belts that border foreland basins and, during uplift, provide much of their sediment. Secondly, a growing knowledge of crustal properties permitted the development of quantitative models relating crustal loading, subsidence, and sedimentation (Jordan and Flemings 1990, 1991).

A significant development during the 1960s and 1970s was the elucidation of the structure of the fold-thrust belts that flank many orogenic uplifts and clearly served as the source for the clastic wedges referred to above. McConnell (1887) was one of the first to emphasize the importance of thrust faulting and crustal shortening in the formation of fold-thrust belts, based on his work in the Rocky Mountains of Alberta. As noted by Berg (1962), the mapping of faults in the Rocky Mountains of the United States and their interpretation in terms of overthrusting became routine in the 1930s. However, as his paper demonstrates, seismic and drilling data available in the early 1960s provided only very limited information about the deep structure of thrust belts. The release of seismic exploration data from the Southern Rocky Mountains of Canada by Shell Canada led to a landmark study by Bally et al. (1966) and set the stage for modern structural analyses of fold-thrust belts. A series of papers by Chevron geologist Clinton Dahlstrom, concluding with a major work in 1970, laid out the major theoretical principles for the understanding of the thrust faulting mechanism.

The final piece of the puzzle was to explain accommodation generation and the occurrence or regional tilts and gentle angular unconformities on cratons hundreds of kilometres from plate margins—the phenomenon termed **epeirogeny**. Modelling of mantle processes indicated the presence of convection currents that caused heating and uplift or cooling and subsidence of the crust. Gurnis (1988, 1990, 1992) termed this **dynamic topography**. Cloetingh (1988) described the process of **intraplate stress** (also termed **in-plane stress**) whereby horizontal stresses exerted on plates, such as the outward-directed compressive stress from sea-floor spreading centres ("ridge push"), may be expressed as intraplate earthquakes that cumulatively develop faults and long-wavelength folds.

## 1.2.9 Unconformities and the Issue of Time in Stratigraphy (1909–1970)

Although some of the ideas discussed in this section have been around for many years, the issue of time in stratigraphy did not begin to have a major influence on the science until Ager's work in the 1970s, and it was not until the full

flowering of sequence stratigraphy in the 1990s that such contributions as Barrell's accommodation diagram (Fig. 5.2) and Wheeler's chronostratigraphic charts (Fig. 8.1) (both discussed below) were fully integrated into the science of stratigraphy. This is why this section is placed here, rather than several pages earlier.

The science of geology began with James Hutton's observations in and around Scotland in the late eighteenth century. His discovery of the angular Silurian-Devonian unconformity at Siccar Point on the coast of southeast Scotland gave rise to Playfair's (1802) famous remark about the "abyss of time."

A predominant strand in geological work during the nineteenth and early twentieth centuries was the gradual documentation of the lithostratigraphy and biostratigraphy of sedimentary basins worldwide. As documented elsewhere (Hancock 1977; Conkin and Conkin 1984; Berry 1968, 1987; Miall 2004), some remarkably refined zonation schemes resulted from this work, and stratigraphic terminology and methods gradually evolved to facilitate description and classification, but until the development of radioisotopic dating by Ernest Rutherford and Arthur Holmes (Holmes first book on the geological time scale was published in 1913) the development of a quantitative understanding of earth processes was limited.

Geological mapping and research in North America during the "frontier" period is usefully summarized by Blackwelder (1909), who discussed the various types of sedimentary break (angular versus structurally conformable) and the duration of the missing time that they represented. His paper contained what is probably the first chronostratigraphic chart for the interior (cratonic) stratigraphy of North America, showing what was then known about the extent of the major Phanerozoic stratigraphic units on this continent and the unconformities that separate them.

In a paper that was remarkably ahead of its time, written shortly after the discovery of the concept of radioisotopic dating, Barrell (1917, pp. 747–748) set out what we now refer to as the concept of **accommodation**:

> In all stratigraphic measures of time, so far as the writer is aware, the rate of deposition of a sedimentary series has been previously regarded as dependent on the type of sediment, whether sandstone, shale, or limestone, combined with the present rate of supply of such sediment to regions of deposition. Here is developed an opposite view: that the deposition of nearly all sediments occurs just below the local baselevel, represented by wave base or river flood level, and is dependent on upward oscillations of baselevel or downward oscillations of the bottom, either of which makes room for sediments below baselevel. According to this control, the rate of vertical thickening is something less that the rate of supply, and the balance is carried farther by the agents of transportation.

Barrell (1917) was probably the first to understand the relationships amongst sedimentation, preservation, and accommodation. He constructed a diagram (Fig. 5.2) showing the "Sedimentary Record made by Harmonic Oscillation in Baselevel" (Barrell 1917, p. 796) that is remarkably similar to diagrams that have appeared in some of the Exxon sequence model publications since the 1980s (e.g., Van Wagoner et al. 1990, Fig. 39; Fig. 5.5 of this book). It shows that when long-term and short-term curves of sea-level change are combined, the oscillations of base level provide only limited time periods when base-level is rising and sediments can accumulate. In his diagram "Only one-sixth of time is recorded" by sediments (Barrell 1917, p. 797). This remarkable diagram (1) anticipated Jervey's (1988) ideas about sedimentary accommodation that became fundamental to models of sequence stratigraphy, (2) it also anticipated Ager's (1981, 1993) point that the sedimentary record is "more gap than record;" and (3) it constitutes the first systematic exploration of the problem of preservation potential.

During the early part of the twentieth century there was much theorizing about the forces at work within the Earth to form mountain ranges and sedimentary basins. This is summarized elsewhere (e.g., Miall 2004) and not dealt with here, because ultimately it did not contribute much to the development of modern stratigraphy. However, the practical work of petroleum exploration did make a difference. The distinguished petroleum geologist A.I. Levorsen was one of the first to describe in detail some examples of the "natural groupings of strata on the North American craton":

> A second principle of geology which has a wide application to petroleum geology is the concept of successive layers of geology in the earth, each separated by an unconformity. They are present in most of the sedimentary regions of the United States and will probably be found to prevail the world over (Levorsen 1943, p. 907).

This principle appears to have been arrived at on the basis of practical experience in the field rather than on the basis of theoretical model building. These unconformity-bounded successions, which are now commonly called "Sloss sequences," for reasons which we mention below, are tens to hundreds of metres thick and, we now know, represent tens to hundreds of millions of years of geologic time. They are therefore of a larger order of magnitude than the cyclothems. Levorsen did not directly credit Grabau, Ulrich, or any of the other contemporary theorists who were at work during this period (see Miall 2004), nor did he cite the description of unconformity-bounded "rock systems" by Blackwelder (1909). Knowledge of these seems to have been simply taken for granted.

The symposium on "Sedimentary facies in geologic history" referred to above in Sect. 1.2.4 contained a lengthy treatment of facies variability in the Paleozoic rocks of the cratonic interior of the United States by Sloss et al. (1949). In this paper much use is made of isopachs and lithofacies maps using Krumbein's (1948) concepts of clastic ratios and sand-shale ratios. The work revealed to the authors the contradictions inherent in current classifications of rock units in North America according to standard geologic time units. The use of the standard time scale (Cambrian, Ordovician, etc.) as a basis for mapping, obscured the fact that the major sedimentary breaks within the succession commonly did not occur at the divisions provided by the time scale, and so they set out to establish "operational units" for mapping purposes. Thus were born the first **sequences** for the North American interior: the Sauk, Tippecanoe, Kaskaskia and Absaroka.

The Sloss et al. (1949) paper in the symposium volume (Longwell 1949) is followed by nearly 50 pages of published discussion by many of the leading American geologists of the day, in which the issues raised by detailed mapping and the concepts and classifications available at the time for their systematization were fully discussed. This broader discussion is dealt with at length elsewhere (Miall 2004; 2010, Chap. 1). For the purpose of this book, the importance of the Sloss et al. (1949) paper and the wider discussion of sedimentary facies contained in the other papers (see Sect. 1.2.4) is that it clearly confirmed, at the time of publication, the need for a systematic differentiation of descriptive terminologies for "time" and for the "rocks." This had been provided by Schenk and Muller (1941), who proposed the following codification of stratigraphic terminology:

Time    division    (for    abstract    concept    of    time)
Time-stratigraphic division (for rock classification)

| Era | – |
|---|---|
| Period | System |
| Epoch | Series |
| Age | Stage |
| Phase | Zone |

Wheeler (1958, p. 1050) argued that a time-rock (chronostratigraphic) unit could not be both a "material rock unit" and one whose boundaries could be extended from the type section as isochronous surfaces, because such isochronous surfaces would in many localities be represented by an unconformity. Wheeler developed the concept of the chronostratigraphic cross-section, in which the vertical dimension in a stratigraphic cross-section is drawn with a time scale instead of a thickness scale (Fig. 8.1). In this way, time gaps (unconformities) become readily apparent, and the nature of time correlation may be accurately indicated. Such diagrams have come to be termed "**Wheeler plots**." Wheeler cited with approval the early work of Sloss and his colleagues, referred to in more detail below:

> As a tangible framework on which to hang pertinent faunal and lithic data, the *sequence* of Sloss et al. (1949, pp. 110–11) generally fulfills these requirements. Paraphrasing these authors'

discussion, *a sequence comprises an assemblage of strata exhibiting similar responses to similar tectonic environments over wide areas, separated by objective horizons without specific time significance* (Wheeler 1958, p. 1050; italics as in original).

Sequences came later to be called simply "**unconformity-bounded units**." However, a brief mention should be made of the concept of the **format**, a term suggested by Forgotson (1957) for laterally equivalent formations enclosed by widely mappable marker beds above and below.

Wheeler's (1958) methods are now universally accepted, although in practice they are still rarely applied. Ager (1973) is famous for his remark that "the sedimentary record is more gap than record." In a later book he expanded on the theme of gaps. Following a description of the major unconformities in the record at the Grand Canyon, he said, (Ager 1993, p. 14):

> We talk about such obvious breaks, but there are also gaps on a much smaller scale, which may add up to vastly more unrecorded time. Every bedding plane is, in effect, an unconformity. It may seem paradoxical, but to me the gaps probably cover most of earth history, not the dirt that happened to accumulate in the moments between. It was during the breaks that most events probably occurred.

Dott (1983, 1996) similarly warned about the episodic nature of sedimentation. However, as discussed in Chap. 7, stratigraphers are still not dealing fully with the issue of time and its representation in the rock record.

The evolution of chronostratigraphic methods and the increasing accuracy and precision with which sedimentary rocks can be dated is discussed in detail elsewhere (Miall 2004, 2010, Chap. 14), and is the focus of Sect. 7.8 and Chap. 8 of this book. A landmark in the development of modern stratigraphy was the adoption in the 1970s of the GSSP principal for the fixing of major chronostratigraphic boundaries. GSSP stands for **Global Stratigraphic Sections and Points**, and is a system for identifying outcrop sections that are accepted by the international community as marking the boundaries of stages and series (McLaren 1970).

## 1.2.10 Sequences and Seismic Stratigraphy (1963–1977)

Building on his earlier work (Sloss et al. 1949), further analysis by Sloss (1963) added two more sequences of Mesozoic-Cenozoic age to the North American suite (Zuni, Tejas) and firmly established the concept of the large-scale control of cratonic stratigraphy by cycles of sea-level change lasting tens of millions of years. In later work Sloss (1972) demonstrated a crude correlation of these sequences with a similar stratigraphy on the Russian Platform, thereby

confirming that global sea-level cycles constituted a major sedimentary control. However, Sloss, unlike his student Peter Vail, was never convinced that global eustasy told the entire story (Sloss 1988, 1991). In his 1963 paper Sloss included a pair of diagrammatic cross-sections of the Sauk and Tippecanoe sequence across the cratonic interior of North America that clearly indicated an angular unconformity between the two sequences, a relationship that could only have been developed as a result of broad warping of the craton before deposition of the Tippecanoe sediments.

Ross (1991) pointed out that all the essential ideas that form the basis for modern sequence stratigraphy were in place by the 1960s. The concept of repetitive episodes of deposition separated by regional unconformities was developed by Wheeler and Sloss in the 1940s and 1950s (Sect. 1.2.9). The concept of the "ideal" or "model" sequence had been developed for the mid-continent cyclothems in the 1930s (Sect. 1.2.2). The hypothesis of glacioeustasy was also widely discussed at that time. Van Siclen (1958) provided a diagram of the stratigraphic response of a continental margin to sea-level change and variations in sediment supply that is very similar to present-day sequence models (Sect. 1.2.3). An important symposium on cyclic sedimentation convened by the Kansas Geological Survey marks a major milestone in the progress of research in this area (Merriam 1964); yet the subject did not "catch on." There are probably two main reasons for this. Firstly, during the 1960s and 1970s sedimentologists were preoccupied mainly by autogenic processes and the process-response model, and by the implications of plate tectonics for large-scale basin architecture (Sect. 1.2.7). Secondly, geologists lacked the right kind of data. It was not until the advent of high-quality seismic-reflection data in the 1970s, and the development of the interpretive skills required to exploit these data, that the value and importance of sequence concepts became widely appreciated. Shell, British Petroleum, and Exxon were all actively developing these skills in their research and development laboratories in the 1970s. The first published use of the term "**seismic stratigraphy**" was in a paper by Fisher et al. (1973) describing a subsurface succession in Brazil (the term appeared in the Portuguese language as estratigrafia sismica). Peter Vail, working with Exxon, was the first to present his ideas in the English-speaking world, at the 1974 annual meeting of the Geological Society of America, but it was his presentation the following year at the American Association of Petroleum Geologists (Vail 1975) that caught the attention of the petroleum geology community. This was the beginning of the modern revolution in the science of stratigraphy.

The key idea that Vail and his colleagues proposed was that large-scale stratigraphic architecture could be reconstructed from reflection-seismic records. Their publication of

Memoir 26 of the American Association of Petroleum Geologists (Vail et al. 1977) was one of the major landmark events in the development of modern stratigraphy. Vail had learned about sequences from his graduate supervisor, Larry Sloss, and added to these his own ideas about global sea-level change (eustasy) as the major allogenic control of sequence development. The debate about global eustasy was long and controversial, and has been amply aired elsewhere (see Miall 2010). However, what emerged from the debate was the critical importance of the "big-picture" in stratigraphic reconstruction, and the predictive value of sequence models. Having once seen a seismic record interpreted in terms of seismic stratigraphy, with its emphasis on seismic terminations and regional unconformities, and the common occurrence of clinoform architectures, old concepts of "layer-cake" stratigraphy were dead forever. The section reproduced here as Fig. 1.3 appeared in Memoir 26, and was one of the first to bring this point home to the geological community.

It also seems likely that, working in the Gulf Coast, Vail learned from the "big-picture" stratigraphers at the Bureau of Economic Geology. The regional view exemplified by work such as the Texas atlas (Fisher et al. 1972) and the seismic interpretation that these individuals were already working on, and which eventually appeared in the same AAPG memoir (Brown and Fisher 1977) were very influential in helping sedimentary geologists understand the large-scale setting and tectonic influences on sedimentary basins at the very time that geophysical basin models were providing the quantitative basis for the plate-tectonic interpretations of these basins (Sect. 1.2.8).

Peter Vail has come to be called the "Father" of sequence stratigraphy, while his graduate supervisor, Larry Sloss, has posthumously earned the title of the "Grandfather" of sequence stratigraphy.

## 1.2.11 Architectural Elements: Sedimentology in Two and Three Dimensions (1983–1990)

Lithofacies maps and isopachs and the reconstruction of regional paleocurrent patterns had become standard tools of the sedimentary geologist (or basin analyst) by the 1970s (the second edition of the Potter and Pettijohn book "*Paleocurrents and basin analysis*" was published in 1977), but they often failed to capture the fine detail of sedimentary processes that were by now emerging from facies studies. As Miall (1984, Sect. 5.3) pointed out, these mapping methods tended to produce generalizations that did not always reflect the rapidly shifting patterns of depositional systems that could now be reconstructed from detailed sedimentological study of outcrops, well records and cores.

There was a scale mismatch. Lithofacies maps deal with large map areas (tens to hundreds of kilometers across) and are essentially two-dimensional. Facies studies at this time (the 1970s to early 1980s) were one-dimensional, focusing on the vertical profile in drill records or outcrops (typically a few metres to tens of metres high). What were clearly needed were the tools to put the observations together. Three-dimensional sedimentological studies provided part of the answer, particularly for outcrop analysis, and sequence studies focused on the larger picture (Sect. 1.2.12).

Work on fluvial systems by Allen (1983) and by Ramos and his colleagues (Ramos and Sopeña 1983; Ramos et al. 1986) led the way. These papers focused on large two-dimensional outcrops of complex fluvial deposits and offered classifications of the lithofacies units that described them in two or three dimensions. Picking up on this early work, Miall (1985, 1988a, b) offered a systematized approach that re-stated the lithofacies classification idea in terms of a limited suite of **architectural elements** that, it was proposed, constitute the basic building blocks of fluvial assemblages. One of the strengths of the approach is the ability to relate paleocurrent observations to the fine detail of the channel and bar complexes, revealing whole new insights into the bar construction and preservation processes. Comparable approaches have subsequently been adopted for other depositional environments. The use of photomosaics as base maps for analysing large outcrops has become standard, and there have been technological developments, such as the use of LIDAR methods for outcrop documentation, facilitating the digitization of observations, corrections for scale problems and perspective effects in ground observations, and so on.

## 1.2.12 Sequence Stratigraphy (1986–1990)

In the decade following the publication of AAPG Memoir 26 (Payton 1977) a wholesale re-evaluation of regional stratigraphy was underway. The significance of this revolution can be exemplified by the first publication that applied the new sequence concepts to an important swath of regional geology, the Cardium Sandstone of Alberta. This loosely defined unit is host to the largest oil field hosted in a clastic reservoir in Canada, the Pembina field, and stratigraphic and sedimentologic studies of the unit had been underway since it was discovered in 1953. The Pembina reservoir was difficult to understand. It consists of locally as much as 9 m of wave- and tide-deposited conglomerate accumulated some 200 km from the assumed contemporary shoreline. How did it get there? The new interpretation by Plint et al. (1986) reconstructed from well-logs a set of seven basin-wide surfaces of erosion and transgression, implying cycles of base-level change lasting about 125 ka. The interpretation was controversial, and was subject to intense discussion at the time (Rine et al. 1987), but the interpretation has stood the test of time, and has led to a complete remapping of Alberta basin stratigraphy using the new sequence concepts (Mossop and Shetsen 1994).

Meanwhile, researchers working with seismic data, particularly in the research laboratories of Shell, BP and Exxon, were applying sequence concepts to basins around the world, yielding many insights into stratigraphic architecture and regional basin controls, particularly the importance of tectonism, even though the global-eustasy paradigm remained dominant throughout the 1980s and early 1990s. Several atlases were published at this time, taking advantage of the large atlas format to display reflection-seismic cross-sections at large scales (Bally 1987). Even more importantly, in 1988 a second major production by the team at Exxon was published (Wilgus et al. 1988), showing in detail how sequence concepts could incorporate facies analysis and could be applied to outcrop studies. The **systems tract** concept reached a full expression in several key papers in this book (Posamentier and Vail 1988; Posamentier et al. 1988), building on experimental models of Jervey (1988) that essentially reinvented Barrell's (1917) ideas about accommodation and its control on sedimentation and developed them further in the light of modern facies concepts.

Another important publication from the Exxon team was that by Van Wagoner et al. (1990) which presented the results of several detailed field mapping projects and extended the reach of sequence concepts further, to regional outcrop and subsurface studies. Largely on the basis of these two publications by the Exxon team, the term seismic stratigraphy began to be replaced in common use by the more general term **sequence stratigraphy**.

### 1.2.13  Reconciling Facies Models with Sequence Stratigraphy (1990)

By the year 1990 a moment of tension had arrived in the evolution of sophisticated stratigraphy. The enormously successful facies model approach, focusing on very detailed local studies, including meticulous analysis of drill cores, had resulted in a proliferation of sedimentological studies and numerous refinements of ideas about how to classify and subdivide sedimentary environments in an ever expanding range of tectonic and climatic settings. Most interpretations dwelt at length on autogenic sedimentary processes. Meanwhile, sequence stratigraphy had introduced an entirely different scale of research, encompassing whole basins, and focusing on allogenic controls, particularly sea-level change. In addition, the architectural-element approach to facies

studies departed from the clean simplicity of the vertical profile by suggesting multi-dimensional assemblages of sedimentary building blocks in patterns difficult to pin down and classify.

The problems may be exemplified by an examination of a paper by Walker (1990), who was attempting to reconcile his facies-model approach to the new concepts and methods. He (Walker 1990, p. 779) complained that the architectural-element approach, which treated elements as building blocks that could be assembled in multiple ways (Miall 1985), constituted "sedimentological anarchy." Walker (1990) conceded that the proliferation of information about environments and facies associations that had resulted from the explosion of facies studies rendered the simple facies-model approach for complex depositional systems (such as submarine fans) inadequate. He referred approvingly to the depositional systems approach exemplified by Fisher and McGowen (1967).

> Future facies modeling must emphasize these contemporaneous, linked depositional environments, and their response to tectonics and changes of relative sea level. This will combine the strengths of classical facies modeling with the recognition that widely spaced and "distinct" geographic environments (summarized as models) can be rapidly superimposed as part of one transgressive or regressive system (Walker 1990, p. 780).

Walker (1990, p. 781) also expressed concern regarding the new concepts of sequence stratigraphy, which were becoming popular at this time. He pointed out the ambiguity in some of the definitions (e.g., that of the **parasequence**; see Sect. 5.2.2), the uncertainty with regard to scale, and the lack of clarity in such expressions as "relatively conformable." The issue of scale arises with reference to such expressions as "genetically related" strata. In facies-model studies, genetically related implies gradational contacts between lithofacies that are related to each other in the sense implied by Walther's Law. In sequence stratigraphy, genetically-related means the deposits formed during a full cycle of base-level change, although, as Walker (1990, p. 784) pointed out, using Galloway's (1989) **genetic stratigraphic sequence model** implies that strata above and below a subaerial erosion surface (the E/T surfaces of Plint et al. 1986), are genetically related, which they are certainly not.

While reluctant to fully embrace the new methods and terminology of sequence stratigraphy, Walker (1990) conceded that the regional patterns and the emphasis on large-scale sedimentary controls that were being revealed by sequence studies were valuable. As a compromise he proposed the adoption of the new system of **allostratigraphy** that had been proposed in 1983 by the North American Commission on Stratigraphic Nomenclature. Allostratigraphy is based on the recognition, mapping and subdivision of unconformity-bounded units. For example, a typical sequence, in the sense implied by Vail et al. (1977) constitutes an alloformation.

## 1.2.14 The Full Flowering of Modern Sequence-Stratigraphic Methods

When sequence stratigraphy was introduced to the geological community through the landmark publications of the Exxon Group (Payton 1977; Wilgus et al. 1988; Van Wagoner et al. 1990) it came with an overriding hypothesis that eustatic sea-level change was the main driver of changes in accommodation, and hence of sequence architecture. Doubts about the universal applicability of this hypothesis began to emerge in the 1980s, and by the mid-1990s most earth scientists had accepted that other factors, including climate change and regional tectonism may play a key role (Miall 1995a). The controversy is described in detail in Miall (2010; Chap. 12).

The realisation that many allogenic processes are at work during the accumulation of a basin fill gave renewed impetus to stratigraphic studies, because it became clear that sequence methods, combining the large scale of reflection-seismic surveying with the facies scale of the outcrop or drill core, could be very powerful tools for the reconstruction of geologic history, as well as provide much more useful predictive stratigraphic models for petroleum exploration and development. The increasing use of horizontal "seiscrop" sections (horizontal sections extracted from three-dimensional data volumes) has led to the development of an entirely new discipline, **seismic geomorphology**, which deals with the analysis of ancient depositional systems based on their preserved landscape architecture and three-dimensional construction (Davies et al. 2007; Hart 2013). Furthermore, the debate about global eustasy placed renewed emphasis on the need for accurate global chronostratigraphic correlations in order to test regional and global correlations, and this also encouraged new work in this field.

The flourishing of sequence stratigraphy as a research topic inevitably led to differences of interpretation and even to differences in the methods for defining sequences. For example, Hunt and Tucker (1992) showed how the Exxon sequence model was quite inadequate in dealing with the falling stage of a base-level cycle. Galloway (1989) proposed defining sequence boundaries at the maximum flooding surface rather than the subaerial erosion surface and its basinward correlative conformity. This and other controversies (discussed in Sect. 7.7) hindered the development of a uniform methodology and common language for dealing with sequences on a formal basis.

Catuneanu (2006), in what has become the standard textbook on sequence stratigraphy addressed these topics topic and showed how different approaches could be reconciled if care is taken with descriptions and definitions. In a series of papers culminating in a review for *Newsletters in Stratigraphy* he and selected colleagues have been leading the way in the work to gain acceptance for sequence stratigraphy as the appropriate formal basis for modern stratigraphic work (Catuneanu et al. 2009, 2010, 2011). More recently, Steel and Milliken (2013) have provided a very useful documentation of the many incremental additions to our knowledge of siliciclastic facies associations and models and their incorporation into sequence-stratigraphic interpretations.

Modern theoretical and experimental work is making substantial contributions to our understanding of processes of sedimentation and sequence generation. The specially-designed experimental facility (eXperimental EarthScape Facility, or XES) described by Paola (2000) and Paola et al. (2001, 2009) is particularly well-equipped to explore what Sheets et al. (2002) termed the stratigraphic "mesoscale", the time scale of years to thousands of years. Within this time frame, "the depositional pattern shifts from reflecting the short-term flow pattern to reflecting long-term basinal accommodation. Individual events are averaged to produce large-scale stratal patterns" (Sheets et al. 2002, p. 288). At this scale, autogenic processes grade into, or are affected by and modified by allogenic forcing. Muto and Steel (2004) demonstrated that, given steady conditions of discharge and sediment supply, prograding deltas will eventually start to "autoincise" over the mesoscale time scale. Strong and Paola (2008) explored the evolving nature of valley incision, terrace formation and valley fill, and demonstrated that the valley-floor surface that ultimately is preserved in the geological record during a cycle of base-level change is an erosion surface that never actually existed in its entirety as a topographic surface in its preserved form, because it undergoes continuous modification by erosion or sedimentation until final burial. Kim and Paola (2007) demonstrated that the autogenic process of delta and channel switching may, under the influence of fault movement, develop cyclothem-like cycles over time periods of $10^5$ years.

Meanwhile, the research theme centred on facies analysis is by no means complete. As discussed in Chap. 4, advances in the understanding of processes and environments continue, aided by the experimental work touched on above and by improved observational methods. Three topics merit note: (1) the increasing recognition of the importance of cool-water environments for carbonate sedimentation, (2) an improved understanding of the development of deep-water turbidite deposits relative to the cycle of sea-level change and sediment delivery patterns, together with a much expanded understanding of the variability and complexity of turbidite systems, due in large measure to developments in marine geology, three-dimensional seismic surveying, and large-scale outcrop work, and (3) the increasing realization that in natural systems mud forms silt- and sand-sized floccules, and most mud is transported and deposited by currents of all kinds. Pelagic settling may be of minor importance as a source of mud deposits.

### 1.2.15 Stratigraphy: The Modern Synthesis

The full flowering of modern stratigraphy represents the amalgamation of the concepts and methods encompassed in all of the separate developments described in the preceding sections. The power of the modern science could not possibly have evolved without the contributions from all of these strands of development. However, for the purpose of education and training, the basic components of modern stratigraphy can be broken down into the following list of seven broad topics. Key references are provided here to some of the main recent reviews and textbooks:

1. Facies analysis methods and facies models (James and Dalrymple 2010).
2. Sequence stratigraphy, concepts, definitions and methods (Catuneanu 2006).
3. Interpretations of the origins of sequences in terms of basin processes (tectonism, eustasy, climate change, etc.) (Miall 2010; Allen and Allen 2013).
4. Modern seismic methods, including seismic geomorphology (Veeken 2007; Davies et al. 2007; Hart 2013).
5. Chronostratigraphy and the Geologic Time Scale (Gradstein et al. 2004a, 2012; see also www.stratigraphy. org).
6. 4. Basin geodynamics: the origins of basins in terms of plate tectonics and crustal behaviour (Miall 1999; Busby and Ingersoll 1995; Allen and Allen 2013).
7. Modern formal stratigraphic methods (Salvador 1994). Updated methods at www.stratigraphy.org.

A specialized branch of stratigraphy deals with the Quaternary record. Specialists include archeologists and anthropologists. Age dating reaches levels of accuracy and precision in the $10^3$–$10^4$-year range, based on dating methods designed specifically for the Recent, including $^{14}$C and U-Th radiometric methods, optically-stimulated luminescence, cosmogenic radionuclides, and amino-acid geochronometry (http://www.inqua-saccom.org/stratigraphic-guide/geochronometry/).

## 1.3 Time in Stratigraphy

Sedimentologists and stratigraphers study sedimentary events over time periods extending through 17 orders of magnitude, from the long-term global changes in plate distribution, atmospheric composition, and oceanic geochemistry that take place over billions of years ($10^9$ year), to the entrainment and displacement of individual sand grains in traction carpets, events that take but a few seconds (3 s = $10^{-7}$ year) (Miall 1991b, 2014b). Different parts of this time spectrum have become the special interest of different scientific groups (and are considered in different parts of this book). Consider this wide variety of observations:

1. Those interested in fluid hydraulics and the development of bedforms are concerned with periods of a few seconds to a few hours ($10^{-7}$–$10^{-3}$ year; Chap. 2).
2. Those who study sedimentary facies and depositional models have to take into account diurnal changes in air and water circulation and tidal effects at one extreme ($10^{-3}$ year), the effects of **dynamic events** such as floods and storms (durations of $10^{-3}$–$10^{-2}$ year, spaced over intervals of $10^{-1}$–$10^1$ year), and, at the other extreme, the growth, migration, and progradation of various subenvironments, such as river channels, tidal flats, delta lobes, and crevasse splays ($10^1$–$10^4$ year; Chap. 3).
3. Conventional stratigraphy—the mapping and formal definition of formations, members and groups, and sequence stratigraphy, deal with sedimentary packages and depositional systems that may take $10^2$–$10^7$ year to accumulate (Chaps. 5, 6 and 7).
4. For those concerned with the establishment of chronostratigraphic relationships and the rates of geological events and processes, chronostratigraphic precision in the dating of stratigraphic events is rarely better than about ± 0.1 m.y. ($10^5$ year), this being the highest level of resolution that the Geologic Time Scale can currently attain for most of the Phanerozoic (Gradstein et al. 2004a; see Chap. 7). For the Neogene the availability of a chemostratigraphic scale based on oxygen isotope variations has provided an even finer degree of resolution, and work is underway to extend a cyclostratigraphic time scale, with a resolution in the $10^5$-year range, back through the Cenozoic (Sect. 7.8.2).
5. There is increasing recognition of the importance of **astronomical (orbital) forcing** as a control on climate. Commonly, in the geologic past, this has been responsible for the growth and decay of continental ice caps, with consequent major effects on climate and global sea level, and hence, on stratigraphic cyclicity, notably the development of the cyclic deposits termed cyclothems.

These are the so-called **Milankovitch processes**, which deal mainly with time periods of $10^4$–$10^5$ year. The study of these cycles constitutes **cyclostratigraphy** and the time scale that may be derived from them is termed **astrochronology** (Chap. 8).

6. A major task for modern stratigraphers is the establishment and correlation of the major sequence framework for the world's sedimentary basins. These have durations ranging from $10^4$ to $10^7$ year. The driving force for these sequences includes eustasy, driven by changes in the global average rate of sea-floor spreading, regional tectonics, and climate change.

7. Geophysical modeling of sedimentary basins has, in recent years, focused on the flexural and rheological behavior of the earth's crust in response to stretching, heating, and loading events, and on the regional (epeirogenic) changes in continental elevation caused by mantle thermal processes (dynamic topography). Subsidence driven by these mechanisms is slow, and the evolution of the large-scale architecture of a basin is a process that takes $10^6$–$10^7$ year to complete.

8. The assembly and dispersion of supercontinents on the earth's surface is a product of the thermal behavior of the core and mantle. Evidence is accumulating for a long-term cyclicity throughout Proterozoic and Phanerozoic time, with a periodicity in the order of $10^8$ year.

The disparities in the time scales that form the background to the work of these various groups have given rise to some tensions in the development of basin analysis. For example, the study of fluid hydraulics in the laboratory has yielded elegant phase diagrams that permit interpretations to be made of instantaneous flow conditions in the development of sequences of hydrodynamic structures (crossbedding, parting lineation, etc.). However, such analyses relate only to the few minutes or hours when the individual bedform was being deposited and can tell us nothing about the usually much longer periods of time (up to several or many years) represented by non-deposition or erosion, and they also tell us nothing about the preservation potential of the various bedforms.

As data have accumulated relating to time intervals and rates of sedimentary processes it has become clear that there are major disparities between the rates derived from the study of modern sedimentary processes and those obtained from analysis of the geological record. Sadler (1981) was the first to document this in detail. Miall (1991b, 2014b) suggested that a hierarchical approach be used to manage discussion of rates and time scales in the geological record. This issue, and the question of preservation potential, are addressed in detail in Chap. 8.

## 1.4  Types of Project and Data Problems

Data collection and analysis procedures are, of course, determined by the nature of the project. The following are some typical basin analysis problems, with a brief discussion of the data collection potential.

### 1.4.1  Regional Surface Stratigraphic Mapping Project

Work of this nature is one of the primary functions of government surveys, intent on providing complete map-sheet coverage of their area of responsibility, both as a service to industry and as a basis for expert advice to government economic planners. Similar regional surveys are commonly undertaken by industry as a preliminary to detailed surface or subsurface exploration, although their studies are rarely as thorough. Many academic theses are also of this type.

Many government surveys are carried out by individuals who are not specifically trained in the analysis of sedimentary basins, the idea being that members of the survey should be generalists, capable of mapping anything. This was an old British tradition, and it is an unfortunate one because it means that the individual survey officer cannot possibly be aware of all the skills that are now available for mapping work in sedimentary rocks, nor are they encouraged to take the time for the specialized observations which would make their work so much more effective. The argument that this is "left for the academics to do later" is not always satisfactory, for it is commonly the case that the stratigraphy of a succession can only be clarified by those who thoroughly understand its sedimentology. Many fruitless arguments about stratigraphic terminology can be avoided if this is realized at the beginning of a mapping endeavor. A team approach to a project is often an ideal solution. For instance the research synthesis edited by Besly and Kelling (1988) is an excellent example of an integrated basin-analysis approach to a broad and complex research problem, in this case the origin of a major series of sedimentary basins in northwest Europe.

Another argument that is sometimes heard about geological mapping is that once it is done it does not need to be done again. Political pressures to defund or disband state geological surveys have been based on this argument, which, of course, ignores the fact that maps constantly need to be updated to incorporate new data, or redrawn to encompass new concepts or to take advantage of new technologies, such as Geographical Information Systems (GIS: digital mapping). It is likely that eventually most standard surface geological maps will be redrawn using sequence concepts instead of lithostratigraphic terminology.

The basis of all surface basin-analysis projects is the careful compilation of vertical stratigraphic sections. These are described by the geologist in the field, who also collects samples for subsequent laboratory analysis, taking care to label each sample according to its position in the section. Where should such sections be located? The choice depends on a variety of factors. First, they should be typical of the area in which they are found and should be as free as possible of structural deformation. Obviously, they should also be well exposed and, ideally, free of chemical or organic weathering, which disturbs or obscures textures and structures.

The best sections for regional correlation are those that include several stratigraphic intervals, but it is rare to find more than a few of these except in exceptionally well exposed areas. Short, partial sections of a stratigraphic unit can provide much valuable sedimentological data, such as facies and paleocurrent information, although it may not be possible to locate them precisely within a stratigraphic framework unless they can be correlated by a marker bed or structural interpretation (Chap. 6). Exposures of this type tend to be ignored by the regional mapper, but they should not be ignored by those intent on producing an integrated basin analysis, because they add to the data base and may provide many useful sedimentological clues.

The great advantage of surface over subsurface studies is the potential, given adequate clean outcrop, to see medium- to large-scale sedimentary features, such as crossbedding, and larger scale architectural elements, including channels, large bar deposits, bioherms, etc., that may be difficult or impossible to identify in a drill hole. Large outcrops can be used to construct lateral profiles. These, of course, add immeasurably to any basin interpretation, particularly paleogeographic aspects, such as the size, geometry, and orientation of depositional elements (e.g., reefs, channels). The geologist should also always be searching for lateral variations in lithologies, fossil content, or sedimentary structures, as these changes may provide crucial control for paleogeographic interpretations.

The disadvantage of studies carried out exclusively at the surface is that most of the rocks in any given basin are buried and may be inaccessible to observation over very large areas. Many basins are depositional basins, in the sense that they preserve at the present day essentially the same outline as during sedimentation. The rocks exposed at the surface, especially around the margins of the basin, may have a quite different thickness and facies to those preserved at the center, and may show erosion surfaces and unconformities not present in the center because of the tendency of basin margins to be affected by a greater degree of tectonic instability. This is illustrated in Fig. 1.1. A basin analysis carried out under such circumstances might therefore produce very incomplete or misleading results.

## 1.4.2 Local Stratigraphic-Sedimentologic Mapping Project

A common incentive for a detailed local study is the occurrence of some highly localized economic deposit, such as an ore body or coal seam. A geologist will examine every available outcrop within a few square kilometers of the deposit, and may also supplement the analysis with logs of diamond drill holes. Another type of local study is an academic thesis project, particularly at the master's level. A small-scale project is chosen because of the time and cost limits imposed by academic requirements.

Many of the comments given in the preceding section apply to local studies, but there are some additional complications that often arise because of the nature of such projects. A common fault is the emphasis on local features to the exclusion of any real consideration of regional implications. The geologist will erect a detailed local stratigraphy and fail to show clearly how it relates to any regional framework that may have been established, or there may be an overemphasis on certain selected parameters chosen, perhaps, as a training exercise, to the exclusion of others.

To set against these problems is the advantage a detailed local project may offer for carrying out very complete, sophisticated paleogeographic reconstruction. Rather than relying on selected stratigraphic sections, the geologist may be able to trace out units on foot or study their variation in a close network of diamond drill holes (Fig. 1.2). In this way, the detailed variations within, for example, an individual reef or channel network or a coal swamp, can be reconstructed. Typically, there will be more time to spend developing architectural data from large outcrops. Such reconstructions are particularly valuable because they provide a mass of three-dimensional data against which to test theoretical facies models and sequence concepts.

Diamond-drill holes are produced by a process of continuous coring. They therefore provide a complete lithologic and stratigraphic sample through the units of interest. The small diameter of the core (2–5 cm) does not permit recognition of any but the smallest sedimentary structures, but the close drill-hole spacing of a few hundred meters or less means that very detailed stratigraphic correlation is usually possible.

Unfortunately, a tradition of retention and curation of diamond-drill-hole core has rarely emerged in government or industrial organizations except, perhaps, for a few key holes in areas of particular interest. The core is normally

**Fig. 1.1** Contrasts between stratigraphic thicknesses and facies at basin center and basin margin. The geology of the deep, hydrocarbon-producing regions of a basin might be quite different from that at the margins, so that surface geology gives little useful information on what lies below. The North Sea Basin is an excellent example. This example is of the Messinian evaporite basin, Sicily (Schreiber et al. 1976)

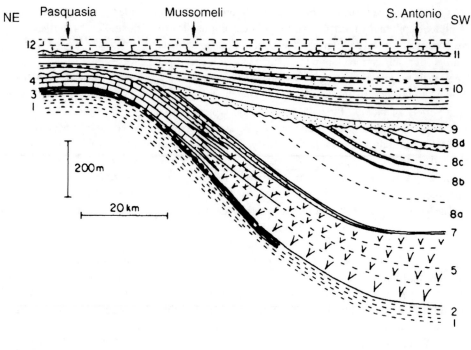

**Fig. 1.2** A typical diamond drill hole (DDH) network across a mining property, a gold prospect in Precambrian metasediments, northern Ontario

discarded once the immediate interest in the area has subsided, and much valuable material is lost to future workers. The mining industry commonly seems to prefer it this way, but what is good for corporate competitiveness is not necessarily the best method for developing a national stratigraphic data base.

### 1.4.3  Regional Subsurface Mapping Project

Exploration activity in the petroleum industry is now mainly of this type. Companies may initially send surface geological mapping parties into the field, but ultimately the most thorough field studies in a given field area are likely to be done by government survey organizations. Much industry activity is now located in offshore regions for which surface

geological information is, in any case, sparse, unobtainable, or irrelevant. As noted previously, the beds exposed at the edge of a basin may bear little relation to those buried near its center (Fig. 1.1).

Regional subsurface work is based initially on geophysical data and subsequently on test drilling. Gravity and aeromagnetic information may provide much useful information on broad structural features, particularly deep crustal structures. Refraction seismic lines may be shot for the same purpose. More detailed structural and stratigraphic data are obtained from reflection-seismic surveys (Fig. 1.3; Chap. 6), and these provide the basis for all exploration drilling in the early phases of basin development. Deep reflection surveys have profoundly changed our ideas about deformed belts during the last decade. Seismic shooting and processing is now a highly sophisticated process, and its practitioners like

**Fig. 1.3** Example of an interpreted seismic line across the continental margin of North Africa (Mitchum et al. 1977a). This is a famous section, having been reproduced in many textbooks as one of the first examples to be published showing the wealth of stratigraphic detail that could be extracted from the data. Note the presence of several angular unconformities, buried paleotopography, the internal architectural details of individual stratigraphic units, and the system of standard coded stratigraphic units employed by the authors. AAPG © 1977, reprinted by permission of the AAPG whose permission is required for further use

to talk about seismic stratigraphy as if it can provide virtually all the answers, not only about structure, but about the stratigraphic subdivision of a basin, regional correlation, and even lithofacies. This is particularly the case where, as is now common, selected areas of a basin are explored using three-dimensional seismic methods. However, the seismic method is only one exploration tool, and its results must be tested against those derived in other ways. For example, test drilling may show that stratigraphic correlations predicted from seismic interpretation are incorrect (Chap. 6). Problems of stratigraphic velocity resolution, the presence of low-angle unconformities, and the obscuring effects of local structure can all introduce errors into seismic interpretation. Resolution of bedding units is still relatively crude, as illustrated in the comparison between typical seismic wave forms and the scale of actual bedding units shown in Fig. 7.1. Dating and correlation of seismic sequences depend heavily on bios-tratigraphic and geophysical log information from explo-ration wells that are used to constrain and calibrate stratigraphic reconstructions made using sequence principles.

Exploration wells, especially the first ones to be drilled in a frontier basin, are as valuable in their own way as space probes sent out to study the planets, and it is unfortunate that in many countries the resulting data are not treated with the same respect afforded space information, but are regarded as the private property of the organization that paid for the drilling. In a competitive world, obviously, a company has the right to benefit from its own expenditures but, in the long-term, knowledge of deep basin structure and stratigra-phy belongs to the people and should eventually be made

available to them. In Canadian frontier areas, well samples and logs must be deposited with the federal government and released for public inspection two years after well comple-tion. Two years competitive advantage is quite long enough for any company in the fast-moving world of the oil industry, and after this time period the well records become part of a national data repository that anybody can use—with obvious national benefits. Seismic records are released after five years.

The nature of the stratigraphic information derived from a well is both better and worse than that derived from surface outcrops. It is better in the sense that there are no covered intervals in a well section, and such sections are generally much longer than anything that can be measured at the surface (perhaps exceeding 6000 m), so that the stratigraphic record is much more complete. The disadvantage of the well section is the very scrappy nature of the actual rock record available for inspection. Three types of sample are normally available:

1. Cuttings. These are produced by the grinding action of the rotary drill bit. These generally are less than 1 cm in length (Fig. 1.4) and can therefore only provide infor-mation on lithology, texture, and microfossil content. The North American practice is to collect from the mud stream and bag for examination samples every 10 ft (3 m) of drilling depth. When drilling in soft lithologies, rock may cave into the mud stream from the side of the hole many meters above the drill bit, so that the samples become contaminated. Also, cuttings of different density

**Fig. 1.4** Typical well cuttings. Photo courtesy of J. Dixon

**Fig. 1.5** Examples of typical core from a petroleum exploration hole (*left*) and a diamond drill hole (*right*)

may rise in the mud stream at different rates, which is another cause of mixing. It is thus necessary to observe the first appearance rule, which states that only the first (highest) appearance of a lithology or fossil type can be plotted with some confidence. Even then, depth distortions can be severe.

2. Full-hole core (Fig. 1.5). The rotary drill bit may be replaced by a coring tool when the well is drilling through an interval of interest, such as a potential or actual reservoir bed. However, this type of core is expensive to obtain, and it is rare to find that more than a few tens of meters, perhaps only a few meters, of core are

available for any given hole.

The advantage of a core is that because of the large diameter (usually on the order of 10 cm) it permits a detailed examination of small- to medium-sized sedimentary structures. Macrofossils may be present, and trace fossils are usually particularly well seen. The amount of sedimentological detail that can be obtained from a core is thus several orders of magnitude greater than is provided by chip samples. However, it is frustrating not to be able to assess the scale significance of a feature seen in a core. An erosion surface, for example, may be the product of a local scour, or it may be a major regional disconformity, but both could look the same in a core. Hydrodynamic sedimentary structures might be present, but it may not be possible to interpret their geometry and, except in rare instances where a core has been oriented in the hole, they provide no paleocurrent information. Orientation can sometimes be deduced if the core shows a structural dip that can be determined from regional structural data or dipmeter (formation microscanner) logs.

3. Side-wall cores. These are small plugs extracted by a special tool from the wall of a hole after drilling has been completed. These cores are rarely available to the geologist because they are used in porosity-permeability tests and are disaggregated for caving-free analysis by biostratigraphers. In any case, their small size limits the amount of sedimentological information that they can yield.

In addition to the samples and core, each exploration hole nowadays is subjected to an extensive series of petrophysical logging methods, which provide records by direct analog tracing or digitization. The description and interpretation of such logs has been the subject of several textbooks and cannot be treated exhaustively here. Log information is discussed further in Chap. 2. The following are a few preliminary remarks discussing the utility of logs in a subsurface data-collection scheme.

Most geophysical tools measure a single physical property of the rock, such as its electrical resistivity, sonic velocity, gamma radioactivity, etc. These properties reflect lithology, and can therefore be used, singly or in combination, to interpret lithology. Because measurements with modern tools have depth resolution of a few centimeters they are of great potential value in deriving accurate, depth-controlled lithologic logs free of the problems of sample caving. The response of a single tool is not unique to each rock type; for example, many different formations will contain rocks with the same electrical resistivity, and so it is not possible to interpret lithology directly from a single log type. However, such interpretations may be possible from a combination of two or more logs, and attempts have been made to automate such interpretations based on computerized calculation routines from digitized log records. Unfortunately, the physical properties of rocks and their formation fluids vary so widely that such automated interpretation procedures can only be successful if they are adapted to the specific conditions of each basin. They thus lose much of their exploration value, but become of considerably greater importance once an initial scatter of exploration wells has become available to calibrate the logs.

A particularly common use to which geophysical logs are put is in stratigraphic correlation (Sect. 6.2). Log records through a given unit may have a distinctive shape, which a skilled geologist can recognize in adjacent holes. Correlation may therefore be possible even if details of lithology are unknown and, indeed, the establishment of correlations in this way is standard practice in subsurface work. Similarly, log shape is a useful tool in environmental interpretation, though it must be used with considerable caution.

Logs are also of considerable importance in calibrating seismic records—seismic velocities are routinely derived from sonic logs and used to improve seismic correlations (e.g., see Fig. 6.20 and discussion in text).

Regional subsurface work may lead to the development of a list of petroleum **plays** for an area. These are conceptual models to explain the local **petroleum system**, including the history of generation, migration and trapping for petroleum pools. For example, a Devonian reef play, which might be based on the occurrence in Devonian strata of porous, dolomitized reef masses enclosed in mudrock, would provide an ideal series of stratigraphic traps. The development of a petroleum play leads to the evolution of an exploration methodology, which summarizes exploration experience in choosing the best combination of exploratory techniques and the type of data required to locate individual pools within the play area. At this stage, exploration moves from the play stage to the **prospect** stage. The first wells drilled in an area might have been drilled on specific prospects, with a specific play in mind, but it is rare for early wells to be successful. Prospect development is the next phase in an exploration program.

It is now common practice in petroleum exploration programs to carry out studies of burial history and thermal maturation of the stratigraphic section in a basin. This is based on what are now called backstripping procedures. Subsidence history is reconstructed by computer programs that reverse the process of sedimentation, stripping off each layer in turn, decompacting the remaining section, and calculating the isostatic balance. In this way, the component of subsidence due to tectonic effects can be isolated from that caused by the weight of sediment, and this, in turn, can be related to the rheological behavior of the crust underlying the basin. Burial history can be related to thermal history in

order to document the organic maturation of the sediments and, in this way, the timing of petroleum generation and release can be evaluated.

The recognition and correlation of stratigraphic sequences (Chaps. 5 and 6) now forms the core of regional stratigraphic work. This framework is then interpreted to assess regional paleogeographic histories and cycles of local to regional sea-level changes. The latter may be compared to regional and global standards, although this is of less importance to the exploration geologist than the production of a useable regional stratigraphic framework.

### 1.4.4  Local Subsurface Mapping Project

A network of closely spaced holes a few kilometers (or less) apart may be drilled to develop a particular petroleum or mineral prospect. These may be in virtually virgin territory or they may be step-out wells from a known pool or deposit in a mature petroleum basin or mine. A three-dimensional seismic survey may be carried out over selected areas of interest. It is at this stage in exploration work that geological skills come most strongly into play. In the early exploration phases, virtually the only information available is geophysical; almost all early petroleum exploration wells are drilled on potential structural traps. Once a hydrocarbon-bearing reservoir is located, the next problem is to understand why it is there, how large and how porous it is, and in what direction it is likely to extend. Interpretations of depositional environments, applications of facies models, and reconstructions of facies architecture and paleogeographic evolution may become crucial in choosing new drilling sites. Similar procedures must be followed in order to develop stratabound ore bodies. In the case of offshore petroleum exploration the exploitation of a new petroleum pool may be achieved by directional drilling of many wells from a central platform, and great pressure may be placed upon the development geologists to design a production strategy based on extremely limited data.

The types of data available for local subsurface mapping are discussed in Sect. 1.4.3. Three-dimensional seismic surveying is becoming an increasingly valuable tool for local structural and stratigraphic mapping, particularly for areas characterized by complex depositional systems, because of the ability to use the techniques of **seismic geomorphology** to map depositional units and their respective lithofacies in great detail (Sect. 6.3.4). At this point, it is pertinent to add a few cautionary notes on the uses of subsurface data. Because of its inherent limitations it is natural that exploration geologists would wish to exploit the data to the full extent but, paradoxically, this can lead to a very limited approach to an exploration problem. Geologists may be so impressed with a particular interpretive technique that they may tend to use it to the exclusion of all others. This has led to many false interpretations. It cannot be overemphasized that, in basin analysis as, no doubt, in every area of geology, every available tool must be brought to bear on a given problem. Examples of techniques in the area of sedimentology that have been overutilized include the use of vertical-profile studies, grain-size analysis, and grain-surface-texture analysis to interpret depositional environments.

## 1.5    Summary of Research and Reporting Procedures

Vail et al. (1991) suggested a working method for the application of sequence methods to the exploration of a new basin. Following a discussion of the application of Exxon sequence analysis methods and concepts to two basin examples, Miall (1997, Sect. 17.3.3) proposed a modified set of research procedures:

1. Develop an allostratigraphic framework based on detailed lithostratigraphic well correlation and/or seismic facies analysis.
2. Develop a suite of possible stratigraphic models that conform with available stratigraphic and facies data.
3. Establish a regional sequence framework with the use of all available chronostratigraphic information.
4. Determine the relationships between sequence boundaries and tectonic events by tracing sequence boundaries into areas of structural deformation, and documenting the architecture of onlap/offlap relationships, fault offsets, unconformable discordancies, etc.
5. Establish the relationship between sequence boundaries and regional tectonic history, based on plate-kinematic reconstructions.
6. Refine the sequence-stratigraphic model. Subdivide sequences into depositional systems tracts and interpret facies.
7. Construct regional structural, isopach, and facies maps, interpret paleogeographic evolution, and develop plays and prospects based on this analysis.
8. Develop detailed subsidence and thermal-maturation history by backstripping/geohistory analysis.

It was suggested that subsidence and maturation analysis (Step 8) be carried out at the conclusion of the detailed analysis, rather than near the beginning, as proposed in the Exxon approach. The reason is that a complete, thorough analysis requires the input of a considerable amount of stratigraphic data. Corrections for changing water depths and

for porosity/lithification characteristics, which are an integral part of such analysis, all require a detailed knowledge of the stratigraphic and paleogeographic evolution of the basin.

Catuneanu (2006, p. 63–71) provided a detailed set of suggestions for completing a regional sequence analysis. His "workflow", which is summarized here is based on "a general understanding [that] the larger-scale tectonic and depositional setting must be achieved first, before the smaller-scale details can be tackled in the most efficient way and in the right geological context." His workflow therefore proceeds from the large-scale through a decreasing scale of observation and an increasing level of detail. These are the basic components of the workflow:

1. Interpret the tectonic setting. This determines the type of sedimentary basin, and therefore controls the subsidence pattern and the general style of stratigraphic architecture. Ideally, the analysis should start from regional seismic lines.
2. Determine the broad regional paleographic setting, including the orientation of regional depositional dip and strike, from which broad predictions may be made about the position and orientation of coastlines, and regional facies architecture.
3. Determine paleodepositional environments, making use of well-log ("well-log motifs"), core data, and 3-D seismic data, as available.
4. From the regional concepts developed by steps 1–3, depositional trends may be predicted, and this provides essential diagnostic clues for the interpretation of sequence architecture. The focus should now be on the recognition and mapping of seismic terminations and major bounding surfaces. For example, coastal onlap may indicate transgression (retrogradation) and downlap indicates regression (progradation). "Only after the depositional trends are constrained, can the sequence stratigraphic surfaces that mark changes in such trends be mapped and labeled accordingly.
5. The last step of a sequence analysis is to identify the systems tracts.

Catuneanu (2006) was not primarily concerned in his book with driving mechanisms, and so his workflow ends with the construction of a detailed sequence stratigraphy. It is only once this has been completed that geohistory (backstripping) analysis can be carried out, a reconstruction of the history of relative sea-level change constructed, and a search for sequence-generating mechanisms be undertaken.

The most recent addition to the geologists' tool kit is three-dimensional reflection seismic data. The concept of extracting seismic reflection data horizontally, or along gently-dipping bedding planes, has been around since at least the mid-1980s (Brown 1985), but it has only been with the steady increase in computer power that 3-D seismic has become a routine exploration tool. Alistair Brown "wrote the book" on this topic in 1986, and his memoir has gone into multiple editions (Brown 2011). At depths of up to a few hundred metres, the resolution of 3-D seismic is adequate to image depositional systems in all their complex detail, providing an essential complement to well data in any project to map potential stratigraphic traps. The ability to map ancient landscapes in this way has given rise to a whole new special research topic, **seismic geomorphology**, pioneered by Henry Posamentier (2000). The subject is explored in detail by Davies et al. (2007) (see Chap. 6).

Formal stratigraphic methods are described at the website of the International Commission on Stratigraphy (www.stratigraphy.org). They do not yet include formal definitions and procedures for the application of sequence stratigraphic methods to the subdivision and naming of units in a basin fill (see Sect. 7.7).

The structure of a final report should also follow a logical order. A summary of the typical contents of a regional report is given below. Elements of this structure commonly are missing in the first drafts of graduate theses and reports submitted for publications in journals, or are not presented in a logical order. Not all components may be required for every individual study.

1. Introduction to the project. Why has this research been undertaken? What regional problem or scientific question has been addressed?
2. A description of the area covered by the report. This is normally illustrated by the use of a map. A project area may be defined geographically (mapping quadrant, state boundary) or geologically (tectonic province, sedimentary basin).
3. The geological setting of the area. The reader must be able to understand the purpose of the report within the context of the regional geological history. If appropriate, a description of the geological setting may include a brief outline of the plate-tectonic history of the region. Significant earlier work on the problem at hand or on the regional geology of the area may be referenced.
4. The stratigraphy of the area, typically including a stratigraphic table. The basis for regional correlations and age designations may be provided. If the purpose of the report is to describe the sequence stratigraphy of the region, earlier lithostratigraphic analyses should be briefly summarized, with references to this earlier work
5. Location of study sections, or wells used, or samples collected. These should be located by annotated points on a map, and other details may be provided if pertinent,

such as GPS data or national map grid references for key sections. If highly detailed location information is required this may be placed in an appendix. Annotated aerial photographs/satellite images/digital elevation models may be used to locate key outcrop Sects.

6. Research methods should be described. This may include, where relevant, sample collection and preparation procedures, field mapping and measurement methods, laboratory procedures, data reduction techniques. Be sure to reference the source of the methods used. Lengthy descriptions of methodology, such as laboratory standardization procedures, may be placed in an appendix (see also point 10, below). Methods that have been used before do not have to be described in detail, but the reader should be able to understand what you have done without needing to read a body of earlier literature.
7. Presentation of observations and/or laboratory results. As far as possible this presentation should be separate from any interpretation. Data should be illustrated without interpretation (e.g., uninterpreted seismic sections, raw stratigraphic sections). This may be difficult, for example, where observations have been guided by a new interpretive hypothesis. The use of systems-tract designations in the description of stratigraphic sections is a good example of a supposedly descriptive approach that has generated many problems.
8. Discussion of results, interpretations, and new hypotheses or theories arising from the project. Suggestions for further work.
9. Summary and conclusions. The major results. What new insights does this research provide? What new research or new problems does the project highlight? Do not introduce new references at this point.
10. Many mainstream journals now offer the opportunity to supplement the information in a published paper with additional documentation accessible on a dedicated journal website. This could include details of research methodologies, calculations, statistical procedures, data tables, etc.

There are also some simple rules to follow regarding illustrations

1. All locations and geological regions referenced in the text must be named on a map, preferably near the beginning of the report.
2. All symbols on all figures must be explained, preferably by legends on the figures themselves. The old European and Russian practice of providing symbols in numbered boxes that are explained by numbered notes in the caption is cumbersome and not to be recommended (this

practice facilitated translations intro different languages while avoiding the need to redraft figures, but now that science is almost universally published in the English language, this need no longer exists).

3. Lettering on figures should not be smaller than 1 mm high when published.

4. Photographs, whether of outcrops, samples, thin-sections, SEM images, etc., must include a scale, and orientation information may also be critical.

5. Figures, Tables, and Plates must be numbered in the order by which they are referenced in the text.

# References

Ager, D. V., 1973, The nature of the stratigraphical record: New York, John Wiley, 114 p.

Ager, D. V., 1981, The nature of the stratigraphical record (second edition): John Wiley, New York, 122 p.

Ager, D. V., 1993, The new catastrophism, Cambridge University Press, 231 p.

Allen, J. R. L., 1962, Petrology, origin and deposition of the highest Lower Old Red Sandstone of Shropshire, England: Journal of Sedimentary Petrology, v. v. 32, p. 657-697.

Allen, J. R. L., 1963a, The classification of cross-stratified units, with notes on their origin: Sedimentology, v. 2, p. 93-114.

Allen, J. R. L., 1963b, Henry Clifton Sorby and the sedimentary structures of sands and sandstones in relation to flow conditions: Geologie en Mijnbouw, v. 42, p. 223-228.

Allen, J. R. L., 1964, Studies in fluviatile sedimentation: six cyclothems from the Lower Old Red Sandstone, Anglo-Welsh basin: Sedimentology, v. 3, p. 163-198.

Allen, J. R. L., 1965, A review of the origin and characteristics of recent alluvial sediments: Sedimentology, v. 5, p. 89-191.

Allen, J. R. L., 1983, Studies in fluviatile sedimentation: bars, bar complexes and sandstone sheets (low-sinuosity braided streams) in the Brownstones (L. Devonian), Welsh Borders: Sedimentary Geology, v. 33, p. 237-293.

Allen, J. R. L., 1993, Sedimentary structures: Sorby and the last decade: Journal of the Geological Society, London, v. 150, p. 417-425.

Allen, P. A., and Allen, J. R., 2013, Basin analysis: Principles and application to petroleum play assessment: Chichester: Wiley-Blackwell, 619 p.

Ball, M. M., 1967, Carbonate sand bodies of Florida and the Bahamas: Journal of Sedimentary Petrology, v. 37, p. 556-591.

Bally, A. W., ed., 1987, Atlas of seismic stratigraphy: American Association of Petroleum Geologists Studies in Geology 27, in 3 vols.

Bally, A. W., 1989, Phanerozoic basins of North America, in Bally, A. W., and Palmer, A. R., eds., The geology of North America—an overview: The geology of North America, Geological Society of America, v. A, p. 397-446.

Bally, A. W., Gordy, P. L., and Stewart, G. A., 1966, Structure, seismic data and orogenic evolution of southern Canadian Rockies: Bulletin of Canadian Petroleum Geology, v. 14, p. 337-381.

Barrell, Joseph, 1917, Rhythms and the measurement of geologic time: Geological Society of America Bulletin, v. 28, p. 745-904.

Beales, F. W., 1958, Ancient sediments of Bahaman type: American Association of Petroleum Geologists, v. 42, p. 1845-1880.

Beaumont, C, 1981, Foreland basins: Geophysical Journal of the Royal Astronomical Society; v. 65, p. 291-329.

Beerbower, J. R., 1964, Cyclothems and cyclic depositional mechanisms in alluvial plain sedimentation: Geological Survey of Kansas Bulletin 169, v. 1, p. 31-42.

Berg, R. R., 1962, Mountain flank thrusting in Rocky Mountain foreland, Wyoming and Colorado: American Association of Petroleum Geologists Bulletin, v. 46, p. 2019-2032.

Berg, R. R., 1968, point-bar origin of Fall River Sandstone reservoirs, northeastern Wyoming: American Association of Petroleum Geologists Bulletin, v. 52, p. 2116-2122.

Berggren, W. A., Kent, D. V., Aubry, M.-P., and Hardenbol, J., eds., 1995, Geochronology, time scales and global stratigraphic correlation: Society for Sedimentary Geology Special Publication 54, 386 p.

Bernard, H. A., Major, C. F. Jr., and Parrott, B. S., 1959, The Galveston Barrier Island and environs - a model for predicting reservoir occurrence and trend: Transactions of the Gulf Coast Association of Geological Societies, v. 9, p. 221-224.

Bernard, H. A., Leblanc, R. J., and Major, C. J., 1962, Recent and Pleistocene geology of southeast Texas, in Rainwater, E. H., and Zingula, R. P., eds., Geology of the Gulf Coast and central Texas: Geological Society of America, Guidebook for 1962 Annual Meeting., p. 175-224.

Berry, W. B. N., 1968, Growth of prehistoric time scale, based on organic evolution: W.H. Freeman and Co., San Francisco, 158 p.

Berry, W. B. N., 1987, Growth of prehistoric time scale based on organic evolution, revised edition: Oxford: Blackwell Science, 202 p.

Besly, B. M., and Kelling, G., eds., 1988, Sedimentation in a synorogenic basin complex: the Upper Carboniferous of northwest Europe: Blackie, Glasgow, 276 p.

Blackwelder, E., 1909, The valuation of unconformities: Journal of Geology, v. 17, p. 289-299.

Blatt, H., Middleton, G. V., and Murray, R. C., 1972, Origin of sedimentary rocks: Englewood Cliffs, New Jersey: Prentice-Hall, 634 p.

Boggs, S., Jr., 1987, Principles of sedimentology and stratigraphy: Prentice Hall, Englewood Cliffs, New Jersey, 784 p.

Bouma, A. H., 1962, Sedimentology of some flysch deposits: Elsevier, Amsterdam, 168 p.

Brouwer, A., 1962, Past and present in Sedimentology: Sedimentology, v. 1, p. 2-6.

Brown, A. R., 1985, The role of horizontal seismic sections in stratigraphic interpretation, in Berg, O. R., and Woolverton, D. G., eds., Seismic stratigraphy II: American Association of Petroleum Geologists Memoir 39, p. 37-47.

Brown, A. R., 2011, Interpretation of three-dimensional seismic data, seventh edition, American Association of Petroleum Geologists Memoir 42, 646 p.

Brown, L. F., Jr., and Fisher, W. L., 1977, Seismic-stratigraphic interpretation of depositional systems: examples from Brazilian rift and pull-apart basins, in Payton, C. E., ed., Seismic stratigraphy — applications to hydrocarbon exploration: American Association of Petroleum Geologists Memoir 26, p. 213-248.

Burk, C. A., and Drake, C. L., eds., 1974, The geology of continental margins, Springer-Verlag, New York, 1009 p.

Busby, C. J., and Ingersoll, R. V., eds., 1995, Tectonics of sedimentary basins: Blackwell Science, Oxford, 579 p.

Catuneanu, O., 2006, Principles of sequence stratigraphy: Elsevier, Amsterdam, 375 p.

Catuneanu, O., Abreu, V., Bhattacharya, J. P., Blum, M. D., Dalrymple, R. W., Eriksson, P. G., Fielding, C. R., Fisher, W. L., Galloway, W. E., Gibling, M. R., Giles, K. A., Holbrook, J. M., Jordan, R., Kendall, C. G. St. C., Macurda, B., Martinsen, O. J., Miall, A. D., Neal, J. E., Nummedal, D., Pomar, L., Posamentier, H. W., Pratt, B. R,. Sarg, J. F., Shanley, K. W., Steel, R. J., Strasser, A., Tucker, M.

E., and Winker, C., 2009, Toward the Standardization of Sequence Stratigraphy: Earth Science Reviews, v. 92, p. 1-33.

Catuneanu, O., Bhattacharya, J. P., Blum, M. D., Dalrymple, R. W., Eriksson, P. G., Fielding, C. R., Fisher, W. L., Galloway, W. E., Gianolla, P., Gibling, M. R., Giles, K. A., Holbrook, J. M., Jordan, R., Kendall, C. G. St. C., Macurda, B., Martinsen, O. J., Miall, A. D., Nummedal, D., Posamentier, H. W., Pratt, B. R,. Shanley, K. W., Steel, R. J., Strasser, A., and Tucker, M. E., 2010, Sequence stratigraphy: common ground after three decades of development: First Break, v. 28, p. 21-34.

Catuneanu, O., Galloway, W.E., Kendall, C.G.St.C., Miall, A.D., Posamentier, H.W., Strasser A., and Tucker M.E., 2011, Sequence Stratigraphy: Methodology and Nomenclature: Report to ISSC: Newsletters on Stratigraphy, v. 4 (3), p. 173-245.

Cloetingh, S., 1988, Intraplate stress: a new element in basin analysis, in Kleinspehn, K., and Paola, C., eds., New Perspectives in basin analysis: Springer-Verlag, New York, p. 205-230.

Conkin, B. M., and Conkin, J. E., eds., 1984, Stratigraphy: foundations and concepts: Benchmark Papers in Geology, New York: Van Nostrand Reinhold, 363 p.

Conybeare, C. E. B., and Crook, K. A. W., 1968, Manual of sedimentary structures: Australian Bureau of Mineral Resources, Geology and Geophysics, Bulletin 102, 327 p.

Curray, J. R., 1956, The analysis of two-dimensional orientation data: Journal of Geology, v. 64, p. 117-131.

Curray, J. R., 1964, Transgressions and regressions, in Miller, R. L., ed., Papers in marine geology, Shepard Commemorative volume: MacMillan Press, New York, p. 175-203.

Curtis, D. M., 1970, Miocene deltaic sedimentation, Louisiana Gulf Coast, in Morgan, J. P., ed., Deltaic sedimentation modern and ancient: Society of Economic Paleontologists and Mineralogists Special Publication 15, p. 293-308.

Dapples, E. C., Krumbein, W. C., and Sloss, L. L., 1948, Tectonic control of lithologic associations: American Association of Petroleum Geologists Bulletin, v. 32, p. 1924-1947.

Davies, R. J., Posamentier, H. W., Wood, L. J., and Cartwright, J. A., eds., 2007, Seismic geomorphology: applications to hydrocarbon exploration and production: Geological Society, London, Special Publication 277, 274 p.

De Raaf, J. F. M., Reading, H. G., and Walker, R. G., 1965, Cyclic sedimentation in the Lower Westphalian of North Devon, England: Sedimentology, v. 4, p. 1-52.

Dewey, J. F., 1982, Plate tectonics and the evolution of the British Isles: Journal of the Geological Society, London, v. 139, p. 371-412.

Dickinson, W. R., 1971, Plate tectonic models of geosynclines: Earth and Planetary Science Letters, v. 10, p. 165-174.

Dickinson, W. R., ed., 1974a, Tectonics and sedimentation: Society of Economic Paleontologists and Mineralogists Special Publication 22. \

Dickinson, W. R., 1974b, Plate tectonics and sedimentation, in Dickinson, W. R., ed., Tectonics and sedimentation: Society of Economic Paleontologists and Mineralogists Special Publication 22, p. 1-27.

Dickinson, W. R., 1976, Sedimentary basins developed during evolution of Mesozoic-Cenozoic arc-trench system in western North America: Canadian Journal of Earth Sciences, v. 13, p. 1268-1287.

Dickinson, W. R., 1980, Plate tectonics and key petrologic associations, in Strangway, D. W., ed., The continental crust and its mineral deposits: Geological Association of Canada Special Paper 20, p. 341-360.

Dickinson, W. R., 1981, Plate tectonics and the continental margin of California, in Ernst, W. G., ed., The geotectonic development of

California: Prentice-Hall Inc., Englewood Cliffs, New Jersey, p. 1-28.

Dickinson, W. R., and Seely, D. R., 1979, Structure and stratigraphy of forearc regions: American Association of Petroleum Geologists Bulletin, v. 63, p. 2-31.

Dietz, R. S., 1963, Collapsing continental rises; an actualistic concept of geosyncline mountain building: Journal of Geology, v. 71, 314-333.

Dietz, R. S., and Holden, J. C., 1974, Collapsing continental rises; actualistic concept of geosynclines—a review, in Dott, R. H., Jr., and Shaver, R. H., eds., Modern and ancient geosynclinal sedimentation: Society of Economic Paleontologists and Mineralogists, Special Publication 19, p. 14-27.

Dott, R. H., Jr., 1983, Episodic sedimentation—how normal is average? How rare is rare? Does it matter? Journal of Sedimentary Petrology, v. 53, p. 5-23.

Dott, R. H., Jr., 1996, Episodic event deposits versus stratigraphic sequences—shall the twain never meet? Sedimentary Geology, v. 104, p. 243-247.

Dott, R. H., Jr., and Shaver, R. H., eds., 1974, Modern and ancient geosynclinal sedimentation: Society of Economic Paleontologists and Mineralogists Special Publication 19, 380 p,.

Doyle, P. and Bennett, M. R., eds., 1998, Unlocking the stratigraphical record: John Wiley and Sons, Chichester, 532 p.

Drake, C. L., Ewing, M., and Sutton, G. H., 1959, Continental margins and geosynclines—the east coast of North America north of Cape Hatteras, in Physics and Chemistry of the Earth: Pergamon Press, London, v. 3, p. 110-198.

Duff, P. M. D., Hallam, A., and Walton, E. K., 1967. Cyclic sedimentation: Developments in Sedimentology, v. 10, Elsevier, Amsterdam, 280 p.

Duff, P. McL. D., and Walton, E. K., 1962, Statistical basis for cyclothems: a quantitative study of the sedimentary succession in the east Pennine Coalfield: Sedimentology, v. 1, p. 235-255.

Dunbar, C. O., and Rodgers, J., 1957. Principles of Stratigraphy: Wiley, New York, 356 p.

Dunham, R. J., 1962, Classification of carbonate rocks according to depositional texture, in Ham, W. E., ed., Classification of carbonate rocks: American Association of Petroleum Geologists Memoir 1, p. 108-121.

Fisher, W. L., and McGowen, J. H., 1967, Depositional systems in the Wilcox Group of Texas and their relationship to occurrence of oil and gas: Transactions of the Gulf Coast Association of Geological Societies, v. 17, p. 105-125.

Fisher, W. L., Brown, L. F., Scott, A. J., and McGowen, J. H., 1969, Delta systems in the exploration for oil and gas: University of Texas Bureau of Economic Geology, 78 p.

Fisher, W. L., McGowen, J. H., Brown, L. F., Jr., and Groat, C. G., 1972, Environmental geologic atlas of the Texas coastal zone — Galveston-Houston area: Bureau of Economic Geology, Texas.

Fisher, W. L., Gama, E., and Ojeda, H. A., 1973, Estratigrafia sismica e sistemas deposicionais da Formaçäo Piaçabuçu, Sociedade Brasileira de Geologia, Anai do XXVII Congresso, v. 3, p. 123-133.

Fisk, H. N., 1944, Geological investigations of the alluvial valley of the lower Mississippi River: U. S. Army Corps of Engineers, Mississippi River Commission, Vicksburg, Mississippi, 78 p.

Folk, R. L., 1962, Spectral subdivision of limestone types, in Ham, W. E., ed., Classification of carbonate rocks: American Association of Petroleum Geologists Memoir 1, p. 62-84.

Forgotson, J. M., 1957, nature, usage and definition of marker-defined vertically segregated rocks units: American Association of Petroleum Geologists Bulletin, v. 41, p. 2108-2113.

Frazier, D. E., 1967; Recent deltaic deposits of the Mississippi River— their development and chronology: Transactions of the Gulf Coast Association of Geological Societies, v. 17, p. 287-315.

Frazier, D. E., 1974, Depositional episodes: their relationship to the Quaternary stratigraphic framework in the northwestern portion of the Gulf Basin: Bureau of Economic Geology, University of Texas, Geological Circular 74-1, 26 p.

Frey, R. W., and Pemberton, S. G., 1984, Trace fossil facies models, in Walker, R. G., ed., Facies models, 2nd edition, Geoscience Canada Reprint Series 1, p. 189-207.

Friedman, G. M., and Sanders, J. E., 1978, Principles of Sedimentology: new York: John Wiley and Sons, 791 p.

Galloway, W. E., 1989, Genetic stratigraphic sequences in basin analysis I: Architecture and genesis of flooding-surface bounded depositional units: American Association of Petroleum Geologists Bulletin, v. 73, p. 125-142.

Gilbert, G. K., 1884, Ripple marks: Science, v. 3, p. 375-376.

Gilbert, G. K., 1899, Ripple marks and cross-bedding: Geological Society of America Bulletin, v. 10, p. 135-140.

Gilbert, G. K., 1914, The transportation of debris by running water: U. S. Geological Survey Professional Paper 86.

Ginsburg, R. N., and Lowenstam, H. A., 1958, The influence of marine bottom communities on the depositional environment of sediments: Journal of Geology, v. 66, p. 310-318.

Gradstein, F. M., Ogg, J. G., and Smith, A. G., eds., 2004a, A geologic time scale: Cambridge University Press, Cambridge, 610 p.

Gradstein, F. M., Ogg, J. G., Schmitz, M. D., and Ogg, G. M., 2012, The Geologic time scale 2012: Elsevier, Amsterdam, 2 vols., 1176 p.

Gressly, A., 1838, Observations geologiques sur le Jura Soleurois: nouv. Mem. Soc. Helv. Sci. Nat. 2: 1-112.

Gurnis, M., 1988, Large-scale mantle convection and the aggregation and dispersal of supercontinents: Nature, v. 332, p. 695-699.

Gurnis, M., 1990, Bounds on global dynamic topography from Phanerozoic flooding of continental platforms: Nature, v. 344, p. 754-756.

Gurnis, M., 1992, Long-term controls on eustatic and epeirogenic motions by mantle convection: GSA Today, v. 2, p. 141-157.

Hancock, J. M., 1977, The historic development of biostratigraphic correlation, in Kauffman, E. G. and Hazel, J. E., eds., Concepts and methods of biostratigraphy: Dowden, Hutchinson and Ross Inc., Stroudsburg, Pennsylvania, p. 3-22.

Harms, J. C., Southard, J. B., Spearing, D. R., and Walker, R. G., 1975, Depositional environments as interpreted from primary sedimentary structures and stratification sequences: Society of Economic Paleontologists and Mineralogists Short Course 2, 161 p.

Hart, B. S., 2013, Whither seismic stratigraphy: Interpretation: A journal of subsurface characterization: Society of Exploration Geophysicists and American Association of Petroleum Geologists, v. 1, p. SA3-SA20.

Hedberg, H. D., 1948, Time stratigraphic classification of sedimentary rocks: Bulletin of the Geological Society of America, v. 59, p. 447-462.

Hunt, D., and Tucker, M. E., 1992, Stranded parasequences and the forced regressive wedge systems tract: deposition during base-level fall: Sedimentary Geology, v. 81, p. 1-9.

Illing, L. V., 1954, Bahaman calcareous sands: American Association of Petroleum Geologists, v. 38, p. 1-95.

Ingersoll, R. V., 1978a, Submarine fan facies of the Upper Cretaceous Great Valley Sequence, northern and central California: Sedimentary Geology, v. 21, p. 205-230.

Ingersoll, R. V., 1978b, Petrofacies and petrologic evolution of the late Cretaceous fore-arc basin, northern and central California: Journal of Geology, v.86, p. 335-352.

Ingersoll, R. V., 1979, Evolution of the Late Cretaceous forearc basin, northern and central California; Geological Society of America Bulletin, v. 90, Pt. 1, p. 813-826.

James, N. P., and Dalrymple, R. W., eds., 2010, Facies Modesl 4: GEOtext 6, Geological Association of Canada, St. John's, Newfoundland, 586 p.

Jervey, M. T., 1988, Quantitative geological modeling of siliciclastic rock sequences and their seismic expression, in Wilgus, C. K., Hastings, B. S., Kendall, C. G. St. C., Posamentier, H. W., Ross, C. A., and Van Wagoner, J. C., eds., Sea level Changes - an integrated approach: Society of Economic Paleontologists and Mineralogists Special Publication 42, p. 47-69.

Jordan, T. E., 1981, Thrust loads and foreland basin evolution, Cretaceous, western United States: American Association of Petroleum Geologists Bulletin, v. 65, p. 2506-2520.

Jordan, T. E., 1995, Retroarc foreland and related basins, in Busby, C. J., and Ingersoll, R. V., eds., Tectonics of sedimentary basins: Blackwell Science, Oxford, p. 331-362.

Jordan, T. E., and Flemings, P. B., 1990, From geodynamic models to basin fill—a stratigraphic perspective, in Cross, T. A., ed., Quantitative dynamic stratigraphy: Prentice-Hall, Englewood Cliffs, p. 149-163.

Jordan, T. E., and Flemings, P. B., 1991, Large-scale stratigraphic architecture, eustatic variation, and unsteady tectonism: a theoretical evaluation: Journal of Geophysical Research, v. 96B, p. 6681-6699.

Kay, M., 1951, North American geosynclines: Geological Society of America Memoir 48.

Kim, W., and Paola, C., 2007, Long-period cyclic sedimentation with constant tectonic forcing in an experimental relay ramp: Geology, v. 35, p. 331-334.

Krumbein, W. C., 1939, Preferred orientation of pebbles in sedimentary deposits: Journal of Geology, v. 47, p. 673-706.

Krumbein, W. C., 1948, Lithofacies maps and regional sedimentary-stratigraphic analysis: American Association of Petroleum Geologists, v. 32, p. 1909-1923.

Krumbein, W. C. and Sloss, L. L., 1951, Stratigraphy and Sedimentation: Freeman, San Francisco, 497 p.

Krumbein, W. C., and Sloss, L. L., 1963, Stratigraphy and sedimentation (2nd. Edition): San Francisco, W. H. Freeman and Co., 660 p.

Kuenen, Ph. H., 1957, Review of Marine Sand-Transporting Mechanisms: Journal of the Alberta Society of Petroleum Geologists, v. 5, No. 4, p. 59-62.

Levorsen, A. I., 1943, Discovery thinking: American Association of Petroleum Geologists Bulletin, v. 27, p. 887-928.

Levorsen, A. I., 1954, Geology of petroleum: San Francisco, W. H. Freeman, 703 p.

Longwell, C. R., ed., 1949, Sedimentary facies in geologic history: Geological Society of America Memoir 39, 171 p.

Lyell, C., 1830-1833, Principles of Geology, 3 vols.: John Murray, London (reprinted by Johnson Reprint Corp., New York, 1969).

Matthews, R. K., 1984a, Oxygen-isotope record of ice-volume history: 100 million years of glacio-isostatic sea-level fluctuation, in Schlee, J. S., ed., Interregional unconformities and hydrocarbon accumulation: American Association of Petroleum Geologists Memoir 36, p. 97-107.

Matthews, R. K., 1984b, Dynamic Stratigraphy, second edition, Englewood Cliffs, New Jersey: Prentice-Hall, 489 p.

McConnell, R. G., 1887, Report of the geological structure of a portion of the Rocky Mountains, accompanied by a section measured near the 51st parallel: Annual Report, 1886, Geological Survey of Canada, p. 1D-41D.

McEachern, J. A., Pemberton, S. G., Gingras, M. K., and Bann, K. L., 2010, Ichnology and facies models, in James, N. P., and Dalrymple, R. W., eds., Facies Models 4: Geotext 6, Geological Association of Canada, p. 19-58.

McKee, E. D., 1957, Flume experiments on the production of stratification and cross-stratification: Journal of Sedimentary Petrology, v. 27, p. 129-134.

McKee, E. D., and Weir, G. W., 1953, Terminology for stratification in sediments: Geological Society of America Bulletin, v. 64, p. 381-389.

McKenzie, D. P., 1978, Some remarks on the development of sedimentary basins: Earth and Planetary Science Letters, v. 40, p. 25-32.

McLaren, D. J., 1970, Presidential address: time, life and boundaries: Journal of Paleontology, v. 44, p. 801-813.

McLaren, D. J., 1978, Dating and correlation: A review, in Cohee, G. V., Glaessner, M. F., and Hedberg, H. D., eds., Contributions to the geologic time scale: American Association of Petroleum Geologists, Studies in Geology 6, p. 1-7.

Merriam, D. F., ed., 1964, Symposium on cyclic sedimentation: Kansas Geological Survey Bulletin 169, 636 p.

Miall, A. D., 1973, Markov chain analysis applied to an ancient alluvial plain succession: Sedimentology, v. 20, p. 347-364.

Miall, A. D., ed., 1980b, Facts and Principles of World Petroleum Occurrence: Canadian Society of Petroleum Geologists Memoir 6, 1003p.

Miall, A. D., 1984, Principles of sedimentary basin analysis: Berlin: Springer-Verlag, 490 p.

Miall, A. D., 1985, Architectural-element analysis: A new method of facies analysis applied to fluvial deposits: Earth Science Reviews, v. 22, p. 261-308.

Miall, A. D., 1988a, Reservoir heterogeneities in fluvial sandstones: lessons from outcrop studies: American Association of Petroleum Geologists Bulletin, v. 72, p. 682-697.

Miall, A. D., 1988b, Facies architecture in clastic sedimentary basins, in Kleinspehn, K., and Paola, C., eds., New perspectives in basin analysis: Springer-Verlag Inc., New York, p. 67-81.

Miall, A. D., 1991b, Hierarchies of architectural units in terrigenous clastic rocks, and their relationship to sedimentation rate, in A. D. Miall and N. Tyler, eds., The three-dimensional facies architecture of terrigenous clastic sediments and its implications for hydrocarbon discovery and recovery: Society of Economic Paleontologists and Mineralogists, Concepts in Sedimentology and Paleontology, v. 3, p. 6-12.

Miall, A. D., 1995a, Whither stratigraphy? Sedimentary Geology, v. 100, p. 5-20.

Miall, A. D., 1997, The geology of stratigraphic sequences, First edition: Springer-Verlag, Berlin, 433 p.

Miall, A. D., 1999, Principles of sedimentary basin analysis, Third edition: New York, N. Y.: Springer-Verlag Inc., 616 p.

Miall, A. D., 2004, Empiricism and model building in stratigraphy: the historical roots of present-day practices. Stratigraphy: American Museum of Natural History, v. 1, p. 3-25.

Miall, A. D., 2010, The geology of stratigraphic sequences, second edition: Springer-Verlag, Berlin, 522 p.

Miall, A. D., 2013, Sophisticated stratigraphy, in Bickford, M. E., ed., The web of geological sciences: Advances, impacts and interactions: Geological Society of America Special Paper 500, p. 169-190.

Miall, A. D., 2014b, The emptiness of the stratigraphic record: A preliminary evaluation of missing time in the Mesaverde Group, Book Cliffs, Utah: Journal of Sedimentary Research, v. 84, p. 457-469.

Middleton, G. V., ed., 1965, Primary sedimentary structures and their hydrodynamic interpretation: Society of Economic Paleontologists and Mineralogists Special Publication 12, 265 p.

Middleton, G. V., 1966, 1967, Experiments on density and turbidity currents, I, II, and III, Canadian Journal of Earth Sciences, v. 3, p. 523-546, p. 627-637, v. 4, p. 475-505.

Middleton, G. V., 1973, "Johannes Walther's law of the correlation of facies": Geological Society of America Bulletin, v. 84, p. 979-988.

Middleton, G. V., 2005, Sedimentology, History, in Middleton, G. V., ed., Encyclopedia of sediments and sedimentary rocks, Berlin: Springer-Verlag, p. 628-635.

Middleton, G. V., and Hampton, M. A., 1976, Subaqueous sediment transport and deposition by sediment gravity flows, in Stanley, D. J., and Swift, D. J. P., eds., Marine sediment transport and environmental management; Wiley, New York, p. 197-218.

Kuenen, Ph. H., and Migliorini, C. I., 1950, Turbidity currents as a cause of graded bedding: Journal of Geology, v. 58, p. 91-127.

Mitchell, A. H., and Reading, H. G., 1969, Continental margins, geosynclines and ocean floor spreading: Journal of Geology, v. 77, p. 629-646.

Mitchum, R. M., Jr., Vail, P. R., and Thompson, S. III, 1977a, Seismic stratigraphy and global changes of sea level, Part 2, The depositional sequence as a basic unit for stratigraphic analysis, in Payton, C. E., ed., Seismic stratigraphy—applications to hydrocarbon exploration: American Association of Petroleum Geologists Memoir 26, p. 53-62.

Moore, R. C., 1949, Meaning of facies, in Longwell, C., ed., Sedimentary facies in geologic history, Geological Society of America Memoir 39, p. 1-34.

Mossop, G. D., and Shetsen, I., compilers, 1994, Geological Atlas of the Western Canada Sedimentary Basin: Canadian Society of Petroleum Geologists, 510 p.

Muto, T., and Steel, R.J., 2004, Autogenic response of fluvial deltas to steady sea-level gall: Implications from flume-tank experiments: Geology, v. 32, p. 401-404.

Nanz, R. H., Jr., 1954, Genesis of Oligocene sandstone reservoir, Seeligson field, Jim Wells and Kleberg Counties, Texas: American Association of Petroleum Geologists Bulletin, v. 38, p. 96-117.

Newell, N. D., and Rigby, J. K., 1957, Geological studies on the Great Bahama Bank: regional aspects of carbonate deposition: Society of Economic Paleontologists and Mineralogists Special Publication 5, p. 15-72.

Oliver, T. A., and Cowper, N. W., 1963, Depositional environments of the Ireton Formation, central Alberta: Bulletin of Canadian petroleum Geology, v. 11, p. 183-202.

Oliver, T. A., and Cowper, N. W., 1965, Depositional environments of the Ireton Formation, central Alberta: Bulletin of the American Association of Petroleum Geologists, v. 49, p. 1410-1425.

Paola, C., 2000, Quantitative models of sedimentary basin filling: Sedimentology, v. 47 (supplement 1) p. 121-178.

Paola, C., Mullin, J., Ellis, C., Mohrig, D. C., Swenson, J. B., Parker, G., Hickson, T., Heller, P. L., Pratson, L., Syvitski, J., Sheets, B., and Strong, N., 2001 Experimental stratigraphy: GSA Today, v. 11, #7, p. 4-9.

Paola, C., Straub, K., Mohrig, D., and Reinhardt, L., 2009, The unreasonable effectiveness of stratigraphic and geomorphic experiments: Earth Science Reviews, v. 97, p. 1-43.

Payton, C. E., ed., 1977, Seismic stratigraphy—applications to hydrocarbon exploration: American Association of Petroleum Geologists Memoir 26, 516 p.

Pettijohn, F. J. 1949, Sedimentary rocks: New York, Harper and Bros, 513 p.

Pettijohn, F. J., 1962, Paleocurrents and paleogeography: American Association of Petroleum Geologists Bulletin, v. 46, p. 1468-1493.

Pettijohn, F. J., and Potter, P. E., 1964, Atlas and glossary of primary sedimentary structures: Springer-Verlag, New York, 370 p.

Playfair, J., 1802, Illustrations of the Huttonian theory of the Earth, Dover Publications, New York 528 p (ed. G.W. White, 1956).

Plint, A. G., Walker, R. G., and Bergman, K. M., 1986, Cardium Formation 6. Stratigraphic framework of the Cardium in subsurface: Bulletin of Canadian Petroleum Geology, v. 34, p. 213-225.

Posamentier, H. W., 2000, Seismic stratigraphy into the next millennium: a focus on 3-D seismic data: American Association of Petroleum Geologists, Annual Conference, New Orleans, p. 16-19.

Posamentier, H. W., and Vail, P. R., 1988, Eustatic controls on clastic deposition II—sequence and systems tract models, in Wilgus, C. K., Hastings, B. S., Kendall, C. G. St. C., Posamentier, H. W., Ross, C. A., and Van Wagoner, J. C., eds., Sea level Changes - an integrated approach: Society of Economic Paleontologists and Mineralogists Special Publication 42, p. 125-154.

Posamentier, H. W., Jervey, M. T., and Vail, P. R., 1988, Eustatic controls on clastic deposition I—Conceptual framework, in Wilgus, C. K., Hastings, B. S., Kendall, C. G. St. C., Posamentier, H. W., Ross, C. A., and Van Wagoner, J. C., eds., Sea level Changes - an integrated approach: Society of Economic Paleontologists and Mineralogists Special Publication 42, p. 109-124.

Potter, P. E., 1959, Facies models conference: Science, v. 129, p. 1292-1294.

Potter, P. E., 1967, Sand bodies and sedimentary environments: A review: American Association of Petroleum Geologists Bulletin, v. 51, p. 337-365.

Potter, P. E., and Pettijohn, F. J., 1963, Paleocurrents and basin analysis: Springer-Verlag, Berlin, 296 p.

Price, R. A., 1973, Large-scale gravitational flow of supracrustal rocks, southern Canadian Rockies, in DeJong, K. A., and Scholten, R. A., eds., Gravity and tectonics: John Wiley, New York, p. 491-502.

Purdy, E. G., 1963, Recent calcium carbonate facies of the Great Bahama Bank: Journal of Geology, v. 71, p. 334-355, p. 472-497.

Ramos, A., and Sopeña, A., 1983, Gravel bars in low-sinuosity streams (Permian and Triassic, central Spain), in Collinson, J. D., and Lewin, J., eds., Modern and ancient fluvial systems: International Association of Sedimentologists Special Publication 6, p. 301-312.

Ramos, A., Sopeña, A., and Perez-Arlucea, M., 1986, Evolution of Buntsandstein fluvial sedimentation in the northwest Iberian Ranges (Central Spain): Journal of Sedimentary Petrology, v. 56, p. 862-875.

Reading, H. G., ed., 1978, Sedimentary environments and facies: Oxford: Blackwell, 557 p.

Reiche, P., 1938, An analysis of cross-lamination: The Coconino Sandstone: Journal of Geology, v. 46, p. 905-932.

Reineck, H. E., and Singh, I. B., 1973, Depositional sedimentary environments—with reference to terrigenous clastics: Berlin: Springer-Verlag, 439 p.

Rich, J. L., 1951, Three critical environments of deposition and criteria for recognition of rocks deposited in each of them: Geological Society of America Bulletin, v. 62, p. 1-20.

Rigby, J. K., and Hamblin, W. K., eds., 1972, recognition of ancient sedimentary environments: Society of Economic Paleontologists and Mineralogists Special Publication 16, 340 p.

Rine, J. M., Helmold, K. P., Bartlett, G. A., Hayes, B. J. R., Smith, D. G., Plint, A. G., Walker, R. G., and Bergman, K. M., 1987, Cardium Formation 6. Stratigraphic framework of the Cardium in subsurface: Discussions and reply: Bulletin of Canadian Petroleum Geology, v. 35, p. 362-374.

Ross, W. C., 1991, Cyclic stratigraphy, sequence stratigraphy, and stratigraphic modeling from 1964 to 1989: twenty-five years of progress?, in Franseen, E. K., Watney, W. L., and Kendall, C. G.. St.C., Sedimentary modeling: computer simulations and methods for improved parameter definition: Kansas Geological Survey Bulletin 233, p. 3-8.

Sadler, P. M., 1981, Sedimentation rates and the completeness of stratigraphic sections: Journal of Geology, v. 89, p. 569-584.

Salvador, A., ed., 1994, International Stratigraphic Guide, Second edition: International Union of Geological Sciences, Trondheim, Norway, and Geological Society of America, Boulder, Colorado, 214 p.

Schenk, H. G., and Muller, S. W., 1941, Stratigraphic terminology: Geological Society of America Bulletin, v. 52, p. 1419-1426.

Schreiber, B. C., Friedman, G. M., Decima, A., and Schreiber, E., 1976, Depositional environments of Upper Miocene (Messinian) evaporite deposits of the Sicilian Basin: Sedimentology, v. 23, p. 729-760.

Seibold, E and Seibold, I., 2002, Sedimentology: from single grains to recent and past environments: some trends in sedimentology in the twentieth century: in Oldroyd, D. R., ed., The Earth Inside and Out: Some major contributions to Geology in the twentieth century: Geological Society, London, Special Publication 192, p. 241-250.

Sheets, B. A., Hickson, T. A., and Paola, C., 2002, Assembling the stratigraphic record: depositional patterns and time-scales in an experimental alluvial basin: Basin Research, v. 14, p. 287-301.

Shepard, F. P., and Wanless, H. R., 1935, Permo-Carboniferous coal series related to southern hemisphere glaciation: Science, v. 81, p. 521-522.

Shepard, F. P., Phleger, F. B., and van Andel, T. H., eds., 1960, Recent sediments, northwest Gulf of Mexico, American Association of Petroleum Geologists, 394 p.

Simons, D. B., and Richardson, E. V., 1961, Forms of bed roughness in alluvial channels: American Society of Civil Engineers Proceedings, v. 87, No. HY3, p. 87-105.

Sloss, L. L., 1963, Sequences in the cratonic interior of North America: Geological Society of America Bulletin, v. 74, p. 93-113.

Sloss, L. L., 1972, Synchrony of Phanerozoic sedimentary-tectonic events of the North American craton and the Russian platform: 24th International Geological Congress, Montreal, Section 6, p. 24-32.

Sloss, L. L., 1988, Tectonic evolution of the craton in Phanerozoic time, in Sloss, L. L., ed., Sedimentary cover—North American Craton: U.S.: The Geology of North America, Boulder, Colorado, Geological Society of America, v. D-2, p. 25-51.

Sloss, L. L., 1991, The tectonic factor in sea level change: a countervailing view: Journal of Geophysical Research, v. 96B, p. 6609-6617.

Sloss, L. L., Krumbein, W. C., and Dapples, E. C., 1949, Integrated facies analysis; in Longwell, C. R., ed., Sedimentary facies in geologic history: Geological Society of America Memoir 39, p. 91-124.

Sorby, H. C., 1852, On the oscillation of the currents drifting the sandstone beds of the southeast of Northumberland, and on their general direction in the coalfield in the neighbourhood of Edinburgh: Proceedings of the West Yorkshire Geological Society, v. 3, p. 232-240.

Sorby, H. C., 1859, On the structures produced by the currents present during the deposition of stratified rocks: The Geologist, v. II, p. 137-147.

Sorby, H. C., 1908, On the application of quantitative methods to the study of the structure and history of rocks. Quarterly Journal of the Geological Society, London, v. 64, p. 171-232.

Steckler, M. S., and Watts, A .B., 1978, Subsidence of the Atlantic-type continental margin off New York: Earth and Planetary Science Letters, v. 41, p. 1-13.

Steel, R. J., and Milliken, K. L., 2013, Major advances in siliciclastic sedimentary geology, 1960-2012, in Bickford, M. E., ed., The web of geological sciences: Advances, impacts and interactions: Geological Society of America Special Paper 500, p. 121-167.

Steno, Nicolaus, 1669, De Solido intra Solidum naturaliter contento dissertationis prodromus, 78 p. Florence.

Stokes, W. L., 1945, Primary lineation in fluvial sandstones: a criterion of current direction: Journal of Geology, v. 45, p. 52-54.

Strangway, D. W., ed., 1980, The continental crust and its mineral deposits: Geological Association of Canada Special Paper 20.

Strong, N., and Paola, C., 2008, Valleys that never were: time surfaces versus stratigraphic surfaces: Journal of Sedimentary Research, v. 78, p. 579-593.

Tankard, A. J., and Welsink, H. J., 1987, Extensional tectonics and stratigraphy of Hibernia oil field, Grand Banks, Newfoundland: American Association of Petroleum Geologists Bulletin, v. 71, p. 1210-1232.

Teichert, C., 1958, Concepts of facies: Bulletin of the American Association of Petroleum Geologists, v. 42, p. 2718-2744.

Torrens, H. S., 2002, Some personal thoughts on stratigraphic precision in the twentieth century, in Oldroyd, D. R., ed., The Earth inside and out: some major contributions to geology in the twentieth century: Geological Society, London, Special Publication 192, p. 251-272.

Vail, P. R., 1975, Eustatic cycles from seismic data for global stratigraphic analysis (abstract): American Association of Petroleum Geologists Bulletin, v. 59, p. 2198-2199.

Vail, P. R., Mitchum, R. M., Jr., Todd, R. G., Widmier, J. M., Thompson, S., III, Sangree, J. B., Bubb, J. N., and Hatlelid, W. G., 1977, Seismic stratigraphy and global changes of sea-level, in Payton, C. E., ed., Seismic stratigraphy - applications to hydrocarbon exploration: American Association of Petroleum Geologists Memoir 26, p. 49-212.

Vail, P. R., Audemard, F., Bowman, S. A., Eisner, P. N., and Perez-Crus, C., 1991, The stratigraphic signatures of tectonics, eustasy and sedimentology—an overview, in Einsele, G., Ricken, W., and Seilacher, A., eds., Cycles and events in stratigraphy: Springer-Verlag, Berlin, p. 617-659.

Van Siclen, D. C., 1958, Depositional topography—examples and theory: American Association of Petroleum Geologists Bulletin, v. 42, p. 1897-1913.

Van Straaten, L.M. J. U., 1954, Composition and structure of recent marine sediments in the Netherlands: Leidse Geol. Mededel., v. 19, p. 1-110.

Van Wagoner, J. C., Mitchum, R. M., Campion, K. M. and Rahmanian, V. D. 1990, Siliciclastic sequence stratigraphy in well logs, cores, and outcrops: American Association of Petroleum Geologists Methods in Exploration Series 7, 55 p.

Veeken, 2007, Seismic stratigraphy, basin analysis and reservoir characterization: Elsevier, Amsterdam, Seismic Exploration, v. 37, 509 p.

Visher, G. S., 1965, Use of vertical profile in environmental reconstruction; American Association of Petroleum Geologists Bulletin: v. 49, p. 41-61.

Waddell, H., 1933, Sedimentation and sedimentology: Science, v. 77, p. 536-537.

Walker, R. G., 1965, The origin and significance of the internal sedimentary structures of turbidites: Proceedings of the Yorkshire Geological Society, v. 35, p. 1-29.

Walker, R. G., 1967, Turbidite sedimentary structures and their relationship to proximal and distal depositional environments: Journal of Sedimentary Petrology, v. 37, p. 25-43.

Walker, R. G., 1973, Mopping up the turbidite mess, in Ginsburg, R. N., ed., Evolving concepts in sedimentology: Johns Hopkins University Press, Baltimore, p. 1-37.

Walker, R. G., 1976, Facies models 1. General Introduction: Geoscience Canada, v. 3, p. 21-24.

Walker, R. G., ed., 1979, Facies models: Geoscience Canada Reprint Series No. 1, 211 p.

Walker, R. G., 1990, Facies modeling and sequence stratigraphy: Journal of Sedimentary Petrology, v. 60, p. 777-786.

Walker, R. G., and James, N. P., ed., 1984, Facies Models, second edition: Geoscience Canada Reprint Series 1, 317 p.

Walker, R. G. and James, N. P., eds., 1992, Facies models: response to sea-level change: Geological Association of Canada, 409 p.

Walther, Johannes, 1893-1894, Einleisung in die Geologie alshistorische Wissemchaft: Jena, Verlag von Gustav Fischer, 3 vols., 1055 p.

Wanless, H. R., and Shepard, E. P., 1936, Sea level and climatic changes related to Late Paleozoic cycles: Geological Society of America Bulletin, v. 47, p. 1177-1206.

Wanless, H. R., and Weller, J. M., 1932, Correlation and extent of Pennsylvanian cyclothems: Geological Society of America Bulletin, v. 43, p. 1003-1016.

Watts, A. B., 1981, The U. S. Atlantic margin: subsidence history, crustal structure and thermal evolution: American Association of Petroleum Geologists, Education Course Notes Series #19, Chap. 2, 75 p.

Watts, A. B., 1989, Lithospheric flexure due to prograding sediment loads: implications for the origin of offlap/onlap patterns in sedimentary basins: Basin Research, v. 2, p. 133-144.

Watts, A. B., and Ryan, W. B. F., 1976, Flexure of the lithosphere and continental margin basins: Tectonophysics, v. 36, p. 24-44.

Weller, J. M., 1960, Stratigraphic Principles and Practice: Harper, New York, 725 p.

Wernicke, B., 1985, Uniform-sense normal simple shear of the continental lithosphere: Canadian Journal of Earth Sciences, v. 22, p. 108-125.

Wheeler, H. E., 1958, Time-stratigraphy: American Association of Petroleum Geologists Bulletin, v. 42, p. 1047-1063.

White, N., and McKenzie, D., 1988, Formation of the "steers' head" geometry of sedimentary basins by differential stretching of the crust and mantle: Geology, v. 16, p. 250-253.

Whitten, E. H. T., 1964, Process-response models in geology: Geological Society of America Bulletin, v. 75, p. 455-464.

Wilgus, C. K., Hastings, B. S., Kendall, C. G. St. C., Posamentier, H. W., Ross, C. A., and Van Wagoner, J. C., eds., 1988, Sea level changes - an integrated approach: Society of Economic Paleontologists and Mineralogists Special Publication 42, 407 p.

Wilson, J. L., 1975, Carbonate facies in geologic history: Springer-Verlag, New York, 471 p.

Woodford, A. O., 1973, Johannes Walther's Law of the Correlation of Facies: Discussion: Geological Society of America Bulletin, v. 84, p. 3737-3740.

## Contents

## 2.1    Introduction

A successful basin analysis requires the collection and integration of several, perhaps many, different kinds of data. Direct observation of the rocks may or may not be fundamental to the study. In the case of a surface geological project, they will be preeminent, though perhaps supplemented by geochemical and geophysical information, plus laboratory analysis of collected samples. For subsurface petroleum studies, actual rock material available for examination may be very limited, consisting of well cuttings from rotary drilling, plus a few short cores. Petrophysical well logs and regional seismic lines may provide at least as important a part of the total data base. Investigations for stratabound ores and minerals typically employ networks of diamond drill holes from which a continuous core normally is available. This provides a wealth of material for analysis, although certain types of observation, such as analysis of sedimentary structures, may be difficult or impossible in such small-diameter cores.

In this chapter we discuss the collection and description of stratigraphic and sedimentologic data from outcrops and drill holes.

## 2.2    Describing Surface Stratigraphic Sections

Vertical stratigraphic sections, whether measured at the surface or derived from subsurface records, constitute the single most important data set that the basin analyst should assemble. Lithostratigraphic or sequence-stratigraphic classification and correlation, and many sedimentological interpretations, depend on the documentation of vertical relationships within and between lithological units.

## 2.2.1  Methods of Measuring and Recording the Data

### 2.2.1.1  Vertical Stratigraphic Sections

The simplest way to record the details of a surface outcrop is by measuring and describing a vertical stratigraphic section. Ideally, the location of the section should be chosen to include important stratigraphic features, such as formation contacts, but, in practice, the location is commonly determined by accessibility, e.g., the presence of bars or beaches allowing us to walk along a river cut, or a negotiable gully cutting through a cliff section. Only those geologists who use their profession as an excuse to practice their favorite sport of mountaineering will be able to apply sound geological principles to the choice of section. The rest of us take what we can reach.

In reconnaissance work, rapid measurement and description techniques are acceptable. For example, a hand-held altimeter (aneroid barometer) may be used in conjunction with dip measurements to reconstruct stratigraphic thicknesses using simple trigonometry. Another method that is commonly described in field handbooks is the pace-and-compass technique, suitable for estimating thicknesses across relatively level ground, given accurate stratigraphic dip. The same distances may be measured from maps or air photographs. Long experience with these methods has shown that they are not very reliable; errors of up to 50 % can be expected.

By far the simplest and most accurate method for measuring a section is the use of a Jacob's staff or "pogo stick." The stick is constructed of a 1.5 m wooden rod, with a clinometer and sighting bar (Fig. 2.1). The clinometer is preset at the measured structural dip and can then be used to measure stratigraphic thickness as fast as the geologist can write down descriptive notes. The best technique is to use two persons. The senior geologist observes the rocks and makes notes, while the junior (who can be an inexperienced student) "pogo's" his or her way up the section recording increments of 1.5 m on a tally counter and collecting samples. The length 1.5 m is convenient for all but the tallest or

shortest persons, although it can be awkward to manipulate on steep slopes. The only skill required by the pogo operator is the ability to visualize the dip of the strata in three dimensions across whatever terrain the geologist may wish to traverse. This is important so that the pogo can at all times be positioned perpendicular to bedding with the line of sight extending from the sighting bar parallel to bedding.

It is far preferable to measure up a stratigraphic section rather than down, even though this often means an arduous climb up steep slopes. Many geologists working by helicopter in rugged terrain have made their traverses physically easier by working downhill wherever possible. But not only is it difficult to manipulate a pogo stick downward in a section, it makes it more difficult for the geologist to comprehend the order of events he or she is observing in the outcrops. It is often convenient to measure a section in separate increments, making use of the most accessible talus slopes or gullies, and traversing laterally along a prominent stratigraphic surface to an adjacent accessible area wherever necessary (Fig. 2.2).

The geologist should search for the cleanest face on which to make observations. Normally, this should be weathered and free of vegetation, talus, or rain wash. Most sedimentary features show up best where they have been etched out by wind or water erosion, or where a face is kept continuously clean and polished by running water, as in a river bed or an intertidal outcrop. Such features rarely show up better on fresh fracture surfaces, so a hammer should only be used for taking samples. Carbonates may benefit from etching with dilute acid. The geologist should methodically examine both vertical cuts and the topside and underside of bedding planes; all may have something to reveal, as described later in this chapter. It is also useful, on larger outcrops, to walk back and examine them from a distance, even from a low-flying helicopter or from a boat offshore, as this may reveal large-scale channels, facies changes, and many other features of interest. A different technique may be used to document such large outcrops, which may conveniently be termed lateral profiling to distinguish it from vertical profiling. We discuss this in the next section.

In the interests of maximum efficiency, the geologist obviously should ensure that all the necessary measurements, observations, and sample collections are made during the first visit to a section. It may be useful to carry along a checklist, for reference each time a lithologic change in the section requires a new bed description. Many geologists have attempted to carry this process one step further by designing the checklist in the form of a computer processable data card, or they record the data in the form of a computer code (e.g., Alexander-Marrack et al. 1970; Friend et al. 1976). There may be two problems with this:

**Fig. 2.1**  Use of pogo stick in section measurement

**Fig. 2.2** Measuring a vertical stratigraphic section in rugged terrain. The section may be offset along a selected stratigraphic surface in order to facilitate ease of access

formation/sequence boundaries

——— segment of measured vertical section

·········· along-strike traverse

1. If attempts are made to record every piece of field information on the computer file, the resulting file is likely to be very large and cumbersome. Storage and retrieval programming may consume far more time than the original field work, unless the geologist can draw on some preexisting program package. This leads to the second problem.

2. If data are to be coded in the field or if a preexisting program system is to be used, it means that decisions will already have been made about what data are to be recorded and how they are to be recorded before the geologist goes into the field. If the geologist knows in advance what is likely to be found, as a result of some previous descriptions or reconnaissance work, this may be satisfactory, but in the case of isolated field areas such foreknowledge may not be available. In this case, there is a certain risk involved in having the observation system designed in advance.

Individual geologists vary in their interests and in the observations they make and, of course, the rocks are highly variable, so that it is not possible to design a single, all-purpose, section-measuring software package. Specialized systems have to be designed for specific projects, and although this leads to expense in programming and debugging, it means that the program can be designed for the specific type of output required. It should not be forgotten, though, that programming, as such, is a technical, not a scientific, skill. Students and other workers who write programs for their research get few points for producing a workable program, only for the interesting scientific ideas their programming allows them to test.

### 2.2.1.2 The Construction of Lateral Profiles

Some stratigraphic units are essentially tabular at the scale of the outcrop, and can be quickly and accurately documented using vertical profiles, in the way outlined in the previous section. However, some types of sedimentary assemblage contain complex facies changes, which may be at a small enough scale to observe in individual outcrops, especially in large outcrops. For example, a reef core, with its reef-front talus slope and back-reef lagoonal deposits, or a large fluvial or submarine-fan channel, with its fill of complex bar deposits, may be spectacularly displayed in a road cut or mountainside. The measurement of a few vertical sections across such an outcrop is a quite inadequate way to document the wealth of facies detail that may be available. Petroleum companies have developed a considerable interest in such large outcrops because of their use as potential analogs of subsurface reservoir units. Internal heterogeneities of these reservoir analogs can be studied and their porosity and permeability characteristics studied, for example by the use of minipermeameters (e.g., Miall and Tyler 1991; Doyle and Sweet 1995).

In order to document the details in a large outcrop it may be necessary to construct a lateral profile, a long section that encompasses the full vertical stratigraphic height of the outcrop, and also extends along strike as far as possible, to illustrate the facies changes. This may be constructed by careful surveying, but a much quicker method is to make use of photographic mosaics of the outcrop. The geologist

**Fig. 2.3** An example of a large outcrop (Upper Cretaceous fluvial Castlegate Sandstone, Utah), with annotation, below. Shown are major bedding surfaces, paleocurrent measurements, and letter codes for major bedding units and bounding surfaces

moves well back from the exposed face, perhaps as much as several hundred meters in the case of a very large outcrop, and carries out a traverse parallel to the face, taking a series of overlapping frames until he or she has covered the entire outcrop. By taking care to remain at the same distance from the face, each frame will be at approximately the same scale. The same end can be achieved by taking photographs from a boat, or even from an aircraft flying low over the outcrop. Unmanned drones are ideal platforms for cameras to take carefully positioned and oriented images.

The images are carefully overlapped in order to construct a mosaic, which can then be used in the field as a kind of topographic map base on which to enter stratigraphic and sedimentologic detail (Fig. 2.3). Some scale distortions inevitably arise. Outcrops are rarely flat, and projections and gullies will not fit together precisely in the mosaic because of the differing perspectives of adjacent frames. Such distortions may be trivial relative to the immense amount of detail that can be shown on such profiles, and if accurate measurements of individual features are required, they should, in any case, be made in the field and not from the photograph.

Modern digital methods, such as the use of Light Detection and Ranging (LIDAR) may make the preparation of the outcrop image for interpretation easier (Hodgetts 2013). This method makes use of laser surveying to record the image, and because the range of each sample point is also recorded, the image can be processed to reduce or eliminate parallax problems, and may then also be rotated to position the image, for example, to remove a structural tilt or to simulate a view parallel to stratigraphic dip.

Techniques for documenting and interpreting lateral profiles are discussed further in Sect. 3.3.4.

## 2.2.2 Types of Field Observation

### 2.2.2.1 Subdivision of the Section into Descriptive Units

This is a subjective operation based on the rock types present, the quality and accessibility of the exposure, and the amount of detail required in the description. Very detailed descriptions may require subdivision into units containing (for example) a single mudstone lens or crossbed set, and will therefore be on the order of a few centimeters or tens of centimeters thick. Thicker units can be defined by grouping similar rock types, but sedimentologically useful detail may be lost thereby. For each unit, the kinds of observations listed in the succeeding paragraphs are made where appropriate.

In the case of lateral profiles (e.g., Fig. 2.3), among the most valuable kinds of observation that can be made is the documentation and classification of the various kinds of bounding surface that separate stratigraphic units. These range from the simple bedding-plane surfaces that separate individual crossbed sets through the surfaces that bound channels and bars to the major (usually horizontal) surfaces that delimit mappable stratigraphic units (formations, members, stratigraphic sequences, etc.). A discussion and classification of bounding surfaces is presented in Sect. 3.5.11.

**Fig. 2.4** Grain-size classification of clastic and carbonate rocks

| PARTICLE DIAMETER | | | CLASTICS | | | CARBONATES | |
|---|---|---|---|---|---|---|---|
| m | mm | Φ | | | | allochems | matrix |
| 10⁰ | 2048 | -11 | v. large | | | | |
| | 1024 | -10 | large | boulders | | v. coarse calcirudite | |
| | 512 | -9 | medium | | | | extremely coarsely crystalline |
| | 256 | -8 | small | | | | |
| 10⁻¹ | 128 | -7 | large | cobbles | | | |
| | 64 | -6 | small | | gravel | | |
| | 32 | -5 | v. coarse | | | coarse calcirudite | |
| 10⁻² | 16 | -4 | coarse | | | | |
| | 8 | -3 | medium | pebbles | | m. calcirudite | |
| | 4 | -2 | fine | | | | |
| | 2 | -1 | v. fine | | | f. calcirudite | v.c.crystalline |
| 10⁻³ | 1 | 0 | v. coarse | | | c. calcarenite | |
| | 0.5 | +1 | coarse | | | m. calcarenite | c. crystalline |
| | 0.25 | +2 | medium | sand | | f. calcarenite | m. crystalline |
| 10⁻⁴ | 0.125 | +3 | fine | | | v. f. calcarenite | |
| | 0.0625 | +4 | v. fine | | | | |
| | 0.0312 | +5 | v. coarse | | | c. calcilutite | f. crystalline |
| | 0.0156 | +6 | coarse | | | m. calcilutite | |
| 10⁻⁵ | 0.0078 | +7 | medium | silt | | f calcilutite | v. f. crystalline |
| | 0.00390 | +8 | fine | | | v. f. calcilutite | |
| | 0.00195 | +9 | v. fine | | | | aphano-crystalline |
| | | | | clay/mud | | | |

micrometres (μm): 500 (+1), 250 (+2), 125 (+3), 62 (+4), 31 (+5), 16 (+6), 8 (+7), 4 (+8), 2 (+9)

**Fig. 2.5** A chart for estimating grain size in the field

## 2.2.2.2 Lithology and Grain Size

Lithologic classification of clastic rocks can usually be done satisfactorily by visual observation in the field, without the necessity of follow-up laboratory work. Classification is based on grain size (Fig. 2.4), which is easily measured on the outcrop. For sand grade rocks, it is useful to take into the field a grain size chart (Fig. 2.5) or a set of sand samples each representing one phi class interval through the sand size ranges. These are used for comparison purposes and permit recognition of the main sand grade subdivisions: very fine,

fine, medium, coarse, and very coarse. Many tests by the author and others have shown that such observations provide adequate, accurate information on the modal size range of the sandstones.

Sorting is also an important descriptive criterion. The description should be modified by appropriate adjectives if sorting characteristics require it; for example, pebbly coarse-grained sandstone, silty mudstone, etc. For the purpose of regional facies analysis, this is usually the only kind of grain size information required. Some examples of the use of sorting

**Fig. 2.6** Sorting and grading in sandstones and conglomerates

adjectives are shown in Fig. 2.6. The distinction between **well-sorted** and **poorly-sorted** refers to the distribution of grain sizes within an individual bed. For example, eolian dune sands (Fig. 2.6a) and many beach sands exhibit a limited range of grains sizes because of current winnowing, whereas many fluvial deposits are rapidly deposited and may consist of mixtures of a range of grain sizes (Fig. 2.6b, d) although this is not invariably the case, as shown by the well-sorted sandstone in Fig. 2.6d. Likewise, debris-flow conglomerates are typically very poorly sorted (Fig. 2.6c, e, f).

Siltstone and mudstone can be distinguished in a hand specimen by the presence or absence of a gritty texture, as felt by the fingers or the tongue. This is, of course, a crude method, and should be checked by making thin sections of selected samples. However, field identifications of this type commonly are adequate for the purpose of facies analysis.

Fine-grained rocks, including those consisting of a mixture of sandstone, siltstone, and mudstone, are difficult to classify and describe. Dean et al. (1985) discussed the methods used in the Deep Sea Drilling Project based on smear slides of soft sediments made on board ship.

For conglomerates, maximum clast size is often a useful parameter to measure. Typically, this is estimated by taking the average of the 10 largest clasts visible within a specified region of an outcrop, such as a given area of a certain bedding plane. In thick conglomerate units, it may be useful to repeat such measurements over regular vertical intervals of the section. It is also important to note the degree of sorting, clast shape and roundness, matrix content, and fabric of conglomerate beds. For example, does the conglomerate consist predominantly of very well-rounded clasts of approximately the same size, or is it composed of angular fragments of varying size and shape (breccia)? Do the clasts "float" in abundant matrix, a rock type termed **matrix-supported conglomerate**, or do the clasts rest on each other with minor amounts of matrix filling the interstices—**clast-supported conglomerate**? These features are discussed at length in Chap. 4.

Carbonate rocks commonly cannot be described adequately or accurately in outcrops, and require description from thin sections or polished sections observed under a low-power microscope. Among the reasons for this are the ready susceptibility of carbonate rocks to fine-scale diagenetic change, and the fact that weathering behavior in many cases obscures rather than amplifies such changes, as seen in outcrops. Another important reason for not relying on outcrop observation is that some of the types of information required for carbonate facies analysis are simply too small to be seen properly with the naked eye. These include mud content, certain sedimentary textures, and biogenic features.

Field geologists traditionally take a dropper bottle of 10 % hydrochloric acid with them to test for carbonate content and to aid in distinguishing limestone from dolomite (on the basis of "fizziness"). However, for research purposes, the test is quite unsatisfactory, and the geologist is advised to abandon the acid bottle (and stop worrying about leakage corroding your field bag). Dolomite commonly can be distinguished from limestone by its yellowish weathering color in the field, but a better field test is to use alizarin red-S in weak acid solution. This reagent stains calcite bright pink but leaves dolomite unstained. In both hand specimens and thin sections use of this reagent can reveal patterns of dolomitization on a microscopic scale.

Because of the problem with carbonate rocks discussed previously the geologist is advised not to rely on field notes for facies analysis of these rocks, but to carry out a rigorous sampling program and supplement (and correct) the field notes using observations made on polished slabs or thin sections. Sampling plans are discussed later in this section. Laboratory techniques for studying carbonates are described by Wilson (1975, Chap. 3).

Evaporites are difficult to study in surface outcrops. They are soft and recessive and commonly poorly exposed, except in arid environments. Like carbonates, they are highly susceptible to diagenetic change, so that field observations must be supplemented by careful laboratory analysis.

Mixed carbonate-clastic sediments are common and are typically dealt with as if they were carbonate or clastic, which may not be the most effective way to emphasize subtle lithologic characteristics. Mount (1985) discussed the problems of classifying these rocks and suggested some methodological approaches.

### 2.2.2.3  Porosity

Porosity and permeability are of particular interest if the rocks are being studied for their petroleum potential. Observations in surface outcrops may be of questionable value because of the effects of surface weathering on texture and composition, but the geologist should always break off a fresh piece of the rock and examine the fracture surface because such observations commonly constitute the only ones made. The geologist should distinguish the various types of porosity, such as intergranular (in detrital rocks), intercrystalline (in chemical rocks), and larger pores, such as vugs, birdseye texture, moulds of allochems, such as oolites or pellets, fossil moulds, fracture porosity, etc. Porosity types should be reported in terms of the estimated percentage they occupy in the bulk volume of the sample.

More accurate observations may be made from thin sections, and samples may be submitted to a commercial laboratory for flow tests if required. Measurements of relative permeability are now routinely made on outcrop profiles of reservoir analogs using minipermeametry equipment.

### 2.2.2.4  Color

Color may or may not be an important parameter in basin analysis. Individual lithologic units may display a very distinctive color, which aids in recognition and mapping. Sometimes it even permits a formation to be mapped almost entirely using helicopter observations from the air, with a minimum of ground checking. However, the sedimentological meaning and interpretation of color may be difficult to resolve.

Some colors are easily interpreted—sandstones and conglomerates commonly take on the combined color of their detrital components, pale grays and white for quartzose sediments, pinks for feldspathic sandstones, and darker colors for lithic rocks. As noted, limestones and dolomites may also be distinguished using color variations. However, color is strongly affected by depositional conditions and diagenesis, particularly the oxidation-reduction balance. Reduced sediments may contain organically derived carbon and $Fe^{2+}$ compounds, such as sulfides, imparting green or drab gray colors. Oxidized sediments may be stained various shades of red, yellow, or brown by the presence of $Fe^{3+}$ compounds such as hematite and limonite. However, local reducing environments, such as those created around decaying organisms, may create localized areas or spots of reduction color. Color can change shortly after deposition, as shown for example by Walker (1967a, b) and Folk (1976). Moberly and Klein (1976) found that oxidation and bacterial action cause permanent color changes when fresh sediments, such as deep sea cores, are exposed to the air. Leaching by groundwaters can also drastically change formation colors.

Thus, the problem is to decide how much time to devote to recording color in the field. Ideally, each descriptive unit in the stratigraphic section should be studied for color using a fresh rock-fracture surface and comparisons to some standard color scheme, such as the U.S. National Research Council Rock-Color Chart (Goddard et al. 1948). In practice, for the purpose of facies and basin analysis, such precision is not required. Simple verbal descriptions, such as pale gray, dark red-brown, etc., are adequate. More precise descriptions may be useful if detailed studies of diagenetic changes are to be undertaken, but research has shown that such studies may give misleading results if carried out exclusively on surface exposures because of the effects of recent weathering (Taylor 1978).

**Fig. 2.7** Bedding and
lamination. See Table 2.1 for key
to abbreviations

### 2.2.2.5 Bedding

An important type of observation, particularly in clastic
rocks, is the thickness of bedding units. Thickness relates to
rate of environmental change and to depositional energy. In
some cases, bed thickness and maximum grain size are
correlated, indicating that both are controlled by the capacity
and competency of single depositional events. Bed-thickness
changes may be an important indicator of cyclic changes in
the environment, and sedimentologists frequently refer to
**thinning-upward** and **fining-upward** or **coarsening-and-
thickening-upward cycles**. It is important to distinguish

bedding from weathering characteristics. For example, a unit
may split into large blocks or slabs upon weathering, but
close examination may reveal faint internal bedding or
lamination not emphasized by weathering. Bedding can be
measured and recorded numerically, or it can be described in
field notes semi-quantitatively (Fig. 2.7) using the descriptive classification given in Table 2.1.

### 2.2.2.6 Inorganic Sedimentary Structures

Sedimentary structures include a wide variety of primary and
post-depositional features (Table 2.2). All individually yield

**Table 2.1** Table of stratification thicknesses

| Descriptor | Range | Figure 2.7 |
|---|---|---|
| Very thickly bedded | >1 m | |
| Thickly bedded | 30–100 cm | THK |
| Medium bedded | 10–30 cm | M |
| Thinly bedded | 3–10 cm | THN |
| Very thinly bedded | 1–3 cm | VTN |
| Thickly laminated | 0.3–1 cm | TKL |
| Thinly laminated | <0.3 cm | TNL |

**Table 2.2** Classification of inorganic sedimentary structures

| I. Hydrodynamic structures |
|---|
| A. By mass gravity transport |
| Graded bedding |
| a. Normal |
| b. Reversed |
| B. By noncohesive flow |
| 1. Lamination |
| 2. Cross-lamination (amplitude <5 cm) |
| 3. Crossbedding (amplitude >5 cm) |
| 4. Clast imbrication |
| 5. Primary current lineation |
| 6. Fossil orientation |
| II. Hydrodynamic erosion of the bed |
| A. Macroscopic |
| 1. Scours |
| 2. Channels |
| 3. Low-relief erosion surfaces |
| B. Mesoscopic |
| 1. Intraformational breccias |
| 2. Hardgrounds |
| 3. Lag concentrates |
| 4. Flutes |
| 5. Tool markings |
| 6. Rain prints |
| III. Liquefaction, load, and fluid loss structures |
| A. Load cast |
| B. Flame structures |
| C. Ball, pillow, or pseudonodule structures |
| D. Convolute bedding |
| E. Syndepositional faults and slumps |
| F. Growth faults |
| G. Deformed crossbedding |
| H. Dish and pillar structures |
| I. Sand volcanoes |
| J. Injection features |
| 1. Dikes |
| 2. Mud lumps |
| 3. Diapirs |
| K. Synaeresis features |
| L. Desiccation cracks |
| M. Ice and evaporite crystal casts |
| N. Gas bubble escape marks |
| O. Teepee structures |
| P. Ptygmatic/enterolithic/chicken wire gypsum/anhydrite |

useful information regarding depositional or diagenetic events in the rocks, and all should be meticulously recorded and described in the context of the lithology and grain size of the bed in which they occur. The assemblage of structures and, in some cases their orientation, can yield vital paleogeographic information.

Inorganic sedimentary structures can be divided into three main genetic classes, as shown in Table 2.2. These are described briefly in order to aid their recognition in the field, but a discussion of their origin and interpretation is deferred to later chapters. Useful texts on this subject include Potter and Pettijohn (1977), Allen (1982), Harms et al. (1982), and Collinson and Thompson (1982). Ashley (1990) carried out an important overview of bedform types and their names, and proposed a unification of concepts and terminology (referred to briefly below) that has received universal acceptance.

Sediment carried in turbulent suspension by mass gravity-transport processes, such as debris flows and turbidity currents, is subjected to internal sorting processes. When the flow slows and ceases, the sorting may be preserved as a distinct texture termed **graded bedding**. Grading commonly consists of an upward decrease in grain size, as illustrated in Fig. 2.6; this is termed **normal grading**. However, certain sedimentary processes result in an upward increase in grain size, termed **inverse grading**.

Clastic grains can be divided into two classes on the basis of their interactive behavior. Cohesive grains are those that are small enough that they tend to be bound by electrostatic forces and thus resist erosion once deposited on a bed. This includes the clay minerals and fine silt particles. A range of erosional sedimentary structures is present in such rocks (Table 2.2), as discussed later. Larger clastic grains, including siliciclastic, evaporite and carbonate fragments, of silt to cobble size, are **noncohesive.** They are moved by flowing water or wind as a traction carpet along the bed, or by intermittent suspension. The dynamics of movement causes the grains to be moulded into a variety of **bedforms,** which are preserved as **crossbedding** within the rock (Figs. 2.8, 2.9).

**Fig. 2.8** Examples of
cross-stratification in outcrop and
drill core. **a** Wave ripples in
shallow-water dolomite; **b** ripples
and climbing ripples in
glaciofluvial outwash; **c** planar
crossbedding in drill core;
**d** trough crossbedding; flow
direction towards the *right*;
**e** low-angle crossbedding; flow
direction towards the *left*

**Fig. 2.9** Examples of types of
crossbedding. **a** Planar
crossbedding in sandstone;
**b** planar crossbedding in modern
river gravel; **c** herringbone
crossbedding (crossbed dip
direction reverses 180° from set
to set); **d** view of the underside of
a bed showing parting lineation
with scour hollows around small
pebbles. Flow direction was from
*right* to *left*

There are three main classes of bedforms and crossbedding found in ancient rocks:

1. Those formed from unidirectional water currents such as are found in rivers and deltas, and oceanic circulation currents in marine shelves and the deep sea.
2. Those formed by oscillatory water-currents, including both wave- and tide-generated features. Although the time scale of current- reversal is, of course, quite different, there are comparable features between the structures generated in these different ways.
3. Those formed by air currents. Such currents may be highly variable, and the structure of the resulting deposits will be correspondingly complex. However, examination of ancient wind-formed (eolian) rocks indicates some consistent and surprisingly simple patterns (Chaps. 4 and 5).

Recognition in an outcrop or in a core of the diagnostic features of these crossbedding classes is an invaluable aid to environmental interpretation (as discussed at length in Chap. 4), and therefore crossbedding structures must be examined and described with great care wherever they are found.

Attributes of crossbedding and examples are illustrated in Figs. 2.8, 2.9, 2.10 and 2.11. A **foreset** represents an avalanche face, down which grains roll or slump or are swept down by air or water currents. Continuous deposition produces repeated foreset bedding or lamination as the bedform accretes laterally, resulting in a crossbed **set** (Fig. 2.10; McKee and Weir 1953). A **coset** is defined as a sedimentary unit made up of two or more sets of strata or crossbedding separated from other cosets by surfaces of erosion, nondeposition, or abrupt change in character (McKee and Weir

1953). Note that a coset can contain more than one type of bedding.

When describing crossbedding, attention must be paid to seven attributes (Fig. 2.11). All but the last of these were first described in detail in an important paper on crossbedding classification by Allen (1963a). In the field, crossbeds are classified first according to whether they are solitary or grouped. Solitary sets are bounded by other types of bedding or crossbedding, grouped sets are cosets consisting entirely of one crossbed type. Scale is the next important attribute. In water-laid strata, it is found that a bedform amplitude of about 5 cm is of hydrodynamic significance (Allen 1982; Ashley 1990) and, accordingly, this amplitude is used to subdivide crossbeds into small- and large-scale forms. An assumption is made that little or none of the top of a bedform is lost to erosion prior to burial; generally, the amount lost seems to increase in approximate proportion to the scale of the bedform or the thickness of the crossbed structure. Forms thinner than 5 cm are termed **ripples**, whereas forms larger than 5 cm have been given a variety of names, reflecting in part a diversity of hydrodynamic causes and in part a considerable terminological confusion. Ashley (1990) proposed several recommendations for a simplification of the terminology that have become widely adopted. Most forms larger than ripples are now termed **dunes**. This will be discussed further in Chap. 4.

Most crossbed sets contain foresets that terminate at the base of the set, in which case the foresets are said to be discordant. In rare cases where the crossbeds are parallel to the lower bounding surface, as occurs in some sets with curved lower surfaces, the crossbeds are described as concordant.

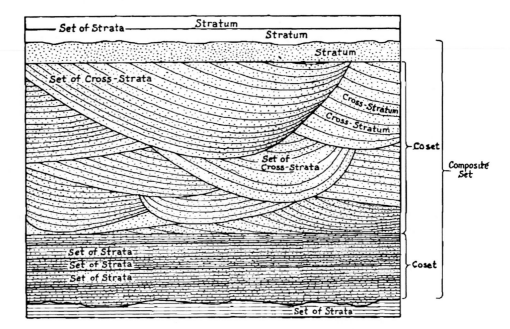

**Fig. 2.10** Terminology for stratified and cross-stratified (crossbedded) units (McKee and Weir 1953)

**Fig. 2.11** Criteria used in the description and definition of crossbedding types (Allen 1963a)

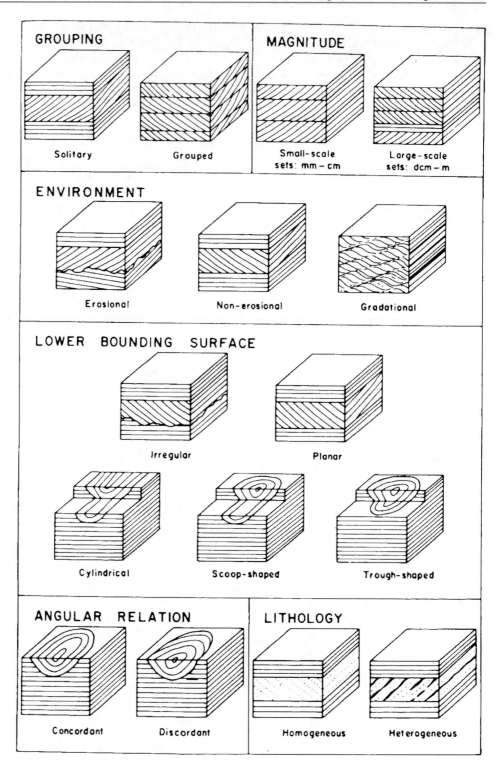

The crossbeds may show either homogeneous or heterogeneous lithology. Homogeneous crossbeds are those composed of foresets whose mean grain size varies by less than two phi classes. Heterogeneous (or heterolithic) crossbeds may contain laminae of widely varying grain size, including interbedded sand and mud or sand and gravel, possibly even including carbonaceous lenses.

The minor internal structures within crossbeds are highly diagnostic of their origin. The dip angle of the foreset relative to the bounding surface is of considerable dynamic significance. Are the foresets curved, linear, or irregular in sections parallel to the dip? Is the direction of dip constant, or are there wide variations or reversals of dip within a set or between sets? Do the sets contain smaller scale hydrodynamic sedimentary structures on the foresets, and if so, what is the dip orientation of their foresets relative to that of the larger structure? What is the small-scale internal geometry of the foresets—are they tabular, lens, or wedge shaped? Do they display other kinds of sedimentary structures, such as trace fossils, synsedimentary faults, or slumps? Are **reactivation surfaces** present? These represent minor erosion surfaces on bedforms that were abandoned by a decrease in flow strength and then reactivated at some later time.

These attributes can be used to classify crossbed sets in the field. It is time-consuming to observe every attribute of every set, but it is usually possible to define a limited range of crossbed types that occur repeatedly within a given stratigraphic unit. These can then be assigned some kind of local unique descriptor, enabling repeated observations to be recorded rapidly in the field notebook. Modern methods of facies classification that encompass variations in crossbed type are discussed in Sect. 3.3.

A vital component of basin analysis is an investigation of sedimentological trends, such as determining the shape and orientation of porous rock units. **Paleocurrent analysis** is one of several techniques for investigating sedimentary trends based, among other attributes, on studying the size, orientation, and relative arrangement of crossbedding structures. Therefore, when describing outcrop sections, it is essential to record the orientation of crossbed sets. The procedures for doing this and the methods of interpretation are described in Sect. 6.7.

Crossbedding represents a macroscopic orientation feature, but each clastic grain is individually affected by a flow system and may take up a specific orientation within a deposit in response to flow dynamics. The longest dimension of elongated particles tends to assume a preferred position parallel or perpendicular to the direction of movement and is commonly inclined upflow, producing an imbricated or shingled fabric. This fabric may be present in sand-sized grains and can be measured optically, in thin section (Martini 1971) or using bulk properties such as dielectric or acoustic anisotropy (Sippel 1971). In recent years, paleomagnetic data have also been recognized to contain much useful information relating to primary sedimentary fabrics. Eyles et al. (1987) discussed magnetic orientation and anisotropy data with reference to the depositional processes of till and till-like diamict deposits. Oriented specimens must be collected in the field for such an analysis (see Sect. 2.2.3.2). In conglomerates, an **imbrication** fabric commonly is visible in an outcrop and can be readily measured by a visual approximation of average orientation or by laborious individual measurements of clasts. Figure 2.12

Force of moving water affects clast differently according to their orientation on the bed: A: clasts pressed into position. Not moved; B: clasts unstable, liable to be flipped over.

**Fig. 2.12** Class imbrication. *Top* imbrication fabric in a modern river. Currents flowed from *left* to *right*. *Bottom* how turbulent currents generate the imbricated fabric

illustrates imbrication in a modern river bed. As discussed in Sect. 6.7.2, it has been found that in nonmarine deposits, in which imbrication is most common, the structure is one of the most accurate of paleocurrent indicators (Rust 1975).

Grain sorting is responsible for generating another type of fabric in sand grade material, which is also an excellent paleocurrent indicator. This is **primary current lineation**, also termed **parting lineation** because it occurs on bedding-plane surfaces of sandstones that are flat bedded and usually readily split along bedding planes. An example is illustrated in Fig. 2.9d. Primary current lineation is the product of a specific style of water turbulence above a bed of cohesionless grains, as are the various bedforms that give rise to crossbedding (Sect. 3.5.4). It therefore has a specific hydrodynamic meaning and is useful in facies analysis as well as paleocurrent analysis.

Rather than the bed itself, objects such as plant fragments, bones, or shells may be oriented on a bedding plane. This should be observed if possible, but interpretation commonly is not easy, as discussed in Sect. 6.7.

### 2.2.2.7 Sedimentary Structures Produced by Hydrodynamic Erosion of the Bed

A wide variety of erosional features is produced by water erosion of newly deposited sediment. These result from changes in water level or water energy in response to floods, storms, tides, or wind-driven waves and currents. They can also result from evolutionary change in a system under steady equilibrium conditions. These processes result in the development of various types of **bounding surfaces** in the rocks (Fig. 2.13). Recognition and plotting of these features in outcrop sections are important components of facies analysis, and with adequate exposure, orientation studies may contribute significantly to the analysis of depositional trends.

These features range in size up to major **river** and **tidal channels**, **submarine canyons**, and **distributary channels** several kilometers across and tens or hundreds of meters deep, but large features such as these can rarely be detected in the average small outcrop. They may be visible in large outcrop sections, where they can be documented using lateral profiles (Sect. 3.5.11) and on seismic sections (Chap. 6), and it may be possible to reconstruct them by careful lithostratigraphic correlation and facies analysis of scattered outcrops, but this is beyond the scope of our immediate discussion. At the outcrop scale, there are two types of small-scale erosional features to discuss, those that truncate one or more bedding units and those that scour or pit the bedding plane without significantly disrupting it.

The first type includes **channels**, **scours**, low-relief **erosion surfaces** and **rill markings**, in decreasing order of scale. These may be classified as macroscopic erosion

**Fig. 2.13** Examples of bounding surfaces exposed in outcrop

**Fig. 2.14** Structures formed during periods of intraformational non-deposition or erosion

features (Table 2.2). Channels and scours are usually filled by sediment that is distinctly different in grain size and bedding characteristics from that into which the channel is cut. Almost invariably the channel fill is coarser than the eroded strata indicating, as might be expected, that the generation of the channel was caused by a local increase in energy level. Exceptions are where a channel was abandoned and subsequently filled by fine sediment and coal under low-energy conditions.

It is a common error to confuse trough crossbedding with channels. Troughs are formed by the migration of trains of dunes or sinuous-crested megaripples (Sect. 3.5.4). They rest on curved scour surfaces, but these are not channels. The scours are formed by vortex erosion in front of the advancing dunes and are filled with sediment almost immediately. Channels, on the other hand, may not be filled with sediment for periods ranging from hours to thousands of years after the erosion surface is cut, and so the cutting and filling of the channel are quite separate events.

Erosion surfaces may exhibit little erosional relief, which may belie their importance. In the nonmarine environment, sheet erosion, wind deflation, and pedimentation can generate virtually planar erosion surfaces. In subaqueous environments, oceanic currents in sediment-starved areas, particularly in abyssal depths, can have the same result. Exposed carbonate terrains may develop **karst** surfaces, with the

formation of extensive cave systems. At the outcrop scale, careful examination of erosion surfaces may reveal a small-scale relief and the presence of features such as infilled **desiccation cracks**, basal intraformational or extrabasinal **lag gravels**, fissures filled with sediment from the overlying bed, zones of bioclastic debris, etc. In some subaerial environments, soil or weathering profiles may have developed, including the development of **caliche** or **calcrete**, and the presence of surfaces of nondeposition. In carbonate environments surfaces of nondeposition commonly develop subaqueously. **Hardgrounds** are organically bored surfaces that may be encrusted with fossils in growth positions (Fig. 2.14). Alternatively, they may be discolored by oxidation, giving a red stain, or blackened by decayed algal matter. Long-continued winnowing of a surface of nondeposition may leave **lag concentrates** or **condensed sections** consisting of larger particles, blackened by algal decay, and possibly including abundant phosphatized fossil material (Wilson 1975, pp. 80–81). Condensed sections may be generated by rapid transgression, which forces coastal sediment sources to undergo retreat (Chap. 5). In continental-slope deposits, giant slumps and slides are common and are particularly well exposed as **intraformational truncation surfaces** in deep water carbonate sediments.

It may be difficult to assess the length of time missing at erosion surfaces. Some may even represent major time

**Fig. 2.15** Sole marks on the underside of a sandstone bed. These are groove casts formed at the base of a turbidity current

breaks detectable by biostratigraphic zonation. In any case, a careful search for and description of such features in the field is an important part of section description. The recognition and mapping of erosion surfaces is a critical element in the study of sequence stratigraphy, as discussed at greater length in Chap. 6. As further discussed in Chap. 7, modern chronostratigraphic documentation of sedimentation rates is indicating that a significant proportion of the elapsed time during which a stratigraphic succession accumulated may be represented by hiatuses and surfaces of non-deposition.

Mesoscopic erosional features fall mainly into a class of structure termed **sole markings**. These are features seen on the underside of bedding planes, usually in sandstones, and they represent the natural casts of erosional features cut into the bed below, which is typically siltstone or mudstone. They attest to the erosive power of the depositional event that formed the sandstone bed, but beyond this, most have little facies or environmental significance. However, they can be invaluable paleocurrent indicators. **Tool markings** are a class of sole structure formed by erosional impact of large objects entrained in the flow, including pebbles, plant fragments, bone, or shell material. The many varieties that have been observed have been assigned names that indicate the interpreted mode of origin. They include **groove, drag, bounce, prod, skip, brush** and **roll markings**. A few examples are illustrated in Fig. 2.15, which show the strongly linear pattern on the bed, providing excellent paleocurrent indicators.

**Flute markings** are formed by vortex erosion, typically at the base of turbidity currents, although they have also been observed at the base of fluvial channels (Fig. 2.16). Erosion is deepest at the up-current end of the scour and decreases down current, so that in a flute cast the high-relief nose of the cast points up current. In rare examples vortex flow lines may be perceived in the walls of the flute. Flutes generally are in the order of a few centimeters deep.

**Fig. 2.16** Flute marks. Diagram at *top* illustrates the formation of flutes by vortex erosion at the base of a high-energy turbid flow. *Middle* view of underside of bed showing flutes and scour around a pebble. *Bottom* flutes at the base of a fluvial channel sandstone

### 2.2.2.8 Liquefaction, Load, and Fluid Loss Structures

Clay deposits saturated with water are characterized by a property termed **thixotropy**: when subjected to a sudden vibration, such as that generated by an earthquake, they tend to liquify and loose all internal strength. This behavior is responsible for generating a variety of structures in clastic rocks. Clay beds commonly are interbedded with sand or silt and, when liquefied, the coarser beds have a higher density than the clay and tend to founder under gravity. This may or may not result in the disruption of the sand units. Where complete disruption does not take place, the sand forms bulbous shapes projecting into the underlying clay, termed **load structures** (Fig. 2.17). These may be well seen in

**Fig. 2.17** Load structures, in the form of balls, pillows, and flame structures of mud trapped between them. Note the deformed lamination. The example (*below*) is from a turbidite succession

ancient rocks by examining the underside of a sandstone bed. They are therefore a class of sole structure, though one produced without current movement. Clay wisps squeezed up between the load masses form pointed shapes termed **flame structures** because of a resemblance to the shape of flames. These are best seen in cross section. Occasionally, loading may take place under a moving current, such as a turbidity flow, and load structures may then be stretched, possibly by shear effects, into linear shapes paralleling the direction of movement.

Commonly, load masses become completely disrupted. The sand bed may break up into a series of ovate or spherical masses that sink into the underlying bed and become surrounded by mud. Lamination in the sand is usually preserved in the form of concave-up folds truncated at the sides or tops of each sand mass, attesting to the fragmentation and sinking of the original layer. Various names have been given to these features, including **ball, pillow,** or **pseudonodule structures** (Fig. 2.17). These structures rarely have a preferred shape or orientation and are not to be confused with slump structures, which are produced primarily by lateral rather than vertical movement.

In many environments, sediment is deposited on a sloping rather than a flat surface (large-scale examples of this are visible on seismic reflection surveys and are called **clinoforms**—see Chaps. 5 and 6), for example, the subaqueous front of a delta, which is something like a very large-scale

foreset built into a standing body of water. The difference is that once deposited such material usually does not move again as individual grains because the angle of the slope is too low. Typically, large-scale submarine fans and deltas exhibit slopes of less than 2°. However, the sediment in such environments is water saturated and has little cohesive strength. Slopes may therefore become oversteepened, and masses of material may be induced to slump and slide downslope by shock-induced failure. Undoubtedly, the thixotropic effects described previously facilitate this process. The result is the production of internal shear or glide surfaces and deformed masses of sediment, termed **slump structures**. Failure surfaces may be preserved as **syndepositional faults**. Some examples of **convolute bedding** may be produced this way, although others are the result of water escape, as discussed below.

Structures produced by failure and lateral movement commonly retain an internal orientation with a simple geometric relationship to the orientation of the depositional slope. This could include the elongation of slump masses and the orientation (strike) of slide surfaces, both parallel to depositional strike, or the asymmetry, even overturning, of folds in convolute beds. Recognition of these geometric properties in an outcrop is important because it helps distinguish the structures from those of different origin, and orientation characteristics obviously have potential as paleoslope indicators.

Very large-scale slumps and syndepositional faults are developed on major deltas. The latter are termed **growth faults**. **Olisthostromes** are giant slumps developed on tectonically active continental slopes.

**Deformed** or **overturned crossbedding** (Fig. 2.18) is developed in saturated sand beds by the shearing action of water or turbid flow across the top of the bedform. The upper few centimeters of the crossbedded unit move down current by a process of intragranular shear, and foreset lamination is overturned as a result, producing an up-current dip. Obviously, to produce this structure the shearing current must have a similar orientation to that of the current that generated the crossbedding. Deformed crossbedding is common in fluvial and deltaic environments. As additional sediment is laid on top of saturated deposits, grains within the substrate begin to settle and pack more tightly. Pore waters are expelled in this process and move upward or laterally to regions of lower hydrostatic pressure. Eventually, they may escape to the surface. This process may take place slowly if sediment is being deposited grain by grain and the fluid movement leaves little or no impression on the sediments.

If loading is rapid, a much more energetic process of fluid loss takes place, and the sediment itself will be moved around in the process. The result, in sand-grade deposits, is a group of features called **dish and pillar structures** (Fig. 2.19c, d, e). Dishes are produced by escaping water

**Undeformed**

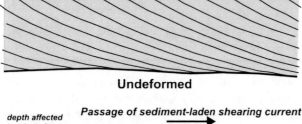

depth affected
by shear force

**Passage of sediment-laden shearing current**

**Deformed**

**Fig. 2.18** The development of deformed or overturned crossbedding

breaking upward through a lamination and turning up the edges; pillars record the vertical path of water flow moving to the surface. Dishes may be up to 50 cm in diameter. These structures are particularly common in the deposits of sediment gravity flows, such as fluidized flows, in which sediment emplacement is rapid. They are produced by water escape as a flow ceases movement, the loss of the lubricating effect of water itself being one of the main reasons why the flow stops. However, dish and pillar structures have also been observed in fluvial and other deposits and are therefore not environmentally diagnostic. Obviously, they can only be seen in deposits containing lamination and will not be present if the sand is uniform in texture.

Fluid movement within a bed is an additional cause of convolute lamination. In this case, the laminations are folded by internal shear and may occasionally be broken through by pipes (Fig. 2.19f). At the sediment-water interface, escaping water may bubble out as a small spring, building up a miniature **sand volcano** (Fig. 2.19g).

The emplacement of relatively more dense material over a lower density layer was discussed previously as the cause of load casts. In addition to the downward movement of the denser material, this situation can be the cause of upward movement of lighter sediment, which is injected, together with contained pore fluids, into the overlying rocks. This can occur on a large scale, producing **diapirs** of evaporite or mud, both of which flow readily under the overburden weight of a few hundred meters of sediment. These diapirs may be several kilometers across and may extend up for several kilometers through overlying deposits. Evaporite diapirs commonly develop on continental margins. Mud diapirs are a characteristic feature of deltas, where coarser deltaic sediment is dumped rapidly on marine mud deposits. Pore fluids may be sealed in the sand beds by this process, with the resulting high overburden pressures leading to overpressuring. Exploration drilling into such deposits must include the use of blow-out preventers to guard against the explosive release of pore fluids and gases when an over-pressured bed is penetrated by the drill.

On a smaller scale, the same injection process can generate **clastic dikes,** consisting of sheets of sandstone or conglomerate (siliciclastic or carbonate) cutting through overlying or underlying beds (Fig. 2.20). The host rocks usually are sharply truncated and not internally deformed, indicating that they were at least partially lithified prior to intrusion. Some dikes intrude along fault planes. Some are intensely folded, suggesting deformation by compaction and further dewatering after injection.

**Desiccation cracks** are readily recognized by even the untrained eye (Fig. 2.21). They are one of the best and most common indicators of subaerial exposure in the rock record (note the spelling of "desiccation", a word that is almost invariably misspelled by students). They may penetrate as deep as a few meters into the underlying rocks (although a few centimeters is more typical) and are normally filled by sediment from the overlying bed. **Teepee structures** are a variety of large desiccation cracks caused by limestone or evaporite expansion on tidal flats. Desiccated beds on tidal flats may peel or curl as they dry, and disrupted fragments commonly are redeposited nearby as an **intraclast breccia** (Fig. 2.14). A subtly different kind of shrinkage feature, termed a **synaeresis structure** may be distinguished from desiccation cracks by two principal differences: unlike desiccation cracks they do not normally form continuous polygonal networks across bedding planes, but may appear as a loose assemblage of small worm-like relief markings on a bedding surface; second, they do not show deep penetration into the substrate, but appear to rest on the bedding plane in which they are found. Synaeresis structures are common in such environments as lakes, lagoons, and tidal pools, where salinity

**Fig. 2.19** **a**, **b** Convolute bedding generated by syndepositional collapse or gravity sliding; **c** mechanism of formation of dish-and-pillar structures; **d**, **e** examples of dish-and-pillar structures; **f** water escape structure formed by sediment loading; **g** sand volcano formed by water escape at the depositional surface

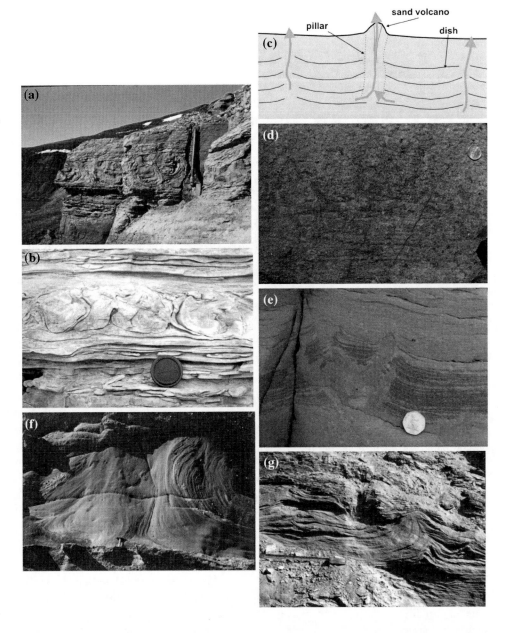

changes may be frequent. As noted in Sect. 3.5.7, these features were thought to be subaqueous shrinkage cracks, but have now been reinterpreted as gypsum pseudomorphs.

Evaporation and freezing may cause the development of crystals of evaporite salts and water, respectively, on the depositional surface, particularly on alluvial floodplains, supratidal flats, and lake margins. Evaporite crystals and nodules may be preserved in the rock record and, of course, major evaporite deposits are common; but individual crystals commonly are replaced by pseudomorphs, or are dissolved and the cavity filled with silt or sand, forming a **crystal cast**. Such structures are a useful indicator of subaerial exposure and desiccation, but do not necessarily imply long-term aridity. Gypsum and halite are the two commonest minerals

to leave such traces. Gypsum forms blade-shaped casts and halite characteristic cubic or "hopper-shaped" structures. Ice casts may be formed in soft sediments during periods of freezing, but have a low preservation potential.

Particularly distinctive evaporite structures on supratidal (sabkha) flats are termed **ptygmatic**, **enterolithic**, and **chicken-wire structures**. These are caused by in-place crystal growth, expansion, and consequent lateral compression of evaporite nodules, possibly aided by slight overburden pressures. Enterolithic structure is so named for a resemblance to intestines; chicken-wire structure is caused by squeezing of carbonate films between the nodules; ptygmatic folds may be caused largely by overburden pressures (Maiklem et al. 1969).

**Fig. 2.20** Clastic dykes

Last, **gas bubble escape structures** should be mentioned. These are produced by carbon dioxide, hydrogen sulfide, or methane escaping from buried, decaying organic matter. Gas passes up through wet, unconsolidated sediment and forms bubbles at the sediment-water interface, leaving small pits on the bedding surface. These structures have been confused with **rain-drop imprints**, but form subaqueously, as may be apparent from associated features preserved in the rocks. True rain-drop imprints probably are very rare.

Mills (1983) provided a discussion and review of this class of structures.

### 2.2.2.9  Fossils

Body fossils are obviously among the most powerful environmental indicators to be found in sedimentary rocks and should be observed and identified with care. Paleontology and paleoecology are specialized subjects, a detailed discussion of which is beyond the scope of this book. However, those engaged in describing outcrop sections for

basin-analysis purposes should be able to make use of such information as they can gather. A complete and thorough paleontological-paleoecological examination of an outcrop section may take several hours, days, or even weeks of work, involving the systematic examination of loose talus and breaking open fresh material or sieving unconsolidated sediment in the search for a complete suite of fossil types. Extensive suites of palynomorphs or microfossils may be extracted by laboratory processing of field samples. Many apparently unfossiliferous or sparsely fossiliferous stratigraphic intervals have been found to contain a rich and varied fauna or flora by work of this kind, but it is the kind of research for which few basin analysts have the time or inclination.

We are concerned in this book with reconstructing depositional environments and paleogeography. Fossils can be preserved in three different ways that yield useful environmental information.

1. In-place life assemblages include invertebrate forms attached to the sea bottom, such as corals, archaeocyathids, rudists, some brachiopods and pelecypods in growth positions, some bryozoa, stromatoporoids (Fig. 2.22a), stromatolites, and trees. In-place preservation usually is easy to recognize by the upright position of the fossil and presence of roots, if originally part of the organism. This type of preservation is the easiest to interpret because it permits the drawing of close analogies with similar modern forms, in the knowledge that the fauna or flora almost certainly is an accurate indicator of the environment in which the rocks now enclosing it were formed.

2. Environmental indicators almost as good as in-place life assemblages are examples of soft-bodied or delicately articulated body fossils preserved intact. These indicate very little transport or agitation after death, and preservation in quiet waters, such as shallow lakes, lagoons, abandoned river channels, or deep oceans. The Cambrian Burgess Shale in the Rocky Mountains near Field, British Columbia, contains one of the most famous examples of such a fossil assemblage, including impressions of soft,

**Fig. 2.21** Desiccation cracks, in bedding plane view (*left*) and cross-section view (*right*)

**Fig. 2.22** Examples of fossils in outcrop. **a** Stromatoporoid "reef", Devonian, Alberta; **b** soft-bodied preservation of arthropod jaws, Burgess Shale, British Columbia; **c** marine reptile bones, Cretaceous, Northeastern British Columbia; **d** trilobites, Burgess Shale, British Columbia; **e** modern shell beach, Florida; **f** pelecypod shell bed, Cretaceous, Arctic Canada

nonskeletal parts of many organisms, such as sponges, that are not found anywhere else (Fig. 2.22b). The Jurassic Solnhofen Limestone of Germany is another good example. These are both examples of what are termed **lagerstätte**, sedimentary units with exceptional, in situ preservation.

The bones of vertebrate animals tend to disarticulate after death, because of the decay of muscle and cartilage and the destructive effects of predators (Fig. 2.22c). Nevertheless, entire skeletons of bony fish, reptiles, and mammals are commonly found in certain rock units, indicating rapid burial under quiet conditions. The Solnhofen Limestone is a well-known example, containing, among other fossils, the entire skeleton and feather impressions of a primitive bird. Such fossil assemblages must, nevertheless, be interpreted with care because a limited amount of transportation is possible from the life environment to the site of eventual burial. Presumably the bird of the Solnhofen Limestone lived in the air, not at the sediment-water interface where it was deposited!

3. Much more common than either of the foregoing are death assemblages of fossils that may have been transported a significant distance, perhaps many kilometers from their life environment. These commonly occur as lag concentrates of shelly debris such as gastropod, pelecypod, brachiopod, or trilobite fragments, and fish bones or scales (Fig. 2.22e). Such concentrations may be abundant enough to be locally rock forming, for example, the famous Silurian Ludlow Bone Bed of the Welsh borderlands. They normally indicate a channel-floor lag concentration or the product of wave winnowing, and can usually be readily recognized by the fact that fragment grain size tends to be relatively uniform.

Transported body fossils may not occur as concentrations but as scattered, individual occurrences, in which case each find must be interpreted with care. Did it live where it is now found or was it transported a significant distance following death? The environmental deduction resulting from such an analysis may be quite different depending on which interpretation is chosen. Evidence of transportation may be obvious and should be sought; for example, broken or abraded fossils may have traveled significant distances. Overturned corals, rolled stromatolites (including oncolites), and uprooted tree trunks, are all obviously transported.

These problems are particularly acute in the case of microfossils and palynomorphs that, on account of their size, are particularly susceptible to being transported long distances from their life environment. For example, modern foraminiferal tests are blown tens of kilometers across the supratidal desert flats of India and Arabia, and shallow water marine forms are commonly carried into the deep sea by sediment gravity flows, such as turbidity currents. Detritus eroded from earlier stratigraphic units may also include derived fossil material. Environmental interpretations based on such fossil types may therefore be very difficult, though it may still be possible if the analysis is carried out in conjunction with the examination of other sedimentary features. In fact, it was the occurrence of sandstones containing shallow-water foraminifera interbedded with mudstones containing deep-water forms in the Cenozoic of the Ventura

Basin, California, that was one of the principal clues leading to the development of the turbidity current theory for the origin of deep water sandstones.

Because of the great variety of life forms preserved as fossils, the subject of paleoecology is a large and complex one. Detailed studies are for the specialist and a complete treatment is beyond the scope of this book. The reader is referred to such texts as Hallam (1973), McKerrow (1978), and Dodd and Stanton (1981). A few examples are discussed in Chap. 4.

### 2.2.2.10  Biogenic Sedimentary Structures

Footprints, burrows, resting, crawling or grazing trails, and escape burrows are examples of what are termed **trace fossils**. They are abundant in some rock units, particularly shallow-marine deposits; all may yield useful environmental information, including water depth, rate of sedimentation, and degree of agitation (Figs. 2.23 and 2.24). The study of trace fossils is termed **ichnology**. Distinctive assemblages of trace fossils Even the nondescript structure **bioturbation**, which is ubiquitous in many shallow marine rocks, can be

**Fig. 2.23** Trace fossils. **a, b.** *Skolithus* (Jurassic) in bedding-plane (**a**) and vertical (**b**) exposure; **c** feeding and crawling trails on a shallow-marine bedding-plane surface (Ordovician); **d, e, f** shallow-marine burrows, all Cretaceous, including *Diplocraterion* (**d**), *Thalassinoides* (**e**) and *Arenicolites* (**f**)

**Fig. 2.24** Example of trace fossils in a drill core, showing how well these features can be seen. They include *Skolithus* (*top*) and *Rosella* (*bottom*)

interpreted usefully by the sedimentologist. Footprints are, of course, best seen on bedding-plane surfaces, as are many types of feeding trails and crawling traces (Fig. 2.25). Burrows are better examined in vertical cross section and are most visible either in very clean, fresh, wetted rock surfaces or in wind- or water-etched weathered outcrops (Fig. 2.23).

Stromatolites are a distinctive component of many carbonate successions, particularly those of Precambrian age, and have been the subject of much detailed study.

Studies of trace-fossils in modern settings and in ancient rocks have led to the recognition of distinctive assemblages, or **ichnofacies**, that are environmentally dependent, reflecting such conditions as the water depths, energy levels and nutrient availability preferred by the organisms that leave the traces. An ecological zonation is shown in Fig. 2.26. Here, for example, can be seen the *Skolithus* assemblage of traces, one which is characteristic of high-energy shoreline environments. Here, a burrowing mode of life permits organisms to live in protected environments while benefiting from the oxygenated and nutrient-rich waters of wave-influenced coastal settings. The *Nereites* assemblage is one composed of organisms that systematically mine the nutrient-poor sediments of deep-water settings. *Glossifungites* organisms like semi-consolidated substrates, and are commonly to be found at major stratigraphic bounding surfaces. And so on.

Figure 2.26 is from McEachern et al. (2010), which is by far the best modern text dealing with the description, classification and interpretation of trace fossils.

### 2.2.3 Sampling Plan

The amount of sampling to be carried out in outcrop section depends on the nature of the problem in hand, as discussed in Sect. 1.4. We are not concerned here with sampling of ore, hydrocarbon source beds, or coal to be analyzed for economic purposes, but the sampling required to perform a satisfactory basin analysis. Sampling is carried out for three basic purposes: (1) to provide a suite of typical lithologic samples illustrating textures, structures, or distinctive fossils on a hand-specimen scale; (2) to provide a set for laboratory analysis of petrography and maturation using polished slabs, the optical microscope, and possibly other tools, such as x-ray diffraction analysis and the scanning electron microscope; and (3) to gather macrofossil and lithologic samples for microfossil or palynological examination to be used for studying biostratigraphy.

#### 2.2.3.1 Illustrative Samples

The choice of such samples is usually simple and can be based on a trade-off between how much it would be useful to take and how much the geologist can physically carry. Large samples showing sedimentary structures or suites of fossils

**Fig. 2.25** Examples of vertebrate footprints. *Left* view of underside of bed, showing a projecting dinosaur footprint; *right* bird footprints

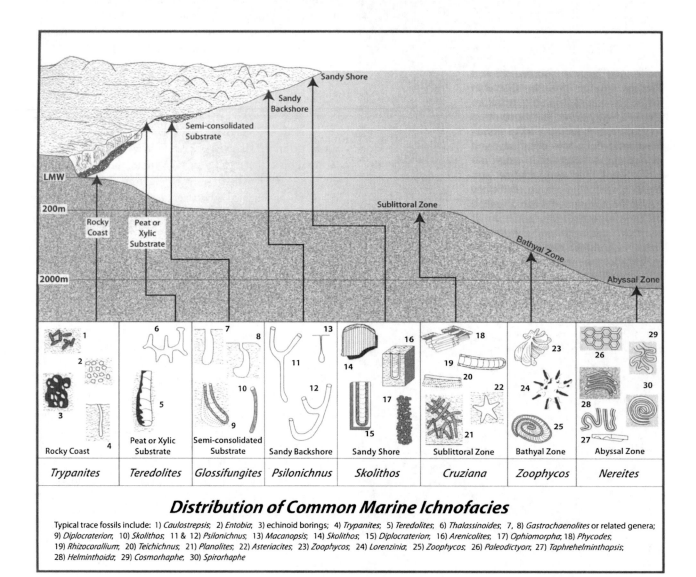

**Fig. 2.26** Ecological zonation of trace-fossil assemblages (McEachern et al. 2010)

may be an invaluable aid in illustrating the geology of the project area to the geologist's supervisor or for practical demonstration at a seminar or a poster display at a conference. Unless field work is done close to a road or is being continuously supported by helicopter, it is rarely possible to collect as much as one would like.

## 2.2.3.2 Petrographic Samples

Before carrying out a detailed sampling program for petrographic work, the geologist should think very carefully about what the samples are intended to demonstrate. Here are some typical research objectives:

1. grain-size analysis of siliciclastic and carbonate clastic rocks, as a descriptive parameter and as an aid to interpreting the depositional environment;
2. petrography of detrital grains, including heavy minerals, as an aid to determining sediment sources;
3. petrography of carbonate grains as an aid to determining the depositional environment;
4. studies of grain interactions, matrix, and cement of both carbonate and clastic grains in order to investigate diagenetic history;
5. studies of detrital grain fabric in order to determine paleocurrent patterns or, in certain cases, as an aid to interpreting the depositional environment;
6. studies of thermal basin-maturity using clay-mineral characteristics, vitrinite reflectance, microfossil color or fluid inclusions; and
7. samples for paleomagnetic study, for use in developing a reversal stratigraphy, a paleopole, or for studying diagenesis.

The geologist must define the scope of the problem before collecting any samples, otherwise it may later be discovered that the collection is unsatisfactory. Work in remote regions is excellent training in such planning exercises, because only rarely is there a chance to return to an outcrop a second time. For the purpose of most regional studies, it is useful to examine petrographic variations vertically through a sequence and areally within a single stratigraphic unit. For example, the composition of detrital grains in a sequence may show progressive vertical changes, recording erosional unroofing of a source area or the switching of source areas. Sampling should be adequate to document this statistically. Samples taken every 10–50 m through a sequence will normally suffice. For diagenetic and fabric studies and for paleomagnetic work, more detailed sampling may be required. Paleomagnetic research normally requires several samples from a single locality in order to permit checks for accuracy.

The most detailed sampling program is required for carbonate sequences, particularly those of shallow marine origin, which show the most facies variation. Laboratory work on polished slabs or thin sections is required for a reliable facies description of most carbonate rocks, and this may call for sampling every meter, or less, through a section.

For certain purposes, it is necessary that the sample be oriented; that is, it should be marked in the field so that its position in space can be reconstructed in the laboratory. For paleomagnetic and fabric studies, this need is obvious, but it may also be useful for petrographic purposes, for example, where it is necessary to examine cavity-filling detrital matrix to determine the time of filling relative to tectonic deformation, as an aid to determining structural top, or for studying microscopic grain-size changes related to bedding (e.g., graded bedding). To collect an oriented sample, the geologist selects a projecting piece of the outcrop that is still in place, not having been moved by frost heave, exfoliation, or other processes, and yet is still removable by hammer and chisel. A flat face on this piece is measured for orientation and marked by felt pen before removal. A more detailed discussion of the methods and purpose of sample orientation is given by Prior et al. (1987).

How much should be collected at each sample station? A few hundred grams is adequate for most purposes. Thin sections can be made from blocks with sides less than 2 cm. Grain mounts of unconsolidated sediment can be made from less than 20 g. Where a particular component is sought, such as disseminated carbonaceous fragments for vitrinite-reflectance measurements, it may be necessary to take a larger sample to ensure that enough fragments are included for a statistical study at each sample station. Samples for paleomagnetic study are collected in a variety of ways, including the use of portable drills, which collect cores about 2 cm in diameter. Alternatively, the geologist may wish to take oriented blocks about 4 × 8 × 15 cm, from which several cores can be drilled in the laboratory. Oriented specimens of unconsolidated sediment are collected by means of small plastic core boxes or tubes pushed into an unweathered face by hand. Textures may be preserved by on-site infiltration with resin (see also Prior et al. 1987).

How do we ensure that samples are truly representative? There is a conflict here between statistically valid experimental design and what is practically possible. Statistical theory requires that we take samples according to some specific plan, such as once every 10 or 30 m (for example) through a vertical section, or by dividing a map area into a square grid and taking one sample from somewhere within each cell of the grid. By these methods we can satisfy the assumptions of statistical theory that our samples are truly representative of the total population of all possible samples. In practice, we can never fully satisfy such assumptions. Parts of any given rock body are eroded or too deeply buried for sampling. Exposures may not be available where sampling design might require them, or a particular interval might be covered by talus. An additional consideration is that the very existence of exposures of a geological unit might be governed by weathering factors related to the parameter the geologist hopes to measure. For example, imagine a sandstone bed formed at the confluence of two river systems, one draining a quartz-rich terrain and one a quartz-poor terrain. The quartzose sandstone intervals may be quartz cemented, much more resistant to erosion, and therefore much more likely to crop out at the surface. Sampling of such a unit might give very biased petrographic results.

In carrying out a basin analysis, we often deal with large areas and considerable thicknesses of strata. Petrographic, textural, and maturity trends are usually strong enough to show through any imperfections in our sampling program. We collect what we can, taking care that our measurements are controlled by the appropriate geological variables, for example, that counts of detrital components are all made on the same grain-size range.

### 2.2.3.3  Biostratigraphic Samples

The study of any fossil group for biostratigraphic purposes is a subject for the appropriate specialists who, ideally, should collect their own material. However, this may not be possible, and the geologist often is required to do the collecting.

Unless a unit is particularly fossiliferous, the collecting of macrofossils can rarely be performed in a fully satisfactory way by the geologist, who is also measuring and describing the section. The search for fossils may take a considerable amount of time, far more than is necessary for the other aspects of the work. In practice, what the geologist usually ends up with are scattered bits and pieces and spot samples of more obviously fossiliferous units, in which it may be fortuitous whether or not any species of biostratigraphic value are present. Given adequate time, for example the two or three field seasons required for dissertation research, the geologist may be able to spend more time on collecting and to familiarize him or herself with the fauna and/or flora, but in reconnaissance mapping exercises, this is usually impracticable.

The increased sophistication of subsurface stratigraphic analysis by the petroleum industry has led to a greatly expanded interest in fossil microorganisms. Groups such as conodonts, acritarchs, foraminifera, palynomorphs, diatoms, and radiolaria have been found to be sensitive stratigraphic indicators and commonly have the inestimable advantage of occurring in large numbers, so that biostratigraphic zonation can be based on statistical studies of taxon distribution. Microfossils are extracted by deflocculation or acid dissolution of suitable host rocks. Most useful fossil forms are pelagic, or are distributed by wind (palynomorphs), so that potentially they may be found in a wide variety of rock types. However, their occurrence is affected by questions of hydrodynamics in the depositional setting and postdepositional preservation. Palynomorphs may be rare in sandstones because they cannot settle out in the turbulent environments in which sand is deposited, but they are abundant in associated silts, mudrocks, and coal. Radiolarians and other siliceous organisms may be entirely absent in mudstones but abundantly preserved in silts, cherts, and volcanic tuffs, because

they are dissolved in the waters of relatively high pH commonly associated with the formation of mud rocks. Conodonts are most commonly preserved in limestones and calcareous mudstones and may be sparse in dolomites, because dolomitization commonly occurs penecontemporaneously in environments inimical to conodonts, such as sabkha flats.

Armed with advice of this kind from the appropriate specialist, the geologist can rapidly collect excellent suites of samples for later biostratigraphic analysis and can cut down on the amount of barren material carried home at great effort and expense, only to be discarded. Advice should be sought on how much material to collect. Normally, a few hundred grams of the appropriate rock type will yield a satisfactory fossil suite, but more may be required for more sparse fossiliferous intervals, for example, several kilograms for conodonts to be extracted from unpromising carbonate units.

Samples should be collected at regular vertical intervals through a section, preferably every 10–50 m. This will vary with rock type, in order to permit more detailed sampling from condensed units or particularly favorable lithologies. Such sample suites permit the biostratigrapher to plot range charts of each taxon and may allow detailed zonation. Scattered or spot samples have to be examined out of context and may not permit very satisfactory age assignments.

## 2.2.4  Plotting the Section

A stratigraphic section can be published in the form of a written description, but this is not a very effective use of the information. It is required for the formal description of type and reference sections of named stratigraphic units, but it is doubtful if much is to be gained by reading such a description. The same information can be conveyed in a much more compact and digestible form as a graphic log, with a central column for lithology and adjacent columns for other features of importance. Such logs have the added advantage that they can be laid side by side, permitting comparisons and correlations between sections from several locations.

It is possible to devise logging techniques with columns and symbols to convey every scrap of lithologic and petrographic information, plus details of fossils, sedimentary structures, paleocurrent measurements, even chemical composition. Many companies and government organizations print blank logging forms of this type for use by their geologists to facilitate the logging process and to standardize the results of different workers. Johnson (1992) developed a comprehensive, all-purpose logging form that he offered as a standard for field workers.

However, such logs have the disadvantage that significant information may be lost in a welter of detail. If a main purpose of a log is to permit visual comparisons between different sections, it is advisable to simplify the logs so that the critical features stand out from the page. For the purpose of basin analysis, the most important data are those that carry paleogeographic information, such as lithology, grain size, sedimentary structures, and fossil content. Paleocurrent data can be added to these, but commonly are treated separately (Chap. 6). As we shall see in Chap. 4, the vertical succession of facies is often of crucial importance in interpreting the depositional environment, and so it is helpful to emphasize this in the logs.

For clastic rocks, one of the most useful techniques is to vary the width of the central lithology column according to grain-size, the wider the column, the coarser the rocks. Examples are illustrated in Fig. 2.27. Many other examples are given by Reading (1986) and James and Dalrymple

**Fig. 2.27** Typical representation of surface stratigraphic sections. Grain-size scale at base of column: *m* mud, *s* silt, *vf* to *vc* very fine to very coarse sand, *c* conglomerate

GSC

(2010). This method imitates the weathering profile of most clastic rocks, as muddy units tend to be less resistant and to form recessive intervals, in contrast to the projecting buttresses and cliffs of coarser sandstones and conglomerates. Drawing the column in this way enables rapid visual comparisons between sections and also permits instant recognition of gradual trends, such as upward fining, and sharp breaks in lithology, as at an erosion surface. Subtleties, such as changes in sorting or a bimodal grain size distribution, cannot be displayed in this way, but are rarely as important from a facies perspective and may, in any case, be accompanied by other kinds of facies change, such as in a sedimentary structure assemblage, which can be readily displayed in visual logs.

The variable column-width technique has not been widely used for carbonate rocks, but there is no reason why it should not be. However, grain size is subject to changes by diagenesis, and this may make interpretation more difficult.

Within the column, various patterns can be used to indicate lithology, including symbols for sedimentary structures and fossils. For siliciclastic sequences consisting of interbedded mudstone, siltstone, sandstone, and conglomerate, the column width conveys most of the necessary lithologic information, and the body of the column can be used primarily for structures and fossils. Some loggers split the column in two, one side for lithology and one for structures, but this is rarely as visually successful. Very little need be placed outside the column, which preserves an uncluttered appearance and increases the graphic impact of the log.

Symbols, abbreviations and other plotting conventions are discussed in Sect. 2.3.4.

At what scale should the logs be drawn? This depends on what it is they are intended to demonstrate. Detailed, local sedimentological logs may require a scale of between 1 cm:1 m and 1 cm:10 m. Regional stratigraphic studies can be illustrated in large foldout diagrams or wall charts at scales in the order of 1 cm:10 m to 1 cm:50 m, whereas page-sized logs of major stratigraphic sequences can be drawn (grossly simplified) at scales as small as 1 cm:1000 m. In the petroleum industry, the scale 1 in.:100 ft has long been a convenient standard for subsurface stratigraphic work. This translates to a convenient approximate metric equivalent of 1 cm:10 m, although it is actually closer to 1 cm:12 m. Metric units should always be used, preferably in multiples of 10.

## 2.3   Describing Subsurface Stratigraphic Sections

### 2.3.1   Methods of Measuring and Recording the Data

Subsurface sections are logged and described using three types of data, well cuttings, core and petrophysical logs (see

Sects. 1.4.3 and 1.4.4). All three are typically available for the large-diameter holes drilled by petroleum exploration companies. Diamond-drill holes (DDH) provide a continuous core but nothing else. The logging techniques and the results obtainable are therefore different.

To reduce costs, corporate practice may make use of outside consultant services to provide wireline-log and core descriptions, or they may limit staff to using photographs of core instead of encouraging them to view the actual core. Key observational detail could be lost as a result, particularly where the investigator is developing an hypothesis that depends on recording critical depositional or diagenetic features.

#### 2.3.1.1   Examination of Well Cuttings

Samples stored in company and government laboratories are of two types, washed and unwashed. Unwashed cuttings consist of samples of all the material that settles out of the mud stream into a settling pit at the drill site. Unconsolidated mudrocks may disperse completely into the mud stream during drilling, in which case little of them will be preserved except as coatings on larger fragments or occasional soft chips. Washed cuttings are those from which all mud has been removed (Fig. 1.4). The washing process makes the cutting examination process easier, but it further biases the distribution of rock types present if the drill penetrated any unconsolidated muddy units. Stratigraphic well logging is normally carried out on the washed cuttings, whereas palynological and micropaleontological sampling is done on the unwashed material. Stratigraphic logging techniques are described later in this chapter. A detailed guide and manual was published by Low (1951) and is well worth reading. Many companies also provide their own manuals. McNeal (1959) has some useful comments, and the American Association of Petroleum Geologists has also issued a logging guide (Swanson 1981).

As described in Sect. 1.4.3, samples are collected at the well site and bagged every 10 ft (3 m). The bag is labeled according to the depth of recovery by the well-site geologist, who makes allowances for the time taken for the mud and cuttings to rise to the surface. Measurements are normally given as "depth below K.B"; K.B. stands for kelly bushing, a convenient measurement location on the drilling platform a few meters above ground level. The altitude above sea level of this point is determined by surveying, so that these drilling depths can be converted to "depths subsea." On offshore rigs K.B. is 25–30 m above sea level.

For various reasons, not all the cuttings in any given bag may be derived from the depth shown. The problems of caving and variable chip density have been referred to in Sect. 1.4.3 (see Fig. 2.28). These problems are particularly acute in soft or unconsolidated rocks, and samples from such a sequence may consist of a heterogeneous mixture bearing

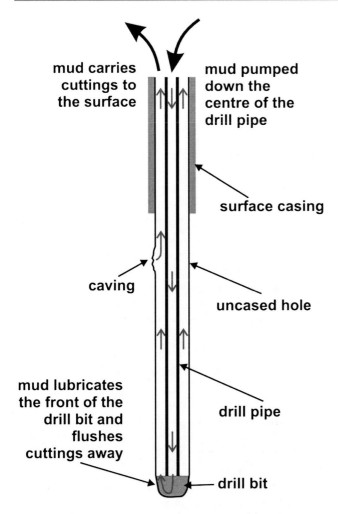

**mud carries cuttings to the surface**

**mud pumped down the centre of the drill pipe**

**surface casing**

**caving**

**uncased hole**

**mud lubricates the front of the drill bit and flushes cuttings away**

**drill pipe**

**drill bit**

**Fig. 2.28** How cavings enter the mud stream in a drill hole

conjunction with the well cuttings. A description of the common petrophysical methods and their uses is given later in this chapter. Once the geologist is familiar with the petrophysical response of the various lithologies in the hole under examination, it is possible to use the logs to adjust or correct the sample description. A lithologic log may therefore be drawn up that bears only a loose relationship to the material actually present in the sample bags. Such a log is an interpretive log, and should be clearly labeled as such. It is likely to be more useful in basin interpretation than a log which simply records the cuttings dogmatically, particularly in the case of poorly consolidated beds or those with rapid vertical lithologic variations. Soft muds will be entirely unrecorded in a straight sample log, which may give the geologist a very inaccurate picture of the subsurface stratigraphy. All this can be allowed for in an interpretive log, but interpretations can be wrong, and the geologist must be aware of it when using this type of record.

Cuttings are observed under a low-power, reflected-light binocular microscope. Immersing the samples in water may aid observation, particularly as dust adhering to the chips can be washed off. The most useful magnification range for such a microscope is from about X5 to X50. The critical petrophysical logs should be unfolded to the appropriate depth interval and placed at one side of the microscope. It is advisable to scan rapidly the samples from several tens of meters of section before beginning the detailed description. Like standing back from an outcrop section, this gives the geologist the opportunity to perceive major lithologic variations. These may be correlated with changes in the petrophysical log response, permitting precise depth control.

As the samples are described, the information may be recorded directly on a preprinted logging form, or written out in note form. If the log is to be an interpretive one, several tens of meters of section should be examined before plotting the graphic log, so as to give the geologist time to digest what is being observed. It may be necessary for publication or other purposes to produce a written sample description, but, as discussed under surface sections, these are difficult for a reader to absorb and are likely to be seldom used. The *Geological Survey of Canada* now publishes them in microfiche form to save paper and space.

Many petroleum and mining exploration companies and service companies (such as the *American Stratigraphic Company, Canadian Stratigraphic Service Ltd.*, and *International Geosystems Corporation*) now use computer processible logging forms. The data are then stored in digital form in data banks. Retrieval programs may be available that can use these data for automated log plotting, and the same data bank can be exploited for automated plotting of maps and sections, as discussed in Chap. 5.

little relation to such stratigraphic detail as thinly interbedded units of contrasting lithology. The loss of the muds from the cutting suite compounds the problem. There are several ways in which these problems can be at least partially resolved. Caved material may be obvious by the large size of the fragments or by its exotic lithology or fossil content. For example, in one well in the Canadian Arctic, I logged Jurassic pelecypods and foraminifera 270 m below the unconformable contact of the Jurassic with the Devonian. The geologist will gradually become familiar with the formations under study, and will then readily recognize such obvious contamination.

A more powerful tool is available to the logger and that is to study the cuttings in conjunction with the suite of petrophysical (wireline) logs from the hole. These logs record various physical properties as a measuring tool is slowly run the length of the hole. Modern petrophysical logging methods are capable of resolving lithologic variations over vertical intervals of a few decimeters or less and can therefore be used, with practice, to interpret lithologies in

### 2.3.1.2  Examination of Core

The large-diameter core produced by petroleum drilling (Fig. 1.5) is stored either in a company laboratory or an official government repository. North American practice is to divide the core into 2.5 ft (0.8 m) lengths, which are stored side by side, two to a box. Top and bottom should be marked on the box. The core usually consists of a series of short pieces, broken from each other by torque during the drilling process. The well-site geologist should number or label each piece with respect to its position in the box and its orientation because, unless this is done, once a piece is moved it may be very difficult to restore it to its correct position, and serious errors may be introduced in reconstructing the vertical lithologic succession. Diamond-drill-hole (DDH) cores are normally stored in 5 ft (1.6 m) lengths, five to a box. The same remarks apply with regard to the position and orientation of core pieces. DDH cores are rarely brought from the field back into the office, except for crucial holes. They are examined and logged in the field, and then are commonly abandoned. They may even be tipped out of the box to prevent competitors from taking a look. This is a great waste of research material, but the practice seems unlikely to change.

Cores are most conveniently examined in a laboratory specifically designed for this purpose (Fig. 2.29). Core boxes are laid on roller tables, so that they can be readily loaded and unloaded using a fork-lift vehicle. A movable platform is positioned above the core boxes, on which are placed the microscope, petrophysical logs, note book, etc.

The grinding action of the core barrel during drilling may smear the core surface and obscure lithologic features; sometimes this can be rectified by washing the core with water or even dilute hydrochloric acid. An even more useful technique is to have the core cut longitudinally with a rock saw, creating a flat section. This should always be wetted with water or etched lightly with dilute acid before examination. The etching technique is particularly useful when examining carbonate rocks, as it tends to generate a fine relief between grains and cement or different carbonate minerals.

Where petrophysical logs are available, it is important to correlate core lithology with log response. This exercise may reveal that parts of the core are missing, perhaps as a result of fragmentation of soft lithologies. Such information is of importance in attempts to reconstruct a detailed vertical lithologic profile.

For a discussion of description and plotting routines, refer to the previous section on well cuttings. Essentially the same methods are used, except that many more features are visible in core, such as bedding features, sedimentary structures and macroscopic trace and body fossils. Orientation of core for geological purposes was discussed by Davison and Haszeldine (1984) and Nelson et al. (1987).

### 2.3.2  Types of Cutting and Core Observation

Large-scale features, including most sedimentary structures and the subtleties of bedding, cannot be identified in well cuttings. They are partly visible in cores, but cores usually provide only frustratingly small snapshots of major features, and core research is rather like trying to describe elephant anatomy by examining a piece of skin with a microscope. The following notes are given in the same format as for surface sections, so that the contrasts with the latter can be emphasized. The description of field observation techniques should be referred to where appropriate.

### 2.3.2.1  Subdivision of the Section into Descriptive Units

This is best carried out by core and sample examination in conjunction with petrophysical logs, in the case of petroleum exploration wells. The combination is a powerful one and yields good generalized stratigraphic subdivisions with precise depth control. For DDH cores the absence of petrophysical logs is compensated by the availability of a continuous core, and stratigraphic subdivision is simple. Because so many features, such as sedimentary structures, cannot be observed in cuttings, descriptive units in the subsurface tend to be thicker and more generalized than those observed in the outcrop. However, for core, the focus on what are really very small samples and the attempt to maximize the use of limited amounts of information tends to lead to very detailed descriptions. Examination of surface sections, sections based on cuttings and those on core, particularly short petroleum cores, require very different

**Fig. 2.29** A core laboratory. Energy Resources Conservation Board, Calgary, Alberta

concepts of scale. These should be borne in mind when a basin-analysis exercise calls for the correlation of all three types of data.

### 2.3.2.2 Lithology and Grain Size

These can be observed satisfactorily in cuttings for all rock types except conglomerates, using the same techniques as described under surface sections. Conglomerates cannot be adequately studied in cuttings where clast size is larger than cutting size. It may not be possible to ascertain which cuttings represent clast fragments and which matrix, and no observations of clast grain size can be made. Remember, also, that unconsolidated silts, muds, and evaporites may not be represented in well cuttings. These may require identification using petrophysical logs.

Well cuttings commonly contain contaminants that the logger should discard or ignore. Many are easy to recognize, such as metal pipe shavings or bit fragments. Oily substances, such as pipe dope or grease, may coat some fragments, but can usually be distinguished from natural oil stains by the fact that they coat the cuttings and do not penetrate them. Drilling mud may also coat the cuttings, particularly poorly washed samples. Casing cement may appear as a flood of cuttings at certain levels, where the hole was reentered following the setting of the casing. Cement can be easily mistaken for sandy, silty, or chalky carbonate. Finally, foreign materials, such as feathers, sacking, seeds, cellophane, perlite, or coarse mica flakes, may be present. These are used in the drilling mud to clog large pores and prevent loss of mud circulation.

Usually there are few restrictions on lithology and grain size determinations in core, except where particularly coarse conglomerates are present.

### 2.3.2.3 Porosity

See the discussion in Sect. 2.2.3. Observations and measurements from subsurface rocks are more reliable than are those from outcrops because of the complications of surface weathering.

### 2.3.2.4 Color

See the discussion under the heading of surface sections (Sect. 2.2.3).

### 2.3.2.5 Bedding

For core, see the discussion under the heading of surface sections (Sect. 2.2.3). Bedding cannot be seen in well cuttings except for fine lamination, whereas core provides good information on bedding variation. Caution should be exercised in attempts to extract any quantitative information about bed thickness from cores, because of the possibility of core loss, as discussed earlier.

### 2.3.2.6 Sedimentary Structures

Very few, if any structures can be observed in well cuttings. However, a wide range of structures is visible in core (e.g., Fig. 2.8c). Large structures, such as major erosion features and giant crossbedding, may be difficult to discern because the small sample of the structure visible in the core may easily be confused with something else. For example, thick crossbed sets could be mistaken for structural dip if the upper or lower termination of the set against a horizontal bedding plane cannot be determined. Dipmeter interpretations and formation microscanner observations may help here, as discussed in Sect. 6.7. Erosion surfaces are practically impossible to interpret in small outcrops or cores. The break may indicate anything from a storm-induced scour surface to an unconformity representing several hundred million years of nondeposition. The presence of soils or regoliths below the erosion surface is about the only reliable indicator of a major sedimentary break (except, of course, where there is structural discordance).

Ripple marks, and crossbed sets up to a few decimeters thick, are more readily recognizable in core, and the reader should turn to the discussion of these in the section on surface exposures for methods of study. The large-scale geometry of crossbed sets is difficult to interpret in core. For example, the difference between the flat shape of a foreset in planar crossbedding and the curved surface of a trough crossbed is practically impossible to detect in core, even the large-diameter petroleum core. Curvature of a typical trough crossbed across such a core amounts to about 2°. Sensitive dipmeter logs have considerable potential for interpreting crossbedding in the subsurface, but the methodology of interpretation has yet to be fully worked out. Dipmeter or formation microscanner logs will eventually permit paleocurrent measurements to be made in the subsurface (Sect. 6.7), something which cannot now be done without an oriented core, and the availability of the latter is practically zero.

Small-scale erosion features such as flutes, tool markings, and rain prints, and other bedding features such as desiccation cracks, and synaeresis markings, are difficult to find in core because bedding plane sections are rare and the geologist is discouraged from creating additional sections by breaking up the core.

Liquefaction, load, and fluid-loss structures are commonly visible in core and, except for the larger features, should be readily interpretable.

### 2.3.2.7 Fossils

Fossil fragments are commonly visible in well cuttings, but are difficult to recognize and interpret. The best solution is to examine them in thin section, when distinctive features of internal structure may be apparent. A program of routine thin-section examination of fossiliferous sequences may be desirable and, particularly if the rocks are carbonates, this can be combined with the lithologic analysis. An excellent textbook by Horowitz and Potter (1971) discusses the petrography of fossils in detail.

Macrofossils can rarely be satisfactorily studied in core, except in the case of highly fossiliferous sections, such as reefs and bioherms. The reason is that the chance sections afforded by core surfaces and longitudinal cuts do not necessarily provide exposures of a representative suite of the forms present, and unlike sparsely fossiliferous surface outcrops, there is no opportunity to break up more rock in a search for additional specimens. Fragments may be studied in thin section, as described previously, but the limitations on the quantity of material still apply.

Both cores and cuttings may be used by biostratigraphic specialists, who extract palynomorphs and microfossils from them by processes of deflocculation or acid dissolution. Much useful ecological information may be obtainable from these suites of fossils. For example, Mesozoic and Cenozoic foraminifera were sensitive to water depth (as are modern forms), and documentation of foram assemblages through a succession can permit a detailed reconstruction of the varying depths of marine depositional environments (Sect. 3.5.8).

### 2.3.2.8 Biogenic Sedimentary Structures

As in the case of most other mesoscopic features, very little can be seen in cuttings, whereas cores commonly contain particularly well-displayed biogenic sedimentary structures. Those confined to bedding planes may not be particularly easy to find, whereas burrows usually are easy to study and can provide invaluable environmental interpretation (Fig. 2.24). Refer to the notes and references in Sect. 2.2.2.10.

### 2.3.3 Sampling Plan

As in the case of surface sections, we are concerned here with sampling for basin analysis purposes in three main categories: illustrative lithologic samples; samples for laboratory petrographic analysis using thin and polished sections, X-ray diffraction, etc.; and samples for biostratigraphic purposes.

Very limited quantities of material are available for any kind of sampling in subsurface sections. The cuttings and core stored in the laboratory are all that will ever be available and once used cannot be replaced. Thus, they should be used with care, and permission should always be sought from the appropriate company or government agency before removing any material for research purposes.

A core may provide excellent illustrative material for demonstrating lithology, sedimentary structures, and facies sequences. The *Canadian Society of Petroleum Geologists* has established a tradition of holding a "core conference" every year in one of the government core laboratories in Calgary (other societies, such as the *Society for Sedimentary Geology* now hold their own workshops). Each contributor to the conference provides a display of a selected core from a producing unit, and uses this as a basis for presenting an interpretation of its geology, with emphasis on depositional environments, diagenesis, and petroleum migration history. The educational value of these conferences is inestimable, and they are always well attended. Some have resulted in the publication of well-illustrated proceedings volumes, for example, Shawa (1974), Lerand (1976), and McIlreath and Harrison (1977). As of Spring 2014 the *Society for Sedimentary Geology (SEPM)* had published 22 "Core Workshop" notes.

The use of well cuttings as illustrative material of this type is clearly limited. However, cuttings may be sampled routinely for petrographic studies, using etching and staining techniques on the raw, unwashed cuttings or preparing polished or thin sections. Fortunately, very small samples are adequate for this kind of work.

Sampling for biostratigraphic purposes should always be carried out on the unwashed rather than the washed cuttings. Depth control is, of course, better for core, but in petroleum wells, there is rarely adequate core for routine biostratigraphic sampling. Very rarely a petroleum exploration well may be drilled by continuous coring for stratigraphic research purposes, and these provide ample sample material free of the problems of sample caving and depth lag. The same is true, of course, for DDH cores. The quantity of material required for biostratigraphic purposes depends on the fossil type under investigation. A few hundred grams is usually adequate for palynological purposes, whereas to extract a representative conodont suite from sparsely fossiliferous carbonate sediments may require several kilograms of material. In the latter case, samples from several depth intervals may have to be combined.

### 2.3.4 Plotting the Section

Well logging is a routine procedure, and most organizations provide standard forms for plotting graphic logs, with a set of standard symbols and abbreviations. That used by the American Stratigraphic Company and Canadian Stratigraphic Services is typical. An example is illustrated in Fig. 2.30. Lithology is shown by color in a column near the center of the log, and accessories, cements, fossil types, and certain structures are shown by symbols. To the left of this column are columns for formation tops, porosity type and

**Fig. 2.30**   Example of part of a log produced by the *American/Canadian Stratigraphic Company*

porosity grade (amount). Formation tops are shown by a formation code. They are interpreted and may be subject to revision. Porosity type, is given in a standard code or symbol and porosity grade by a crude graph based on visual estimates. The depth column may be used for symbols to indicate hydrocarbon shows or stains. To the right of the lithology column is a column for crystal or detrital grain size, based on visual estimates or measurements against a grain-size comparison set (Fig. 2.5). Both to the left and right of the lithology column are spaces for selected petrophysical logs. On the right-hand side of the log there, is a space for typing in an abbreviated description of each lithologic interval. The American/Canadian Stratigraphic Company system uses a list of more than 450 abbreviations, covering almost every conceivable descriptive parameter. Remaining columns are used for grain rounding and sorting and for engineering data.

Much of this detail is not necessary for basin analysis purposes. Figure 2.31 illustrates the more limited range of symbols and codes used by Tassonyi (1969) in a study of the subsurface stratigraphy of the Mackenzie Basin in northern Canada. Figure 2.32 illustrates one of his graphic logs. This style emphasizes lithology and other important stratigraphic variables.

## 2.4   Petrophysical Logs

A wide range of physical parameters can be measured using tools lowered down a petroleum exploration hole. Because of the method of data acquisition these are commonly called by the alternative name **wireline logs**. These tools provide information on lithology, porosity, and oil and water saturation (Table 2.3). In many cases, the measurements are not

LEGEND

*ROCK TYPES. LITHOLOGICAL COLUMN*

Limestone

Dolomite

Limestone and dolomite, interbedded or intergrading

Limestone, dolomite with shale partings and interbeds

Shale

Shale, dark chocolate or black, bituminous

Shale, dark, slightly bituminous

Sandstone

Calcareous sandstone, argillaceous sandstone

Siltstone

Calcareous siltstone, argillaceous siltstone

Sandy shale

Silty shale

Shale, interbedded with sandstone

Shale interbedded with siltstone

Anhydrite (very rarely gypsum)

Salt

Surface deposits, gravels, clay including Pleistocene

Covered interval

No samples available

Unconformity

| | |
|---|---|
| Pyrite, pyritic . . . . . . . . . . . . . . . . . . . . . . . | ■ |
| Glauconite . . . . . . . . . . . . . . . . . . . . . . . | ■' |
| Bentonite . . . . . . . . . . . . . . . . . . . . . . . | ▸ |
| Authigenic quartz, in carbonates, conspicuous under 12X . . . . . . . . . . | ◦ |
| Vein quartz . . . . . . . . . . . . . . . . . . . . . . | ◂ |
| Quartz crystal, vug lining . . . . . . . . . . . . . . | ♦ |
| Clear, large calcite or dolomite crystals . . . . . . . . | . |
| Brecciated or flow texture . . . . . . . . . . . | ▬ |
| Oolitic . . . . . . . . . . . . . . . . . . . . . . | ◇ |
| Pelletoid . . . . . . . . . . . . . . . . . . . . . . | ◊ |
| Obscure pellet, lump, grapestone texture . . . . . . . . | ▱ |
| Algal texture, oogonia . . . . . . . . . . . . . . . | ▲ |
| Bioclastic . . . . . . . . . . . . . . . . . . . . . | ⟩ |
| Red . . . . . . . . . . . . . . . . . . . . . . . . | , |
| Pink . . . . . . . . . . . . . . . . . . . . . . . . | ▸ |
| Variegated . . . . . . . . . . . . . . . . . . . . | ' |
| Green, used exclusively for smooth or waxy shales of pre-Hume strata . . . . . . . . . . | ● |
| Lignite . . . . . . . . . . . . . . . . . . . . . . | L |
| Fossil fragment, indeterminate . . . . . . . . . . | f |
| Microforaminifera, visible under 12X . . . . . . . . | ⋈ |
| Plant fragment . . . . . . . . . . . . . . . . . . | ⋔ |
| Spores (megasporangia) . . . . . . . . . . . . . . | ⊛ |
| Brachiopods . . . . . . . . . . . . . . . . . . . . | ■ |
| Crinoids . . . . . . . . . . . . . . . . . . . . . . | c |
| Gastopods . . . . . . . . . . . . . . . . . . . . . | ɢ |
| Ostracods . . . . . . . . . . . . . . . . . . . . . | o |
| Trilobites . . . . . . . . . . . . . . . . . . . . . | Tᵣ |
| Tentaculites . . . . . . . . . . . . . . . . . . . . | T |
| Styliolina . . . . . . . . . . . . . . . . . . . . . | s |
| Fish scale or fragment . . . . . . . . . . . . . . . | fₛ |
| Inoceramus (prismatic fragments) . . . . . . . . . . | I |

*SYMBOLS USED OUTSIDE LITHOLOGICAL COLUMN*

| | |
|---|---|
| Gas (in test only) . . . . . . . . . . . . . . . . . . | ○ |
| Oil saturation . . . . . . . . . . . . . . . . . . | ● |
| Oil stain . . . . . . . . . . . . . . . . . . . . | ● |
| Formation boundary . . . . . . . . . . . . . | ─── |
| Member boundary . . . . . . . . . . . . . . | ─── |
| Zone boundary, correlation line . . . . . . | - - - - - |

*MECHANICAL LOGS*
Gamma ray
Self-potential        Resistivity

Corad interval

*SYMBOLS USED WITHIN LITHOLOGICAL COLUMN*

| | |
|---|---|
| Calcareous . . . . . . . . . . . . . . . . . . . . . . . | − |
| Dolomitic . . . . . . . . . . . . . . . . . . . . . . | ╱ |
| Argillaceous . . . . . . . . . . . . . . . . . . . . . | − |
| Sandy . . . . . . . . . . . . . . . . . . . . . . . | ... |
| Silty . . . . . . . . . . . . . . . . . . . . . . . . | ... |
| Anhydritic or gypsiferous . . . . . . . . . . . . . . | ⁄ |
| Coralline, stromatoporoids, amphiporoidal, may include rubbly clastics . . . . . . . . | ^ |
| Salt casts . . . . . . . . . . . . . . . . . . . . . | ⊞ |
| Granules or larger grains (used also for 1± mm limestone grains in shales in the Gossage Formation . . . . . . | • |
| Floating sand grains in carbonates . . . . . . . . . | . |
| Ironstone beds or concretions . . . . . . . . . . | ⌒ |
| Chert or chertification, conspicuous . . . . . . . . | ▲ |
| Chert or chertification, minor . . . . . . . . . . | △ |
| Siliceous . . . . . . . . . . . . . . . . . . . . . | ∟ |
| Chert pebbles or granules . . . . . . . . . . . . . . | ∧ |

Approximate Vertical Scale: 1 inch to 250 feet

250          0          250          500

**Fig. 2.31** Codes and symbols used in a regional subsurface study of Paleozoic and Mesozoic stratigraphy in the Mackenzie Valley region, Northwest Territories (Tassonyi 1969)

direct, but require interpretation by analogy or by correlating values between two or more logs run in the same hole. The subject of petrophysics is a highly advanced one and is beyond the scope of this book. Some approaches are described by Pirson (1977); others are given in the various interpretation manuals issued by the logging companies. Some excellent texts that discuss geological interpretations are those by Cant (1992) and Doveton (1994). Whittaker (1998) provided a more recent review of petrophysical logging methods.

In this section, some of the principal log types are described briefly, and some demonstrations of their utility are given. Petrophysical logs are used routinely by stratigraphers and basin analysts to provide information on lithology and to aid facies analysis, as discussed in Sects. 1.

4.3 and 3.5.10. The use of the gamma ray and other log types in stratigraphic correlation is discussed in Sect. 6.2.

## 2.4.1    Gamma Ray Log (GR)

This log measures the natural radioactivity of the formation, and therefore finds economic application in the evaluation of radioactive minerals, such as potash or uranium deposits. In sedimentary rocks, radioactive elements (potassium, thorium) tend to concentrate in clay minerals, and therefore the log provides a measurement of the muddiness of a unit. Texturally and mineralogically mature clastic lithologies, such as quartz arenites and clean carbonate sediments, give a

**Fig. 2.32** A subsurface log drawn to emphasize stratigraphically important details, using the codes and symbols given in Fig. 2.31. Devonian, Mackenzie Valley, Northwest Territories. SP and resistivity logs have been redrawn beside the lithologic column (Tassonyi 1969)

low log response, whereas mudstones and certain special sediment types, such as volcanic ash and granite wash (which has a high feldspar content), give a high log response. Absolute values and quantitative calculations of radioactivity are not necessary for the stratigrapher. The shape of the log trace is a sensitive lithostratigraphic indicator, and the gamma ray log is commonly used in correlation and facies studies. The log has the advantage that gamma radiation penetrates steel, and so the log can be run in cased holes. Low GR readings (deflection to the left) correspond to cleaner, sandy parts of the succession. For clastic rocks, the variable-width column defined by the two log traces for each well provides a graphical portrayal of grain-size variations analogous to the variable-width log-plotting methods recommended for drawing surface sections (Sect. 2.2.4). The gamma ray log is therefore widely used for plotting interpretive sample logs of subsurface clastic sections, and for log-shape studies. An example is illustrated in Fig. 2.33, showing how "upward-fining" and "upward-coarsening" cycles commonly yield a distinctive signature on gamma ray logs (see also Fig. 3.61).

### 2.4.2 Spontaneous Potential Log (SP)

The curve is a recording of the potential difference between a movable electrode in the borehole and the fixed potential of a surface electrode. SP readings mainly record currents of electrochemical origin, in millivolts, and are either positive or negative. There are two separate electrochemical effects, as illustrated in Fig. 2.34. When the movable electrode is opposite a muddy unit, a positive current is recorded as a result of the membrane potential of the mudstone. The latter is permeable to most cations, particularly the $Na^+$ of saline formation waters, as a result of ion exchange processes, but is impermeable to anions. A flow of cations therefore proceeds toward the least saturated fluid, which in most cases is the drilling mud in the hole (log deflection to the right).

Opposite permeable units such as porous limestones and sandstones there is a liquid junction potential that generates a negative potential in the movable electrode. Both anions and cations are free to diffuse into the drilling mud from the more concentrated saline formation waters, but $Cl^-$ ions have the greater mobility, and the net effect is a negative charge (deflection to the left).

The SP log trace, particularly in clastic sequences, is very similar in shape to that of the GR log (examples are given in Figs. 2.34, 2.36, 3.63 and 3.64), and the two may be used alternatively for correlation purposes, as illustrated in Fig. 6.3. The SP curve is normally smoother, and this log type does not offer the same facility for identifying thin beds. SP deflections are small where the saltiness and resistivities of the formation fluids and the drilling mud are similar.

**Table 2.3** The major petrophysical log types and their uses

| Log | Property measured | Unit | Geological uses |
|---|---|---|---|
| Spontaneous potential | Natural electric potential (compared to drilling mud) | Millivolts | Lithology (in some cases), correlation, curve shape analysis, identification of porous zones |
| Resistivity | Resistance to electric current flow | Ohm-metres | Identification of coals, bentonites, fluid evaluation |
| Gamma-ray | Natural radioactivity-related to K, Th, U | API units | Lithology (shaliness), correlation, curve shape analysis |
| Sonic | Velocity of compressional sound wave | Microseconds/metre | Identification of porous zones, coal, tightly cemented zones. |
| Caliper | Size of hole | Centimetres | Evaluate hole conditions and reliability of other logs |
| Neutron | Concentrations of hydrogen (water and hydrocarbons) in pores | Percent porosity | Identification of porous zones, cross-plots with sonic, density logs for empirical separation of lithologies |
| Density | Bulk density (electron density) includes pore fluid in measurement | Kilograms per cubic metre ($gm/cm^3$) | Identification of some lithologies such as anhydrite, halite, non-porous carbonates |
| Dipmeter | Orientation of dipping surfaces by resistivity changes | Degrees (and direction) | Structural analysis, stratigraphic analysis |

## 2.4.3    Resistivity Logs

Most rock types, in a dry state, do not transmit electric currents and are therefore highly resistive. The main exception consists of those rocks with abundant clay minerals. These transmit electricity by ion exchange of the cations in the clay lattice. In the natural state, rocks in the subsurface are saturated with water or hydrocarbons in pore spaces. Formation waters are normally saline and thus act as electrolytes. The resistivity therefore depends on the salinity and the continuity of the formation waters. The latter depends, in turn, on porosity and permeability, so that resistivity is lowest in units such as clean sandstone and vuggy dolomite and highest in impermeable rocks, for example, poorly sorted, dirty, silty sandstones and tight carbonates. Evaporites and coal are also highly resistive. Metallic ores have very low resistivity. Oil is highly resistive and so, under certain conditions, resistivity tools may be used to detect oil-saturated intervals.

A wide variety of resistivity measurement tools has been devised, as listed below. Published logs are commonly identified by the appropriate abbreviation, such as shown in Fig. 6.3.

Electrical survey (ES)
Laterolog[1] (LL)
Induction-electrical survey (IES)
Dual induction laterolog (DIL)
Microlog
Microlaterolog

The conventional electrical survey was, together with the SP log, the only logging tool available for many years, during the early days of petrophysical logging before the war. An electrical current is passed into the formation via an electrode, and this sets up spherical equipotential surfaces centered on the electrode-rock contact. Three additional electrodes are positioned on the tool to intersect these surfaces at set distances from the first electrode. The standard spacing is

short-normal electrode: 40.6 cm (16 in.)
medium-normal electrode: 1.63 m (64 in.)
lateral electrode: 5.69 m (18 ft 8 in.)

The wider the spacing the deeper the penetration of the current into the formation. This permits comparisons between zones close to the hole, permeated by drilling mud, and uninvaded zones further out (Fig. 2.35). The wide spacing of the electrodes also means that the ES tool is not sensitive to thin beds, and so it is not a very satisfactory device for stratigraphic studies.

The laterolog uses an arrangement of several electrodes designed to force an electrical current to flow horizontally out from the borehole as a thin sheet. A monitoring electrode measures a variable current that is automatically adjusted to maintain this pattern as the tool passes through lithologies of variable resistivity. This device is much more responsive to thin beds. Examples, run together with a SP survey and plotted on a logarithmic scale, are shown in Fig. 2.36.

Induction logs were developed for use with oil-based drilling muds which, because they are nonconductive, make the use of electrodes unsatisfactory. A high-frequency alternating current is passed through a transmitting coil, creating a magnetic field that induces a secondary current to flow in the surrounding rocks. This, in turn, creates a

---

[1]These are registered trade names for tools developed by Schlumberger Limited.

**GR count increases**

"sand line"

"shale line"

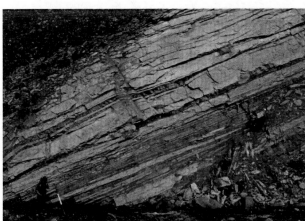

**Fig. 2.33** The gamma ray log. Radioactive traces of uranium, radium, thorium and potassium are concentrated in clay minerals. Gamma-ray readings are therefore a measure of the "clayeyness" of the sediments. The gamma ray log is therefore an excellent recorder of sand-shale variations, as in this "upward-coarsening" or "funnel-shaped" profile, characteristic of a deltaic succession. The outcrop example is a deltaic profile in the Tertiary Eureka Sound Group of the Canadian Arctic Islands

magnetic field that induces a current in a receiver coil. The strength of the induced current is proportional to formation conductivity.

The DIL survey is a combination of two induction devices and a laterolog device, with different formation penetration characteristics. It is normally run with a SP tool. An example is illustrated in Fig. 2.36. Separation between the three resistivity curves occurs opposite permeable units, where the low resistivity of the saline formation waters contrasts with the higher resistivity of the zone close to the borehole, which has been invaded by low-salinity drilling mud (Fig. 2.35). The deep-penetration induction log (ILd) therefore gives the lowest resistivity reading and the shallow penetration laterolog (LL8) the highest. The presence of these permeable zones is confirmed by the SP curve, which has a pattern very similar to that of the induction logs in the example shown.

The Microlog is the most sensitive device for studying lithologic variations in thin bedded sequences. Its primary petrophysical purpose is to measure the resistivity of the invaded zone. The principle is as follows: during the drilling process mud enters permeable beds and hardens on the surface as a mud cake up to about 1 cm thick. No mud cake is formed opposite impermeable units. The microlog tool consists of two closely spaced electrodes. Opposite the mud cake, they record the low resistivity of the mud itself, whereas opposite impermeable units, they record the generally higher resistivities of uninvaded rocks. The presence of mud cake is confirmed by a caliper log, which is sensitive to the slight reduction in hole diameter when a mud cake is present. Muddy units commonly cave and give a very erratic caliper log. The microlog readings are also likely to be erratic because of the poor electrode contact. All these responses are illustrated in Fig. 2.37. The Microlaterolog is a more sensitive version of the Laterolog. Use of both the microresistivity devices permits accurate determinations of permeable sandstone and carbonate thickness, of considerable use in regional subsurface facies studies (Chaps. 5 and 6).

### 2.4.4 Sonic Log

The sonic tool consists of a set of transmitters for emitting sound pulses and a set of receivers. The fastest path for sound waves to travel between transmitters and receivers is along the surface of the hole, in the rock itself rather than through the mud or the actual tool. The time of first arrival of the sound pulses is therefore a measure of formation density, which depends on lithology and porosity. The sonic log is normally run with a GR tool.

The sonic tool has two main uses, the estimation of porosity where lithology is known, and the calibration of regional seismic data (Sect. 6.3.1). For the latter purpose,

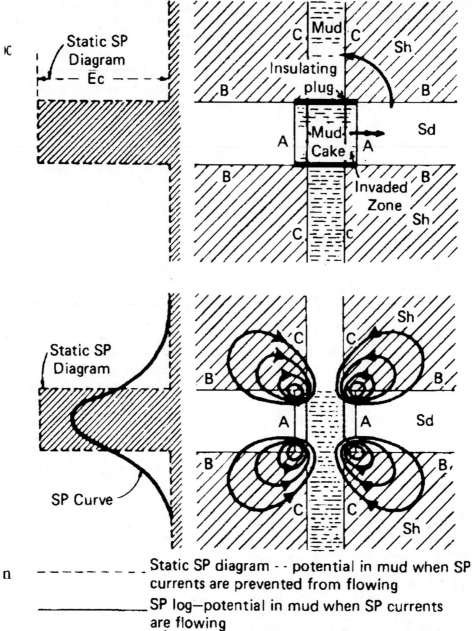

**Fig. 2.34**   The principal of the SP log

a computer in the recording truck at the wellhead integrates the travel time over each increment of depth as the survey is run. Every time this amounts to 1 ms a small pip appears on a track down the center of the log.

scattering is dependent on the density of the electrons in the formation with which the gamma rays collide. This, in turn, depends on rock density, porosity, and composition of the formation fluids.

## 2.4.5   Formation Density Log

The tool contains a radioactive source emitting gamma rays. These penetrate the formation and collide with it, in a process known as Compton scattering. The deflected gamma rays are recorded at a detector on the tool. The rate of

## 2.4.6   Neutron Log

For this log, a radioactive source emitting neutrons is used. These collide with the nuclei of the formation material, with a consequent loss of energy. The greatest loss of energy occurs when the neutron collides with a hydrogen nucleus,

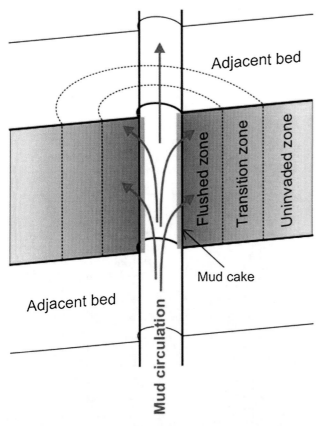

**Fig. 2.35** Porous beds adjacent to a drill hole are flushed and filled with the drilling mud. This also leaves a "mud cake" adhering to the wall of the hole

and so the total loss of energy depends mainly on the amount of hydrogen present, either in formation waters, hydrocarbons, or bound water in clay minerals, gypsum, etc. The detector measures either the amount of scattered low-energy neutrons or the gamma rays emitted when these neutrons are captured by other nuclei.

### 2.4.7 Crossplots

Where the formation is known to consist of only two or three rock types, such as sandstone-siltstone-mudstone or limestone-dolomite, combinations of two or more logs can be used to determine lithology, porosity, and hydrocarbon content. These crossplots are therefore of considerable stratigraphic use where only generalized lithologic information is available from well cuttings.

For clean, nonmuddy formations, combinations of the sonic, formation density, and neutron logs are the most useful. For example, Figs. 2.38 shows the neutron-density combination. These logs are commonly calibrated in terms of apparent limestone porosity; that is, if the rock is indeed limestone, its porosity has the value indicated. For other lithologies, the porosity estimate will be in error, but by reading values for both logs, it is possible to determine both lithology and correct porosity. The curves in Fig. 2.38 give ranges of actual porosity readings for four principal rock types. To show how this graph can be used, compare it to the neutron-density overlay given in Fig. 2.38. Such overlays may be provided on a routine basis by the logging company or can be redrawn on request. The thick sandstone interval at the top can be recognized by the distinctly higher readings on the density curve. Values range from about 4 to 12, whereas those on the neutron curve are mostly close to zero. Examination of Fig. 2.38 shows that only sandstone can give this combination. The thick limestone at the bottom of the overlay in Fig. 2.38 is indicated by the near coincidence of the two curves. Dolomite or anhydrite would be suggested by relatively higher neutron readings. Mudstones can commonly be recognized by very high neutron readings relative to density, because of the abundant water in the clay mineral lattice. This would be confirmed by the gamma ray response or the SP or caliper log, if available.

**Fig. 2.36** Examples of SP and resistivity logs. Impermeable beds, such as shales, are indicated where the Laterolog curves combine. This is confirmed by the deflection of the SP curve towards the shale line. Permeable beds are indicated where the Laterolog curves diverge, indicating the different resistivity response between the invaded and uninvaded zone

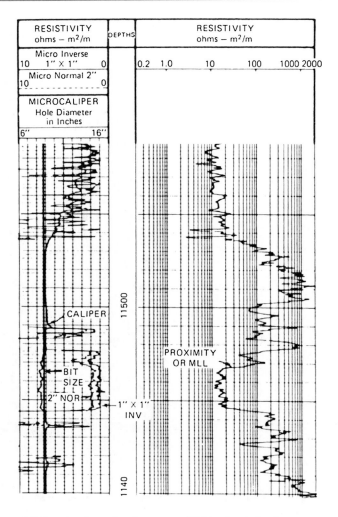

Fig. 2.37 Example of a Microlog and Caliper log (reproduced courtesy of Schlumberger Inc.)

**Fig. 2.38** *Left* a neutron (ON): density (OD) crossplot, showing how readings of the two logs can be used to determine lithology and porosity (reproduced courtesy of Schlumberger Inc.). *Right* example of a neutron: density overlay, illustrating how curve separation and deflections can be used to determine lithology (reproduced courtesy of Schlumberger Inc.)

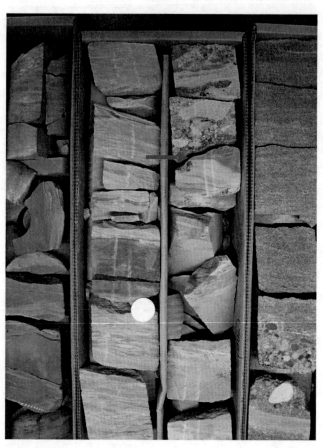

**Fig. 2.39** Integrating core and log data is a very important and useful step in the documentation of subsurface stratigraphy and sedimentology. This example is form the Alberta Basin. The cored interval in the well is shown by the *black bar* down the center of the wireline log. The contact between Fernie shales and siltstones and Cadomin conglomerates and sandstones is indicated on the core by a *red arrow*

## 2.4.8    Integrating Cores and Wireline Logs

The combination of core and petrophysical log data is a powerful one. An essential step is to locate the core on the petrophysical strip log by referring to the core depth information on the core box, as shown in Fig. 2.39. In this example, an important regional bounding surface and disconformity that can be traced for hundreds of kilometres shows up as a sharp discontinuity in the log and can be identified by a sharp lithologic break in the core (red arrow). Typically, regional cross-sections constructed from petrophysical logs (e.g., Figs. 5.4 and 5.5) are used to document the regional stratigraphic variability, while the core is used to highlight local facies characteristics, particularly vertical facies changes, contact relationships and sedimentary structures, including crossbedding and trace fossils.

## References

Alexander-Marrack, P. D., Friend, P. F., and Yeats, A. K., 1970, Mark sensing for recording and analysis of sedimentological data, in Cutbill, J. L., ed., Data processing in biology and geology: Systematics Association Special Volume 3, Academic Press, London, p. 1-16.

Allen, J. R. L., 1963a, The classification of cross-stratified units, with notes on their origin: Sedimentology, v. 2, p. 93-114.

Allen, J. R. L., 1982, Sedimentary structures: their character and physical basis: Elsevier Scientific Publications, Developments in Sedimentology 30, 663 p.

Ashley, G. M., 1990, Classification of large-scale subaqueous bedforms: a new look at an old problem: Journal of Sedimentary Petrology, v. 60, p. 160-172.

Cant, D. J., 1992, Subsurface facies analysis, in Walker, R. G., and James, N. P., eds., Facies models: response to sea level change: Geological Association of Canada, St. John's, Newfoundland, p. 27-45.

Collinson, J. D., and Thompson, D. B., 1982, Sedimentary structures: George Allen and Unwin, London, 194 p.

Davison, I., and Haszeldine, R. S., 1984, Orienting conventional cores for geological purposes: a review of methods: Journal of Petroleum Geology, v. 7, p. 461-466.

Dean, W. E., Leinen, M., and Stow, D. A. V., 1985, Classification of deep-sea, fine-grained sediments: Journal of Sedimentary Petrology, v. 55, p. 250-256.

Dodd, J. R., and Stanton, R. J., Jr., 1981, Paleoecology, concepts and applications: Wiley. New York.

Doveton, J. H., 1994, Geologic log interpretation: Society for Sedimentary Geology (SEPM) Short Course Notes, #29, 184 p.

Doyle, J. D., and Sweet, M. L., 1995, Three-dimensional distribution of lithofacies, bounding surfaces, porosity, and permeability in a fluvial sandstone: Gypsy Sandstone of northern Oklahoma: American Association of Petroleum Geologists Bulletin, v. 79, p. 70-96.

Eyles, N., Day, T. E., and Gavican, A., 1987, Depositional controls on the magnetic characteristics of lodgement tills and other glacial diamict facies: Canadian Journal of Earth Sciences, v. 24, p. 2436-2458.

Folk, R. L., 1976, Reddening of desert sands: Simpson Desert, N.T, Australia: Journal of Sedimentary Petrology, v. 46, p. 604-615.

Friend, P. F., Alexander-Marrack, P. D., Nicholson, J., and Yeats, A. K., 1976, Devonian sediments of east Greenland I: introduction, classification of sequences, petrographic notes: Meddelelser om Gronland Bd. 206, nr. 1.

Goddard, E. N., Trask, P. D., De Ford, R. K., Rove, O. N., Singewald, J. T., Jr., and Overbeck, R. M., 1948, Rock-color chart: Geological Society of America, 16 p.

Hallam, A., ed., 1973, Atlas of palaeobiogeography: Elsevier, Amsterdam.

Harms, J. C., Southard, J. B., and Walker, R. G., 1982, Structures and sequences in clastic rocks: Society of Economic Paleontologists and Mineralogists Short Course 9, Calgary.

Hodgetts, D., 2013, laser scanning and digital outcrop geology in the petroleum industry: A review: Marine and Petroleum Geology, v. 46, p. 335-354.

Horowitz, A. S., and Potter, P. E., 1971, Introductory Petrography of Fossils: Springer-Verlag, New York, 302 p.

James, N. P., and Dalrymple, R. W., eds., 2010, Facies Modesl 4: GEOtext 6, Geological Association of Canada, St. John's, Newfoundland, 586 p.

Johnson, M. R., 1992, A proposed format for general-purpose comprehensive graphic logs: Sedimentary Geology, v. 81, p. 289-298.

Lerand, M. M., ed., 1976, The sedimentology of selected clastic oil and gas reservoirs in Alberta: Canadian Society of Petroleum Geologists, 125 p.

Low, J. W., 1951, Examination of well cuttings: Quarterly of Colorado School of Mines, v. 46, no. 4, 47 p.

Maiklem, W. R., Bebout, D. G., and Glaister, R. P., 1969, Classification of anhydrite— a practical approach: Bulletin of Canadian Petroleum Geology, v. 17, p. 194-233.

Martini, I. P., 1971, A test of validity of quartz grain orientation as a paleocurrent and paleoenvironmental indicator: Journal of Sedimentary Petrology, v. 41, p. 60-68.

McEachern, J. A., Pemberton, S. G., Gingras, M. K., and Bann, K. L., 2010, Ichnology and facies models, in James, N. P., and Dalrymple, R. W., eds., Facies Models 4: Geotext 6, Geological Association of Canada, p. 19-58.

McIlreath, I. A., and Harrison, R. D., eds., 1977, The geology of selected carbonate oil, gas and lead-zinc reservoirs in Western Canada: Canadian Society of Petroleum Geologists, 124 p.

McKee, E. D., and Weir, G. W., 1953, Terminology for stratification in sediments: Geological Society of America Bulletin, v. 64, p. 381-389.

McKerrow, W. S., ed., 1978, The ecology of fossils, an illustrated guide: Duckworth & Co. Ltd., London, 384 p.

McNeal, R. P., 1959, Lithologic analysis of sedimentary rocks: American Association of Petroleum Geologists Bulletin, v. 43, p. 854-879.

Miall, A. D., and Tyler, N., eds., 1991, The three-dimensional facies architecture of terrigenous clastic sediments, and its implications for hydrocarbon discovery and recovery: Society of Economic Paleontologists and Mineralogists Concepts and Models Series, v. 3, 309 p.

Mills, P. C., 1983, Genesis and diagnostic value of soft-sediment deformation structures - a review; Sedimentary Geology, v. 35, p. 83-104.

Moberly, R., and Klein, G., deV., 1976, Ephemeral color in deep-sea cores: Journal of Sedimentary Petrology, v. 46, p. 216-225.

Mount, J., 1985, Mixed siliciclastic and carbonate sediments: a proposed first-order textural and compositional classification: Sedimentology, v. 32, p. 435-442.

Nelson, R. A., Lenox, L. C., and Ward, B. J., Jr., 1987, Oriented core: its use, error, and uncertainty: American Association of Petroleum Geologists Bulletin, v. 71, p. 357-367.

Pirson, S. J., 1977, Geologic well log analysis, 2nd ed.: Gulf Publishing Co., Houston, 377 p.

Potter, P. E., and Pettijohn, F. J., 1977, Paleocurrents and basin analysis, 2nd edition: Academic Press, San Diego, California, 296 p.

Prior, D. J., Knipe, R. J., Bates, M. P., Grant, N. T., Law, R. D., Lloyd, G. E., Welbon, A., Agar, S. M., Brodie, K. H., Maddock, R. H., Rutter, E. H., White, S. H., Bell, T. H., Ferguson, C. C., and Wheeler, J., 1987, Orientation of specimens: essential data for all fields of geology: Geology, v. 15, p. 829-831.

Reading, H. G., ed., 1986, Sedimentary environments and facies, second ed.: Blackwell, Oxford, 615 p.

Rust, B. R., 1975, Fabric and structure in glaciofluvial gravels, in Jopling, A. V., and McDonald, B. C., eds., Glaciofluvial and glaciolacustrine sedimentation: Society of Economic Paleontologists and Mineralogists Special Publication 23, p. 238-248.

Shawa, M. S., ed., 1974, Use of sedimentary structures for recognition of clastic environments: Canadian Society of Petroleum Geologists, 66 p.

Sippel, R. F., 1971, Quartz grain orientation - 1 (the photometric method): Journal of Sedimentary Petrology, v. 41, p. 38-59.

Swanson, R. G., 1981, Sample examination manual: American Association of Petroleum Geologists.

Tassonyi, E. J., 1969, Subsurface geology, Lower Mackenzie River and Anderson River area, District of Mackenzie: Geological Survey of Canada Paper 68-25.

Taylor, J. C. M., 1978, Introduction to state of the art meeting, 1977, on sandstone diagenesis: Journal of the Geological Society, London, v. 135, p. 3-6.

Walker, R. G., 1967, Turbidite sedimentary structures and their relationship to proximal and distal depositional environments: Journal of Sedimentary Petrology, v. 37, p. 25-43.

Walker, R. G., and James, N. P., eds., 1992, Facies models: response to sea-level change: Geological Association of Canada, 409 p.

Walker, T. R., 1967, Formation of redbeds in modern and ancient deserts: Geological Society of America Bulletin, v. 78, p. 353-368.

Whittaker, A., 1998, Borehole data and geophysical log stratigraphy, in Doyle, P., and Bennett, M. R., eds., Unlocking the stratigraphic record: advances in modern stratigraphy: John Wiley and Sons, Chichester, p. 243-273.

Wilson, J. L., 1975, Carbonate facies in geologic history: Springer-Verlag, New York, 471 p.

# Facies Analysis

## Contents

## 3.1   Introduction

The study and interpretation of the textures, sedimentary structures, fossils, and lithologic associations of sedimentary rocks on the scale of an outcrop, well section, or small segment of a basin (the subject of Chap. 2) comprise the subject of facies analysis. A very large literature has grown since the 1960s that synthesizes this information in the form of facies models of varying complexity and sophistication. Several excellent texts deal extensively with facies analysis methods and process-response facies models (Reading 1996; Boggs 2012; James and Dalrymple 2010). The purpose of this chapter is to focus on a discussion of analytical methods and reviews of the kinds of practical information relating to depositional processes and sedimentary environments that can be obtained from sediments, based on the observations described in Chap. 2. It is hoped that this material will provide students with an introduction to modern facies-modeling methods. Practice on specific ancient examples can then be carried out using the summaries of facies models set out in Chap. 4 (and perhaps also including one of the advanced texts mentioned above).

As explained in Sect. 1.2, since the 1990s, facies analysis methods have become an integral component of the broader study of sequence stratigraphy. The transition from the local-scale focus of facies modeling to the basin-scale approach of sequence stratigraphy is exemplified by the review paper of Walker (1990), which is examined in Sect. 1.2.13. Sequence analysis is aided by various mapping methods (Chap. 6) and by the separate and independent skills involved in correlation and dating (Chap. 7).

## 3.2   The Meaning of Facies

The meaning of the word facies has been much debated in geology (e.g., Longwell 1949; Teichert 1958; Krumbein and Sloss 1963; Middleton 2005; see Sect. 1.2.4). It is widely used in sedimentary geology, but also has a somewhat different meaning in the area of metamorphic petrology (Fawcett 1982). Anderton (1985), Reading (1996, Chap. 2), Pirrie (1998) and Dalrymple (2010a, b) provided excellent discussions of the modern sedimentological uses of the term and the methods of interpreting individual facies and facies relationships.

The word facies is now used in both a descriptive or an interpretive sense, and the word itself may have either a singular or plural meaning. Descriptive facies include **lithofacies** and **biofacies**, both of which are terms used to refer to certain observable attributes of sedimentary rock bodies that can be interpreted in terms of depositional or biological processes. (When used without a prefix in this book, the word facies is intended to mean either lithofacies or biofacies.) An individual lithofacies is a rock unit defined on the basis of its distinctive lithologic features, including composition, grain size, bedding characteristics, and sedimentary structures. Each lithofacies represents an individual depositional event. Lithofacies may be grouped into **lithofacies associations** or **assemblages,** which are characteristic of particular depositional environments. These assemblages form the basis for defining **lithofacies models**; they commonly are repetitive and may be fully cyclic. A biofacies is defined on the basis of fossil components, including either body fossils or trace fossils. The term biofacies is normally used in the sense of an assemblage of such components. For the purpose of sedimentological study, a deposit may be divided into a series of lithofacies units, each of which displays a distinctive assemblage of lithologic or biologic features. These units may be single beds a few millimeters thick or a succession of beds tens to hundreds of meters thick. For example, a river deposit may consist of decimeter thick beds of a conglomerate lithofacies interbedded with a crossbedded sandstone facies. Contrast this with the biofacies terms used to describe the fill of many major early Paleozoic basins. Commonly, this may be divided into units hundreds of meters thick comprising a shelly biofacies, containing such fossils as brachiopods and trilobites, and a graptolitic biofacies. At the other extreme, Wilson (1975) recommended the use of microfacies in studying thin sections of carbonate rocks and defined 24 standard types.

The scale of an individual lithofacies or biofacies unit depends on the level of detail incorporated in its definition. It is determined by the variability of the succession, by the nature of the research undertaken (basin-wide reconnaissance versus a detailed local study), or by the availability of rock material for examination. Facies units defined on the basis of outcrop, core, well-cutting, or geophysical criteria tend to refer to quite different scales and levels of detail. Geophysicists in the petroleum industry refer to seismic facies, but this is not comparable to the small-scale type of facies discussed in this chapter. (Seismic facies are referred to in Chaps. 4 and 5.) Modern, high-resolution shallow

seismic surveying coupled with side-scan sonar imaging is providing a powerful tool for the analysis of facies compositions and geometries in modern environments and has had a major impact on the understanding of shelf and slope sedimentary environments (Sect. 6.3). More attention is being paid to the three-dimensional geometry of facies units, particularly in outcrop studies and subsurface studies involving reservoir development, where three-dimensional seismic-reflection methods are increasingly being employed. To a large extent, the scales at which facies units are defined reflect criteria of convenience. The term is thus a very flexible and convenient one for descriptive purposes.

The term facies can also be used, usually for lithofacies assemblages, in an interpretive sense for groups of rocks that are thought to have been formed under similar conditions. This usage may emphasize specific depositional processes, such as till facies or turbidite facies. Alternatively, it may refer to a particular depositional environment, such as shelf carbonate facies or fluvial facies, encompassing a wide range of depositional processes (see also Dalrymple 2010a for additional discussion of the meaning and usage of the term facies).

Widespread use is still made of two nineteenth-century Swiss stratigraphic terms that have acquired a generalized facies meaning encompassing lithologic characteristics, depositional environment and tectonic setting. The first of these is the *flysch facies*, comprising marine sediments, typically turbidites and other sediment gravity flow deposits, formed on tectonically active continental margins. The *molasse facies* consists of nonmarine and shallow marine sediments, mainly sandstones and conglomerates, formed within and flanking fold belts during and following their elevation into mountain ranges. Both of these facies types may make up major stratigraphic units hundreds or thousands of meters thick and extending for hundreds or thousands of kilometers. Continued use of the terms flysch and molasse is not recommended because of ambiguities about their implications for tectonic setting. The vague association with orogenic belts is inadequate now that we have the theory of plate tectonics to assist us in interpreting the relationship between sedimentation and tectonics (Allen and Allen 2013).

Lithostratigraphy and lithofacies analysis are two contrasting approaches to the study of sedimentary rocks. The first is the traditional descriptive approach. The second is based on detailed facies descriptions, which provide the basis for the genetic study of sediments using facies models. Lithofacies and biofacies analysis must be used to assist in stratigraphic studies, because by understanding the depositional environments and paleogeography existing at the time a rock unit was formed we are much better placed to make predictions and extrapolations about lateral changes in thickness and composition. Obviously, this will be invaluable for correlation purposes, and can make for a much more logical definition of formal lithostratigraphic units. Biofacies analysis is crucial in the definition and comprehension of biostratigraphic units. Lithostratigraphy and facies analysis are essential components of the study of **sequence stratigraphy**.

## 3.3 Recognition and Definition of Facies Types

### 3.3.1 Philosophy and Methods

In order to make sense of the lithologic variability present in most sedimentary basins, it is necessary to generalize, categorize and simplify what we see in well sections and outcrops. It is necessary because the human brain can only absorb and process a limited amount of detail at any one time. In sedimentology, we find that much of the variability disguises a limited range of basic lithofacies and biofacies types, and that variations between these types represent minor random environmental fluctuations or are even the result of accidental exposure or the position of a thin-section cut. The existence of this natural pattern is what makes facies studies, facies modeling, and paleogeographic reconstruction possible. For example, for many years sedimentologists categorized most deep-sea deposits in terms of only five basic lithofacies. These were the $T_a$, $T_b$, $T_c$, $T_d$, and $T_e$ divisions of the classic Bouma turbidite sequence (Bouma 1962). This Bouma model served sedimentologists well, and was used as the type example in Walker's (1976; and subsequently repeated in Walker 1979, 1984; Walker and James 1992) definition of facies models. But it has always been known that some beds do not fit the pattern, as discussed further below. The standard submarine fan model, that evolved in the 1970s (Mutti and Ricci-Lucchi 1972; Walker 1978) interpreted Bouma sequences as occurring mainly on the outer, non-channelized part of a fan. The lithofacies spectrum for submarine fans and for other types of deep-sea sand deposit is now rather more complex, as discussed in Sect. 4.2.9.

The deposits of many other depositional environments have proved to be amenable to some simple, empirical classifications. For example, Miall (1977, 1978c, 1996) showed that most river deposits could be described using about 20 lithofacies types. Wilson (1975), in a major review of Phanerozoic carbonate rocks, concluded that most could be described satisfactorily by drawing from a list of only 24 standard microfacies. Eyles et al. (1983) proposed a simple lithofacies classification scheme for the description of glacial-marine and glacial-lacustrine deposits. Farrell et al. (2012) provided the basis for a universal set of codes for describing clastic sediments, which they suggest (p. 377)

"supports process-based interpretations of stratigraphic sequences."

Facies modelers have yet to categorize all sediment types in every variety of depositional environment; for example, no one has yet attempted to erect a universal lithofacies scheme for clastic rocks comparable to that of J.L. Wilson for carbonates. Similarly, many fossil groups await detailed biofacies analysis. For this reason, every basin analyst should study each basin with a fresh eye and, at least in the preliminary stages of the research, erect a local facies scheme without too much dependence on previously published work. Slavish adherence to such readily available research keys may result in minor but critical lithofacies types and lithofacies relationships being missed or forced to fit an inappropriate mould. New facies models or a better definition of old ones will never come about if geologists are content merely with such replication studies. Bridge (1993) expressed this concern forcefully with respect to the facies scheme now in widespread use for fluvial deposits (see Sect. 4.2.1), and Shanmugam (1997) argued strongly that the Bouma-sequence concept had been seriously misused in the study of deep-sea sandstones. Others, such as Walker (1990, 1992), Reading and Levell (1996), and Dalrymple (2010a, b) are strongly in favor of the simplifying and categorizing methods of facies analysis. The problem that concerned Bridge (1993) and Shanmugam (1997) is that hasty or lazy field observation can too easily lead to incorrect facies classification. For example, Shanmugam (1997) argued that for too long sedimentologists have been applying the Bouma sequence concept and its five-fold subdivisions to deposits that exhibit few of the key characteristics of the original sequence. Many successions contain no graded basal unit, they contain internal breaks in sedimentation indicating that they are not the product of single depositional events, they contain evidence of deposition from other types of sediment-gravity flow, such as rafted clasts, that would indicate debris-flow deposition, and so on. In fact such practices have become so common, in Shanmugam's (1997) view, that he refers to a phenomenon that he termed the "turbidite mind set."

Hummocky cross-stratification is an excellent example of a structure that sedimentologists had been looking at without seeing for many years, until focused upon and given a name by Harms et al. (1975). Suddenly, it was realized that this distinctive style of crossbedding is characteristic of many storm-generated ancient deposits. Its recognition led to the reappraisal of numerous shallow marine sequences, to the extent that the term became something of a cliché in the early 1980s (e.g. see Byers and Dott 1981). This is a good example of critical advances in sedimentological interpretation being missed because of the widespread lack of critical, independent observation.

How, then, does the basin analyst perform a facies analysis of an undescribed rock sequence or succession of fossils? The method for measuring and describing stratigraphic sections is discussed in detail in Chap. 2. For the purpose of lithofacies analysis, the focus must be on recognizing associations of attributes that are repeated through the section (or parts of the entire basin). Lithofacies may be distinguished by the presence of bedding units with a characteristic sedimentary structure or structures, a limited grain size range, a certain bed thickness, and perhaps a distinctive texture or color (color is subject to diagenetic change and should not be used as a primary criterion in definition; see Sect. 2.2.2.4). Biofacies represent associations within the same stratigraphic interval of a limited suite of genera or species. Biofacies studies may be carried out on single taxonomic groups, because many of these, if examined by specialists, can yield highly detailed paleoecological (and hence depositional) information. The definition of biofacies is an exercise in defining assemblages and will be discussed in the next section.

In order to recognize these associations of lithologic attributes or fossil types, it may be useful to set up a checklist and tabulate their occurrence. Statistical analysis is commonly used by paleoecologists to establish biofacies from such tabulated data. It is less used for erecting individual lithofacies because many of the common associations of lithologic attributes are well enough known to be recognized without the aid of statistics. But the key to success is very careful, empirical observation.

### 3.3.2  Field Examples of Facies Schemes

Two examples of lithofacies schemes that have been erected from field data are given in this section to illustrate the methods. A simple biofacies association is also described.

Cant and Walker (1976) described a Devonian fluvial section in eastern Quebec and subdivided it into eight lithofacies. The section is shown in Fig. 3.1, in which lithologic symbols are keyed to the lithofacies scheme. Note the use in this illustration of the variable-width column plotting technique for drawing stratigraphic sections that is discussed in Sect. 2.2.4. Here are some examples of their lithofacies descriptions, edited in order to focus on salient features.

Well defined trough crossbedded facies (B): This facies is composed of well-defined sets of trough crossbedding..., with trough depths averaging 15 to 20 cm (range 10 to 45 cm). The troughs are regularly stacked on top of each other, but in some individual occurrences of the facies, trough depths decrease upward... The sets are composed of well-sorted medium sand... A few of the coarser sets have granules and pebbles concentrated at their bases.

Asymmetrical scour facies (E): This facies consists of large, asymmetrical scours and scour fillings, up to 45 cm deep and 3 m wide... The scours cut into each other and into underlying troughed facies (A and B), and occurrences of the asymmetrical

scour facies have a flat, erosionally truncated top... The main difference between the scour fillings and the two troughed facies (A and B) lies in the geometry of the infilling strata. In the asymmetrical scours, the layers are not at the angle of repose, but are parallel to the lower bounding surface.

Rippled sandstone and mudstone Facies (F): This facies includes cross-laminated sandstones..., and alternating cross-laminated sandstones and mudstones. [An example of the latter] is 1.5 m thick and consists of three coarsening-upward sequences, which grade from basal mudstones into trough cross-laminated fine sandstone and finally into granule sandstone. The sandstones capping each coarsening-upward sequence have sharp, bioturbated tops.

A variety of criteria has been used in defining these lithofacies, but the authors refrained, at this stage, from making any environmental interpretation.

A carbonate example was described by Wilson (1975), who studied the Upper Jurassic Oolite Series (actually a formation, but an incorrect, older nomenclature is still in use) of Dorset, England. He recognized four lithofacies, as shown in Fig. 3.2. Note in this illustration the relationship between lithostratigraphic and lithofacies units. The four lithofacies are as follows (described using the carbonate classification of Folk 1962):

1. Coarsening-upward units shown by two beds (1) the *Chlamys qualicosta* bed of intramicrite-oomicrite-oosparite-poorly washed biosparite and (2) pisolite, consisting of quartz sands and phyllosilicate clay-intramicrite-oomicrite-oosparite-oncolites.
2. Crossbedded sets of oosparite showing 20–250 dips and sharp contacts either with phyllosilicate clays with nodular micrites or bioturbated oolite. Some minor flaser bedding and clay drapes over current ripples also occur.
3. Association of *Rhaxella* biomicrites.
4. Sheet deposits (5–10 cm) and large accretion sets (30 cm) of oomicrite and biomicrite with subsidiary oosparite and biosparite. Shell debris often shows imbricate structure, and the oomicrites are texturally inverted sediments, being a mixture of extremely well-sorted oolites in a micrite matrix. Some sets showing alternating current directions occur.

The careful reader will note two points about these descriptions. First, a mixture of lithofacies and biofacies criteria is used. This commonly is desirable in fossiliferous successions where the fossils are distinctive rock-forming components. Second, several of the lithofacies consist of different carbonate rock types interbedded on a small scale. Each of these could be described as a lithofacies in its own right, in which case Wilson's four facies become facies assemblages. It is a question of the scale of description that is the most suitable for the purpose at hand. Those engaged in analyzing entire basins or major stratigraphic intervals in a

**Fig. 3.1** A stratigraphic section plotted using a standardized facies scheme and the variable-width column technique. Fluvial cycles of the Battery Point Formation (Devonian), Quebec (Cant and Walker 1976)

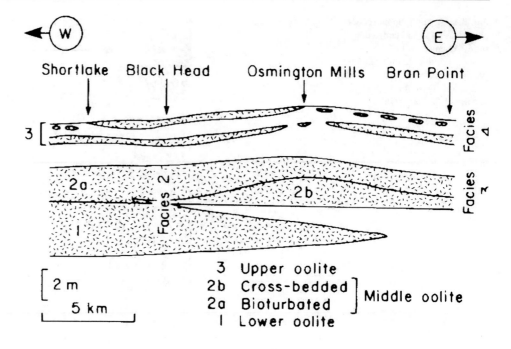

**Fig. 3.2** Carbonate facies and their distribution in the Osmington Oolite (Jurassic), southern England (Wilson 1975)

basin cannot afford to spend too much time on fine detail. Decisions must be made with the core or outcrop in front of the analyst about how thick the descriptive units are going to be and how to combine thin beds for the purpose of lithofacies description. Generally, lithofacies thicknesses on the order of a few decimeters to a few meters have been found to be the most useful. Smaller-scale subdivisions may be erected for selected examples of well-exposed sections or continuous cores if desired, to illustrate particular points in a facies description.

Wilson suspected that his Oolite Series beds were tidal in origin, and he tabulated various features of each of the four facies using a range of criteria suggested by Klein (1971) for recognizing tidal deposits (Fig. 3.3). Many of these make specialized use of paleocurrent evidence, which is discussed in Sect. 6.7. Figure 3.3 illustrates the use of a checklist for lithofacies description. This can be a useful technique for field observation if the geologist suspects in advance what it is he or she is about to study. It is then possible to incorporate all the special observational detail and environmental criteria available from studies of the appropriate modern environment or other ancient rock units. In doing so there is, of course, the danger that other features might be missed and interpretation becomes a self-fulfilling prophecy. Tables of this kind are not recommended for illustration of written or oral presentation of a lithofacies study except, perhaps, as an appendix. They are visually difficult to absorb, and the geologist does better to provide summary descriptions, such as those given previously, plus photographs and graphic logs.

### 3.3.3    Establishing a Facies Scheme

The examples in the previous section serve to illustrate the methods used in three different kinds of basin problems to define facies types. The objective of the basin analyst should be to erect a facies scheme that can encompass all the rock types present in his or her project area. If the project encompasses several or many stratigraphic units deposited under widely varying environments, it may make more sense to erect a separate scheme for each lithostratigraphic unit, because the lithofacies and, most likely, the biofacies assemblages will be different for each such unit. However, this can lead to repetition and overlap, so some judgment is required.

Facies schemes should be kept as simple as possible, otherwise they defeat the whole purpose of carrying out a facies analysis. Some authors have subdivided their rocks into 20 or 30 lithofacies and erected subclasses of some of these. This gives the appearance of meticulous research and great analytical precision, but such schemes are difficult to understand or absorb. Remember that the purpose of a facies analysis is to aid in interpretation, and in basin analysis, this is often best accomplished by judicious simplification. It is rare that many more than half a dozen distinct lithofacies occur together in intimate stratigraphic association. Commonly, two to four lithofacies occur in repeated, monotonous succession through tens or hundreds of meters of section. In defining only a few facies to cover thick and varied successions, the problem of what to group together and what to separate is always present. What should be done about thin beds? Should units that show gradational contacts

**Fig. 3.3** Distribution of selected lithofacies attributes in the Osmington Oolite (Wilson 1975)

| | Facies of Osmington Oolites | | | |
|---|:-:|:-:|:-:|:-:|
| | 1 | 2 | 3 | 4 |
| A. 1. Cross stratification with sharp set boundaries | | ■ | | ■ |
| 2. Herringbone cross-stratification | | | ■ | ■ |
| 4. Parallel laminae | ■ | | | |
| 5. Complex internal organization of dunes, etc. | | ■ | ■ | ■ |
| B. 6. Reactivation surfaces | | ■ | | ■ |
| 7. Bimodal distribution of set thicknesses | | ■ | | |
| 9. Unimodal distribution of dip direction of cross strata | | ■ | ■ | |
| 10. Orientation of cross-strata parallels sand-body trend | | | ? | |
| C. 13. Small current ripples to larger current ripples | | | ■ | |
| 19. Symmetrical ripples | | | ■ | |
| D. Flaser bedding | | | ■ | |
| F. 35 & 36 channel log deposits | | | ■ | |
| H. Burrowing, etc. | ■ | ■ | ■ | |

be split, and if so, how? The only answer is to do what works. One approach is to compile a very detailed description of the first well or section of a new project and to use the resulting notes to erect a preliminary facies scheme; this is then employed for describing each new section. It has the advantage that, instead of writing down basic descriptions for every bedding unit, it may be adequate simply to assign it to the appropriate facies and record in one's notes any additional observations that seem necessary, such as grain-size differences or additional sedimentary structures relative to the original descriptive scheme. The scheme itself may thus be modified as one goes along. This approach has worked well for the author for many years of studying clastic sedimentary sequences in outcrops and cores, in the Canadian Arctic, the southwest United States, and elsewhere. It is more difficult to apply this method to carbonate rocks, because of their finer scale variability and the difficulty of seeing all the necessary features in weathered outcrops. One approach is to take into the field a rock saw and other necessary equipment for constructing etched slabs or peels. These can then be examined and described as field work proceeds. (Wilson 1975, pp. 56–60).

Friend et al. (1976) avoided these problems entirely by recording only attributes of a sequence, not lithofacies as such. They then used statistical techniques to group these attributes into lithofacies assemblages (Sect. 2.2.1).

Anderton (1985) suggested that facies criteria be ranked in order of priority, permitting an increasingly refined subdivision of the facies units as the work proceeds. His suggested ranking for channelized clastic deposits (as an example) is

1. bases of major channels or scoured surfaces;
2. tops and bases of nonclastic lithologies (e.g., coal, limestone);
3. abrupt changes from section composed predominantly of one grain size to another (e.g., sandstone to shale, fine sandstone to coarse); and
4. changes of internal structure within units of similar grain size.

This approach, if extended regionally, can form the basis for a practical means of recognizing and defining sequences, of which the facies assemblages are distinct components.

With continual study of the same kind of rocks, a standard set of facies classes will gradually emerge and can be used for rapid field or laboratory description. For the writer, this led to the erection of a generalized lithofacies scheme for describing the deposits of braided rivers based on field studies of various ancient deposits and a review of literature on both ancient deposits and modern rivers. The initial scheme contained 10 lithofacies types (Miall 1977). Further research by the writer and use by other workers led to the addition of about 10 more lithofacies types (Miall 1978c), but the original 10 include most of the common ones. This scheme has now been applied to a wide variety of fluvial deposits and to the fluvial component of deltaic successions by numerous workers around the world. Table 3.1 lists these lithofacies, showing the codes used for note taking and a sedimentological interpretation of each. The lithofacies codes consist of two parts, a capital letter for modal grain size (G, gravel; S, sand; F, fines) and a lowercase letter or letters chosen as a mnemonic of a distinctive texture or structure of each lithofacies. The three lithofacies B, E, and F of Cant and Walker (1976), discussed in the previous section, are St, Ss, and Fl in this scheme. Le Blanc Smith (1980) and Uba et al. (2005) developed this fluvial facies scheme still further by incorporating additional structures and information on grain size (see also Miall 1996, for

further discussion of code schemes). However, as Dalrymple (2010a, p. 6) noted, this scheme does not contain categories and codes for the tidal features that are common in some coastal (especially estuarine) fluvial systems.

J.L. Wilson's microfacies scheme for carbonates contains 24 types. Figure 3.4 illustrates his standard legend, facies numbers, and abbreviated description. Figures 3.5 and 3.6 illustrate the use of these two schemes in drawing stratigraphic sections.

### 3.3.4    Facies Architecture

Many of the facies-analysis techniques discussed in this chapter focus on the use of vertical facies relationships, as derived from the measurement of stratigraphic sections (Sects. 2.3, 3.5.10). Although section measuring has evolved

**Table 3.1**  Facies classification (from Miall 1996)

| Facies code | Facies | Sedimentary structures | Interpretation |
|---|---|---|---|
| Gmm | Matrix-supported, massive gravel | Weak grading | Plastic debris flow (high-strength, viscous) |
| Gmg | Matrix-supported gravel | Inverse to normal grading | Pseudoplastic debris flow (low strength, viscous) |
| Gci | Clast-supported gravel | Inverse grading | Clast-rich debris flow (high strength), or pseudoplastic debris flow (low strength) |
| Gcm | Clast-supported massive gravel | – | Pseudoplastic debris flow (inertial bedload, turbulent flow) |
| Gh | Clast-supported, crudely bedded gravel | Horizontal bedding, imbrication | Longitudinal bedforms, lag deposits, sieve deposits |
| Gt | Gravel, stratified | Trough crossbeds | Minor channel fills |
| Gp | Gravel, stratified | Planar crossbeds | Transverse bedforms, deltaic growths from older bar remnants |
| St | Sand, fine to v. coarse, may be pebbly | Solitary or grouped trough crossbeds | Sinuous-crested and linguoid (3-D) dunes |
| Sp | Sand, fine to v. coarse, may be pebbly | Solitary or grouped planar crossbeds | Transverse and linguoid bedforms (2-D dunes) |
| Sr | Sand, very fine to coarse | Ripple cross-lamination | Ripples (lower flow regime) |
| Sh | Sand, v. fine to coarse, may be pebbly | Horizontal lamination parting or streaming lineation | Plane-bed flow (critical flow) |
| Sl | Sand, v. fine to coarse, may be pebbly | Low-angle (<15°) crossbeds | Scour fills, humpback or washed-out dunes, antidunes |
| Ss | Sand, fine to v. coarse, may be pebbly | Broad, shallow scours | Scour fill |
| Sm | Sand, fine to coarse | Massive, or faint lamination | Sediment-gravity flow deposits |
| Fl | Sand, silt, mud | Fine lamination, v. small ripples | Overbank, abandoned channel, or waning flood deposits |
| Fsm | Silt, mud | Massive | Back swamp or abandoned channel deposits |
| Fm | Mud, silt | Massive, desiccation cracks | Overbank, abandoned channel, or drape deposits |
| Fr | Mud, silt | Massive, roots, bioturbation | Root bed, incipient soil |
| C | Coal, carbon-aceous mud | Plant, mud films | Vegetated swamp deposits |
| P | Paleosol carbonate (calcite, siderite) | Pedogenic features: nodules, filaments | Soil with chemical precipitation |

SMF-1   Basinal spiculite, black lime mudstone

SMF-2   Microbioclastic and peloid

SMF-3   Pelagic forams, nannoplankton

SMF-4   Talus, lithoclastic conglomerate

SMF-4   Lithoclastic, bioclastic microbreccia

SMF-5   Foreslope calcarenite, bioclastic grainstone, packstone-floatstone

SMF-6   Reef rudstone

SMF-7   Bindstone, lamellar, encrusting

SMF-7   Platy algal baffelstone

SMF-7   Tubiphytes bindstone

SMF-7   Dendroid coral, stromatoporoid boundstone

SMF-7   Massive coral, stromatoporoid boundstone

SMF-8   Whole fossil wackestone

SMF-9   Foreslope, neritic open marine bioclastic wackestone

SMF-10   Rounded bioclasts in micrite, textural inversion

SMF-11 & 12   Rounded, coated, bioclastic calcarenite grainstone to packstone

SMF-12   Encrinite grainstone

SMF-12   Rounded bioclastic calcarenite with dasycladaceans

SMF-22   Onkoidal-large foram wackestone

SMF-14   Carbonate conglomeratic lag deposit

SMF-15   Oolite-pisolite

SMF-16   Peloids, pelsparite

SMF-17-18   Peloidal grapestone

SMF-19   Restricted marine peloidal wackestone, pelleted mudstone

SMF-20-21   Algal stromatolitic and spongiostrome micrite

SMF-24   Flat pebble conglomerate, lithoclastic grainstone

SMF-23   Pure micrite and micrite matrix

Basinal shale or silt, commonly siliceous

Shale or marl of shelf

Sandy silty limestone

**Fig. 3.4** The standard carbonate microfacies scheme of Wilson (1975)

into a sophisticated art, it is essentially a one-dimensional approach to the study of what commonly are very complex three-dimensional rock bodies. Increasingly, it has been realized in recent years that vertical sections are inadequate for the description of facies variations in many environments, and techniques are evolving for the description and classification of facies units in three dimensions. At the surface, this can be accomplished by the search for and the description of large, laterally extensive outcrops. In the subsurface, lithostratigraphic correlation of closely spaced cores, coupled with high-resolution seismic data (especially 3-D seismic, if available), may serve the purpose. The results are important not only to facilitate the development of more quantitatively accurate reconstructions of depositional environments, but to provide a usable data base for development geologists and engineers to model fluid flow in reservoir units.

Outcrop techniques were described by Anderton (1985) and Miall (1988a, b, 1996), and field examples were given by Allen (1983), Miall and Tyler (1991), and Miall (1988b, 1996). Ideally, the geologist should search for long natural or artificial cliff faces in areas of minimal structural disturbance. The entire outcrop should be photographed to permit the construction of a lateral profile based on a photomosaic or LIDAR profile, using the techniques described in Sect. 2.2.1. Hodgetts (2013) provided a review of LIDAR techniques.

Once printed and mounted on card or board, the mosaic, with a transparent overlay, can be taken back to the field and used as a base map on which to plot detailed sedimentological observations. Alternatively, hand-held digital devices, such as tablets, are now flexible enough that they may be taken into the field and used for the direct entry of field data at the time of observation. The geologist should exploit the detail available in locations of excellent outcrop to measure and describe as many facies criteria as possible, taking care to enter these on the mosaic with as precise as possible a record of their locations. Locations and orientations of hydrodynamic sedimentary structures and the position and attitude of various bounding surfaces between sedimentary units are among the most important kinds of

**Fig. 3.5** Use of the fluvial lithofacies scheme of Table 3.1 for plotting a section in braided fluvial deposits (Miall 1978)

Sl

St

Se

Sr

Sp

Gm

Fl

Sh

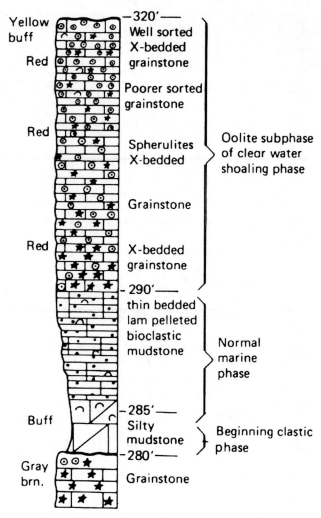

**Fig. 3.6** Use of the carbonate microfacies scheme for plotting a limestone section (Wilson 1975)

large outcrops, it may be useful to employ the service of an assistant to carry out the actual plotting, because once one is standing on a large outcrop it may be difficult to see precisely where one is on the mosaic. The assistant can base him or herself as closely as possible at the location from which the photograph was taken and enter measurements on the profile that are communicated by voice (or radio) from the geologist on the outcrop. Some highly detailed work is now being done by working across vertical cliff faces using climbing gear. Although introducing a significant logistic complexity to field work, in some cases this is the only way to ensure that critical fine detail is collected at the exact spot that observations are needed.

If necessary, vertical stratigraphic sections can be constructed in the office from such lateral profile data. However, the wealth of detail yielded by a good quality lateral profile may make such section drawing superfluous. As noted by Anderton (1985), "Photographic techniques are sufficiently important in facies description that if one ever has to choose

observations to make. For average road cuts or river cliffs a hundred meters or so long and a few tens of meters high, this fieldwork may take as long as a day or two to complete. For between taking a camera and a hammer into the field, take the camera."

Figure 2.3 is an example of a lateral profile constructed using these techniques; a brief discussion of the kinds of facies observations that can be made from profiles is presented in Sect. 3.5.11. When combined with paleocurrent data, detailed architectural reconstructions may be made from good outcrop data (Fig. 6.64).

## 3.4 Facies Associations and Models

### 3.4.1 The Association and Ordering of Facies

The term **facies association** was defined by Potter (1959) as "a collection of commonly associated sedimentary attributes" including "gross geometry (thickness and areal extent); continuity and shape of lithologic units; rock types …, sedimentary structures, and fauna (types and abundances)." A facies association (or assemblage) is therefore based on observation, perhaps with some simplification. It is expressed in the form of a table, a statistical summary, or a diagram of typical stratigraphic occurrences (e.g., a vertical profile). A **facies model** is an interpretive device, which is erected by a geologist to explain the observed facies association. A facies model may be developed at first to explain only a single stratigraphic unit, and similar units may then be studied in order to derive generalized models.

Lithofacies group together into assemblages because they represent various types of depositional events that frequently occur together in the same overall depositional environment. For example, a submarine fan environment typically contains canyon, channel, levee, overbank, and proximal slope subenvironments, each of which produces a distinctive lithofacies. These lithofacies become stacked into stratigraphic units because the environments shift through time, permitting different lithofacies to accumulate along any given vertical axis. The nature of these environmental shifts is often predictable, which means that the resulting lithofacies successions are equally predictable. For example, on the submarine fan, channels may shift in position and subenvironments within a turbidity current may move (very rapidly) down slope. Deltas and tidal flats prograde seaward, river channels and tidal inlets migrate, and so on. These processes may repeat themselves many times. This is the basis for the principle of cyclic sedimentation. It was clearly understood by one of the founders of modern sedimentary geology, Johannes Walther. He enunciated what has come to be known as Walther's law, the "rule of succession of facies," which states

The various deposits of the same facies area and, similarly, the sum of the rocks of different facies areas were formed beside

each other in space, but in a crustal profile we see them lying on top of each other … it is a basic statement of far-reaching significance that only those facies and facies-areas can be superimposed, primarily, that can be observed beside each other at the present time (Walther 1893–1894, as translated from the German by Middleton 1973).

The value of this rule to basin analysts is that it means lateral facies relationships can be predicted by studying them in vertical succession. Stratigraphic sections, such as outcrops measured up cliffs or drill hole sections, comprise most of our data base. Lateral facies relationships are much more difficult to study because they tend to be gradual and stretched out over considerable distances. Few outcrops are long enough to reveal such features in their entirety. Paleogeographic reconstruction therefore relies heavily on the examination of vertical profiles.

It must be stressed that Walther's law only applies to continuous sections without significant stratigraphic breaks. An erosion surface may mark the removal of the evidence of several or many subenvironments. It is therefore important to observe the nature of contacts between lithofacies units when describing sections. Contacts may be sharp or gradational. Sharp contacts may or may not imply significant sedimentary breaks. Periods of erosion or nondeposition may be revealed by truncated sedimentary structures, extensive boring, penecontemporaneous deformation, hardgrounds, or submarine diagenesis. Walther's Law does not apply in some settings. For example, shallow-water carbonate platforms, lakes, and evaporite environments may undergo rapid, basin-wide changes in water chemistry and physical conditions (temperature, water clarity) in response to climate change or modest sea-level change, with the sedimentary response being basin-wide. The vertical succession of facies is then not a reflection of a lateral shift in environments.

Given these constraints, the study of cyclic sedimentation has been one of the most popular and fruitful in sedimentology because, as Reading (1986, p. 5) stated, "it enabled geologists to bring order out of apparent chaos, and to describe concisely a thick pile of complexly interbedded sedimentary rocks." Interpretations of cyclic sedimentation form the basis of most facies models.

Statistical methods have been used for defining and describing lithofacies and biofacies associations. Cluster analysis methods were used by Imbrie and Purdy (1962) and Purdy (1963) to document and describe the associations of facies components on the modern Bahama Platform (Figs. 3.7 and 3.8). Multivariate statistical methods have also been used in the erection of biofacies associations (Mellow and Buzas 1968; Ludvigsen 1978). These methods were described at some length in Miall (1999), from which this chapter has been updated, but this material has been omitted here, because such techniques, having served their purpose, have now fallen into disuse. After more than forty years of intensive research the major facies associations in the most important sedimentary environments have now been recognized (Reading 1996; Galloway and Hobday 1996; James and Dalrymple 2010), and such exploratory techniques are no longer needed. Attention has shifted to the investigation of the three-dimensional attributes of sedimentary successions, including architectural-element analysis, at the outcrop scale (Sect. 3.5.11), and sequence stratigraphy, at the basin scale (Chap. 6). The following is a brief summary of two statistical methods that were popular in the 1960s and 1970s.

Imbrie and Purdy (1962) and Purdy (1963) carried out a landmark study of carbonate facies that involved a statistical analysis of data from modern sediments of the Great Bahama Bank. Purdy performed point count analyses under the microscope of 218 samples from the northern part of this bank. He recognized 12 major organic and inorganic components: coralline algae, *Halimeda* (calcareous alga), Peneroplidae (a family of Foraminifera), other foraminifera, corals, molluscs, fecal pellets, mud aggregates, grapestone, oolite, cryptocrystalline grains, and silt grains <1/8 mm in size. Cluster analysis was then used to determine which components tended to occur together; the results are shown in Fig. 3.7. From this information, five major lithofacies

were defined, and by referring back to the sample data, these could be mapped as shown in Fig. 3.8. Purdy (1963) interpreted the lithofacies data in terms of physical and chemical processes operating over the bank. The work has been widely used and quoted as an example of a lithofacies study in carbonate rocks.

There has been considerable controversy over the methods used to define sedimentary cycles. Early attempts were subjective and qualitative. The culmination of the analysis was commonly the erection of a single model or ideal cycle (e.g., Visher 1965). This approach was criticized by Duff and Walton (1962) on the grounds that, in fact, this ideal cycle may rarely be represented in the section. Random environmental changes, local reversals of the sequence, or the effect of some external event, such as a flash flood or tectonic pulse, commonly intercede to modify the succession. The many analyses of the Carboniferous Coal Measure cyclothems prior to about 1960 illustrate this difficulty well, but cannot be entered into here for reasons of space (modern ideas about these cycles are discussed in Miall 2010).

As noted in Sect. 1.2.7, a step forward in the analysis of cyclicity was introduced by de Raaf et al. (1965) in the form of the *facies relationship diagram*, which is a semi-quantitative illustration of all possible vertical facies transitions, highlighting those that are more common.

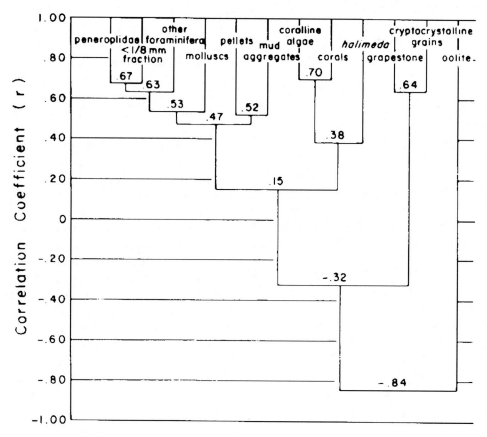

**Fig. 3.7** Cluster analysis (R-mode) of facies components on the modern Bahama Platform (Purdy 1963)

**Fig. 3.8**  Distribution of lithofacies of the Bahama Platform (Purdy 1963)

The technique of Markov chain analysis, which was popular in the 1970s, added a certain statistical rigor to this type of analysis. Like the multivariate techniques described above, it results in a grouping of lithofacies into an assemblage, but it also has the additional property of revealing the order in which the lithofacies tend to occur. It can reveal not only a single-model cycle, but also the several or many statistically most probable variations on this theme. A Markov processes is one "in which the probability of the process being in a given state at a particular time may be deduced from knowledge of the immediately preceding state" (Harbaugh and Bonham-Carter 1970, p. 98). The methods were described by Miall (1973) and Harper (1984), but are no longer used, because the extensive body of empirical research that has accumulated since the 1970s has thoroughly documented cyclicity in the sedimentary record, and attention has turned to architectural analysis and sequence stratigraphy.

### 3.4.2  The Theory of Facies Models

The concept of a facies model as a summary of a depositional environment and its products is a recent one in sedimentology. The term was first discussed at a conference reported by Potter (1959), but was used in the sense we now imply by the term facies assemblage. The difference is critical. A facies assemblage is essentially descriptive, whereas a facies model attempts to provide an interpretation of a particular type of facies assemblage in terms of depositional environments. Much use is usually made of comparisons with actual modern environments, and so it is now common practice to refer to these mental constructs as *actualistic facies models* (although Shea 1982, has criticized the use of the word "actualistic", partly on etymological grounds).

As noted in Sect. 1.2.7, the modern usage of the term facies model commenced with the classic paper by Walker (1976). The most recent expression of this concept is given by Dalrymple (2010a), drawing on Walker (1992), who described the process of constructing a facies model, using submarine fans as an example. Data from various sources (modern sediments, ancient deposits) are collected, and the wealth of information sifted to determine the important features that they have in common, The local variability is "distilled" away (Fig. 3.9). Various statistical procedures may be used to achieve this, such as those discussed in the previous section. The objective is to produce a general summary of submarine fans. But, as Walker (1992) stated, "what constitutes local detail, and what is general? Which aspects do we dismiss, and which do we extract and consider important?"

**Fig. 3.9** The development of a
facies model, using
submarine-fan deposits as an
example (Dalrymple 2010a;
Modified from Walker (1976)

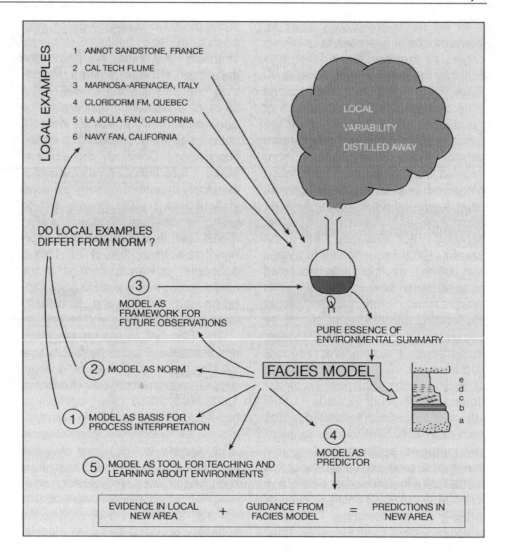

Walker (1976) stated that a facies model should fulfill
four functions:

1. it must act as a **norm**, for purposes of comparison;
2. it must act as a **framework** and **guide** for future
   observations;
3. it must act as a **predictor** in new geological situations;
   and
4. it must act as a basis for **environmental interpretation**.

This is illustrated in Fig. 3.9. Facies models are a powerful tool for the interpretation of poorly exposed sediments
because they suggest certain critical observations or clues for
interpreting the sedimentary record. As with all such summaries and models, however, uncritical use may lead to loss
of information or misinterpretation, because it is tempting to
observe strata in terms of a preconceived model. If a facies
model is being used properly in new field situations, each
use may generate a refinement of the model or it may lead to
the recognition of situations where the particular model is

inappropriate. The development of a new model could then
follow.

What does a facies model look like? The answer depends
on how the model was constructed. Paleogeographic sket-
ches, block diagrams, and vertical profile logs are typical
components of a published facies model. Some models are
based primarily on geomorphology (e.g., deltas, alluvial
fans), others on transport processes (e.g., turbidite, glacial
till, and eolian dune models), and others on organic processes (e.g., reefs). Many are governed by climatic and
tectonic variables, which have not, in all cases, been
exhaustively explored.

Many models are based primarily on the study of modern
environments, for example, the three-fold subdivision of
deltas erected by Galloway (1975) used maps of typical
deltas to convey information on lithofacies geometry. For
continental environments, there is no problem seeing and
describing a wealth of detail in modern settings, and this has
led to the recognition that for fluvial deposits at least sixteen
distinct fluvial styles can be described (Miall 1996). Rapid

evolution in eolian facies models came about by improved understanding of eolian bedforms (Sect. 3.5.4.7).

Facies models for subaqueous environments are less dependent on studies of modern settings, because the techniques of marine geology (geophysical profiling, shallow coring, sonar, satellite navigation) only began to enter general use in the early 1980s and did not, at first, permit very precise descriptions of lithofacies geometry and relationships. The first comprehensive submarine-fan model of Mutti and Ricci-Lucchi (1972) was based mainly on well-exposed Cenozoic fan deposits in Italy, and interpretations of other ancient deposits figure prominently in the first detailed discussions of this environment (e.g., Walker 1978). Models for storm-dominated shelf deposits incorporate hummocky cross-stratification (e.g., Hamblin and Walker 1979), a structure not reliably documented from modern environments until the 1980s. Howard (in Byers and Dott 1981, p. 342) explained how it may be obscured in cores of recent sediments by our methods of sampling the sea bottom and offered what may be the first observation from a modern offshore region (Howard and Reineck 1981). Application of side-scan sonar techniques to the study of submarine depositional systems brought about a revolution in our knowledge of shelf and deep marine environments, yielding a knowledge of channels, bedforms, and other depositional features that had a major impact on facies models for these environments (e.g., Bouma et al. 1985).

Many of these points are reviewed at greater length in Chap. 4, but this summary should serve to illustrate the varied state of the art of facies modeling at the present day. It is very much a case of work in progress, and basin analysts can potentially add much to our fund of knowledge by careful studies of newly explored stratigraphic units.

## 3.4.3 The Present as the Key to the Past, and Vice Versa

We have all heard the cliche "the present is the key to the past." It was one of the great generalizations to emerge from James Hutton's enunciation of the principle of uniformitarianism toward the end of the eighteenth century. Charles Lyell's work a half century later seemed to nail it down forever as a cornerstone of the still unborn science of sedimentology. Yet it is only true in a limited sense. Undoubtedly the study of modern depositional environments provides the essential basis for modern facies studies, but there are at least a dozen major problems that emerge to confuse this work. It turns out that the past is also a very important key to the present, that many geomorphological processes can best be understood by adopting a geological perspective and looking at the ancient record. The art of facies modeling is therefore a two-way process.

The greatest advantage in studying modern environments is that we can observe and measure sedimentary processes in action. We can measure current strength in the rivers and oceans; we can observe at least the smaller bedforms moving and evolving; we can measure temperatures and salinities and study the physics and chemistry of carbonate sedimentation and the critical effects of organic activity in the photic zone (shallow depth zone in the sea affected by light penetration). We can sample evaporite brines in inland lakes and deep oceans, such as the Red Sea, and deduce the processes of concentration and precipitation. We can put down shallow core holes and study the evolution of the sedimentary environment through the Recent to the present day. Many physical, chemical, thermodynamic, and hydrodynamic sedimentary models have resulted from this type of work since the 1960s. As noted in the previous section, for some sedimentary environments, work of this type has constituted the major advance in our understanding. Examples of early facies models that depended largely on studies of modern environments include those for deltas (Coleman and Wright 1975; Galloway 1975), barrier islands (Bernard et al. 1959; Dickinson et al. 1972), tidal inlets (Kumar and Sanders 1974), clastic tidal flats (Evans 1965; van Straaten 1951, 1954; Reineck and Singh 1980), tidal deltas (Hayes 1976), and sabkhas (Shearman 1966; Kinsman 1966, 1969).

Some of the most difficult problems to resolve when using actualistic models stem from the inadequate time scale available to us for observation purposes. A very persistent geologist may be stubborn enough to pursue the same field project for 10, perhaps even 20, years. Aerial photographs might push observations of surface form as far back as about 1920 at the earliest. Old maps may go back for another 100 years or more, but become increasingly unreliable. Weather records may be available back into the nineteenth century; stream gauge data have been collected for only a few decades. It is difficult to assess the relevance of a 100-year record to a geological unit that may have taken a million or more years to accumulate. Have the last 100 or so years been typical? Were the same sorts of processes occurring at the same rates in the distant geological past? As discussed in the following paragraphs the answer is a qualified maybe.

The most important aspects of this question are the difficulty of assessing the importance of ephemeral events and judging the preservation potential of deposits we can see forming at the present day. Dott (1983) and Reading and Levell (1996) have called attention to the need to distinguish between normal and catastrophic sedimentation. Normal processes persist for the greater proportion of time. Net sedimentation is usually slow. It may be nil or even negative if erosion predominates. Normal processes include pelagic settling, organic growth, diagenesis, tidal and fluvial currents (Reading and Levell 1996). Reading and Levell (1996) distinguished catastrophic processes as those that "occur almost instantaneously. They frequently involve 'energy' levels several orders of magnitude greater than those operating during normal sedimentation." They may deposit a small proportion of the total rock and give rise to only an occasional bed, or they may deposit a large proportion of the total rock and so become the dominant process of deposition. Examples of catastrophic processes include flash floods in rivers, hurricanes, and sediment gravity flows. Although geologists have studied many modern flash flood deposits and the effects of several recent hurricanes (e.g., Hayes 1967), we cannot be sure that their magnitude and frequency at the present day are the same as that in some past period of interest without attempting to obtain some geological perspective from studying the ancient record. The most violent and geologically important event may be one that only occurs every 200 or 500 years and has not yet been seen. Sediment gravity flows are thought to be the chief agent of erosion of submarine canyons, but in spite of years of oceanographic observation, no major flows have actually been observed there (Shepard et al. 1979).

Ager (1973, 1981) and Dott (1983) argued that many stratigraphic sequences contain more gaps than record, and that significant proportions of the sedimentary record are deposited in a very short time by particularly violent dynamic events. Such events (hurricanes, sediment gravity flows, flash floods, etc.) are rare and difficult to study in action. Most of our energy as sedimentologists is expended in studying the less violent processes that occupy most of geological time but may contribute volumetrically far less sediment to the total record. Ager (1986) re-examined a well-known basal conglomerate in the Jurassic Lias of Wales, and concluded that it was a storm deposit formed very rapidly. In fact, he facetiously suggested that (p. 35) "it all happened one Tuesday afternoon," a remark that was deliberately designed to draw attention to the difference between "normal" and "event" sedimentation (we return to this topic in Chap. 8).

Studies of modern environments also suffer from the fact that many of the deposits we see forming at the present day are lost to erosion and never preserved. Thus, the geological record may be biased. The bias may be in favor of more deeply buried sediments, which the geologist, scratching the surface with trenches and box cores, never sees. For example, Picard and High (1973) published a detailed study of the sedimentary structures of modern ephemeral streams, but many of the structures are sufficiently ephemeral to be rarely, if ever, found in the ancient record. Fluvial flash flood deposits commonly are preferentially preserved because they infill deep scours below the normal level of fluvial erosion (Miall 1996, Chap. 8). Many of our fluvial facies models are based on studies of modern rivers in upland regions undergoing net degradation. How relevant are they to research in some of the great ancient alluvial basin that by the thickness of preserved deposits demonstrate a long history of aggradation? Facies models for barrier and shoreface environments are subject to similar constraints. For years, the classic Galveston Island (Texas) model of coarsening-upward beach accretion cycles dominated geological thinking (Bernard et al. 1959), but more recently it was realized that the barrier sediments may be removed by lateral migration of deep tidal inlets and that many barriers consist of superimposed inlet deposits with a superficial skin of wave-formed shoreline sediments (Kumar and Sanders 1974). Hunter et al. (1979) argued that many shallow subtidal deposits are systematically removed by rip currents and are never preserved in the geological record.

Some of the best studies of modern environments are those that use shallow drill cores to penetrate into pre-Recent deposits. Such sediments can be said to have "made it" into the geological record, and yet they can be placed in the context of a still extant and presumably little modified modern environment. Some work on modern turbidites (Bennetts and Pilkey 1976), anastomosed rivers (Smith and Smith 1980), and reefs (Adey 1975; Adey et al. 1977; Shinn et al. 1979) is of this type.

In two important ways, the present is quite unlike the past, and is therefore an unreliable laboratory for reconstructing sedimentary environments. The Pleistocene ice age generated several geologically rapid changes of sea-level culminating in a major rise and transgression since about 12,000 years ago. Second, the present configuration of continents and oceans is a unique pattern, different from any in the past because of the long history of sea-floor spreading, rifting, subduction, and suturing. This plate movement has had an important effect on some of the broader aspects of facies models.

Because of the recent sea-level rise, modern continental shelf, shallow-marine, and coastal plain deposits around the world have been formed under transgressive conditions. Sea level was approximately 150 m lower during the Plio-Pleistocene glacial phases, rivers graded their profile to mouths located near the edge of present continental shelves, and carbonate platforms, such as the Bahamas, were exposed to subaerial erosion and may have developed extensive karst systems. Submarine canyons were deeply entrenched by active subaerial and submarine erosion (Shepard 1981; McGregor 1981). During the rapid transgressions that followed in interglacial phases, shoreline sands were continually reworked. On the Atlantic shelf off North America, extensive barrier islands were formed and receded into their present position (Swift 1975a). River valleys were drowned and filled with estuarine and deltaic deposits. Submarine canyons were commonly deprived of their abundant supply of river-borne detritus, as this was now deposited on the landward side of a widening and deepening continental shelf. This had a drastic effect on the rate of growth of some submarine fans. Carbonate sedimentation began afresh in warm, detritus-free waters, over resubmerged platforms, but the local water depths, circulation patterns, and facies distribution may have been partly controlled by erosional topography (Purdy 1974; although this is disputed by Adey 1978). We therefore have excellent modern analogues for studies of rapid transgressions in the geological record, but few for rapid regressions or for periods of still stand. The Mississippi and other large deltas are good examples of regressive deposits built since the last ice age, but they represent an environment that may be characterized by unusually rapid progradation. Some have maintained that most modern continental shelves are covered by relict sediments, implying that their study may not be of much geological relevance (Emery 1968), but more recent work has shown that the dynamic effects of tidal currents and storms do in fact result in continual change (Swift et al. 1971).

Miall (2014a, 2015) has argued that basing stratigraphic interpretations on comparisons with the post-glacial record is a mis-application of the principal of uniformitarianism for two important reasons: (1) the issue of preservation—the present post-glacial record exemplifies the unfinished nature of the geological preservation machine. A future glacioeustatic fall in sea level would remove much of the sedimentary record along continental margins and within estuaries, and subsequent (future) cycles of rise and fall would generate a long-term geological record by the superimposition of the lowermost fragments of successive cycle. (2) Rates of processes calculated from study of the post-glacial record (e.g., fluvial and deltaic channel avulsion; lobe switching on deltas and submarine fans) are orders of magnitude greater that the rates that can be calculated from apparently similar processes in the geological past, suggesting that uniformitarianist comparisons of processes may be incorrect. For example, Miall (2014a) suggested that changes in fluvial channel stacking patterns in the ancient record are not reflections of changing rates of sedimentary accommodation but record proximal-distal shifts in fluvial styles in response to allogenic forcing.

Rapid transgressions may have occurred commonly during other ice ages in the geological past (Miall 2010), but most other changes in the geological past were somewhat slower. For two lengthy periods in the Phanerozoic, our suite of modern analogues is particularly inadequate. These were the times of high global sea-level stand during the Ordovician-Silurian and again in the Cretaceous, when vast areas of the world's continents were covered by shelf seas. We simply do not have modern equivalents of these huge inland seas, and many of the large shelf seas that do exist, such as the Bering Sea and Yellow Sea, have yet to be studied in detail. Likewise, there are no modern analogues for the very large evaporite basins that exist in the geological record (James et al. 2010).

Ginsburg (in Byers and Dott 1981) discussed the problem of interpreting Cambro-Ordovician carbonate banks of the North American craton. These are up to hundreds of kilometers long and thousands of kilometers wide. "The dilemma is how the vast extent of the banks, most of which suggest carbonate production and deposition in but a few meters or less of water, could all be bathed in normal marine or slightly restricted water. Would not such banks form major circulation barriers?" (Ginsburg, as reported by L.C. Pray, in Byers and Dott 1981). It has been suggested that the banks are actually diachronous, and developed by seaward progradation, or they may have been crossed by "irregular to channelized deeper water areas facilitating water circulation." Tidal currents certainly would have been able to assist with the latter. Careful biostratigraphic work has now demonstrated the diachronous nature of these tidal deposits. Runkel et al. (2007) showed that the Sauk sequence in Wisconsin consists of suites of superimposed, extremely low-angle, clinoform sets.

Climatic patterns and the network of oceanic currents are controlled by global plate configurations. In many respects, our present geography is unique. Therefore, we have modern analogues for situations that may not have existed in the past and, conversely, we cannot replicate certain conditions that did exist. For example, the Mesozoic Tethyan Ocean and the Pangea supercontinent had profound effects on climate and water circulation and hence on sedimentation patterns, and we have only generalized models for interpreting them. The study of paleoclimates is beyond the scope of this book, but is considered in an excellent review by Frakes et al. (1992).

The last group of problems with the actualistic modeling method arises from the important effects organisms have on sedimentary processes. Plants and animals have, of course, both evolved profoundly since Archean time. Therefore, in many environments, sedimentation is controlled or modified by sediment-organism relations that did not exist, or were different, in the geological past (James et al. 2010). Many authors, beginning with Schumm (1968), have speculated on the implications of the evolution of land plants on fluvial patterns. Vegetation has a crucial effect on stabilizing channel banks, colonizing bars and islands, as controlling chemical weathering, sediment yield, and discharge fluctuations. Our typical braided and meandering fluvial facies models reflect these effects as they have been studied in temperate, humid, and hot and cold arid climates. Going back in time, the evolution of grasses in Mesozoic time must have changed geomorphic patterns profoundly. Abundant land vegetation is thought to have first appeared in the Devonian, and prior to that time the majority of river channels may have been unconfined, ephemeral, and braided, as in modern arid regions (Davies and Gibling 2010a, b). Modern deserts are commonly used as analogues for pre-Devonian rivers, but there is no reason why they had to be arid. However, we have no modern analogue for a humid, vegetation-free environment with which to study the pre-Devonian except, possibly, the south coast of Iceland.

Turning to shallow marine environments, the same kinds of difficulties apply. H.E. Clifton (in Byers and Dott 1981) pointed out that salt marsh vegetation, so important on modern tidal flats, did not exist prior to the Cenozoic. Similarly, foraminifera, which provide sensitive bathymetric indicators for younger sediments, particularly those of Cenozoic age, did not exist in the Paleozoic. James et al. (2010) discussed the implications of these various evolutionary developments on changes in carbonate depositional environments.

The ecology of forms that are now extinct may be difficult to interpret, which makes them less useful in facies studies (Sect. 3.5.8). Instead of providing independent, unambiguous environmental information, as do still living forms, it may be necessary to interpret them with reference to their sedimentary context, which itself may be of uncertain origin. Functional morphology is studied to determine probable habits, but many uncertainties may remain. In the Precambrian, most sedimentary units are entirely devoid of fossils, and here problems of lithofacies interpretation without supporting fossil evidence may become acute. Many Precambrian units have been reinterpreted several times for this reason, as different environmental criteria are brought to bear on particular problems. Long (1978) discussed many of these difficulties with reference to the recognition of fluvial deposits in the Proterozoic.

Reviews of Precambrian clastic sedimentation systems by Eriksson et al. (1998, 2013) documented several major differences in the preserved Precambrian record relative to that of the Phanerozoic. Marine shelf deposits are characterized by very uniform suites of sediments lacking distinctive vertical trends. Fluvial deposits are dominated by those of braided style. Foreshore deposits are rare, and eolian dune deposits appear to be absent in rocks older than about 1.8 Ga.

### 3.4.4  To Classify and Codify, or Not?

We now know, more than a decade into the twenty-first century, a great deal about the origins of sedimentary rocks, and have amassed a considerable published record of case studies documenting sedimentary processes, facies and facies assemblages in a wide variety of modern settings, plus a wealth of detailed descriptions and interpretations of ancient units (e.g., James and Dalrymple 2010). These data are in danger of overwhelming the practitioner, and summaries and classifications are more needed than ever.

Stephen Jay Gould (1989) stated: "Classifications are theories about the basis of natural order, not dull catalogues complied only to avoid chaos." In a descriptive science like geology, there will always be a need for classifications. These serve two purposes, they attempt to create order out of apparent chaos, as aids to memory, and they are an attempt to understand genesis. The best classifications are those that highlight differences in origins of geological features, and are therefore particularly useful for purposes of prediction and extrapolation, for example in geological exploration.

There will always be a creative tension between those who focus on the minute differences amongst geological attributes and those who like to simplify and generalize. This is the tension between those, the purists, who seek constantly to perfect our understanding of descriptive detail and of processes, and those, the synthesizers, who think they know enough details for the time being, and want to move onto larger problems. It is the same as the debate amongst taxonomists between the "lumpers" and the "splitters". Facies classification and facies models offer plenty of fuel for debate between the advocates of both sides of this argument. Anderton (1985) provided an excellent discussion of the various levels of complexity that are embraced by the modeling process (see also Pirrie 1998).

The purists like to point to classifications that have failed, as proof of the failure of the idea of classification. For example, Bridge (1993) used Mike Leeder's (pers. com.) example of the geochemical classification of igneous rocks proposed by Shand (1947), a complex system that did not catch on and was not used, no doubt because it was not based on any understanding of genetic reality that withstood the test of time. Likewise, Kay's (1951) classification of geosynclines was widely used until the 1960s, at which time it virtually vanished, to be replaced by classifications based on plate tectonics. Ingersoll and Busby (1995) proposed 27 varieties of sedimentary basin, based mainly on plate-tectonic principles—more types than there were geosynclines. Is this being objected to? No, in fact the contrary (e.g., see Dickinson 1997), because it is now recognized that, at least according to current plate-tectonic concepts, this is a genetically-based classification and therefore has deep meaning, much like a phylogenetic classification of organisms.

Replacement and abandonment is always likely to be the fate of classifications. But does this mean that we should avoid the attempt to classify? At the very least, classifications represent way stations on the road to perfect understanding. At best, they offer a common language for description and interpretation. This was why I proposed my classification and code scheme for fluvial deposits in 1977. Numerous descriptions of modern and ancient deposits were accumulating in the literature at that time, but most were described using unique systems of nomenclature, rendering comparative studies almost impossible. Reinterpretation of the descriptions in terms of a standard classification revealed a number of common themes in terms of facies assemblages and vertical profiles (Miall 1977), and this began the work of recognizing and systematizing the variety of processes in braided fluvial systems that is still continuing. The classification of fluvial sandstone facies is based in part on fluvial hydraulics. It has been demonstrated that common bedforms are predictable in form and are generated under predictable conditions (Ashley 1990; Sect. 3.5.4). This justifies the erection of a simple classification of those lithofacies that represent the preserved deposits of these bedforms. Revisions and improvements to this classification have not ceased. An updated version was offered by Miall (1996), but other workers have developed their own, and in fact the system has been adopted and adapted several times by others without reference to the original work (e.g., Uba et al. 2005), indicating that the ideas have, to a considerable extent, developed a life of their own.

Turning to fluvial facies models: the explosion of descriptive work in fluvial sedimentology has led to the discovery of numerous variants on the original simple suite of models, and this has led to a variety of attacks on the concept of the model itself, rather than a recognition that the models were only intended as mental concepts, as teaching aids, and as temporary fixed points to aid in the comprehension of nature. The proliferation of facies studies led Dott and Bourgeois (1983) to remark that by the early 1980s fluvial facies models had "multiplied like rabbits so that every real-world example now seems to require a new model. Such proliferation defeats the whole purpose of the conceptual model by encouraging excessive pigeon-holing, which obscures rather than reveals whatever unity may exist among the variants". It was largely to manage this problem that I began to explore the concept of the architectural element as an object capable of simplification and classification at a sub-facies model level (Miall 1985).

Some textbooks detailing simple classifications have been very successful. The Geological Association of Canada "Facies Models" volume, now in its fourth edition (James and Dalrymple 2010), has sold more than 70,000 copies. This work is popular because it is useful in teaching and preliminary interpretation, but nobody ever claimed that the book represents the last word in description and interpretation.

Warnings about the dangers of oversimplification in the use of classifications and models are common in the literature describing them (e.g., see Miall 1980a, 1985, 1996, on my own fluvial models, and the general remarks of Bridge 1993), but this does not prevent such misuse. Don Gorsline (Personal communication 1984) has told the story of a graduate student of his who came to him deeply upset one day and about to quit the program, because he could not make his rocks fit any of the standard models, and felt that he had failed as a sedimentologist. Gorsline had to gently point out to him the purpose of research, which is to do original things.

The debate continues. In Shanmugam's (1997) fascinating and provocative review of the use of the Bouma sequence model he states: "Miall (1995b, p. 379) asks, '… who would now object to the use of Bouma's (1962) five divisions (A–E) as a framework for the field description of turbidites?' I, for one, would." And then, in the Geotimes (February, 1988, p. 32) review of advances in clastic sedimentology Neil Wells wrote, with reference to the book "The geology of fluvial deposits" (Miall 1996) the opinion that the "codification of facies and architectural elements [in this book] risks promoting rigid classification, superficial observation, and simplistic interpretation."

The use of predetermined classifications in clastic sedimentology was debated at length by Bridge (1993, 1995) and Miall (1995b, 1996, p. 78, 89) with regard to the specific application of classification techniques to fluvial deposits. Such classifications are widely used, but their existence causes unease to some researchers. Shanmugam's discussion of the Bouma sequence concept and the turbidite mind-set shows why this is so. He has demonstrated rather thoroughly the misuse of the original concept by subsequent workers. The misuse of the turbidite model by workers on deep-marine sandstones had even led to the description of something called the "non-turbulent turbidity current", which Shanmugam rightly called the "ultimate example of an oxymoron".

What has happened here, and is there a systemic problem with certain sedimentological methods? I do not think so. Shanmugam's critique of the Bouma model does not mean we should abandon the model, only use it more critically. Shanmugam's work is, in fact, an application of Walker's approach that we use a model as a guide to future observations in order to determine whether they fit or not. He also refers to Anderton's (1985) critique of modeling techniques and methods. But it is precisely because we have in our minds a good concept of what a "true" Bouma turbidite should look like, that we can readily appreciate how far off the track many sedimentological descriptions and interpretations have strayed, when someone like Shanmugam comes along and brings us up short with careful, detailed observation. It turns out that many deep-marine sands may not be turbidites at all. But this does not prove that Bouma was wrong or that the concept of the facies model is wrong, only that these elegant concepts have been applied too carelessly.

Sequence stratigraphy, both as a body of concepts for the teaching of stratigraphy, and as a research tool, has been heavily dependent on models, especially the by-now virtually classic papers of Posamentier et al. (1988), Posamentier and Vail (1988), and van Wagoner et al. (1990). Yet this area, too, has been bedeviled by the misunderstanding and misuse of models, to the extent that the original proponents of the models have had to become apologists for them, pointing out in such overview papers as Posamentier and James (1993) and Weimer and Posamentier (1993) that the models were never intended to be of universal application. These controversies have now largely been resolved by the comparative, integrative work of Catuneanu (2006; see Chap. 5).

There is no question but that classifications and models in sedimentology have been used in too facile a manner as research tools by many workers, but this does not mean that the approach is wrong, only that all due caution is required in the application of the models. Classifications and models that genuinely reflect variations in the genesis of sedimentary units are likely to survive, thrive and be used, possibly with numerous revisions and accretions. Those that do not will be ignored.

### 3.4.5  Facies Analysis and Sequence Stratigraphy

As noted by Nummedal (in Nummedal et al. 1987, p. iii),

> The interpretation of the sedimentary geological record has been greatly stimulated over the past few years by rapid conceptual advances in "sequence stratigraphy", i.e., the attempt to analyze stratigraphic successions in terms of genetically related packages of strata. The value of the concept of a "depositional sequence" lies both in the recognition of a consistent three-dimensional arrangement of facies within the sequence, the facies architecture, and the regional (and inter-regional) correlation of the sequence boundaries. It has also been argued that many sequence boundaries are correlatable globally, and that they reflect periods of sea-level low stand, i.e., sequence boundaries are subaerial erosion surfaces.

Sequence stratigraphy is not a radically new concept (Sect. 1.2.10). For example, many of the sedimentological ideas were discussed by Frazier (1974), and the first assemblage of sequences, in the modern sense, was established by Sloss et al. (1949). But the study of sequences has received considerable impetus since developments in seismic stratigraphy in the late 1970s reminded geologists of the ubiquity of sea-level changes throughout the geologic past.

Sequence stratigraphy is a combination of lithostratigraphy, allostratigraphy, and facies analysis. The vertical arrangement of facies is interpreted in terms of Walther's law, leading to reconstructions of the lateral movement through time of depositional environments (Chap. 5). Sequences typically consist of transgressive and regressive half-sequences that together comprise repeated cyclic successions. Up to this point the study of sequences does not differ greatly from the analysis of vertical profiles for the purpose of documenting autogenic depositional events, for example, the progradation and lobe switching of a deltaic coastal plain complex. However, as noted by Nummedal, careful regional lithostratigraphic, allostratigraphic, and chronostratigraphic correlation demonstrates that, typically, the sequences extend laterally for much greater distances than could be attributed to autogenic controls. The influence of a regional or global allogenic control, such as changes in relative sea level, may then be suspected (Miall 2010).

## 3.5  Review of Environmental Criteria

In Chap. 2, a summary of major environmental indicators is given under the heading of what to look for in outcrop and subsurface sections. This list is by no means exhaustive; for example, it concentrates almost exclusively on what can be seen with the naked eye. However, it includes many of the sedimentary features vital to a generalized environmental interpretation and hence to an effective basin analysis. In this section, some indications are given of how to use this information. Beginners at the art of basin analysis, or those who wish to understand what their specialist colleagues are up to, may find this section both enlightening and confusing. The numerical precision or well-defined statistical error of laboratory-oriented geological studies is not to be found in the area of facies interpretation. Statistical methods may be used to aid in analyzing the composition of facies assemblages, but the business of interpreting their meaning depends heavily on qualitative study. The sedimentologist must be aware of the meaning and limitations of all the facies criteria visible and must be able to weigh the evidence of all these against each other. A good knowledge of published facies models and modern analogues is, of course, essential.

A common basin-analysis problem is that the geologist is faced with a new outcrop or core showing certain assemblages of lithologies, textures, structures, and, perhaps, fossils and may have no idea which facies model to turn to for assistance in making an environmental interpretation. The intent of this section is to review very briefly the types of interpretation that can be made from the principal kinds of sedimentological observation in order to provide some environmental clues and an entry into some of the crucial literature. Discussions of the physical and chemical conditions of formation of most sediment types are beyond the scope of this book.

In some of the notes that follow, siliciclastics and carbonates are treated separately because they require a different approach. This is particularly the case with interpretations of grain size and texture. For others, such as hydrodynamic sedimentary structures, there is no reason not to consider all rock types under the same heading. Structures such as crossbedding and ripple marks are commonly regarded as the domain of the sandstone specialists, but they also occur in carbonates and evaporites. Chemical sediments are often studied by geologists whose first interest is in the chemistry of their formation and diagenesis, but could usually benefit from the approach taken to sedimentology by clastic sedimentologists.

The emphasis throughout is on features that are most useful for constructing depositional environments and paleogeography. Diagenesis and geochemistry are not dealt with in this book.

It cannot be overemphasized that very few sedimentological criteria have an unambiguous environmental interpretation. Many years ago it was thought that ripple marks only occurred in shallow water, until oceanographers started taking photographs of the bottom of the oceans. More recently dish structures were thought to be indicators of submarine grain flows, but have now been found in fluvial and other deposits (Nilsen et al. 1977). The common trace fossil *Ophiomorpha* is a good indicator of shallow marine environments but J. Coleman (Personal communication 1976) reported finding it many miles inland in the deposits of the modern Mekong River. Marine it certainly is, but the Mekong River has an extensive marine salt wedge that flows far upstream during high tide. *Ophiomorpha* has also been found in outer shelf environments (Weimer and Hoyt 1964). Reliance should never be placed on a single structure or feature of the rocks or on a single method of analysis for making environmental interpretations. The geologist must review and assess all the evidence available.

## 3.5.1 Grain Size and Texture

In a general sense, the grain-size of a clastic sediment indicates the relative amount of energy required to emplace the grains in their final resting place. This energy might have been derived from the force transmitted by air or water movement or it may represent downward movement under gravity. Most clastic sediments represent a combination of both of these processes.

Grain-size interpretation in siliciclastic rocks (Fig. 2.4) tends to be much simpler than it does in carbonates and evaporites because diagenesis in these chemical sediments frequently obscures original grain relationships. The grain size of carbonate sediments may be related entirely to in-place primary or diagenetic crystal growth and not at all to transport processes. The size of allochems in carbonate sediments is commonly determined by their organic origins. However, some chemical sediments show evidence of having behaved as clastic detritus at some stage in their formation, so that considerations of grain dynamics might provide useful environmental information.

There are two ways in which general grain-size data yield environmental information in siliciclastic sediments. Local vertical variations in mean or maximum grain size are frequently cyclic or rhythmic, and these, coupled with variations in sedimentary structures, are powerfully diagnostic. Analysis of vertical profiles for cyclic patterns has been

discussed earlier in this chapter (Sect. 3.4.1) and is dealt with in greater detail in Sect. 3.5.10. The second aspect of grain-size information is the environmental information contained in the size distribution of individual samples. This subject has received an enormous amount of attention from sedimentologists, who have proposed a wide variety of statistical techniques for distinguishing the deposits of various depositional environments on the basis of some supposed environmental signature retained in the sample. Most of this attention has been directed toward the study of sandstones. It has long been realized that the hydraulic sorting effects of waves, wind, and unidirectionally flowing water result in the movement of different populations of grains. Most sandstones are mixtures of several populations, but it has long been the hope that the right analytical methods would infallibly recognize these subpopulations based on some size or sorting criteria, enabling the depositional environment to be recognized. All such techniques depend on careful laboratory size analyses, preferably carried out using sieves or a setting tube or, failing this in the event of lithified samples, counts of grains in thin section. Various methods were discussed by Visher (1969), Solohub and Klovan (1970), Glaister and Nelson (1974), Friedman and Sanders (1978), and Friedman (1979). None of these methods is fully reliable because of the problem of inherited size distributions in the case of second or multicycle sands, and the effects of diagenesis (cementation, secondary porosity development) on lithified sandstones. Extensive laboratory work and data analysis are required to complete a rigorous grain-size analysis. Even then the results are usually ambiguous and, in this writer's opinion, do not justify the great effort expended. Accordingly, the use of this tool is not normally recommended. The method does, perhaps, have some uses in the analysis of well cuttings or sidewall cores, where these are the only samples available, because the technique can be applied to small samples, whereas most of the observational techniques discussed here cannot. Even here, visual inspection of petrophysical logs may provide equally reliable environmental information in a fraction of the time. A new, comprehensive, descriptive classification of clastic sediments, including those consisting of very mixed grain-size populations, was offered by Farrell et al. (2012).

A useful technique in the study of conglomerates is the measurement of maximum clast size. Normally, this is determined by averaging the intermediate diameter of the 10 largest clasts present at the sample level. Such measurements have been used in the study of grading and cyclic changes in subaerial alluvial fans (Gloppen and Steel 1981; Steel 1974) and in subaqueous resedimented conglomerates (Nemec et al. 1980). It is found that there is a direct relationship between maximum particle size and bed thickness for deposits formed by subaerial or subaqueous debris flows, indicating that the beds were formed by single depositional events without subsequent reworking (Bluck 1967; Nemec et al. 1980).

Grain size, fabric, and texture are, together with sedimentary structures and vertical profiles, important criteria for distinguishing flow type in the study of sediment gravity flows and their deposits. The recognition of particular grain-size populations in a deposit, the internal sorting and grading of each population, and the vertical and lateral relationships of the populations to each other have been interpreted in terms of grain-support mechanisms and their evolution during the passage of flow events (Middleton and Hampton 1976; Lowe 1979, 1982; Postma 1986). Magnetic fabrics have been used to help distinguish sediment gravity flow deposits from other types of deposits. For example, Shor et al. (1984) used this technique in an attempt to discriminate turbidites from contourites. Gravenor and Wong (1987) and Eyles et al. (1987) employed magnetic-fabric data in studies of glacigenic deposits.

Grain-size and textural studies in conglomerates were used by Walker (1975) to suggest models of deposition on submarine fans (Fig. 3.10). The disorganized-bed model is thought to represent rapid, clastic deposition on steep slopes, perhaps in the feeder submarine canyon at the head of a fan. Debris flows passing to the inner fan are thought to exhibit first an inverse- to normally graded texture, passing down current into a graded-stratified type. Inverse grading develops as a result of dispersive pressure, possibly including a "kinetic sieve" mechanism whereby smaller clasts fall down between the larger ones. Hein (1982) extended this work to pebbly and massive sandstones.

Similar studies of texture and fabric in glacigenic deposits have helped to distinguish the various mechanisms of formation of poorly sorted conglomerates. Early workers referred to all glacigenic conglomerates as **till** (tillite is the lithified variety), a deposit formed by the accumulation of unsorted debris below or in front of grounded glacial ice. However, careful facies analysis of many Quaternary and more ancient glacial deposits has shown that most are water laid (Eyles and Eyles 2010). Analyses of clast fabric and the magnetic fabric of the matrix have been used as a supplement to outcrop facies studies in order to determine depositional processes (Eyles et al. 1987). Glacial conglomerates may have originated as tills, but have moved down local depositional slopes as slumps or sediment gravity flows or by the accumulation of material moved into the basin by ice rafting. Fabric studies can help distinguish these processes. Eyles et al. (1983) suggested referring to all poorly sorted conglomerates by the nongenetic term **diamictite** (unlithified variety: diamict or diamicton), and they erected a lithofacies classification for diamictites that places primary emphasis on the presence or absence of grading and stratification, and the density of clast packing (Sect. 4.2.4). Schultz (1984) applied this same classification to the analysis of diamictites formed in a subaerial alluvial fan setting.

In the conglomerate deposits of alluvial fans and other gravelly rivers, grain-size variations commonly reveal crude stratification and large-scale crossbedding (Fig. 2.9b). The stratification is the product of longitudinal-bar growth or superimposition of debris flows. Individual bar deposits may show an upward decrease in grain size (Miall 1996). True grading is rare in fluvial gravels formed by traction transport.

Many resedimented conglomerates and pebbly sandstones contain a clast fabric with the long (a) axis dipping upstream, indicating that they were deposited from a dispersed sediment mass without bedload rolling (Walker 1975;

**Fig. 3.10** Stratigraphic-textural-fabric models for resedimented conglomerates (Walker 1975)

Hein 1982). This contrasts with the common imbrication of flat clasts in conglomerates deposited by traction transport, such as in gravelly rivers (Rust 1972). The fabric is therefore a useful indicator of transport mode.

For most carbonate sediments, a different approach must be taken to grain size and texture. Most grains, both fine and coarse, are locally produced by organic activity. The mean grain size or the size of the largest grains may mean little in terms of local hydraulics. Dunham (1962) has pointed out that it is more useful to focus on the fine material in the rock, because this is an accurate measure of the strength of winnowing currents. Dunham (1962) devised a system for classifying the texture of carbonate rocks based on whether the sediment is a self-supporting grain framework or whether the grains are enclosed in a micrite matrix. The spectrum of packing categories erected by Dunham is shown in Fig. 3.11. The terms **mudstone**, **wackestone**, **packstone**, **grainstone**, and **boundstone**, are widely used for describing the texture of carbonates. In general, the amount of current winnowing energy implied by the rock type name increases from left to right. A similar carbonate textural spectrum was devised by Folk (1962), who used a two-part terminological system (Fig. 3.11). The suffixes **micrite** and **sparite** are used for the predominant matrix type. Sparry rocks are packstones and grainstones from which the micrite matrix has been removed

by winnowing and replaced by coarse calcite cement. Documenting textural relations in this way is an important first step in subdividing the various subenvironments of carbonate platforms. Wilson (1975, Chap. 1) discussed the subject at greater length.

Schreiber et al. (1976) and Schreiber (1981, 1986) have shown that manyevaporite deposits, particularly those formed in marginal marine environments, can be treated as clastic deposits in that they show similar bedding and sedimentary structures. The same is true for some carbonate rocks, particularly oolite sand shoals (Ball 1967) and carbonate turbidites (Mountjoy et al. 1972), which are formed by clastic sorting and redistribution processes although consisting of carbonate particles. Considerations of grain size (cyclic or lateral changes, grading) and texture (packing) might provide useful environmental information in these cases.

### 3.5.2   Petrology

The composition of the major rock-forming constituents of siliciclastic sediments (including mudrocks) is not directly diagnostic of environment. Krynine (1942) suggested that many sandstone types were characteristic of particular tectonic-sedimentary environments, for example **arkoses**

**Fig. 3.11** The textural spectrum in limestones and the two basic classifications used for limestones (Dunham 1962; Folk 1962)

**LIMESTONE CLASSIFICATION**
**1. The "Dunham" classification**

| Less than 10% components > 2 mm | | | | More than 10% components > 2 mm |
|---|---|---|---|---|
| Contains lime mud (<0.3 mm) | | | no lime mud | |
| mud supported | | grain supported | | |
| Less than 10% grains >0.03 mm | More than 10% grains >0.03 mm | | | |
| MUDSTONE | WACKESTONE | PACKSTONE | GRAINSTONE | RUDSTONE |

micrite                                                                          sparite

**2. The "Folk" classification**

| micrite | sparse- to packed biomicrite | poorly washed biosparite | biosparite | coarse biosparite |
|---|---|---|---|---|

(feldspathic sandstones) supposedly represent nonmarine sediments derived from granitic orogenic complexes and **greywackes** (lithic sandstones) represent early "geosynclinal" sedimentation. These interpretations have been discarded, although Dickinson and Suczek (1979) and Schwab (1981) have shown that sandstone composition may closely reflect the plate-tectonic setting of the basin. However, this does not necessarily translate into depositional environment. For example, volcaniclastic forearc sediments may be deposited in fluvial, lacustrine, marginal-marine, shelf, or submarine-fan settings, depending on the continental-margin configuration. Dalrymple (2010b) provided an updated summary of these ideas.

Certain minor components of sandstones may be strongly suggestive of the depositional environment. For example, glauconite pellets form only in shallow marine environments (Odin and Matter 1981) and are rare as detrital or resedimented grains. Carbonaceous debris from plants is typical of nonmarine environments. Paleosol development is indicated by lenses of calcium carbonate (caliche, calcrete) and other minerals. Abundant red iron staining indicates oxygenated environments, typically either the preservation of oxidized states in detrital particles (van Houten 1973) or the production of oxidized colors during early diagenesis (Walker 1967). Red beds are therefore mostly indicative of nonmarine or high intertidal environments (Turner 1980), although there are exceptions (e.g., Franke and Paul 1980).

Within a given basin, detrital composition may reflect variations in source area or depositional environment. These data may therefore be used as a paleocurrent indicator (Sect. 6.5.1) and may assist in the stratigraphic correlation of units formed under the same hydraulic conditions. Davies and Ethridge (1975) showed that detrital composition varied between fluvial, deltaic, beach, and shallow marine environments on the Gulf Coast and in various ancient rock units as a result of hydraulic sorting and winnowing processes and chemical destruction. Thus, although composition is not environmentally diagnostic, it may be useful in extending interpretations from areas of good outcrop or core control into areas where only well cuttings are available.

For carbonate sediments, in contrast to siliciclastics, petrographic composition is one of the most powerful environmental indicators (James et al. 2010). Carbonate facies studies, therefore, require routine thin-section petrographic analysis, nowadays supplemented by the use of the scanning electron microscope and cathodoluminescence (Wilson 1975; Scholle 1978; Flügel 1982; Machel 1985). Most carbonate grains are organic in origin, including micrite mud derived from organic decay or mechanical attrition, sand-sized and larger particles consisting of organic fragments, fecal pellets, grapestone, and ooliths, all of which are produced in part by organic cementation processes, and boundstones or biolithites, formed by framework-building

organisms. Most carbonate particles are autochthonous; therefore, an examination of the composition of a carbonate sediment is of crucial importance.

These and other differences between carbonate and siliciclastic sediments are summarized in Fig. 3.12. Textural classifications of carbonate sediments were discussed in the previous section (Fig. 3.11) and have been expanded to include compositional details of framework-building organisms by Embry and Klovan (1971) and Cuffey (1985). Interpretation of ancient carbonate sediments is complicated by the fact that the organisms that generate carbonate particles have changed with time (James et al. 2010, Fig. 12). However, there are many similarities in form and behavior between modern and extinct groups, so that actualistic modeling can usually be carried out with caution (see also Sect. 4.3). Another problem is that diagenetic change is almost ubiquitous in carbonate rocks and may obscure primary petrographic features (this is briefly discussed in Sect. 4.3.1). Ginsburg and Schroeder (1973) showed that some carbonates are converted contemporaneously from reef boundstones or grainstones to wackestones by continual boring, followed by infill of fine-grained sediment and cement. Mountjoy (1980) suggested that many carbonate mud mounds may owe their texture to this process.

Most environmental interpretation of carbonates is based on thin-section examination using a microfacies description system such as that erected by Wilson (1975; Fig. 3.4 of this book). Wilson was able to define a set of standard facies belts for the subenvironments of a carbonate platform and slope (Fig. 3.13), each characterized by a limited suite of

## DIFFERENCES BETWEEN SEDIMENT TYPES

| CARBONATES | CLASTICS |
|---|---|
| Mainly shallow marine | Occur in all environments |
| Grain size commonly reflects size of organic components | Grain size reflects hydraulic energy of environment |
| Mud (micrite) reflects prolific organic activity | Mud reflects settling from suspension |
| Shallow-water carbonate sands reflect localized fixation of carbonate | Shallow-water sands reflect currents and waves |
| Carbonate buildups alter hydraulic regime | Changes in sedimentary environment reflect hydraulic regime |
| Sediments commonly cemented in place | Sediments remain un-cemented until burial |
| Sediments very susceptible to diagenesis | Sediments relatively resistant to diagenesis |

**Fig. 3.12** The differences between carbonate and clastic sediments. Adapted from James (1984a)

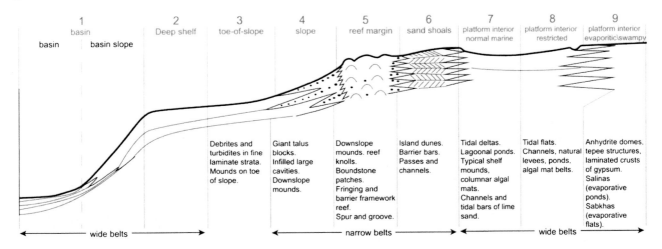

**Fig. 3.13** The standard facies belts of Wilson (1975), as modified by Schlager (2005, Fig. 4.3)

microfacies reflecting the variations in water depth, water movement, oxygenation, and light penetration. A classic example of a carbonate petrology study, the analysis of modern sediments of the Bahama Platform, is described briefly in Sect. 4.3 (see Figs. 3.7 and 3.8). Each facies assemblage can readily be related to environmental variables, such as the quiet-water pelletoidal facies and the high-energy skeletal sand and oolite lithofacies.

Many carbonates consist of dolomite rather than limestone. In most cases, the dolomite is clearly a replacement, as indicated by the presence of dolomite rhombs penetrating allochemical particles such as shell fragments or ooliths. In other cases, there may be evidence for a primary or penecontemporaneous origin, which may be environmentally useful information. Primary dolomite crusts form in association with evaporites in areas where seawater is evaporated at high rates, such as on supratidal flats and in shallow tidal lagoons. Modern examples include parts of Bonaire, sabkhas on the Persian Gulf, Deep Springs Lake, California, and Coorong Lagoon, Australia (Blatt et al. 1980, pp. 512–522; Friedman 1980). The dolomite-evaporite association, together with the evidence of certain evaporite or desiccation textures (Sect. 3.5.7) is strongly environmentally diagnostic. The dolomitization processes depends on "seepage refluxion" or "evaporative pumping" of sea water to the surface under hot, arid conditions (Adams and Rhodes 1960). However, this is only one mechanism for the production of dolomite. Most dolomite probably forms at depth by a diagenetic mixing of sea water and fresh water, a process known as the Dorag model (Badiozamani 1973; Land 1973). In rare cases dolomite can also occur as detrital grains.

The composition of evaporite minerals is not a good guide to the depositional environment of evaporites. A sample of normal seawater if evaporated to dryness yields a sequence of precipitates in the following order: calcite,

gypsum, halite, epsomite, sylvite, and bischofite. However, the composition of the final deposit in nature may vary considerably because of the effects of temperature, the availability of earlier formed components for later reaction, and the rate and nature of replenishment of the water supply. Evaporites may form in a variety of marine and nonmarine environments (Schreiber et al. 1976; Schreiber 1981; Kendall 2010), and it is not their composition as much as their internal structure and lithologic associations that are the best clues to the depositional environment (e.g. the association with penecontemporaneous dolomite mentioned previously).

The formation of many chemical sediments, such as chalk, chert, phosphates, and glauconites depends on factors such as organic activity and ocean water oxygenation and temperature (Gorsline 1984). These factors are controlled in part by large-scale oceanic circulation patterns, which are being studied in order to develop predictive models of facies development (Parrish 1983).

Certain other chemical deposits contain useful environmental information. Chert is common as a replacement mineral in carbonate sediments, where it forms nodules and bedded layers commonly containing replacement casts of fossils, ooliths, etc. Knauth (1979) suggested that such chert was formed by the mixing of fresh and marine waters in a shallow subsurface, marginal-marine setting. Chert also occurs in abyssal oceanic sediments in association with mafic and ultramafic igneous rocks. Radiolarians, sponge spicules, and diatoms are common. These are some of the typical components of **ophiolites**, which are remnants of oceanic crust and indicate a former deep water environment (Grunau 1965; Barrett 1982).

Iron-rich rocks occur in a variety of settings. Their chemistry is controlled partly by Eh and pH conditions. Pyrite and siderite are common as early-diagenetic crystals and nodules in the reduced environment of organic-rich muds, particularly in fluvial or coastal (deltaic, lagoonal) swamps. Occasionally, such deposits may be present in rock-forming abundance, as in the pisolitic bog iron ores. Plant remains, impressions, and replaced (petrified) wood are commonly associated with these forms of iron. Pyrite is also associated with unoxidized, disseminated organic particles in the black muds of anoxic lake and ocean basins.

Organic-carbon-rich black shales are of considerable importance as petroleum source beds. Arthur et al. (1984) reviewed their origins and significance in terms of oceanic organic productivity, sedimentation rate, and oxygenation of ocean bottom waters (Fig. 3.14). Ultimate controls on sedimentation are plate-tectonic configurations, which determine global sea level and climate, and which control the nature of oceanic circulation patterns. Such shales cannot, therefore, be interpreted strictly on a local basis, but must be considered in a regional or even a global context.

The increasing importance of shale-gas as an energy source has focused considerable attention on the facies analysis and petrology of mudrocks. Contrary to common assumptions, most mudrocks contain significant proportions of non-mud components, particularly quartz and calcium carbonate, with clay minerals comprising less than 50 % by volume, based on thin-section point counts (Shaw and Weaver 1965; McQuaker and Adams 2003; Fig. 3.15). When examined in detail, shales are not homogeneous, but

**Fig. 3.14** Oxygen content at the sediment-water interface and generalized relationships between types of benthic organisms, sedimentary structures, water chemistry, mineralogy, organic content, and sediment color. Notes: *1**, certain polychaete organisms and foraminifera may inhabit surficial environments where benthic metazoans are excluded; *2**, degree of lamination depends in part on seasonal variations in clastic input and productivity. (Arthur et al. 1984, reproduced by permission of the Geological Society, London)

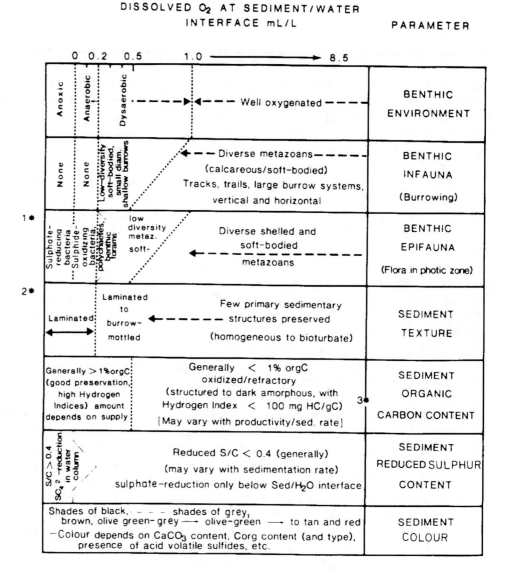

GENERAL CRITERIA FOR RECOGNITION OF
BOTTOM-WATER DISSOLVED OXYGEN
LEVELS IN MUDDY MARINE ENVIRONMENTS

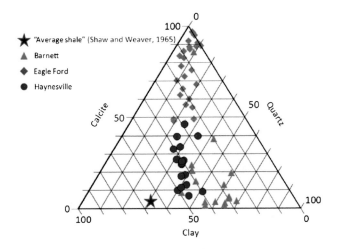

**Fig. 3.15** Composition of typical mudrocks. Note that clay mineral composition is rarely more than 50 % (Hart et al. 2013, after Shaw and Weaver 1965). AAPG © 2013, reprinted by permission of the AAPG whose permission is required for further use

vary significantly in petrographic composition and internal structures, which, in turn, significantly affects their reservoir properties. It is increasingly being recognized that mud deposits are in many cases not simply passive pelagic accumulations in quiet, deep-water environments, but are the deposits of mud floccules transported by traction currents or sediment gravity flows, and commonly exhibiting bedforms comparable to those formed in sand-bed deposits (Plint 2010; Shieber et al. 2013).

Hematite and chamosite iron ores are locally important in the Phanerozoic. Most are oolitic and display typical shallow-water sedimentary structures. As van Houten (1985) noted, these ironstones are particularly abundant in the Ordovician and Jurassic record, but display widely varying paleogeographic and paleoclimatic settings, and so their use as facies indicators is uncertain at present. The iron is probably an early diagenetic replacement. Precambrian iron formations are widespread in rocks between about 1.8 and 2.6 Ga. Their mineralogy is varied and unusual (Eichler 1976); it may reflect formation in an anoxic environment (Cloud 1973; Pufhal 2010).

Manganese- and phosphate-rich rocks are locally important. Both commonly occur as crusts and replacements on disconformity and hardground surfaces, where they are taken as indicators of nondeposition or very slow sedimentation. Phosphates are particularly common on continental margins in regions of upwelling oceanic currents. More detailed discussions of these sediments are given by Blatt et al. (1980) and Pufhal (2010).

Coal always indicates subaerial swamp conditions, usually on a delta plain, river floodplain or raised swamp. However, lacustrine coals and coal formed in barrier-lagoon settings have also been described. Calcrete (or caliche) occurs in alluvial and coastal environments and is an

excellent indicator of subaerial exposure (Bown and Kraus 1981; James 1972; Wright 1986a).

### 3.5.3 Bedding

Are the various rock types present in a stratigraphic unit interbedded in major packages several or many meters thick, or are they interlaminated on a scale of a few centimeters or millimeters (Fig. 2.7)? Is the bedding fine or coarse? Is it flat, disturbed, undulatory, distinct, or indistinct (gradational)? These questions, while rarely providing answers that are uniquely diagnostic of a depositional environment, may provide important supplementary information. In a general way the bed thickness is proportional to the depositional energy level.

Finely laminated sediments are mostly formed in quiet-water environments. X-radiography and scanning electron microscopy are commonly used to define lamination and very small-scale sedimentary structures in fine-grained rocks (e.g., see several papers in Stow and Piper 1984; O'Brien 1990). Such sediments may include laminated pelagic mudrocks, prodelta deposits, deep-water evaporites, thin-bedded basin-plain turbidites, and delta plain lagoonal and fluvial overbank muds. Laminated fluvial sheetflood sandstones are an exception to this pattern (see next section). Thicker beds form in a variety of high-energy wave- or current-dominated environments. Reef rocks, formed in extremely high energy conditions, may lack bedding entirely.

An interbedding of contrasting lithofacies may be environmentally indicative. For example wavy, flaser, and lenticular bedding (**tidal bedding**: Reineck and Wunderlich 1968; Fig. 3.16) record the alternation of quiet-water mud sedimentation and higher energy flow conditions under which rippled sand is deposited. This can occur during tidal reversals on exposed mudflats, on fluvial floodplains, or below normal wave base on the shelf, at depths affected by infrequent storm waves. Another common bedding association is that produced by the alternation of storm- and fair-weather processes on the shoreface. A sequence, basal gravel → laminated sand → bioturbated or rippled sand, indicates storm suspension followed by decreasing energy levels and then a return to low-energy wave activity and bioturbation during periods of fair weather (Kumar and Sanders 1976).

As noted in Sect. 3.5.1 maximum grain size and bed thickness are commonly correlated in deposits formed by individual sediment gravity flow events. This can provide invaluable interpretive data.

Bedding surfaces that bound depositional units have a particular importance in facies analysis, as discussed in Sect. 3.5.11.

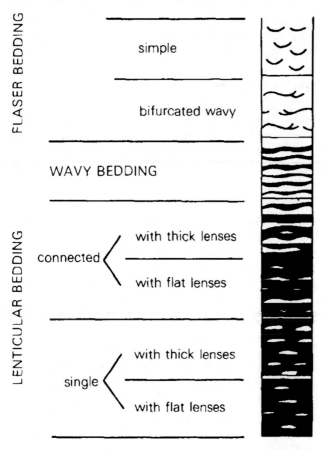

FLASER BEDDING

simple

bifurcated wavy

WAVY BEDDING

LENTICULAR BEDDING

connected
- with thick lenses
- with flat lenses

single
- with thick lenses
- with flat lenses

**Fig. 3.16** Flaser, wavy and lenticular bedding Reineck and Singh (1973)

### 3.5.4 Hydrodynamic Sedimentary Structures

Many environmental deductions can be made from the details of internal structure of hydrodynamic sedimentary structures and from orientation (paleocurrent) information. Three general groups of structures can be distinguished:

1. structures formed by unimodal water currents in gravel, sand and mud in rivers, deltas, parts of ebb and flood tidal deltas in inlets, submarine fans, and continental slopes (contour currents);
2. structures formed by reversing (bimodal) water currents, such as tides and wave oscillation in sand and mud in shelf and marginal-marine environments and in lakes; and
3. structures formed by eolian currents in coastal dune complexes, inland sand seas, and some alluvial-lacustrine environments.

Unimodal currents are readily recognized from unimodal foreset orientations (e.g., Fig. 2.8), but such patterns are not necessarily environmentally diagnostic. For example, as discussed below, it has been found that in areas of strongly

reversing currents such as tidal inlets and their associated deltas, ebb and flood currents are segregated into different parts of the system. Structures in a single outcrop of a tidal delta may therefore be misinterpreted as fluvial in origin, based on structure type and paleocurrent patterns. Simple paleocurrent models, such as those of Selley (1968), should therefore be used with caution. Other evidence, such as fauna, might yield clues as to the correct interpretation. Crossbedding structures may contain evidence of stage fluctuation, in the form of **reactivation surfaces** (Collinson 1970). These are erosion surfaces formed during a fall in the water level, but again, they are not environmentally diagnostic as water levels rise and fall in rivers, deltas, and tidal environments.

Reversing currents can be recognized from such structures as herringbone crossbedding (Fig. 2.9c) or wave-ripple cross-lamination, in which foreset dip directions are at angles of up to 180° to each other. Herringbone crossbedding is a classic indicator of reversing tidal currents, but it can also form under oscillatory wave-generated flow conditions (Clifton et al. 1971), and even in fluvial environments, where bars migrate toward each other across a channel. Because of the segregation of ebb and flood currents in estuaries and inlets, herringbone crossbedding is, in fact, not common in many marginal marine deposits. Some examples of this structure may result from the deposition from the dominant and subordinate currents that generate tidal dunes (Fig. 3.27; see Allen 1980; Dalrymple 2010c, p. 210). Some examples of herringbone cross-bedding as observed in flat outcrops may, in fact, be oblique exposures of unimodally-oriented, cross-cutting trough cross-sets.

The reversing ripples, chevron ripples, lenticular foresets, and variable symmetry and orientation of wave-formed ripple cross-lamination, as discussed below, are strongly diagnostic of a low-energy wave environment, such as a gently shelving marine beach or a lake margin (Fig. 3.29; see de Raaf et al. 1977). Similar structures could also form in abandoned meanders or floodplain ponds in an alluvial environment, but would comprise a less conspicuous part of the overall succession. In many marginal-marine environments, crossbedding will be formed by both waves and tides, resulting in very complex paleocurrent patterns (e.g., Klein 1970). Careful documentation of structure types and their orientations may be necessary to distinguish the precise environment and mode of origin, but such work may also yield invaluable information on sand-body geometry, shoreline orientation, beach and barrier configuration, etc. (see further discussion on Sect. 3.5.4.4).

Work is underway to document and classify the hydrodynamic structures that are increasingly being recognized in mudrocks as a result of the almost ubiquitous tendency of mud to flocculate and then respond to shear stress much as a bed of loose sand (Schieber et al. 2013). It has been realized that most mud rocks are probably deposited from moving water, not by pelagic settling. The lack of recognition of this important interpretive point is partly explained by the fact that bedforms are flattened out and almost disappear during burial, generating what appears to be simple lamination. Newly deposited mud is highly porous (70–80 %), and loses most of this porosity during burial compaction, with almost complete loss of the vertical dimension in preserved sedimentary structures. This can readily be corrected by directional stretching of digital images of clean mudrock surfaces, such as core slabs.

### 3.5.4.1 The Flow-Regime Concept

The interpretation of hydrodynamic sedimentary structures is one of the most important components of facies analysis, particularly in siliciclastic sediments. Certain carbonate and evaporite deposits contain similar structures and can be studied in the same way.

The basis for interpreting structures formed in aqueous environments is the **flow-regime concept**. This fundamental theory states that the flow of a given depth and velocity over a given bed of noncohesive grains will always produce the same type of bed configuration and therefore the same internal stratification. If such structures are predictable, their presence can be used to interpret flow conditions. These fundamental ideas were first enunciated following an extensive series of flume experiments by Simons and Richardson (1961) and were developed further for use by geologists by Simons et al. (1965), Harms and Fahnestock (1965), Southard (1971), and Harms et al. (1975, 1982) (Sect. 1.2.5). Valuable syntheses of hydrodynamic sedimentary structures were provided by Collinson and Thompson (1982) and Allen (1982), and reviews of later developments were provided by Leeder (1983), Ashley (1990) and Allen (1993).

It is now realized that bedforms are controlled mainly by three parameters, sediment grain size, flow depth, and flow velocity, and a series of experiments has demonstrated the sequences of bedforms produced as these parameters are varied. For example, Fig. 3.17 shows the stability fields of ripples, dunes, and other bedforms for flow depths of about 20 cm. A welter of terms for various types of bedforms in different environments was clarified by Ashley (1990), following lengthy debate by a research group under the auspices of the Society for Sedimentary Geology (SEPM). Two broad classes of flow-transverse bedforms are now defined, two-dimensional (2-D) forms which occur at lower flow velocities, and three-dimensional (3-D) forms, which occur at higher velocities for a given grain size. There is a continuum of sizes of these bedforms from small-scale ripples to very large dunes more than 10 m in height. A natural discontinuity in size occurs at a spacing (wavelength) of 0.5–1.0 m, which reflects the structure of the turbulence that generates these

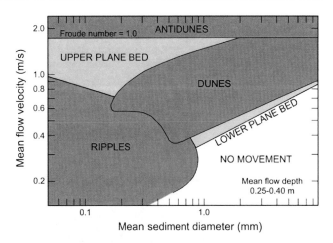

**Fig. 3.17** Stability fields of bedforms in sand and silt in a flow depth of 25–40 cm. Adapted from Ashley (1990)

bedforms (Leeder 1983). Forms smaller than this are termed ripples; Ashley (1990) recommended that those larger should all be termed **dunes**. Small-scale ripples, dunes, and sand waves are forms out of phase with surface water movement; indeed, their form may bear no relation to surface water patterns at all. These bedforms have traditionally been classified as **lower flow-regime** forms (Fig. 3.18).

Increasing the flow velocity over a field of dunes causes changes in the turbulence in the boundary layer, and this has important consequences for the geometry of the bedform that results. In particular, the separation eddies become flatter, and skewed more and more downstream. Grains are swept over the tops of dunes from which the crests are eroded in the strong currents, and distinct avalanche faces may not form. Bedforms assume a "humpback" or "washed-out" shape (Fig. 3.18). The crossbed deposit that is preserved may be characterized by the low dip of the foresets (<10°), as shown in the example of low-angle crossbedding illustrated in Fig. 2.8e.

The **upper flow regime** is characterized by **antidunes** and standing waves (Fig. 3.19), which are in phase with surface water waves. The transition from lower to upper flow regime is marked by a streaming out of transverse turbulent eddies into longitudinal eddies. An intermediate **upper flat-bed condition** is marked by streaming flow, which aligns the sand grains and produces primary current lineation (parting lineation; Fig. 2.9d).

How can these flume data be used to interpret ancient sediments? First, Allen (1968) and Harms et al. (1975) demonstrated the relationships between bedforms and sedimentary structures. For example, planar tabular crossbedding is produced by the migration of straight-crested megaripples, such as sand waves (what are now termed 2-D dunes; Fig. 3.20a), whereas trough crossbedding develops from the migration of 3-D dunes (Fig. 3.20b). Allen (1968) demonstrated the dependence of dune and ripple shape on water depth (Fig. 3.21). Second, the flow-regime concept

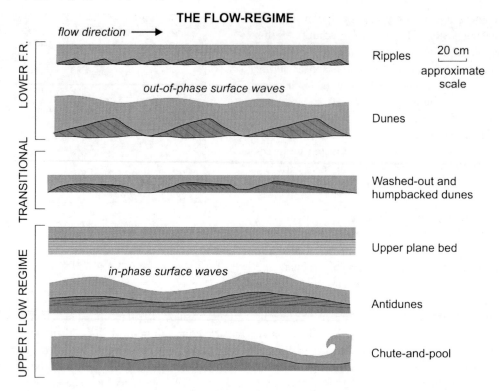

**THE FLOW-REGIME**

Fig. 3.18 The flow-regime concept, illustrating the general succession of bedforms that develops with increasing flow velocity. Dashed lines indicate areas of flow separation. Note internal stratification. Adapted from Simons et al. (1965, Blatt et al. 1980)

Fig. 3.19 *Left* Standing waves in a modern river, Alaska (*photo* N. Smith); *right* antidunes on a modern beach, South Carolina (*photo* M. Hayes)

may be used to interpret ordered sequences of sedimentary structures in terms of gradations in flow conditions. Examples of applications to fluvial point-bar deposits, Bouma turbidite sequences, and wave-formed sedimentary structures are discussed below. This is by no means an exhaustive listing. For example, the concepts have been adapted by Dott and Bourgeois (1982) to the interpretation of hummocky crossbedding, a product of storm wave activity (Fig. 3.32; see below and Sect. 4.2.8).

The size of a bedform depends on the depth of the system in which it forms and the amount of sediment available. In large, energetic, channels with an abundant sand supply, very large dunes may form, possibly exceeding 10 m in height (e.g., Fig. 3.22). Large dunes also occur on the continental shelf, in areas where there are strong tidal or oceanic currents. Numerous observations made in a range of channel types shows that there is a continuum in sizes of dunes, from those about 10 cm high, up to forms two orders of magnitude

**Fig. 3.20** Relationships between bedforms and sedimentary structures. **a** Sand waves (2-D dunes) and planar crossbedding; **b** Linguoid (3-D) dunes and trough crossbedding. Adapted from Harms et al. (1975)

**Fig. 3.21** Variations in bedform morphology with depth and velocity (Allen 1968)

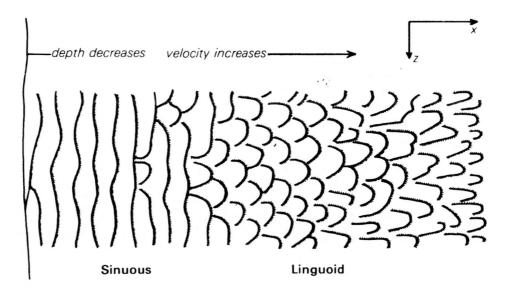

larger (Fig. 3.23). Note, however, the gap in this diagram corresponding to heights of about 4–8 cm. Bedforms of this scale are rare, because of the change in turbulence structure that takes place as flow velocity increases from the ripple to the dune stage.

There is a fairly consistent relationship between dune heights and wavelength with flow depth. This may be useful in estimating water depths or channel size from preserved dune deposits in the ancient record.

### 3.5.4.2  Bedform Preservation

In nature, bedform conditions change continually in response to changing flow conditions. What finally gets preserved into the rock record is an important consideration, because it is the final preserved result that we need to be able to understand in order to interpret the sedimentary record. For example, the gradual fill of a river or tidal channel is commonly recorded by a succession showing an upward decrease in grain size and an upward change in bedform type

**Fig. 3.22** An example of a large, simple, planar-crossbed set formed in a deep channel. Devonian Old Red Sandstone, Scotland

**Fig. 3.23** The range of ripple and dune sizes observed in all depositional setting. The two darker areas indicate clusters of particularly common dune sizes (adapted from Ashely 1990)

from forms higher to ones lower in the flow-regime suite, reflecting the gradual reduction in flow velocity and depth in the channel. It is common for the base of a channel fill to be characterized by dune deposits, with the upper part of the fill showing abundant ripples. The changing suite of crossbed structures is part of the description of the standard fining-upward cycle of channel-fill deposits (Sect. 3.5.10), and similar suites occur in other cycle types.

The actual act of preservation may come about because of a sudden local switch in flow directions, which leaves the last-formed deposit abandoned. Antidune and chute-and-pool

structures are rare in the rock record, because typically they are modified back into a lower-energy structure, such as the plane bed condition, as flow velocities drop. Their preservation may, therefore, indicate a sudden change of this type. Consider, for a moment, how bedforms advance along a bed. Given a space on the floor of a channel or an area of the continental shelf, so long as the sediment supply is constant and current strength remains the same, each bedform will migrate in the direction of the current, to be replaced by the one behind it. Under these conditions, how do bedforms ever stack upon each other to form successions that may consist of several or many individual crossbed sets one above the other? Usually, the scour hollow in the lee of the bedform crests is slightly erosive. As each bedform advances, the top of the one in front is stripped away. Complete bedforms are rarely preserved. Stacking to form multiple crossbed sets may be simply a case of successive trains of bedforms filling available accommodation space.

Under certain conditions, successive bedforms climb up the backs of each other without intervening erosion. This occurs when sediment is settling from suspension in addition to the sediment rolled or bounced along the bed by traction. The structure so produced is called climbing ripples (Fig. 2.8 b). The angle of climb increases with the amount of suspended sediment in the mix (Fig. 3.24). The right-hand image in Fig. 3.24 shows deposition and preservation of the upcurrent (stoss) sides of the ripples, because of the addition of suspended sediment to each ripple as it forms. As a result, the angle of climb is high.

Examples of planar and trough crossbedding are illustrated in Fig. 3.25. Illustration A shows five virtually identical sets of planar-tabular cross-bedding, with almost identical foreset orientations, indicating a consistency in channel orientation. Diagram B shows several large-scale crossbed sets more than 1 m thick. Diagram C shows intersecting trough sets in cross-section view, while Diagram D is a bedding-plane view of trough crossbbeding.

### 3.5.4.3   Bedforms and Crossbedding in Gravels

Most of what we have been describing up to this point concerns sand systems. However, gravel also forms bedforms, and these are preserved as crossbedding in many conglomerate deposits formed in rivers (Fig. 2.9b), on beaches, and in some submarine deposits that are otherwise dominated by sediment-gravity flows.

Individual clasts are oriented by flow (Fig. 2.12). We have not discussed this in the case of sand grains, because it requires specially oriented samples and careful laboratory examination to measure and document grain orientation. Nevertheless, such orientation measurements have been made on a number of sandstone bodies, and provide a useful supplementary form of paleocurrent measurement. In the

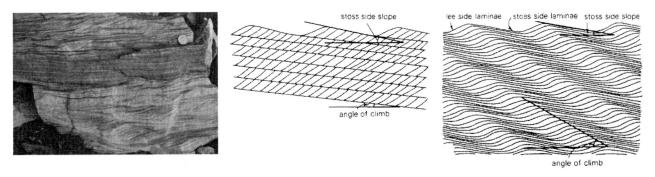

**Fig. 3.24** The formation of climbing ripples. Flow is from left to right in all three images. Photo at left is of a Paleogene unit in Ellesmere Island, Arctic Canada; Diagrams are from Collinson and Thompson (1982)

**Fig. 3.25** Examples of planar (**a**, **b**) and trough (**c**, **d**) crossbedding

case of gravels, the orientation of individual particles is much more obvious, and more readily measured in outcrop (Fig. 2. 12). The reasons for imbricated orientations are explained in Fig. 2.12, which makes it clear that this is simply an issue of clast stability under the force of moving water. Once a clast assumes a position dipping upstream it is more likely to remain in this position, and become buried and preserved into the sedimentary record holding such an orientation.

### 3.5.4.4 Structures Formed by Reversing (Tidal) Currents

Flow-regime concepts apply to tidal settings, with the difference that currents directions reverse on a diurnal (daily) or semi-diurnal (twelve-hour) timetable, as tides flood and ebb. The details of where and how this occurs, and the relationship to other processes, including river influx, wave action, and so on, are discussed in Sect. 4.2.5. Here we focus on the specifics of how reversing tidal currents generate distinctive bedforms and crossbed structures.

The key to understanding the distinctiveness of tidal structures is to realize the importance of the process of current reversal. During the flood or the ebb, currents in tidal channels are typically comparable in strength to those occurring in rivers of comparable channel dimensions. Bedforms and crossbed sets of similar size and geometry are formed. However, fluvial currents are unidirectional, and vary in strength mainly on a seasonal basis. Tidal currents slow and stop altogether every twelve hours (diurnal currents) or every six hours (semi-diurnal currents). During the slackwater phase, mud may be deposited, and then current strength picks up again, but in the reverse direction. These processes are shown in diagrammatic form in Fig. 3.26. The "first mud drape" formed during a slackwater phase may be partially re-eroded when the tidal current resumes in the opposite direction. Commonly, this current will plane off the top of the earlier dune deposit, forming a curved **reactivation surface**, while ripples may form at the base of the set, where current strength is lower. What commonly characterizes tidal dune deposits are the rhythmically-spaced reactivation surfaces denoting regular intervals of current cessation and/or erosion as the dune forms.

## THE DEVELOPMENT OF TIDAL MUD-SAND COUPLETS

Fig. 3.26 The succession of events as tidal currents pass through a slackwater condition and then are renewed in the opposite direction (adapted from Visser 1980)

Sand waves that develop in tidal environments range from 1 to 15 m in thickness and have characteristic internal structures, as described by Allen (1980, 1982). In detail, the geometry of tidal bedforms depends on the relative strengths of the flood and ebb currents. As we show in Sect. 4.2.5, these are usually not equal locally, with one being dominant over the other in terms of current strength, sediment transport, and water depths during the most active phase of sedimentation. The result can be a complex construction of reversing crossbedding, the relative size of the sets corresponding to each phase of the tide providing some indication of the relative strengths of the tidal phases (Fig. 3.27). It is not at all uncommon for some areas of tidal deposition, particularly in the middle of tidal deltas, at the inner or outer mouths of tidal channels, to be dominated by one phase of the tidal current, with the result that, locally, the crossbed structures formed and preserved into the rock record appear unimodal and quite fluvial in character (Fig. 3.27). It may be necessary to carefully search for other evidence of marine influence, such as fossil evidence, to arrive at the correct environmental interpretation.

Another distinctive character of many tidal deposits is the variation in current strength and resulting deposit thickness that reflects the monthly evolution from spring to neap tides and back again. Spring tides are the stronger tidal currents and higher tidal range that occur when the earth and moon are acting together, whereas the weaker, neap tides occur when the sun and the moon exert their gravitational force in divergent directions, reducing the scale of the global tidal circulation. The result may be rhythmic variations in the thickness of tidal bed sets (Fig. 3.28). Meticulous work on modern tidal flats in the Netherlands and some ancient deposits has shown how the thickness variations may be related to the specifics of the sun-moon system at the time of deposition (Visser 1980; Williams 1989). Many other subtle indicators of tidal influence were described by Reineck and Singh (1980) and Nio and Yang (1991).

The alternation of strong and weak currents is shown at a smaller scale, within the ripple deposits that form on most tidal flats. A terminology for these various forms of so-called **tidal bedding** was evolved by Dutch workers, and is shown Fig. 3.16, and examples are illustrated in Fig. 3.29. **Flasers** are the small mud drapes that develop over ripples during slackwater phases. Wavy and lenticular bedding occur when there is a limited sand supply on a tidal flat, and ripples may not join together to form continuous beds, but occur as isolated ripple crests.

A word of caution: tidal bedding is a very distinctive lithofacies in tidal settings, but its presence is not definitive proof of a tidal environment. The alternation of sand and mud indicates fluctuating current strengths, but this can also occur on river floodplains on a seasonal or more irregular basis, and can occur on the deeper parts of continental

**Fig. 3.27** Tidal deposits formed under the influence of reversing flow. *Left* Ebb and flood have comparable strengths. The result may be herringbone crossbedding. *Right* The more unimodal appearance of crossbedding formed when one phase of the tide is strongly dominant, resulting in affectively unimodal current strengths. Diagrams adapted from Allen (1980)

shelves, where occasional storm waves may stir the sediment on a sea floor that is otherwise a tranquil environment of mud deposition. In making any kind of environmental interpretation from the rock record, the geologist should use all the facies indicators that he or she can observe.

### 3.5.4.5 Structures Formed by Oscillating Currents (Waves)

Clifton et al. (1971) carried out one of the first detailed studies of the sedimentary structures that form on coastlines under breaking waves. They recognized a direct relationship between wave type, resulting water motion, and structure type (Fig. 3.30). Waves are generated by wind shear across a water surface. The greater the distance over which this takes place (called wave "fetch") and the stronger the winds, the greater the amplitude of the waves that are generated. The water motion that is expressed as surface waves is circular, with the diameter of the circle increasing with wave height. The circular motion is transmitted to the water column below (Fig. 3.30). Under average conditions this motion extends downwards for 5–15 m, but under storm conditions, it exceeds 20 m, and may reach as much as 200 m. Where the waves touch bottom, the circular motion is expressed as oscillation back and forth, the flow velocity usually being

adequate to move sediment of sand grade. Waves are, therefore, a significant factor in the generation of sediments on continental shelves and the bottom of lakes. Their effects are even more pronounced at shorelines, where the waves shoal, steepen and break, with the familiar rush of water up the beach slope and then back.

The nature of the deposits formed beneath shoaling and breaking waves was first studied by SCUBA divers on the Pacific coast of Oregon. The divers anchored themselves to the sea bottom so that they could make observations in this very energetic environment. The results of their studies are shown in Fig. 3.30. It was found that there are distinct zones of bedforms in the nearshore to offshore, corresponding to the evolving state of the waves, as they approach the beach. Flow regime concepts can be adapted to understand the formation of these zones.

In offshore areas, as the oscillatory motion of the waves begin to touch bottom, sand may be moved into ripples. With continued approach of the wave toward the beach, the velocity of the oscillation increases as frictional shear at the base of wave increases. A zone of three-dimensional dunes is formed. The zone of highest energy is that beneath the breaking wave, and extending immediately to the landward of that. Here, the characteristic bedform is a plane bed.

**NEAP-SPRING BUNDLING
OF TIDAL DUNE CROSSBEDDING**

Neap bundles    Spring bundles    Neap bundles

**Fig. 3.28** Current speed during individual flood-ebb cycles is shown by the successive peaks in the *upper diagram*. The peaks are joined by the smooth curve at the *top*, which traces out the rise and fall in maximum current strength from neap to spring and back to the neap condition. Areas coloured in *brown* indicate those periods when sand transport is possible. These show a maximum duration around the time of the spring tides, as reflected by the increased bundle thicknesses formed at such times (*lower diagram*). Adapted from Yang and Nio 1989)

These three zones (Fig. 3.30) correspond to the progression through the lower flow regime to the beginning of the upper flow regime, as flow velocity is increased in a flume.

The "inner rough zone" is an area of rapid change in flow direction and energy. Lower flow-regime bedforms are the result. Finally, the swash zone, where the water from the breaking wave rushes up onto the beach and then flows back or is infiltrated, is a zone where flow velocity may be high for a few seconds, but water depths are small. This is called the "inner planar zone," and is an area where flat bed conditions may develop, but on gently shelving beaches this is commonly an area of wave-ripple formation (Fig. 3.31).

Wave ripples are typically symmetrical in external form, and have distinctive inner structures, reflecting the rapidly alternating flow directions. The points noted in the upper diagram of Fig. 3.31 may be used to rapidly identify wave ripples in outcrop and drill core. These are small structures, and close examination is necessary.

In the rock record it is not expected that the zones identified in Fig. 3.30 would be perfectly preserved exactly in their original configuration, because the zones normally move up and down the shore with the rise and fall of the tides. As a beach deposit accumulates by regressive sedimentation, it could be predicted that there would be some overlap and mixing of the bedforms representing adjacent zones. As shown in Sect. 4.2.5, however, the overall succession of

**Fig. 3.29** Examples of tidal bedding. *Top* ripples on a modern tidal flat, near Seattle; *bottom* wavy bedding in a Cretaceous unit, Utah

lithofacies generated by beach regression can readily be explained with reference to the general zonation of transport energy and bedform assemblages shown in Fig. 3.30.

### 3.5.4.6 Storm Sedimentation and Geostrophic Flow

In the mid-1970s, once the description of what came to be known as **flow-regime bedforms** (and their preserved equivalents) became routine, sedimentologists studying the ancient record began to notice types of crossbedded structures that did *not* fit the standard descriptions. Amongst these, a particularly distinctive type was a variety of low-angle crossbedding with both concave- and convex-up curvature. In plan view, on bedding planes, these structures had the appearance of low domes, or mounds (Fig. 3.32). The context of these structures—the presence of shallow marine fossils, the stratigraphic bracketing between coastal and deep-water deposits—all suggested that these bedforms were formed on the continental shelf, but nobody had observed anything of this type actually present, or forming, on modern continental shelves. This presented something of a problem, until marine geologists, going back over their records, and carrying out new, targeted surveys of the sea

**Fig. 3.30** The formation of bedforms beneath breaking waves. Adapted from Clifton et al. (1971)

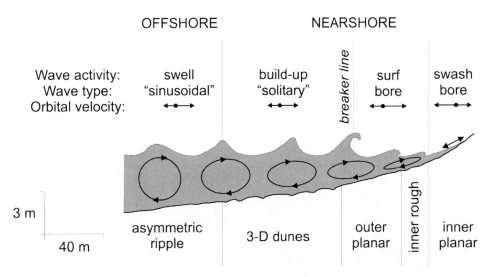

# WAVE BUILD-UP AND BEDFORM ZONES ON A SANDY COASTLINE

floor, were able to identify the structure and confirm that, indeed, this structure is a characteristic feature of the continental shelf. The descriptive name **hummocky cross-stratification** was coined for this new structure (Byers and Dott 1981; Dott and Bourgeois 1982). The acronym HCS is often used as a convenient shorthand (Harms et al. 1975; Fig. 3.36).

It is now known that HCS is the product of what is called **combined flow**, that is, both waves and ocean currents are involved. This combination occurs during storms, when wind pressures lead to the generation of temporary **geostrophic currents**. These, combined with the effects of deep scour by storm waves, cause a considerable amount of sand transport across the shelf (Swift et al. 1983; Nottvedt and Kreisa 1987; Allen 1993; Plint 2010).

The principle of the geostrophic current is illustrated in Fig. 3.33. Storm winds approaching the shore push coastal waters onland. This coastal **set-up** may raise local water levels by as much as 5 m. The set-up is part of a temporary overturning of water, with a return flow at depth, across the continental shelf. The return flow is influenced by the Coriolis effect, and may be diverted to flow obliquely across the continental shelf, as shown in Fig. 3.33. At the same time, the oscillatory sediment motion induced by large storm waves reaches deep into the water column, and adds a back-and-forth component to the transport of individual sand grains. The overall result of this complex sand movement pattern is the development of the mounds, or hummocks, that we now recognize as HCS. The local pattern of growth of the hummocks is indicated by the progradation and truncation of the crossbedding within the hummocks.

### 3.5.4.7  Eolian Bedforms

Eolian sand dunes are photogenic. They are often very large, ranging up to several hundred metres high, and forming dune fields many kilometers across. Images taken from space and from the ground of Earth's various deserts, and now, also, of the dunes on Mars, are commonplace. In certain parts of the stratigraphic record eolian dunes make up significant thicknesses, particularly in the Mesozoic of the American southwest and the Permian of the North Sea Basin (where they form a very important gas reservoir) (Brookfield 2010).

Two important developments since the mid-1970s led to a revolution in the study of eolian strata in the rock record. First Brookfield (1977) and Gradzinski et al. (1979) examined the mechanics of dune construction and migration and presented some useful ideas on the nature of large-scale crossbed bounding surfaces. Second Hunter (1977a, b, 1980, 1981) and Kocurek and Dott (1981) studied the details of sand movement by wind on modern and ancient dunes and showed that several distinctive crossbedding and lamination patterns are invariably produced. In addition, Walker and Harms (1972) and Steidtmann (1974) carried out useful detailed facies studies of ancient eolian units. All this work has brought us to the point where eolian crossbedding should now be relatively simple to recognize, even in small outcrops.

Figure 3.34 is adapted from a global study of sand seas carried out in the late 1970s that utilized early "Landsat" images, and shows four of the main types of large sand dune. Barchans (Fig. 3.35d) tend to occur in areas of limited sand supply. They are commonly seen migrating across stony,

## OSCILLATION RIPPLES

variable direction and degree of ripple asymmetry,
commonly inconsistent with internal structure,
adjacent sets dissimilar

unidirectional cross-lamination,          draping foresets
commonly opposed                                              bundled upbuilding
chevron upbuilding

irregular, undulating lower
set boundary

**Fig. 3.31** *Top* The fine-scale structure of wave-formed (oscillation) ripples. These typically are symmetrical and consist of bundles of laminae formed by short-lived bursts of flow in opposing directions. *Middle* Modern wave-formed beach, South Carolina; *bottom* wave ripples in Ordovician sediments, near Ottawa (*diagram* de Raaf et al. 1977; *middle photo* M. O. Hayes)

**Fig. 3.32** Hummocky cross-stratification. The outcrop example is from the Jurassic of Yorkshire, UK. Diagram adapted from Harms et al. (1975)

sand-free, desert surfaces. Transverse dunes are comparable to water-lain planar-tabular crossbedding in terms of their internal structure. The other dune types (and only a few are shown in Figs. 3.34 and 3.35) reflect the product of complex wind patterns.

The pattern of air flow over a bed of loose sand is superficially similar to that of aqueous traction currents. Sand grains are moved by traction, saltation (bouncing) and intermittent suspension; flow separates at the dune crest and

there is turbulence in the lee of the dune. In detail, of course, flow patterns are distinctly different. For example, turbulence patterns commonly result in air flowing horizontally *across* the dune foreset. Air currents are less "competent" than aqueous traction currents, in the sense that the lower density of air means that the size range of particles that can be moved by wind is small, typically not larger than sand grade. Pebbles may be rolled along the foot of the dune, but larger material cannot be moved by the wind. Silt and clay are transported in suspension and may settle when the wind drops. Interdune areas (e.g., the dark areas in Fig. 3.35c, which are dark because they are damp) may be affected by rare flash floods and are therefore environments where temporary ponds and erosional river channels may occur, as discussed on more detail in Sect. 4.2.2.

Sand is transported and deposited on dunes by three processes, which generate distinctive textures and structures that readily enable the processes to be identified in small outcrops or drill core. These processes are:

1. Formation of **wind ripples**. Long-wavelength, low-amplitude ripples commonly form as a result of airflow along the active front face of a dune (Fig. 3.36a). Because of their small size, ripple foresets are rarely visible. An example is shown in Fig. 3.36b.
2. **Grain-fall lamination**. Sand blown over a dune crest and settling out of suspension in the lee, forms compact,

**Fig. 3.33** Storm-driven return flow across the continental shelf is diverted by the Coriolis effects, resulting in sediment transport obliquely across the shelf. Adapted from Walker, in Walker and James (1992)

**Fig. 3.34** The main types of large-scale eolian dune (adapted from McKee 1979)

commonly laminated units on the dune face (Fig. 3.36e). The carpet of sand moving close to the dune surface as it passes over the crest causes the "fuzzy" appearance of the dune crest behind the observer in Fig. 3.36c.

3. **Grain-flow lobes**. These are small avalanches of sand triggered whenever the depositional surface becomes too steep (Fig. 3.36c, d). They are most easily identified in strike-parallel cross-section, where their lenticular shape is quite distinctive (Fig. 3.36f).

Figure 3.37 shows an exposed dune foreset in the Permian Canyon de Chelly Sandstone, Arizona (at the point of the arrow). The enlarged view of this face (at right) shows

the linear pattern of wind-ripple crests oriented down the dune face, exactly as in the photograph of the modern dune surface shown in Fig. 3.36a. Figure 3.38 illustrates the three "orders" of bounding surface that may be identified in large eolian crossbed sets.

### 3.5.5  Sediment Gravity Flows

Sediment gravity flows are initiated by a failure of some kind on a sloping basin margin. Sedimentary over steepening on a delta slope may cause this, or an unstable slope may undergo collapse as a result of an earthquake. Sediment

**Fig. 3.35** Modern eolian dunes. **a** Great Sand Dunes National Monument, Colorado; **b, c** aerial views of the Gobi Desert, western China; **d** a barchan in the Gobi Desert

gravity flows can also occur where a sediment-laden water body flows into a standing body of water, either a lake or the sea.

Turbidity currents were the first type of sediment-gravity flow to be identified. After the construction of the Hoover Dam on the Colorado River in the 1930s, it was noticed that as the turbid water of the river entered Lake Mead (the water body created by damming the river) it descended beneath the surface as underflows, because of the greater density of the turbid water relative to the clear water of the lake (Grover and Howard 1938). Engineers noticed that sediment was being transported well into the lake before being deposited (a present-day example of this process is shown in Fig. 3.45). At about the same time, marine geologists suggested that submarine currents might be responsible for eroding out submarine canyons (Daly 1936). Then, in the 1950s, turbidity currents were modeled in laboratory flumes (Kuenen and Migliorini 1950). This classic paper clearly established subaqueous sediment gravity flows as the cause of graded bedding and other features in the rock record.

The ability of turbidity currents to carry sediment from shallow into deep water explained a puzzle that had bothered petroleum geologists working in the Ventura Basin of California for some time. There, sandstones containing a shallow-water fauna of foraminifera are interbedded with shales containing a deep-water fauna (Natland and Kuenen 1951). How could this be? We now know that turbidity currents can carry large quantities of sediment for thousands of kilometers under water (see Sect. 4.2.9). In the early 1970s, data relating to a major earthquake that had occurred on the Grand Banks off Newfoundland was interpreted in terms of turbidity current activity, and became a classic object lesson in the power and scale of turbidity current events (Heezen and Hollister 1971; see Figs. 3.46 and 3.47 of this book). Middleton carried out an important set of experiments on turbidity currents in the 1960s (Middleton 1966, 1967; see Sect. 1.2.5 for additional information concerning the origins of our knowledge about sediment-gravity flows).

In order to understand the several different kinds of sediment-gravity flow and the variety of textures and fabrics that they display, it is necessary to make a short diversion into some elements of fluid hydraulics (Middleton and Hampton 1976; Lowe 1979). Fluids such as water, including water that contains a dilute suspension of dispersed particles, have low viscosity, and are said to be **Newtonian fluids**, because they obey Newton's laws of viscosity. Fluids that contain a modest concentration of sediment show increased viscosity, a viscosity that may vary with the rate of shear. Such fluids are called **non-Newtonian**, and are characterized by concentrations of greater than 30 % sand, or somewhat less if the sediment consists of cohesive clay particles. If the concentration of sediment is large, viscosity is also large,

**Fig. 3.36** Types of eolian stratification. **a** Wind ripples formed by air currents passing across the front of a dune; **b** example of wind ripples in a Permian eolian sandstone, Arizona; **c, d** grain-fall deposits (beneath the observer's feet) and grain-flow deposits (the sand lobes oriented down the dune slope); **e, f** Permian sandstone in northern England showing grain-fall (dark, laminated beds) and grain-flow lenses (lenticular units seen in strike-parallel cross-section in **f**)

and the flow then has a certain mechanical strength. In fact, its strength may hold it in place on a sloping basin margin until an applied shear force exceeds a certain critical strength. Such a substance has the properties of a plastic. If the viscosity remains constant after the failing shear stress is exceeded, the flow is called a **Bingham plastic**. If the viscosity varies, it is called a pseudoplastic. Actual sediment gravity flows, at different times and under different conditions, display all of these conditions. Careful study and experimentation has led to the classification shown in Fig. 3.39. Column three lists the five main types of sediment-gravity flow. The boxes to the left indicate how these correspond to the types of flow discussed in the preceding paragraph, and the column on the right lists the

**Fig. 3.37** Exposure of active dune face (*at point of arrow*). Close up (*right*) shows wind-ripples oriented down the dune face. Permian, Arizona

**Fig. 3.38** Structure of typical ancient eolian dune deposit. *MBS* main bounding surface; *SS* second order bounding surface; *TS* third order bounding surface; *ER* eolian ripples; *FC* fluvial channel or sheetflood deposits (after Gradzinski et al. 1979). Example at *right* is from the Schnebly Hills Formation, near Flagstaff, Arizona

Main (first-order) bounding surfaces - MBS
(Major surfaces of wind deflation)

Second-order surfaces - SS
(Surfaces formed by changes
in wind direction)

Third-order surfaces - TS
(Minor dune reactivation surfaces)

specific processes that mobilize the sediment and keep it mobilized during transportation. Some of these processes are remarkably persistent. Notably, the sediment transported by turbidity currents may remain in suspension for days to weeks, and major flows can travel across the ocean floor for hundreds to several thousands of kilometers. All these types of flow are common on the ocean floor. The two "plastic" types of flow, grain flows and mudflows/debris flows, may also occur on land, under subaerial conditions.

Flow types can also evolve with time (Pickering et al. 1986). As a flow passes down slope, a loss of sediment from the base of the flow, or turbulence causing incorporation of additional fluid or sediment, may cause a change in the flow dynamics. The evolution from one flow type to another is shown in Fig. 3.40. Basic "support" mechanisms of the major sediment-gravity flow types are shown in Fig. 3.41, and are discussed below.

### 3.5.5.1 Debris Flow

Debris flows consist of poorly sorted mixtures of clay, silt, sand and larger clasts, possibly including large boulders (examples are illustrated in Fig. 3.42). *Mudflows* are a type of debris flow dominated by a mud matrix, with floating clasts. Debris flows may evolve from slumps or slides that become lubricated by infiltrated water. Subaerial flows are characteristic of arid areas, where weathered debris may accumulate on hillsides for many years, until mobilized by a sudden downpour. They flow at speeds of 1–3 m/s. Witnesses report that flows are very noisy, sounding like the onrush of several large freight trains. Debris flows may also occur in humid, vegetated areas when exceptionally heavy rains occur after periods of several or many decades (Fig. 3.42a). They are common on the slopes of arc volcanoes, where loose ash is easily mobilized by the rain. Debris flows also occur under water, on continental slopes, as a

**Fig. 3.39** A classification of sediment-gravity flows. Adapted from Lowe (1979)

| Flow behaviour | Flow type | | Sediment support mechanism |
|---|---|---|---|
| Fluid | Fluidal flow | Turbidity current | Fluid turbulence |
| | | Fluidized flow | Escaping pore fluid (full support) |
| | | Liquefied flow | Escaping pore fluid (Partial support) |
| Plastic (Bingham) | Debris flow | Grain flow | Dispersive pressure |
| | | Mudflow or cohesive debris flow | Matrix strength Matrix density |

**Fig. 3.40** The evolution of flow types with time, showing the changes in sediment-support mechanisms, and the resulting flow types (Pickering et al. 1986)

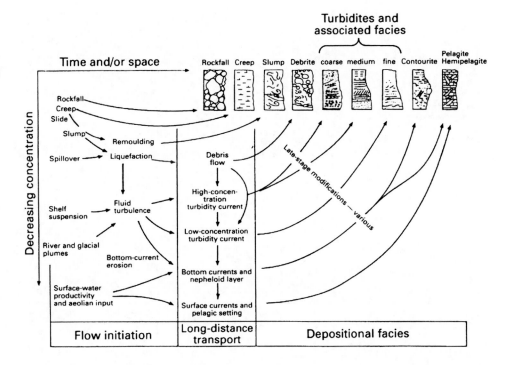

result of sedimentary over steepening. They may be triggered by an earthquake. Subaerial flows may travel for 10–20 km out across a flat basin floor. Subaqueous flows may travel for several hundred kilometers under water (Fig. 3.43).

The concentrated sediment-water mixture of a debris flow gives it a certain mechanical strength, which enables the flow to suspend and transport large boulders for significant distances (Fig. 3.42a). Loss of water as a flow moves down slope may lead to the development of a "rigid plug" of sediment, whereas within the remainder of the flow, sediment and water continue to interact with each other, until enough water is lost or gravitational energy is dissipated on a flat surface. At this point, the internal frictional forces absorb the flow momentum and it ceases to move. Internal shear may develop a recognizable fabric, with elongate grains all tending to be rotated into vertical or imbricated fabrics (e.g., Figure 3.42c). Large clasts may sink to the base of the flow and settle out earlier than smaller clasts, resulting in a

## Sediment gravity flows

Rheological flow classes

Flow types

Dominant clast-support mechanism

Fluidal flows (viscous behavior)

Debris flows (plastic behavior)

Turbidity currents | concentrated suspensions | Liquified flows | Grain flows (non-cohesive debris flows) | Cohesive debris flows

*Turbulence* | *Escaping pore fluid (due to elevated pore pressure)* | *Grain interaction* | *Matrix strength*

Buoyancy acts in all cases

**Fig. 3.41** Types of sediment gravity flow and principal "support" mechanisms (Middleton and Hampton 1976)

**graded fabric** (Fig. 3.42e). Alternatively, agitation within the flow may cause smaller clasts to move downwards between the larger ones (a process called a kinetic sieve), resulting in a **reverse-graded** fabric, in which the large clasts are at the top (Fig. 3.42d).

### 3.5.5.2   Grain-Flow

Grain flows are a rather specialized form of sediment-gravity flow; one we already encountered in the discussion of eolian bedforms. The grain-flow lobes that form as small avalanches on the front of active dunes are grain flows (Fig. 3.36c, d and f). They are categorized this way because they constitute a form of mass sediment movement. In this case, the "support" mechanism that keeps the grains in suspension so that gravity overcomes friction, is what is called **dispersive pressure**. This is the force of grains "bouncing" against each other as a result of their elasticity. This process only works effectively where the sediment mass consists of well-sorted grains, and flow requires a slope in the range of 30° to provide the necessary gravitational energy.

### 3.5.5.3   Liquified/Fluidized Flow

These flows are kept in suspension by the movement of pore fluids between the grains, which thereby overcomes the internal grain-on-grain friction. This mechanism is important mainly in the last stages of movement of the flow, as the

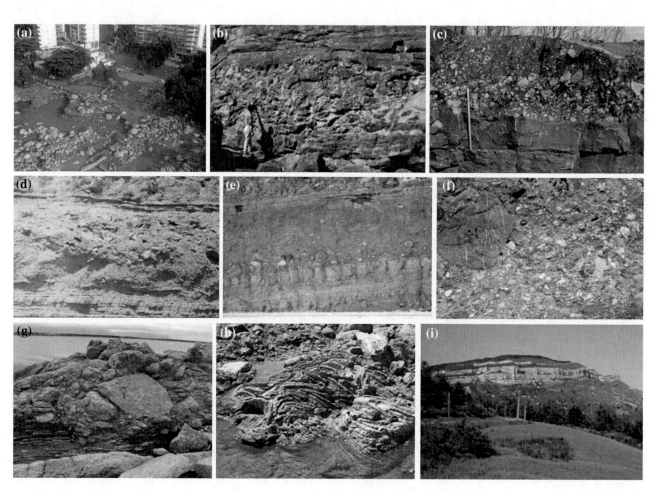

**Fig. 3.42** Examples of debris flow deposits. **a** Modern debris flow that swept though an apartment complex, coastal Venezuela, February 2000. **b** Subaerial flow, with large logs; Cenozoic, Japan. **c** Marine debris flow consisting of reworked glacial debris, Proterozoic, Ontario. **d**, **e** Miocene deposits, Turkey. **f** Cenozoic deposit, California. **g**, **h** Carbonate debris flows, Cambrian-Ordovician, Newfoundland. Note deformed clast in **h**. **i** Distant view of carbonate debris flow lenses, Cretaceous, France (*photos* A. O. Colmenares, D, E: W. Nemec)

**Fig. 3.43** A large slump on the continental margin of NW Africa triggered a muddy debris flow that traveled for several hundred kilometers across the floor of the Atlantic Ocean (after Embley 1976)

grains begin to settle and pore fluids are extruded. The upward movement of the pore fluids typically forms dish-and-pillar structures (Fig. 2.19c, d, e and f), which constitutes the most distinctive criterion for recognizing this category of sediment-gravity flow. Examples are shown in Fig. 3.44.

### 3.5.5.4 Turbidity Current

The first systematic exploitation of the turbidity-current concept to the interpretation of the ancient record was by a young Dutch geologist, by the name of A.H. Bouma, who was completing his Ph.D. thesis work on some Cretaceous sandstones in the south of France in the early 1960s (Bouma 1962). He recognized the repeated occurrence of a type of sedimentary cycle a few decimetres to a few metres in thickness, and attributed these cycles to the action of turbidity currents. The publication containing his results generated widespread interest, and his name rapidly became attached to the cycle model he had proposed. Nowadays it is commonplace to refer to the typical or classical turbidity current deposit as a **Bouma sequence**. An example of one of these sequences, observed in a drill core, is shown in Fig. 3.48, together with a modern interpretation of the cycle in terms of the flow-regime concept.

The succession of five divisions constituting the standard Bouma sequence provide a good example of how sediment support and transport mechanisms change with time within

**Fig. 3.44** Fluidized/liquefied flows. **a** Cretaceous, south of France; **b** Cenozoic, Riverdale, California

an individual flow event (Fig. 3.40). Initially, the sediment is maintained in suspension by the turbulence within the flow. The suspended sediment is what provides the flow with its increased density and hence the gravitational momentum to keep flowing, which in turns generates more turbulence. This process, known as "autosuspension", is what allows the flows to keep traveling under water, across the floor of an ocean, for hundreds of kilometers. Sediment gradually drops out of suspension as the momentum of the flow decreases, the coarser particles being deposited first, resulting in a graded bed (Bouma division A). This basal unit usually rests on a bed marked by sole markings (flutes, tool marks; Figs. 2.15 and 2.16), which are evidence of the erosive power of the flow (Figs. 3.45, 3.46 and 3.47).

The deposition of the A division corresponds to a "quick bed" phase of the upper flow regime. A decrease in flow power configures the flow as a plane-bed, from which is deposited the laminated sands of the B division (Fig. 3.48). The flow is here acting as a traction current, with loose sediment particles moved along the bed by the frictional force of the flow.

In the final phases of the flow, at its most distal end, the frictional force of the flow on the bed corresponds to the lower flow regime, forming the ripples or small dunes of the C division. The remaining very fine sands and silts being transported by the flow then settle out as the D division.

The A–D divisions of the flow deposit are formed rapidly, the A division perhaps within a few minutes, the remainder of the deposit within hours. The E division is not a product of turbidity-current deposition, but represents the background **pelagic settling** of fine muds that continues at all the time on a quiet ocean floor. This division represents the tens to thousands of years between flow events.

The composition of each turbidity-current deposit changes in a down-current direction, as shown in Fig. 3.49. The wedge-shaped cross-section of a flow shown at the centre of this diagram represents an individual flow unit, perhaps hundreds of kilometers long, and varying from a couple of metres thick, at most, near source, to a few centimetres, at its most distant end.

### 3.5.6 Sedimentary Structures Produced by Hydrodynamic Erosion of the Bed

Few of these are environmentally diagnostic, although their presence may add weight to an interpretation made from other evidence.

As discussed in Chap. 2, there are two main classes of erosional structure at the outcrop scale, macroscopic and mesoscopic. Macroscopic structures comprise channels, scours and low-relief erosion surfaces. Examples are illustrated in Fig. 2.13. They can occur in practically any environment as a result of current activity. More diagnostic than the channel itself is the channel fill, which can be analyzed in terms of vertical profile, lithology, sedimentary structures, etc. Subordinate features on an erosion surface may also

**Fig. 3.45** Turbid stream waters draining into Lake Louise, Alberta, plunge beneath the clear lake waters as a turbidity current

**Fig. 3.46** Cross-section of the continental margin off the Grand Banks of Newfoundland, showing data relating to a major earthquake that occurred there in 1929. Slumps triggered by the quake generated a turbidity current that flowed down the slope and continued for more than 1000 km out across the Atlantic Ocean. The times at which successive submarine cables were broken by the current enabled geologists to calculate the velocity of the flow (*graph at top*). Adaped from Heezen and Hollister (1971)

provide some environmental information. Lag concentrates of pebbles, bioclastic debris, or phosphatized fossils may be interpreted in terms of such processes as condensed sedimentation or wind deflation, storm erosion and deposition, or submarine nondeposition (hardgrounds). Desiccated surfaces may break up, yielding breccias of partly lithified mud, silt or carbonate clasts, which can then be interpreted as the result of subaerial exposure.

Another term for macroscopic erosion surfaces is **bounding surfaces**. Their physical extent, geometry, and facies associations have become important components of the architectural subdivision and classification of clastic deposits (Sects. 2.2.2, 3.5.11), and also feature in the definition and subdivision of stratigraphic sequences (Chap. 5).

Mesoscopic erosional features include the variety of sole markings described and illustrated in Chap. 2 (Fig. 2.15). These are abundant in submarine fan deposits and submarine basin plain sediments, particularly in sandy turbidites. Flute marks, in particular, indicate a style of vortex turbulence that is common at the base of turbidity currents (Fig. 2.16). However, they have been observed in a wide range of other settings, including fluvial traction current deposits. Turbidity currents are not restricted to the deep oceans, but are common in lakes and subaqueous ice-margin environments.

### 3.5.7  Liquefaction, Load and Fluid Loss Structures

A few of these are environmentally diagnostic, others are practically ubiquitous in occurrence. Load and slump structures, convolute bedding, and other products of thixotropy, can occur wherever sediments are saturated, which is to say practically anywhere.

Water escape features, such as dish and pillar structures (Fig. 2.19c, d, e and f), are also found in the deposits of many environments (Nilsen et al. 1977), although they are particularly abundant in fluidized sediment gravity flows.

As noted in Chap. 2, subtly different sedimentary structures have been attributed to desiccation and synaeresis. Desiccation cracks are an indicator of subaerial exposure, synaeresis has been attributed to volume changes resulting from salinity fluctuations in tidal, lagoonal, or lacustrine settings (e.g., Donovan 1973), but a reexamination of one of the classic locations of synaeresis cracks, the Orcadian Basin in Scotland, suggested that the "cracks" are gypsum pseudomorphs (Astin and Rogers 1991). Desiccation may be accompanied by the break-up of the uppermost beds and resedimentation as an intraformational breccia. Shinn (1983) provided a useful review of the terminology of some of the small-scale structures that develop on tidal flats and the kinds of interpretations that can be made from them.

Displacive evaporite growth in sabkhas produces a variety of distinctive structures that are useful diagnostic indicators.

### 3.5.8  Paleoecology of Body Fossils

In Chap. 2, we discuss briefly the three main types of body fossil preservation, those in growth positions, soft-bodied organisms that would not survive significant transportation, and the far more common death assemblage. In Sect. 2.2.2.9, we touch on the problems of biostratigraphic interpretation

**Fig. 3.47** The extent of the turbidite deposited following the 1929 quake. Laurentian Channel at top left (Walker, in Walker and James 1992)

Paleoecology is the study of the relationships between fossil organisms and their environment. In most paleoecological research and in the resulting books and papers, the primary focus is on the animals, with the use of such criteria as species interrelationships and fossil-sediment characteristics in the analysis of the ecology of some fossil species or community. Such research is typically carried out by specialists whose background is usually paleontology or biology, and these individuals may have little or no interest in regional stratigraphy or basin history. Here we wish to reverse the process, to use fossils as paleogeographic indicators. In the Phanerozoic record, fossils are among the most powerful environmental indicators available, but of course, their distribution and usefulness in the Precambrian is very limited. Useful texts on this subject have been provided by Ager (1963), Boucot (1981) and Dodd and Stanton (1981).

Even the simplest paleoecological observations can be invaluable. The presence of abundant bioturbation or (for example) a brachiopod fauna may be the key evidence in distinguishing the deposits of tidal channels and their point bars, tidal inlets, and deltas from certain fluvial and delta plain facies, which can be sedimentologically very similar (Barwis 1978). At the other extreme are sophisticated statistical studies of particular communities or fossil groups, such as the trilobite population analysis of Ludvigsen (1978). Most paleoecological studies by stratigraphers and basin analysts fall somewhere between these extremes.

Fossil data are subject to two biases, which may distort our observations. First there is the bias of unskilled or incomplete collection, which is particularly likely to arise when paleoecological analysis is attempted by the nonspecialist or when it is rushed through, as in a reconnaissance survey, by a prospecting team. Ager (1963) demonstrated this with data assembled by B.W. Sparks, who used two methods to collect molluscs from Quaternary deposits in southern Britain (Fig. 3.50). A bulk sample was analyzed by sieving, and the distribution was compared with that of a collection made by hand picking from the outcrop surface. The bias in the second collection in favor of larger or more brightly colored species is startling.

The second bias is that introduced by the geological obstacle courses organisms are put through before they end up under the geologist's hammer. The various hazards and possible paths through them are summarized in Fig. 3.51. The study of this flow diagram is itself the subject of a specialized science called **taphonomy**. Many of the processes of transportation, removal, or break-up of a fossil depend on sedimentological processes and are therefore of interest to basin analysts. Figure 3.52 illustrates the fate of dinosaur remains in Cretaceous fluvial deposits of the Red Deer Valley, Alberta (from Dodson 1971). The state of preservation may yield much information on local transport energy and hence on the depositional environment.

of death assemblages and reworked fossils and the difficulties of correlating faunas and floras limited in their distribution by factors of ecology or biogeography (not necessarily the same thing). All these points are relevant when we focus on the use of body fossils as indicators of depositional environment.

**Fig. 3.48** The Bouma sequence model. An example of a complete sequence, from the Precambrian of northern Ontario, is shown at *left*. The five divisions (**a–e**) of the standard sequence are interpreted at *right* (diagram from Blatt et al. 1980. *Core photo* Gowganda Formation, northern Ontario)

**Fig. 3.49** The down-current evolution of turbidity-current deposits. As shown by the much-shortened cross-section of an individual flow (at centre of diagram), near the source of the flow, the deposit may consist mainly of the *A* division, whereas at the distal end of the flow, turbidites may begin with the *B* or *C* division (diagram from Reading 1986)

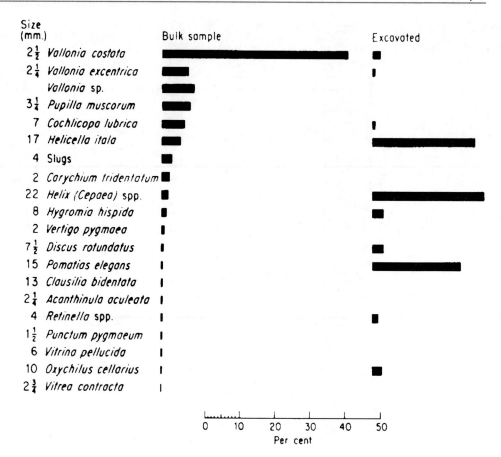

**Fig. 3.50** Distribution of species in two collections made from the same outcrop by two different methods (from Ager (1963), reproduced with permission of the publisher)

As shown by a thematic set of papers in the journal Palaios (v.1, #3), the study of taphonomy is growing into a useful subdiscipline. Although valuable in a negative sense, revealing the role of postmortem processes in biasing the fossil record, taphonomic studies can also provide useful information on the physical and chemical processes occurring after death and during burial (**biostratinomy**) and those that occur during early diagenesis (fossil diagenesis)(Brett and Baird 1986). Speyer and Brett (1986) showed that the varying state of preservation of fossil trilobites (disarticulation, fragmentation) records useful information regarding energy levels in the depositional environment and proposed that these variations be formalized into taphonomic facies or **taphofacies**.

One of the first and most obvious observations that must be made in a paleoecological study is to relate biofacies to lithofacies. The relationships may be obvious, including marked contrasts in facies, such as that shown in Fig. 3.53 (from Hecker 1965). Here, there is a red-bed lithofacies characterized by fish, a brackish lithofacies with Lingula, and marine deposits with corals, stromatoporoids and abundant bioturbation. This pattern of interbedded sediments and fossil types is characteristic of Early Paleozoic continental-marine transitions around the world.

Many excellent paleoecological-sedimentological studies have been made of carbonate reefs, such as the Permian reef complex of the Guadalupe Mountains in the southwestern United States (Newell et al. 1953), Silurian reefs of the Great Lakes (Lowenstam 1950, 1957), and Devonian reefs of Alberta (Andrichuk 1958; Klovan 1964). Figure 3.54 illustrates the lithofacies and biofacies subdivisions of some Silurian carbonate banks (Wilson 1975; Sellwood 1986). More subtle fossil-sediment relationships may only be apparent after careful statistical studies, such as the brachiopod analysis carried out by Jones (1977). In some cases, biofacies are not obviously related to lithofacies, as in the Welsh Paleozoic brachiopod assemblages discussed in Sect. 7.5.1 (see Fig. 7.8).

How are these observations on biofacies actually used to indicate the depositional environment? In four main ways: (1) by comparison with the ecology of living relatives, (2) by deductions about functional morphology, (3) by deductions from position and preservation, and (4) by deductions from assemblages and associations. These lines of reasoning are not intended to be mutually exclusive, and indeed, in many ancient deposits, there is an overlap in their application.

To start with the first point, comparisons with modern relatives become more difficult the further back in time we

**Fig. 3.51** The various possibilities for the preservation or destruction of a living animal community and its eventual collection as a fossil assemblage (from Ager (1963), reproduced with permission of the publisher)

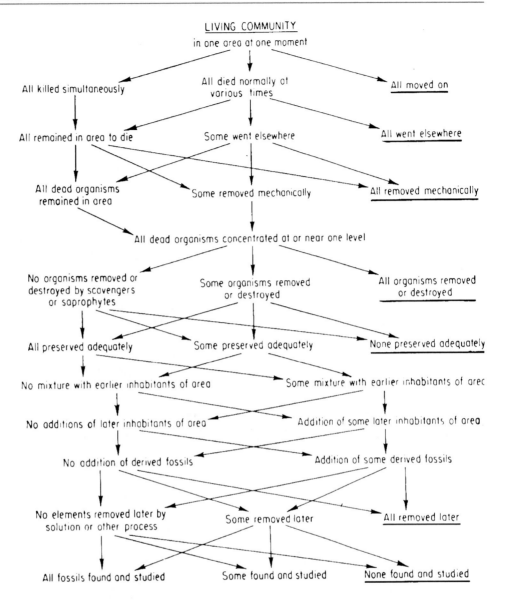

recede, as we trace evolutionary lines back to more and more distant ancestors, for whom there is no guarantee of unchanged habits. This is where examination of functional morphology, position, and associations might provide invaluable supportive evidence. The corals are a classic example of a group of organisms that has undergone marked evolutionary change during the Phanerozoic, yet seems to have continued to inhabit rather similar ecological niches throughout this time. Modern hermatypic (colonial, reef forming) corals inhabit depths of water down to about 100 m, but are only abundant above about 15 m. They are best adapted to grow in waters between 18 and 36 °C, with an optimum of 25 to 29 °C. Strong sunlight, a continuous nutrient supply, minimal salinity variations, and clean, sediment-free water are also necessary. These requirements restrict colonial corals to a present-day latitudinal range of 30°S to 30°N, with variations depending on the extent of cold

or warm oceanic currents, sediment input from large rivers, etc. (Schlager 2005; James and Wood 2010). For Mesozoic and Cenozoic corals, which are closely related to modern forms (orders Scleractinia and Alcyonaria), environmental deductions based on modern ecology seem safe. Paleozoic corals belong to two different orders, Rugosa and Tabulata, yet their gross appearance, lithofacies and biofacies associations (particularly the symbiotic association with algae) remain similar, as is evident from the various Paleozoic reef studies quoted previously. Heckel (1972a) suggested, however, that they may have occupied less agitated environments, with the now extinct stromatoporoids filling part of this niche. Table 3.2 summarizes some other modern-ancient comparisons for carbonate-producing organisms.

A group of organisms that has become extremely important in recent years for unraveling the history of continental margins is the Foraminifera. Comparisons of ancient

**Fig. 3.52** The taphonomy of
dinosaur skeletons based on
studies of the Cretaceous rocks of
Red Deer Valley, Alberta
(Dodson 1971)

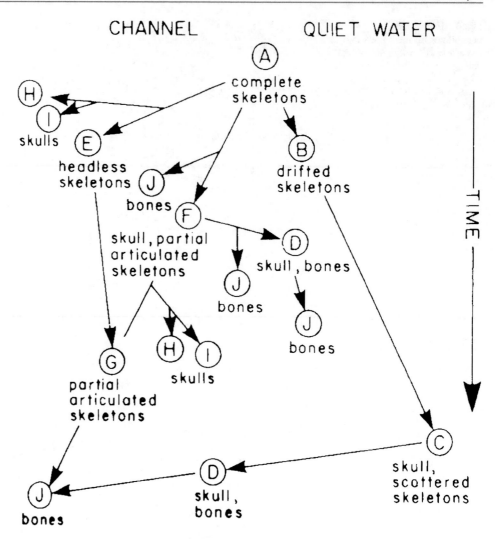

and modern assemblages yield invaluable paleobathymetric data for Cenozoic and later Mesozoic time, as discussed below.

Many ancient groups, such as the trilobites, graptolites, and many vertebrates, including early fish, amphibians, and reptiles, are now extinct, as are numerous genera and species of still thriving phyla. Therefore, comparisons with the ecology of modern relatives becomes tenuous or impossible. Ecological and environmental interpretations may then depend partly or wholly on deductions made from functional morphology. For example, Nichols (1959) was able to demonstrate the change in burrowing habit of the Cretaceous echinoid *Micraster* by careful examination of the functional morphology and comparison with similar features in modern echinoids. Micraster tests show an increase with time in the number of pores required for respiratory tube feet, an obvious adaptation to deeper burial.

There has been much debate about the habitats of trilobites, which show considerable variation in morphology. Some have large eyes, suggesting dim light (deep water?),

others are blind and may have lived below the photic zone. Long horizontal spines may have been a device to inhibit sinking in a soft substrate, and the same effect may have been achieved by the broad, flat exoskeletons of some species. Others were burrowers, for example *Bathyrus*, which has a concave-up, wedge shape adapted to wriggling into the substrate. Ludvigsen (1978) discussed these and other morphological features and related them to trilobite biofacies zones and their position on the Ordovician platform of Western Canada.

A widely used climatic indicator is the shape of tree leaves. It is known from present-day distributions that in tropical regions the leaves of dicotyledonous trees are entire (smooth), whereas in temperate regions they are more commonly dentate. Edwards (1936) showed that this characteristic could be used to interpret fossil floras, even for extinct or unidentifiable forms. Examples are illustrated in Fig. 3.55.

Factors of abundance, state of preservation, and position may yield important environmental information. Fossil forms

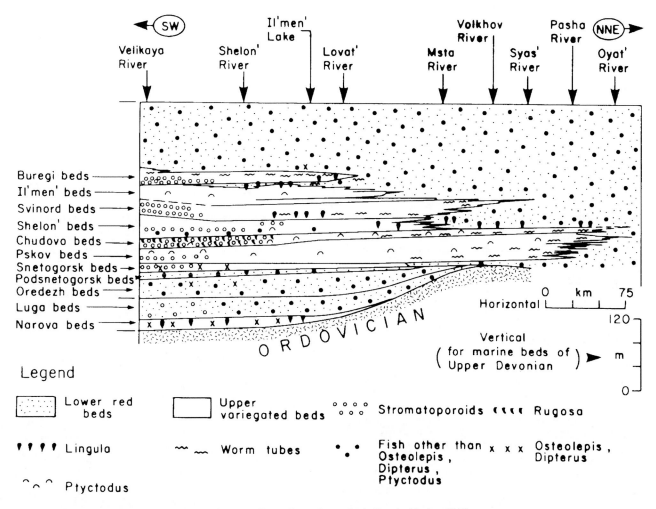

**Fig. 3.53** Lithofacies-biofacies relationships for some Upper Devonian rocks in Russia (Hecker 1965)

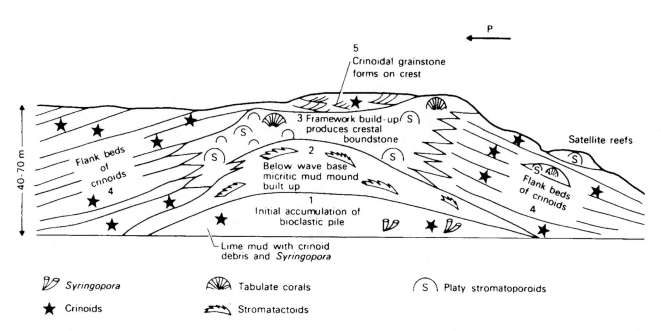

**Fig. 3.54** Lithofacies-biofacies composition of carbonate buildups in the mid-Silurian shelf of the American Mid-west (Sellwood 1986; after Wilson 1975. Reproduced with permission from Wilson 1975)

**Table 3.2** Modern and ancient carbonate-producing organisms and their sedimentary products (James 1984a)

| Modern organism | Ancient counterpart | Sedimentary aspect |
| --- | --- | --- |
| Corals | Archaeocyathids, corals, stromatoporoids, bryozoan, rudistid bivalves, hydrozoans | The large components (often in place) of reefs and mounds |
| Bivalves | Bivalves, brachiopods, cephalopods, trilobites, and other arthropods | Remain whole or break apart in several pieces to form sand- and gravel-sized particles |
| Gastropods, benthic foraminifers | Gastropods, tintinitids, tentactulitids, salterelids, benthic foraminifera, brachiopods | Whole skeletons that form sand-and gravel-sized particles |
| Codiacian algae *Halimeda*, sponges | Crinoids and other pelmatozoans, sponges | Spontaneously disintegrate upon death into many sand-sized particles |
| Planktonic foraminifers | Planktonic foraminifera, coccoliths (post-Jurassic) | Medium sand-sized and smaller particles in basinal deposits |
| Encrusting foraminifers and coralline algae | Coralline algae, phylloid algae, renalcids, encrusting foraminifers | Encrust on or inside hard substrates, build up thick deposits, or fall off on death to form lime-sand particles |
| Codiacean algae *Penicillus* | Codiacean algae, *Penicillus*-like forms | Spontaneously disintegrate upon death to form lime mud |
| Blue–green algae | Blue–green algae (especially in pre-Ordovician) | Trap and bind fine-grained sediments to form mats and stromatolites |

**Fig. 3.55** Leaf shape as a reflection of climate. **a–d** Tropical plants typically have entire margins; assemblage from Lower Eocene, southern England. **e–h** Temperate plants typically have dentate margins; assemblage from Miocene, southern France. (Edwards 1936; from Ager (1963), reproduced with permission from the publisher)

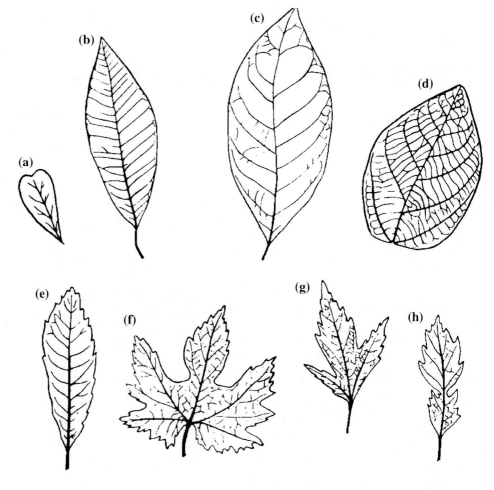

that normally grew attached to a substrate may occur as disoriented, broken, and abraded individuals or as hydraulically accumulated shell beds, suggesting storm wave activity. Isolated individuals may be found out of context; for example, I have found rare abraded orthocone cephalopods and compound corals in laminated stromatolitic dolomites formed on a Silurian tidal flat in the Canadian Arctic. Ballance et al. (1981) reported finding coconuts in Miocene turbidites and suggested that they may have been transported offshore bytsunamis, the same events that probably triggered the turbidity currents. Fossils found in place, such as coral colonies, pelecypods in their burrows, and upright trees attached to their roots, are particularly diagnostic of depositional environment (so long as their habits are known) because such occurrences are usually devoid of any ambiguity about possible transportation into an unrelated environment. However, Jeletzky (1975) and Cameron (1975) have engaged in heated controversy over the significance of shallow-water invertebrate fossils found in nodules in a turbidite succession on Vancouver Island. Are the nodules still in place where they were formed around the fossils, or is there evidence of abrasion indicating transportation to deeper water? The geologist should be prepared for the most unlikely event. For example upright trees, weighed down by boulders in their roots, have been found in the flood deposits of Mt. St. Helens and offshore near islands in the South Pacific, where they were emplaced by storms.

Preferred orientations of attached or transported fossils may have paleocurrent significance. Corals, crinoids, stromatolites, and other immobile organisms may grow in particular directions in response to current conditions, and accumulations of elongated fossils such as graptolites, gastropods, logs, or belemnites may be aligned hydraulically. Deductions based on preservation or position require particularly careful field observations and normally cannot be made on loose talus.

Perhaps the most reliable paleoecological interpretations are those based on assemblages and associations in an entire fossil community. Deductions made from extinct groups can be checked against information yielded by those with living relatives, and a picture of ecological relationships within the community can be attempted. Some forms may yield very generalized environmental information, including many pelagic groups such as the graptolites, whose occurrence is governed more by energy conditions at the site of postmortem deposition than by habitat during life. Interpretations based on tolerant groups may be finely tuned if other, more selective species can be found. Particularly accurate interpretations may be possible if species with different but slightly overlapping habitat ranges can be found together. For example, in the Lower Cretaceous of northern Texas, arenaceous foraminifera are associated with the oyster *Ostrea carinata* in deposits assumed to be close to the shoreline, but the association only occurs in limestones and marls, not in mudstones. Modern arenaceous foraminifera are known to be tolerant of brackish water. *O. carinata* therefore appears to be tolerant of reduced salinities, but not of turbid water. By contrast, the oyster *Gryphaea washitaensis* occurs in all rock types, but never in association with arenaceous foraminifera. It therefore seems to have

preferred open marine conditions, but had no preference for either turbid or non turbid water (Laughbaum 1960; Ager 1963). Another example is the association of ostracodes, gastropods, and the brachiopod *Lingula* in some Silurian dolomite beds in the Canadian Arctic. This is interpreted as a brackish-water tidal-flat assemblage and contrasts strongly with the rich coral, brachiopod, crinoid, and trilobite assemblage of interbedded open marine limestones (Miall and Gibling 1978). Heckel (1972b) provided tables relating the modern distribution of fossilizable invertebrate groups to water salinity, turbidity, and depth (e.g., Fig. 3.56), and within each group individual genera or species may have much more restricted tolerance ranges, as in the various examples given above.

Some of the most detailed assemblage studies carried out for basin-interpretation purposes are those on ancient reefs of all ages and on Cenozoic foraminifera. For example, Fig. 3.57 shows the vertical changes in assemblage through Niagaran reefs of the Great Lakes (Lowenstam et al. 1956). The reefs were built on a soft substratum in quiet water and gradually extended up into the zone of breaking waves.

Studies of foraminifera paleoecology began in the Cenozoic of the Gulf Coast and California in the 1930s (Natland 1933). The initial stimulus for the work was the need to correlate petroleum-bearing sediments, and this remains one of its most important applications. However, the same data are now finding increasing application in the interpretation of the subsidence history of continental margins, a subject of considerable relevance to our attempts at understanding the mechanisms of plate tectonics. An excellent example of this is the work on subsurface Cenozoic strata of the continental margin off Labrador and Baffin Island by Gradstein and Srivastava (1980). They distinguished four biofacies and interpreted their environments based on associated faunas and comparisons with other modern and ancient foraminifera assemblages. Considerable use was made of DSDP data. The biofacies are as follows:

1. Nonmarine: spores and pollen, no foraminifera.
2. Shallow neritic: marginal marine to inner shelf (<100 m deep). Diagnostic foraminiferal assemblages are of very low generic and specific diversity, with rare or no planktonics. In the Late Neogene sections *Cibicidiodes, Elphidium, Cassidulina, Bulimina, Melonis*; gastropods and bryozoans occur in low numbers.
3. Deep neritic: 100–200 m water depths. Foraminiferal generic and specific diversity varies, planktonics occur locally. Cenozoic benthonic genera include *Uvigerina, Pullenia, Gyroidina, Alabamina*. Paleogene assemblages are impoverished.
4. Bathyal: upper slope, water depths 200–1000 m. Foraminiferal generic and specific diversity is high, and the

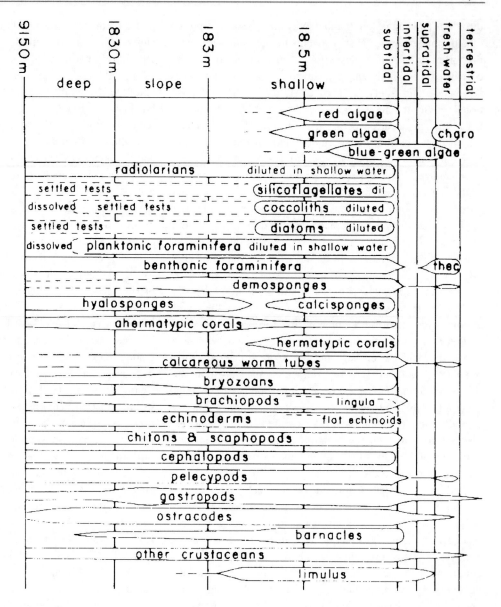

**Fig. 3.56** Depth distribution of common types of fossil invertebrates (Heckel 1972b)

assemblages are dominated by coarse, often large-sized agglutinated taxa. Over 50 species, including rich *Cyclammina* spp. Locally, planktonics are abundant (mostly turborotaliid and globigerinid forms). In a few wells, Paleogene calcareous benthonics occur in low numbers, including *Pleurostomella, Osangularia, Stilostomella*, and *Nuttalides*.

### 3.5.9    Ichnology

Trace fossils are widely used environmental indicators in marine sediments, where they are most abundant. They have one advantage over body fossils in that they are almost invariably in place; derived and transported trace fossils are very rare. They are well preserved and readily studied in drill core, and have therefore made a substantial contribution to subsurface environmental analysis Trace fossils may be used in two ways: (1) to indicate local sedimentation and erosion patterns and (2) to indicate the depositional environment. Trace-fossil assemblages have become valuable tools in the analysis of discontinuities and their allostratigraphic significance. The study of ichnofacies has become a specialized form of paleoecology. It is widely used in subsurface studies because of the ease with which trace fossils may be studied in cores. Many recent review articles and textbooks have been written on this subject, among the most useful of which is that by MacEachern et al. (2010).

There are various ways in which trace fossils may be used to evaluate local erosion and sedimentation. The density of bioturbation in a bed varies inversely with the rate of

**Fig. 3.57** Vertical variations in fossil assemblages of Niagaran reefs, Great Lakes (Lowenstam et al. 1956)

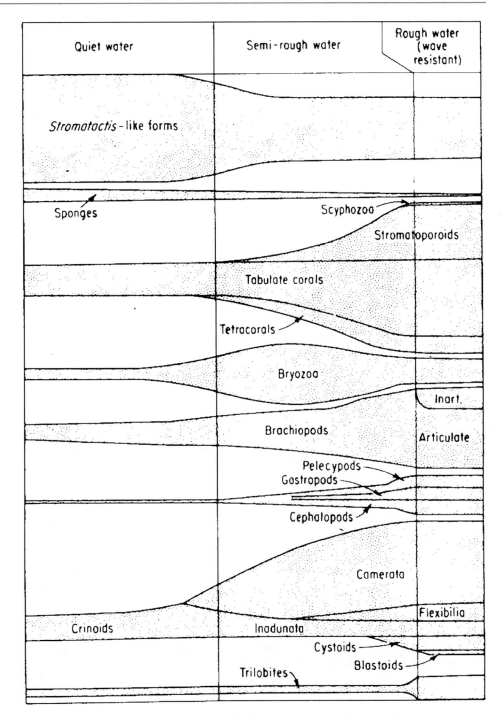

sedimentation, so that where sediment supply is low and invertebrate life abundant primary structures such as bedding may be completely destroyed MacEachern et al. (2010, Fig. 3) illustrated a **bioturbation index**, which is an attempt to quantify the intensity of alteration. Beds formed by rapid sedimentation, such as the passage of a turbidity current or a storm, may contain little or no penetration by organisms, except perhaps for escape burrows made vertically through the bed. Types of bioturbation may vary from one lithology to the next, because of different behavior patterns of the organisms in response to different sediment types. Ichnofacies are useful in the examination of cores and outcrops for the analysis of discontinuities, including the various bounding surfaces used in sequence-stratigraphic analysis (MacEachern et al. 2010, p. 45).

Important and widely quoted work by Seilacher (1967) showed how trace fossils could be used to interpret paleo-bathymetry. A set of five biofacies (now termed ichnofacies) were erected. These have now been expanded to nine (MacEachern et al. 2010; see Fig. 2.26 in this book). In the shallowest waters waves and currents keep nutrients in suspension. Animals are subject to violent conditions and build deep vertical burrows such as *Skolithus*. In less turbulent waters faunal diversity increases, sediment feeding becomes more important and a variety of crawling and grazing feeding trails appear (*Cruziana* biofacies). Below wave base sedimentation is slower, the oxygen content of the sediments is lower and nutrient supply more sparse. Sediment mining organisms of the *Zoophycos* and *Nereites* biofacies are characteristic. The latter, with its complex but highly systematic grazing patterns, is particularly typical of abyssal submarine-fan deposits. Other assemblages include a nonmarine *Scoyenia* biofacies, a "softground" *Psilonichnus* assemblage, in addition to the *Skolithus* and *Cruziana* assemblages, and the *Glossifungites* biofacies, which develops on firm but unlithified substrates. The details of the biofacies distributions varies from basin to basin and depend on sediment type, water temperature, salinity and circulation patterns. Frey and Pemberton (1984) pointed out that

local sets of environmental factors are most important in controlling the distribution of tracemakers, whether or not these parameters occur at specific water depths. For instance, many of the estuarine point bars … exhibit a high-energy, channelward side typified by a *Skolithus* association and a low-energy bankward side typified by a *Cruziana* association… The respective associations occur in close proximity, at the same stratigraphic or bathymetric datum.

Also, as Howard (in Basan 1978) pointed out, different organisms may make very similar structures, and conversely, the same species may make different structures when engaged in different activities or when interacting with substrates of different composition. Interpretations must therefore always be made in conjunction with other facies studies, and the geologist would be wise to consult published work on trace fossil assemblages in other basins of a similar age. Given these caveats, "ichnofacies are facies models that address animal-sediment responses in the depositional environment" (MacEachern et al. 2010, p. 29). The term facies model is used here with explicit reference to the term as defined by Walker (1976).

In carbonate environments, algal mats and stromatolites generally are excellent indicators of intertidal deposition, although recent discoveries of stromatolites in subtidal settings has indicated that an environment hostile to browsing organisms may be the critical control. According to James (1984b), their "upper limit is controlled by climate; in arid areas they cannot grow above the high intertidal into the supratidal zone, whereas in areas of high rainfall where the supratidal zone is moist or flooded for days at a time, mats are prolific. The lower limit is more variable and appears to be controlled by the presence of gastropods that eat algae."

In hypersaline zones, gastropods cannot survive and mats grow into the subtidal zone. After deposition and burial, mats commonly rot away, but they leave voids as a result of the disappearance of organic materials, entrapped gas, or shrinkage. The resulting distinctive type of porosity has been referred to as laminoid fenestrate, loferite, or birds-eye. Stromatolites are present well back into the Precambrian and may have lived in similar environments throughout this time (see also Pratt 2010).

MacEachern et al. (2010, p. 31) report on recent advances in the analysis and understanding of ichnofacies in continental environments.

## 3.5.10  Vertical Profiles

The history of cyclic sedimentation and the early significance of vertical profiles is discussed in Sect. 1.2.2. The importance of Walther's law and of the vertical profile in modern facies analysis is discussed in Sect. 3.4.1. Methods of statistical analysis of cyclic sedimentation are also described briefly in this section. The recognition of cyclic sequences has become one of the most widely used tools for reconstructing depositional environments in the subsurface. One of the reasons for this is that cyclic changes are commonly readily recognizable in petrophysical logs and are often interpreted without access to core or well cuttings. Whether such interpretations are always correct is another question. The sedimentological literature is full of shorthand references to **fining-upward** or **coarsening-upward** cycles or to **fining-and-thinning** or **coarsening-and-thickening upward**. Grain size, bed thickness and scale of sedimentary structures are commonly correlated in clastic rocks, so that the cyclicity may be apparent from several types of observation. French sedimentologists tend to use the terms positive and negative cycles; dipmeter analysts recommend coding cyclic changes diagrammatically in red and blue, but I am embarrassed to have to admit that I am not sure if fining-upward cycles are positive and red or negative and blue or the opposite and (or) vice versa. Such terms are obviously not helpful if one cannot remember which way they are used, and a simple descriptive terminology seems preferable.

An important warning: cyclicity may be in the eye of the beholder. Many facies interpretations in sedimentology place considerable emphasis on the recognition of repeated, ordered vertical change in facies characteristics, and many well-known facies models have been built upon such interpretations. But, as Hiscott (1981) demonstrated, with respect to submarine fan deposits, the presence of cyclicity may be based on very subjective interpretations. Anderton (1995) argued that most submarine fan deposits are non-cyclic. He stated "the human brain is very good at pattern recognition.

It seldom misses a real pattern and can often see patterns in random noise. Thus, objective correlation tends to be over-correlation." Wilkinson et al. (1998) took issue with the interpretations of cyclicity in certain shelf-carbonate successions, descriptions of which have formed the basis for some important studies in sequence stratigraphy. They argued that other natural, random processes can be demonstrated to generate very similar patterns. The solution may be to apply rigorous statistical tests to the data. Harper (1998) provided such tests for the examination of thickening- and thinning-upward patterns.

There are two common basic types of cycle, those indicating an increase in transport energy upward and those demonstrating a decrease. Both types can be caused by a variety of sedimentary, climatic and tectonic mechanisms. Beerbower (1964) divided these into **autocyclic** and **allocyclic** controls. The terms **autogenic** and **allogenic** are now preferred, in order to avoid the connotation of cyclicity. Autocyclic mechanisms are those that result in the natural redistribution of energy within a depositional system. Examples include the meandering of a channel in a river, tidal creek, or submarine fan; subaerial flood events; subaqueous sediment gravity flows; channel switching on a subaerial or submarine fan or a delta (avulsion); storms; and tidal ebb and flood. All of these can potentially produce cyclic sequences. Allocyclic mechanisms are those in which change in the sedimentary system is generated by some external cause. Tectonic control of basin subsidence, sediment supply, and paleoslope tilt, eustatic sea-level change, and climatic change, are the principal types of allocyclic mechanisms. These are large-scale basinal sedimentary controls. A sedimentary basin may be affected by several of these processes at the same time, so that it is not uncommon to find that there are two or three scales of cyclicity nested in a vertical profile. Allocyclic cycles tend to be thicker and more widespread in their distribution than autocyclic cycles. The latter are generally formed only within the confines of the subenvironment affected by the particular autocyclic process. This assists the geologist in distinguishing and interpreting sedimentary cycles, a matter of some importance in the definition of various scales of sequences and parasequences, but such interpretations may be far from easy.

Vertical profiles formed a crucial component of the first facies models, including the point-bar model of Bernard and Major (1963) and Allen (1964), the barrier island model of Bernard et al. (1959), and the Bouma (1962) turbidite sequence (Sects. 1.2.2, 1.2.4, and 1.2.7). They were the focus of a classic paper on facies models by Visher (1965), and their recognition in the subsurface has been covered in many articles and textbooks (e.g., Pirson 1977; Fisher et al. 1969; Sneider et al. 1978; Selley 1979). They still provide an essential basis of many modern facies models (Reading 1996; James and Dalrymple 2010), because they provide a simple way of synthesizing many different types of observation, including all the environmental criteria described in this chapter.

Vertical profiles have found an important use in recent years in the definition and documentation of stratigraphic sequences. As sea-level rises and falls, coastal and shallow-marine depositional environments may migrate landward and seaward, respectively. The result is a corresponding vertical succession of facies assemblages recording the drowning and reestablishment of the various coastal environments. Coarsening-upward and fining-upward cycles may demonstrate the presence of deltaic, estuarine, or beach-barrier systems at the shoreline, depending on the local paleogeography. Shoaling-upward successions bounded by marine flooding surfaces have been termed **parasequences** by sequence stratigraphers, and the nature of their vertical stacking is an important component of sequence analysis (van Wagoner et al. 1987; Catuneanu 2006). Where younger successions extend successively further into the basin they are said to exhibit a **progradational stacking pattern**, and this is interpreted to indicate a rate of sediment input more rapid than the rate at which depositional space (**accommodation**) is being generated by basin subsidence or sea-level rise. **Aggradational** (vertically stacked) and **retrogradational** (**backstepping**) patterns indicate a balance between sediment delivery and accommodation, and a deficit in sediment delivery, respectively. The latter condition is particularly common during transgression (Sect. 5.3.1). Lithostratigraphic and chronostratigraphic techniques may be used to correlate sequences and their component facies successions across a basin in order to determine the regional history of sea-level change (Chaps. 5 and 6).

Coarsening-and-thickening-upward cycles are the most varied in their origins. Several distinct types are produced by coastal regression and progradation (lateral accretion), where there is a gradation from low-energy environments offshore to higher energy in the shoaling wave and intertidal zones. Examples are illustrated in Fig. 3.58. Other types are formed where there is a steep slope and abundant sediment supply, and the flow system that develops over it attempts to grade itself to a balance by filling in the basin margin with a wedge of sediment. Coarsening-upward cycles are formed under these circumstances by prograding submarine fans and alluvial fans, particularly where the relief is maintained or even increased by active tectonic uplift. Other examples of coarsening-upward cycles (not illustrated) are those produced by crevasse splays in fluvial and deltaic settings, washover fans building from a barrier landward into a lagoon, and fluvioglacial sequences formed in front of an advancing continental ice sheet.

Fining-and-thinning-upward cycles commonly occur in fluvial environments as a result of lateral channel migration (point-bar succession) or vertical channel aggradation.

**Fig. 3.58** Typical examples of thickening-and-coarsening-upward profiles from James and Dalrymple (2010). A,B,C: Examples of delta-front successions (Bhattacharya 2010, Fig. 19, p. 246); D: Shoreface-offshore profile (Dalrymple 2010a, Fig. 5, p. 9)

Alluvial fans may also show fining-upward cycles where they form under conditions of tectonic stability. These three types are shown in Fig. 3.59. Other examples are the tidal-creek point-bar and intertidal beach progradation sequences. Sediments deposited by catastrophic runoff events, including fluvial flash floods and debris flows and many types of subaqueous sediment gravity flows, also show a fining-upward character.

Many of these cycles are superficially similar, and it may require careful facies and paleocurrent studies to distinguish them. Information on lateral variability may be crucial but this may be difficult to obtain in the case of subsurface studies.

**Fig. 3.59** Typical examples of thinning-and-fining-upward profiles from James and Dalrymple (2010); Vertical scales approximately the same; **a** tidal-inlet channel-fill succession (Boyd 2010, Fig. 23, p. 277); **b** tidal channel or tidal bar succession (Dalrymple 2010c, Fig. 20, p. 213); **c** meandering-river point bar or channel-fill (Dalrymple 2010b, Fig. 6, p. 64)

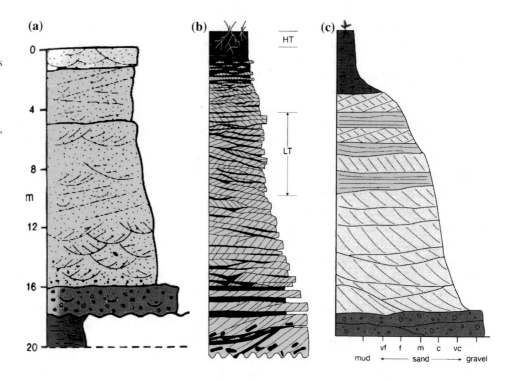

For carbonate environments, less emphasis has traditionally been placed on the vertical facies succession or profile, and more on the grain type, fauna, and structures of individual beds. Assemblages of such attributes are commonly environmentally diagnostic, whereas in the case of siliciclastic sediments, much ambiguity may be attached to their interpretation, and such additional features as vertical profile and lateral facies relationships assume a greater importance. The range of environments in which carbonates are formed is much narrower than that of siliciclastics; they are mainly confined to shallow continental shelves, platforms, or banks and adjacent shorelines and continental margin environments. Yet enormous variability is apparent in these various settings, particularly in shallow-water and coastal regions, and this is another reason why standard vertical profile models have not become as popular as they have with clastic sedimentologists.

Ginsburg (1975), James (1984b) and Pratt (2010) discussed shoaling-upward succession formed in shallow-subtidal to supratidal settings. These are common in

the ancient record, reflecting the fact that the rate of carbonate sedimentation is generally much greater than the rate of subsidence. Shallowing-up sequences therefore repeatedly build up to sea level and prograde seaward. Lateral shifts in the various subenvironments are common. Pratt (2010) offered three generalized sequences as models of vertical profiles that could develop under different climatic and energy conditions (Fig. 3.60). Ginsburg and Hardie (1975) and Ginsburg et al. (1977) developed an exposure index representing the percent of the year an environmental zone is exposed by low tides. By studying tide gauges and careful surveying of part of the modern Andros Island tidal flat, they were able to demonstrate that a variety of physical and organic sedimentary structures is each present over a surprisingly narrow tidal exposure zone. This idea has considerable potential for interpreting shoaling-upward sequences, as demonstrated by Smosna and Warshauer (1981).

Carbonate buildups or reefs may contain an internal cyclicity that is the result of upward reef growth. James and

**Shallowing-upward Cycles**

**Fig. 3.60** Examples of meter-scale shoaling-upward carbonate succession formed in shallow subtidal to intertidal environments (Pratt 2010, Fig. 18, p. 416)

Bourque (1992) suggested that the vertical profile may show an upward transition from an initial pioneer or stabilization phase to colonization, diversification and domination phases, characterized by distinctive textures and faunas. James and Wood (2010) described the response of reefs to sea-level change in terms of **keep-up**, **catch-up** or **give-up** modes, reflecting the rate of accumulation relative to the rate of sea-level change. In practice, most ancient reefs are the product of numerous sedimentation episodes separated by diastems or disconformities, attesting to fluctuating water levels (e.g., Upper Devonian reefs of Alberta: Mountjoy 1980). Analysis of vertical profiles of repeated cyclic patterns may therefore not be very helpful for basin-analysis purposes, although such work may be useful for documenting small-scale patterns of reef growth (e.g., Wong and Oldershaw 1980).

Deep-water carbonates comprise a variety of allochthonous, shelf-derived breccias and graded calcarenites, contourite calcarenites and hemipelagic mudstones, cut by numerous intraformational truncation (slide) surfaces (Cook and Enos 1977; McIlreath and James 1984; Coniglio and Dix 1992; Playton et al. 2010). Slope sedimentation is commonly most rapid during times of high sea level, when abundant carbonate material is being generated on the platform and shed from the margins, the process termed **highstand shedding** (Schlager 1991). The lithofacies assemblages are distinctive, but variations in slope topography and the random occurrence of sediment gravity flows seem to preclude the development of any typical vertical profile. Organic stabilization and submarine cementation of carbonate particles probably prevent the development of carbonate submarine fans comparable to those formed by siliciclastic sediments, with their distinctive channel and lobe morphology and characteristic vertical profile.

Cyclic sequences are common in evaporite-bearing sediments, reflecting a sensitive response of evaporite environments to climatic change, brine level, or water chemistry. Vertical profile models are therefore of considerable use in environmental interpretation. One of the most well known of these is the coastal sabkha, based on studies of modern arid intertidal to supratidal flats on the south coast of the Persian Gulf (Shearman 1966; Kinsman 1969; Kendall 1992, 2010). Coastal progradation and growth of displacive nodular anhydrite results in a distinctive vertical profile that has been widely applied (indeed, overapplied) to ancient evaporite-bearing rocks. Kendall (1992) discussed variations in this profile model, reflecting differences in climate and water chemistry that arise in other coastal and playa lake margin settings.

As noted elsewhere, evaporites can occur in a variety of other lacustrine and hypersaline marine settings. They mimic many kinds of shallow- to deep-marine carbonate and siliciclastic facies, and a range of sedimentary criteria is required to demonstrate origin. The vertical profile is only one of these, but may be useful particularly when examining subsurface deposits in cores. For example, sulphates that accumulate below wave base commonly display a millimeter-scale lamination interbedded with carbonate and organic matter and possibly including evaporitic sediment gravity flow deposits (Kendall 1992). The latter may even display Bouma sequences (Schreiber et al. 1976). Shoaling-upward intertidal to supratidal cycles have been described by Schreiber et al. (1976) and Vai and Ricci-Lucchi (1977) in Messinian (Upper Miocene) evaporites of the Mediterranean basin (Fig. 3.52). Caution is necessary in interpreting these cycles because they may not indicate a buildup or progradation under stable water levels, but instead they may be the product of brine evaporation and falling water levels. Many cycles of recharge and evaporation have been proposed for major evaporite basins such as the Mediterranean (Hsü et al. 1973).

Lacustrine environments are in general characterized by a wide variety of vertical profiles, reflecting many cyclic processes involving changes in water level and water chemistry. Many of these contain a chemical-sediment component. Lakes are highly sensitive to climate change, and their sediments have, therefore, become important in the investigation of orbital forcing mechanisms (Miall 2010). van Houten (1964) described a shoaling-upward, coarsening-upward type of cycle in the Lockatong Formation (Triassic) of New Jersey. The cycles are about 5 m thick and consist, in upward order, of black, pyritic mudstone, laminated dolomitic mudstone, and massive dolomitic mudstone with desiccation cracks and bioturbation. Chemical cycles which have an upper member of analcime-rich mudstone are also present. The cycles are interpreted as the product of short-term climatic change, with differences between the two types of cycle reflecting a greater tendency toward humidity or aridity, respectively. Eugster and Hardie (1975) described transgressive-regressive playa-margin cycles in the oil-shale-bearing Green River Formation of the Rocky Mountains. Donovan (1975) erected five profile models for cycles that occur as a result of changes in lake level and fluvial-deltaic lake-margin progradation in the Devonian Orcadian Basin of northern Scotland. Numerous other examples could be quoted.

Much has been written on the recognition of cyclic sedimentation from petrophysical logs (Fisher et al. 1969; Skipper 1976; Pirson 1977; Cant 1984, 1992). At present, the technique is best suited to the study of clastic cycles. As explained in Sect. 2.4, the gamma ray, spontaneous potential, and resistivity logs are sensitive indicators of sand-mud variations and so are ideally suited to the identification of fining- and coarsening-upward cycles. These appear as **bell-shaped** and **funnel-shaped** log curves, respectively, and various subtleties of environmental change may be detected by observing the convexity or concavity of the curves, smooth versus serrated curves, the presence of nested cycles of different thicknesses, and so on. The technique has been referred to as **log-shape analysis** (Fig. 3.61). Fisher et al. (1969) published a series of typical profiles for coastal and marginal-marine clastic environments based on examples from the Gulf Coast (Fig. 3.62). Other examples from the Beaufort-Mackenzie Basin (from Young et al. 1976) and elsewhere are illustrated in Figs. 3.63, 3.64, 3.65 and 3.66. These curves are commonly interpreted in the absence of cores or cuttings. As should be apparent from the preceding pages similar cycles may be produced in different environments, and so this is a risky procedure. However, by paying close attention to appropriate facies models and scale considerations (cycle thickness, well spacing) good paleogeographic reconstructions can be attempted. The availability of cores in a few crucial holes may make all the difference.

Figure 3.65 illustrates wireline logs showing the typical bell-shaped log profile, together with two examples. These are both fluvial examples. Figure 3.66a illustrates a variation on the channel-fill log shape, the so-called **blocky** log response, indicating a sand-fill with little vertical variation in mud content or grain size. Finally, Fig. 3.66c, d illustrates a log response (and example) of a clastic succession showing no cyclicity, in this case a suite of turbidite sandstones.

### 3.5.11   Architectural Elements and Bounding Surfaces

One of the fundamental objectives of stratigraphy is to describe and interpret the three-dimensional architecture of the sediments filling sedimentary basins. On the largest scale, this involves the disciplines of lithostratigraphy, biostratigraphy, allostratigraphy and chronostratigraphy (Chap. 7), the reconstruction of major depositional sequences and the sequence-stratigraphic architecture, perhaps with the aid of regional seismic sections (Chaps. 5 and 6), an assessment of the roles of sea-level change, and the various mechanisms of basin subsidence (Miall 2010; Allen and Allen 2013). At an intermediate to small scale, basin architecture depends on the interplay among subsidence rates, sea-level, and the autocyclic processes governing the distribution and accumulation of sediments within a particular

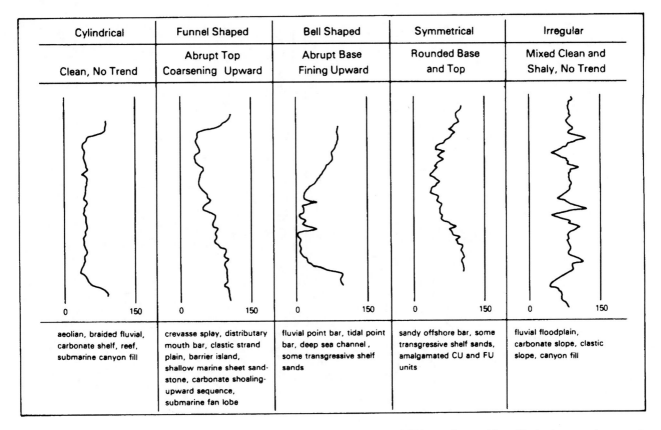

**Fig. 3.61** "Log-shape analysis". Certain petrophysical logs, particularly gamma ray and SP logs, when working with clastic successions, may be very useful indicators of cyclic processes in vertical profiles (Cant 1992)

**Fig. 3.62** Examples of typical petrophysical log profiles through coastal plain and shelf clastic sequences based on examples from the Gulf Coast. *Left log* gamma ray or S.P.; *right log* resistivity. Center bar shows scale subdivisions of 30 m (100 ft) (Fisher et al. 1969)

**Fig. 3.63** Some examples of actual log profiles from the Beaufort-Mackenzie Basin (Young et al. 1976)

range of depositional systems. Exploration for stratigraphic petroleum traps and for many types of stratabound ore bodies require close attention to basin architecture at the intermediate scale. For example, the search for and the interpretation of subsurface trends in potential reservoir rocks is a major preoccupation of petroleum geologists. Since the late 1950s many improvements have been made in techniques of basin analysis that facilitate this task.

For some time, we have been quite good at predicting the location and trend of such major depositional entities as alluvial fans, barrier reefs, various types of delta, and barrier island sand bodies (to take some examples at random). But one of the outcomes of modern sedimentological research on depositional processes has been a much improved understanding of the composition and architecture of the various facies assemblages formed within each type of depositional system. The predictive value of sequence models has also

proved its usefulness. We now have the ability, therefore, to construct much better models of the heterogeneities that exist within clastic reservoirs. Conventional techniques of sedimentological analysis and facies modeling place primary emphasis on the use of Walther's law and the interpretation of vertical profiles, particularly for the study of clastic sediments (Reading 1996; James and Dalrymple 2010). Meanwhile, the study of sequence stratigraphy has reached a level of maturity that terminologies and techniques have largely stabilized (Catuneanu 2006; Catuneanu et al. 2011).

A better understanding of the types of depositional units or elements that comprise stratigraphic sequences, including data describing in three dimensions their scales, lithofacies compositions, length of time required for accumulation, preservation potential, and their spacing in time and space, facilitates quantitative modeling of clastic depositional systems. A more immediate practical application of these data is

**Fig. 3.64** Wireline log showing the typical "funnel-shaped" profile, with three examples: **b**, **d** tertiary delta-front cycles, Arctic Canada; **c** deltaic bay-fill succession, Carboniferous, Kentucky (log diagram from Pirson 1977)

to improve the information used by reservoir engineers and geologists in their models of fluid flow in clastic reservoirs. It has been estimated that complex internal architectures are responsible for intrareservoir stratigraphic entrapment of an average of 30 % of the original oil in place, amounting to as much as 100 billion barrels of movable, nonresidual oil in the United States alone (Tyler et al. 1984). A better understanding of architectural complexities of petroleum reservoirs would facilitate improved primary production and would increase the success rate of enhanced recovery projects.

There are two important interrelated ideas (Miall 1988a, b, c):

1. The first is the concept of **architectural scale**. Deposits consist of assemblages of lithofacies and structures over a wide range of physical scales, from the individual small-scale ripple mark to the assemblage produced by an entire depositional system. Recent work, particularly in eolian, fluvial, tidal, and turbidite environments, suggests that it is possible to formalize a hierarchy of scales. Depositional units at each size scale originate in response

to processes occurring over a particular time scale and are physically separable from each other by a hierarchy of internal **bounding surfaces**.

2. The second is the concept of the **architectural element**. An architectural element is a lithosome characterized by its geometry, facies composition, and scale, and it is the depositional product of a particular process or suite of processes occurring within a depositional system.

### 3.5.11.1  Architectural Scale and Bounding Surface Hierarchies

As noted by Allen (1983, p. 249)

> The idea that sandstone bodies are divisible internally into "packets" of genetically related strata by an hierarchically ordered set of bedding contacts has been exploited sedimentologically for many years, although not always in an explicit manner. For example, McKee and Weir (1953) distinguished the hierarchy of the stratum, the set of strata, and the coset of sets of strata, bedding contacts being used implicitly to separate these entities.

Allen (1966) showed that flow fields in such environments as rivers and deltas could be classified into a

**Fig. 3.65 a, c** Wireline logs showing the typical "bell-shaped" profile of a fining-upward succession. **b** Point-bar, Carboniferous, Alabama; **d** fluvial channel-fill, Tertiary, Arctic Canada (log diagram from Pirson 1977)

hierarchical order. His hierarchy was designed as an aid to the interpretation of variance in paleocurrent data collected over various areal scales, from the individual bed to large outcrops or outcrop groups. The hierarchy consists of five categories, small-scale ripples, large-scale ripples, dunes, channels, and the integrated system, meaning the sum of the variances over the four scales. Miall (1974) added the scale of the entire river system to this idea, and compiled some data illustrating the validity of the concept (Sect. 6.7.4).

Brookfield (1977) discussed the concept of an eolian bedform hierarchy and tabulated the characteristics of four orders of eolian bedform elements: draas, dunes, aerodynamic ripples, and impact ripples. These four orders occur simultaneously, superimposed on each other. Brookfield showed that this superimposition resulted in the formation of three types of internal bounding surface. His first-order

surfaces are major, laterally extensive, flat-lying, or convex-up bedding planes between draas (**macroforms**, in the terminology of Jackson 1975). Second-order surfaces are low- to moderately-dipping surfaces bounding sets of cross-strata formed by the passage of dunes across draas (**mesoforms**). Third-order surfaces are reactivation surfaces bounding bundles of laminae within crossbed sets and are caused by localized changes in wind direction or velocity (mesoforms to **microforms**).

Similar hierarchies of internal bounding surfaces have been recognized in some subaqueous bedforms. Shurr (1984) developed a fivefold hierarchy of morphological elements of shelf sandstone bodies, ranging from the lithosome (widespread stratigraphic units) to individual facies packages. Yang and Nio (1989) carried out a similar type of study on some ancient tidal sandstone bodies.

**(a)**

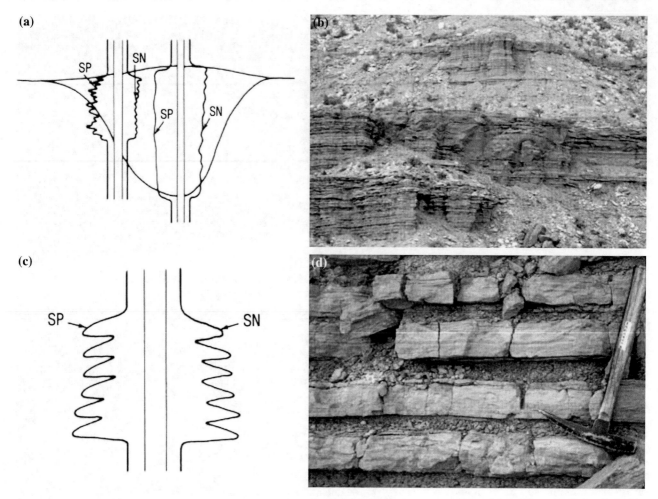

**Fig. 3.66 a** Blocky log response characteristic of some channel-fills, with an example (**b**) of a channel in the Cutler Group (Permian), New Mexico. **c** Serrated log (from Pirson 1977) and an example (**d**), turbidite sandstones, France

Brookfield's (1977) development of the relationship among the time duration of a depositional event, the physical scale of the depositional product, and the geometry of the resulting lithosome was a major step forward that has been of considerable use in the analysis of eolian deposits. Brookfield (1977), Gradzinski et al. (1979), and Kocurek (1981) showed how these ideas could be applied to the interpretation of ancient eolian deposits. Kocurek (1988) has found that first-order surfaces include two types of surfaces, the most laterally extensive of which he terms **super-surfaces**. Characterization of eolian deposits therefore now requires a four-fold hierarchy of bounding surfaces.

Several workers have attempted to develop a breakdown of the range of physical scales present in fluvial deposits. Allen's (1983) study of the Devonian Brownstones of the Welsh Borders represents the first explicit attempt to formalize the concept of a hierarchy of bounding surfaces in fluvial deposits and makes reference to Brookfield's work in eolian strata as a point of comparison. Allen described three types of bounding surfaces. He reversed the order of numbering from that used by Brookfield (1977), such that the surfaces with the highest number are the most laterally extensive. No reason was offered for this reversal, but the result is an open-ended numbering scheme that can readily accommodate developments in our understanding of larger scale depositional units, as discussed below. First-order contacts, in Allen's scheme, are set boundaries, in the sense of McKee and Weir (1953). Second-order contacts "bound clusters of sedimentation units of the kinds delineated by first-order contacts." They are comparable to the coset boundaries of McKee and Weir (1953), except that more than one type of lithofacies may comprise a cluster. Allen (1983) stated that "these groupings, here termed complexes, comprise sedimentation units that are genetically related by facies and/or paleocurrent direction." Many of the complexes in the Brownstones are macroforms, in the sense defined by Jackson (1975). Third-order surfaces are comparable to the major surfaces of Bridge and Diemer (1983). No direct relationship is implied between Allen's three orders of surfaces and those of Brookfield because of the different hydraulic behavior and depositional patterns of eolian and aqueous currents.

Miall (1988a, b, 1996) found it useful to expand Allen's classification to a seven-fold hierarchy to facilitate the definition of fluvial macroform architecture and to include the largest, basin-scale heterogeneities in the classification (Fig. 3.67). This hierarchy is summarized in Table 3.3 and is compared with other examples of architectural hierarchies erected for eolian deposits (Brookfield 1977; Kocurek 1988), coastal-estuarine sand waves (Allen 1980), hummocky cross-stratification cycles (Dott and Bourgeois 1982), shelf deposits (Shurr 1984), and turbidite depositional systems (Mutti and Normark 1987). The various rankings are correlated mainly on the basis of the time duration represented by the deposits or the length of time stored in the bounding surfaces between them (column 2 in Table 3.3). The numbers in the first column refer to a hierarchy of

**Sedimentation Rate Scales** that is discussed in Chap. 8. The formal bounding-surface classification shown in Fig. 3.67 is provided in column 3 of Table 3.3. An example of an outcrop lateral profile showing the application of the bounding-surface classification is shown in Fig. 2.3. Note that the depositional units listed in Table 3.3 represent a range of 12 orders of magnitude in time scale. They also represent at least 14 orders of magnitude in size (area). Note also that although this hierarchy has, at the time of writing, only been applied to fluvial deposits (several examples are given in Miall and Tyler 1991), there is considerable potential for similar subdivision of all clastic deposits, including an adaptation of the other hierarchical systems illustrated in the table. Miall (1988c) noted the similarities of fluvial deposits to some coarse-grained submarine-fan deposits, particularly the channel fills, and suggested ways of applying the architectural concepts described here to these

**Fig. 3.67** A six-fold hierarchy of depositional elements and bounding surfaces for fluvial deposits. Diagrams **a–e** represent successive enlargements of part of a fluvial unit in which ranks of bounding surfaces are indicated by circled numbers. A similar hierarchy could probably be developed for channelized deposits in other environments (Miall 1988a)

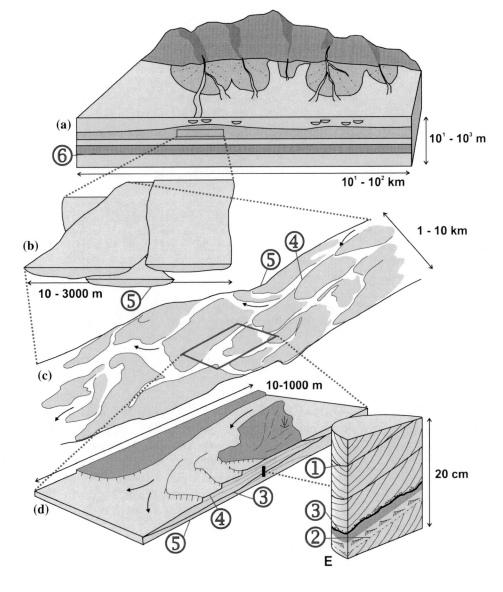

**Table 3.3** Hierarchy of depositional units

| SRS | Time scale (years) | Inst. sed. rate (m/ka) | Bound surf. | Sedimentary process | Eolian architecture (Brookfield) | Coastal-estuarine architecture (Allen) | Shelf architecture (Dott and Bourgeois, Shurr) | Submarine fan (Mutti and Normark) |
|---|---|---|---|---|---|---|---|---|
| 1 | $10^{-6}$ [30 s] | $10^6$ | 0 | Burst-sweep cycle | | | | |
| 2 | $10^{-5}$ to $10^{-4}$ [5–50 min] | $10^5$ | 0 | Ripple migration | | Ripple [E3 surface] | 3-surface in HCS | |
| 3 | $10^{-3}$ [8 h] | $10^5$ | 1 | Dune migration, foreset bundles | | Tidal bundle [E2 surface] | 2 surface in HCS | |
| 4 | $10^{-2}$ to $10^{-1}$ [3–36 days] | $10^4$ | 2 | Diurnal variability to normal meteorological floods (dynamic events) | | Neap-Spring bundle [E2] storm layer | HCS sequence [1 surface] | |
| 5 | $10^0$ to $10^1$ | $10^2$ to $10^3$ | 3 | Seasonal to 10-year flood: macroform growth increment | Reactivation [3rd order surface] | | | |
| 6 | $10^2$ to $10^3$ | $10^2$ to $10^3$ | 3, 4 | Long-term (100-year) flood: macroform, point bar, splay | Dune [2nd order surface] | Sand-wave field, wash over fan | Facies package (V) | Macroform [5] |
| 7 | $10^3$ to $10^4$ | $10^0$ to $10^1$ | 5 | Long-term geomorphic process: channel, delta lobe, coal seam | Draa [1st order surface] | Sand ridge, tidal channel, barrier island | Elongate lens (IV) | Minor lobe, channel-levee [4] |
| 8 | $10^4$ to $10^5$ | $10^{-1}$ | 6 | Channel belt, delta, orbital cycle [5th order] | Erg [supersurface] | | Regional lentil (III) | Major lobe [turbidite stage 3] |
| 9 | $10^5$ to $10^6$ | $10^{-2}$ to $10^{-1}$ | 7 | Depositional system, alluvial fan, major delta complex, orbital cycle [4th order] | Erg [supersurface] | | Sandstone sheet (II) | Depositional system |
| 10 | $10^6$ | $10^{-1}$ to $10^0$ | 7 | Rapid subsidence of convergent-margin basins, syntectonic clastic progradation, growth strata | | | Lithosome (I) | |
| 11 | $10^6$ to $10^7$ | $10^{-2}$ to $10^{-1}$ | | Basin-fill complex, tectonic cyclothem (e.g., "clastic wedge") | | Coastal-plain complex | Lithosome (I) | Fan complex |
| 12 | $10^6$ to $10^7$ | $10^{-3}$ to $10^{-2}$ | | Very low-accommodation cratons | | | | |

*SRS* Sedimentary rate scale; *HCS* Hummocky cross-stratification
Information in square brackets in columns 6–9 refers to classifications used by the authors named at the head of the column

rocks. These ideas were picked up by Clark and Pickering (1996), who provided numerous illustrations of modern and ancient deep-marine channel systems.

First- and second-order surfaces in the sixfold fluvial hierarchy record boundaries within microform and mesoform deposits (Fig. 3.67). The definition of *first-order surfaces* is unchanged from Allen (1983). *Second-order surfaces* are simple coset bounding surfaces, in the sense of McKee and Weir (1953). Third- and fourth-order surfaces are defined when architectural reconstruction indicates the presence of macroforms, such as point bars or sand flats (Fig. 3.67). *Third-order surfaces* represent growth increments of macroforms, such as the reactivation surfaces of

Jones and McCabe (1980). Individual depositional units (storeys or architectural elements) are bounded by surfaces of *fourth-order* or higher rank. *Fifth-order surfaces* are surfaces bounding major sand sheets, such as channel-fill complexes (Fig. 3.67). They are generally flat to slightly concave upward, but may be marked by local cut-and-fill relief and by basal lag gravels (these were termed third-order surfaces by Allen 1983, and are the major surfaces of Bridge and Diemer 1983). *Sixth-order surfaces* are surfaces defining groups of channels, or paleovalleys. Mappable stratigraphic units, such as members or submembers, and sequence boundaries, are bounded by surfaces of sixth-order and higher (Fig. 3.67; not defined by Allen 1983). Miall (1997,

Table 15.3) suggested that sixth-order surfaces, as defined here, are equivalent to the "F" surfaces of Nio and Yang (1991) and to the bounding unconformities of allomembers or submembers, in the sense used by some workers in the area of sequence stratigraphy (Chap. 6). *Seventh-order surfaces* define individual depositional systems. *Eighth-order surfaces* bound major basin-fill complexes. Stratigraphic sequences are bounded by surfaces of sixth- and higher-order rank.

Surfaces of fifth and higher order are potentially the easiest to map in the subsurface because of their wide lateral extent and essentially simple, flat, or gently curved, channelized geometry. Many examples of the mapping of channel and paleovalley surfaces have, in fact, been reported in the literature (e.g., Busch 1974). Doyle and Sweet (1995) reported on an attempt to apply Miall's hierarchical system of bounding surfaces to a subsurface study of a fluvial sandstone succession. Considerable potential now exists for mapping these surfaces with three-dimensional seismic data,

as described by Brown (2011; see Chap. 6). All higher order surfaces may appear very similar in cores. They are best differentiated by careful stratigraphic correlation between closely spaced cores, an objective that is best achieved in the most intensively developed fields, where well spacing may be a few hundred meters, or less. Third- and fourth-order surfaces may be recognized by gentle depositional dips (usually <10°).

Identification and correlation of these various bounding surfaces can clearly make a major contribution to the unraveling of the complexities of a channelized depositional system. They are likely to be particularly useful in the recognition and documentation of macroforms, about which much remains to be learned. Further discussion and illustration of the fluvial bounding surface hierarchy as applied to outcrop examples of fluvial deposits is given by Miall (1996).

The basis for a universal scheme with which to subdivide complex three-dimensional rock bodies is beginning to

**Fig. 3.68** Principal sandstone and/or conglomerate architectural elements in fluvial deposits (modified from Miall 1985)

emerge, as suggested by Table 3.3, and the development of a common terminology may become necessary. Thus, the eolian super surfaces of Kocurek (1988) are equivalent to Miall's (1988a) sixth-order surfaces, whereas Brookfield's (1977) first-order eolian surfaces could perhaps be equated with fifth-order surfaces. Similarly, in the shelf environment, tidal sand waves and sand banks also contain a hierarchy of facies units and internal bounding surfaces (Allen 1980; Shurr 1984; Berné et al. 1988; Harris 1988). The basal E1 surfaces of tidal sand waves (Allen 1980) are equivalent to Miall's third-order surfaces in terms of their time significance, and E2 surfaces are of first or second order, depending on their origin as diurnal or spring-neap erosion surfaces. The basal erosion surfaces of sand banks (Harris 1988) are of fifth order, whereas the internal master bedding planes are equivalent to fourth-order surfaces. Note that these rankings are based on time significance, not on geometry. E2 surfaces of sand waves and master bedding surfaces of sand banks both show depositional dips and appear superficially similar to Miall's third-order surfaces (e.g., compare Fig. 3.67 with Fig. 3.27), but represent

different processes acting over different time periods. These differences make the erection of a common terminology difficult.

### 3.5.11.2  Architectural Elements

Application of the bounding surface concept permits the subdivision of a clastic succession into a hierarchy of three-dimensional rock units. This facilitates description and also makes it easier to visualize the appropriate physical extent and time duration of the processes that controlled sedimentation at each level of the hierarchy. Rock description involves the definition of lithofacies and the recognition of facies assemblages.

Most deposits may be subdivided into several or many types of three-dimensional bodies characterized by distinctive lithofacies assemblages, external geometries, and orientations (many of which are macroforms). Allen (1983) coined the term **architectural element** for these depositional units, and Miall (1985, 1996) attempted a summary and classification of the current state of knowledge of these elements as they occur in fluvial deposits, suggesting that

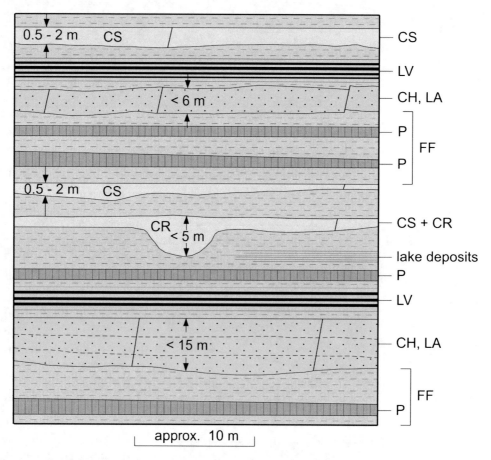

**Fig. 3.69** Principal architectural elements in the overbank or floodplain of river systems, based on an architectural diagram of a floodplain succession in the Lower Freshwater Molasse of Switzerland (Platt and Keller 1992), showing the range of elements to be expected in a floodplain setting (Miall 1996, Fig. 7.3). Element codes: *CH* channel, *CR* crevasse channel, *CS* crevasse splay, *FF* floodplain fines, *LA* lateral accretion element, *LV* levee. *P* pedogenic (paleosoil) unit

**Fig. 3.70** A diagram to explain a system for the documentation of architectural elements and bounding surfaces in a fluvial deposit. The architecture of a simple, hypothetical mid-channel macroform that shows lateral accretion (*LA*), oblique accretion, and downstream accretion (*DA*) in different parts of the unit. Arbitrary cutoff between these accretion geometries is indicated by dashed line. *Numerals in brackets* refer to ranks of bounding surfaces. *Capital letters* indicate numbering system for bounding surfaces as they would be labeled in outcrop. Based on the sand-flat model of Allen (1983, Fig. 19a) and the Brahmaputra bar mapped by Bristow (1987, Fig. 12)

there about eight basic architectural elements comprising the coarse units in fluvial depositional systems (Fig. 3.68). A comparable number of fine-grained deposits make up the overbank or floodplain suite (Fig. 3.69).

Two interpretive processes are involved simultaneously in the analysis of outcrops that contain a range of scales of depositional units and bounding surfaces: (1) the definition of the various types and scales of bounding surfaces and (2) the subdivision of the succession into its constituent lithofacies assemblages, with the recognition and definition of macroforms and any other large features that may be present. Figure 3.70 illustrates a hypothetical set of cross-sections through an assemblage of architectural elements, showing how a system of labeling and nomenclature may be used to clarify the sedimentology of the deposit.

In general, the most distinctive characteristic of a macroform is that it consists of genetically related lithofacies, with sedimentary structures showing similar orientations and internal minor bounding surfaces (first- to third-order of the classification given previously) that extend from the top to the bottom of the element, indicating that it developed by long-term lateral, oblique, or downstream accretion. A macroform is comparable in height to the depth of the channel in which it formed and in width and length is of similar order of magnitude to the width of the channel. However, independent confirmation of these

**Fig. 3.71** Channel hierarchies in the Brahmaputra River, *A*, and the Donjek River, *B* (after Williams and Rust 1969). Numbers in circles refer to bars, other numbers refer to channels. The first-order channel comprises the whole river, which includes several second-order channels. Bars scale within the channels in which they occur. In the Brahmaputra River, third-order channels modify higher-order bars but still have bars within them, which cannot be shown at this scale (Bristow 1987)

dimensions is difficult in multistory sandstone bodies, where channel margins are rarely preserved and the storeys commonly have erosional relationships with each other.

The definition of a macroform in any given outcrop is in part an interpretive process. Some types of macroforms, such as the lateral accretion deposits that comprise the typical point bar, are by now so well known that their recognition in outcrops would be classified by some workers as a descriptive rather than an interpretive exercise. Other types of macroform are less well known. The recognition of a macroform may depend in part on the type of bounding surface that encloses it. Conversely, the appropriate classification of a bounding surface may depend on a description of the lithofacies assemblage and geometry of the beds above and below it. For these reasons, description,

classification, and interpretation cannot always be completely separate exercises.

It may be possible to document the range in scales and the potential mappability of the hierarchy of depositional units and their bounding surfaces in any given stratigraphic unit (e.g., Miall 1988b). This type of subdivision is of importance in reservoir development. However, hierarchies of this type do not have a fixed range of scales (Fig. 3.71). For example, in parts of the Donjek River, Yukon, large compound bars (the first-order bars of Williams and Rust 1969, using an earlier classification; fourth-order macroforms in the present scheme) are in the order of 200–400 m in downstream length. Comparable features in the Brahmaputra River, Bangladesh, are several kilometers in length (Coleman 1969)—an order of magnitude larger (Bristow and Best 1993).

**Table 3.4** Scales of sedimentological analysis

| 1. Scales | Facies analysis | Depositional systems analysis |
| | | Sequence stratigraphy |
| 2. General definitions | Outcrop or local-scale facies assemblages and environmental interpretations, within-basin (autogenic) cyclic mechanisms, facies models | Basin-wide paleogeography and stratigraphic architecture, allogenic process: contemporaneous tectonics, sea-level change, climate change, sequence stratigraphy |
| 3. Selected analytical methods | Facies analysis scale | Depositional systems scale |
| 3a. Sedimentary structures and paleocurrents | Bar and bedform geometry and relative arrangement, local hydraulics, identification of river, wave and tide influence | Basin dispersal patterns and relationship to tectonic elements |
| 3b. Vertical profile studies | Environmentally diagnostic, autogenic mechanisms | Allogenic mechanisms |
| 3c. Petrology (siliciclastic) | Authigenic components may be environmentally diagnostic | Basin dispersal patterns, plate-tectonic setting of source rocks |
| 3d. Petrology (carbonate) | Basis of most lithofacies identifications and paleogeographic reconstructions at all scales | |
| 3e. Paleoecology | Local environmental interpretations | Regional paleobathymetry, gross salinity and temperature changes |
| 4. Selected environments | Facies-analysis scale | Depositional-systems scale environments |
| 4a. Fluvial | Bar type, channel geometry, local fluvial style | Areal and stratigraphic variations of fluvial style, relation of dispersal patterns to contemporaneous tectonic elements |
| 4b. Eolian | Dune type and geometry, inter-dune environments | Erg paleogeography, relation to playa and fluvial systems |
| 4c. Deltaic | Discrimination of river, wave and tide influence is main objective at all levels of analysis as these are main controls on all scales of cyclicity and facies geometry | |
| 4d. Shelf | Types of HCS and storm sequence, tidal sand wave evolution | Relative dominance of tides versus storms |
| 4e. Continental rise | Fan channel geometry and facies, channel migration and lobe progradation | Canyon-fan-basin plain relations, paleogeography of fan complexes |
| 4f. Carbonate platform | Local environments, tidal exposure, reef paleoecology | Discrimination of windward and leeward shelf margins, reef distribution |
| 4g. Evaporite basins | Recognition of deep or shallow water origin | Large scale cyclicity and relation to water-level fluctuations |

# 3.6 Conclusions and Scale Considerations

The focus of this chapter has been on analytical methods. Details of all the major facies models are summarized in Chap. 4, but this material is also covered in several excellent textbooks, to which reference is made elsewhere.

At this point, it would be useful to draw together a number of ideas appearing in different parts of this book that touch on stratigraphic scale.

In Sect. 1.4, we discuss different types and scales of basin-analysis projects, the kinds of data collection typically undertaken, and the problems and opportunities each offers. Elsewhere in this chapter a distinction has been drawn between facies analysis and facies models on the one hand and depositional systems analysis on the other hand; the latter now forming an essential element of the study of sequence stratigraphy. These represent different levels of the stratigraphic hierarchy and corresponding analytical complexity. This point is illustrated in Table 3.4. Selected analytical methods are listed with an indication of the kinds of information obtained at the smaller, facies analysis scale and the larger, depositional systems scale. Similarly, each depositional environment can be analyzed at the two different scales, and some examples are given to demonstrate this idea. The table is not exhaustive, but is offered as an illustration of these scale considerations. To tackle the larger scale, including the mapping of systems tracts and stratigraphic sequences, many different basin mapping techniques are required, which is why the discussion of sequence stratigraphy (Chap. 5) needs to be complemented by the mapping methods described in Chap. 6.

To some extent, the difference between facies models and depositional systems is artificial; the boundary between the two is vague, and for some environments, such as deltas, the distinction is all but impossible to define. Yet, it is a useful approach to take, because it helps to distinguish and clarify the purpose of a number of different procedures we perform more or less simultaneously as a basin analysis is carried out.

# References

Adams, J. E., and Rhodes, M. L., 1960, Dolomitization by seepage refluction: American Association of Petroleum Geologists Bulletin, v. 44, p. 1912-1920.

Adey, W.H., 1975, The algal ridges and coral reefs of St. Croix; their structure and Holocene development: Atoll Research Bulletin, No. 187, 67 p.

Adey, W. H., 1978, Coral reef morphologenesis: a multidimensional model: Science, v. 202, p. 831-837.

Adey, W. H., Macintyre, R., and Stuckenrath, R. D., 1977, Relict barrier reef system off St. Croix: its implications with respect to late Cenozoic coral reef development in the Western Atlantic: Proceedings of the 3rd International Coral Reef Symposium, v. 2, p. 15-21.

Ager, D. V., 1963, Principles of palaeoecology: McGraw-Hill, New York, 371 p.

Ager, D. V., 1973, The nature of the stratigraphical record: New York, John Wiley, 114 p.

Ager, D. V., 1981, The nature of the stratigraphical record (second edition): John Wiley, New York, 122 p.

Ager, D., 1986, A reinterpretation of the basal 'Littoral Lias' of the Vale of Glamorgan, Proceedings of the Geologists Association, v. 97, p. 29-35.

Allen, J. R. L., 1964, Studies in fluviatile sedimentation: six cyclothems from the Lower Old Red Sandstone, Anglo-Welsh basin: Sedimentology, v. 3, p. 163-198.

Allen, J. R. L., 1966, On bed forms and paleocurrents: Sedimentology, v. 6, p. 153-190.

Allen, J. R. L., 1968, Current ripples: North Holland Publishing Company, Amsterdam, 433 p.

Allen, J. R. L., 1980, Sand waves: a model of origin and internal structure: Sedimentary Geology, v. 26, p. 281-328.

Allen, J. R. L., 1982, Sedimentary structures: their character and physical basis: Elsevier Scientific Publications, Developments in Sedimentology 30, 663 p.

Allen, J. R. L., 1983, Studies in fluviatile sedimentation: bars, bar complexes and sandstone sheets (low-sinuosity braided streams) in the Brownstones (L. Devonian), Welsh Borders: Sedimentary Geology, v. 33, p. 237-293.

Allen, J. R. L., 1993, Sedimentary structures: Sorby and the last decade: Journal of the Geological Society, London, v. 150, p. 417-425.

Allen, P. A., and Allen, J. R., 2013, Basin analysis: Principles and application to petroleum play assessment: Chichester: Wiley-Blackwell, 619 p.

Anderton, R., 1985, Clastic facies models and facies analysis, in Brenchley P. J., and Williams, B. P. J., eds., Sedimentology: Recent developments and applied aspects: Blackwell Scientific Publications, Oxford, p. 31-47.

Anderton, R., 1995, Sequences, cycles and other nonsense: are submarine fan models any use in reservoir geology? in Hartley, A. J., and Prosser, D. J., eds, Characterization of deep marine clastic systems, Geological Society, London Special Publication 94, p. 5-11.

Andrichuk, J. M., 1958, Stratigraphy and facies analysis of Upper Devonian reefs in Leduc, Stettler and Redwater area, Alberta: American Association of Petroleum Geologists Bulletin, v. 42, p. 1-93.

Arthur, M. A., Dean, W. E., and Stow, D. A. V., 1984, Models for the deposition of Mesozoic-Cenozoic fine-grained organic-carbon-rich sediment in the deep sea, in Stow, D. A. V., and Piper, D. J. W., eds., Fine-grained sediments: deep-water processes and facies: Geological Society of London Special Publication 15, p. 527-560.

Ashley, G. M., 1990, Classification of large-scale subaqueous bedforms: a new look at an old problem: Journal of Sedimentary Petrology, v. 60, p. 160-172.

Astin, T. R., and Rogers, D. A., 1991, "Subaqueous shrinkage cracks" in the Devonian of Scotland reinterpreted: Journal of Sedimentary Petrology, v. 61, p. 850-859.

Badiozamani, K., 1973, The dorag dolomitization model—application to the Middle Ordovician of Wisconsin: Journal of Sedimentary Petrology, v. 43, p. 965-984.

Ball, M. M., 1967, Carbonate sand bodies of Florida and the Bahamas: Journal of Sedimentary Petrology, v. 37, p. 556-591.

Ballance, P. F., Gregory, M. R., and Gibson, G. W., 1981, Coconuts in Miocene turbidites in New Zealand: possible evidence for tsunami origin of some turbidity currents: Geology, v. 9, p. 592-595.

Barrett, T. J., 1982, Stratigraphy and sedimentology of Jurassic bedded chert overlying ophiolites in the North Apennines, Italy: Sedimentolgy, v. 29, p. 353-373.

Barwis, J. H., 1978, Sedimentology of some South Carolina tidal-creek point bars, and a comparison with their fluvial counterparts, in Miall, A. D., ed., Fluvial sedimentology: Canadian Society of Petroleum Geologists Memoir 5, p. 129-160.

Basan, P. B., ed., 1978, Trace fossil concepts: Society of Economic Paleontologists and Mineralogists Short Course No. 5, 201 p.

Beerbower, J. R., 1964, Cyclothems and cyclic depositional mechanisms in alluvial plain sedimentation: Geological Survey of Kansas Bulletin 169, v. 1, p. 31-42.

Bennetts, K. R. W., and Pilkey, O. H., 1976, Characteristics of three turbidites, Hispaniola-Caicos Basin: Geological Society of America Bulletin, v. 87, p. 1291-1300.

Bernard, H. A., and Major, C. J., 1963, Recent meander belt deposits of the Brazos River; an alluvial "sand" model (abs): American Association of Petroleum Geologists Bulletin, v. 47, p. 350.

Bernard, H. A., Major, C. F. Jr., and Parrott, B. S., 1959, The Galveston Barrier Island and environs - a model for predicting reservoir occurrence and trend: Transactions of the Gulf Coast Association of Geological Societies, v. 9, p. 221-224.

Berné, S., Auffret, J.-P., and Walker, P., 1988, Internal structure of subtidal sandwaves revealed by high-resolution seismic reflection: Sedimentology, v. 35, p. 5-20.

Bhattacharya, J., 2010, Deltas, in James, N. P., and Dalrymple, R. W., eds., Facies Models 4: GEOtext 6, Geological Association of Canada, St. John's, Newfoundland, p. 233-264.

Blatt, H., Middleton, G. V., and Murray, R., 1980, Origin of sedimentary rocks; 2nd edition: Prentice-Hall Inc., Englewood Cliffs, New Jersey, 782 p.

Bluck, B. J., 1967, Deposition of some Upper Old Red Sandstone conglomerates in the Clyde area: a study in the significance of bedding: Scottish Journal of Geology, v. 3, p. 139-167.

Boggs, S., Jr., 2012, Principles of sedimentology and stratigraphy: Fifth edition, Prentice Hall, Englewood Cliffs, New Jersey, 585 p.

Boucot, A. J., 1981, Principles of marine benthic paleoecology: Academic Press, New York,463 p.

Bouma, A. H., 1962, Sedimentology of some flysch deposits: Elsevier, Amsterdam, 168 p.

Bouma, A. H., Normark, W. R., and Barnes, N. E., eds., 1985, Submarine fans and related turbidite systems: Springer-Verlag Inc., Berlin and New York, 351 p.

Bown, T. M., and Kraus, M. J., 1981, Lower Eocene alluvial paleosols (Willwood Formation, northwest Wyoming, U.S.A.) and their significance for paleoecology, paleoclimatology, and basin analysis: Palaeogeography, Palaeoclimatology, Palaeoecology, v. 34, p. 1-30.

Boyd, R., 2010, Transgressive wave-dominated coasts, in James, N. P., and Dalrymple, R. W., eds., Facies Models 4: GEOtext 6, Geological Association of Canada, St. John's, Newfoundland, p. 265-294.

Brett, C. E., and Baird, G. C., 1986, Comparative taphonomy: a key to paleoenvironmental interpretation based on fossil preservation: Palaios, v. 1, p. 207-227.

Bridge, J. S., 1993, Description and interpretation of fluvial deposits: a critical perspective: Sedimentology, v. 40, p. 801-810.

Bridge, J. S., 1995, Description and interpretation of fluvial deposits: a critical perspective: Reply to Discussion: Sedimentology, v. 42, p. 384-389.

Bridge, J. S., and Diemer, J. A., 1983: Quantitative interpretation of an evolving ancient river system: Sedimentology, v. 30, p. 599-623.

Bristow, C. S., 1987, Brahmaputra River: Channel migration and deposition, in Ethridge, F. G., Flores, R. M., and Harvey, M. D., eds., Recent developments in fluvial sedimentology: Society of Economic Paleontologists and Mineralogists Special Publication 39, p. 63-74.

Bristow, C. S., and Best, J. L., 1993, Braided rivers: perspectives and problems, in Best, J. L., and Bristow, C. S., eds., Braided rivers: Geological Society, London, Special Publication 75, p. 1-11.

Brookfield, M. E., 1977, The origin of bounding surfaces in ancient aeolian sandstones: Sedimentology, v. 24, p. 303-332.

Brookfield, M. E., and Silvestro, S., 2010, Eolian systems, in James, N. P., and Dalrymple, R. W., eds., Facies Models 4: GEOtext 6, Geological Association of Canada, St. John's, Newfoundland, p. 139-166.

Brown, A. R., 2011, Interpretation of three-dimensional seismic data, seventh edition, American Association of Petroleum Geologists Memoir 42, 646 p.

Busby, C. J., and Ingersoll, R. V., eds., 1995, Tectonics of sedimentary basins: Blackwell Science, Oxford, 579 p.

Busch, D. A., 1974, Stratigraphic traps in sandstones—exploration techniques: American Association of Petroleum Geologists Memoir 21.

Byers, C. W., and Dott, R. J., Jr., 1981, SEPM Research Conference on modern shelf and ancient cratonic sedimentation—the orthoquartzite-carbonate suite revisited: Journal of Sedimentary Petrology, v. 51, p. 329-347.

Cameron, B. E. B., 1975, Geology of the Tertiary rocks north of Latitude 49°, west coast of Vancouver Island: Geological Survey of Canada Paper 75-1A, p. 17-19.

Cant, D. J., 1984, Subsurface facies analysis, in Walker, R. G., ed., Facies models, 2nd edition: Geoscience Canada Reprint Series 1, p. 297-310.

Cant, D. J., 1992, Subsurface facies analysis, in Walker, R. G., and James, N. P., eds., Facies models: response to sea level change: Geological Association of Canada, St. John's, Newfoundland, p. 27-45.

Cant, D. J., and Walker, R. G., 1976, Development of a braided-fluvial facies model for the Devonian Battery Point Sandstone, Quebec; Canadian Journal of Earth Sciences, v. 13, p. 102-119.

Catuneanu, O., 2006, Principles of sequence stratigraphy: Elsevier, Amsterdam, 375 p.

Catuneanu, O., Galloway, W.E., Kendall, C.G.St.C., Miall, A.D., Posamentier, H.W., Strasser A., and Tucker M.E., 2011, Sequence Stratigraphy: Methodology and Nomenclature: Report to ISSC: Newsletters on Stratigraphy, v. 4 (3), p. 173-245.

Clark, J. D., and Pickering, K. T., 1996, Architectural elements and growth patterns of submarine channels: application to hydrocarbon exploration: American Association of Petroleum Geologists Bulletin, v. 80, p. 194-221.

Clifton, H. E., Hunter, R. E., and Phillips, R. L., 1971, Depositional structures and processes in the non-barred, high energy nearshore: Journal of Sedimentary Petrology, v. 41, p. 651-670.

Coleman, J. M., 1969, Brahmaputra River: Channel processes and sedimentation: Sedimentary Geology, v. 3, p. 129-239.

Coleman, J. M., and Wright, L. D., 1975, Modern river deltas: variability of processes and sand bodies, in Broussard, M. L., ed., Deltas, models for exploration: Houston Geological Society, Houston, p. 99-149.

Collinson, J. D., 1970, Bedforms of the Tana River, Norway: Geografisker Annaler, v. 52A, p. 31-55.

Collinson, J. D., and Thompson, D. B., 1982, Sedimentary structures: George Allen and Unwin, London, 194 p.

Coniglio, M., and Dix, G. R., 1992, Carbonate slopes, in Walker, R. G. and James, N. P., eds., Facies models: response to sea-level change: Geological Association of Canada, Geotext 1, p. 349-373.

Cook, H. E., and Enos, P., eds., 1977, Deep-water carbonate environments: Society of Economic Paleontologists and Mineralogists Special Publication 25, 336 p.

Cuffey, R. J., 1985, Expanded reef-rock textural classification and the geologic history of bryozoan reefs: Geology, v. 13, p. 307-310.

Dalrymple, R. W., 2010a, Interpreting sedimentary successions: facies, facies analysis and facies models, in James, N. P., and Dalrymple, R. W., eds., Facies Models 4: GEOtext 6, Geological Association of Canada, St. John's, Newfoundland, p. 3-18.

Dalrymple, R. W., 2010b, Introduction to siliciclastic facies models, in James, N. P., and Dalrymple, R. W., eds., Facies Models 4: GEOtext 6, Geological Association of Canada, St. John's, Newfoundland, p. 59-72.

Dalrymple, R. W., 2010c, Tidal depositional systems, in James, N. P., and Dalrymple, R. W., eds., Facies Models 4: GEOtext 6, Geological Association of Canada, St. John's, Newfoundland, p. 201-231.

Daly, R. A., 1936, Origin of submarine canyons: American Journal of Science, v. 31, p. 401-420.

Davies, D. K., and Ethridge, F. G., 1975, Sandstone composition and depositional environments: American Association of Petroleum Geologists Bulletin, v. 59, p. 239-264.

Davies, N. S. and Gibling, M. R., 2010a, Cambrian to Devonian evolution of alluvial systems: The sedimentological impact of the earliest land plants: Earth-Science Reviews, v. 98, p. 171–200.

Davies, N. S., and Gibling, M. R., 2010b, Paleozoic vegetation and the Siluro-Devonian rise of fluvial lateral accretion sets: Geology, v. 38, p. 51-54.

De Raaf, J. F. M., Reading, H. G., and Walker, R. G., 1965, Cyclic sedimentation in the Lower Westphalian of North Devon, England: Sedimentology, v. 4, p. 1-52.

De Raaf, J. F. M., Boersma, J. R., and Van Gelder, A., 1977, Wave generated structures and sequences from a shallow marine succession, Lower Carboniferous, County Cork, Ireland: Sedimentology, v. 24, p. 451-483.

Dickinson, W. R., and Suczek, C. A., 1979, Plate tectonics and sandstone compositions: American Association of Petroleum Geologists Bulletin, v. 63, p. 2164-2182.

Dickinson, W. R. (Chair, U.S. Geodynamics Committee), 1997, The dynamics of sedimentary basins: National Research Council, National Academy Press, 43 p.

Dickinson, K. A., Berryhill, H. L., Jr., and Holmes, C. W., 1972, Criteria for recognizing ancient barrier coastlines, in Rigby, J. K., and Hamblin, W. K., eds., Recognition of ancient sedimentary environments: Society of Economic Paleontologists and Mineralogists Special Publication 16, p. 192-214.

Dodd, J. R., and Stanton, R. J., Jr., 1981, Paleoecology, concepts and applications: Wiley. New York.

Dodson, P., 1971, Sedimentology and taphonomy of the Oldman Formation (Campanian), Dinosaur Provincial Park, Alberta (Canada): Palaeogeography, Palaeoclimatology, Palaeoecology, v. 10, p. 21-74.

Donovan, R. N., 1973, Basin margin deposits of the Middle Old Red Sandstone at Dirlot, Caithness: Scottish Journal of Geology, v. 9, p. 203-212.

Donovan, R. N., 1975, Devonian lacustrine limestones at the margin of the Orcadian Basin, Scotland: Journal of the Geological Society of London, v. 131, p. 489-510.

Dott, R. H., Jr., 1983, Episodic sedimentation—how normal is average? How rare is rare? Does it matter? Journal of Sedimentary Petrology, v. 53, p. 5-23.

Dott, R. H., Jr., and Bourgeois, J., 1982, Hummocky stratification: significance of its variable bedding sequences: Geological Society of America Bulletin, v. 93, p. 663-680.

Dott, R. H., Jr., and Bourgeois, J., 1983, Hummocky stratification: significance of its variable bedding sequences. Reply to Discussion by R. G. Waker et al.: Geological Society of America Bulletin, v. 94, p. 1245-1251.

Doyle, J. D., and Sweet, M. L., 1995, Three-dimensional distribution of lithofacies, bounding surfaces, porosity, and permeability in a fluvial sandstone: Gypsy Sandstone of northern Oklahoma: American Association of Petroleum Geologists Bulletin, v. 79, p. 70-96.

Duff, P. McL. D., and Walton, E. K., 1962, Statistical basis for cyclothems: a quantitative study of the sedimentary succession in the east Pennine Coalfield: Sedimentology, v. 1, p. 235-255.

Dunham, R. J., 1962, Classification of carbonate rocks according to depositional texture, in Ham, W. E., ed., Classification of carbonate rocks: American Association of Petroleum Geologists Memoir 1, p. 108-121.

Edwards, W. N., 1936, The flora of the London Clay: Proceedings of the Geologists' Association of London, v. 47, p. 22-31.

Eichler, J., 1976, Origin of the Precambrian banded iron formations, in Wolf, K. H., ed., Handbook of strata-bound and stratiform ore deposits: Elsevier, Amsterdam, v. 7, p. 157-201.

Embley, R. W., 1976, New evidence for occurrence of debris flow deposits in the deep sea: Geology, v. 4, 371-374.

Embry, A. F., and Klovan, J. E., 1971, A Late Devonian reef tract on northeastern Banks Island, N.W.T.: Bulletin of Canadian Petroleum Geology, v. 19, p. 730-781.

Emery, K. O., 1968, Relict sediments on continental shelves of the world: American Association of Petroleum Geologists Bulletin, v. 52, p. 445-464.

Eriksson, P. G., Condie, K. C., Tirsgaard, H., Mueller, W. U., Alterman, W., Miall, A. D., Aspler, L. B., Catuneanu, O., and Chiarenzelli, J. R., 1998, Precambrian clastic sedimentation systems, in Eriksson, P. G., ed., Precambrian clastic sedimentation systems: Sedimentary Geology, Special Issue, v. 120, p. 5-53.

Eriksson, P. G., Banerjee, S., Catuneanu, O, Corcoran, P. L., Eriksson, K. A., Hiatt, E. E, Laflamme, M., Lenhardt, N., Long, D. G. F., Miall, A. D., Mints, M. V., Pufahl, P. K., Sarkar, S., Simpson, E. L., Williams, G. E., 2013, Secular changes in sedimentation systems and sequence stratigraphy, in Kusky, T., Stern, R., and Dewey, J., eds., Secular Changes in Geologic and Tectonic Processes: Gondwana Research, Special issue, v. 24, issue #2, p. 468-489.

Eugster, H. P., and Hardie, L. A., 1975, Sedimentation in an ancient playa-lake complex: The Wilkins Peak Member of the Green River Formation of Wyoming: Geological Society of America Bulletin, v. 86, p. 319-334.

Evans, G., 1965, Intertidal flat sediments and their environments of deposition in the Wash: Quarterly Journal of the Geological Society of London, v. 121, p. 209-245.

Eyles, C. H., and Eyles, N., 2010, Glacial deposits, in James, N. P., and Dalrymple, R. W., eds., Facies Models 4: GEOtext 6, Geological Association of Canada, St. John's, Newfoundland, p. 73-104.

Eyles, N., Eyles, C. H., and Miall, A. D., 1983, Lithofacies types and vertical profile models; an alternative approach to the description and environmental interpretation of glacial diamict sequences: Sedimentology, v. 30, p. 393-410.

Eyles, N., Day, T. E., and Gavican, A., 1987, Depositional controls on the magnetic characteristics of lodgement tills and other glacial diamict facies: Canadian Journal of Earth Sciences, v. 24, p. 2436-2458.

Farrell, K. M., Harris, W. B., Mallinson, D. J., Culver, S. J., Riggs, S. R., Pierson, J., Self-Trail, J. M., and Lautier, J. C., 2012, Standardizing texture and facies codes for a process-based classification of clastic sediment and rock: Journal of Sedimentary Research, v. 82, p. 364-378.

Fawcett, J. J., 1982, Facies (metamorphic facies), in McGraw-Hill Encyclopedia of Science and Technology, 5th ed.: McGraw-Hill, New York, p. 303.

Fisher, W. L., Brown, L. F., Scott, A. J., and McGowen, J. H., 1969, Delta systems in the exploration for oil and gas: University of Texas Bureau of Economic Geology, 78 p.

Flügel, E., 1982, Microfacies analysis of limestones: Springer-Verlag, Berlin and New York, 633 p.

Folk, R. L., 1962, Spectral subdivision of limestone types, in Ham, W. E., ed., Classification of carbonate rocks: American Association of Petroleum Geologists Memoir 1, p. 62-84.

Frakes, L. A., Francis, J. E., and Syktus, J. I., 1992, Climate modes of the Phanerozoic: Cambridge University Press, Cambridge, 274 p.

Franke, W., and Paul, J., 1980, Pelagic redbeds in the Devonian of Germany—deposition and diagenesis: Sedimentary Geology, v. 25, p. 231-256.

Frazier, D. E., 1974, Depositional episodes: their relationship to the Quaternary stratigraphic framework in the northwestern portion of the Gulf Basin: Bureau of Economic Geology, University of Texas, Geological Circular 74-1, 26 p.

Frey, R. W., and Pemberton, S. G., 1984, Trace fossil facies models, in Walker, R. G., ed., Facies models, 2nd edition, Geoscience Canada Reprint Series 1, p. 189-207.

Friedman, G. M., 1979, Address of the retiring President of the International Association of Sedimentologists: Differences in size distributions of populations of particles among sands of various origins: Sedimentology, v. 26, p. 3-32.

Friedman, G. M., 1980, Dolomite is an evaporite mineral: evidence from the rock record and from sea-marginal ponds of the Red Sea: Society of Economic Paleontologists and Mineralogists Special Publication 28, p. 69-80.

Friedman, G. M., and Sanders, J. E., 1978, Principles of Sedimentology: new York: John Wiley and Sons, 791 p.

Friend, P. F., Alexander-Marrack, P. D., Nicholson, J., and Yeats, A. K., 1976, Devonian sediments of east Greenland I: introduction, classification of sequences, petrographic notes: Meddelelser om Gronland Bd. 206, nr. 1.

Galloway, W. E., 1975, Process framework for describing the morphologic and stratigraphic evolution of the deltaic depositional systems, in Broussard, M. L., ed., Deltas, models for exploration: Houston Geological Society, Houston, p. 87-98.

Galloway, W. E., and Hobday, D. K., 1996, Terrigenous clastic depositional systems, second edition: Springer-Verlag, Berlin, 489 p.

Ginsburg, R. N., ed., 1975, Tidal deposits: a casebook of recent examples and fossil counterparts: Springer-Verlag, Berlin and New York, 428 p.

Ginsburg, R. N. and Hardie, L. A., 1975, Tidal and storm deposits, northeastern Andros Island, Bahamas, in Ginsburg, R. N., ed., Tidal deposits: a casebook of recent examples and fossil counterparts: Springer-Verlag, Berlin and New York, p. 201-208.

Ginsburg, R. N. and Schroeder, J. H., 1973, Growth and submarine fossilization of algal cup reefs, Bermuda: Sedimentology, v. 20, p. 575-614.

Ginsburg, R. N., Hardie, L .A., Bricker, O. P., Garrett, P., and Wanless, H. R., 1977, Exposure index: a quantitative approach to defining position within the tidal zone, in Hardie, L. A., ed., Sedimentation on the modern carbonate tidal flats of northwestern Andros Island, Bahamas: Johns Hopkins University Studies in Geology, v. 22, p. 7-11.

Glaister, R. P., and Nelson, H. W., 1974, Grain-size distributions, an aid in facies identification: Bulletin of Canadian Petroleum Geology, v. 22, p. 203-240.

Gloppen, T. G., and Steel, R. J., 1981, The deposits, internal structure and geometry in six alluvial fan-fan delta bodies (Devonian-Norway)—A study in the significance of bedding sequences in conglomerates, in Ethridge, F. G., and Flores, R. M., eds., Recent and ancient nonmarine depositional environments: models for exploration: Society of Economic Paleontologists and Mineralogists Special Publication 31, p. 49-70.

Gorsline, D. S., 1984, A review of fine-grained sediment origins, characteristics, transport and deposition, in Stow, D. A. V., and Piper, D. J. W., eds., Fine-grained sediments: deep-water processes and facies: Geological Society of London Special Publication 15, p. 17-34.

Gould, S. J., 1989, Wonderful life. The Burgess Shale and the nature of history: W. W. Norton, New York.

Gradstein, F. M. and Srivastava, S. P., 1980, Aspects of Cenozoic stratigraphy and paleoceanography of the Labrador Sea and Baffin Bay: Palaeogeography, Palaeoclimatology, Palaeoecology, v. 30, p. 261-295.

Gradzinski, R., Gagol, J., and Slaczka, A., 1979, The Tumlin Sandstone (Holy Cross Mts., central Poland): Lower Triassic deposits of aeolian dunes and interdune areas: Acta Geologica Polonica, v. 29, p. 151-175.

Gravenor, C. P., and Wong, T., 1987, Magnetic and pebble fabrics and origin of the Sunnybrook Till, Scarborough, Ontario, Canada: Canadian Journal of Earth Sciences, v. 24, p. 2038-2046.

Grover, N. C., and Howard, C. S., 1938, The passage of turbid water through Lake Mead: American Society of Civil Engineers, v. 103, p. 720-732.

Grunau, H. R., 1965, Radiolarian cherts and associated rocks in space and time: Eclogae Geologica Helvetica, v. 58, p. 157-208.

Hamblin, A. P., and Walker, R. G., 1979, Storm-dominated shallow marine deposits; the Fernie-Kootenay (Jurassic) transition, southern Rocky Mountains: Canadian Journal of Earth Sciences, v. 16, p. 1673-1690.

Harbaugh, J. W., and Bonham-Carter, G., 1970, Computer simulation in geology: Wiley-Interscience, New York, 98 p.

Harms, J. C., and Fahnestock, R. K., 1965, Stratification, bed forms, and flow phenomena (with an example from the Rio Grande), in Middleton, G. V., ed., Primary sedimentary structures and their hydrodynamic interpretation: Society of Economic Paleontologists and Mineralogists Special Publication 12, p. 84-115.

Harms, J. C., Southard, J. B., Spearing, D. R., and Walker, R. G., 1975, Depositional environments as interpreted from primary sedimentary structures and stratification sequences: Society of Economic Paleontologists and Mineralogists Short Course 2, 161 p.

Harms, J. C., Southard, J. B., and Walker, R. G., 1982, Structures and sequences in clastic rocks: Society of Economic Paleontologists and Mineralogists Short Course 9, Calgary.

Harper, C. W., Jr., 1984, Improved methods of facies sequence analysis, in Walker, R. G., ed., Facies models, 2nd ed.: Geoscience Canada Reprint Series 1, p. 11-13.

Harper, C. W., Jr., 1998, Thickening and/or thinning upward patterns in sequences of strata: tests of significance: Sedimentology, v. 45, p. 657-696.

Harris, P. T., 1988, Large-scale bedforms as indicators of mutually evasive sand transport and the sequential infilling of wide-mouthed estuaries: Sedimentary Geology, v. 273-298.

Hart, B. S., Macquaker, J. H. S., and Taylor, K. G., 2013, Mudstone ("shale") depositional and diagenetic processes: Implications for seismic analysis of source-rock reservoirs: Interpretation, A journal of subsurface characterization: Society of Exploration Geophysicists and American Association of Petroleum Geologists, v. 1, p. B7-B26.

Hayes, M. O., 1967, Hurricanes as geological agents: Case studies of Hurricanes Carla, 1961, and Cindy, 1963: University of Texas, Bureau of Economic Geology, Austin, Texas, Report of Investigations No. 61.

Hayes, M. O., 1976, Morphology of sand accumulation in estuaries; an introduction to the symposium, in Cronin, L. E., ed., Estuarine

Research, v. 2, Geology and Engineering: Academic Press, London, p. 3-22.

Heckel, P. H., 1972a, Pennsylvanian stratigraphic reefs in Kansas, some modern comparisons and implications: Geologische Rundschau, v. 61, p. 584-598.

Heckel, P. H., 1972b, Recognition of ancient shallow marine environments, in Rigby, J. K., and Hamblin, W. K., eds., Recognition of ancient sedimentary environments: Society of Economic Paleontologists and Mineralogists Special Publication 16, p. 226-296.

Hecker, R. F., 1965, Introduction to paleoecology (translated from Russian): American Elsevier Publishing Company, New York, 166 p.

Heezen, B. C. and Hollister, C. D., 1971, The face of the deep: Oxford University Press, New York, 659 p.

Hein, F .J., 1982, Depositional mechanisms of deep-sea coarse clastic sediments, Cap Enragé Formation, Quebec: Canadian Journal of Earth Sciences, v. 19, p. 267-287.

Hiscott, R. N., 1981, Deep-sea fan deposits in the Macigno Formation (Middle-Upper Oligocene) of the Gordana valley, northern Apennines, Italy — Discussion: Journal of Sedimentary Petrology, v. 51, p. 1015-1033.

Hodgetts, D., 2013, laser scanning and digital outcrop geology in the petroleum industry: A review: Marine and Petroleum Geology, v. 46, p. 335-354.

Howard, J. D., and Reineck, R.-E., 1981, Depositional facies of high-energy beach-to-offshore sequence: comparison with low energy sequence: American Association of Petroleum Geologists Bulletin, v. 65, p. 807-830.

Hsü, K. J., Cita, M. B., and Ryan, W. B. F., 1973, The origin of the Mediterranean evaporites, in Ryan, W. B. F., Hsü, K. J., et al., Initial Reports of the Deep Sea Drilling Project: v. 13, p. 1203-1231.

Hunter, R. E., 1977a, Basic types of stratification in small eolian dunes: Sedimentology, v. 24, p. 361-388.

Hunter, R. E., 1977b, Terminology of cross-stratified sedimentary layers and climbing-ripple structures: Journal of Sedimentary Petrology, v. 47, p. 697-706.

Hunter, R. E., 1980, Depositional environments of some Pleistocene coastal terrace deposits, southwestern Oregon—case history of a progradational beach and dune sequence: Sedimentary Geology, v. 27, p. 241-262.

Hunter, R. E., 1981, Stratification styles in eolian sandstones: some Pennsylvanian to Jurassic examples from the Western Interior U.S. A., in Ethridge, F. G., and Flores, R., eds., Recent and ancient nonmarine depositional environments: models for exploration: Society of Economic Paleontologists and Mineralogists Special Publication 31, p. 315-329.

Hunter, R. E., Clifton, H. E., And Phillips, R. L., 1979, Depositional processes, sedimentary structures and predicted vertical sequences in barred nearshore systems, southern Oregon Coast: Journal of Sedimentary Petrology, v. 49, p. 711-726.

Imbrie, J., and Purdy, E. G., 1962, Classification of modern Bahamian carbonate sediments, in Ham, W. E., ed., Classification of carbonate rocks: American Association of Petroleum Geologists Memoir 1, p. 253-272.

Jackson, R. G., II, 1975, Hierarchical attributes and a unifying model of bed forms composed of cohesionless material and produced by shearing flow: Geological Society of America Bulletin, v. 86, p. 1523-1533.

James, N. P., 1972,: Holocene and Pleistocene calcareous crust (caliche) profiles: criteria for subaerial exposure: Journal of Sedimentary Petrology, v. 42, p. 817-836.

James, N. P., 1984a, Introduction to carbonate facies models, in Walker, R. G., ed., Facies models, second edition, Geoscience Canada Reprint Series 1, p. 209-211.

James, N. P., 1984b, Shallowing-upward sequences in carbonates, in Walker, R. G., ed., Facies models, second edition, Geoscience Canada Reprint Series 1, p. 213-228.

James, N. P., and Bourque, P.-A., 1992, Reefs and mounds, in Walker, R. G. and James, N. P., eds., Facies models: response to sea-level change: Geological Association of Canada, Geotext 1, p. 323-347.

James, N. P., and Dalrymple, R. W., eds., 2010, Facies Modesl 4: GEOtext 6, Geological Association of Canada, St. John's, Newfoundland, 586 p.

James, N. P., Kendall, A. C., and Pufahl, P. K., 2010, Introduction to biological and biochemical facies models, in James, N. P., and Dalrymple, R. W., eds., Facies Models 4: GEOtext 6, Geological Association of Canada, St. John's, Newfoundland, p. 323-340.

James, N. P., and Wood, R., 2010, Reefs, in James, N. P., and Dalrymple, R. W., eds., Facies Models 4: GEOtext 6, Geological Association of Canada, St. John's, Newfoundland, p. 421-447.

Jeletzky, J. A., 1975, Hesquiat Formation (new): a neritic channel and interchannel deposit of Oligocene age, western Vancouver Island, British Columbia (92E): Geological Survey of Canada Paper 75-32.

Jones, B., 1977, Variations in the Upper Silurian brachiopod *Atrypella phoca* (Salter) from Somerset and Prince of Wales Islands, Arctic Canada: Journal of Paleontology, v. 51, p. 459-479.

Jones, C. M., and McCabe, P. J., 1980, Erosion surfaces within giant fluvial cross-beds of the Carboniferous in northern England: Journal of Sedimentary Petrology, v. 50, p. 613-620.

Kay, M., 1951, North American geosynclines: Geological Society of America Memoir 48.

Kendall, A. C., 1992, Evaporites, in Walker, R. G. and James, N. P., eds., Facies models: response to sea-level change: Geological Association of Canada, Geotext 1, p. 375-409.

Kendall, A. C., 2010, Marine evaporites, in James, N. P., and Dalrymple, R. W., eds., Facies Models 4: GEOtext 6, Geological Association of Canada, St. John's, Newfoundland, p. 505-539.

Kinsman, D. J. J., 1966, Gypsum and anhydrite of recent age, Trucial Coast, Persian Gulf, in Rau, J. L., ed., Second Symposium on Salt: Northern Ohio Geological Society, Cleveland, Ohio, v. 1, p. 302-326.

Kinsman, D. J. J., 1969, Modes of formation, sedimentary associations, and diagenetic features of shallow-water supratidal evaporites: American Association of Petroleum Geologists Bulletin, v. 53, p. 830-840.

Klein, G. deV., 1970, Depositional and dispersal dynamics of intertidal sand bars: Journal of Sedimentary Petrology, v. 40, p. 1095-1127.

Klein, G. deV., 1971, A sedimentary model for determining paleotidal range: Geological Society of America Bulletin, v. 82, p. 2585-2592.

Klovan, J. E., 1964, Facies analysis of the Redwater Reef complex, Alberta, Canada: Bulletin of Canadian Petroleum Geology, v. 12, p. 1-100.

Knauth, L. P., 1979, A model for the origin of chert in limestone: Geology, v. 7, p. 274-277.

Kocurek, G. A., 1981, Significance of interdune deposits and bounding surfaces in aeolian dune sands: Sedimentology, v. 28, p. 753-780.

Kocurek, G. A., 1988, First-order and super bounding surfaces in eolian sequences - bounding surfaces revisited: Sedimentary Geology, v. 56, p. 193-206.

Krumbein, W. C., and Sloss, L. L., 1963, Stratigraphy and sedimentation (2nd. Edition): San Francisco, W. H. Freeman and Co., 660 p.

Krynine, P. D., 1942, Differential sedimentation and its products during one complete geosynclinal cycle: Proceedings of the 1st. Pan American Congress on Mining and Engineering Geology, Pt. 1, v. 2, p. 537-560.

Kocurek, G., and Dott, R. H., Jr., 1981, Distinctions and uses of stratification types in the interpretation of eolian sand: Journal of Sedimentary Petrology, v. 51, p. 579-595.

Kumar, N., and Sanders, J. E., 1974, Inlet sequence: a vertical succession of sedimentary structures and textures created by the lateral migration of tidal inlets: Sedimentology, v. 21, p. 491-532.

Kumar, N., and Sanders, J. E., 1976, Characteristics of shoreface deposits; modern and ancient: Journal of Sedimentary Petrology, v. 46, p. 145-162.

Land, L. S., 1973, Holocene meteoric dolomitization of Pleistocene limestones, North Jamaica: Sedimentology, v. 20, p. 411-424.

Laughbaum, L. R., 1960, A paleoecologic study of the Upper Denton Formation, Tarrant, Denton, and Cooke Counties, Texas: Journal of Paleontology, v. 34, p. 1183-1197.

Le Blanc Smith, G., 1980, Logical-letter-coding system for facies nomenclature: Witbank Coalfield; Transactions of the Geological Society of South Africa, v. 83, p. 301-312.

Leeder, M. R., 1983, On the interactions between turbulent flow, sediment transport and bedform mechanics in channelized flows, in Collinson, J. D., and Lewin, J., eds., Modern and ancient fluvial systems: International Association of Sedimentologists Special Publication 6, p. 5-18.

Long, D. G. F., 1978, Proterozoic stream deposits: some problems of recognition and interpretation of ancient sandy fluvial systems, in Miall, A. D., ed., Fluvial sedimentology: Canadian Society of Petroleum Geologists Memoir 5, p. 313-341.

Longwell, C. R., ed., 1949, Sedimentary facies in geologic history: Geological Society of America Memoir 39, 171 p.

Lowe, D .R., 1979, Sediment gravity flows: their classification and some problems of application to natural flows, in Doyle, L. J., and Pilkey, O. H., Jr., eds., Geology of Continental Slopes: Society of Economic Paleontologists and Mineralogists Special Publication 27, p. 75-82.

Lowe, D. R., 1982, Sediment gravity flows II. Depositional models with special reference to the deposits of high-density turbidity currents: Journal of Sedimentary Petrology, v. 52, p. 279-297.

Lowenstam, H. A., 1950, Niagaran reefs in the Great Lakes area: Journal of Geology, v. 58, p. 430-487.

Lowenstam, H. A., 1957, Niagaran reefs in the Great Lakes area: Geological Society of America Memoir 67, v. 2, p. 215-248.

Lowenstam, H. A., Willman, H. B., and Swann, D. H., 1956, The Niagaran reef at Thornton, Illinois: American Association of Petroleum Geologists and Society of Economic Paleontologists and Mineralogists Guidebook, p. 1-19.

Ludvigsen, R., 1978, Middle Ordovician trilobite biofacies, southern Mackenzie Mountains, in Stelck, C. R., and Chatterton, B. D. E., eds., Western and Arctic Canadian biostratigraphy: Geological Association of Canada Special Paper 18, p. 1-37.

MacEachern, J. A., Pemberton, S. G., Gingras, M. K., and Bann, K. L., 2010, Ichnology and facies models, in James, N. P., and Dalrymple, R. W., eds., Facies Models 4: Geotext 6, Geological Association of Canada, p. 19-58.

Machel, H.-G., 1985, Cathodoluminescence in calcite and dolomite and its chemical interpretation: Geoscience Canada, v. 12, p. 139-147.

Macquaker, J. H. S., and Adams, A. E., 2003, Maximizing information from fine-grained sedimentary rocks: An inclusive nomenclature for mudstones: Journal of Sedimentary Research, v. 73, p. 735–744.

McGregor, B. A., 1981, Ancestral head of Wilmington Canyon: Geology, v. 9, p. 254-257.

McIlreath, I. A., and James, N. P., 1984, Carbonate slopes, in Walker, ed., R. G., Facies Models, 2nd edition: Geoscience Canada Reprint Series 1, p. 245-257.

McKee, E. D., ed., 1979, A study of global sand seas: U.S. Geological Survey Professional Paper 1052.

McKee, E. D., and Weir, G. W., 1953, Terminology for stratification in sediments: Geological Society of America Bulletin, v. 64, p. 381-389.

Mello, J. F., and Buzas, M. A., 1968, An application of cluster analysis as a method of determining biofacies: Journal of Paleontology, v. 42, p. 747-758.

Miall, A. D., 1973, Markov chain analysis applied to an ancient alluvial plain succession: Sedimentology, v. 20, p. 347-364.

Miall, A. D., 1974, Paleocurrent analysis of alluvial sediments: a discussion of directional variance and vector magnitude: Journal of Sedimentary Petrology, v. 44, p. 1174-1185.

Miall, A. D., 1977, A review of the braided river depositional environment: Earth Science Reviews, v. 13, p. 1-62.

Miall, A. D., 1978c, Lithofacies types and vertical profile models in braided river deposits: a summary, in Miall, A. D., ed., Fluvial sedimentology: Canadian Society of Petroleum Geologists Memoir 5, p. 597-604.

Miall, A. D., 1980a, Cyclicity and the facies model concept in geology: Bulletin of Canadian Petroleum Geology, v. 28, p. 59-80.

Miall, A. D., 1985, Architectural-element analysis: A new method of facies analysis applied to fluvial deposits: Earth Science Reviews, v. 22, p. 261-308.

Miall, A. D., 1988a, Reservoir heterogeneities in fluvial sandstones: lessons from outcrop studies: American Association of Petroleum Geologists Bulletin, v. 72, p. 682-697.

Miall, A. D., 1988b, Facies architecture in clastic sedimentary basins, in Kleinspehn, K., and Paola, C., eds., New perspectives in basin analysis: Springer-Verlag Inc., New York, p. 67-81.

Miall, A. D., 1988c, Architectural elements and bounding surfaces in channelized clastic deposits: notes on comparisons between fluvial and turbidite systems, in Taira, A., and Masuda, F., eds., Sedimentary facies in the active plate margin: Terra Scientific Publishing Company, Tokyo, Japan, p. 3-15.

Miall, A. D., 1995b, Description and interpretation of fluvial deposits: A critical perspective: Discussion: Sedimentology, v. 42, p. 379-384.

Miall, A. D., 1996, The geology of fluvial deposits: sedimentary facies, basin analysis and petroleum geology: Springer-Verlag Inc., Heidelberg, 582 p.

Miall, A. D., 1997, The geology of stratigraphic sequences, First edition: Springer-Verlag, Berlin, 433 p.

Miall, A. D., 1999, Principles of Sedimentary Basin Analysis, Third edition: Springer-Verlag, Berlin, 616 p.

Miall, A. D., 2010, The geology of stratigraphic sequences, second edition: Springer-Verlag, Berlin, 522 p.

Miall, A. D., 2014a, Fluvial depositional systems: Springer-Verlag, Berlin 316 p.

Miall, A. D., 2015, Updating uniformitarianism: stratigraphy as just a set of "frozen accidents", in Smith, D. G., Bailey, R., J., Burgess, P., and Fraser, A., eds., Strata and time: Geological Society, London, Special Publication 404, p. 11-36.

Miall, A. D., and Gibling, M. R., 1978, The Siluro-Devonian clastic wedge of Somerset Island, Arctic Canada, and some regional paleogeographic implications: Sedimentary Geology, v. 21, p. 85-127.

Miall, A. D., and Tyler, N., eds., 1991, The three-dimensional facies architecture of terrigenous clastic sediments, and its implications for hydrocarbon discovery and recovery: Society of Economic Paleontologists and Mineralogists Concepts and Models Series, v. 3, 309 p.

Middleton, G. V., 1966, 1967, Experiments on density and turbidity currents, I, II, and III, Canadian Journal of Earth Sciences, v. 3, p. 523-546, p. 627-637, v. 4, p. 475-505.

Middleton, G. V., 1973, "Johannes Walther's law of the correlation of facies": Geological Society of America Bulletin, v. 84, p. 979-988.

Middleton, G. V., 2005, Sedimentology, History, in Middleton, G. V., ed., Encyclopedia of sediments and sedimentary rocks, Berlin: Springer-Verlag, p. 628-635.

Middleton, G. V., and Hampton, M. A., 1976, Subaqueous sediment transport and deposition by sediment gravity flows, in Stanley, D. J., and Swift, D. J. P., eds., Marine sediment transport and environmental management; Wiley, New York, p. 197-218.

Kuenen, Ph. H., and Migliorini, C. I., 1950, Turbidity currents as a cause of graded bedding: Journal of Geology, v. 58, p. 91-127.

Mountjoy, E. W., 1980, Some questions about the development of Upper Devonian carbonate buildups (reefs), Western Canada: Bulletin of Canadian Petroleum Geology, v. 28, p. 315-344.

Mountjoy, E. W., Cook, H. E., Pray, L. C., and McDaniel, P. N., 1972, Allochthonous carbonate debris flows—worldwide indicators of reef complexes, banks or shelf margins: 24th International Geological Congress, Montreal, Section 6, p. 172-189.

Mutti, E., and Normark, W. R., 1987, Comparing examples of modern and ancient turbidite systems: problems and concepts; in Leggett, J. K., and Zuffa, G. G., eds., Marine clastic sedimentology: concepts and case studies: Graham and Trotman, London, p. 1-38.

Mutti, E., and Ricci-Lucchi, F., 1972, Le turbiditi dell'Appennino settentrionale: introduzione all' analisi di facies: Soc. Geol. Ital. Mem. 11, p. 161-199.

Natland, M. L., 1933, The temperature- and depth-distribution of some Recent and fossil foraminifera in the southern California region: Bull. Scripps Institute of Oceanography, v. 3, p. 225-230.

Natland, M. L., and Kuenen, Ph.D., 1951, Sedimentary history of the Ventura Basin, California, and the action of turbidity currents: Society of Economic Paleontologists and Mineralogists Special Publication 2, p. 76-107.

Nemec, W., Porebski, S. J., and Steel, R. J., 1980: Texture and structure of resedimented conglomerates: examples from Ksiaz Formation (Fammenian-Tournaisian), southwestern Poland: Sedimentology, v. 27, p. 519-538.

Newell, N. D., Rigby, J. K., Fischer, A. G., Whiteman, A. J., Hickox, J. E., and Bradley, J. S., 1953, The Permian reef complex of the Guadalupe Mountains Region, Texas and New Mexico: W. H. Freeman, San Francisco, 236 p.

Nichols, D., 1959, Changes in the Chalk heart-urchin *Micraster* interpreted in relation to living forms: Philosophical Transactions of the Royal Society, London, Series B., v. 242, p. 347-437.

Nilsen, T., Bartow, J. A., Stump, E., And Link, M. H., 1977, New occurrences of dish structure in the stratigraphic record: Journal of Sedimentary Petrology, v. 47, p. 1299-1304.

Yang Chang-shu, and Nio, S.-D., 1989, An ebb-tide delta depositional model - a comparison between the modern Eastern Scheldt tidal basin (southwest Netherlands) and the Lower Eocene Roda Sandstone in the southern Pyrenees (Spain): Sedimentary Geology, v. 64, p. 175-196.

Nio, S.-D., and Yang, C., 1991, Diagnostic attributes of clastic tidal deposits: a review, in Smith, D. G., Reinson, G. E., Zaitlin, B. A., and Rahmani, R. A., eds., Clastic tidal sedimentology: Canadian Society of Petroleum Geologists memoir 16, p. 3-27.

Nottvedt, A., and Kreisa, R. D., 1987, model for the combined-flow origin of hummocky cross-stratification: Geology, v. 15, p. 357-361.

Nummedal, D., Pilkey, O. H., and Howard, J. D., eds., 1987, Sea-level fluctuation and coastal evolution: Society of Economic Paleontologists and Mineralogists Special Publication 41, 267 p.

O'Brien, N. R., 1990, Significance of lamination in Toarcian (Lower Jurassic) shales from Yorkshire, Great Britain: Sedimentary Geology, v. 67, p. 25-34.

Odin, G. S., and Matter, A., 1981, De glauconiarum origine: Sedimentology, v. 28, p. 611-642.

Parrish, J. T., 1983, Upwelling deposits: nature of association of organic-rich rock, chert, chalk, phosphorite and glauconite: American Association of Petroleum Geologists Bulletin, v. 67, p. 529.

Picard, M. D., and High, L. R., Jr., 1973, Sedimentary structures of ephemeral streams: Developments in Sedimentology 17, Elsevier, Amsterdam, 223 p.

Pickering, K. T., Stow, D. A. V., Watson, M. P., and Hiscott, R. N., 1986, Deep-water facies, processes and models: a review and classification scheme for modern and ancient sediments: Earth Science Reviews, v. 23, p. 75-174.

Pirrie, D., 1998, Interpreting the record, facies analysis, in Blundell, D. J., and Scott, A. C., eds., Lyell: the past is the key to the present: Geological Society of London, Special Publication 143, p. 395-420.

Pirson, S. J., 1977, Geologic well log analysis, 2nd ed.: Gulf Publishing Co., Houston, 377 p.

Platt, N. H., and Keller, B., 1992, Distal alluvial deposits in a foreland basin setting — the Lower Freshwater Molasse (Lower Miocene), Switzerland: sedimentology, architecture and palaeosols: Sedimentology, v. 39, p. 545-565.

Playton, T. E., Janson, X., and Kerans, C., 2010, Carbonate slopes, in James, N. P., and Dalrymple, R. W., eds., Facies Models 4: GEOtext 6, Geological Association of Canada, St. John's, Newfoundland, p. 449-476.

Plint, A. G., 2010, Wave- and storm-dominated shoreline and shallow-marine systems, in James, N. P., and Dalrymple, R. W., eds., Facies Models 4: GEOtext 6, Geological Association of Canada, St. John's, Newfoundland, p. 167-199.

Posamentier, H. W., and James, D. P., 1993, An overview of sequence stratigraphic concepts: uses and abuses, in Posamentier, H. W., Summerhayes, C. P., Haq, B. U., and Allen, G. P., eds., Sequence stratigraphy and facies associations: International Association of Sedimentologists Special Publication 18, p. 3-18.

Posamentier, H. W., and Vail, P. R., 1988, Eustatic controls on clastic deposition II—sequence and systems tract models, in Wilgus, C. K., Hastings, B. S., Kendall, C. G. St. C., Posamentier, H. W., Ross, C. A., and Van Wagoner, J. C., eds., Sea level Changes - an integrated approach: Society of Economic Paleontologists and Mineralogists Special Publication 42, p. 125-154.

Posamentier, H. W., Jervey, M. T., and Vail, P. R., 1988, Eustatic controls on clastic deposition I—Conceptual framework, in Wilgus, C. K., Hastings, B. S., Kendall, C. G. St. C., Posamentier, H. W., Ross, C. A., and Van Wagoner, J. C., eds., Sea level Changes - an integrated approach: Society of Economic Paleontologists and Mineralogists Special Publication 42, p. 109-124.

Postma, G., 1986, Classification for sediment gravity-flow deposits based on flow conditions during sedimentation: Geology, v. 14, p. 291-294.

Potter, P. E., 1959, Facies models conference: Science, v. 129, p. 1292-1294.

Pratt, B. R., 2010, Peritidal carbonates, in James, N. P., and Dalrymple, R. W., eds., Facies Models 4: GEOtext 6, Geological Association of Canada, St. John's, Newfoundland, p. 401-420.

Pufhal, P. K., 2010, Bioelemental sediments, in James, N. P., and Dalrymple, R. W., eds., Facies Models 4: GEOtext 6, Geological Association of Canada, St. John's, Newfoundland, p. 477-503.

Purdy, E. G., 1963, Recent calcium carbonate facies of the Great Bahama Bank: Journal of Geology, v. 71, p. 334-355, p. 472-497.

Purdy, E. G., 1974, Karst-determined facies patterns in British Honduras: Holocene carbonate sedimentation model: American Association of Petroleum Geologists Bulletin, v. 58, p. 825-855.

Reading, H. G., ed., 1986, Sedimentary environments and facies, second ed.: Blackwell, Oxford, 615 p.

Reading, H. G., ed., 1996, Sedimentary environments: processes, facies and stratigraphy, third edition: Blackwell Science, Oxford, 688 p.

Reading, H. G., and Levell, B. K., 1996, Controls on the sedimentary rock record, in Reading, H. G., ed., Sedimentary environments: processes, facies and stratigraphy, third edition: Blackwell Science, Oxford, p. 5-36.

Reineck, H. E., and Singh, I. B., 1973, Depositional sedimentary environments—with reference to terrigenous clastics: Berlin: Springer-Verlag, 439 p.

Reineck, H. E. and Singh, I. B., 1980, Depositional sedimentary environments - with reference to terrigenous clastics: 2nd edition, Springer-Verlag, New York, 549 p.

Reineck, H. E., and Wunderlich, R., 1968, Classification and origin of flaser and lenticular bedding: Sedimentology, v. 11, p. 99-104.

Runkel, A. C., Miller, J. F., McKay, R. M., Palmer, A. R., and Taylor, J. F., 2007, High-resolution sequence stratigraphy of lower Paleozoic sheet sandstones in central North America: the role of special conditions of cratonic interiors in development of stratal architecture: Geological Society of America Bulletin, v. 119, p. 860-881.

Rust, B. R., 1972, Pebble orientation in fluviatile sediments: Journal of Sedimentary Petrology, v. 42, p. 384-388.

Schieber, J., Southard, J. B., Kissling, P., Rossman, B., and Ginsburg, R., 2013, Experimental Deposition of Carbonate Mud From Moving Suspensions: Importance of Flocculation and Implications For Modern and Ancient Carbonate Mud Deposition: Journal of Sedimentary Research, v. 83, p. 1025-1031.

Schlager, W., 1991, Depositional bias and environmental change—important factors in sequence stratigraphy: Sedimentary Geology, v. 70, p. 109-130.

Schlager, W., 2005, Carbonate sedimentology and sequence stratigraphy: SEPM Concepts in Sedimentology and Paleontology #8, 200p.

Scholle, P. A., 1978, A color illustrated guide to carbonate rock constituents, textures, cements and porosities: American Association of Petroleum Geologists Memoir 27, 241 p.

Schreiber, B. C., 1981, Marine evaporites: facies development and relation to hydrocarbons and mineral genesis: American Association of Petroleum Geologists Fall Education Conference, 44 p.

Schreiber, B. C., 1986, Arid shorelines and evaporites, in Reading, H. G., ed., Sedimentary environments and facies, 2nd ed.; Blackwell Scientific Publications, Oxford, p. 189-228.

Schreiber, B. C., Friedman, G. M., Decima, A., and Schreiber, E., 1976, Depositional environments of Upper Miocene (Messinian) evaporite deposits of the Sicilian Basin: Sedimentology, v. 23, p. 729-760.

Schultz, A. W., 1984, Subaerial debris flow deposition in the Upper Paleozoic Cutler Formation, western Colorado: Journal of Sedimentary Petrology, v. 54, p. 759-772.

Schumm, S. A., 1968: Speculations concerning paleohydrologic control of terrestrial sedimentation: Geological Society of America Bulletin, v. 79, p. 1573-1588.

Schwab, F. L., 1981, Evolution of the western continental margin, French-Italian Alps: sandstone mineralogy as an index of plate tectonic setting: Journal of Geology, v. 89, p. 349-368.

Seilacher, A., 1967, Bathymetry of trace fossils: Marine Geology, v. 5, p. 413-428.

Selley, R. C., 1968, A classification of paleocurrent models: Journal of Geology, v. 76, p. 99-110.

Selley, R. C., 1979, Dipmeter and log motifs in North Sea submarine-fan sands: American Association of Petroleum Geologists Bulletin, v. 63, p. 905-917.

Sellwood, B. W., 1986, Shallow-marine carbonate environments, in Reading, H. G., ed., Sedimentary environments and facies, 2nd ed.: Blackwell, Oxford, p. 283-342.

Shand, S. J., 1947, Eruptive rocks, 3rd edition: John Wiley and Sons, New York.

Shanmugam, G., 1997, The Bouma sequence and the turbidite mind set: Earth Science Reviews, v. 42, p. 201-229.

Shaw, D. B., and Weaver, C. E., 1965, The mineralogical composition of shales: Journal of Sedimentary Petrology, v. 35, p. 213–222.

Shea, J. H., 1982, Twelve fallacies of uniformitarianism: Geology, v. 10, p. 449-496.

Shearman, D. J., 1966, Origin of marine evaporites by diagenesis: Transactions of the Institute of Mining and Metallurgy Bulletin, v. 75, p. 208-215.

Shepard, F. P., 1981, Submarine canyons: multiple causes and long-time persistence: American Association of Petroleum Geologists Bulletin, v. 65, p. 1062-1077.

Shepard, F. P., Marshall, N. F., McLoughlin, P. A., and Sullivan, G. G., 1979, Currents in submarine canyons and other sea valleys: American Association of Petroleum Geologists Studies in Geology 8.

Shinn, E. A., 1983, Birdseyes, fenestrae, shrinkage pores, and loferites: a reevaluation: Journal of Sedimentary Petrology, v. 53, p. 619-628.

Shinn, E. A., Halley, R. B., Hudson, J. H., Lidz, B., and Robbin, D. M., 1979, Three dimensional aspects of Belize patch reefs: American Association of Petroleum Geologists Bulletin, v. 63, p. 528.

Shor, A. N., Kent, D. V., and Flood, R. D., 1984, Contourite or turbidite?: magnetic fabric of fine-grained Quaternary sediments, Nova Scotia continental rise, in Stow, D. A. V., and., Piper, D J. W, eds., Fine-grained sediments: deep-water processes and facies: Geological Society of London Special Publication 15, p. 257-273.

Shurr, G. W., 1984, Geometry of shelf-sandstone bodies in the Shannon Sandstone of southeastern Montana, in Tillman, R. W., and Siemers, C. T., eds., Siliciclastic shelf sediments: Society of Economic Paleontologists and Mineralogists Special Publication 34, p. 63-83.

Simons, D. B., and Richardson, E. V., 1961, Forms of bed roughness in alluvial channels: American Society of Civil Engineers Proceedings, v. 87, No. HY3, p. 87-105.

Simons, D. B., Richardson, E. V., and Nordin, C. F., 1965, Sedimentary structures generated by flow in alluvial channels, in Middleton, G. V., ed., Primary sedimentary structures and their hydrodynamic interpretation: Society of Economic Paleontologists and Mineralogists Special Publication 12, p. 34-52.

Skipper, K., 1976, Use of geophysical wireline logs for interpreting depositional processes: Geoscience Canada, v. 3, p. 279-280.

Sloss, L. L., Krumbein, W. C., and Dapples, E. C., 1949, Integrated facies analysis; in Longwell, C. R., ed., Sedimentary facies in geologic history: Geological Society of America Memoir 39, p. 91-124.

Smith, D. G., and Smith, N. D., 1980, Sedimentation in anastomosed river systems: examples from alluvial valleys near Banff, Alberta: Journal of Sedimentary Petrology, v. 50, p. 157-164.

Smosna, R., and Warshauer, S. M., 1981, Rank exposure index on a Silurian carbonate tidal flat: Sedimentology, v. 28, p. 723-731.

Sneider, R. M., Tinker, C. N., and Meckel, L. D., 1978, Deltaic environment reservoir types and their characteristics: Journal of Petroleum Technology, v. 20, p. 1538-1546.

Solohub, J. T., and Klovan, J. E., 1970, Evaluation of grain-size parameters in lacustrine sediments: Journal of Sedimentary Petrology, v. 40, p. 81-101.

Southard, J. B., 1971, Representation of bed configurations in depth-velocity-size diagrams: Journal of Sedimentary Petrology, v. 41, p. 903-915.

Speyer, S. E., and Brett, C. E., 1986, Trilobite taphonomy and Middle Devonian taphofacies: Palaios, v. 1, p. 312-327.

Steel, R. J., 1974, New Red Sandstone floodplain and piedmont sedimentation in the Hebridean Province: Journal of Sedimentary Petrology, v. 44, p. 336-357.

Steidtmann, J. R., 1974, Evidence for eolian origin of cross-stratification in sandstone of the Casper Formation, southernmost Laramie Basin, Wyoming: Geological Society of America Bulletin, v. 85, p. 1835-1842.

Stow, D. A. V., and Piper, D. J. W., eds., 1984, Fine-grained sediments: deep-water processes and facies: Geological Society of London Special Publication 15, Blackwell Scientific Publications, Oxford, 659 p.

Swift, D. J. P., 1975a, Barrier island genesis: evidence from the Middle Atlantic shelf of North America: Sedimentary Geology, v. 14, p. 1-43.

Swift, D. J. P., Stanley, D. J., and Curray, J. R., 1971, Relict sediments on continental shelves: a reconsideration: Journal of Geology, v. 79, p. 322-346.

Swift, D. J. P., Figueiredo, A. G., Jr., Freeland, G. L., and Oertel, G. F., 1983, Hummocky cross-stratification and megaripples: a geological double standard? Journal of Sedimentary Petrology, v. 53, p. 1295-1318.

Teichert, C., 1958, Concepts of facies: Bulletin of the American Association of Petroleum Geologists, v. 42, p. 2718-2744.

Turner, P., 1980, Continental red beds: Developments in sedimentology 29, Elsevier, Amsterdam, 562 p.

Tyler, N., Galloway, W. E., Garrett, C. M., Jr., and Ewing, T. E., 1984, Oil accumulation, production characteristics, and targets for additional recovery in major oil reservoirs of Texas: Bureau of Economic Geology, Texas, Geologic Circular 84-2, 31 p.

Uba, E., Heubeck, C., and Hulka, C., 2005, Facies analysis and basin architecture of the Neogene Subandean synorogenic wedge, southern Bolivia: Sedimentary Geology, v. 180, p. 91-123.

Vai, G. B., and Ricci-Lucchi, F., 1977, Algal crusts, autochthonous and clastic gypsum in a cannibalistic evaporite basin: a case history from the Messinian of Northern Apennines: Sedimentology, v. 24, p. 211-244.

Van Houten, F. B., 1964, Cyclic lacustrine sedimentation, Upper Triassic Lockatong Formation, central New Jersey and adjacent Pennsylvania, in Merriam, D. F., ed., Symposium on cyclic sedimentation: Geological Survey of Kansas Bulletin 169, p. 495-531.

Van Houten, F. B., 1973, Origin of red beds: a review—1961-1972: Annual Review of Earth and Planetary Sciences, v. 1, p. 39-62.

Van Houten, F. B., 1985, Oolitic ironstones and contrasting Ordovician and Jurassic paleogeography: Geology, v. 13, p. 722-724.

Van Straaten, L. M. J. U., 1951, Texture and genesis of Dutch Wadden Sea sediments: Proceedings of the 3rd International Congress on Sedimentology, Netherlands, p. 225-255.

Van Straaten, L.M. J. U., 1954, Composition and structure of recent marine sediments in the Netherlands: Leidse Geol. Mededel., v. 19, p. 1-110.

Van Wagoner, J. C., Mitchum, R. M., Jr., Posamentier, H. W., and Vail, P. R., 1987, Seismic stratigraphy interpretation using sequence stratigraphy, Part 2: key definitions of sequence stratigraphy, in Bally, A. W., ed., Atlas of seismic stratigraphy: American Association of Petroleum Geologists Studies in Geology 27, v. 1, p. 11-14.

Van Wagoner, J. C., Mitchum, R. M., Campion, K. M. and Rahmanian, V. D. 1990, Siliciclastic sequence stratigraphy in well logs, cores, and outcrops: American Association of Petroleum Geologists Methods in Exploration Series 7, 55 p.

Visher, G. S., 1965, Use of vertical profile in environmental reconstruction; American Association of Petroleum Geologists Bulletin: v. 49, p. 41-61.

Visher, G. S., 1969, Grain size distributions and depositional processes: Journal of Sedimentary Petrology, v. 39, p. 1074-1106.

Visser, M. J., 1980, Neap-spring cycles reflected in Holocene subtidal large-scale bedform deposits: a preliminary note: Geology, v. 8, p. 543-546.

Walker, R. G., 1967, Turbidite sedimentary structures and their relationship to proximal and distal depositional environments: Journal of Sedimentary Petrology, v. 37, p. 25-43.

Walker, R. G., 1975, Generalized facies models for resedimented conglomerates of turbidite association: Geological Society of America Bulletin, v. 86, p. 737-748.

Walker, R. G., 1976, Facies models 1. General Introduction: Geoscience Canada, v. 3, p. 21-24.

Walker, R. G., 1978, Deep-water sandstone facies and ancient submarine fans: models for exploration for stratigraphic traps: American Association of Petroleum Geologists Bulletin, v. 62, p. 932-966.

Walker, R. G., ed., 1979, Facies models: Geoscience Canada Reprint Series No. 1, 211 p.

Walker, R. G., 1990, Facies modeling and sequence stratigraphy: Journal of Sedimentary Petrology, v. 60, p. 777-786.

Walker, R. G., 1992, Facies, facies models and modern stratigraphic concepts, in Walker, R. G. and James, N. P., eds., Facies models: response to sea-level change: Geological Association of Canada, p. 1-14.

Walker, T. R. and Harms, J. C., 1972, Eolian origin of Flagstone beds, Lyons Sanstone (Permian), type area, Boulder County, Colorado: Mountain Geologist, v. 9, p. 279-288.

Walker, R. G., and James, N. P., ed., 1984, Facies Models, second edition: Geoscience Canada Reprint Series 1, 317 p.

Walker, R. G. and James, N. P., eds., 1992, Facies models: response to sea-level change: Geological Association of Canada, 409 p.

Walther, Johannes, 1893-1894, Einleisung in die Geologie alshistorische Wissemchaft: Jena, Verlag von Gustav Fischer, 3 vols., 1055 p.

Weimer, R. J., and Hoyt, H. J., 1964, Burrows of Callianassa major Say, geologic indicators of littoral and shallow neritic environments: Journal of Paleontology v. 38, p. 761-767.

Weimer, P., and Posamentier, H. W., 1993, Recent developments and applications in siliciclastic sequence stratigraphy, in Weimer, P., and Posamentier, H. W., eds., Siliciclastic sequence stratigraphy: American Association of Petroleum Geologists Memoir 58, p. 3-12.

Wilkinson, B. H., Diedrich, N. W, Drummond, C. N., and Rothman, E. D., 1998, Michigan hockey, meteoric precipitation, and rhythmicity of accumulation on peritidal carbonate platforms: Geological Society of America Bulletin, v. 110, p. 1075-1093.

Williams, G. E., 1989, Late Precambrian tidal rhythmites in South Australia and the history of the Earth's rotation: Journal of the Geological Society, London, v. 146, p. 97-111.

Williams, P. F. and Rust, B. R., 1969, The sedimentology of a braided river: Journal of Sedimentary Petrology, v. 39, p. 649-679.

Wilson, J. L., 1975, Carbonate facies in geologic history: Springer-Verlag, New York, 471 p.

Wong, P. K., and Oldershaw, A. E., 1980, Causes of cyclicity in reef interior sediments, Kaybob Reef, Alberta: Bulletin of Canadian Petroleum Geology, v. 28, p. 411-424.

Wright, V. P., 1986a, Facies sequences on a carbonate ramp: the Carboniferous Limestone of South Wales: Sedimentology, v. 33, p. 221-242.

Young, F. G., Myhr, D. W., and Yorath, C. J., 1976, Geology of the Beaufort-Mackenzie Basin: Geological Survey of Canada Paper 76-11.

# Facies Models

**4**

## Contents

## 4.1 Introduction

As we describe in Chaps. 2 and 3, there are two broad classes of sedimentary rock, clastic sediments and chemical sediments. **Clastic sediments** are generated by the transport of detrital particles by currents of water or air; **chemical sediments** are created in place by chemical and biochemical processes, although current reworking may be an important part of the final set of processes that develops a given facies assemblage. Most clastic detritus is derived from the erosion and weathering, or recycling, of siliceous igneous, metamorphic or sedimentary rocks, and is referred to as **siliciclastic detritus**. Chemical sediments, particularly carbonate sediments, may be reworked as clastic detritus by marine currents or in sediment gravity flows.

Somewhat different techniques are needed to study these two groups (Chap. 3). The importance of currents of all types in the formation of clastic sediments—unimodal, bimodal and oscillating traction currents, eolian currents, and sediment-gravity flows—means that identifying the lithofacies and structures associated with each type of current process represents a substantial part of the work carried out to interpret clastic depositional environments. The processes of chemical sedimentation are identified by the petrological products of various chemical and biochemical processes, including the allochems of carbonate sediments and the precipitates of evaporate sediments. Current reworking is, of course, common in these environments, too.

Modern techniques to identify sedimentary environments based on sedimentary successions, fossils and structures date back to the early 1960s (see Chap. 1). The development of the flow-regime concept was a critical step forward (Sect. 1.2.5). Another was the realization of the significance of vertical profiles (Sects. 1.2.2, 1.2.7), a concept that began to be exploited by several geologists more or less simultaneously in Britain and the southern United States. In both cases, much of this early work was concerned with river (fluvial) deposits. This is no accident; it obviously reflects the accessibility of rivers for easy examination.

© Springer International Publishing Switzerland 2016
A.D. Miall, *Stratigraphy: A Modern Synthesis*, DOI 10.1007/978-3-319-24304-7_4

The documentation of marine deposits took longer, especially those of the deep marine environment, notably submarine fans and shelf deposits. The major breakthrough in the evolution of facies studies for these environments came in the early 1980s with the development of side-scan sonar methods for imaging the sea floor (Sect. 1.2.6). Prior to that time, interpretations of shelf and deep-marine sediments relied on bathymetric charts of variable accuracy, coupled with limited bottom coring.

Nowadays we can not only confidently identify the products of a wide range of environments, but we can detect and describe small differences between them. For example, in the 1960s, fluvial deposits were identified as either of the "meandering" or the "braided" variety. A modern synthesis of fluvial environments (Miall 1996) identified at least sixteen distinct varieties of fluvial environment, and the point was made that there exists intermediate conditions between most of any pair of the sixteen types. Similarly, so many actual modern submarine fans have now been studied and documented in detail that sedimentologists are no longer able to agree on how to classify them—there are simply too many different kinds (Bouma et al. 1985).

Amongst the most important developments in facies analysis in recent years has been the recognition of the dynamic nature of mud deposition. The long-standing assumption that most mud is deposited in static water by pelagic settling has been shown to be largely incorrect. Mud typically flocculates in most natural environments, and is subject to distribution by ambient currents, just as is loose sand. This opens a rich field for the reinterpretation of muddy deposits, some of which is touched on in this chapter. Readers may also wish to turn directly to some of the research and review literature that focuses on this topic, e.g., Macquaker et al. (2007), Plint (2010), Schieber et al. (2013), Trabucho-Alexandre (2015).

The intent of this chapter is to provide a brief overview of what is, in fact, a very complex topic. Facies models, or "summaries of the environment" (Sect. 3.4.2) are essential mental constructs that stratigraphers must build as they develop sequence-stratigraphic and chronostratigraphic interpretations of their subject rocks. This chapter is designed to provide a flavor of the type of work that needs to be done. For a more complete understanding of depositional systems and facies models the reader should turn to James and Dalrymple (2010), which is the fourth edition of the famous Canadian "facies models volume." Steel and Milliken (2013) provided a review of major advances in siliciclastic facies studies since 1960. Schlager (2005) published a comprehensive overview of carbonate facies and sequence stratigraphy.

## 4.2 Clastic Environments

### 4.2.1 Fluvial Environments

All siliciclastic detritus is transported at some point, by a river system. First-cycle debris eroded from a mountainous source area enters torrential mountain streams and is carried into mountain basins, or lakes, deposited at the edges of a basin in coarse alluvial fans, or transported out onto a coastal plain. The work of the sedimentologist involves recognizing various depositional settings (subenvironments) within a river (Fig. 4.1), and accumulating enough information about these to be able to reconstruct the local fluvial style (Fig. 4.2). Three examples of fluvial styles are illustrated in Figs. 4.3, 4.4 and 4.5.

The facies classification and coding system introduced by Miall (1977, 1978c) and updated by Miall (1996; see Table 3.1) is now virtually in standard use for fluvial studies, although as noted in Sect. 3.3.1, not all workers agree with the standardized approach to classification implicit in this system (e.g., Bridge 1993). Conversely, Farrell et al. (2012) have developed what they offer as a universal code system for use with all clastic rocks. An increasing number of workers are employing architectural methods for the three-dimensional documentation and classification of fluvial deposits in the surface and subsurface, particularly for channel-fill sandstone and conglomerate bodies (e.g., Miall and Tyler 1991; Miall 1996, Chap. 6). The classification of bounding surfaces proposed by Miall (1988a, b; see Sect. 3.5.11.1) has also found considerable use, although it has, as expected, proved difficult to make much use of the classification in subsurface work (Doyle and Sweet 1995). A simplified classification of channel style, showing their dependency on sediment grain size and other factors, and the range of typical architectural elements present in each type, is shown in Fig. 4.2. However, researchers are encouraged to devise their own classifications to suit the particular needs of their research.

Detailed work on modern floodplains, such as the study of the Mississippi by Farrell (1987), and classification studies, such as that by Nanson and Croke (1992), have contributed to a much better understanding of floodplain processes and sedimentation during the last decade. Description and classification of ancient floodplain deposits in terms of facies assemblages and architectural elements was attempted by Platt and Keller (1992), and the importance of this environment for coal development has been examined by Flores (1981). Important work on the genesis of paleosols and their stratigraphic context has been carried out by Kraus and Bown (1988) and Wright (1986b, 1992).

**Fig. 4.1** Terminology of the subenvironments of a river

**Fig. 4.2** Relationship between grain size and fluvial channel pattern. Architectural elements are those of Miall (1985; see Fig. 4.49). Note that FM has now been replaced by DA: downstream-accretion element (Orton and Reading 1993)

**Fig. 4.3** The point-bar model, showing its derivation either from a modern meandering river (*top*) or from preserved deposits (*bottom*)

From the point of view of basin analysis probably the most interesting study of overbank deposits is that of Willis and Behrensmeyer (1994), who used an extremely detailed stratigraphic study of part of the Miocene Siwalik succession in Pakistan as a basis for the building and testing of various models for the development of floodplain stratigraphies.

Only the four main fluvial styles are discussed briefly here. **Meandering rivers** were the subject of one of the first modern facies models, based on the recognition of point bars from their distinctive vertical profile (Bernard and Major

1963; Bernard et al. 1962; Allen 1964). In fact, the point-bar model is an excellent one for the purpose of illustrating the principle of the facies model and the key role played by its characteristic vertical profile (Fig. 4.3). Sediments in meandering rivers may be gravel-, sand- or silt-dominated. The lateral accretion of the point bar is the most distinctive feature of meandering-river deposits, but some bars in braided rivers also exhibit lateral accretion, so a certain amount of caution is needed in interpreting fluvial style from this one characteristic.

**Fig. 4.4** A model for a
sandy-braided river, based on the
South Saskatchewan River,
Canada (Miall 1996)

The other common fluvial style is that of the **braided river**, characterized by a wide active valley with several or many component channels of varying sinuosity (Fig. 4.4). The main sediment load may be sand or gravel. Individual channels undergo rapid migration, switches in position and abandonment. The deposits of braided rives are typically composed of an assemblage of crossbedded fragments of bar deposits, exhibiting accretion directions oriented in almost any direction, the most common being downstream and oblique to channel orientation.

**Anastomosed rivers** occur in areas of active vertical aggradation, such as coastal systems during a time of rapidly rising sea level, and in the case of inland systems where a river valley is backfilling behind a constriction. Many channels of variable sinuosity occur. They are relatively stable in position, showing little or no evidence of the lateral migration common to the meandering and braided styles (Fig. 4.5). Broad floodplains occur between the channels. These may be wetlands, peat swamps or desiccated flats, depending on the climate.

**Straight rivers** occur in areas of very low slope, such as near base level. As such, they are most common as the distributaries of certain deltas. Although the river itself may be nearly straight, there is typically a slightly sinuous deep channel, and long, narrow bars, called alternate bars,

accumulate along the banks inside the meanders defined by this channel.

Some examples of channel cutbanks are shown in Fig. 2. 13, but preservation of these features is, in fact, rather rare. In most outcrops the geologist will encounter tabular assemblages of variably crossbedded sandstone or conglomerate, and an interpretation of fluvial style must be made from this limited evidence. The orientation of crossbedding may be a useful clue (see Sect. 6.7). Many, but by no means all fluvial successions are characterized by fining-upward vertical profiles (Fig. 3.59c). In many cases, the fine-grained deposits forming the top of the cycle are truncated by erosion at the base of the succeeding channel system. In some sand-laden rivers the entire channel fill is composed of a similar lithology, and there is no discernible upward fining. On petrophysical logs this is commonly referred to as a "blocky" or "cylindrical" log response (Fig. 3.61).

The nature of floodplain deposits in fluvial systems depends largely on the climate. Figure 4.6 illustrates a variety of floodplain deposits formed in arid to semi-arid climates. In more humid climates, coals with or without soil horizons, may be present.

Caliche (Fig. 4.6a, b) is characteristic of semi-arid climates, where dissolved carbonate accumulates by capillary

**Fig. 4.5** A model for an
anastomosed river in a humid
environment. Based on the work
of Smith and Smith (1980) in
Banff National Park Canada

**Fig. 4.6** Fluvial floodplain deposits. **a**, **b** caliche, Triassic, Spain; **c** interbedded floodplain siltstones and mudstones. Note small channel filled with fine-grained deposits near the base of the section; **d** thin floodplain muds form the recessive unit at the top of a succession of sheet sandstone formed by flash floods in an ephemeral system; Devonian, Arctic Canada

action near the sediment surface. This example shows large nodules that have coalesced to form an extensive bed of nodular limestone. Such deposits take many thousands of years to develop.

Fossils in fluvial deposits may include remains of land vertebrates (footprints, bones), and a range of invertebrate trace fossils. Plant remains (logs, leaves, pollen) may be abundant in rocks of Devonian age and younger.

### 4.2.2 Eolian Environments

The publication of satellite imagery of some of the large and inaccessible deserts of Africa and Asia (McKee 1979) demonstrated a complexity of dune types that has not been recognized in the ancient record. However, it has become apparent that many of these dune fields are not currently active, except that minor modifications of the giant bedforms (**draas**) are taking place. The major episodes of sand movement were during the Pleistocene ice age. This curtails the usefulness of modern-process studies of giant dune formation. Fossil dune deposits demonstrate a surprising similarity and relative simplicity of form, as demonstrated by the useful summary papers in McKee (1979). We are only now beginning to understand why.

At the present day, large eolian sand systems tend to be clustered along the Tropic of Cancer and the Tropic of Capricorn (Fig. 4.7), because these belts, centering on 22.5° N and S, are characterized by zones of dry, high-pressure air masses, with steady winds but minimal rainfall. Major sand

accumulations are termed **ergs**. These have the superimposed deposits of large dunes at the centre, with other facies, such as the deposits of ephemeral rivers, typically occurring closer to the margins (Fig. 4.8). Smaller dune systems are common in coastal areas, where they may form a cap to barrier islands. These occur where winds pick up dry beach sands and blow them back above the high-tide mark. Such dune systems are often prominent features of barrier coastlines, but they have very low preservation potential.

Eolian systems are also present at higher latitudes in continental interiors, such as the Gobi, Taklamakam and Turkestan deserts of central Asia. Their presence reflects the tendency towards extreme aridity of high-elevation continental interiors. An ancient analog of the continental interior eolian setting is that of the Rotliegende Sandstone of the North Sea basin (an important gas reservoir: Glennie 1972, 1983).

The formation of eolian dunes and the larger stratigraphic features of erg systems have been clarified by the recognition of the importance of the different types of **bounding surface** that may be distinguished within the deposits (Fig. 3.38). Second- and third-order surfaces reflect changing wind patterns within the dunes. First-order surfaces record the migration of the major dunes across the desert surface. As illustrated below, these are commonly surfaces of wind deflation, and are marked by distinctive lithofacies (Fig. 4.9). Most important of all are what are termed **supersurfaces**, which are major stratigraphic surfaces extending for many kilometers through an erg (Kocurek 1988). These may be the product of climate changes or, in the case of coastal ergs, changes in sea level, which affects the level of the water

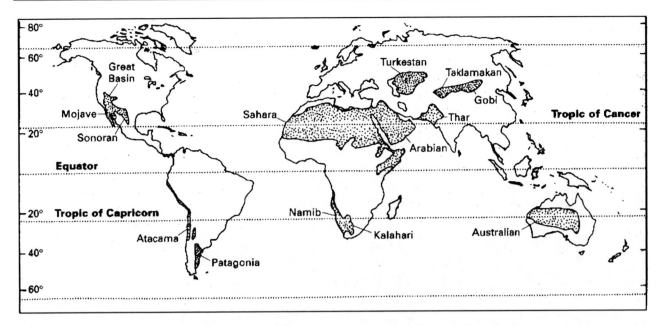

**Fig. 4.7** The distribution of major eolian deserts at the present day. Note the clustering along the latitudes 22.5°N and S (Reading 1996)

**Fig. 4.8** Facies model for an erg (Porter 1986)

table in the dune deposits (Brookfield and Silvestro 2010; see Fig. 5.21 of this book).

The characteristics of eolian bedforms are described and illustrated in Sect. 3.5.4.7. Here, in Fig. 4.9 we can see some of the other distinctive features of eolian systems. The interbedding of eolian and ephemeral fluvial deposits is common (Fig. 4.9a). Ephemeral interdune ponds may appear after flash floods, and become the site for the accumulation of thin silt and mud lenses as the ponds dry out (Fig. 4.9b). These may be blown away by the wind before they are buried by dune sands, and may not have a very high preservation potential. Temporary saturation may also be the cause of the slumping of dune faces, developing deformed crossbedding (Fig. 4.9c).

Surfaces of deflation are particularly characteristic of eolian deserts (Fig. 4.9d–f), and may, stratigraphically, mark

**Fig. 4.9** Characteristics of eolian environments and deposits. **a** A stream cutting through dune sands; **b** dried-out interdune pond deposit; **c** slumped face of dune sandstone; **d** tire tracks disappearing beneath a barchan; **e**, **f** surface of deflation. **e** Shows a surface covered with ventifacts; **g** close-up of a ventifact; **h**, **i** cross-sections of ancient eolian deposits showing surfaces of deflation with ventifacts. Locations: **a**, Colorado; **b**, **d**, **e**, **f**, Gobi desert, China; **c** Navajo Sandstone, Arizona; **h** Triassic, Spain; **i** Permian, Scotland

the position of first-order surfaces or supersurfaces. Figure 4.9d illustrates the rapid movement of barchan dunes across a deflation surface—note the disappearance of modern tire tracks beneath the arms of the dune.

A particularly useful indicator of eolian processes is the **ventifact**, which is a large clast that has been sand-blasted by the wind, and periodically turned over (possibly by flash floods), so that it has developed a series of flat faces (Fig. 4.9g). The term **dreikanter** is also used for these objects. Figure 4.9e illustrates a particularly common feature of the desert floor—a deflation surface covered by sand-blasted pebbles. Ancient equivalents of these surfaces are shown in Figs. 4.9h, i.

### 4.2.3 Lacustrine Environments

Lakes are among the most varied of all depositional environments, even though they occupy a relatively small percentage of the earth's surface at the present day (about 1 % according to Talbot and Allen 1996). There are no universal

facies models for the lacustrine environment, and recent books and review articles on this subject are not so much syntheses of ideas as catalogs of case studies, each example varying markedly from the next in terms of lithofacies, geochemistry, thickness, and extent (Matter and Tucker 1978; Ryder 1980; Picard and High 1981; Katz 1990; Anadón et al. 1991; Talbot and Allen 1996; Renaut and Gierlowski-Kordesch 2010). The documentation of this variability is perhaps the most important achievement of recent facies studies of lake sediments. This is by no means an academic subject. Vast petroleum reserves occur as oil shales in the Green River Formation of the Uinta Basin and as liquid petroleum in China (e.g., Daqing Oil Field), the western United States, West Africa, and elsewhere (Ryder 1980).

Most major lake deposits owe their origin to tectonic isolation of a sedimentary basin from the sea. The advent of plate tectonics has helped explain the origins of these basins, many of which are surprisingly broad and deep. Typical tectonic settings for lakes include rifts (Lake Baikal, East African Rift System), transform plate margins (Dead Sea,

**Fig. 4.10** Lacustrine shales of the Eocene Green River Formation, Utah

**Fig. 4.11** Laminated lacustrine muds and silts formed in a glaciolacustrine setting, and containing dropstones deposited from floating ice. Pleistocene, Ontario

Cenozoic basins of California), remnant ocean basins (Black Sea?), some foreland basins (Cenozoic Uinta Basin), and intracratonic depressions (Lake Eyre). Other lakes owe their origins to ice damming or volcanic damming or to glacial scour. Temporary lakes form in various fluvial and coastal settings and may be the site for minor but distinctive deposits.

Another major sediment control is water chemistry. This can vary widely, reflecting variations in the inflow/evaporation balance, nature, and quantity of dissolved riverine sediment load, temperature, water level, and water body structure (presence of stratification, seasonal overturning, etc.). As Talbot and Allen (1996) noted, lakes are very sensitive to climatic change. They provided a useful review of this topic and suggested a broad classification of lacustrine deposits into those formed in dilute lakes of hydrologically open basins and those formed in saline lakes of closed basins. However, lakes may change from one of these states to the other with variations in climate or tectonic setting. Hardie et al. (1978) discussed subenvironments and subfacies that exist in saline lakes and showed how lacustrine brines evolve during evaporative concentration.

In many ways, lakes represent a microcosm of almost every other environment, usually on a smaller scale than in most marine basins (however, lacustrine carbonates are rather rare). The lake margin may exhibit coastal beach-barrier or delta systems, or there may be coarse fan-deltas prograding into the deep basin, with submarine fans forming in the deeper parts of the basin. The lake centre may be a quiet water environment characterized by muds (Fig. 4.10). Storm deposits, including hummocky cross-stratification, have been identified in some ancient lake deposits.

Because of their sensitivity to climate change and tectonism, and the commonly rapid lateral and vertical facies changes, lacustrine sediments have been much studies by those interested in the processes of allogenic cyclic sedimentation. The processes of **orbital forcing** of climate change (the so-called **Milankovitch mechanisms**) have been studied in lacustrine sediments since their influence was first suspected in the Eocene Green River Shales of Wyoming (Bradley 1929). Recent studies of such cyclicity (reviewed in Miall 2010) include work on Miocene deposits of northeast Spain (Anadón et al. 1991) and the Triassic Newark Group of the eastern United States (Van Houten 1964; Olsen 1984, 1986, 1990). In the latter, there is considerable variation between humid-climate cycles containing abundant fluvial sandstone deposits, and arid-climate cycles with evaporates. In pro-glacial settings, lakes may contain many of the same facies as those occurring in glaciomarine settings, including sediment-gravity flow deposits from the reworking of glacial tills, and dropstones deposited from floating ice (Fig. 4.11).

Figure 4.12 provides a facies model and hydrological model for arid, intermontane basins, of the type that characterizes the present-day arid southwest of the United States. These basins (especially Death Valley) are commonly cited as analogues of the numerous lacustrine basins of Permian-Triassic age that formed all across what is now eastern North America and western Europe as Pangea was beginning to break up.

### 4.2.4 Glacial Environments

Glacigenic deposits were among the last to be studied using facies analysis methods. The analogue method of facies interpretation, in this case the use of Quaternary environmental and facies data, did not at first work well for glacigenic deposits. There are several reasons for this. A considerable body of specialized knowledge had built up from the study of Quaternary sediments, but much of this is stratigraphic in nature, rather than sedimentologic, and therefore offers little assistance in the interpretation of the ancient record. A thorough understanding of ancient

**(a)**  Mountains (source area)   Alluvial fans (bajada)   Dry mudflat   Saline mudflat   Sandflat   Perennial saline lake or saline pan

**(b)**

M   Meteoric waters
F   Formation waters
H   Hydrothermal waters
D   Density (salinity-driven) circulation
S   Springs (ambient or hydrothermal)

Evaporation
Surface runoff

**Fig. 4.12** Playa-lake model. **a** General model for closed-basin sedimentation in an arid climate, based on the intermontane basins of the western United States. **b** Hydrology of an intermontane closed basin (Renaut and Gierlowski-Kordesch 2010, Fig. 18, p. 562)

glacigenic deposits was also hindered by the tendency to describe all coarse-grained, poorly sorted glacigenic deposits as "**till**". While this term means primarily deposition directly from grounded ice, some authorities have used a much broader definition. For example, Dreimanis (1985) stated that till "is formed and deposited by primary glacial and also by closely associated penecontemporaneous secondary non-glacial processes." It required much sedimentological investigation to document those secondary processes and to demonstrate that in many basins they were dominant. Most ancient glacigenic deposits formed by the dispersal and redeposition of glacigenic materials by a wide variety of processes in almost every conceivable environment, and it has taken some time for sedimentologists to sort out these various processes, to recognize their importance, and to devise criteria for their recognition in the rock record.

Eyles et al. (1983) were amongst the first to develop systematic facies criteria for this purpose. Most authors now recommend use of the nongenetic term **diamict** (or **diamicton** or **mixton**) for the description of poorly sorted gravel-sand-silt-mud mixtures and **diamictite** (or **mixtite**) for the lithified variety. The term till is then restricted for use in an interpretive sense to mean only those deposits that it can be demonstrated were laid down directly by the action of grounded ice and not subsequently reworked (lodgement

till). Texturally similar, poorly sorted diamicts also occur in volcaniclastic settings (Eyles and Eyles 2010, p. 78).

The bulk of Quaternary deposits are continental, and tills (in the restricted sense of a primary glacial deposit) are common. However, it is now becoming clear that in the ancient record subaqueous resedimented glacigenic deposits are much more important, including ice-rafted material and the deposits of sediment-gravity flows. The preservation potential of subaqueous sediments deposited along a continental margin (Fig. 4.13) or in a basin center is much higher than that of subaerial sediments formed on continental interiors, and in subaqueous environments tills are rare. Glacigenic deposits form in almost every clastic environment, and it could be argued that there is no need for a separate class of clastic sedimentary facies labeled "glacigenic". However, there are several unique features about the type of sediment supply, transport processes, and facies associations of glacigenic deposits that justify treating them as a separate category.

Several comprehensive syntheses of glacial sedimentary processes and facies are now available, including those by Boulton and Deynoux (1981), Andrews and Matsch (1983; includes an extensive annotated bibliography), Eyles (1984), Eyles and Miall (1984), Eyles et al. (1985), Ashley et al. (1985), Edwards (1986), Brodzikowski and van Loon (1987), Eyles and Eyles (1992, 2010), and Miller (1996). Criteria for the recognition of tills were discussed by Hambrey and Harland (1981a) and Anderson (1983); a comprehensive classification of glacial lithofacies was offered by Eyles et al. (1983); and Earth's glacial record was considered in detail by Frakes (1979), Hambrey and Harland (1981b), Anderson (1983), Chumakov (1985), Frakes et al. (1992), and Eyles (1993, 2008). Interpretation of glacigenic deposits has been facilitated by several major studies of modern subaqueous glacial environments, including the Antarctic shelf and deep sea (Anderson et al. 1983; Wright et al. 1983), the Gulf of Alaska (Molnia 1983a), and the fiords of southern Alaska (Molnia 1983a; Powell 1981, 1983, 1984). Detailed facies studies of ancient glacigenic sequences are now too numerous to list. Several are contained in the invaluable research syntheses edited by Molnia (1983b) and Deynoux (1985).

In studying ancient glacial deposits, it must be recognized that while many facies and other criteria may indicate a glacial origin for the detritus, the material may have been redeposited several to many kilometers from the ice front in a **periglacial** setting and still preserve the evidence of its glacial origins. Criteria that indicate glacial origins include particle shape and grain-surface textures, including the presence of striae on clasts. True tills (in the restricted sense of the word) may be distinguished from other diamicts, including ice-rafted deposits and sediment gravity flows, by such sedimentologic and stratigraphic criteria as internal grain-size sorting (e.g., grading) and sedimentary structures, clast and matrix fabrics, bed thickness, geometry and lateral

**Fig. 4.13** The main environments, processes and deposits of a typical glaciated continental margin. Representative vertical profiles are shown *below*. Eyles and Eyles (2010, Fig. 3, p. 75) estimate that as much as 90 % of the glacigenic deposits that accumulated during the Pleistocene were deposited in environments such as these, and that glaciomarine deposits dominate the record of older glaciations because the terrestrial record is easily eroded

extent, and facies associations. Magnetic characteristics, including the anisotropy of magnetic susceptibility (AMS) and the orientation of natural remanent magnetism (NRM), have become useful tools for the investigation of matrix fabrics, because the results can be related to final depositional processes. For example, lodgement tills may have distorted NRM distributions as a result of post-depositional subglacial shear. Rain-out deposits, which may appear similar in outcrops, have tightly clustered NRM orientations similar to those of other fine-grained, non-glacial, subaqueous deposits (Eyles et al. 1987).

One of the best lines of evidence for glacial environments is the dropstone or lonestone facies—dispersed clasts several millimeters to several meters in diameter, interbedded with sediments of much smaller grain size, commonly mudstones (Fig. 4.11). This very distinctive facies can only be formed

in abundance where there are numerous drifting icebergs transporting glacial debris out to sea until dropped by gradual melting. This facies may be formed several thousands of kilometers from the ice front. For example, it is forming at the present day off Newfoundland, from the melting of icebergs that originate at the terminus of valley glaciers in northwest Greenland. Seasonal land-fast ice cannot produce this facies in the same volume. The only significant nonglacial origin for such a facies is the transport of debris by floating plants, but the volume is likely to be insignificant compared with that forming in polar waters close to floating ice margins.

Striated pavements are a spectacular demonstration of grounded ice, and many examples have been documented in the ancient record; but, as Anderson (1983) warned, they cannot be taken as indicating continental glacial conditions.

The Antarctic ice cap is in places grounded at depths of up to 2 km below sea level, and large icebergs are known to form deeply gouged grooves at depths of hundreds of meters on, for example, the continental shelf off eastern Canada, and the floor of Beaufort Sea.

Other traditional glacial indicators must also be used with caution. Rhythmically-laminated fine-grained muds and silts formed in glacial settings are routinely described as varves, and interpreted as products of seasonal freeze-thaw cycles. However, similar deposits may form as a result of low-viscosity density interflows and underflows and from slump-generated surge currents, and the number of rhythms does not necessarily bear any relationship to the number of years during which the deposit accumulated. Ashley et al. (1985) described the variations between these different types of rhythm in some detail, with many excellent photographs. Mackiewicz et al. (1984) coined the term **cyclopel** for varve-like laminated interflow and underflow deposits.

Sand-wedge structures may indicate subaerial permafrost conditions, but Eyles and Clark (1986) demonstrated that superficially similar structures are produced by soft-sediment gravity loading in subaqueous settings. The correct interpretation may be a critical factor in discriminating between terrestrial and subaqueous environments for the beds in which the structures occur, as they demonstrated in the case of a glacigenic unit of Upper Proterozoic age in northwest Scotland.

Despite their low preservation potential, continental glacial deposits have been recorded from the ancient glacial record. Probably the most well preserved are those of Upper Ordovician age in North Africa, which formed around a short-lived continental ice cap in the center of a craton and subsequently remained undisturbed to the present day. Terrestrial tills are well preserved, as are many glacial landforms typical of Quaternary glacial landscapes, such as ice-push structures, kettles, sand-wedge casts, polygonal ground, drumlins, pingos, and eskers (Beuf et al. 1971). Marwick and Rowley (1998) reviewed the evidence for glaciation between the Triassic and the Pleistocene, citing the criteria discussed in this section, and concluded that the evidence is very limited, and some indicators, such as scattered dropstones, can be interpreted as detritus carried in the roots of floating trees.

Some ancient glacigenic deposits are interbedded with carbonate sediments, and for many years this has raised interpretive problems because of the assumption that carbonates indicate warm waters and low latitudes. However, it can be demonstrated that some interbedded carbonates are detrital in origin and formed as, for example, continental-slope sediment-gravity-flow deposits (e.g., Eyles et al. 1985). Anderson (1983) pointed out that the critical factor in carbonate sedimentation is not water temperature but an absence of detritus. Biogenic carbonate production takes place, albeit slowly, in waters of extremely low

temperature. For example, the Antarctic continent is ringed with carbonate oozes and shell hash formed in waters of 0–8 °C.

A variety of glacial environments has now been identified and described. Eyles (1984) and Ashley et al. (1985) synthesized our knowledge of terrestrial glacial landforms and associated terrestrial depositional settings (valley glaciers, outwash plains, glacial lakes). Fiord sedimentation has been described by Powell (1981, 1983, 1984) based on his work in Alaska. Shelf sedimentation has been described by Molnia (1983a) and Anderson et al. (1983). Three main ice-margin models have evolved: that where the ice is grounded on land and the detritus is carried into the sea by outwash rivers, that where the ice is grounded on the shelf below sea level, and that where the ice margin floats on tidewater. Distal glacial environments are those characterized mainly by ice-rafted sediments. Glacial marine sedimentation is controlled by a variety of factors, including the rate of berg calving, glacier advance and retreat behavior, the position of debris on the glacier, the position of meltwater streams, and oceanographic factors, such as the control of bottom currents on berg drift tracks and sediment plumes, bottom currents, tidal currents, and vertical water mixing. Sedimentation rates in front of ice margins can be astonishingly high. Short-term rates up to 9 m/a have been measured in bays near the termini of glaciers in southern Alaska, while shelf sedimentation rates in the Gulf of Alaska typically range from 2 to 10 mm/a, with an average of 4.5 mm/a (Molnia 1983a).

A current focus of sedimentological research is the attempt to distinguish glacial thermal regimes using evidence from the depositional record. In subarctic regions, such as southern Alaska, basal glacier ice slides over a substrate lubricated by meltwater, and abundant meltwater streams move in tunnels through the ice. In frigid polar environments, such as that along most of the Antarctic ice margin, the glacial ice is frozen to the substrate and moves by internal shear and flow. The delivery of detritus to the ice margin is several orders of magnitude slower, and basin sedimentation rates are considerably less (Eyles et al. 1985, Table 1). Most ice margins are probably characterized by both types of thermal regime over time, and sedimentologists have not yet succeeded in developing criteria (other than sedimentation rates) to distinguish the resulting depositional products.

## 4.2.5 Coastal Wave- and Tide-Dominated Environments

Coastlines, where the land meets the sea, may be either erosional or depositional environments. Erosional coastlines are rarely preserved, although in some cases the architecture of an unconformity may indicate the presence of buried cliffs (e.g., see Fig. 1.3. Just such a feature is visible between units

**Fig. 4.14** beach-barrier-lagoon depositional environments. **a** Hypothetical coastline, showing the interaction of waves and tides; **b, c** aerial views of South Carolina coastline; **c** Galveston Island and Matagordia Bay, Texas; **e** the beach at South Padre Island, Texas

J3.2 and K1.1). Unconformity surfaces are sometimes covered by a deposit of large boulders emplaced by waves at the commencement of a new depositional cycle. Coastlines that are tectonically stable or undergoing subsidence are places of net sediment accumulation.

Recent textbooks treat clastic shorelines in different ways. All types of shoreline environment are included in a single chapter on this topic in the third edition of the classic "Reading" textbook (Reading and Collinson 1996), whereas in the later editions of the "Facies models" book the subject matter is subdivided into several chapters. In Walker and James (1992) there are three chapters: "Deltas", "Transgressive barrier island and estuarine systems" and "Tidal depositional systems." In James and Dalrymple (2010), shoreline and shallow-marine systems are treated in two chapters, "Transgressive wave-dominated coasts" and "Wave- and storm-dominated shoreline and shallow-marine systems." This leads to a certain amount of overlap in the case of estuarine environments. An excellent compendium of recent research on tidal sedimentation was compiled by Smith et al. (1991).

This is a highly dynamic environment, where sediment is moved by waves and tides, and episodically affected by storms (Figs. 4.14, 4.15). Rivers may deposit large volumes of sediment at the coast, in the form of deltas, which may or may not be modified by marine processes. However, rising sea-levels may cause a river valley to be drowned, forming an estuary. Deltas and estuaries are discussed separately in the next section. Here we describe coastlines characterized by beaches, and beach-barrier-lagoon systems (Fig. 4.14).

The subdivision of coasts into domains on the basis of tide range (**microtidal, mesotidal,** and **macrotidal**; Hayes 1976) has been a very useful concept, but Hayes (1979) and Davis and Hayes (1984) showed that it is the relative effectiveness of waves and tides, not their absolute energy levels (wave height, tide range), that is important Thus, it is possible for a microtidal coast to be tide dominated if wave energy is low enough. Boothroyd et al. (1985) showed how this idea could be used to develop a modified classification of various coastal regions around North America and elsewhere.

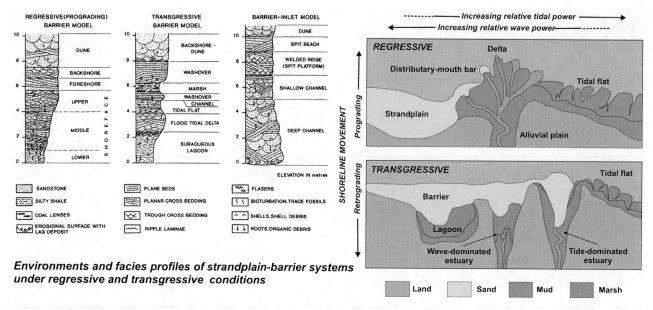

**Environments and facies profiles of strandplain-barrier systems under regressive and transgressive conditions**

**Fig. 4.15** Environments and facies profiles of strandplain-barried systems under regressive and transgressive conditions. Vertical profiles at left from Reinson (1992); Diagrams at *right* adapted from Reading (1996)

Wave- and tide-influenced environments may be recognized in small outcrops or drill core by the distinctive styles of hydrodynamic sedimentary structures (Figs. 3.26, 3.27, 3.28, 3.29, 3.30 and 3.31). Wave-formed structures are distinguished by the evidence of oscillating currents (Fig. 3.31); tidal environments by reversing bedforms, and the presence of sand-mud couplets (Fig. 3.27).

Typical beach profiles are shown in Fig. 4.15. The profile for a **regressing coastline** is that of a coarsening-upward succession, reflecting the increase in wave energy from lower shoreface to foreshore. The same energy trend is shown in a beach profile in Fig. 3.30. The vertical profile in Fig. 4.15 (see also Fig. 3.58d) was first developed from studies on the Galveston Island barrier, and was one of the first vertical-profile models to be established, beginning in the late 1950s (Bernard et al. 1959; see Sect. 1.2.6). Regression by the addition of sediment to the front of a beach under conditions of stable sea level is called **normal regression**, to distinguish this from the regression caused by falling sea level, which has been termed **forced regression** (Plint 1988; see Sect. 5.5.1).

Much of the recent work in the area of sequence stratigraphy has focused on shoreline deposits because of their obvious sensitivity to sea-level change, and this has had a new dimension to our knowledge of their long-term response to environmental change. For example, a long-standing controversy regarding the origins of barrier systems and their preservation was largely resolved by Swift at al. (1985) and Niederoda et al. (1985), in a pair of detailed papers dealing with U.S. Atlantic coast barriers, that demonstrated rather conclusively the ability of barriers to migrate landward during transgression, feeding on their own

sand to maintain volume. Sequence-stratigraphic reconstructions clearly confirm this process.

There are differences between the geography and sedimentology of coastlines undergoing normal regression versus those undergoing transgression (Fig. 4.15). Transgression across a gently sloping shelf may cause the main wave energy to be dissipated offshore, with the beach deposits isolated from the actual shoreline by a lagoon. Barrier-lagoon systems are mainly the product of transgression, such as the extensive system of such landforms along the southeast Atlantic and Gulf coasts of the United States (Fig. 4.14). Tidal rise and fall causes marine waters to enter and leave the lagoon, and this is accomplished by the maintenance of **tidal inlets** through the barrier system (Fig. 4.14a, d). These are not necessarily associated with river mouths, but may be initiated by storm erosion, which is discussed in greater length below. Tidal inlets may accumulate channel-fill sediment if the inlet is abandoned. Such a deposit is similar to a fluvial channel fill, but may be distinguished by the presence of marine fossils, and crossbedding evidence of reversing currents.

Current patterns on wave- and tide-dominated coastlines may be very complex. The pattern of prevailing winds may be responsible for establishing a dominant angle of attack by waves (Fig. 4.14a). Waves are refracted by friction on the sea bottom as they approach the shore, but still may tend to strike the coast obliquely, driving sediment obliquely up the foreshore. The backwash rolls the grains down the beach slope, and then they are driven obliquely up again by the next wave. In this way, sediment is gradually caused to move along the shore, a process called **longshore drift**. Sediment is driven into the mouths of tidal inlets, which may cause the tidal inflow and outflow to impinge erosively on the opposite bank. In this way, the position of the inlet may gradually shift laterally along the barrier. Sand will also be driven as a growing **spit** into open areas of the sea. The "growth" lines of the spit are commonly visible from the air (e.g., Fig. 4.14b).

Sediment is flushed in and out of the lagoon by the rise and fall of the tides. Where the flow expands into the lagoon (during the flood) or out into the sea (during the ebb) much of this sediment is deposited as **tidal deltas**. An ebb-tidal delta is shown in the diagram of Fig. 4.14a and a South Carolina example in Fig. 4.14d. Current patterns over these deltas are complex, as temporary distributary channels migrate and switch position with the changing weather and seasons, and bedforms migrate and are eroded. Parts of these deltas are commonly dominated by either the ebb or the flood phase of the tides, because of accidents of local topography, the result being that crossbedded tidal delta deposits may be surprisingly unimodal, as in the example shown in Fig. 3.27 (right-hand example).

**Lagoons** are sites of fine-grained sedimentation. They are typically characterized by broad tidal flats, colonized by various salt-resistant grasses around the margins. The waters may be brackish or, in hot climates, hypersaline. Tidal flood and ebb currents are focused along a network of tidal creeks, that function like reversing fluvial channels, complete with meanders and point bars.

Plint (2010) has drawn attention to the importance of mud deposition on continental shelves. Many modern shelves are covered with coarse, relict (**palimpsest**) sediment left by post-glacial sea-level rise, but stratigraphic studies, particularly Plint's own work on the Cretaceous stratigraphy of the Western Interior Basin, has demonstrated that significant thicknesses of mud may be trapped on continental shelves. The long-prevailing idea that mud deposition is characteristic of low-energy, pelagic environments, has been demonstrated to be an incomplete interpretation of fine-grained facies. While some mud is deposited in such environments, for example, in lake basins and the deep oceans, mud particles commonly aggregate in saline (marine) water and are redistributed by coastal and geostrophic currents on continental shelves and at their margins, where they may form gently-dipping clinoform deposits. For example, huge volumes of mud are currently accumulating on the continental shelf adjacent to the mouth of the Amazon River. Application of these new ideas to the interpretation of the preserved record of mud deposits is creating some entirely new approaches to the facies modeling of fine-grained deposits (Plint 2010).

Studies of Cretaceous units in the Western Interior Plains of North America, especially in the Alberta Basin, led to the recognition of a distinctive facies assemblage for which Thomas et al. (1987) coined the term **inclined heterolithic stratification,** and the acronym IHS. This facies consists of

sand-mud couplets developed under tidal influence on the point bars of tidal creeks and tidally-influenced rivers. Depositional dips reflect the formation of this facies by lateral accretion. Particularly spectacular examples of IHS occur in the Athabasca Oil Sands, near Fort McMurray, Alberta, where tidally-influenced point bars up to 25 m thick are exposed (Smith 1987, 1988a, b). The review by Thomas et al. (1987) provided a useful summation of ideas regarding estuarine channel sedimentation, many of which had been known since the Dutch work of the 1950s, but required reanalysis in the light of modern sedimentological concepts.

Many studies of tidal bedforms have now appeared. Allen's (1980) model of tidal sand waves was an invaluable theoretical study (Fig. 3.27), the results of which have been applied to several ancient deposits (e.g., Blakey 1984; Teyssen 1984). Modern estuarine examples of sand waves have been described by Dalrymple (1984). Harris (1988) reviewed the formation of larger estuarine bedforms, including sand ribbons and sand banks (tidal current ridges of other authors), and confirmed earlier ideas regarding the evolution of the latter by tidally-driven racetrack circulation of sand around them. Superb images of tidal sandwaves in a shelf setting were obtained by high-resolution seismic surveying, echo sounding, and side-scan sonar surveying off the coast of Normandy, by Berné et al. (1988, 1991). Studies of this type will provide the basis for the next generation of facies modeling of the ancient record, because of the detailed picture of surface bedform patterns and three-dimensional facies architecture which they provide.

As illustrated in Chap. 5, the transition from lowstand to transgression is accompanied in many cases by backstepping or retrogradation of coastal depositional systems, and the partial stripping away of strandplain and barrier systems by the process of **ravinement** (Nummedal and Swift 1987) with the sediment so released forming new beach systems further landward (Fig. 4.16). Wave scour of the sea floor is at its greatest in the shallow offshore, over depths of up to perhaps 10–15 m below sea level. At such depths, erosion typically predominates over deposition. Given enough time, and a stable sea level, waves may develop a flat **wave-cut platform** somewhat below the low-tide water level. Under conditions of transgression, the zone of maximum waves scour is gradually translated obliquely upward, as the shoreline is pushed landward. Beach and back-barrier lagoonal deposits are truncated by this process, while shelf sediments begin to cover the surface as water depths increase with the rising sea level. The end result is the development of a major erosion surface that juxtaposes lagoonal and shelf deposits. But it is a time-transgressive surface (as shown by the dashed correlation line in Fig. 4.16). It commonly forms a sequence boundary in the rock record.

Episodic storms can dramatically change the configurations of the beach-barrier systems just described. Hurricanes

**Fig. 4.16** The process of
ravinement. The *dashed red line*
indicates the time correlation
between shelf sediments
deposited on the ravinement
surface and contemporaneous
lagoon deposits that are truncated
by ravinement erosion as sea level
rises (after Nummedal and Swift
1987)

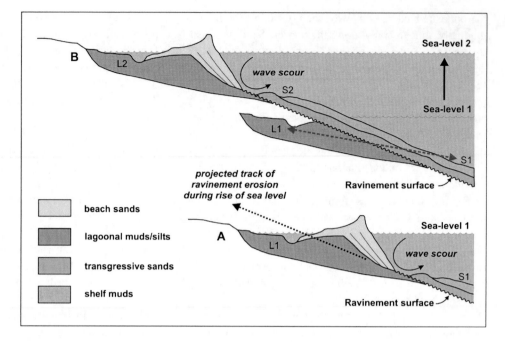

approaching the shore generate wind pressures that can push
the sea onshore, raising local water levels by as much as 5 m.
Coupled with the high waves generated by the winds, such
storm surges can do enormous damage and completely
change the landscape. Figure 4.17 before-and-after map of a
portion of the Galveston Island barrier system, which was
substantially modified by Hurricane Carla in 1964 (Hayes
1967). Moribund channels are opened up to permit the storm
surge to pass over the barrier, and considerable volumes of
sediment are carried landward, forming **washover fans** on
the lagoon side of the barrier. The storm channels may
gradually be filled again during subsequent quiet weather,
and this can lull property developers into a false sense that
these lands are safe to develop. US citizens will be familiar
with these processes as a result of the cathurricanes Katrina
(Coastal Louisiana 2005) and Sandy (New Jersey 2012).

### 4.2.6 Deltas

Deltas typically represent the single largest repository of
sediment in many sedimentary basins. Some deltaic complexes are known to have existed for tens of millions of
years and have accumulated many kilometers of sediment.
Recent work on deltaic sedimentation has been summarized
by Bhattacharya and Walker (1992), Orton and Reading
(1993), Reading and Collinson (1996) and Bhattacharya
(2010). A collection of papers on coarse-grained deltas was
edited by Colella and Prior (1990).

The most well-known delta is that of the Mississippi
(Fig. 4.18). It is one of the world's largest, and has been the
subject of detailed study since the 1930s, partly because of

**Fig. 4.17** The effects of Hurricane Carla on Galveston Island 1964
(Hayes 1967)

its importance to coastal navigation in the Gulf Coast region
of the United States, and partly because it was recognized

**Fig. 4.18** The classic delta—that of the Mississippi river, as seen from space. Note the turbid river water mixing with the clear sea-water of the Gulf

many years ago that it serves as an excellent modern laboratory for the study of geological processes in deltas (Blum and Roberts 2012). However, it represents only one of three broad classes of deltas (Fig. 4.19). The Mississippi drains into a semi-enclosed sea in which tidal range is low, and wave energy is also typically low (except during the rare hurricanes that affect the coastline). Sediment carried into the Gulf by the river has been deposited alongside the distributary channels as long fingers extending out into the sea. In other settings, where marine processes are more important, much deltaic sediment is redistributed by powerful

waves and tides, and the resulting architecture of the delta can be very different (Fisher et al. 1969; Coleman and Wright 1975; Galloway 1975). A widely used classification of deltas based on the relative importance of fluvial input and wave and tide reworking was published by Galloway (1975). The various styles of delta shown in Fig. 4.19 are summarized and classified in Fig. 4.20.

New classifications of deltas were offered by Postma (1990) and Orton and Reading (1993). As noted by Orton and Reading (1993), in addition to recognizing the importance of the three major processes governing sediment distribution on deltas, the grain size of the sediment supply is also a critical factor in determining the configuration of the delta and the architecture of the resulting deposits. McPherson et al. (1987) were amongst the first to attempt to incorporate grain size into delta classification, with the recognition of coarse-grained types termed **fan deltas**, and finer-grained varieties **termed braid-deltas** and **common deltas**. These ideas were developed further and generalized in the classifications by Postma (1990) and Orton and Reading (1993). Postma (1990) focused on the nature of the fluvial feeder system and the water depth in the receiving basin, which reflects the setting of the delta at the margin of a broad continental shelf or on the shelf-margin or slope. This classification departs from Galloways' emphasis on the processes of redistribution of sediment at the delta front by not incorporating data dealing with wave- and tide-generated sediment redistribution. The classification of Orton and Reading (1993) builds on that of Galloway to incorporate data on the grain size of the sediment load. The latter

**Fig. 4.19** Deltas as seen from space

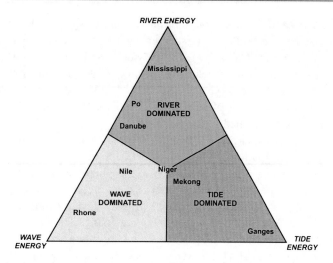

**Fig. 4.20** Processes influencing deltas, and a delta classification (Galloway 1975)

authors also provide a thorough discussion of the importance of other variables such as delta plain slope, and the control of grain size and wave height on the tendency of wave energy to be dissipated or reflected at the shoreline, and the effects of these variables on the resulting sedimentary deposit.

Along coastlines where there are strong prevailing winds, there may be a dominant direction of wave attack, and sediment deposited at the mouths of deltaic distributaries is redistributed along the coast, forming arcuate beach deposits. The delta of the Rhone, in the Mediterranean Sea, is commonly cited as a typical example of a wave-influenced delta. The Nile delta and the Niger delta also show the smoothly curving stretches of sand-beach coastline that characterizes a wave-influenced delta.

Where tidal influence is strong, deltaic distributaries are shaped by the constant rush of the flood and ebb currents. The volume of water rushing into and out of the delta, in such situations, requires a greater than usual number of distributaries. Such situations occur where the tidal range is high (macrotidal regimes), such as at the head of wide, but narrowing, bays in the ocean. The Bay of Bengal, with the delta of the Ganges/Brahmaputra River at its head, is a good example. Most deltas experience a combination of influences, by rivers, waves and tides. A summary of the processes at work on deltas and a classification developed from this summary is shown in Fig. 4.20. As indicated in this

classification, there are rivers such as the Mississippi, Ganges and Rhone, that appear to be dominated by just one of the three main processes, whereas the geography and architecture of others suggests a combination of forces. The Niger, in particular, shows evidence of all three at work. The delta itself has a lobate outline, which is consistent with active seaward progradation by rivers. The delta is fronted by long, narrow beaches, indicating some wave reworking, and the beaches are cut by numerous distributaries that are backed by tidal flats, indicating continual tide-driven flux of water into and out of the delta.

The history of the Mississippi delta complex provides a good illustration of how deltas work. The modern delta shown in Fig. 4.18 is only the most recent of a series of deposits formed along this part of the coastline since the last ice age. Based on numerous drill cores, geologists have mapped seven previous deltas comparable in size to the modern lobe, labeled "6" in Fig. 4.21. It appears that each lobe gradually grows, and extends out into the sea, for several thousand years, until it becomes unstable, at which point the main river switches into a different course. The process is illustrated in Fig. 4.22. Delta switching is a process called **avulsion**. It may be triggered during episodes of unusually large runoff, when the flow overtops its banks and finds a shorter, steeper route to the sea.

After abandonment, the deposits continue to compact slowly and to subside, allowing the sea to re-advance. This is well seen in the case of lobe 4, the St. Bernard lobe, where the lobe has undergone transgression, and much of the sand has been reworked into a series of offshore beach shoals called the Chandaleur Islands. The Atchafalaya lobe is a potential successor to the present, modern lobe. The Atchafalaya River is an old course of the lower Mississippi (located where the word "Atchafalaya" is in Fig. 4.21). When modern river control and navigation works began in the Mississippi River, late in the nineteenth century, the Mississippi was in the process of switching back to this course, which, as can be seen from the map in Fig. 4.21, is a considerably shorter route to the sea than the modern river. Dams and locks were installed to prevent this happening, because of the importance of the flow through New Orleans, a major sea port. The flow that is permitted to divert down the Atchafalaya is nonetheless developing an active delta at its mouth (Fig. 4.19).

**Fig. 4.21**  The seven postglacial lobes of the Mississippi delta (Penland et al. 1988; after Frazier 1967)

**Fig. 4.22**  Oblique view of a delta, demonstrating why delta lobes switch in position. (1) As a delta builds out from the coastline the river channel aggrades above sea level (vertical height beneath point A); (2) the path from point A to point B becomes steadily longer and the slope gentler; (3) the path from point A to point C remains shorter and steeper; (4) If the river finds the route to point C by overbank flooding it will result in switching of the river to the shorter course, leading to abandonment of the delta lobe and initiation of a new lobe starting from point C

The modern delta lobe is an example of what has been called the **birdsfoot** type of delta, narrow, branching fingers of land extending seaward, flanked by bays. The bays themselves gradually fill with sediment deposited in a series of prograding mini-deltas, as shown in Fig. 4.23. Initiation of each of these begins with a **crevasse**, or break in the bank of the main channel. Diversion of sediment and water and filling of the bay is achieved quite rapidly.

Figure 4.24 shows the processes that occur at the mouths of distributaries. Expansion of the flow occurs as it passes

**Fig. 4.23** The development of small deltas infilling the bay areas flanking the main distributaries of the modern Mississippi. The amount of sediment in the water is apparent from its turbid appearance in the space photograph. The date of initiation of each of the deltas is shown in the map at the *left* (diagram from Coleman and Gagliano 1964)

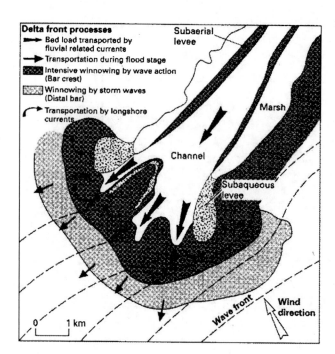

**Fig. 4.24** A mouth bar forming at the end of one of the Mississippi delta distributaries (Coleman and Gagliano 1964)

out through the mouth. Most of the coarser sediment load is deposited here to form a mouth bar deposit. This may be worked and reworked by fluvial and marine processes,

resulting in a well-sorted sand characterized by much internal scouring. Fine sediment stays in suspension a little longer, and is carried further away from the mouth, much of it being deposited eventually to form the distal portion of the mouth bar. However, the water remains turbid for a considerable distance out from the distributary mouths, as can be seen in the two space photographs (Figs. 4.18, 4.23). Sediment deposited at the margins of the channel forms a subaqueous extension of the levee system. The incremental additions to the levee and the mouth bar cause the mouth to prograde. The mouth bar therefore develops by a process of continual offlapping, as the mouth is shifted seaward. The mouth bar develops a long, narrow ("**shoestring**") geometry and has an upward coarsening grain-size trend, reflecting the process whereby coarser sediments, formed on the upper bar, are gradually extended out over the finer sediments that form the lower or more distal bar.

A very useful depositional model for Mississippi-type mouth bars was developed at the Bureau of Economic Geology in Texas, and is shown here as Fig. 4.25. This helps to explain the classic shoestring geometry, a term that has long been used by petroleum geologists in their search for sandstone reservoirs in deltaic systems. Note, also, the various sub-environments illustrated in this diagram, including the interdistributary areas of the delta plain, characterized by fine-grained deposits, possibly peat swamps, that may ultimately become coal deposits.

**Fig. 4.25** The classic shoestring-sand facies model (Frazier 1967)

(a) INITIAL PROGRADATION

NATURAL LEVEE

FRESH-WATER STREAM-MOUTH MARSH

DELTA-FRONT DISTRIBUTARY-MOUTH-BAR SILTY SAND

DELTA-FRONT SILTY SAND AND SILTY CLAY

PRODELTA SILTY CLAY

(b) ENLARGEMENT BY FURTHER PROGRADATION

DELTA-PLAIN PEAT

DELTA-PLAIN INORGANIC SILTY CLAY

DELTA-PLAIN NATURAL-LEVEE CLAYEY SILT AND SILTY CLAY

(c) DISTRIBUTARY ABANDONMENT AND TRANSGRESSION

MORIBUND DISTRIBUTARY

NEW DISTRIBUTARY RESULTING FROM STREAM DIVERSION

TRANSGRESSIVE DELTA-MARGIN-ISLAND SAND

TRANSGRESSIVE BAY DEPOSITS

Up to this point we have been looking primarily at the shape and surface evolution of deltas. Figure 4.26 helps to explain how deltas develop geologically significant thicknesses of sediment. This diagram was drawn following a study of deltaic sedimentation in the Gulf of Mexico, but has a universal applicability. A vertical drill hole placed anywhere along this hypothetical cross-section will penetrate successively more distal and fine-grained deposits the lower it goes. This explains the coarsening-upward nature of most deltaic deposits (Fig. 6.15).

Figure 4.27 illustrates some examples of ancient deltaic deposits.

Coal is commonly associated with deltas, but not all deltaic peat swamps become coal. Much depends on the clastic influx into the swamp. Frequent overbank floods carry silt and clay into the swamp, and the resulting deposit is then a carbonaceous shale rather than coal. Such a deposit is of too poor a quality to be used as coal because of the large amount of non-burnable sediment ("ash") left after combustion. The best coals develop where a peat swamp accumulates plant material at a rate rapid enough that the swamp become slightly raised above the level of the surrounding channel system. This elevated position then serves to keep clastic influx out of the swamp, and the peat can then accumulate as a potentially cleaner form of carbon.

Careful subsurface mapping using well data may help to clarify the style of delta that formed a given deposit. The distribution of channel sands, in particular, reveals key information about the transport and reworking of the coarser portion of the sediment load by fluvial and marine processes (Fig. 4.28). The isopachs shown in this figure are net-sand isopachs, which are constructed by careful mapping and correlation of individual deltas in the subsurface. Clearly, maps of this type require good well control.

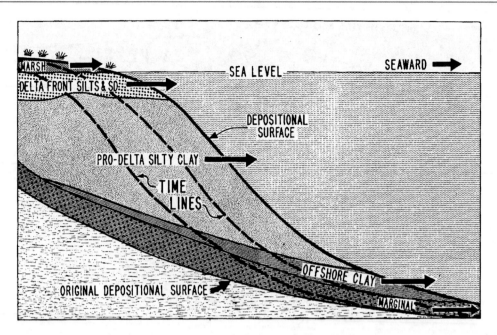

**Fig. 4.26** The prograding front of a delta defines three main subenvironments: (1) the delta front, which is the seaward margin of the delta plain and is characterized by channel, overbank and interdistributary bay environments; (2) the prodelta, which is the environment of the outbuilding distributary. *Sloping lines* are successive depositional surfaces, and are the clinoforms of some reflection-seismic sections. The offshore is the deep-marine environment beyond the influence of the delta (Scruton 1960). AAPG © 1960, reprinted by permission of the AAPG whose permission is required for further use

**Fig. 4.27** Examples of ancient deltaic deposits. **a** Channel sandstone over coal seam. Jurassic, Alberta; **b** channel cut into delta-front sandstone, Paleogene, Arctic Canada. Base of channel indicated by dotted red line; **c–f** carboniferous deposits, Kentucky; **c** overbank and crevasse-splay deposits; **e, f** coals overlain by crevasse-splay deposits

## 4.2.7   Estuaries

Estuaries form at the mouths of rivers undergoing flooding or transgression. They are areas of open water, and modest sediment input. Facies models and/or collections of papers detailing estuarine deposition have been provided by Dalrymple et al. (1994, 2006) and Boyd (2010).

Sedimentary processes may vary considerably in an estuary. Wave processes, including the formation of a barrier beach deposit, are common at the mouth (Fig. 4.29). At the head, fluvial processes may be important. Tidal influence may be significant throughout the length of an estuary, but are most likely to be most important in controlling sedimentary processes near the middle of the estuary, downstream from the main fluvial input and back from the full force of the waves near the mouth. All these various influences may be studies using the indicators of fluvial, wave and tidal processes discussed in Chap. 3. Figure 4.29 provides many of these sedimentological details, and stitches them together into two diagrams, a plan-view and a

**Fig. 4.28** Isopach patterns of typical deltas, reflecting various combinations of sedimentary processes. Adapted from Coleman and Wright (1975)

**River dominated:**
(seaward-extending sand fingers. Local thick patches indicate pauses in distributary progradation)

**Tide-dominated:**
(linear sand bodies and offshore sand shoals indicating strong tidal currents)

**Wave-influenced river-dominated:**
(seaward-extending distributaries, with reworked sands forming beaches parallel to shore)

**Strongly wave modified:**
(Delta front sands reworked into a barrier-beach complex)

**River dominated:**
(Delta-topset sheet sand reflects shifting of braided distributaries)

**Wave and longshore-current influence:**
(stacked beach ridges skewed along-shore from river input point)

*River input point* ↘   **Net sand isopachs, increasing thickness with darker colour**   *Scale: each delta few km to tens of km across*

cross-section of typical estuaries. The panels showing sedimentary structures and the vertical profiles provide the kinds of practical indicators that field geologists need to assign deposits to the appropriate facies model.

Interest in estuarine sedimentation received a tremendous impetus when it was realized that many producing sandstone reservoirs in the Western Interior basin, particularly in the Lower Cretaceous Mannville Group of Alberta, occur in broad linear "channels" and contain evidence of tidal influence during sedimentation. Careful mapping revealed that these channels are deeply incised valleys that form part of regional erosion surfaces, and that they originated as estuaries at the margins of the Western Interior Seaway (e.g., Rosenthal 1988). Incised valleys in the sedimentary record are important indicators of episodes of low relative sea level, and as such may provide important evidence in the mapping of sequences and their component systems tracts (Chap. 5).

### 4.2.8 Continental Shelf Environment

Most shelves are either tide dominated (e.g., North Sea, Georges Bank) or storm dominated (e.g., U. S. Atlantic shelf, Gulf Coast). A few areas of the continental shelf, such as the Agulhas Bank off southeast Africa, are dominated by strong unidirectional oceanic currents that spill up onto the shelf (e.g., Flemming 1980; Martin and Flemming 1986). In tide-dominated shelves, the principal sedimentary processes

are the development of sand waves and sand banks (ridges), whereas in storm-dominated shelves, the main process is geostrophic flow (Fig. 3.33), leading to the development of storm cycles containing HCS (Walker and James 1992; Johnson and Baldwin 1996) and to storm-generated tidal sand ridges. Plint (2010) provided a review of modern work on wave- and storm-dominated shoreline and continental-shelf deposition. Dalrymple (2010c) discussed tidally-dominated shelf and coastal systems. We now realize that significant volumes of mud are transported across and deposited within shelf settings.

Tidal currents may reach speeds of 1 m/s, quite adequate to transport sand and mud (the latter in the form of floccules). Sand waves, which may be as much as 15 m high, and large sand ridges up to 50 m high, occur in areas of particularly vigorous tidal currents, such as the continental shelf around the British Isles (Figs. 4.30, 4.31). In the North Sea, the tides are semi-diurnal, with separate tidal "waves" entering this sea from the English Channel and around northern Scotland directly from the Atlantic Ocean.

Storms may be a particularly important sedimentary process, scouring the sea floor to depths perhaps as much as 100 m below sea level. The main sedimentary effect of storms is to generate hummocky cross-stratification (HCS) by the combined flow process described in Chap. 3 (see Figs. 3.32, 3.33).

Hummocky cross-stratification is the name given to the main, large-scale crossbedded structure formed by storms,

**Fig. 4.29** Sedimentary models for estuaries. *Top* Reinson (1992); *bottom* Dalrymple et al. (1994)

but it is commonly only part of a sequence of deposits formed during storm activity. A block diagram model of a **storm cycle** is provided in Fig. 4.32, variations on the storm theme are shown in Fig. 4.33, and some examples of HCS in drill core from Alberta are illustrated in Fig. 4.34.

The storm cycle typically begins with a scoured base. This reflects the build-up of the storm to maximum strength, during which gutter casts may be eroded and a basal lag of coarse particles transported. The deposition of HCS reflects the beginning of the waning of the storm, with the rest of the cycle recording the completion of the waning phase. In nearshore environments it is common for only the bases of the HCS domes to be preserved (Fig. 4.33, column A). The cycle fines upward through plane-laminated and rippled sand, not unlike a Bouma sequence (compare to Fig. 3.48). In deeper water settings of the mid- to outer shelf, only the finer-grained upper portions of the cycle may be deposited.

Fig. 4.30 Giant sand ridges, up to 50 km long, in the North Sea, an area characterized by strong semi-diurnal currents (Reading 1986)

SHELF TIDAL SAND WAVE
(interpreted from seismic-reflection data)

Fig. 4.31 Tidal sand wave on the continental shelf off Brittany, France (diagram drawn from reflection-seismic record). Adapted from Berné et al. (1988)

### 4.2.9 Continental Slope and Deep Basin Environment

The ultimate fate of clastic debris is to be shed off the continental shelf and down the continental slope to the deep basin floor of the ocean, to depths of up to 5 km (to even greater depths in the trenches above subduction zones). The most important transport paths are submarine canyons, but detritus may also be swept over the continental shelf by

storms and by oceanic currents, forming sediment prisms on the brink of the continental slope that periodically fail by slumping (Fig. 4.35). Such events may or may not evolve into sediment gravity flows, as described in Sect. 3.5.5. The most volumetrically significant deposits in deep oceanic basins are accumulations of sediment-gravity flow deposits. The traditions term for these is **submarine fans**, the term implying deposition on radiating patterns of distributaries. However, given the wide variety of depositional patterns that have been identified on modern continental slopes, many workers now refer to these deposits by the more general term **turbidite systems** (Bouma et al. 1985; Arnott 2010). This is the characteristic mode of sedimentation at the mouths of submarine canyons, which commonly serve as point sources for much of the sediment. Giant submarine fans have developed at the mouths of many of the world's large rivers, as described below. Owing largely to the ability of turbidity currents to travel for thousands of kilometres across the sea floor, these fans include the largest depositional systems on earth, including the Amazon, Indus and Bengal fans (Bouma et al. 1985).

Deposition of large sand sheets also occurs, especially where the deep basin is fed by multiple sediment sources. An example of this would be the floor of the Labrador Sea, between Labrador and Greenland. The continental shelf on both sides of this sea is covered with a thick blanket of glacial debris, which underwent repeated slope failure during and following multiple phases of glaciation and glacioeustatic sea-level change. This resulted in gully erosion and numerous individual sources of sediment, feeding a network of tributaries to a few major channels flowing towards the centre and along the axis of the sea, and ultimately out into the Atlantic Ocean (Chough and Hesse 1976).

Early ideas about the architecture of submarine fans were based almost entirely on ancient fan deposits in California and Italy (Mutti and Ricchi Lucchi 1972; Walker 1978). The Navy fan, off California, was one of the first modern fan to be studied in detail, back in the 1970s, but at that time navigational and remote sensing instrumentation was much more primitive than at present, and sedimentary details were based on very limited piston coring and grab sampling. The scale problem has bedeviled the study of submarine fans, because typical outcrops and the range of sizes of features resolvable by marine photography is much too small to encompass the main architectural features of most fan systems (Fig. 4.36). Modern fans range through three orders of magnitude in scale, from a few tens of kilometers across, to a few giants several thousands of kilometres across. The scale problem has now been considerably ameliorated by the use of side-scan sonar methods for mapping modern fans, and 3-D seismic for the analysis of ancient fan systems (Sect. 6.3). A modern review of sedimentary processes has been

| Structure | Orientation Relative to Shoreline |
|---|---|
| 1  Gutter Cast | ⊥ |
| 2  Sole Marks | ⊥  (Offshore) |
| 3  Parting Lineation | ⊥ |
| 4  Wave-Ripple Crests | // |
| 5  Combined-Flow Ripple Cross Lamination | ⊥  (Offshore) |
| 6  Current-Ripple Cross Lamination | ⊥  (Offshore) |

**Fig. 4.32** Block diagram model of a storm cycle. Leckie and Krystinick (1989, Fig. 4, p. 868)

provided by Arnott (2010). Stow and Mayall (2000) presented an overview of fan architecture and models. Nilsen et al. (2007) assembled an atlas of outcrops of deep-marine deposits.

Sedimentary lobes on the Navy fan are small, and the fan is confined by the topography created by extensional and strike-slip faulting on the continental borderland of California (Fig. 4.37). The Monterey fan is more typical of submarine fan physiography, in its radiating wedge of sediment, and the presence of one active distributary channel, with remnants of earlier, inactive channels. The main channel is up to several hundred metres deep, where it is incised across the upper fan, and is up to nearly 4 km wide (Fig. 4.37). This is much larger than most outcrops of ancient fan deposits.

Figure 4.38 illustrates two giant fans. The Mississippi fan occupies most of the eastern Gulf of Mexico, its eastern margin coinciding with the foot of the continental slope off

western Florida, and its southwest flank lapping up onto the base of slope off the Yucatan Peninsula of Mexico. This fan has one main channel. Shallow-seismic exploration has identified numerous ancient, buried channels. Currently, the fan is not active because its main feeder channel, the canyon that cuts across the continental shelf off Louisiana, does not line up with the distributaries of the Mississippi delta. As we showed earlier in this chapter, the Mississippi delta lobes have switched in position several times in the last few thousand years. A few hundred years from now the prograding delta may reach the edge of the continental shelf, and then sediment will once again be shed directly down the continental slope, and a new fan channel may develop dispersing sediment across the Mississippi fan. Until then the fan is simply, slowly, accumulating a thin blanket of mud.

The surface of the Amazon fan is crossed by numerous old channels. These have a surprisingly "fluvial" appearance to them. Figure 4.39 shows a side-scan sonar image and a

**Fig. 4.33** Variations in the composition of storm cycles with depth, across the continental shelf from shoreface to storm wave base. Abbreviatons: *WS* wave scour, *CS* current scour, *B* basal lag, *G* grading, *P* planar lamination, *H* hummocky cross-stratification, *WX* wave cross-lamination, *CX* current cross-lamination, *M* bioturbation, *A* amalgamation Brenchley (1985)

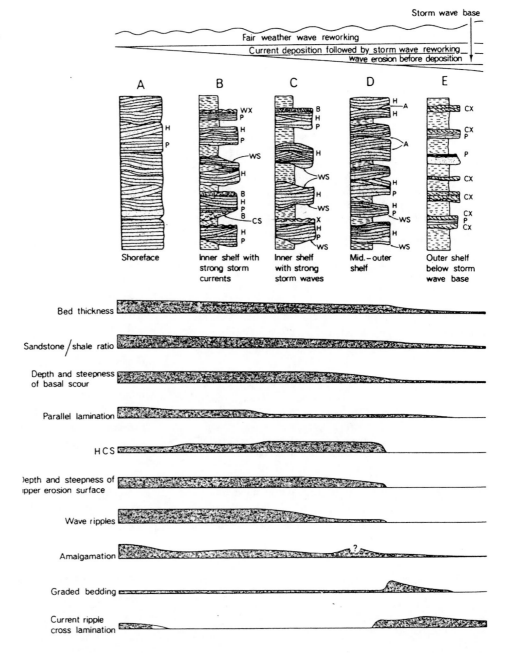

seismic cross-section through one of the channel-levee complexes of the Amazon. It is now known that in some environments, the flow of a turbidity current may last a long time because of an almost continuous supply of sediment at the upstream source. The hydraulic behaviour of the flows, including internal turbulence and the tendency to meander, is therefore similar to that of the traction currents that characterize fluvial flow. In the case of the Amazon, the fan is fed by a very deep canyon, linked directly to the mouth of the Amazon River. The sediment supply was, therefore, not affected by the late Cenozoic fluvio-glacial sea-level changes, and has been essentially continuous for millions of years.

The Bengal fan is the world's largest depositional system (Fig. 4.40). This giant depository consists of sediment eroded from the rising Himalayan mountains, and carried to the sea by the Ganges and Brahmaputra/Jamuna rivers. It is nearly 3000 km long, and more than 10 km thick. Deposition commenced in the Eocene following the initial collision of India with Asia (Curray 2014).

Whereas the first facies models for submarine fans were based on largely hypothetical vertical profiles (e.g., Walker 1978), modern marine surveying and 3-D seismic data have indicated that fans have a highly variable composition and architecture. Stow and Mayall (2000) subdivided the larger

**Fig. 4.34** Examples of hummocky cross-stratification in Cretaceous oil sands, Alberta

components of deep-marine systems into a series of architectural elements, including fan elements, contourite mounds and bounding surfaces (Fig. 4.41). They noted (p. 130).

> These elements are commonly a few hundred metres to several kilometres in width, a few metres to a few hundred metres in relief or thickness and may be irregular in outline, approximately equidimensional or markedly elongate. Each element may occur at a range of scales and within a hierarchy of similar features The sedimentary composition, including facies associations and vertical sequences, that make up any one element, can vary, typically within a limited range.

Stow and Mayall (2000, p. 132) commented that "The natural variability of channels (sensu lato) is enormous: they include steep-sided v-shaped canyons, deeply incised -

shaped delta-front troughs, simple straight to slightly sinuous slope gullies, braided to meandering channel complexes, shallow ephemeral high-gradient fan-delta channels, extremely low-gradient ocean basin channels, and so on."

The large-scale architecture of deep-marine systems is highly variable. As Arnott (2010, p. 302) has argued, this variability in shape and composition renders the terms "submarine fan" misleading. As noted above, his preference is for the term **turbidite system**, following Bouma et al. (1985). A primary control on large-scale architecture is grain size, and the nature of the sediment feeder system is also critical. A classification of deep-marine systems, based on these two main variables, was developed by Reading and Richards (1994) and was also used by Stow and Mayall (2000) and Arnott (2010) (Fig. 4.42). A subdivision of submarine fans into upper, middle and lower fans was suggested by Walker (1978), and is still in use. The upper fan is where the main feeder channel is flanked by levees and is an area of sediment bypass. In some cases the main channel may be incised into earlier fan deposits within this area. The middle fan is an area of flow expansion and net deposition, typically from a set of radiating distributary channels. Beyond the mouths of these channels is a nearly flat area, where flow is unconfined and deposition is predominantly sheet-like in geometry; this is defined as the lower fan.

Turbidite systems are an important component of many continental-margin stratigraphic sequences. Controversy has surrounded their appropriate place in the development of systems tracts (Sect. 5.3.1), but it is now recognized that their development may occur at any stage in the base-level cycle, depending on the local physiography and the control of sediment supply (e.g., see Fig. 5.13).

**Fig. 4.35** The range of processes that influence the sorting, transport and deposition of clastic detritus at a continental margin (Stow and Mayall 2000, Fig. 4, p. 128)

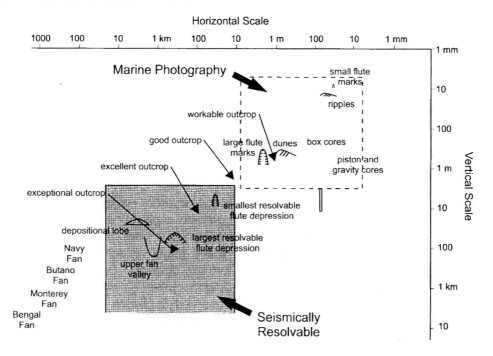

**Fig. 4.36** Vertical and horizontal scales of deep-marine systems, modern and ancient, illustrating the scale problem that arises when using available observational methods to reconstruct submarine-fan architecture (Barnes and Normark 1985)

**Fig. 4.37** The Navy (*left*) and Monterey (*right*) fans, on the California borderland (Bouma et al. 1985)

**Fig. 4.38** The Amazon (*Below side-scan sonar image of several of the old*) and Mississippi (*above*) submarine fans (Bouma et al. 1985)

## 4.3    Carbonate Environments

### 4.3.1    Conditions of Carbonate Sedimentation

The facies analysis of carbonate sediments must, of necessity, take into account some differences in the ways siliciclastic and carbonate sediments accumulate (Fig. 3.12). Carbonate sediments are primarily biochemical in origin. Light intensity, temperature, salinity and nutrient supply are critical controls, and limit the zone of rapid carbonate sedimentation to depths of less than about 20 m (Fig. 4.43a). Warm-water carbonates are generated within the latitudinal range 30°N to 30°S (Jones 2010). Sedimentation follows what Schlager (1992) called the **law of sigmoidal growth** (Fig. 4.43b):

> Growth is slow at first, then accelerates, often exceeding the rate of change in the forcing function; finally, growth decreases as the system reaches the limits of the newly formed niche. In carbonate sedimentology, this pattern is known as the start-up, catch-up, keep-up stages of growth … (Schlager 1992)

The interaction of the controls illustrated in Fig. 4.43 is the reason why most carbonate sedimentation takes place in

the so-called **carbonate factory** of the shallow-marine continental shelf. Sedimentation is most rapid in the **photic zone,** which extends to depths of between 80 and 100 m, with maximum productivity at depths of less than 20 m (Fig. 4.44). Most carbonate grains are originally biogenic, but may be modified by contemporary marine processes (e.g., ooliths, intraclasts). Under conditions of warm, clear water, organic productivity is very rapid (Schlager 2005; Jones 2010). In fact, carbonate generation and accumulation under ideal conditions can keep up with almost any geologically-imposed rate of subsidence or sea-level change. Near the coast, carbonate sedimentation is inhibited by clastic influxes, and on the continental margin, slower sedimentation under conditions of deeper water gradually exaggerates the differences between shelf and slope as subsidence takes place. This gradually results in the development of a **rimmed platform**, as shown in Fig. 4.44.

Changes in environmental conditions (temperature, nutrient supply, salinity, etc.) may bring carbonate sedimentation to a halt, with the subsequent development of a **hardground**. These are areas of lithified substrate,

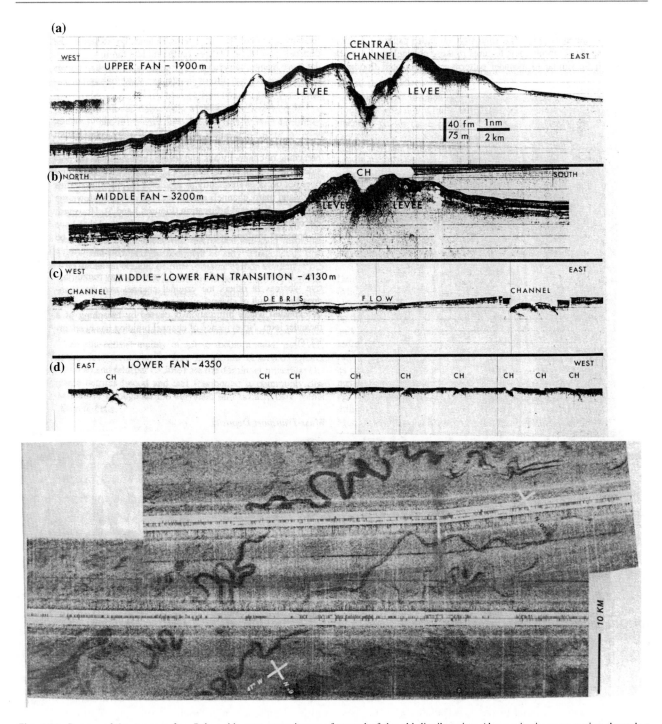

**Fig. 4.39** Images of the Amazon fan. *Below* side-scan sonar image of several of the old distributaries; *Above* seismic cross-section through a channel-levee complex (Bouma et al. 1985)

characterized by submarine cementation, typically with distinctive paleoecological communities, ichnofacies and mineralized crusts (Fig. 2.14). They may extend over substantial distances, and can provide important correlatable surfaces in the rock record (Jones 2010).

The study of carbonate facies cannot be divorced from considerations of diagenesis. In many environments, carbonate sediments start to undergo cementation and alteration as soon as they are deposited. The ubiquity of organic processes in carbonate seas introduces a significant complication, and many carbonate deposits are modified penecontemporaneously by submarine cementation and other types of early diagenesis. For example, reef and carbonate slope build-ups commonly have steeply dipping

**Fig. 4.40** The Bengal fan, between India and Myanmar. *Left* bathymetry, *right* physiography (Bouma et al. 1985)

sediment-water interfaces, much steeper than the angle of repose of loose talus, sometimes even vertical or over-hanging. This is due to organic binding and early cementation. Most research now makes extensive use of special microscopic techniques, such as cathodoluminescence and scanning-electron microscopy, plus a range of geochemical techniques, including the analysis of stable isotopes and trace elements, in order to track these diagenetic changes (Dickson 1985). Cathodoluminescence is widely used to see through diagenetic modifications to primary depositional fabrics (Machel 1985). Analyses of oxygen and carbon isotopes are used to determine depositional temperature and to distinguish early from late diagenetic cementation (Arthur et al. 1983). In many other respects, carbonate facies studies can be pursued in a similar manner to siliciclastic studies, in that many carbonate deposits are accumulated by clastic processes controlled by waves, tides, and gravity.

One of the most important contributions to the study of carbonate facies models was the book by Wilson (1975) on Phanerozoic carbonates. Wilson demonstrated that carbonate sediments could be described using a standard set of 24 microfacies types (Fig. 3.4). He showed that lithofacies, structures, and biota defined nine standard facies assemblages, which could be used to describe practically all

carbonate rocks in the ancient record, although not all nine assemblages are present in every rock unit. Wilson's models are widely used in interpreting the ancient record. More recent syntheses and reviews of carbonate sedimentation have been provided by Scholle et al. (1983), Tucker (1985), Walker and James (1992), Wright and Burchette (1996) and James and Dalrymple (2010). Pomar and Hallock (2008) documented the changes in the "carbonate factory" with time, as biogenic carbonate-producing environments changed in response to plate-tectonic movements (the opening and closing of seaways, changes in oceanic currents), climate change and extinctions events (which at critical times, such as at the Permian-Triassic extinction, led to wholesale changes in the assemblage of carbonate-producing organisms).

Sellwood (1986) suggested that five major environmental zones could be recognized in shallow-marine settings: supratidal flats, intertidal flats, the marine shelf, the carbonate sand belt, and the reef belt. To these should be added the continental slope and pelagic environments, most of which are deep marine (Jenkyns 1986). Schlager (2005), following Wilson (1975), broadened the subdivision into nine subenvironments (Fig. 3.13). In many respects, shelf and slope carbonate environments are similar to their siliciclastic counterparts. Thus, the sedimentology of carbonate

**Fig. 4.41** The principal architectural elements in deep-marine sedimentary systems (Stow and Mayall 2000, Fig. 7, p. 131)

**Fig. 4.42** A classification of deep-marine sedimentary systems, based on sediment grain size and the nature of the sediment sources (Stow and Mayall 2000, Fig. 9, p. 133; based on Reading and Richards 1994)

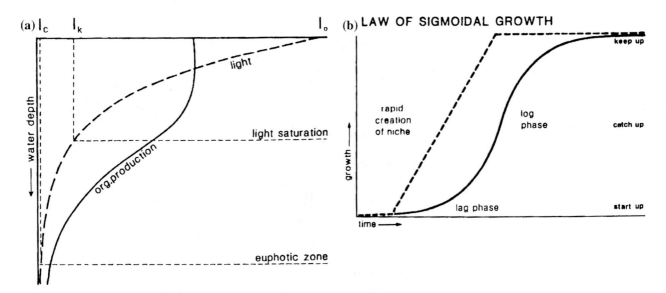

**Fig. 4.43 a** Growth rates for carbonate sediments, as dependent on light intensity. **b** The law of sigmoidal growth in carbonate sediments. Populations of organisms respond to the opening up of new living space slowly at first. In most situations sediment production can exceed the rate of increase of new space but is limited by the rate of growth of space (Schlager 1992). AAPG © 1992, reprinted by permission of the AAPG whose permission is required for further use

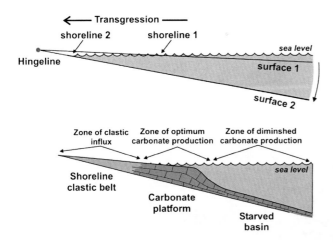

**Fig. 4.44** The zone of optimal carbonate production is located between the clastic-contaminated shore zone and the rapidly subsiding continental margin. Adapted from Wilson (1975)

shelves involves the study of storm deposits, some containing HCS (Aigner 1985; Handford 1986) and carbonate

sand shoals similar to siliciclastic sand waves and banks (Hine 1977; Cook et al. 1983). Carbonates also occur as eolianites, most of which originated as coastal dune deposits (McKee and Ward 1983).

By far the most important environments for carbonate sedimentation are warm, clear, shelf (platform) seas. Carbonate generation is most rapid in tropical settings, such as on the Bahamas Banks or Platform (Fig. 4.45), around South Pacific atolls, and along the Great Barrier Reef of Australia (Fig. 4.46). The Great Barrier Reef, at 344,000 km$^2$, is the largest area of the earth's surface currently undergoing predominantly carbonate sedimentation. However, recent studies have shown that some cooler-water settings, such as the southern continental margin of Australia and the west coast of Ireland, are also sites of carbonate sediment accumulation, although here the processes of sediment generation and accumulation are proceeding somewhat more slowly (James and Lukasik 2010). There are no modern analogues for the vast epeiric seas that covered the North American cratonic interior for much of the early Paleozoic.

**Fig. 4.45** The Bahamas Banks. Areas of shallow water are distinguished by paler colours. The seismic section is Fig. 4.49. From Google Earth

**Fig. 4.46**  The Great Barrier reef, Queensland, Australia. Diagram at *right* from Davies et al. (1989)

**Fig. 4.47**  The continental shelf
of Nicaragua, showing the
composition of surface sediments.
Note the belt of clastic-dominated
sediment along the coast, and the
fringing reef at the edge of the
shelf (Roberts 1987)

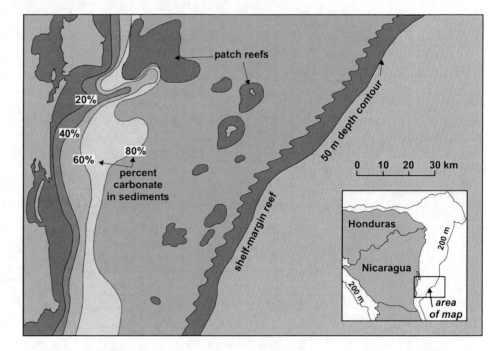

### 4.3.2  Platforms and Reefs

An excellent example of the range of environments shown in
Fig. 4.44 is illustrated in Fig. 4.47. The Nicaragua continental shelf has a coastal, mud-dominated zone and a marginal barrier reef system. The wide carbonate platform is
dotted with small reefs or bioherms (small organic buildups).

Many other examples, including a suite of facies models, are
provided by Jones (2010).

Satellite images showing two other examples of carbonate environments are shown in Figs. 4.46 and 4.48. The
Florida platform has a fringing reef (the Keys) and a
back-barrier area that merges gradually into the mangrove
swamp of the Everglades. Carbonate production takes place

**Fig. 4.48** The Florida keys, showing the shallow waters behind the reef, merging with the mangrove swamps of the everglades. The bar next to the compass diagram is 20 km long

today on the seaward side of the barrier. The Great Barrier Reef is a belt of patch reefs and a fringing reef that extends for more than 1600 km along the Queensland coast of Australia (Fig. 4.46).

The Bahamas Banks (Fig. 4.45) constitutes a broad area of very shallow water (much of it less than 10 m), but with a steeply dipping slopes on the margins. Slopes on the northeast, windward, margin, are particularly steep. On the leeward, western margin, carbonate sediment production has exceeded the space created for sediment by subsidence, and much of this material has ended up being transported by tides and currents off the edge of the platform, where it has contributed to a zone of westward-prograding clinoforms (Fig. 4.49). These have extended the bank margin by nearly 30 km since sedimentation on the banks commenced in the mid-Cretaceous.

The Nicaragua shelf and the Great Barrier Reef are examples of what are termed **rimmed platforms,** whereas the Bahamas Banks exemplify **isolated platforms**, which are surrounded by deep-water environments on all sides. In the ancient record, rimmed platforms are more commonly located on extensional continental margins. Isolated platforms typically develop over fault blocks. Those forming the basement to the Bahamas developed during the initial

break-up of Pangea early in the development of the central Atlantic Ocean.

We touch on the importance of petrology in the facies analysis of carbonate sediments in Sect. 3.5.2. The Bahamas Platform was the site of a classic study of modern carbonate sediments (Purdy 1963). This work demonstrated that carbonate sediment composition is dependent mainly on the water depths and energy levels of the local setting. In high-energy environments, such as the windward margin of the Bahamas Bank, framework-building corals have constructed wave-resistant reefs. On the northern and western margins of the bank, reef debris and oolitic accumulations form belts of carbonate sand, while in the interior of the lagoon, micrite muds accumulate, together with allochems such as invertebrate shells and related fecal pellets (Fig. 4.50). A generalized model for the subenvironments that develop around a modern reef are shown in Fig. 4.51, together with a diagram constructed from a detailed core analysis of a Devonian reef in Alberta (Klovan 1964). While corals form the main reef core at the present day, in the geological past other types of organisms fulfilled this function. In the Devonian, for example, stromatoporoids were particularly important.

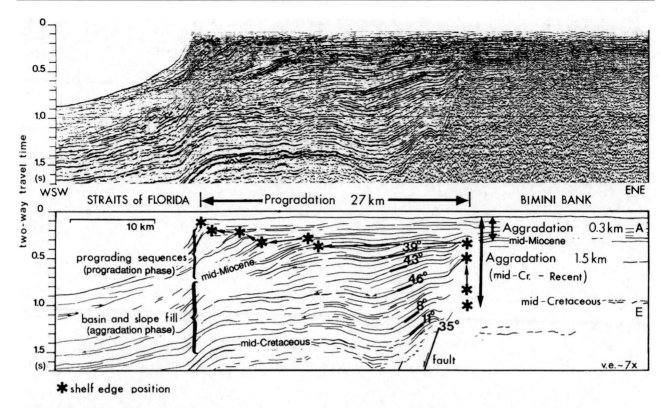

**Fig. 4.49** Seismic cross-section through the leeward margin of the Bahamas Bank. The location is shown in Fig. 4.45. Note the extensive lateral progradation of the platform as a series of clinoforms (Eberli and Ginsburg 1989)

**Fig. 4.50** Facies analysis of the surface sediments accumulating over part of the Great Bahamas Bank (after Purdy 1963). The image at *right* shows the same area as seen on Google Earth

**Fig. 4.51** *Top* a cross-section through a typical modern reef. In this diagram the open sea is to the *left*, and the reef is backed by a lagoon (adapted from James and Bourque, in Walker and James 1992); *bottom* a cross-section constructed from core analysis of a Devonian reef north of Edmonton, Alberta. The fore reef slope is to the *right*, and is characterized by talus deposits. The main reef core does not form the structural high in this example, because of post-depositional structural disturbance (Klovan 1964)

**Fig. 4.52** Examples of reef deposits. **a–e** are Devonian, **f** is Ordovician. **a** Gyrfalcon Bluff, Banks Island, Arctic Canada, **b** Crinoidal debris forming the talus slope to a small reef at Princess Royal Islands, Arctic Canada; **c, d, e** fossiliferous limestones near Norman Wells, NWT; **c** shows Kee Scarp, which exposes the reservoir rock that produces from fractures in the Norman Wells field. **f** The cavity exposes the rubbly beds of a small bioherm near Niagara Falls, Ontario

Examples of Devonian reef limestones and dolomites from Alberta and northern Canada are illustrated in Figs. 4.52 and 4.53. Gyrfalcon Bluff, in Banks Island in the Canadian Arctic, is an example of a small reef that has been "exhumed", that is, modern erosion has caused the original reef to re-appear with approximately its original outline, following removal of the sediments that covered it following deposition in the mid-Devonian. The Grassi Lakes outcrops, in Alberta, (Fig. 4.53) are classic examples of stromatoporoid reefs. The rocks here have been pervasively dolomitised, and exhibit large-scale vuggy porosity.

Given the importance of these reefs to the petroleum economy of Alberta, it is not surprising that there has been considerable exploration work carried out on these rocks in the subsurface. Figure 4.54 shows two examples of drill cores through domomitised stromatoporoid reefs rocks. In

these examples, much of the vuggy porosity has been filled with calcite cement, reducing their value as reservoir rocks, in contrast to the highly porous example shown in Fig. 4.53.

Reefs constructed by framework-building organisms span all Phanerozoic periods. From the Proterozoic through the Lower Paleozoic, limestones were also constructed by a process of sediment-trapping by algae, building fossils called stromatolites. Figure 4.55 illustrates some examples of these. These were not true reefs, in the sense that this word is used today. Stromatolites are still being formed today, for example, in Shark Bay in Western Australia. But the algal mats provide food for some browsing organisms, and they tend, therefore, to grow and survive mainly in environments that are hostile to other forms of life, such as in saline pools.

**Fig. 4.53** The dolomitised, stromatoporoid reef rocks of the Cairns Formation, at Grassi Lakes, near Canmore, Alberta. *Blue dotted line* in *top* view outlines stromatoporoid reef bodies. Note the abundant vuggy porosity in the close-up image

### 4.3.3  Tidal Sedimentation

The Paleozoic craton of North America was covered by shallow tropical to subtropical shelf seas for hundreds of millions of years (Pratt and Holden 2008; Pratt 2010). Through most of this time, water circulation by tides and storms was able to maintain an open marine environment. The result was hundreds of metres of platform carbonates constituting units containing facies that, in some cases, can be traced for hundreds of kilometers. There are no modern analogies for depositional systems of this gigantic scale, but modern settings such as the Bahamas Banks can serve to illustrate the major processes and types of sediment that were deposited. Another good analog is the southern margin of the Persian Gulf. Figure 4.56 is a model of coastal, tidal flat and lagoonal sedimentation generalized from studies of this hot, arid region. Figure 4.57 illustrates some examples of carbonate sediments deposited in environments of this type.

There are some analogies between the Persian Gulf tidal coastline and clastic barrier systems, such as those characterizing the Atlantic continental margins of the southeastern United States. The main point of comparison is the significant sedimentary influence of reversing tidal currents, and their role in forming barrier systems cut by regular tidal channels, through which strong tidal currents reverse direction, typically twice a day. Detrital carbonate material is transported by such currents, and the resulting deposits show abundant evidence of current action, in the form of scour surfaces and cross bedding (Fig. 4.57a, c, f). Another analogy is the presence of tidal flats, which are exposed to the air during the falling tide, and therefore subject to desiccation. Here, however, lies a major difference with the Atlantic barriers, in the form of the extreme heat and aridity of the Persian Gulf environment, which leads to intense evaporation, and to drying out and incipient lithification of carbonate tidal sediments. Embryonic tidal-flat limestone beds formed in such environments are commonly broken up by currents during subsequent high tides, and form fractured or brecciated beds, in which the fragmented pieces may still be almost in their original position, or broken up and dispersed to form an intraclast breccia (Fig. 4.57b). Burrowing (Fig. 4.57e) may lead to pervasive mottling texture, as in some of the Ordovician limestones of the Ottawa area (Fig. 4.57d).

In the supratidal zone, high evaporation rates leads to precipitation of evaporite nodules form the groundwaters, that are continually drawn to the surface by evaporation and capillary action. This environment is termed a **sabkha**, based on the Arabic word for "salt-encrusted desert." Limestones formed here are also prone to early dolomitization, although this is not the primary mode of dolomitization that has affected vast swaths of Paleozoic carbonates across North America. These appear to have been dolomitized by large-scale groundwater circulation millions of years after deposition.

Tidal deposits identified in the rock record are commonly cyclic, consisting of repeated shoaling-up successions (Pratt

**Fig. 4.54** Drill cores through stromotoporoid dolomites from the Middle Devonian Swan Hills formation of northern Alberta. The core in the box at *left* shows former "vuggy" porosity, now filled with calcite

**Fig. 4.55** Algal mats were important rock-forming organisms during the Proterozoic and early Paleozoic. Here are examples of stromatolites from the Proterozoic (Dismal Lakes, NWT, *top*) and the Silurian (Somerset Island, NWT, *bottom*)

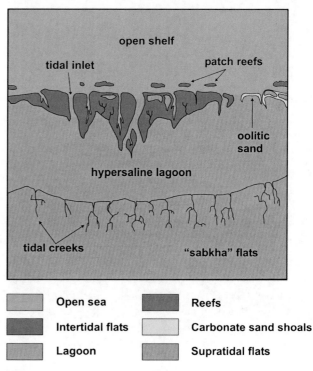

**Fig. 4.56** Depositional model for a hot, arid, tidal flat environment, based on the modern Persian Gulf (adapted from Purser 1973)

2010). The successions commonly begin with subtidal carbonate mudstones or oolitic grainstones, passing up into a range of intertidal facies, including peloidal, bioturbated, stromatolitic or bioclastic wackestones to packstones, and capped by supratidal carbonate muds, commonly containing

**Fig. 4.57** Examples of ancient limestones and dolomites formed in shallow, tidal environments. **a** Limestone with crossbedding, in which the bedding is emphasized by chert layers; Lisburne limestone (Mississippian), North Slope, Yukon; **b** intraclast dolomite breccia, Ordovician, Somerset Island NWT.; **c, f** rippled dolomites with small scour surfaces, Silurian, Somerset Island, NWT; **d** mottled limestone, a texture formed by burrowing, Ordovician, Ottawa; **e** mottled, bioturbated dolomite, Red River Formation, Tyndall Stone, Ordovician, Garson Quarry, near Selkirk, Manitoba

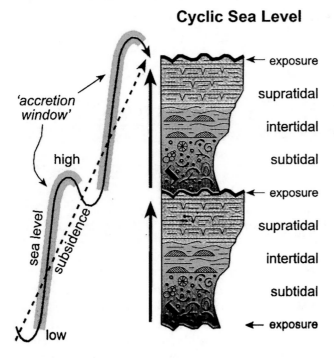

**Fig. 4.58** The development of peritidal carbonate cycles by periodic rises in sea level superimposed on a background of uniform subsidence (Pratt 2010, Fig. 19, p. 416)

evidence of desiccation in the form of shrinkage cracks and evaporite nodules or casts (Fig. 4.58). There are several mechanisms that can generate such cycles (Fig. 4.59), including local island growth on the carbonate platform, progradation of the shelf, or simple aggradation of an initially subtidal setting. Regional tracing of the cycles may be necessary to determine which of these mechanisms is appropriate in any given case.

### 4.3.4 Carbonate Slopes

Active carbonate platforms where the "carbonate factory" is operating at maximum productivity typically generate carbonate sedimentary material at a rate faster than new space can accommodate it (hence the term **accommodation** for sedimentary space: see Sect. 5.2.1 for a discussion of this term). Much of this sediment eventually is deposited on the continental slope, where it is responsible for the lateral progradation of the platform margin (Fig. 4.60). The carbonate factory typically achieves maximum productivity during high stands of sea level, which has given rise to the term **highstand shedding** for this process.

**Fig. 4.59** Three contrasting models for the development of peritidal carbonate cycles (Pratt 2010, Fig. 20, p. 417)

**Origin of Peritidal Cycles**

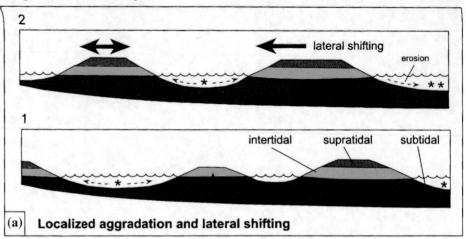

(a) **Localized aggradation and lateral shifting**

(b) **Regional progradation**

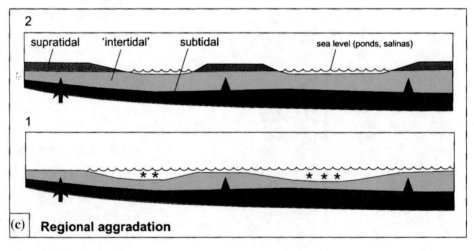

(c) **Regional aggradation**

Carbonate slopes vary in the processes and the deposits formed there (Fig. 4.60). Deep-water carbonates comprise a variety of allochthonous, shelf-derived breccias and graded calcarenites, contourite calcarenites and hemipelagic mudstones, cut by numerous intraformational truncation (slide) surfaces (Cook and Enos 1977; McIlreath and James 1984;

Coniglio and Dix 1992; Playton et al. 2010). On windward slopes, currents may be strong enough to sweep away much or all of the sediment deposited down the slope, resulting in thin sedimentary accumulations, or even disconformity surfaces. More commonly, carbonate debris accumulates and periodically slides down the slope, triggering sediment

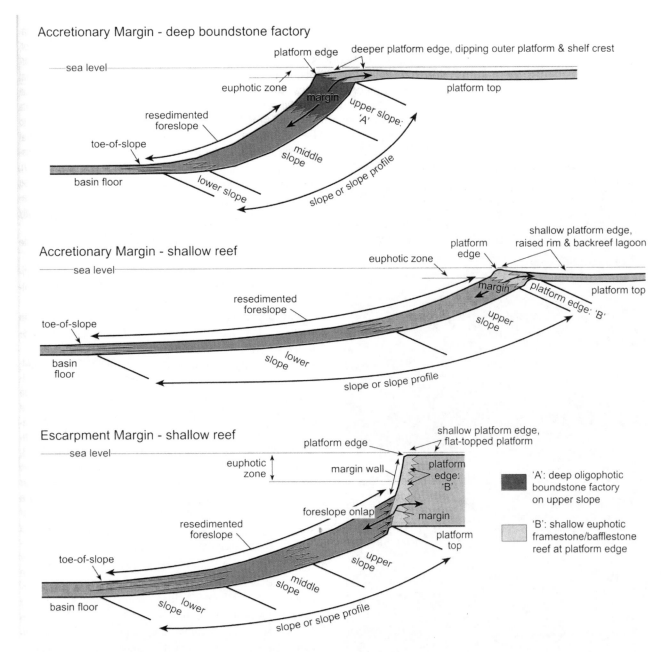

**Fig. 4.60** Typical carbonate shelf-to-slope depositional models (Playton et al. 2010, Fig. 7, p. 453)

gravity flows. These range from thin-bedded turbidites formed from the finest-grained material, to debris flows, in which boulders metres across may be present. Extreme examples of the slumping of giant pieces of carbonate platform-carbonate rocks are termed **olistoliths**.

One of the most well-known examples of carbonate-slope deposition is the Cow Head Breccia of Newfoundland. This was deposited on the extensional continental margin of the Iapetus (proto-Atlantic) Ocean during the Ordovician, and is

now well-exposed within Gros Morne National Park in the western part of the Province (Fig. 4.61).

In very exceptional circumstances, the transition from platform to slope or from slope to basin may be exposed. Such is the case in a few very large outcrops in the Canadian Arctic Islands. Figure 8.63 illustrates one such example, where carbonate platform and slope deposits of the Upper Paleozoic Nansen Formation pass down-slope into the slope deposits of the Hare Fiord Formation. The basin centre

**Fig. 4.61** Examples of carbonate-slope deposits, Cow Head Breccia, Newfoundland. **a** Turbidites, **b** debris flows, **c** a deformed clast within a debris flow

(Sverdrup Basin) is occupied by deep-water carbonates and by evaporates (Otto Fiord Formation: see next section). Note, in this illustration, the clinoform architecture of the slope deposits, something that is rarely seen in outcrops.

## 4.4   Evaporites

Evaporites are the deposits generated by the evaporation of saline waters. Some vast evaporite deposits, hundreds of metres thick and extending for thousands of kilometers, have been formed at different times during the geological past by the total evaporation of geographically enclosed water bodies. The formation of a new ocean by sea-floor spreading can generate a water body that is partially enclosed early in its history. If this sea is located in a low-latitude setting especially if it straddles the belt of high-pressure dry air at 30°N or 30°S, evaporation can result in the formation of a major evaporite deposit at an early stage in the development of an extensional continental margin. Examples include the Jurassic evaporites of the Gulf of Mexico and the margins of the central Atlantic Ocean, off Nova Scotia, and the Cretaceous evaporites of the central Atlantic deep basin, off Brazil and Angola (Fig. 4.62).

Extensive deposits have also been formed by the evaporation of cratonic seas within large continents, such as North America during the Devonian (the Prairies Evaporites of Alberta and Saskatchewan) and northwest Europe in the late Paleozoic (Zechstein evaporates of the North Sea Basin). The geological record also contains many examples of evaporates formed in Lacustrine settings. Lakes are particularly susceptible to climate change, and cyclic alternations between evaporites and clastic intervals (formed during the humid phases) are distinctive components of some series formed in cratonic, rift and intramontane settings (e.g., many of the Triassic rift basins bordering the North Atlantic Ocean).

The major evaporite minerals are listed in Table 4.1, and Table 4.2 shows the relative abundance of the major constituents of sea water, from which most evaporite deposits are formed. The abundance of the sodium and chlorine ions would suggest that sodium chloride—the mineral halite—should be the commonest evaporite deposit, but it is a highly soluble mineral and is only deposited under conditions of extreme evaporation, as indicated in Fig. 4.63. Gypsum and anhydrite are more common and, in fact, are the predominant evaporite minerals in the ancient record (Fig. 4.64).

As noted in the previous section, salt crystals may form on the supratidal flats of very hot, arid environments, such as the

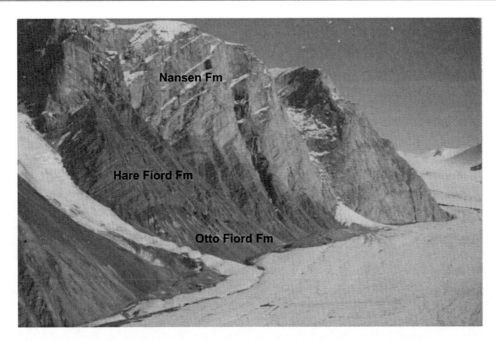

**Fig. 4.62** A dramatic facies transition from platform to basin. The Nansen formation consists of a bioclastic-carbonate debris slope, consisting partly of carbonate turbidites. The Otto Fiord Formation consists of interbedded anhydrite, shales and detrital carbonates. Sverdrup Basin (Carboniferous), Ellesmere Island, Arctic Canada (*photo* G. R. Davies)

**Table 4.1** Major evaporite minerals

| Mineral | Chemical formula |
|---------|------------------|
| Chlorides | |
| Halite | $NaCl$ |
| Sylvite | $KCl$ |
| Carnallite | $KMgCl_3 \cdot 6H_2O$ |
| Sulphates | |
| Anhydrite | $CaSO_4$ |
| Langbeinite | $K_2Mg_2(SO_4)_3$ |
| Polyhalite | $K_2Ca_2Mg(SO_4)_4 \cdot 2H_2O$ |
| Kieserite | $MgSO_4 \cdot H_2O$ |
| Gypsum | $CaSO_4 \cdot 2H_2O$ |
| Kainite | $KMg(SO_4)Cl \cdot 3H_2O$ |

**Table 4.2** Relative amounts of the major dissolved components of sea water

| | |
|---|---|
| $Cl^-$ | 48.70 % |
| $Na^+$ | 41.8 |
| $Mg^{2+}$ | 4.7 |
| $SO_4^{2-}$ | 2.5 |
| $Ca^{2+}$ | 0.9 |
| All others | 1.3 |

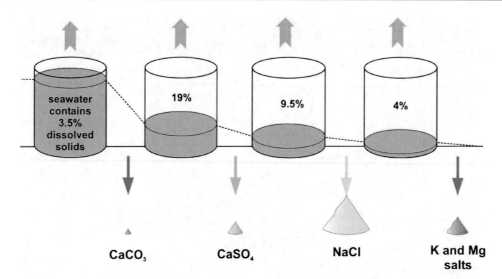

**Fig. 4.63** The succession of evaporite salts precipitated from seawater. On evaporation, CaCO₃ is precipitated first. When evaporation has reduced the volume to 19 % of the original amount, CaSO₄ begins to precipitate; at 9.5 % of the original volume, NaCl starts to precipitate, and so on. The volume of the piles represents the relative proportions of the precipitated salts (adapted from http://www.geo.utexas.edu/courses)

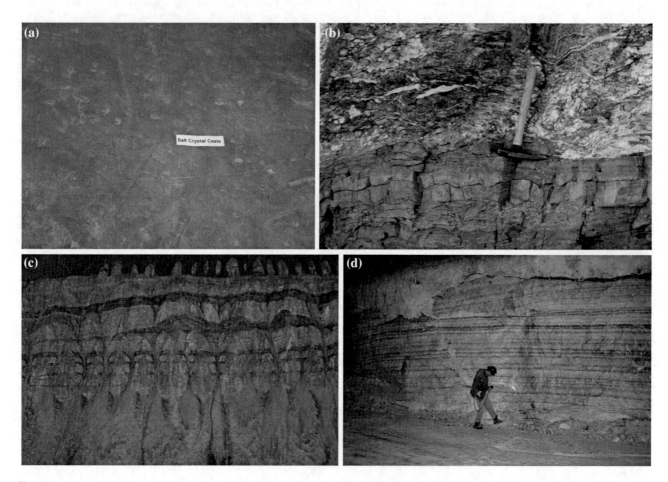

**Fig. 4.64** Evaporites in outcrop. **a** Salt pseudomorphs in sandstone, Jurassic, Utah; **b** interbedded gypsum and calcisiltite, Silurian, Somerset Island, Arctic Canada; **c** interbedded anhydrite and shale, Carboniferous, Ellesmere Island, Arctic Canada; **d** cyclic deposits of halite in mine, Devonian, Goderich, Ontario (photos **c**, by G. Davies; **d**, by N. Eyles)

**Initial stage**

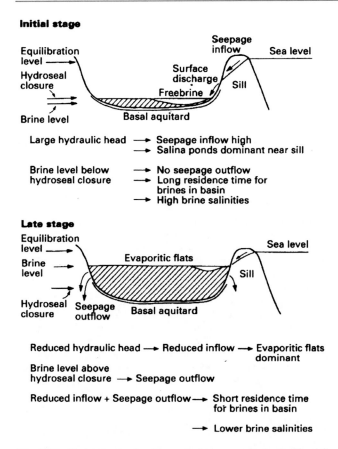

Large hydraulic head → Seepage inflow high
→ Salina ponds dominant near sill

Brine level below → No seepage outflow
hydroseal closure → Long residence time for
brines in basin
→ High brine salinities

**Late stage**

Reduced hydraulic head → Reduced inflow → Evaporitic flats
dominant

Brine level above
hydroseal closure → Seepage outflow

Reduced inflow + Seepage outflow → Short residence time
for brines in basin

→ Lower brine salinities

**Fig. 4.65** Model for the development of an evaporite basin (Kendall 1992)

modern south coast of the Persian Gulf—the type of environment geologists refer to as a **sabkha**. These commonly dissolve away, and their crystal form replaced by sand, mud or some other mineral acting as a pseudomorph (Fig. 4.64a).

The major evaporite deposits are formed within deep, oceanic basins or on continental shelves, in the basins that may develop behind barriers, such as reefs, that restrict water circulation. Calculations of the amounts of seawater needed to generate the evaporite accumulations in some ancient basins indicate that simple evaporation of a volume of water equivalent to that of the basin is entirely inadequate. Multiple basin-volumes of water typically are indicated. This may reflect cyclic drying out and refilling of the basin as a result of climate cyclicity, but it could also simply represent long-continued seepage of groundwaters into a basin through the basin-margin rocks, with continual evaporation, the groundwaters being replenished by recharge in a distant upland area within a more humid climatic setting.

A model for the development of evaporates in a sedimentary basin is shown in Fig. 4.65. Evaporation may lead to drawdown of the basin waters to well below normal sea level, as is speculated to have occurred during the Mediterranean Sea during the Miocene (Messinian age). Water depths during evaporite sedimentation may vary from very shallow to below wave base, depending on the salinities and rate of replenishment of the basin waters.

Not all evaporite beds represent simple precipitation and crystallization in place. Once deposited, an evaporite crystal, like any other sediment, is prone to re-sedimentation as a detrital particle by wind erosion and transportation, by rivers, or by marine processes, so long as water concentrations do not drop low enough to permit dissolution of the clastic material. Outcrops of evaporite areas in present-day arid areas, where the surface textures have not been destroyed by rain solution, exhibit a remarkable variety of sedimentary structures and textures. Figure 4.66 illustrates a range of these, as observed in the Messinian evaporates of the Mediterranean basin. Current transportation is indicated by lamination and ripples, and mass movement downslope is revealed by grain-size cycles and internal structures identical to those of thin-bedded turbidites and debris flows.

**Fig. 4.66** Textures and structures in re-sedimented evaporates, Messinian, Italy (after Schreiber et al. 1976)

|  | Sulphate | | Salt |
|---|---|---|---|
|  | Increasing turbulence ⟶ | | |
| Shallow | Laminated | Cross-laminated & rippled | Chevron-halite beds |
| Intermediate | Crystalline with carbonate | Wavy, anastomosing beds | Laminae composed of 'hopper' xl rafts |
| Deep | Crystalline | Debris flows    Turbidites | Laminated cyclic salts |

# References

Aigner, T., 1985, Storm depositional systems: Springer-Verlag Inc., New York, Lecture Notes in Earth Sciences 3, 174 p.

Allen, J. R. L., 1964, Studies in fluviatile sedimentation: six cyclothems from the Lower Old Red Sandstone, Anglo-Welsh basin: Sedimentology, v. 3, p. 163-198.

Allen, J. R. L., 1980, Sand waves: a model of origin and internal structure: Sedimentary Geology, v. 26, p. 281-328.

Anadón, P., Cabrera, L., and Kelts, K., eds., 1991, Lacustrine facies analysis: International Association of Sedimentologists Special Publication 13.

Anderson, J. B., 1983, Ancient glacial-marine deposits: their spatial and temporal distribution, in Molnia, B. F., ed., Glacial-marine sedimentation: Plenum Press, New York, p. 3-92.

Anderson, J. B., Brake, C., Domack, E., Meyers, N., and Wright, R., 1983, Development of a polar glacial-marine sedimentation model from Antarctic Quaternary deposits and glaciological information, in Molnia, B. F., ed., Glacial-marine sedimentation: Plenum Press, New York, p. 233-264.

Andrews, J. T., and Matsch, C. L., 1983, Glacial marine sediments and sedimentation: an annotated bibliography: Geo Abstracts, Norwich, Bibliography II, 227 p.

Arnott, R. W. C., 2010, Deep-marine sediments and sedimentary systems, in James, N. P., and Dalrymple, R. W., eds., Facies Models 4: GEOtext 6, Geological Association of Canada, St. John's, Newfoundland, p. 295-322.

Arthur, M. A., Anderson, T. F., Kaplan, I. R., Veizer, J., and Land, L. S., 1983, Stable isotopes in sedimentary geology: Society of Economic Paleontologists and Mineralogists Short Course 10, 435 p.

Ashley, G. M., Shaw, J., and Smith, N. D., 1985, Glacial sedimentary environments: Society of Economic Paleontologists and Mineralogists Short Course 16, 246 p.

Barnes, N. E., and Normark, W. R., 1985, Diagnostic parameters for comparing modern fans and ancient turbidite systems, in Bouma, A. H., Normark, W. R., and Barnes, N. E., eds., Submarine fans and related turbidite systems: Frontiers in Sedimentary Geology, Springer-Verlag, New York, p. 13-14.

Bernard, H. A., and Major, C. J., 1963, Recent meander belt deposits of the Brazos River; an alluvial "sand" model (abs): American Association of Petroleum Geologists Bulletin, v. 47, p. 350.

Bernard, H. A., Major, C. F. Jr., and Parrott, B. S., 1959, The Galveston Barrier Island and environs - a model for predicting reservoir occurrence and trend: Transactions of the Gulf Coast Association of Geological Societies, v. 9, p. 221-224.

Bernard, H. A., Leblanc, R. J., and Major, C. J., 1962, Recent and Pleistocene geology of southeast Texas, in Rainwater, E. H., and Zingula, R. P., eds., Geology of the Gulf Coast and central Texas: Geological Society of America, Guidebook for 1962 Annual Meeting., p. 175-224.

Berné, S., Auffret, J.-P., and Walker, P., 1988, Internal structure of subtidal sandwaves revealed by high-resolution seismic reflection: Sedimentology, v. 35, p. 5-20.

Berné, S., Durand, J., and Weber, O., 1991, Architecture of modern subtidal dunes (sand waves), Bay of Bourgneuf, France, in Miall, A. D., and Tyler, N., eds., The three-dimensional facies architecture of terrigenous clastic sediments, and its implications for hydrocarbon discovery and recovery: Society of Economic Paleontologists and Mineralogists, Concepts in Sedimentology and Paleontology, v. 3, p. 245-260.

Beuf, S., Biju-Duval, B., De Charpal, O., Rognon, P., Gariel, O., and Bennacef, A., 1971, Les Grès du Paléozoïque Inférieur au Sahara. Sédimentaion et discontinuités, évolution structurale d'un craton: Technip, Paris, 464 p.

Bhattacharya, J., 2010, Deltas, in James, N. P., and Dalrymple, R. W., eds., Facies Models 4: GEOtext 6, Geological Association of Canada, St. John's, Newfoundland, p. 233-264.

Bhattacharya, J., and Walker, R. G., 1992, Deltas, in Walker, R. G., and James, N. P., eds., Facies models: response to sea level change: Geological Association of Canada, Geotext 1, p. 157-177.

Blakey, R. C., 1984: Marine sand-wave complex in the Permian of central Arizona: Journal of Sedimentary Petrology, v. 54, p. 29-51.

Blum, M. D., and Roberts, H. H., 2012, The Mississippi delta region: Past, present, and future: Annual Review of Earth and Planetary Sciences, v. 40, p. 655-683.

Boothroyd, J. C., Friedrich, N. E., and McGinn, S. R., 1985, Geology of microtidal coastal logoons: Rhode Island, in Oertel, G. F., and Leatherman, S. P. eds., Barrier Islands: Marine Geology, Special Issue, v. 63, p. 35-76.

Boulton, G. S., and Deynoux, M., 1981, Sedimentation in glacial environments and the identification of tills and tillites in ancient sedimentary sequences: Precambrian Research, v. 15, p. 397-422.

Bouma, A. H., Normark, W. R., and Barnes, N. E., eds., 1985, Submarine fans and related turbidite systems: Springer-Verlag Inc., Berlin and New York, 351 p.

Boyd, R., 2010, Transgressive wave-dominated coasts, in James, N. P., and Dalrymple, R. W., eds., Facies Models 4: GEOtext 6, Geological Association of Canada, St. John's, Newfoundland, p. 265-294.

Bradley, W. H., 1929, The Varves and Climate of the Green River Epoch: U.S. Geological Survey Professional Paper 158-E, 110 p.

Brenchley, P J., 1985, Storm influenced sandstone beds: Modern Geology, v. 9, p. 369-396.

Bridge, J. S., 1993, Description and interpretation of fluvial deposits: a critical perspective: Sedimentology, v. 40, p. 801-810.

Brodzikowski, K., and Van Loon, A. J., 1987, A systematic classification of glacial and periglacial environments, facies and deposits: Earth Science Reviews, v. 24, p. 297-381.

Brookfield, M. E., and Silvestro, S., 2010, Eolian systems, in James, N. P., and Dalrymple, R. W., eds., Facies Models 4: GEOtext 6, Geological Association of Canada, St. John's, Newfoundland, p. 139-166.

Chough, S., and Hesse, R., 1976, Submarine meandering talweg and turbidity currents flowing for 4,000 km in the Northwest Atlantic Mid-Ocean Channel, Labrador Sea: Geology, v. 4, p. 529-533.

Chumakov, N. M., 1985, Glacial events of the past and their geological significance: Palaeogeography, Palaeoclimatology, Palaeoecology, v. 319-346.

Colella, A., and Prior, D. B., eds., 1990, Coarse-grained deltas: International Association of Sedimentologists Special Publication 10, 357 p.

Coleman, J. M., and Gagliano, S. W., 1964, Cyclic sedimentation in the Mississippi River deltaic plain: Transactions of the Gulf Coast Association of Geological Societies, v. 14, p. 67-80.

Coleman, J. M., and Wright, L. D., 1975, Modern river deltas: variability of processes and sand bodies, in Broussard, M. L., ed., Deltas, models for exploration: Houston Geological Society, Houston, p. 99-149.

Coniglio, M., and Dix, G. R., 1992, Carbonate slopes, in Walker, R. G. and James, N. P., eds., Facies models: response to sea-level change: Geological Association of Canada, Geotext 1, p. 349-373.

Cook, H. E., and Enos, P., eds., 1977, Deep-water carbonate environments: Society of Economic Paleontologists and Mineralogists Special Publication 25, 336 p.

Cook, H. E., Hine, A. C., and Mullins, H. T., 1983, Platform margin and deep water carbonates: Society of Economic Paleontologists and Mineralogists Short Course 12, 573 p.

Curray, J. R., 2014, The Bengal depositional system: from rift to orogeny: Marine Geology, v. 352, p. 59-69.

Dalrymple, R. W., 1984, Morphology and internal structure of sandwaves in the Bay of Fundy: Sedimentology, v. 31, p. 365-382.

Dalrymple, R. W., 2010c, Tidal depositional systems, in James, N. P., and Dalrymple, R. W., eds., Facies Models 4: GEOtext 6, Geological Association of Canada, St. John's, Newfoundland, p. 201-231.

Dalrymple, R. W., Boyd, R., and Zaitlin, B. A., eds., 1994, Incised-valley systems: origin and sedimentary sequences: SEPM (Society for Sedimentary Geology) Special Publication 51, 391 p.

Dalrymple, R. W., Leckie, D. A., and Tillman, R. W., eds., 2006, Incised valleys in space and time: SEPM Special Publication 85, 348 p.

Davies, P. J., Symonds, P. A., Feary, D. A., and Pigram, C. J., 1989, The evolution of the carbonate platforms of northeast Australia, in Crevello, P. D., Wilson, J. D., Sarg, J. F., and Read, J. F., eds., Controls on carbonate platform and basin development: Society for Sedimentary Geology (SEPM) Special Publication 44, p. 233-258.

Davis, R. A., Jr., and Hayes, M. O., 1984, What is a wave-dominated coast? Marine Geology, v. 60, p. 313-329.

Deynoux, M., ed., 1985, Glacial record: Palaeogeography, Palaeoclimatology, Palaeoecology, v. 51, 451 p. (special issue).

Dickson, J. A. D., 1985, Diagenesis of shallow-marine carbonates, in Brenchley, P. J., and Williams, B. P. J. eds., Sedimentology, recent developments and applied aspects: Blackwell Scientific Publications, Oxford, p. 173-188.

Doyle, J. D., and Sweet, M. L., 1995, Three-dimensional distribution of lithofacies, bounding surfaces, porosity, and permeability in a fluvial sandstone: Gypsy Sandstone of northern Oklahoma: American Association of Petroleum Geologists Bulletin, v. 79, p. 70-96.

Dreimanis, A., 1985, Field criteria for the recognition of till or tillite: Palaeogeography, Palaeoclimatology, Palaeoecology, v. 51, p. 7-14.

Eberli, G., and Ginsburg, R. N., 1989, Cenozoic progradation of northwestern Great Bahama Bank, a record of lateral platform growth and sea-level fluctuations, in Crevello, P. D., Wilson, J. L., Sarg, J. F., and Read, J. F., eds., Controls on carbonate platform and basin development: Society of Economic Paleontologists and Mineralogists Special Publication 44, p. 339-351.

Edwards, M. B., 1986, Glacial environments, in Reading, H. G., ed., Sedimentary environments and facies, 2nd edition: Blackwell Scientific Publications, Oxford, p. 445-470.

Eyles, N., 1984, Glacial geology: an introduction for engineers and earth scientists: Pergamon Press, Oxford, 409 p.

Eyles, N., 1993, Earth's glacial record and its tectonic setting: Earth-Science Reviews, v. 35, p. 1-248.

Eyles, N., 2008, Glacio-epochs and the supercontinent cycle after ~ 3.0 Ga: Tectonic boundary conditions for glaciation: Palaeogeography, Palaeoclimatology, Palaeoecology, v. 258, p. 89-129.

Eyles, N., and Eyles, C. H., 1992, Glacial depositional systems, in Walker, R. G., and James, N. P., eds., Facies models: response to sea level change: Geological Association of Canada Geotext 1, p. 73-100.

Eyles, C. H., and Eyles, N., 2010, Glacial deposits, in James, N. P., and Dalrymple, R. W., eds., Facies Models 4: GEOtext 6, Geological Association of Canada, St. John's, Newfoundland, p. 73-104.

Eyles, N., and Clark, B. M., 1986, Significance of hummocky and swaley cross-stratification in late Pleistocene lacustrine sediments of the Ontario basin, Canada: Geology, v. 14, p. 679-682.

Eyles, N., and Miall, A. D., 1984, Glacial facies, in Walker, R. G., ed., Facies models, 2nd edition: Geoscience Canada Reprint Series 1, p. 15-38.

Eyles, N., Eyles, C. H., and Miall, A. D., 1983, Lithofacies types and vertical profile models; an alternative approach to the description and environmental interpretation of glacial diamict sequences: Sedimentology, v. 30, p. 393-410.

Eyles, C. H., Eyles, N., and Miall, A. D., 1985, Models of glaciomarine sedimentation and their application to the interpretation of ancient glacial sequences: Palaeogeography, Palaeoclimatology, Palaeoecology, v. 51, p. 15-84.

Eyles, N., Day, T. E., and Gavican, A., 1987, Depositional controls on the magnetic characteristics of lodgement tills and other glacial diamict facies: Canadian Journal of Earth Sciences, v. 24, p. 2436-2458.

Farrell, K. M., 1987, Sedimentology and facies architecture of overbank deposits of the Mississippi River, False River region, Louisiana, in Ethridge, F. G., Flores, R. M., and Harvey, M. D., eds., Recent developments in fluvial sedimentology: Society of Economic Paleontologists and Mineralogists Special Publication 39, p. 111-120.

Farrell, K. M., Harris, W. B., Mallinson, D. J., Culver, S. J., Riggs, S. R., Pierson, J., Self-Trail, J. M., and Lautier, J. C., 2012, Standardizing texture and facies codes for a process-based classification of clastic sediment and rock: Journal of Sedimentary Research, v. 82, p. 364-378.

Fisher, W. L., Brown, L. F., Scott, A. J., and McGowen, J. H., 1969, Delta systems in the exploration for oil and gas: University of Texas Bureau of Economic Geology, 78 p.

Flemming, B. W., 1980, Sand transport and bedform patterns on the continental shelf between Durban and Port Elizabeth (southeast Africa continental margin): Sedimentary Geology, v. 26, p. 179-205.

Flores, R. M., 1981, Coal deposition in fluvial paleoenvironments of the Paleocene Tongue River Member of the Fort Union Formation, Powder River area, Powder River basin, Wyoming and Montana, in Ethridge, F. G., and Flores, R. M., eds., Recent and ancient nonmarine depositional environments: models for exploration: Society of Economic Paleontologists and Mineralogists Special Publication 31, p. 161-190.

Frakes, L. A., 1979, Climates throughout geologic time: Elsevier, Amsterdam, 310 p.

Frakes, L. A., Francis, J. E., and Syktus, J. I., 1992, Climate modes of the Phanerozoic: Cambridge University Press, Cambridge, 274 p.

Frazier, D. E., 1967, Recent deltaic deposits of the Mississippi River—their development and chronology: Transactions of the Gulf Coast Association of Geological Societies, v. 17, p. 287-315.

Galloway, W. E., 1975, Process framework for describing the morphologic and stratigraphic evolution of the deltaic depositional systems, in Broussard, M. L., ed., Deltas, models for exploration: Houston Geological Society, Houston, p. 87-98.

Glennie, K. W., 1972, Permian Rotliegendes of northwest Europe interpreted in light of modern desert sedimentation studies: American Association of Petroleum Geologists Bulletin, v. 56, p. 1048-1071.

Glennie, K. W., 1983, Lower Permian Rotliegend desert sedimentation in the North Sea area, in Brookfield, M. E., and Ahlbrandt, T. S., eds., Eolian sediments and processes: Elsevier, Amsterdam, Developments in Sedimentology, v. 38, p. 521-541.

Hambrey, M. J., and Harland, W. B., 1981a, Criteria for the identification of glacigenic deposits, in Hambrey, M. J., and Harland, W. B., eds., Earth's Pre-Pleistocene glacial record: Cambridge University Press, Cambridge, p. 14-27.

Hambrey, M. J., and Harland, W. B., eds., 1981b, Earth's Pre-Pleistocene glacial record: Cambridge University Press, Cambridge, 1004 p.

Handford, C. R., 1986, Facies and bedding sequences in shelf-storm-deposited carbonates - Fayettville Shale and Pitkin Limestone: Journal of Sedimentary Petrology, v. 56, p. 123-137.

Hardie, L. A., Smoot, J. P., and Eugster, H. P., 1978, Saline lakes and their deposits: a sedimentological approach, in Matter, A., and Tucker, M. E., eds., Modern and ancient lake sediments:

International Association of Sedimentologists Special Publication 2, p. 7-41.

Harris, P. T., 1988, Large-scale bedforms as indicators of mutually evasive sand transport and the sequential infilling of wide-mouthed estuaries: Sedimentary Geology, v. 273-298.

Hayes, M. O., 1967, Hurricanes as geological agents: Case studies of Hurricanes Carla, 1961, and Cindy, 1963: University of Texas, Bureau of Economic Geology, Austin, Texas, Report of Investigations No. 61.

Hayes, M. O., 1976, Morphology of sand accumulation in estuaries; an introduction to the symposium, in Cronin, L. E., ed., Estuarine Research, v. 2, Geology and Engineering: Academic Press, London, p. 3-22.

Hayes, M. O., 1979, Barrier island morphology as a function of tidal and wave regime, in Leatherman, S. P. ed., Barrier islands—from the Gulf of St. Lawrence to the Gulf of Mexico: Academic Press, New York, p. 1-27.

Hine, A. C., 1977, Lily Bank, Bahamas: history of an active oolite sand shoal: Journal of Sedimentary Petrology, v. 47, p. 1554-1581.

James, N. P., and Dalrymple, R. W., eds., 2010, Facies Modesl 4: GEOtext 6, Geological Association of Canada, St. John's, Newfoundland, 586 p.

James, N. P., and Lukasik, J., 2010, Cool- and cold-water neritic carbonates, in James, N. P., and Dalrymple, R. W., eds., Facies Models 4: GEOtext 6, Geological Association of Canada, St. John's, Newfoundland, p. 371-420.

Jenkyns, H. C., 1986, Pelagic environments, in Reading, H. G., ed., Sedimentary environments and facies, 2$^{nd}$ ed., Blackwell Scientific Publications, Oxford, p. 343-397.

Johnson, H. D., and Baldwin, C. T., 1996, Shallow clastic seas, in Reading. H. G., ed., Sedimentary environments: processes, facies and stratigraphy: Blackwell Science, Oxford, p. 232-280.

Jones, B., 2010, Warm-water neritic carbonates, in James, N. P., and Dalrymple, R. W., eds., Facies Models 4: GEOtext 6, Geological Association of Canada, St. John's, Newfoundland, p. 341-369.

Katz, B. J., ed., 1990, Lacustrine basin exploration: case studies and modern analogs: American Association of Petroleum Geologists Memoir 50, 340 p.

Kendall, A. C., 1992, Evaporites, in Walker, R. G. and James, N. P., eds., Facies models: response to sea-level change: Geological Association of Canada, Geotext 1, p. 375-409.

Klovan, J. E., 1964, Facies analysis of the Redwater Reef complex, Alberta, Canada: Bulletin of Canadian Petroleum Geology, v. 12, p. 1-100.

Kocurek, G. A., 1988, First-order and super bounding surfaces in eolian sequences - bounding surfaces revisited: Sedimentary Geology, v. 56, p. 193-206.

Kraus, M. J., and Bown, T. M., 1988, Pedofacies analysis; a new approach to reconstructing ancient fluvial sequences: Geological Society of America Special Paper 216, p. 143-152.

Leckie, D. A., and Krystinick, L. F., 1989, Is there evidence for geostrophic currents preserved in the sedimentary record of inner to middle shelf deposits? Journal of Sedimentary Petrology, v. 59, p. 862-870.

Machel, H.-G., 1985, Cathodoluminescence in calcite and dolomite and its chemical interpretation: Geoscience Canada, v. 12, p. 139-147.

Mackiewicz, N. E., Powell, R. D., Carlson, P. R., and Molnia, B. F., 1984, Interlaminated ice-proximal glacimarine sediments in Muir Inlet, Alaska: Marine Geology, v. 57, p. 113-148.

Macquaker, J. H. S., Taylor, K. G., and Gawthorpe, R. L. 2007, High-resolution facies analyses of mudstones: implications for paleoenvironmental and sequence stratigraphic interpretations of offshore ancient mud-dominated successions: Journal of Sedimentary Research, v. 77, p. 324–339.

Martin, A. K., and Flemming, B. W., 1986, The Holocene shelf sediment wedge off the south and east coast of South Africa; in Knight, R. J., and McLean, J. R., eds., Shelf sands and sandstones: Canadian Society of Petroleum Geologists Memoir 11, p. 27-44.

Matter, A., and Tucker, M., eds., 1978, Modern and Ancient lake sediments: International Association of Sedimentologists Special Publication 2, 290 p.

Marwick, P. J., and Rowley, D. B., 1998, The geological evidence for Triassic to Pleistocene glaciation: implications for eustasy, in Pindell, J. L., and Drake, C. L., eds., Paleogeographic evolution and non-glacial eustasy, northern South America: Society for Sedimentary Geology (SEPM) Special Publication 58, p. 17-43.

McIlreath, I. A., and James, N. P., 1984, Carbonate slopes, in Walker, ed., R. G., Facies Models, 2nd edition: Geoscience Canada Reprint Series 1, p. 245-257.

McKee, E. D., ed., 1979, A study of global sand seas: U.S. Geological Survey Professional Paper 1052.

McKee, E. D., and Ward, W. C., 1983, Eolian environment, in Scholle, P. A., Bebout, D. G., and Moore, C. H., eds., Carbonate depositional environments: American Association of Petroleum Geologists Memoir 33, p. 131-170.

McPherson, J. G., Shanmugam, G., and Moiola, R. J., 1987, Fan-deltas and braid deltas: varieties of coarse-grained deltas: Geological Society of America Bulletin, v. 99, p. 331-340.

Miall, A. D., 1977, A review of the braided river depositional environment: Earth Science Reviews, v. 13, p. 1-62.

Miall, A. D., 1978c, Lithofacies types and vertical profile models in braided river deposits: a summary, in Miall, A. D., ed., Fluvial sedimentology: Canadian Society of Petroleum Geologists Memoir 5, p. 597-604.

Miall, A. D., 1985, Architectural-element analysis: A new method of facies analysis applied to fluvial deposits: Earth Science Reviews, v. 22, p. 261-308.

Miall, A. D., 1988a, Reservoir heterogeneities in fluvial sandstones: lessons from outcrop studies: American Association of Petroleum Geologists Bulletin, v. 72, p. 682-697.

Miall, A. D., 1988b, Facies architecture in clastic sedimentary basins, in Kleinspehn, K., and Paola, C., eds., New perspectives in basin analysis: Springer-Verlag Inc., New York, p. 67-81.

Miall, A. D., 1996, The geology of fluvial deposits: sedimentary facies, basin analysis and petroleum geology: Springer-Verlag Inc., Heidelberg, 582 p.

Miall, A. D., 2010, The geology of stratigraphic sequences, second edition: Springer-Verlag, Berlin, 522 p.

Miall, A. D., and Tyler, N., eds., 1991, The three-dimensional facies architecture of terrigenous clastic sediments, and its implications for hydrocarbon discovery and recovery: Society of Economic Paleontologists and Mineralogists Concepts and Models Series, v. 3, 309 p.

Miller, J. N. G. 1996, Glacial sediments, in Reading, H. G., ed., Sedimentary environments: processes, facies and stratigraphy: Blackwell Science, Oxford, p. 454-484.

Molnia, B. F., 1983a, Subarctic glacial-marine sedimentation: a model, in Molnia, B. F., ed., Glacial-marine sedimentation: Plenum Press, New York, p. 95-144.

Molnia, B. F., ed., 1983b, Glacial-marine sedimentation: Plenum Press, New York, 844 p.

Mutti, E., and Ricci-Lucchi, F., 1972, Le turbiditi dell'Appennino settentrionale: introduzione all' analisi di facies: Soc. Geol. Ital. Mem. 11, p. 161-199.

Nanson, G. C., and Croke, J. C., 1992, A genetic classification of floodplains: Geomorphology, v. 4, p. 459-486.

Niederoda, A. W., Swift, D. J. P., Figueiredo, A. G., Jr., and Freeland, G. L., 1985, Barrier island evolution, middle Atlantic shelf, U.S.A. Part II: Evidence from the shelf floor, in Oertel, G. F. and

Leatherman, S. P. eds., Barrier islands: Marine Geology Special Issue, v. 63, p. 363-396.

Nilsen, T., Shew, R., Steffens, G., and Studlick, J., eds., 2007, Atlas of deep-water outcrops: American Association of Petroleum Geologists, Studies in Geology 56, 504 p.

Nummedal, D., and Swift, D. J. P., 1987, Transgressive stratigraphy at sequence-bounding unconformities: some principles derived from Holocene and Cretaceous examples, in Nummedal, D., Pilkey, O. H., and Howard, J. D., eds., Sea-level fluctuation and coastal evolution; Society of Economic Paleontologists and Mineralogists Special Publication 41, p. 241-260.

Olsen, P. E., 1984, Periodicity of lake-level cycles in the Late Triassic Lockatong Formation of the Newark Basin (Newark Supergroup, New Jersey and Pennsylvania), in Berger, A., Imbrie, J., Hays, J., Kukla, G., and Saltzman, B., eds., Milankovitch and climate: NATO ASI Series, D. Reidel Publishing Company, Dordrecht, Part 1, p. 129-146.

Olsen, P. E., 1986, A 40-million year lake record of Early Mesozoic orbital climatic forcing: Science, v. 234, p. 842-848.

Olsen, P. E., 1990, Tectonic, climatic, and biotic modulation of lacustrine ecosystems—examples from Newark Supergroup of eastern North America, in Katz, B. J., ed., Lacustrine basin exploration: case studies and modern analogs: American Association of Petroleum Geologists Memoir 50, p. 209-224.

Orton, G. J., and Reading, H. G., 1993, Variability of deltaic processes in terms of sediment supply, with particular emphasis on grain size: Sedimentology, v. 40, p. 475-512.

Penland, S., Boyd, R., and Suter, J. R., 1988, Transgressive depositional systems of the Mississippi delta plain: a model for barrier shoreline and shelf sand development: Journal of Sedimentary Petrology, v. 58, p. 932-949.

Picard, M. D., and High, L. R., Jr., 1981, Physical stratigraphy of ancient lacustrine deposits, in Ethridge, F. G., and Flores, R. M., eds., Recent and ancient nonmarine depositional environments: models for exploration: Society of Economic Paleontologists and Mineralogists Special Publication 31, p. 233-259.

Platt, N. H., and Keller, B., 1992, Distal alluvial deposits in a foreland basin setting — the Lower Freshwater Molasse (Lower Miocene), Switzerland: sedimentology, architecture and palaeosols: Sedimentology, v. 39, p. 545-565.

Playton, T. E., Janson, X., and Kerans, C., 2010, Carbonate slopes, in James, N. P., and Dalrymple, R. W., eds., Facies Models 4: GEOtext 6, Geological Association of Canada, St. John's, Newfoundland, p. 449-476.

Plint, A. G., 1988, Sharp-based shoreface sequences and "offshore bars" in the Cardium Formation of Alberta: their relationship to relative changes in sea level, in Wilgus, C. K., Hastings, B. S., Kendall, C. G. St. C., Posamentier, H. W., Ross, C. A., and Van Wagoner, J. C., eds., Sea-level Changes: an integrated approach: Society of Economic Paleontologists and Mineralogists Special Publication 42, p. 357-370.

Plint, A. G., 2010, Wave- and storm-dominated shoreline and shallow-marine systems, in James, N. P., and Dalrymple, R. W., eds., Facies Models 4: GEOtext 6, Geological Association of Canada, St. John's, Newfoundland, p. 167-199.

Pomar, L., and Hallock, P., 2008, Carbonate factories: a conundrum in sedimentary geology: Earth Science Reviews, v. 87, p. 134-169.

Porter, M. L., 1986, Sedimentary record of erg migration: Geology, v. 14, p. 497-500.

Postma, G., 1990, Depositional architecture and facies of river and fan deltas: a synthesis, in Colella, A., and Prior, D. B., eds., Coarse-grained deltas: International Association of Sedimentologists Special Publication 10, p. 13-27.

Powell, R. D., 1981, A model for sedimentation by tidewater glaciers: Annals of Glaciology, v. 2, p. 129-134.

Powell, R. D., 1983, Glacial marine sedimentation processes and lithofacies of temperate tidewater glaciers, Glacier Bay, Alaska, in Molnia, B. F., ed., Glacial-marine sedimentation: Plenum Press, New York, p. 185-232.

Powell, R. D., 1984, Glaciomarine processes and inductive lithofacies modelling of ice shelf and tidewater glacier sediments based on Quaternary examples: Marine Geology, v. 57, p. 1-52.

Pratt, B. R., 2010, Peritidal carbonates, in James, N. P., and Dalrymple, R. W., eds., Facies Models 4: GEOtext 6, Geological Association of Canada, St. John's, Newfoundland, p. 401-420.

Pratt, B. R., and Holmden, C., eds., 2008, Dynamics of epeiric seas: Geological Association of Canada Special Paper 48.

Purdy, E. G., 1963, Recent calcium carbonate facies of the Great Bahama Bank: Journal of Geology, v. 71, p. 334-355, p. 472-497.

Purser, B. H., ed., 1973, The Persian Gulf: Holocene carbonate sedimentation and diagenesis in a shallow epicontinental sea: Springer-Verlag, Berlin and New York, 471 p.

Reading, H. G., ed., 1986, Sedimentary environments and facies, second ed.: Blackwell, Oxford, 615 p.

Reading, H. G., ed., 1996, Sedimentary environments: processes, facies and stratigraphy, third edition: Blackwell Science, Oxford, 688 p.

Reading, H. G., and Collinson, J. D., (1996) Clastic coasts, in Reading, H. G., ed., Sedimentary environments: processes, facies and stratigraphy, third edition: Blackwell Science, Oxford, p. 154-231.

Reading, H. G., and Richards, M., 1994, Turbidite systems in deep-water basin margins classified by grain-size and feeder system: American Association of Petroleum Geologists Bulletin, v. 78, p. 792-822.

Reinson, G. E., 1992, Transgressive barrier island and estuarine systems, in Walker, R. G. and James, N. P., eds., Facies models: response to sea-level change: Geological Association of Canada, Geotext 1, p. 179-194.

Renaut, R. W., and Gierlowski-Kordesch, E. H., 2010, Lakes, in James, N. P., and Dalrymple, R. W., eds., Facies Models 4: GEOtext 6, Geological Association of Canada, St. John's, Newfoundland, p. 541-575.

Roberts, H. H., 1987, Modern carbonate-siliciclastic transitions: humid and arid tropical examples: Sedimentary Geology, v. 50, p. 25-66.

Rosenthal, L., 1988, Wave dominated shorelines and incised channel trends: Lower Cretaceous Glauconite Formation, west-central Alberta, in James, D. P., and Leckie, D. A., eds., 1988, Sequences, stratigraphy, sedimentology: surface and subsurface: Canadian Society of Petroleum Geologists Memoir 15, p. 207-230.

Ryder, R. T., 1980, Lacustrine sedimentation and hydrocarbon occurrences: a review, in American Association of Petroleum Geologists Fall Education Conference, 103 p.

Schieber, J., Southard, J. B., Kissling, P., Rossman, B., and Ginsburg, R., 2013, Experimental Deposition of Carbonate Mud From Moving Suspensions: Importance of Flocculation and Implications For Modern and Ancient Carbonate Mud Deposition: Journal of Sedimentary Research, v. 83, p. 1025-1031.

Schlager, W., 1992, Sedimentology and sequence stratigraphy of reefs and carbonate platforms: American Association of Petroleum Geologists Continuing Education Course Notes Series 34, 71 p.

Schlager, W., 2005, Carbonate sedimentology and sequence stratigraphy: SEPM Concepts in Sedimentology and Paleontology #8, 200p.

Scholle, P. A., Bebout, D. G., and Moore C. H., eds., 1983, Carbonate depositional environments: American Association of Petroleum Geologists Memoir 33, 708 p.

Schreiber, B. C., Friedman, G. M., Decima, A., and Schreiber, E., 1976, Depositional environments of Upper Miocene (Messinian) evaporite deposits of the Sicilian Basin: Sedimentology, v. 23, p. 729-760.

Scruton, P. C., 1960, Delta building and the delta sequence, in F. P. Shepard, F. B. Phleger, and T. H. van Andel, eds., Recent sediments, northwest Gulf of Mexico, American Association of Petroleum Geologists, p. 82-102.

Sellwood, B. W., 1986, Shallow-marine carbonate environments, in Reading, H. G., ed., Sedimentary environments and facies, 2nd ed.: Blackwell, Oxford, p. 283-342.

Smith, D. G., 1987, Meandering river point bar lithofacies models: modern and ancient examples compared, in Ethridge, F. G., Flores, R. M., and Harvey, M. D., eds., Recent developments in fluvial sedimentology: Society of Economic Paleontologists and Mineralogists Special Publication 39, p. 83-91.

Smith, D. G., 1988a, Modern point bar deposits analogous to the Athabasca Oil Sands, Alberta, Canada, in de Boer, P. L., van Gelder, A., and Nio, S. D., eds., Tide-influenced sedimentary environments and facies: Reidel Publishing Co., Netherlands, p. 417-432.

Smith, D. G., 1988b, Tidal bundles and mud couplets in the McMurray Formation, northeastern Alberta, Canada: Bulletin of Canadian Petroleum Geology, v. 36, p. 216-219.

Smith, D. G., and Smith, N. D., 1980, Sedimentation in anastomosed river systems: examples from alluvial valleys near Banff, Alberta: Journal of Sedimentary Petrology, v. 50, p. 157-164.

Smith, D. G., Reinson, G. E., Zaitlin, B. A., and Rahmani, R. A., eds., 1991, Clastic tidal sedimentology: Canadian Society of Petroleum Geologists memoir 16, 387 p.

Steel, R. J., and Milliken, K. L., 2013, Major advances in siliciclastic sedimentary geology, 1960-2012, in Bickford, M. E., ed., The web of geological sciences: Advances, impacts and interactions: Geological Society of America Special Paper 500, p. 121-167.

Stow, D. A. V. and Mayall, M., 2000, Deep-water sedimentary systems: new models for the 21$^{st}$ century: Marine and Petroleum Geology, v. 17, p. 125-135.

Swift, D. J. P., Niederoda, A. W., Vincent, C. E., and Hopkins, T. S., 1985, Barrier island evolution, middle Atlantic shelf, U.S.A. Part I: shoreface dynamics: marine Geology, v. 63, p. 331-361.

Talbot, M. R., and Allen, P. A., 1996, Lakes, in Reading, H. G., ed., Sedimentary environments, processes, facies and stratigraphy, third edition, Blackwell Science, Oxford, p. 83-124.

Teyssen, T. A. L., 1984, Sedimentology of the Minette oolitic ironstones of Luxembourg and Lorraine: a Jurassic subtidal sandwave complex: Sedimentology, v. 31, p. 195-212.

Thomas, R. G., Smith, D. G., Wood, J. M., Visser, J., Calverley-Range, E. A., and Koster, E. H., 1987, Inclined heterolithic stratification - terminology, description, interpretation and significance: Sedimentary Geology, v. 53, p. 123-179.

Trabucho-Alexandre, J. 2015. More gaps than shale: erosion of mud and its effect on preserved geochemical and palaeobiological signals. in: Smith, D. G., Bailey, R. J., Burgess, P. M., and Fraser, A. J., eds., Strata and Time: Probing the Gaps in Our Understanding: Geological Society, London, Special Publications, 404, p. 251-270.

Tucker, M. E., 1985, Shallow-marine carbonate facies and facies models, in Brenchley P. J., and Williams, B. P. J., eds., Sedimentology: Recent developments and applied aspects: Blackwell Scientific Publications, Oxford, p. 147-169.

Van Houten, F. B., 1964, Cyclic lacustrine sedimentation, Upper Triassic Lockatong Formation, central New Jersey and adjacent Pennsylvania, in Merriam, D. F., ed., Symposium on cyclic sedimentation: Geological Survey of Kansas Bulletin 169, p. 495-531.

Walker, R. G., 1978, Deep-water sandstone facies and ancient submarine fans: models for exploration for stratigraphic traps: American Association of Petroleum Geologists Bulletin, v. 62, p. 932-966.

Walker, R. G. and James, N. P., eds., 1992, Facies models: response to sea-level change: Geological Association of Canada, Geotext 1, 409 p.

Willis, B. J., and Behrensmeyer, A. K., 1994, Architecture of Miocene overbank deposits in northern Pakistan: Journal of Sedimentary Research, v. B64, p. 60-67.

Wilson, J. L., 1975, Carbonate facies in geologic history: Springer-Verlag, New York, 471 p.

Wright, V. P., ed. 1986b, Paleosols: their recognition and interpretation: Blackwell Scientific Publications, Oxford, 315 p.

Wright, V. P., 1992, Paleopedology, stratigraphic relationships and empirical models, in Martini, I. P., and Chesworth, W., eds., 1992, Weathering, soils and paleosols: Developments in Earth Surface Processes, No. 2, Elsevier, Amsterdam, p. 475-499.

Wright, V. P., and Burchette, T. P., 1996, Shallow-water carbonate environments, in Reading, H. G., ed., Sedimentary environments: processes, facies and stratigraphy, 3$^{rd}$ edition: Blackwell Science, p. 325-394.

Wright, R., Anderson, J. B., and Fisco, P. P., 1983, Distribution and association of sediment gravity flow deposits and glacial/glacial-marine sediments around the continental margin of Antarctica, in Molnia, B. F., ed., Glacial-marine sedimentation: Plenum Press, New York, p. 265-300.

# Sequence Stratigraphy

**5**

## Contents

## 5.1  Introduction

The purpose of this chapter is to present a succinct summary of sequence concepts, focusing on what has become the standard, or most typical, sequence model, and making only brief references to exceptions and complexities. These are described more completely elsewhere, most notably by Catuneanu (2006).

The nomenclature for sequence architecture and systems tracts was established initially for use in analyzing and interpreting seismic records by the Exxon group, led by Vail et al. (1977), and included much new terminology, new concepts and new interpretive methods (Sects. 1.2.10, 1.2.14). It is important to be familiar with these, as the terms and concepts came to be used by virtually all stratigraphers. Application of the methods to different types of data (outcrop, well-log, 2-D seismic, 3-D seismic) from around the world permitted a quantum leap in understanding of basin architectures (e.g., Van Wagoner et al. 1990), but also revealed some problems with the Exxon concepts and terminology that have stimulated vigorous debate over the last two decades. The purpose of this chapter is to bring the debate up to date by presenting current concepts, with enough historical background that readers familiar with the debates will be able to comprehend the need for revisions and new approaches.

The history of sequence concepts is presented briefly in Chap. 1. An excellent history is also provided by Nystuen (1998). Current concepts and definitions are presented by Catuneanu (2006) and Catuneanu et al. (2009, 2011), from which much of this chapter is summarized. The reader is referred to these various sources which, at the time of publication of this book, could be considered the most up to date and definitive sources. The introduction to Chap. 5 in Catuneanu (2006, pp. 165–171) is particularly useful in providing a history of the controversies, the main players and publications, and the concepts and terminologies about which there have been disputes.

Sequence stratigraphic concepts are, to a considerable extent, independent of scale (Schlager 2004; Catuneanu

© Springer International Publishing Switzerland 2016
A.D. Miall, *Stratigraphy: A Modern Synthesis*, DOI 10.1007/978-3-319-24304-7_5

2006, p. 9). Posamentier and Vail (1988) and Posamentier et al. (1992) gave examples of sequence architecture evolving from base-level changes in small natural systems, and much experimental work has successfully simulated sequence development in small laboratory tanks (e.g., Wood et al. 1993; Paola 2000). Interpretations of the temporal significance of sequences can, therefore, be difficult. However, as discussed later in this book, some aspects of the driving process, notably tectonism and climate change, generate facies and architecture that are sufficiently distinctive to yield reliable interpretations.

In this book I do not discuss the concept of global eustasy, which was an integral part of the sequence stratigraphic model when first introduced by Vail et al. (1977). The second edition of my book *The Geology of Stratigraphic Sequences* (Miall 2010) contains an extensive discussion of this issue, and it is assumed here that sequence-generating mechanisms are in accord with what Miall and Miall (2001) termed the **complexity paradigm**; that is, that there are multiple causes of the accommodation changes that lead to the generation of sequences.

## 5.2    Elements of the Model

A practical, working geologist faces two successive questions: firstly, is his/her stratigraphy subdivisible into stratigraphic sequences? And, secondly, what generated these sequences: regional tectonism, global eustasy, orbital forcing, or some other cause? The first part of this chapter deals with the methods for analyzing the sequence record. These include the following:

- The mapping of unconformities as a first step in identifying unconformity-bounded sequences.
- The mapping of onlap, offlap and other stratigraphic terminations in order to provide information about the internal architectural development of each sequence.
- Clarifying the relationship between regional structural geology and the large-scale configuration of sequences.
- The mapping of cyclic vertical facies changes in outcrop or well records in order to subdivide a stratigraphic succession into its component sequences and depositional-systems tracts, and as an indicator of changes in accommodation, including changes in relative sea-level.

The first three steps may be based on seismic-reflection data, well records or outcrops (Chap. 6); the last step cannot be accomplished using seismic data alone. This fourth step depends on the use of facies-cycle and other data, as described in Chaps. 3 and 4 (see also Dalrymple 2010b).

Sequence stratigraphy is based on the recognition of **unconformity-bounded units**, which may be formally defined and named using the methods of **allostratigraphy**, as noted in Sect. 1.2.13. Formal definitions and procedures for the employment of allostratigraphy are provided by the North American Commission on Stratigraphic Nomenclature (2005) and are touched on briefly in Sects. 7.6 and 7.7. Mitchum et al. (1977b, p. 53) defined a depositional sequence as "a stratigraphic unit composed of a relatively conformable succession of genetically related strata and bounded at its top and base by unconformities or their correlative conformities." An unconformity may be traced laterally into the deposits of deep marine environments, where it may be represented by a **correlative conformity**. However, recognizing the bounding contacts in a conformable succession might, in practice, be difficult. Documentation of the two- and three-dimensional architecture of sequences was one of the most important breakthroughs of the seismic method, as explained in the Sect. 5.2.2.

Sequences reflect the sedimentary response to **base-level cycles**—the rise and fall in sea level relative to the shoreline, and changes in the sediment supply. Change in sea-level relative to the shoreline may result from **eustasy** (absolute changes in sea-level elevation relative to the centre of Earth) or from vertical movements of the basin floor as a result of tectonism. Because of the difficulty in distinguishing between these two different processes, the term **relative sea-level change** is normally used in order to encompass the uncertainty. These basic controls are explained below. In nonmarine settings, upstream controls (tectonism and climate change) are the major determinant of sequence architecture (Sect. 5.3.2).

The cycle of rise and fall of base level generates predictable responses in a sedimentary system, such as the transgressions that occur during rising relative sea level, and the widespread subaerial erosion and delivery of clastic detritus to the continental shelf, slope, and deep basin during a fall in relative sea level. The depositional systems that result, and their vertical and lateral relationships, provide the basis for subdividing sequences into systems tracts. These are described and explained in Sect. 5.2.2.

Unconformities provided the basis for the first definitions of sequences, by Blackwelder, Levorsen, Sloss and Vail (Sect. 1.2.10). The unconformities that are the key to sequence recognition are those that develop as a result of subaerial exposure. Unconformities may develop below sea level as a result of submarine erosion, but are not used as the basis for sequence definition. Where subaerial unconformities are present, as in nonmarine and coastal settings, sequence definition is relatively straightforward. Carrying a correlation into the offshore, including recognition of a correlative conformity, is not necessarily simple; in fact this has been the cause of considerable debate and controversy, as discussed in Sect. 5.2.3. Several methods of defining sequences evolved from these controversies, and this has inhibited further progress in the establishment of sequence terminology and classification as a formal part of what might be called the official language of stratigraphy. Catuneanu (2006) and Catuneanu et al. (2009, 2011) have gone a long way towards resolving these problems, and their proposals are summarized in Sect. 7.7.

## 5.2.1  Accommodation and Supply

Sequence stratigraphy is all about **accommodation** (Fig. 5.1). Accommodation is defined as "the space available for sedimentation". Jervey (1988) explained the concept this way:

> In order for sediments to accumulate, there must be space available below base level (the level above which erosion will occur). On the continental margin, base level is controlled by sea level and, at first approximation, is equivalent to sea level…. This space made available for potential sediment accumulation is referred to as accommodation.

In marine basins this is equivalent to the space between sea level and the sea floor. In nonmarine basins, a river's graded profile functions as sedimentary base level (Holbrook et al. 2006). Sequences are a record of the balance between accommodation change and sediment supply. As accommodation is filled by sediment, the remaining space is measured by the depth of water from the sea surface to the sediment-water interface at the bottom of the sea. Total accommodation increases when the basin floor subsides or sea level rises faster than the supply of sediment to fill the available space. Barrell (1917) understood this decades

before geologists were in a position to appreciate its significance (Sect. 1.2.9; Fig. 5.2). Where supply > accommodation, progradation results. Where supply < accommodation, retrogradation results. These contrasting scenarios were recognized many years ago, and are illustrated in Fig. 5.3 with reference to the stacking patterns of deltas on a continental margin. Figure 5.4 illustrates the initial Exxon concept of how "sediment accommodation potential" is created and modified by the integration of a curve of sea-level change with subsidence. A diagram very similar to that of Barrell's was provided by Van Wagoner et al. (1990; see Fig. 5.5) and used to illustrate the deposition of shoaling-upward successions (**parasequences**).

The three major controls on basin architecture, subsidence/uplift (tectonism), sea-level change and sediment supply are themselves affected by a range of allogenic mechanisms. Crustal extension, crustal loading, and other regional tectonic processes provide the ultimate control on the size and architecture of sedimentary basins (Allen and Allen 2013). Sea level change is driven by a range of low- and high-frequency processes, as discussed at length by Miall (2010). Sediment supply is affected by the tectonic elevation of the source area, which controls rates of erosion, and by climate, which affects such factors as rates of erosion, the calibre, volume and type of erosional detrital product, and the rates of subaqueous biogenic carbonate production.

## 5.2.2  Stratigraphic Architecture

The recognition of unconformities and **stratigraphic terminations** is a key part of the method of **seismic stratigraphy**, as introduced by Vail et al. (1977). This helps to explain why modern sequence stratigraphy developed initially from the study of seismic data, because conventional basin analysis based on outcrop and well data provides little direct information on stratigraphic terminations, whereas these are readily, and sometimes spectacularly, displayed on seismic-reflection cross-sections (e.g., Fig. 1.3).

Unconformities may be used to define stratigraphic sequences because of two key, interrelated characteristics: (1) The break in sedimentation that they represent has a constant maximum time range, although parts of that time range may be represented by sedimentation within parts of the areal range of the unconformity; (2) The sediments lying above an unconformity are everywhere entirely younger than those lying below the unconformity.

There are a few exceptions to the second rule, that of the age relationships that characterize unconformities. There are at least three situations in which diachronous unconformities may develop, such that beds below the unconformity are locally younger than certain beds lying above the unconformity. The first case is that where the unconformity is

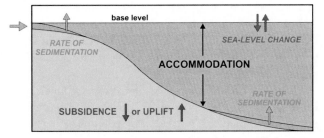

**Fig. 5.1** Accommodation, and the major allogenic sedimentary controls. Total accommodation in a basin is that generated by the subsidence of the basin floor (measured by backstripping methods). At any moment in time, the remaining accommodation in the basin represents that not yet filled by sediment, and is measured by water depth (the space between sea level and the sediment-water interface). Changes in accommodation (eustasy + tectonism) almost never correlate with bathymetric changes (eustasy + tectonism + sedimentation). Water depth reflects the balance between simultaneous creation (eustasy + tectonics) and consumption (sedimentation) of accommodation

**Fig. 5.2** Barrell's (1917) explanation of how oscillatory variations in base level control the timing of deposition. Sedimentation can only occur when base level is actively rising. These short intervals are indicated by the *black bars* in the *top* diagram. The resulting stratigraphic column, shown at the *left*, is full of disconformities, but appears to be the result of continuous sedimentation

FIGURE 5.—*Sedimentary Record made by harmonic Oscillations in Baselevel*

A-A. Primary curve of rising baselevel.
B-B. Diastrophic oscillations, giving disconformities D-D.
C-C. Minor oscillations, exaggerated and simplified, due largely to climatic rhythms.
    Equation of curve C-C: $y = \sin x - .25 \cos 8 x - .05 \cos 64 x$.

generated by marine erosion caused by deep ocean currents (Christie-Blick et al. 1990, 2007). These can shift in position across the sea floor as a result of changes in topography brought about by tectonism or sedimentation. Christie-Blick et al. (1990) cited the case of the Western Boundary Undercurrent that flows along the continental slope of the Atlantic Ocean off the United States. This current is erosive where it impinges on the continental slope, but deposition of entrained fine clastic material takes place at the margins of the main current, and the growth of this blanket is causing the current to gradually shift up the slope. The result is onlapping of the deposits onto the slope below the current, and erosional truncation of the upslope deposits.

The second type of diachronous unconformity is that which develops at basin margins as a result of syndepositional tectonism. Continuous deformation during sedimentation may lead to migration of a surface of erosion, and subsequent rapid onlap of the erosion surface by alluvial sediments (Riba 1976; Anadón et al. 1986). Typically these unconformities are associated with coarse conglomeratic sediments and die out rapidly into the basin. There is, therefore, little danger of their presence leading to the development of erroneous sequence stratigraphies.

The third case is that of incised alluvial valleys at continental margins (Strong and Paola 2008). Deepening and widening of an alluvial valley initiated during base-level fall may continue during the final stages of the evolution of a subaerial surface, even while a turn-around in the base-level cycle has begun to transgress and bury the surface during the beginning of a transgression. Channel or overbank deposits that are preserved as terrace remnants, resting on the basal erosion surface, could therefore predate the coastal deposits formed during the final stages of base-level fall, and would therefore be older than the sequence boundary at the coast, although resting on it.

As discussed in Sect. 5.3.3, breaks in sedimentation, called **drowning unconformities**, occur in carbonate sedimentary environments, as a result of environmental change, having nothing to do with changes in relative sea level.

A **correlative conformity** (defined by Mitchum et al. 1977a, b) is the deep-offshore equivalent of a subaerial unconformity. In practice, as noted later in this section, in many cases this surface is conceptual or hypothetical, occurring within a continuous section bearing no indication of the key stratigraphic processes and events taking place contemporaneously in shallow-marine and nonmarine environments. It may be possible to determine an approximate

**Fig. 5.3** Relationship between rate of deposition (Rd) and rate of subsidence (Rs) in a delta complex. **a** Progradational, **b** aggradational, **c** retrogradational. Maps at *right show* successive positions of delta fronts (Curtis Curtis 1970)

Delta complex: Dip section    Delta complex: Stratigraphic map

Rd/Rs > 1

Rd/Rs = 1

Rd/Rs < 1

position of the correlative conformity by tracing seismic reflections, but this may be quite inadequate for the purpose of formal sequence documentation and classification. As pointed out by Christie-Blick et al. (2007, p. 222), given that an unconformity represents a span of time, not an instant, "at some scale, unconformities pass laterally not into correlative conformities but into correlative intervals. Such considerations begin to be important as the resolution of the geological timescale improves at a global scale." The issue of time in stratigraphy is discussed at length in Chap. 8.

Various architectural, or geometric, characteristics record the lateral shift in depositional environments in response to sea-level change and subsidence (Figs. 5.6 and 5.7). **Onlap**

typically takes place at the base of the succession, recording the beginning of a cycle of sedimentation. **Offlap** develops when the rate of sedimentation exceeds the rate of accommodation generation. An offlap architecture may predominate in settings of high sediment supply. **Toplap** represents the abrupt pinch-out of offlapping units at the shelf-slope break. This develops when there is a major difference in sedimentary accumulation between the shelf and slope, for example when wave, tide, or storm processes inhibit or prevent accumulation on the shelf. Sediment transported across the shelf is eventually delivered to the slope, a process termed **sediment bypass**. Toplap may represent abrupt thinning rather than truncation, with a thick slope unit

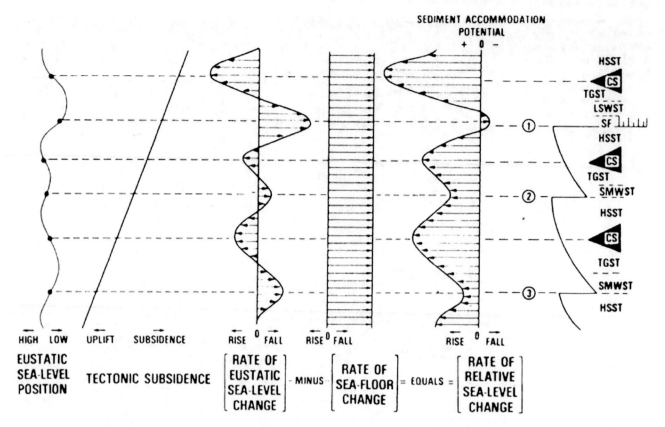

**Fig. 5.4** The standard Exxon diagram illustrating the relationship between eustasy and tectonism and the creation of "sediment accommodation potential". Integrating the two curves produces a curve of relative change of sea level, from which the timing of sequence boundaries can be derived (events 1-3 in right-hand column). However, changing the shape of the tectonic subsidence curve will change the shape and position of highs and lows in the relative sea-level curve, a point not acknowledged in the Exxon work. This version of the diagram is from Loutit et al. (1988). *CS* condensed section, *HSST* highstand systems tract, *LSWST* lowstand wedge systems tract, *SF* submarine fan, *SMWST* shelf-margin wedge systems tract, *TGST* transgressive systems tract

passing laterally into a condensed section on the shelf. Discrimination between truncation and condensation may then depend on seismic resolution. **Downlap surfaces** may develop as a result of progradation across a basin floor, and they also develop during a transition from onlap to offlap. They typically develop above flooding surfaces, as basin-margin depositional systems begin to prograde seaward following the time of maximum flooding. The dipping, prograding units are called **clinoforms** (after Rich 1951), and they lap out downward onto the downlap surface as lateral progradation takes place. The word **lapout** is used as a general term for all these types of stratigraphic termination.

The broad internal characteristics of stratigraphic units may be determined from their **seismic facies**, defined to mean an areally restricted group of seismic reflections whose appearance and characteristics are distinguishable from those of adjacent groups (Sangree and Widmier 1977). Various attributes may be used to define facies: reflection configuration, continuity, amplitude and frequence spectra, internal velocity, internal geometrical relations, and external three-dimensional form (Sect. 6.3.3).

Figure 5.8 illustrates the main styles of seismic-facies reflection patterns (Mitchum et al. 1977a), and a further discussion and illustrations of seismic facies are presented in Sect. 6.3.3. Most of these are best seen in sections parallel to depositional dip. Parallel or subparallel reflections indicate uniform rates of deposition; divergent reflections result from differential subsidence rates, such as in a half-graben or across a shelf-margin hinge zone. Clinoform reflections comprise an important class of seismic-facies patterns. They are particularly common on continental margins, where they commonly represent prograded deltaic or continental-slope outgrowth. Variations in clinoform architecture reflect different combinations of depositional energy, subsidence rates, sediment supply, water depth and sea-level position. **Sigmoid clinoforms** tend to have low depositional dips, typically less than 1°, whereas **oblique clinoforms** may show depositional dips up to 10°. **Parallel-oblique clinoform** patterns show no topsets. This usually implies shallow water depths with wave or current scour and sediment bypass to deeper water, perhaps down a submarine canyon that may be revealed on an adjacent seismic cross section. Many seismic sequences show

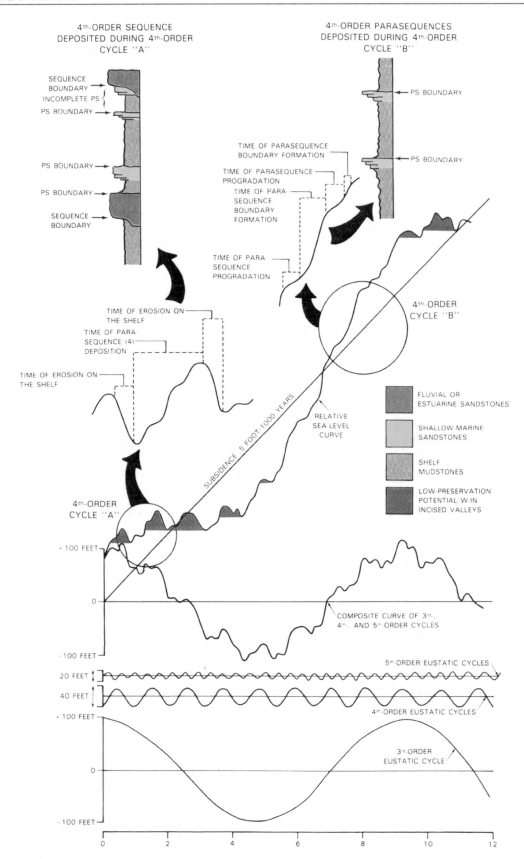

**Fig. 5.5** A modern version of Barrell's diagram (Fig. 1.3), showing the relationship between accommodation changes and sedimentation. The "relative sea-level curve" is a composite of three "eustatic" sea-level curves (although this could include other, non-eustatic mechanisms, as discussed by Miall 2010), integrated with a smooth subsidence curve. The *coloured areas* of the composite curve indicate intervals of time when accommodation is being generated, and parasequences are deposited. Examples of parasequences are indicated by the *arrows* (Van Wagoner et al. 1990). AAPG © 1990, reprinted by permission of the AAPG whose permission is required for further use

**Fig. 5.6** Sequence architecture, showing common characteristics of "seismic reflection terminations" (redrawn from Vail et al. 1977)

**Fig. 5.7** A later diagram of sequence architecture (from Vail 1987), which incorporates the concept of initial onlap followed by progradation across a downlap surface. AAPG © 1987, reprinted by permission of the AAPG whose permission is required for further use

complex offlapping stratigraphy, of which the complex sigmoid-oblique clinoform pattern in Fig. 5.8 is a simple example. This diagram illustrates periods of sea-level still-stand, with the development of truncated topsets (toplap) alternating with periods of sea-level rise (or more rapid basin subsidence), which allowed the lip of the prograding sequence to build upward as well as outward. Mitchum et al. (1977a) described the **hummocky clinoform** pattern as consisting of "irregular discontinuous subparallel reflection segments forming a practically random hummocky pattern marked by nonsystematic reflection terminations and splits. Relief on the hummocks typically is low, approaching the limits of seismic resolution. The reflection pattern is generally interpreted as strata forming small, interfingering clinoform lobes building into shallow water," such as the upbuilding or offlapping lobes of a delta undergoing distributary switching. Submarine fans may show the same hummocky reflections. **Shingled clinoform** patterns typically reflect offlapping sediment bodies on a continental shelf.

**Chaotic reflections** may reflect slumped or contorted sediment masses or those with abundant channels or cut-and-fill structures, such as submarine fan systems. Many carbonate reefs also yield chaotic reflections. Disrupted reflections are usually caused by faults. Lenticular patterns are likely to be most common in sections oriented perpendicular to depositional dip. They represent the depositional lobes of deltas or submarine fans.

A **marine flooding surface** is a surface that separates older from younger strata, across which there is evidence of an abrupt increase in water depth. These surfaces are typically prominent and readily recognizable and mappable in the stratigraphic record. Each of the heavy, arrowed lines within the lower, retrogradational part of the sequence shown in Fig. 5.7 are marine flooding surfaces, as are the heavy lines in Fig. 5.9b. The **maximum flooding surface** records the maximum extent of marine drowning, and separates transgressive units below from regressive units above (the dashed line extending obliquely across the centre of the cross-section in Fig. 5.7 is a maximum flooding surface). It commonly is a surface of considerable regional stratigraphic prominence and significance. It may be marked by a widespread shale, or by a condensed section, indicating slow sedimentation at a time of sediment starvation on the continental shelf (Loutit et al. 1988), and may correspond to a downlap surface, as noted above. The prominence of these surfaces led Galloway (1989) to propose that sequences be defined by the maximum flooding surface rather than the subaerial erosion surface. We discuss this, and other alternative concepts, in Sect. 7.7.

Sequences may consist of stacked facies successions, each of which shows a gradual upward change in facies character, indicating a progressive shift in local depositional environments. The small packages of strata contained between the heavy lines in Fig. 5.9a are examples of these component packages of strata. Van Wagoner et al. (1987) erected the term **parasequence** to encompass "a relatively conformable succession of genetically related beds or bedsets bounded by marine flooding surfaces and their correlative surfaces... Parasequences are progradational and therefore the beds within parasequences shoal upward." As Walker (1992) pointed out, "parasequences and facies successions... are essentially the same thing, except that the concept of facies succession is broader." However, other types of facies succession occur within sequences (e.g., channel-fill fining-upward successions), and the term parasequence is therefore unnecessarily restrictive. Many such successions are generated by autogenic processes, such as delta-lobe switching, and channel migration, that have nothing to do with sequence controls, and to include them in a term that has the word "sequence" within it may be misleading. Walker (1992) recommended that the term parasequence not be used. Catuneanu (2006, pp. 243–245) pointed out numerous problems with the concept of the parasequence, including the imprecise meaning of the term "flooding surface" (which it is now recognized, may have several different meanings) and the potential confusion with surfaces generated by autogenic processes. He recommended using the term only in the context of progradational units in coastal settings. I suggest

**Fig. 5.8** Typical seismic reflection patterns, illustrating the concept of seismic facies (Mitchum et al. 1977b). AAPG © 1977, reprinted by permission of the AAPG whose permission is required for further use

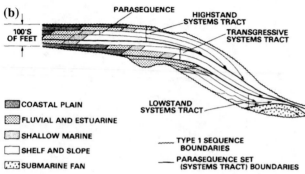

**Fig. 5.9** Diagram of sequences, sequence sets, and composite sequences. **a** Parasequences are the shoaling-upward successions, bounded by flooding marine surfaces (the *heavy lines*). **b** Sequences are composed of parasequences, which stack into lowstand, transgressive, and highstand sequence sets to form composite sequences (Mitchum and Van Wagoner 1991)

that the term be abandoned altogether. We return to this point at the end of the chapter.

### 5.2.3 Depositional Systems and Systems Tracts

The concept of the **depositional-system** and basin-analysis methods based on it were developed largely in the Gulf Coast region as a means of analyzing and interpreting the immense thicknesses of Mesozoic-Cenozoic sediment there that are so rich in oil and gas (Sect. 1.2.3). A depositional system is defined in the Schlumberger online Oilfied Glossary as

> The three-dimensional array of sediments or lithofacies that fills a basin. Depositional systems vary according to the types of sediments available for deposition as well as the depositional processes and environments in which they are deposited. The dominant depositional systems are alluvial, fluvial, deltaic, marine, lacustrine and eolian systems.

The principles of depositional-systems analysis have never been formally stated, but have been widely used, particularly by geologists of the Bureau of Economic

Geology at the University of Texas (notably W.L. Fisher, L. F. Brown Jr., J.H. McGowen, W.E. Galloway and D.E. Frazier). Useful papers on the topic are those by Fisher and McGowen (1967) and Brown and Fisher (1977). Textbook discussions are given by Miall (1999, Chap. 6) and Walker (1992). The concept of depositional episode was developed by Frazier (1974) to explain the construction of Mississippi delta by progradation of successive delta lobes.

Posamentier et al. (1988, p. 110) defined a **depositional system** as "a three-dimensional assemblage of lithofacies, genetically linked by active (modern) or inferred (ancient) processes and environments". A **systems tract** is defined as

> A linkage of contemporaneous depositional systems... Each is defined objectively by stratal geometries at bounding surfaces, position within the sequence, and internal parasequence stacking patterns. Each is interpreted to be associated with a specific segment of the eustatic curve (i.e., eustatic lowstand-lowstand wedge; eustatic rise-transgressive; rapid eustatic fall-lowstand fan, and so on), although not defined on the basis of this association (Posamentier et al. 1988).

Elsewhere, Van Wagoner et al. (1987) stated that "when referring to systems tracts, the terms lowstand and highstand are not meant to imply a unique period of time or position on a cycle of eustatic or relative change of sea level. The actual time of initiation of a systems tract is interpreted to be a function of the interaction between eustasy, sediment supply, and tectonics." There is clearly an inherent, or built-in contradiction here, that results from the use in a descriptive sense of terminology that has a genetic connotation (e.g., transgressive systems tract implies transgression). We return to this problem below.

Systems tracts are named with reference to their assumed position within the sea-level cycle, and these names incorporate ideas about the expected response of a basin to the changing balance between the major sedimentary controls (accommodation and sediment supply) during a base-level cycle. There are four standard systems tracts. These are the **highstand, falling-stage, lowstand,** and **transgressive systems tracts.** Each is illustrated here by a block diagram model with summary remarks outlining the major sedimentary controls and depositional patterns prevailing at that stage of sequence development (Fig. 5.10). Other terms have been used by different workers, but these four systems tracts and their bounding surfaces provide a useful, easy-to-understand model from which to build interpretive concepts.

The linking of systems tracts to stages of the sea-level cycle is unfortunate, because it has now been repeatedly demonstrated that the geometric and behavioural features that supposedly characterize each systems tract and its link to the sea-level cycle are not necessarily diagnostic of sea level, but may reflect combinations of several factors. For example, on the east coast of South Island, New Zealand,

**Fig. 5.10**  The evolution of a sequence through a complete base-level cycle, showing the development of the major systems tracts

different stretches of coast are simultaneously undergoing coastal progradation, reflecting a large sediment supply, and coastal retreat and transgression, because of locally high wave energy (Leckie 1994). Andros Island, in the Bahamas, currently exhibits three different systems-tract conditions. Lowstand conditions characterize the eastern (windward) margin of the island, facing the deep-water channel, the Tongue of the Ocean. Transgressive conditions occur along the northwest margin, where tidal flats are undergoing erosion, whereas on the more sheltered leeward margin, along the southwest edge of the Andros coastline, highstand conditions are suggested by tidal-flat progradation (Schlager 2005, Fig. 7.6).

The use of seismic-stratigraphic data for the construction of regional stratigraphies is discussed in Sect. 6.3.

## 5.3   Sequence Models in Clastic and Carbonate Settings

In this section, a brief overview of sequence models is provided for the main areas and styles of deposition, marine and nonmarine clastics, and carbonates. A much more complete treatment of this topic forms the core of the book by Catuneanu (2006). In addition, many useful review articles and books provide additional insights into particular areas or themes (e.g., Emery and Myers 1996; Schlager 1992, 1993, 2005; Posamentier and Allen 1999; Hunt and Gawthorpe 2000; Yoshida et al. 2007). What follows corresponds to the model Catuneanu (2006) described as "Depositional Sequence IV" (Fig. 7.26). Variants on this model, and other models, are described briefly in Sect. 7.7.

### 5.3.1   Marine Clastic Depositional Systems and Systems Tracts

In this section a description is provided of the most typical, or "model" cycle of sequence development, in which it is assumed that the main sediment transport mode is normal to the coastline. Sediment transport and deposition out of this line of section may introduce significant complications, and is the reason why sequence analysis should be a completely three-dimensional process. An example of how tectonic and sedimentary processes change dramatically along strike is described briefly in Sect. 6.3.3 (Fig. 6.29).

The acronyms provided in brackets for many of the terms are the standard abbreviations that are now in common use in research literature and in many textbooks.

*The highstand systems tract (HST)* corresponds to a period when little new accommodation is being added to the depositional environment. As shown in the relative sea-level curve at the lower right corner of Fig. 5.10, base-level rise is in the process of slowing down as it reaches its highest point, immediately prior to commencement of a slow fall. If the sediment supply remains more or less constant, then Rd > Rs at this point (the upper of the three conditions shown in Fig. 5.3). The most characteristic feature of this systems tract is the lateral progradation of coastal sedimentary environments. Major coastal barrier-lagoon and deltaic complexes are the result. **Normal regression** is the term used to describe the seaward advance of the coastline as a result of the progressive addition of sediment to the front of the beach or the delta systems, developing a broad **topset** environment (the **undathem** of Rich 1951). This is in contrast to the condition of forced regression, which is described below.

**Fig. 5.11** The Dunvegan alloformation of northwest Alberta is dominated by highstand deposits, reflecting its origins as a deltaic complex. Each shingle represents an individual delta lobe, terminated by a flooding surface (indicated by a downlapping *half-arrow*). Adapted from Bhattacharya (1991)

Where the terrigenous sediment supply is high, delta systems may largely dominate the resulting sedimentary succession, as shown in the accompanying example of the Dunvegan delta, Alberta (Fig. 5.11). The allomember boundaries in this diagram indicate times of relative low sea level, followed by flooding. Each allomember boundary is overlain by a mudstone representing the maximum flooding surface, over which delta complexes prograded. Sedimentary environments characteristically include coastal mangrove swamps, and may include significant peat swamps, the sites of future coal development. The numbered subdivisions of each allomember indicate individual deltaic shingles. Subsurface mapping may indicate that shingles of this type shifted laterally as a result of delta switching, in a manner similar to the Mississippi delta and the Yellow River delta. This points to potential confusion in terminology, because upward-shoaling successions, such as those illustrated in Fig. 5.11 correspond to the Van Wagoner et al. (1987) definition of parasequence. We return to this point elsewhere (Sect. 7.7).

The thickness of highstand shelf deposits depends on the accommodation generated by marine transgression across the shelf, typically a few tens of metres, up to a maximum of about 200 m. Where the shelf is narrow or the sediment supply is large, deltas may prograde to the shelf-slope break, at which point deltaic sedimentation may extend down slope into the deep basin (Porebski and Steel 2003). High-amplitude clinoforms may result, including significant volumes of sediment-gravity-flow deposits.

*The falling-stage systems tract (FSST).* A fall in base level from the highstand position exposes the coastal plain and then the continental shelf, to subaerial erosion. River mouths migrate seaward, and under most conditions, river valleys incise themselves as they continually grade downward to progressively lower sea levels. Deeply incised paleovalleys may result. In the rock record, many of these show evidence of multiple erosional events (Korus et al. 2008), indicating repeated responses to autogenic threshold triggers (Schumm 1979) or perhaps to minor cycles of base-level change. Significant volumes of sediment are eroded from the coastal plain and the shelf, and are fed through the coastal fluvial systems and onto the shelf or down submarine canyons. Eventually, these large sediment volumes may be tipped directly over the edge of the shelf onto the continental slope, triggering submarine landslides, debris flows, and turbidity currents. These have a powerful erosive effect, and may initiate development of submarine erosional valleys at the mouths of the major rivers or offshore from major delta distributaries. Many submarine canyons are initiated by this process, and remain as major routes for sediment dispersal through successive cycles of base-level change. The FSST is typically a major period of growth of submarine fans.

The falling base level causes basinward retreat of the shoreline, a process termed **forced regression** by Plint (1991) (Fig. 5.12). The occurrence of forced regression, as distinct from normal regression, may be detected by careful mapping of coastal shoreline sandstone complexes. Fall of

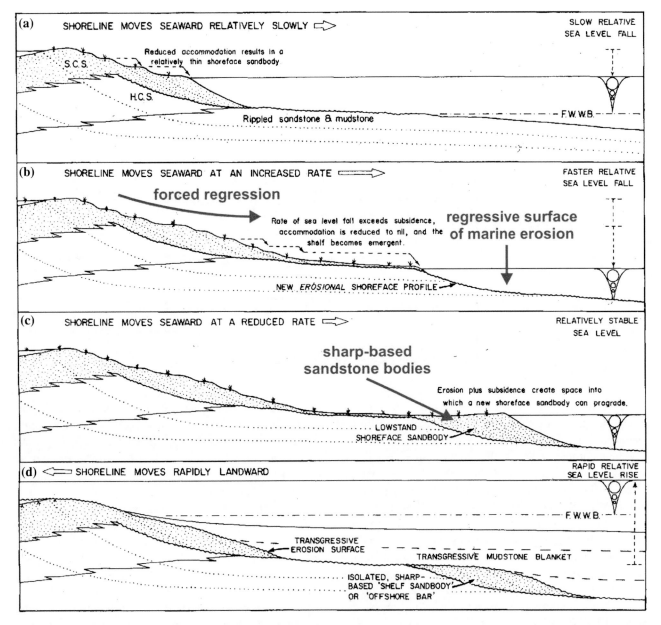

**Fig. 5.12** The process of forced regression, and the development of the regressive surface of marine erosion and "sharp-based sand bodies". Original diagram from Plint (1988)

sea level causes water depths over the shelf to decrease, increasing the erosive power of waves and tides. This typically leads to the development of a surface called the **regressive surface of marine erosion (RSME)**, which truncates shelf and distal coastal (e.g., deltaic) deposits that had been formed during the preceding highstand phase (Fig. 5.12b). The first such surface to form, at the commencement of a phase of sea-level fall, is termed the **basal surface of forced regression**. Some specialists used this surface as the basis for sequence definition, as discussed in Sect. 7.7. Given an adequate sediment supply, especially if there are pauses during the fall of sea level (Fig. 5.12c), shoreface sand accumulates above the RSME, forming what have come to be informally termed **sharp-based sandstone bodies** (Plint 1988). These are internally identical to other coastal, regressive sandstone bodies, except that they rest on an erosion surface instead of grading up from the fine-grained shelf sediments, as in the initial coastal sands shown in Fig. 5.12a (which are the product of normal regression). Repeated pulses of sea-level fall punctuated by stillstand may develop several offlapping surfaces of marine erosion. Shelf-margin deltas may form where the mouths of major river systems regress to the shelf-slope break during forced regression (Porebski and Steel 2003).

**Fig. 5.13** Sediment supply and depositional processes on a continental margin, in relationship to a cycle of sea-level change (Arnott 2010)

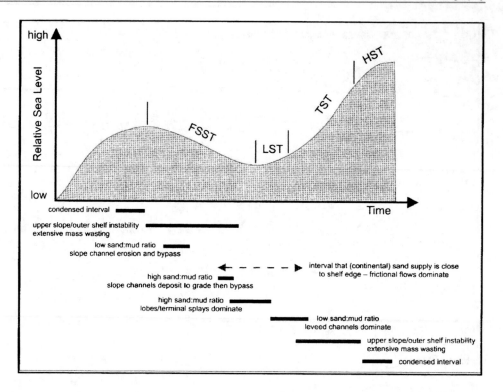

As noted above, the falling-stage is typically the interval during the sea-level cycle when the sediment supply to the continental shelf is at its greatest. However, the timing and magnitude of the sediment flux to the continental slope depends on many factors, including the width and depth of the continental shelf (Fig. 5.13). Most of the early sequence models (e.g., Posamentier et al. 1988; Posamentier and Vail 1988) showed submarine fans resting on a basal sequence boundary, but this configuration now seems unlikely. Given a low-accommodation shelf and adequate sediment supply, submarine fans (or turbidite systems, to use the term that many now prefer) may develop at the base of the continental slope at any time during a sea-level cycle (e.g., Burgess and Hovius 1998; Carvajal and Steel 2006).

On the continental shelf and coastal plain, the sequence boundary is an erosion surface representing the lowest point to which erosion cuts during the falling stage of the base-level cycle. As sea-level fall slows to its lowest point, sediment delivery from the newly exposed coastal plain and shelf will gradually diminish. Sedimentation on submarine fans may slow down, but much depends on the processes that lead to sediment-bypass on the shelf and mass wasting of the slope (Fig. 5.13). The sequence boundary, therefore, may be located anywhere within a submarine-fan succession. However, there is unlikely to be an actual mappable break in sedimentation at this level, and it may be difficult to impossible to locate the position in the section corresponding to the turn-around from falling to rising sea level. This

horizon is, therefore, what Vail et al. (1977) called a **correlative conformity**, although his original application of the term was to the fine-grained sediments formed in deep water beyond the submarine-fan wedge, out in the deep basin where it was assumed sedimentation would be continuous throughout a sea-level cycle.

*The sequence boundary (SB)* marks the lowest point reached by erosion during the falling stage of the sea-level cycle. On land this is represented by a subaerial erosion surface, which may extend far onto the continental shelf, depending on how far sea level falls. The sequence boundary cuts into the deposits of the highstand systems tract and is overlain by the deposits of the lowstand or transgressive systems tracts. It is therefore typically a surface where a marked facies change takes place, usually from a relatively lower-energy deposit below to a high-energy deposit above. Mapping of such a surface in outcrop or in the subsurface, using well logs, is facilitated by this facies change, except where the boundary juxtaposes fluvial on fluvial facies. In such cases, distinguishing the sequence boundary from other large-scale channel scours may be a difficult undertaking.

The *lowstand systems tract (LST)*: This systems tract represents the interval of time when sea-level has bottomed out, and depositional trends undergo a shift from seaward-directed (e.g., progradational) to landward-directed (e.g., retrogradational). Within most depositional systems there is little that may be confidently assigned to the lowstand systems tract. The initial basal fill of incised river

**Fig. 5.14** Depositional model for estuaries (Reinson 1992; Dalrymple et al. 1994)

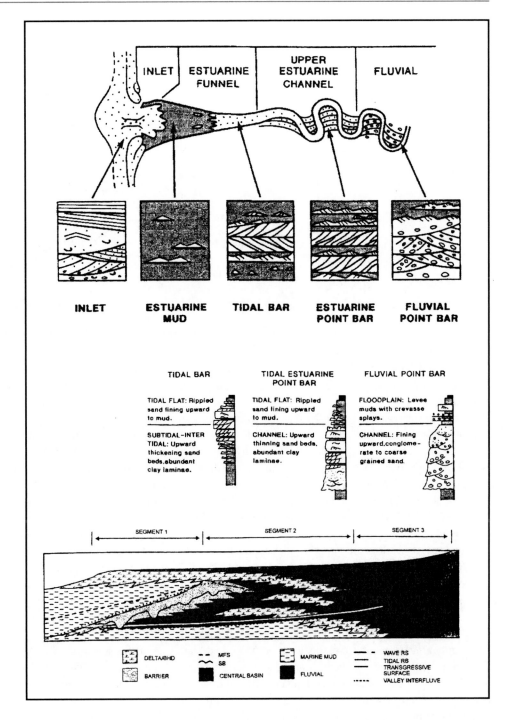

valleys, and some of the fill of submarine canyons are deposited during this phase. Volumetrically they are usually of minor importance, but they may be of a coarser grain size than succeeding transgressive deposits. In parts of the incised valley of the Mississippi River, for example (the valley formed during Pleistocene glacioeustatic sea-level lowstands), the basal fill formed during the initial post-glacial transgression is a coarse braided stream deposit, in contrast to the sandy meandering river deposits that form the bulk of the Mississippi river sediments. The episode of

active submarine-fan sedimentation on the continental slope and deep basin may persist through the lowstand phase.

There may be a phase of normal regressive sedimentation at the lowstand coastline. On coastal plains, the lowstand is a time of stillstand, when little erosion or sedimentation takes place. Between the major rivers, on the interfluve uplands, this may therefore be a place where long-established plant growth and soil development takes place. Peat is unlikely to accumulate because of the lack of accommodation, but soils, corresponding in time to the sequence boundary, may be

**Fig. 5.15** The process of transgressive ravinement. The *dashed red line* indicates the time correlation between shelf sediments deposited on the ravinement surface and contemporaneous lagoon deposits that are truncated by ravinement erosion as sea-level rises. Adapted from Nummedal and Swift (1987)

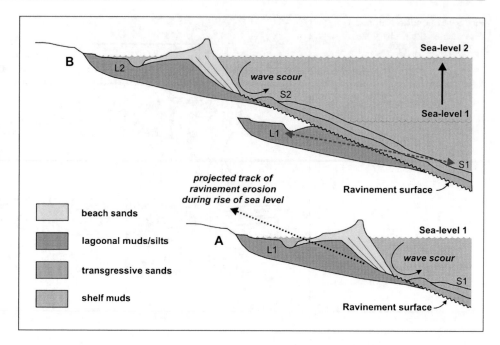

extensive, and the resulting paleosols may therefore be employed for mapping purposes (e.g., McCarthy et al. 1999; Plint et al. 2001).

*Transgressive systems tract (TST)*: A rise in base level is typically accompanied by flooding of incised valleys and transgression across the continental shelf (Fig. 5.10). Base-level rise may exceed sediment supply, leading to retrogradation of depositional systems (Rd < Rs in Fig. 5.3), although at the mouths of the largest rivers sediment supply may be sufficiently large that deltas may continue to aggrade or prograde.

Flooded river valleys are estuaries; they typically provide ample accommodation for sedimentary accumulation. In estuarine successions, the upward transition from lowstand to transgressive systems tract in estuaries and other coastal river systems is commonly marked by the development of wave- or tide-influenced fluvial facies, such as tidal sand bars containing sigmoidal crossbedding or flaser bedding. The sedimentology of this environment has received much attention (Fig. 5.14), because of the potential for the development of stratigraphic sandstone traps, in the form of valley-fill ribbon sands. Studies of ancient paleovalley fills have shown that many are complex, indicating repeated cycles of base level change and/or autogenic changes in sediment dispersal (Korus et al. 2008).

On the continental shelf the most distinctive feature of most transgressive systems tracts is the development of a widespread **transgressive surface (TS)**, a flooding surface covered with an equally widespread marine mudstone. A transgressive conglomeratic or sandy lag may blanket the flooding surface. Offshore, rapid transgression may cut the deep-water environment off from its sediment source,

leading to slow sedimentation, and the formation of a **condensed section**. This is commonly a distinctive facies, consisting of concentrated shell or fish fragments, amalgamated biozones, and a "hot" (high gamma-ray) response on well-logs, reflecting a concentration of radioactive clays (Loutit et al. 1988; Bohacs et al. 2002). Significant volumes of clastic sediment deposited on the shelf may be reworked during transgression, and mass wasting of the continental slope, with delivery of large volumes of detritus to submarine fan complexes at the base of the slope is also common. Posamentier (2002) documented numerous complexes of shelf sand ridges constituting parts of shelf transgressive systems tracts that were formed by vigorous wave and tide action. Offshore, limestones may be deposited, such as the several Jurassic and Cretaceous limestones and chalks in the Western Interior Seaway (Greenhorn Limestone, Austin Chalk).

In the nearshore setting, wave erosion during transgression is usually the cause of ravinement, with the development of a diachronous **ravinement surface** (Fig. 5.15). The juxtaposition of marine shelf sediments, above, over coastal shoreline or lagoonal sediments below, creates a prominent surface which should not be confused with a sequence boundary. A ravinement surface marks an upward deepening, the opposite of the facies relationships at most sequence boundaries. However, in some cases, ravinement erosion may cut down through lowstand deposits and into the underlying highstand systems tract, and in such cases the ravinement surface becomes the sequence boundary (Nummedal and Swift 1987).

Peat may be deposited on the coastal plain and in deltaic settings at any time during a cycle of base-level change.

**Fig. 5.16** The dependency of the lateral extent and thickness of coal seams on the rate of change of base level (Bohacs and Suter 1997). AAPG © 1997, reprinted by permission of the AAPG whose permission is required for further use

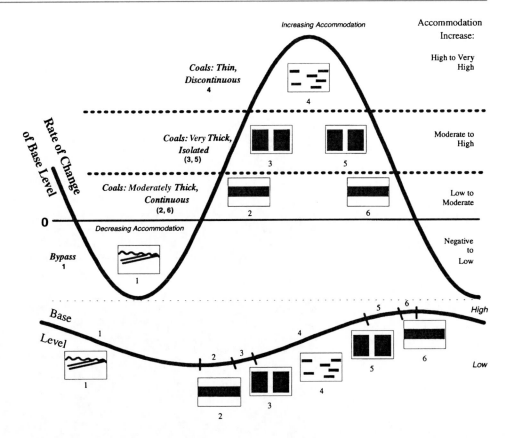

However, the thickest and most widespread coals are now known to be those formed from peat accumulated during transgression, because of the accommodation provided by rising base level, during a time when clastic influx into the coastal plain is "held back" by the landward-advancing shoreline (Bohacs and Suter 1997; see Fig. 5.16).

*The maximum flooding surface (MFS)* marks the end of the phase during which the difference between the rate of sea-level rise and the rate of sediment supply is at its greatest (Figs. 5.10). Sea-level rise continues beyond this point, but as the rate of rise slows, sediment input begins to re-establish progradation at the shoreline, and this defines the transition into the highstand systems tract. The offshore shale formed around the time of the MFS is an excellent mapping marker, because of its widespread nature and distinctive facies. In areas distant from the shoreline, where clastic sediment supply is at a minimum, the MFS is commonly marked by calcareous shale, marl or limestone. In some studies, sequence mapping is accomplished using this surface in preference to the sequence boundary, because of its more predictable facies and its consistent horizontality (Sect. 7.7).

The preceding paragraphs constitute a set of useful generalizations. However, there are many exceptions and special cases. For example, consider the ultimate fate of the clastic sediment flux on continental margins during cycles of sea level change. In the traditional model (Posamentier et al. 1988), on which this section is largely based, coastal plain complexes, including deltas, typically accumulate during highstand phases, following a period of coastal plain transgression and flooding, and basin slope and plain deposits, including submarine fans, accumulate during the sea-level falling stage and lowstand. However, these generalizations do not necessarily apply to all continental margins. As Carvajal and Steel (2006, p. 665) pointed out,

This model has been challenged using examples from narrow shelf settings (e.g., fans in the California Borderland, Gulf of Corinth, and Mediterranean Sea; see Piper and Normark 2001; Ito and Masuda 1988) or extremely high supply systems (e.g., Bengal Fan; Weber et al. 1997). In these cases slope canyons extending to almost the shoreline may receive sand from littoral drift or shelf currents during rising sea level. In addition, deltas may easily cross narrow shelves and provide sand for deep-water deposits under normal supply conditions during relative sea-level highstand. It has also been postulated that in moderately wide (tens of kilometers) to wide shelf (hundreds of kilometers) settings, significant volumes of sand can be bypassed to deep-water areas at highstand through shelf-edge deltas (Burgess and Hovius 1998; Porebski and Steel 2006). Nonetheless, documenting such delivery either in the modern or ancient has been difficult (except for suggestions from studies at the third order time scale, e.g., McMillen and Winn 1991), biasing researchers to interpret ancient deep-water deposits preferentially following the lowstand model. Thus, focus on this lowstand model has tended to cause us to overlook (1) the dominant role that sediment supply may play in deep-water sediment delivery, and (2) how such supply-dominated shelf margins can generate deep-water fans even during periods of rising relative sea level.

**Fig. 5.17** The application of a sequence model to coastal glaciomarine sedimentation (Pedersen 2012, Fig. 1, p. 30)

Covault et al. (2007) similarly noted the development of submarine fans on the California borderland at times of sea-level highstand. The connection of canyon and fan dispersal systems to the littoral sediment supply is the key control on the timing of deposition in this setting.

In addition to the physiographic variations noted here, which complicate the relationship between the base level cycle and systems-tract architecture and development, it is quite possible for episodic changes in systems-tract development at continental margins to have nothing to do with sea-level change at all. To cite two examples, in the case of the modern Amazon fan, the marked facies variations mapped by the ODP bear no relation to Neogene sea-level changes, but reflect autogenic avulsion processes on the upper fan (Christie-Blick et al. 2007). White and Lovell (1997) demonstrated that in the North Sea basin, peaks in submarine fan sedimentation occurred at times of regional uplift of the crust underlying the British Isles, as a

consequence of episodes of magma underplating, resulting in increased sediment delivery to the marine realm.

Given the enormous increase in interest in fine-grained rocks, because of their value as self-sourced gas reservoirs, a special focused study on the sequence stratigraphy of these deposits by Bohacs et al. (2002) is referred to briefly here. Shales and mudstones are typically assumed to be deep-water deposits, or to represent continental-shelf deposits formed during the flooding phase. However, as this study points out, they may be formed at any stage in the sequence cycle. Some offshore successions are composed entirely of mudstone-dominated sequences. Bohacs et al. (2002, p. 322) noted that:

> sequence boundaries and flooding surfaces continue as recognizable entities well beyond the distal limit of coarse grained lithologies. We also recognize that the correlative "conformity" that is the distal portion of a sequence boundary may exhibit significant erosion, even within completely fine-grained strata

and that truly continuous sedimentation at even a fine scale is relatively scarce.

Careful study of some classic black shales, such as the Chattanooga, have shown that they may be subdivided using sequence methods, and that many of the sequence boundaries, some traceable for hundreds of kilometres, are clearly erosional.

Recent studies have successfully developed a model for the application of sequence concepts to glaciomarine environments (Fig. 5.17). The falling stage is the stage of glacial freeze-up and subglacial erosion. Most glaciomarine sediments are deposited during the subsequent glacial-retreat phase, which is during the transgressive systems tract, when sea level is rising.

Note, in closing, the caveats at the end of Sect. 5.2.3 regarding the possible confusion between the terminology of systems tracts (highstand, falling stage, etc.) and the actual state of the sea-level cycle which they represent.

## 5.3.2  Nonmarine Depositional Systems

The early sequence model of Posamentier et al. (1988) and Posamentier and Vail (1988) emphasized the accumulation of fluvial deposits during the late highstand phase of the sea-level cycle, based on the graphical models of Jervey (1988). The model suggested that the longitudinal profile of rivers that are graded to sea level would shift seaward during a fall in base level, and that this would generate accommodation for the accumulation of nonmarine sediments. This idea was examined critically by Miall (1991a), who described scenarios where this may and may not occur. The response of fluvial systems to changes in base level was examined from a geomorphic perspective in greater detail by Wescott (1993) and Schumm (1993), and the sequence stratigraphy of nonmarine deposits was critically reviewed in an important paper by Shanley and McCabe (1994). An extensive discussion of the sequence stratigraphy of fluvial deposits was given by Miall (2014a, Chap. 6), and only a brief, updated summary of this material is given here.

Shanley and McCabe (1994) discussed the relative importance of downstream base-level controls versus upstream tectonic controls in the development of fluvial architecture. In general, the importance of base-level change diminishes upstream. In large rivers, such as the Mississippi, the evidence from the Quaternary record indicates that sea-level changes affect aggradation and degradation as far upstream as the region of Natchez, Mississippi, about 220 km upstream from the present mouth. Farther upstream than this, source-area effects, including changes in discharge and sediment supply, resulting from tectonism and climate change, are much more important. In the Colorado River of Texas, base-level influence extends about 90 km upstream,

beyond which the river has been affected primarily by the climate changes of Neogene glaciations. Blum (1994, p. 275), based on his detailed work on the Gulf Coast rivers, stated "At some point upstream rivers become completely independent of higher order relative changes in base level, and are responding to a tectonically controlled long-term average base level of erosion." The response of river systems to climate change is complex. As summarized by Miall (1996, Sect. 12.13.2), cycles of aggradation and degradation in inland areas may be driven by changes in discharge and sediment load, which are in part climate dependent. These cycles may be completely out of phase with those driven primarily by base-level change.

Holbrook et al. (2006) introduced the useful concepts of **buttresses** and **buffers** to account for longitudinal changes in fluvial facies and architecture upstream from a coastline. A buttress is some fixed point that constitutes the downstream control on a fluvial graded profile. In marine basins this will be marine base level (sea level). In inland basins it will be the lip or edge of a basin through which the trunk river flows out of the basin. The buffer is the zone of space above and below the current graded profile which represents the range of reactions that the profile may exhibit absent changes in upstream controls, such as tectonism or climate change, that govern the discharge and sediment load of the river. Tectonic uplift may raise the graded profile and the upper and lower limits of the buffer. A drop in the buttress, for example as a result of a fall in sea level, may result in incision of the river system, but if the continental shelf newly exposed by the fall in sea level has a similar slope to that of the river profile, there may be little change in the fluvial style of the river. In any of these cases, the response of the river system is to erode or aggrade towards a new dynamically maintained equilibrium profile that balances out the water and sediment flux and the rate of change in accommodation. The zone between the upper and lower limits is the buffer zone, and represents the available (potential) **preservation space** for the fluvial system.

Two sequence models developed specifically for fluvial systems were proposed in the early 1990s, those of Wright and Marriott (1993) and Shanley and McCabe (1994), referred to hereafter as the WMSM models (Fig. 5.18). These two papers drew extensively on the concepts relating fluvial architecture to accommodation that had been explored in the numerical simulation model of Bridge and Leeder (1979). That model, in turn, drew on concepts developed earlier by J.R.L. Allen, and was further developed in a series of papers referred to collectively as the LAB models, after the authors Leeder, Allen and Bridge.

Both of the WMSM models highlight the architecture and stacking pattern of channel sand bodies. It was suggested that the lowstand and highstand systems tracts are

**Fig. 5.18** Sequence model for
fluvial deposits. **a** Wright and
Marriott (1993), **b** Shanley and
McCabe (1994)

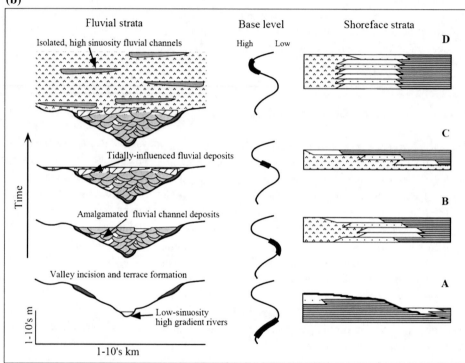

characterized by laterally-amalgamated sand bodies, whereas
the mid-portion of each sequence, including the transgressive systems tract, consists of relatively more isolated sand
bodies separated by thicker intervals of floodplain fines.
These variations were attributed to changes in the rate of
accommodation generation. Amalgamated sand bodies
indicate slow rates of accommodation generation, so that
channels have time to wander freely within the floodplain
and erode laterally into each other, whereas intervals of
isolated sand bodies develop at times of high rates of
accommodation generation, which creates space for floodplain deposits as channels switch in position within the
floodplain.

However, the LAB and WMSM models are based primarily on sedimentary processes studied in modern rivers
and the post-glacial record. Because such studies are focused

on relatively high-frequency processes and short time scales
(typically $10^3$–$10^5$ years), they are of limited value in
interpretations of the ancient rock record. Most examples of
fluvial systems in the rock record that have been interpreted
using the WMSM models span time intervals in the $10^6$-year
range (several examples were described by Miall 2014a,
Chap. 6; see also Colombera et al. 2015) indicating that
accommodation generation and sedimentation rates would
be at least an order of magnitude slower than those systems
on which the LAB and WMSM models were based.

It is suggested that systematic stratigraphic changes in
alluvial architecture in the ancient rock record are not the
product of changing avulsion rates and changes in fluvial
style under the influence of variable rates of accommodation,
as in the LAB models, but reflect regional shifts in facies
belts, that themselves are a response to tectonism and to

changes in accommodation and other variables (climate, discharge, sediment supply, etc.) at much slower rates, by one to several orders of magnitude, than those assumed for the LAB models (Miall 2014a, Sect. 6.2.4). The WMSM models may be appropriate for high-frequency fluvial cycles generated by orbital forcing, but further work is needed to test this hypothesis.

Blum (1994), who studied post-glacial fluvial systems deposited over the last few thousands of years, demonstrated that nowhere within coastal fluvial systems is there a single erosion surface that can be related to lowstand erosion. Such surfaces are continually modified by channel scour, even during transgression, because episodes of channel incision may reflect climatically controlled times of low sediment load, which are not synchronous with changes in base level. This is particularly evident landward of the limit of base-level influence. Post-glacial terraces within inland river valleys reveal a history of alternating aggradation and channel incision reflecting climate changes, all of which occurred during the last post-glacial rise in sea level. A major episode of valley incision occurred in Texas not during the time of glacioeustatic sea-level lowstand, but at the beginning of the postglacial sea-level rise, which commenced at about 15 ka. (Blum 1994). The implications of this have yet to be resolved for inland basins where aggradation occurs (because of tectonic subsidence), rather than incision and terrace formation. However, it would seem to suggest that no simple relationship between major bounding surfaces and base level change should be expected.

Figure 5.19 shows a model of fluvial processes in relationship to glacially controlled changes in climate and vegetation, based on Dutch work. These studies, and those in Texas, deal with periglacial regions, where climate change was pronounced, but the areas were not directly affected by glaciation. Vandenberghe (1993) and Vandenberghe et al. (1994) demonstrated that a major period of incision occurred during the transition from cold to warm phases because runoff

increased while sediment yield remained low. Vegetation was quickly able to stabilize river banks, reducing sediment delivery, while evapotranspiration remained low, so that the runoff was high. Fluvial styles in aggrading valleys tend to change from braided during glacial phases to meandering during interglacials (Vandenberghe et al. 1994). With increasing warmth, and consequently increasing vegetation density, rivers of anastomosed or meandering style tend to develop, the former particularly in coastal areas where the rate of generation of new accommodation space is high during the period of rapidly rising base level (Törnqvist 1993; Törnqvist et al. 1993). Vandenberghe (1993) also demonstrated that valley incision tends to occur during the transition from warm to cold phases. Reduced evapotranspiration consequent upon the cooling temperatures occurs while the vegetation cover is still substantial. Therefore runoff increases, while sediment yield remains low. With reduction in vegetation cover as the cold phase becomes established, sediment deliveries increase, and fluvial aggradation is reestablished.

It is apparent that fluvial processes inland and those along the coast may be completely out of phase during the climatic and base-level changes accompanying glacial to interglacial cycles. Within a few tens of kilometres of the sea, valley incision occurs at times of base level lowstand, during cold phases, but the surface may be modified and deepened during the subsequent transgression until it is finally buried. Inland, major erosional bounding surfaces correlate to times of climatic transition, from cold to warm and from warm to cold, that is to say during times of rising and falling sea level, respectively.

The Dakota Group of northeast New Mexico and southeast Colorado provides a good example of an internally architecturally complex fluvial unit generated by a combination of upstream tectonic controls and downstream sea-level cycles (Fig. 5.20; Holbrook et al. 2006). At the coastline, progradation and retrogradation creating three sequences were caused by sea-level cycles on a $10^5$-year time scale. Each of

**Fig. 5.19** Relationship among temperature, vegetation density, evapotranspiration, precipitation, and sedimentary processes in river systems during glacial and interglacial phases, and the relationship to the contemporaneous marine cycle. Based on work in the modern Rhine-Meuse system (Vandenberghe 1993), with the marine cycle added

**Fig. 5.20** Longitudinal NW-SE section, approximately 250 km in length, through the mid-Cretaceous Dakota Group, from Colorado into the northeast corner of New Mexico. The internal architecture consists of a series of unconformity-bounded sandstone sheets that that reflect "deposition during repetitive valley-scale cycles of aggradation and incision." (Holbrook et al. 2006, p. 164)

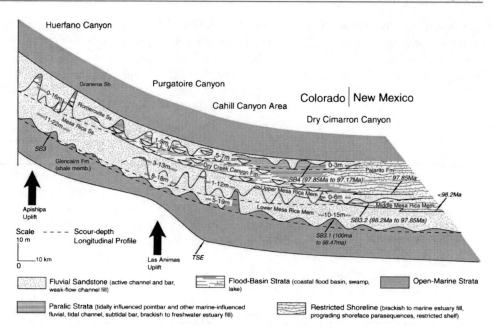

**Fig. 5.21** A sequence model for a coastal eolian succession. *Heavy arrows* within the cross-section point to supersurfaces (Blakey and Middleton 1983)

these sequences can be traced updip towards the west, where they are composed of repeated cycles of aggradational valley-fill successions and mutually incised scour surfaces. These cycles reflect autogenic channel shifting within the limited preservation space available under conditions of modest, tectonically-generated accommodation. This space is defined by a lower buffer (in the Holbrook et al. 2006 terminology) set by maximum local channel scour, and an upper buffer set by the ability of the river to aggrade under the prevailing conditions of discharge and sediment load.

Certain eolian environments and deposits can be interpreted using a sequence model. A climate model for erg

generation and the formation of the major inter-erg bounding surfaces was first proposed by Talbot (1985). These surfaces were termed **super bounding surfaces** (or **supersurfaces**) by Kocurek (1988). Application of the model, with the discussion of a range of examples, was provided by Clemmensen and Hegner (1991), Kocurek et al. (1991), Brookfield (1992), Kocurek and Havholm (1993), Kocurek (1996) and Brookfield and Silvestro (2010).

Figure 5.21 shows the sequence model developed for the Coconino sandstone of Arizona. A cyclic repetition of facies has been mapped, commencing with the sequence boundary, which corresponds to a supersurface in the sense of Kocurek

(1988). Migrating facies succession above this surface indicate a gradual seaward progradation of successively more inland facies, starting with a supratidal sabkha environment, and continuing through coastal and then large, inland dunes. As noted by Brookfield and Silvestro (2010, pp. 159–160), supersurfaces, that may be used to delineate eolian sequences, may reflect climatic or tectonic change, or change in sea level.

A brief note on the importance of orbital forcing as a major influence in the lacustrine environment is provided in Sect. 4.2.3. Cyclic repetition of local depositional environments, from coastal fans and fluvial plains to deep lacustrine settings is a common theme.

### 5.3.3 Carbonate Depositional Systems

Carbonate and clastic depositional systems respond very differently to sea-level change (Sarg 1988; James and Kendall 1992; Schlager 1992, 1993, 2005; James et al. 2010). The differences between carbonates and clastics were not understood at the time the original Exxon sequences models were developed by Vail et al. (1977) and Vail (1987).

Most carbonate sediments accumulate in shallow water, typically on continental-margin **platforms** or isolated **banks**. Large **epeiric platorms** covering much of the North American craton developed during the early Paleozoic during an extended period of high global sea level—a paleogeographic configuration that has not been replicated since. Figure 4.47 illustrates a typical **rimmed carbonate platform**, consisting of a wide, carbonate-dominated shelf with a fringing barrier reef, scattered bioherms or patch reefs in the platform interior, and a marginal clastic belt, the width of which depends on the clastic supply delivered to the coast from river mouths, and the strength of the waves and tides redistributing it along the coast. Unrimmed platform margins, lacking a marginal barrier reef, occur on the leeward sides of platforms. Carbonate **ramps** are unrimmed shelves sloping at about 1°.

Figure 5.22 illustrates the differences between the responses of carbonate and clastic systems to sea-level change, which are described briefly here, based on a standard sequence model. At times of sea-level lowstand, terrigenous clastics bypass the continental shelf, leading to exposure, erosion, and the development of incised valleys. Submarine canyons are deepened, and sand-rich turbidites systems develop submarine-fan complexes on the slope and the basin floor. Carbonate systems essentially shut down at times of lowstand, because the main "carbonate factory", the continental shelf, is exposed, and commonly undergoes **karstification,** which is the process of solution, and the formation of caves and below-surface drainages characteristic of limestone terranes. A narrow shelf-edge belt of reefs or sand shoals may occur, while the deep-water basin is starved of sediment or possibly subjected to hyperconcentration, with the development of evaporite deposits. Evaporites may also develop on the continental shelf during episodes of sea-level fall, when reef barriers serve to block marine circulation through to the shelf.

During transgression of a clastic system, incised valleys fill with estuarine deposits and eventually are blanketed with marine shale. There may be a rapid landward translation of facies belts, leaving the continental shelf starved of sediment, so that condensed successions are deposited. By contrast, transgression of a carbonate shelf serves to "turn on" the carbonate factory, with the flooding of the shelf with warm, shallow seas. This may be preceded by a short period of shallow waters in which evaporites may develop. Thick platform carbonate successions are expected to develop, reefs, in particular, being able to grow vertically at extremely rapid rates as sea level rises. At times of maximum transgression, the deepest part of the shelf may pass below the photic zone, leading to cessation of carbonate sedimentation and development of a condensed section or hardground. The resulting surface is termed a **drowning unconformity** (Schlager 1989).

Carbonate and clastic shelves are most alike during times of highstand. The rate of addition of sedimentary accommodation space is low, and lateral progradation will take place if sediment generation exceeds accommodation generation, with the development of clinoform slope architectures. Autogenic shoaling-upward cycles are common in both types of environment (e.g., terrigenous deltaic lobes, tidal carbonate cycles). Schlager (1992) stated:

> Prograding [carbonate] margins dominated by offshore sediment transport most closely resemble the classical [siliciclastic] sequence model. They are controlled by loose sediment accumulation and approach the geometry of siliciclastic systems (e.g., leeward margins...).

Carbonate platforms generate sediment at the highest rate during highstands of sea level, when platforms are flooded and the carbonate factory is at maximum productivity. The sediment volume commonly exceeds available accommodation, and the excess is delivered to the continental slope, where it may provide the sediment for large-scale progradation by carbonate talus and sediment-gravity flows. This process, termed **highstand shedding,** is exemplified by the architecture of the Bahamas Platform (Schlager 1992; see Fig. 4.49 of this book). It is the converse of the pattern of siliciclastic sedimentation, within which, as already noted, sediment is fed to the continental slope most rapidly at times of low sea level.

Many ancient shelf deposits are mixed carbonate-clastic successions, containing thin sand banks or deltaic sand sheets interbedded with carbonate platform deposits (Dolan 1989; Southgate et al. 1993). Galloway and Brown (1973)

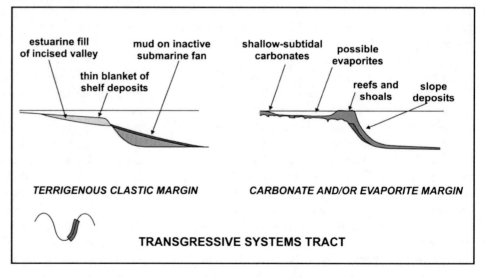

described an example from the Pennsylvanian of northern central Texas, in which a deltaic system prograded onto a stable carbonate shelf (Fig. 5.23). The term **reciprocal** **sedimentation** has been used for depositional systems in which carbonates and clastics alternate (Wilson 1967). In the example given in Fig. 5.23, deltaic distributary channels are

**Fig. 5.23** The stratigraphic architecture of the late Paleozoic continental margin, central Texas, an example of "reciprocal sedimentation" (Galloway and Brown 1973)

**Highstand carbonate platform and coastal clastic systems**

— sequence boundaries

Deltaic channels

Deltaic sandstone

Carbonate platform deposits

Slope sandstone/mudstone

Black shale/dark limestone

Lowstand clastic pathways (shelf-edge canyons)

**Lowstand slope fans, basinal muds and carbonate turbidites**

50 m

10 km

incised into the underlying shelf carbonate deposits. Widespread shelf limestones alternate with the clastic sheets and also occur in some inter-deltaic embayments. Carbonate banks occur on the outer shelf edge, beyond which the sediments thicken dramatically into a clinoform slope-clastics system. This association of carbonates and clastics reflects regular changes in sea-level, with the carbonate phase representing high sea-level and the clastic phase low sea-level. The deltaic and shelf-sand sheets and the slope clinoform deposits represent lowstand systems tracts, while the carbonate deposits are highstand deposits. During episodes of high sea-level, clastics are trapped in nearshore deltas, while during low stands much of the detritus bypasses the shelf and is deposited on the slope (arrows in Fig. 5.23).

The reciprocal-sedimentation model should, however, be used with caution. Dunbar et al. (2000) reported on the results of a detailed sampling program across the Great Barrier Reef, from which they concluded that through the glacioeustatic sea-level cycles of the last 300 ka, the maximum rate of siliciclastic sedimentation on the continental slope occurred during transgression, not during the falling stage. They explained this as the result of very low fluvial slopes across the shelf that became exposed during the sea-level falling stage and lowstand, resulting in sediment accumulation and storage on the shelf during this phase of the sea-level cycle. The sediment stored there was mobilized and transported seaward by vigorous marine processes during the subsequent transgressive phases.

### 5.3.3.1 Breaks in Sedimentation in Carbonate Environments

Submarine erosion, and other processes, can generate breaks in sedimentation without any change in sea level. This is particularly the case in carbonate sediments, which are very sensitive to environmental change, and may develop drowning unconformities (Schlager 1989, 1992). These may be architecturally similar to lowstand unconformities, and care must be taken to interpret them correctly. They may, in fact, represent an interval of slow sedimentation, represented by thin condensed sections containing many small hiatuses. The stratigraphic details may be indistinguishable on the seismic record from actual unconformable breaks because of limited seismic resolution. Schlager (1992) stated:

> Drowning requires that the reef or platform be submerged to subphotic depths by a relative rise that exceeds the growth potential of the carbonate system. The race between sea level and platform growth goes over a short distance, the thickness of the photic zone. Holocene systems indicate that their short-term growth potential is an order of magnitude higher than the rates of long-term subsidence or of third-order sea level cycles… This implies that drowning events must be caused by unusually rapid pulses of sea level or by environmental change that reduced the growth potential of platforms. With growth reduced, drowning may occur at normal rates of rise.

Schlager (1992) pointed to such environmental changes as the shifts in the El Niño current, which bring about sudden rises in water temperature, beyond the tolerance of many corals. Drowning can also occur when sea-level rise invades flat bank tops, creating shallow lagoons with highly

**Fig. 5.24** The types of sequence that develop, depending on variations in the rate of relative sea-level rise and carbonate sediment production (Jones and Desrochers 1992)

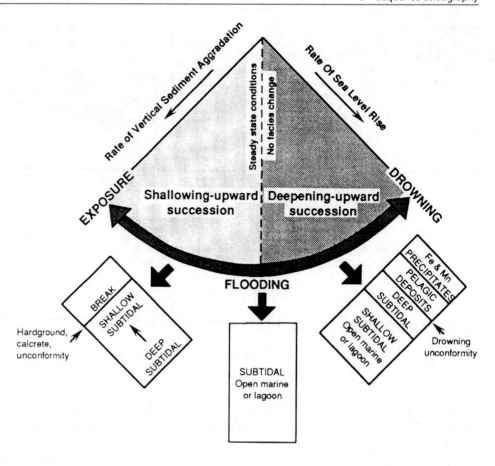

variable temperatures and salinities, plus high suspended-sediment loads due to coastal soil erosion. Oceanic anoxic events, particularly in the Cretaceous, are also known to have caused reef drowning. Schlager (1992) suggested that two Valanginian sequence boundaries in the Haq et al. (1987, 1988) global cycle chart may actually be drowning unconformities that have been misinterpreted as lowstand events. He also noted the erosive effects of submarine currents, and their ability to generate unconformities that may be mapped as sequence boundaries but that have nothing to do with sea-level change. The Cenozoic sequence stratigraphy of the Blake Plateau, off the eastern United States, is dominated by such breaks in sedimentation that do not correlate with the Exxon global cycle chart, but have been interpreted as the result of erosion by the meandering Gulf Stream.

### 5.3.3.2  Platform Carbonates: Catch-up Versus Keep-up

The style of carbonate sequence-stratigraphy on the platform is mainly a reflection of the balance between sea-level rise and carbonate production, as summarized in reviews by Jones and Desrochers (1992) and Jones (2010). Where sea-level fall exceeds subsidence rates, exposure occurs, and karst surfaces may develop (Figs. 5.24 and 5.25). Rapid transgression, on the other hand, leads to the development of condensed sections, and may shut down the carbonate

factory, resulting in a drowning unconformity. Nutrient poisoning and choking by siliciclastic detritus may also shut down the carbonate factory at times of high sea level, and this can also lead to the development of drowning unconformities, which may be mistaken for sequence boundaries (e.g., Erlich et al. 1993). James and Bourque (1992) argued that poisoning and choking were the processes most likely to cause a shut down of reef sedimentation, because studies have indicated that under ideal conditions vertical reef growth is capable of keeping pace with the most rapid of sea-level rises.

Figure 5.25 illustrates the main variations in platform architecture that develop in response to changes in the controls noted above. Drowning during a rapid rise in relative sea level is typically followed by backstepping. A slightly less rapid transgression may lead to **catch-up** architecture. Here the sea-floor remains a site of carbonate production, and as sea-level rise slows, sedimentation is able to catch up to the new sea level. Vertical aggradation characterizes the first stage of the catch-up, but lateral progradation may occur late in the cycle, when the rate of generation of new accommodation space decreases. Shoaling-upward sequences are the result.

A balance between sea-level rise and sediment-production rates will lead to **keep-up** successions, in which cyclicity is poorly developed. Eventually, typically at the close of a cycle of sea-level rise, carbonate production

**Fig. 5.25** The styles of carbonate-platform architecture and their dependence on the balance between the rate of sea-level rise and carbonate productivity (Jones and Desrocher 1992)

may exceed the rate of generation of accommodation space. This can lead to lateral progradation, and highstand shedding of carbonate detritus onto the continental slope.

As with coastal detrital sedimentation, autogenic processes may generate successions that are similar to those that are inferred to have formed in direct response to sea-level change. Pratt et al. (1992) and Pratt (2010) summarized the processes of autogenesis in peritidal environments, where shallowing-upward cycles are characteristic, as a result of short-distance transport and accumulation of carbonate detritus. Lateral progradation and vertical aggradation both may occur. Rapid filling of the available accommodation space may lead to a shut-down of the carbonate factory until relative sea-level rises enough to stimulate its reactivation. The alternative to autogenesis is the invocation of a form of rhythmic tectonic movement to generate the required sea-level change, or Milankovitch mechanisms. As Pratt et al. (1992) noted, the scale and periodicity of tectonism required for metre-scale cycles has not yet been demonstrated.

## 5.4    Conclusions

This chapter has attempted to provide an introduction to standard sequence models. Many complications have been omitted, and the reader is recommended to turn to such works as Catuneanu (2006) and Catuneanu et al. (2011) for additional details. Modern developments include the development of protocols for the integration of sequence stratigraphy into the formal structure of stratigraphy, and in order for this to be accomplished agreement must be reached about how to reconcile the various approaches to sequence mapping and definition that have emerged since the work of Vail et al. (1977). This topic is addressed in Sect. 7.7.

## References

Allen, P. A., and Allen, J. R., 2013, Basin analysis: Principles and application to petroleum play assessment: Chichester: Wiley-Blackwell, 619 p.

Anadón, P., Cabrera, L., Colombo, F., Marzo, M., and Riba, O, 1986, Syntectonic intraformational unconformities in alluvial fan deposits, eastern Ebro basin margins (NE Spain), in Allen, P. A., and Homewood, P., eds., Foreland basins: International Association of Sedimentologists Special Publication 8, p. 259-271.

Arnott, R. W. C., 2010, Deep-marine sediments and sedimentary systems, in James, N. P., and Dalrymple, R. W., eds., Facies Models 4: GEOtext 6, Geological Association of Canada, St. John's, Newfoundland, p. 295-322.

Barrell, Joseph, 1917, Rhythms and the measurement of geologic time: Geological Society of America Bulletin, v. 28, p. 745-904.

Bhattacharya, J., 1991, Regional to sub-regional facies architecture of river-dominated deltas, Upper Cretaceous Dunvegan Formation, Alberta subsurface, in Miall, A. D., and Tyler, N., eds., The three-dimensional facies architecture of terrigenous clastic sediments and its implications for hydrocarbon discovery and recovery, Society of Economic Paleontologists and Mineralogists, Concepts in Sedimentology and Paleontology, v. 3, p. 189-206.

Blakey, R. C., and Middleton, L. T., 1983, Permian shoreline aeolian complexes in central Arizona; dune changes in response to cyclic sea level changes, in Brookfield, M. E., and Ahlbrandt, T. S., eds., Eolian sediments and processes: Elsevier, Amsterdam, p. 551-581.

Blum, M. D., 1994, Genesis and architecture of incised valley fill sequences: a late Quaternary example from the Colorado River, Gulf Coastal Plain of Texas, in Weimer, P., and Posamentier, H. W., eds., Siliciclastic sequence stratigraphy: Recent developments and applications: American Association of Petroleum Geologists Memoir 58, p. 259-283.

Bohacs, K., and Suter, J., 1997, Sequence stratigraphic distribution of coaly rocks: fundamental controls and paralic examples: American Association of Petroleum Geologists Bulletin, v. 81, p. 1612-1639.

Bohacs, K. M., Neal, J. E., and Grabowski, G. J., Jr., 2002, Sequence stratigraphy in fine-grained rocks: Beyond the correlative conformity: 22nd Annual Gulf Coast Section SEPM Foundation Bob F. Perkins Research Conference, p. 321-347.

Bridge, J. S. and Leeder, M. R., 1979, A simulation model of alluvial stratigraphy: Sedimentology, v. 26, p. 617-644.

Brookfield, M. E., 1992, Eolian systems, in Walker, R. G. and James, N. P., eds., Facies models: response to sea-level change: Geological Association of Canada, Geotext 1, p. 143-156.

Brookfield, M. E., and Silvestro, S., 2010, Eolian systems, in James, N. P., and Dalrymple, R. W., eds., Facies Models 4: GEOtext 6, Geological Association of Canada, St. John's, Newfoundland, p. 139-166.

Brown, L. F., Jr., and Fisher, W. L., 1977, Seismic-stratigraphic interpretation of depositional systems: examples from Brazilian rift and pull-apart basins, in Payton, C. E., ed., Seismic stratigraphy — applications to hydrocarbon exploration: American Association of Petroleum Geologists Memoir 26, p. 213-248.

Burgess, P.M., and Hovius, N., 1998, Rates of delta progradation during highstands; consequences for timing of deposition in deep-marine systems: Geological Society, London, Journal, v. 155, p. 217–222.

Carvajal, C. R., and Steel, R. J., 2006, Thick turbidite successions from supply-dominated shelves during sea-level highstand: Geology, v. 34, p. 665-668.

Catuneanu, O., 2006, Principles of sequence stratigraphy: Elsevier, Amsterdam, 375 p.

Catuneanu, O., Abreu, V., Bhattacharya, J. P., Blum, M. D., Dalrymple, R. W., Eriksson, P. G., Fielding, C. R., Fisher, W. L., Galloway, W. E., Gibling, M. R., Giles, K. A., Holbrook, J. M., Jordan, R., Kendall, C. G. St. C., Macurda, B., Martinsen, O. J., Miall, A. D., Neal, J. E., Nummedal, D., Pomar, L., Posamentier, H. W., Pratt, B. R,. Sarg, J. F., Shanley, K. W., Steel, R. J., Strasser, A., Tucker, M. E., and Winker, C., 2009, Toward the Standardization of Sequence Stratigraphy: Earth Science Reviews, v. 92, p. 1-33.

Catuneanu, O., Galloway, W.E., Kendall, C.G.St.C., Miall, A.D., Posamentier, H.W., Strasser A., and Tucker M.E., 2011, Sequence Stratigraphy: Methodology and Nomenclature: Report to ISSC: Newsletters on Stratigraphy, v. 4 (3), p. 173-245.

Christie-Blick, N., Mountain, G. S., and Miller, K. G., 1990, Seismic stratigraphy: record of sea-level change, in Revelle, R., ed., Sea-level change: National Research Council, Studies in Geophysics, Washington, National Academy Press, p. 116-140.

Christie-Blick, N., Pekar, S. F., and Madof, A. S., 2007, Is there a role for sequence stratigraphy in chronostratigraphy? Stratigraphy, v. 4, p. 217-229.

Clemmensen, L. B., and Hegner, J., 1991, Eolian sequence and erg dynamics: the Permian Corrie Sandstone, Scotland: Journal of Sedimentary Petrology, v. 61, p. 768-774.

Colombera, L., Mountney, N. P., and McCaffrey, W. D., 2015, A meta-study of relationships between fluvial channel-body stacking pattern and aggradation rate: implications for sequence stratigraphy: Geology,

Covault, J. A., Normark, W. R., Romans, B. W., and Graham, S. A., 2007, Highstand fans in the California borderland: the overlooked deep-water depositional systems: Geology, v. 35, p. 783-786.

Curtis, D. M., 1970, Miocene deltaic sedimentation, Louisiana Gulf Coast, in Morgan, J. P., ed., Deltaic sedimentation modern and ancient: Society of Economic Paleontologists and Mineralogists Special Publication 15, p. 293-308.

Dalrymple, R. W., 2010b, Introduction to siliciclastic facies models, in James, N. P., and Dalrymple, R. W., eds., Facies Models 4: GEOtext 6, Geological Association of Canada, St. John's, Newfoundland, p. 59-72.

Dalrymple, R. W., Boyd, R., and Zaitlin, B. A., eds., 1994, Incised-valley systems: origin and sedimentary sequences: SEPM (Society for Sedimentary Geology) Special Publication 51, 391 p.

Dolan, J. F., 1989, Eustatic and tectonic controls on deposition of hybrid siliciclastic/carbonate basinal cycles: American Association of Petroleum Geologists Bulletin, v. 73, p. 1233-1246.

Dunbar, G. B., Dickens, G. R. and Carter, R. M., 2000, Sediment flux across the Great Barrier reef to the Queensland Trough over the last 300 ky: Sedimentary Geology, v. 133, p. 49-92.

Emery, D., and Myers, K. J., 1996, Sequence stratigraphy: Blackwell, Oxford, 297 p.

Erlich, R. N., Longo, A. P., Jr., and Hyare, S., 1993, Response of carbonate platform margins to drowning: evidence of environmental collapse: in Loucks, R. G., and Sarg, J. F., Jr., eds., Carbonate sequence stratigraphy – recent developments and applications: American Association of Petroleum Geologists, memoir 57, p. 241-266.

Fisher, W. L., and McGowen, J. H., 1967, Depositional systems in the Wilcox Group of Texas and their relationship to occurrence of oil and gas: Transactions of the Gulf Coast Association of Geological Societies, v. 17, p. 105-125.

Frazier, D. E., 1974, Depositional episodes: their relationship to the Quaternary stratigraphic framework in the northwestern portion of the Gulf Basin: Bureau of Economic Geology, University of Texas, Geological Circular 74-1, 26 p.

Galloway, W. E., 1989, Genetic stratigraphic sequences in basin analysis I: Architecture and genesis of flooding-surface bounded depositional units: American Association of Petroleum Geologists Bulletin, v. 73, p. 125-142.

Galloway, W. E., and Brown, L. F., Jr., 1973, Depositional systems and shelf-slope relations on cratonic basin margin, uppermost Pennsylvanian of north-central Texas: American Association of Petroleum Geologists Bulletin, v. 57, p. 1185-1218.

Haq, B. U., Hardenbol, J., and Vail, P. R., 1987, Chronology of fluctuating sea levels since the Triassic (250 million years ago to present): Science, v. 235, p. 1156-1167.

Haq, B. U., Hardenbol, J., and Vail, P. R., 1988, Mesozoic and Cenozoic chronostratigraphy and cycles of sea-level change, in Wilgus, C. K., Hastings, B. S., Kendall, C. G. St. C., Posamentier, H. W., Ross, C. A., and Van Wagoner, J. C., eds., Sea-level Changes: an integrated approach: Society of Economic Paleontologists and Mineralogists Special Publication 42, p. 71-108.

Holbrook, J., Scott, R. W., and Oboh-Ikuenobe, F. E., 2006, Base-level buffers and buttresses: a model for upstream versus downstream control on fluvial geometry and architecture within sequences: Journal of Sedimentary Research, v. 76, p. 162-174.

Hunt, D., and Gawthorpe, R. L., eds., 2000, Sedimentary response to forced regressions: Geological Society of London Special Publication 172, 383 p.

Ito, M., and Masuda, F., 1988, Late Cenozoic deep-sea to fan-delta sedimentation in an arc-arc collision zone, central Honshu, Japan; sedimentary response to varying plate-tectonic regime, in Nemec, W., and Steel, R.J., eds., Fan deltas; sedimentology and tectonic settings: Glasgow, UK, Blackie and Son, p. 400–418.

James, N. P., and Bourque, P.-A., 1992, Reefs and mounds, in Walker, R. G. and James, N. P., eds., Facies models: response to sea-level change: Geological Association of Canada, Geotext 1, p. 323-347.

James, N. P., and Kendall, A. C., 1992, Introduction to carbonate and evaporite facies models, in Walker, R. G., and James, N. P., eds., Facies models: response to sea level change: Geological Association of Canada, Geotext 1, p. 265-275.

James, N. P., Kendall, A. C., and Pufahl, P. K., 2010, Introduction to biological and biochemical facies models, in James, N. P., and Dalrymple, R. W., eds., Facies Models 4: GEOtext 6, Geological Association of Canada, St. John's, Newfoundland, p. 323-340.

Jervey, M. T., 1988, Quantitative geological modeling of siliciclastic rock sequences and their seismic expression, in Wilgus, C. K., Hastings, B. S., Kendall, C. G. St. C., Posamentier, H. W., Ross, C. A., and Van Wagoner, J. C., eds., Sea level Changes - an integrated approach: Society of Economic Paleontologists and Mineralogists Special Publication 42, p. 47-69.

Jones, B., 2010, Warm-water neritic carbonates, in James, N. P., and Dalrymple, R. W., eds., Facies Models 4: GEOtext 6, Geological Association of Canada, St. John's, Newfoundland, p. 341-369.

Jones, B., and Desrochers, A., 1992, Shallow platform carbonates, in Walker, R. G., and James, N. P., eds., Facies models: response to sea level change: Geological Association of Canada, Geotext 1, p. 277-301.

Kocurek, G. A., 1988, First-order and super bounding surfaces in eolian sequences - bounding surfaces revisited: Sedimentary Geology, v. 56, p. 193-206.

Kocurek, G. A., 1996, Desert aeolian systems, in Reading, H. G., ed., Sedimentary environments, processes, facies, stratigraphy, third edition: Blackwell Science, Oxford, p. 125-153.

Kocurek, G., and Havholm, K. G., 1993, Eolian sequence stratigraphy —a conceptual framework, in Weimer, P., and Posamentier, H. W., eds., Siliciclastic sequence stratigraphy: American Association of Petroleum Geologists Memoir 58, p. 393-409.

Kocurek, G. A., Havholm, K. G., Deynoux, M., and Blakey, R. C., 1991, Amalgamated accumulations resulting from climatic and eustatic changes: Akchar Erg, Mauritania: Sedimentology, v. 38, p. 751-772.

Korus, J. T., Kvale, E. P., Eriksson, K. A., and Joeckel, R. M., 2008, Compound paleovalleys fills in the Lower Pennyslvanian New River Formation, West Virginia, USA: Sedimentary Geology, v. 208, p. 15-26.

Leckie, D. A., 1994, Canterbury Plains, New Zealand—implications for sequence stratigraphic models: American Association of Petroleum Geologists Bulletin, v. 78, p. 1240-1256.

Loutit, T. S., Hardenbol, J., Vail, P. R., and Baum, G. R., 1988, Condensed sections: the key to age dating and correlation of continental margin sequences, in Wilgus, C. K., Hastings, B. S., Kendall, C. G. St. C., Posamentier, H. W., Ross, C. A., and Van Wagoner, J. C., eds., Sea-level Changes: an integrated approach: Society of Economic Paleontologists and Mineralogists Special Publication 42, p. 183-213.

McCarthy, P. J., Faccini, U. F., and Plint, A. G., 1999, Evolution of an ancient coastal plain: palaeosols, interfluves and alluvial architecture in a sequence stratigraphic framework, Cenomanian Dunvegan Formation, NE British Columbia, Canada: Sedimentology, v. 46, p. 861-891.

McMillen, K.M., and Winn, R.D., Jr., 1991, Seismic facies of shelf, slope, and submarine fan environments of the Lewis Shale, Upper Cretaceous, Wyoming, in Weimer, P., and Link, M.H., eds., Seismic facies and sedimentary processes of submarine fans and turbidite systems: New York, Springer-Verlag, p. 273–287.

Miall, A. D., 1991a, Stratigraphic sequences and their chronostratigraphic correlation: Journal of Sedimentary Petrology, v. 61, p. 497-505.

Miall, A. D., 1996, The geology of fluvial deposits: sedimentary facies, basin analysis and petroleum geology: Springer-Verlag Inc., Heidelberg, 582 p.

Miall, A. D., 1999, Principles of sedimentary basin analysis, Third edition: New York, N. Y.: Springer-Verlag Inc., 616 p.

Miall, A. D., 2010, The geology of stratigraphic sequences, second edition: Springer-Verlag, Berlin, 522 p.

Miall, A. D., 2014a, Fluvial depositional systems: Springer-Verlag, Berlin 316 p.

Miall, A. D., and Miall, C. E., 2001, Sequence stratigraphy as a scientific enterprise: the evolution and persistence of conflicting paradigms: Earth Science Reviews, v. 54, #4, p. 321-348.

Mitchum, R. M., Jr., and Van Wagoner, J. C., 1991, High-frequency sequences and their stacking patterns: sequence-stratigraphic evidence of high-frequency eustatic cycles: Sedimentary Geology, v. 70, 131-160.

Mitchum, R. M., Jr., Vail, P. R., and Thompson, S. III, 1977a, Seismic stratigraphy and global changes of sea level, Part 2, The depositional sequence as a basic unit for stratigraphic analysis, in Payton, C. E., ed., Seismic stratigraphy—applications to hydrocarbon exploration: American Association of Petroleum Geologists Memoir 26, p. 53-62.

Mitchum, R. M., Jr., Vail, P. R., and Sangree, J. B., 1977b, Seismic stratigraphy and global changes of sea level, Part six: Stratigraphic interpretation of seismic reflection patterns in depositional sequences, in Payton, C. E., ed., Seismic stratigraphy—applications to hydrocarbon exploration; American Association of Petroleum Geologists Memoir 26, p. 117-133.

North American Commission on Stratigraphic Nomenclature (NACSN), 2005, North American Stratigraphic Code: American Association of Petroleum Geologists Bulletin, v. 89, p. 1547-1591.

Nummedal, D., and Swift, D. J. P., 1987, Transgressive stratigraphy at sequence-bounding unconformities: some principles derived from Holocene and Cretaceous examples, in Nummedal, D., Pilkey, O. H., and Howard, J. D., eds., Sea-level fluctuation and coastal evolution; Society of Economic Paleontologists and Mineralogists Special Publication 41, p. 241-260.

Nystuen, J. P., 1998, History and development of sequence stratigraphy, in Gradstein, F. M., Sandvik, K. O. and Milton, N. J., eds., Sequence stratigraphy — concepts and applications: Norwegian Petroleum Society Special Publication 8, p. 31-116.

Paola, C., 2000, Quantitative models of sedimentary basin filling: Sedimentology, v. 47 (supplement 1) p. 121-178.

Pedersen, S. A. S., 2012, Glaciodynamic sequence stratigraphy, in Huuse, M., Redfern, J., LeHeron, D. P., Dixon, R. J., Moscariello, A., and Craig, J., eds., Glaciogenic reservoirs and hydrocarbon systems, Geological Society, London, Special Publication 368, p. 29-51.

Piper, D.J.W., and Normark, W.R., 2001, Sandy fans; from Amazon to Hueneme and beyond: American Association of Petroleum Geologists Bulletin, v. 85, p. 1407–1438.

Plint, A. G., 1988, Sharp-based shoreface sequences and "offshore bars" in the Cardium Formation of Alberta: their relationship to relative changes in sea level, in Wilgus, C. K., Hastings, B. S., Kendall, C. G. St. C., Posamentier, H. W., Ross, C. A., and Van Wagoner, J. C., eds., Sea-level Changes: an integrated approach: Society of Economic Paleontologists and Mineralogists Special Publication 42, p. 357-370.

Plint, A. G., 1991, High-frequency relative sea-level oscillations in Upper Cretaceous shelf clastics of the Alberta foreland basin: possible evidence for a glacio-eustatic control? in Macdonald, D. I. M., ed., Sedimentation, tectonics and eustasy: sea-level changes at active margins: International Association of Sedimentologists Special Publication 12, p. 409-428.

Plint, A. G., McCarthy, P. J., and Faccini, U. F., 2001, Nonmarine sequence stratigraphy: updip expression of sequence boundaries and systems tracts in a high-resolution framework: Cenomanian Dunvegan Formation, Alberta foreland basin, Canada: American Association of Petroleum Geologists Bulletin, v. 85, p. 1967-2001.

Pore¸bski, S.J., and Steel, R.J., 2006, Deltas and sea-level change: Journal of Sedimentary Research, v. 76, p. 390–403.

Porebski, S. J., and Steel, R. J., 2003, Shelf-margin deltas: their stratigraphic significance and relation to deepwater sands: Earth-Science Reviews, v. 62 p. 283–326.

Posamentier, H. W., 2002, Ancient shelf ridges—a potentially significant component of the transgressive systems tract: case study from offshore northwest Java: American Association of Petroleum Geologists Bulletin, v. 86, p. 75-106.

Posamentier, H. W., and Allen, G. P., 1999, Siliciclastic sequence stratigraphy—concepts and applications: Society for Sedimentary Geology (SEPM), Concepts in sedimentology and paleontology 7, 210 p.

Posamentier, H. W., and Vail, P. R., 1988, Eustatic controls on clastic deposition II—sequence and systems tract models, in Wilgus, C. K., Hastings, B. S., Kendall, C. G. St. C.,

Posamentier, H. W., Jervey, M. T., and Vail, P. R., 1988, Eustatic controls on clastic deposition I—Conceptual framework, in Wilgus, C. K., Hastings, B. S., Kendall, C. G. St. C., Posamentier, H. W., Ross, C. A., and Van Wagoner, J. C., eds., Sea level Changes - an integrated approach: Society of Economic Paleontologists and Mineralogists Special Publication 42, p. 109-124.

Posamentier, H. W., Allan, G. P., and James, D. P., 1992, High-resolution sequence stratigraphy – the East Coulee Delta, Alberta: Journal of Sedimentary Petrology, v. 62, p. 310-317.

Pratt, B. R., James, N. P., and Cowan, C. A., 1992, Peritidal carbonates, in Walker, R. G. and James, N. P., eds., Facies models: response to sea-level change: Geological Association of Canada, Geotext 1, p. 303–322.

Pratt, B. R., 2010, Peritidal carbonates, in James, N. P., and Dalrymple, R. W., eds., Facies Models 4: GEOtext 6, Geological Association of Canada, St. John's, Newfoundland, p. 401-420.

Reinson, G. E., 1992, Transgressive barrier island and estuarine systems, in Walker, R. G. and James, N. P., eds., Facies models: response to sea-level change: Geological Association of Canada, Geotext 1, p. 179-194.

Riba, O., 1976, Syntectonic unconformities of the Alto Cardener, Spanish Pyrenees, a genetic interpretation: Sedimentary Geology, v. 15, p. 213-233.

Rich, J. L., 1951, Three critical environments of deposition and criteria for recognition of rocks deposited in each of them: Geological Society of America Bulletin, v. 62, p. 1-20.

Sangree, J. B., and Widmier, J. M., 1977, Seismic stratigraphy and global changes of sea level, part 9: Seismic interpretation of clastic depositional facies, in Payton, C. E., ed., Seismic stratigraphy — applications to hydrocarbon exploration: American Association of Petroleum Geologists Memoir 26, p. 165-184.

Sarg, J. F., 1988, Carbonate sequence stratigraphy, in Wilgus, C. K., Hastings, B. S., Kendall, C. G. St. C., Posamentier, H. W., Ross, C. A., and Van Wagoner, J. C., eds., Sea level Changes - an integrated approach: Society of Economic Paleontologists and Mineralogists Special Publication 42, p. 155-181.

Schlager, W., 1989, Drowning unconformities on carbonate platforms, in Crevello, P. D., Wilson, J. L., Sarg, J. F., and Read, J. F., eds., Controls on carbonate platforms and basin development: Society of Economic Paleontologists and Mineralogists Special Publication 44, p. 15-25.

Schlager, W., 1992, Sedimentology and sequence stratigraphy of reefs and carbonate platforms: American Association of Petroleum Geologists Continuing Education Course Notes Series 34, 71 p.

Schlager, W., 1993, Accommodation and supply—a dual control on stratigraphic sequences: Sedimentary Geology, v. 86, p. 111–136.

Schlager, W., 2004, Fractal nature of stratigraphic sequences, Geology, v. 32, p. 185-188.

Schlager, W., 2005, Carbonate sedimentology and sequence stratigraphy: SEPM Concepts in Sedimentology and Paleontology #8, 200p.

Schumm, S. A., 1979, Geomorphic thresholds: the concept and its applications: Transactions of the Institute of British Geographers, v. 4, p. 485-515.

Schumm, S. A., 1993, River response to baselevel change: Implications for sequence stratigraphy: Journal of Geology, v. 101, p. 279–294.

Shanley, K. W., and McCabe, P. J., 1994, Perspectives on the sequence stratigraphy of continental strata: American Association of Petroleum Geologists Bulletin, v. 78, p. 544–568.

Southgate, P. N., Kennard, J. M., Jackson, M. J., O'Brien, P. E., and Sexton, M. J., 1993, Reciprocal lowstand clastic and highstand carbonate sedimentation, subsurface Devonian reef complex, Canning Basin, Western Australia, in Loucks, R. G., and Sarg, J. F., eds., Carbonate sequence stratigraphy: American Association of Petroleum Geologists Memoir 57, p. 157–179.

Strong, N., and Paola, C., 2008, Valleys that never were: time surfaces versus stratigraphic surfaces: Journal of Sedimentary Research, v. 78, p. 579–593.

Talbot, M. R., 1985, Major bounding surfaces in aeolian sandstones—a climatic model: Sedimentology, v. 32, p. 257-265.

Törnqvist, T. E., 1993, Holocene alternation of meandering and anastomosing fluvial systems in the Rhine-Meuse delta (central Netherlands) controlled by sea-level rise and subsoil erodibility: Journal of Sedimentary Petrology, v. 63, p. 683-693.

Törnqvist, T. E., van Ree, M. H. M., and Faessen, E. L. J. H., 1993, Longitudinal facies architectural changes of a Middle Holocene anastomosing distributary system (Rhine-Meuse delta, central Netherlands): Sedimentary Geology, v. 85, p. 203-219.

Vail, P. R., 1987, Seismic stratigraphy interpretation using sequence stratigraphy, Part 1: seismic stratigraphy interpretation procedure, in Bally, A. W., ed., Atlas of seismic stratigraphy: American Association of Petroleum Geologists Studies in Geology 27, v. 1, p. 1-10.

Vail, P. R., Mitchum, R. M., Jr., Todd, R. G., Widmier, J. M., Thompson, S., III, Sangree, J. B., Bubb, J. N., and Hatlelid, W. G., 1977, Seismic stratigraphy and global changes of sea-level, in Payton, C. E., ed., Seismic stratigraphy - applications to hydrocarbon exploration: American Association of Petroleum Geologists Memoir 26, p. 49-212.

Van Wagoner, J. C., Mitchum, R. M., Jr., Posamentier, H. W., and Vail, P. R., 1987, Seismic stratigraphy interpretation using sequence stratigraphy, Part 2: key definitions of sequence stratigraphy, in Bally, A. W., ed., Atlas of seismic stratigraphy: American Association of Petroleum Geologists Studies in Geology 27, v. 1, p. 11-14.

Van Wagoner, J. C., Mitchum, R. M., Campion, K. M. and Rahmanian, V. D. 1990, Siliciclastic sequence stratigraphy in well logs, cores, and outcrops: American Association of Petroleum Geologists Methods in Exploration Series 7, 55 p.

Vandenberghe, J., 1993, Changing fluvial processes under changing periglacial conditions: Z. Geomorph, N.F., v. 88, p. 17-28.

Vandenberghe, J., Kasse, C., Bohnke, S., and Kozarski, S., 1994, Climate-related river activity at the Weichselian-Holocene transition: a comparative study of the Warta and Maas rivers: Terra Nova, v. 6, p. 476-485.

Walker, R. G., 1992, Facies, facies models and modern stratigraphic concepts, in Walker, R. G. and James, N. P., eds., Facies models: response to sea-level change: Geological Association of Canada, p. 1-14.

Weber, M.E., Wiedicke, M.H., Kudrass, H.R., Huebscher, C., and Erlenkeuser, H., 1997, Active growth of the Bengal Fan during sea-level rise and highstand: Geology, v. 25, p. 315–318.

Wescott, W. A., 1993, Geomorphic thresholds and complex response of fluvial systems—some implications for sequence stratigraphy: American Association of Petroleum Geologists Bulletin, v. 77, p. 1208-1218.

White, N., and Lovell, B., 1997, Measuring the pulse of a plume with the sedimentary record: Nature, v. 387, p. 888-891.

Wilson, J. L., 1967, Cyclic and reciprocal sedimentation in Virgilian strata of southern New Mexico: Geological Society of America Bulletin, v. 78, p. 805-818.

Wood, L. J., Ethridge, F. G., and Schumm, S. A., 1993, The effects of rate of base-level fluctuations on coastal plain, shelf and slope depositional systems: an experimental approach, in Posamentier, H. W., Summerhayes, C. P., Haq, B. U., and Allen, G. P., eds., Sequence stratigraphy and facies associations: International Association of Sedimentologists Special Publication 18, p. 43-53.

Wright, V. P., and Marriott, S. B., 1993, The sequence stratigraphy of fluvial depositional systems: the role of floodplain sediment storage: Sedimentary Geology, v. 86, p. 203-210.

Yoshida, S., Steel, R. S, and Dalrymple, R. W., 2007, Changes in depositional processes – an ingredient in a new generation of sequence stratigraphic models: Journal of Sedimentary Research, v. 77, p. 447-460.

## Contents

## 6.1   Introduction

In the introduction to this chapter in the 2000 edition of *"Principles of sedimentary basin analysis"* I said (p. 249):

> The art of basin mapping is one of reconstructing paleogeography and fill geometry from very limited evidence. Being skilled at data synthesis, interpretation, and extrapolation is of prime importance. Maps drawn only from outcrop data are rarely sufficiently accurate or useful for subsurface predictions, because they do not contain an adequate, three-dimensional distribution of data points.

This statement now qualifies as a historically interesting comment, because the maturation of sequence stratigraphy and developments in reflection-seismic methods have caused a complete change in the practice of basin mapping. In 2000, seismic stratigraphy had assumed considerable importance, but a reliance on scattered exploration wells was still essential. Sequence methods have advanced substantially since then, and now comprise a comprehensive set of predictive tools for the tracking of stratigraphic units in the surface and subsurface. The basic methods, including an introduction to the topic of seismic faces, are described in Chap. 5. Modern, sophisticated, digital methods for the processing and display of seismic data have made the work of the subsurface geologist immeasurably easier. These new methods include an increasingly common use of three-dimensional seismic data volumes, from which cross-sections in any orientation, and stratigraphically horizontal seismic maps can readily be constructed, enabling the three-dimensional visualization of complex stratigraphic objects. Such objects reflect the depositional, structural and diagenetic processes by which stratigraphy is constructed, and this new ability for three-dimensional visualization has led to the growth of an entirely new tool, that of **seismic geomorphology**—the study of ancient landscapes and depositional systems based on 3-D seismic data. Petrophysical and core data are essential in providing the "ground truthing" of seismic interpretations, such as confirming complex correlations and providing necessary lithofacies and biostratigraphic detail, but in many cases they no longer

© Springer International Publishing Switzerland 2016
A.D. Miall, *Stratigraphy: A Modern Synthesis*, DOI 10.1007/978-3-319-24304-7_6

provide the primary data set from which the subsurface geophysicists and geologists do their work.

Seismic data are used to generate and extend a stratigraphic synthesis across a broad project area. Older methods, such as facies and isopach mapping, and paleocurrent analysis (mainly useful for surface work) do not provide this level of detail, and are useful mainly in order to explore regional basinal trends, including the direction of paleoslope, differential subsidence patterns, and regional facies trends. These methods are included in this chapter for completeness, because not all mapping projects have access to seismic data.

## 6.2   Stratigraphic Mapping with Petrophysical Logs

### 6.2.1   Log Shape and Electrofacies

Well-log data are used for stratigraphic correlation and for lithologic interpretation. They may also be used directly for facies mapping purposes. As noted briefly in Sect. 2.4, various combinations of logs can be employed in crossplots to yield lithologic information. These relationships can be converted into computer algorithms, and where log data are digitized and stored in data banks, a powerful automated mapping technique becomes available. Digital log data can also be displayed and manipulated using interactive computer graphics routines, a facility that permits easy comparisons for stratigraphic correlation purposes, plus checking for accuracy, and for the normalizing of logs made at different times using different hole conditions, and many other purposes. Well-log service companies have devoted considerable energy to devising and marketing automated processing and display techniques for use in basin analysis and petroleum development, but all such techniques suffer from the limitation that petrophysical responses are very location-specific. The techniques cannot be used without much initial careful calibration to local petrographic and groundwater conditions.

An early example of the use of petrophysical log data for mapping purposes was given by McCrossan (1961). He constructed a resistivity map for the lower and middle Ireton Formation of central Alberta that can be used to predict the proximity to reefs (Fig. 6.1). The Ireton is an Upper Devonian basinal shale that drapes against the reefs. Calcareous content and thus resistivity increase close to the reefs because of the presence of reef talus. The resistivity data therefore provide a useful analogue paleogeographic map.

The recognition of characteristic log "shapes" and distinctive vertical-profile character in siliciclastic deposits (**electrofacies**) may be a useful tool for correlation purposes (Figs. 6.2 and 6.3). Regardless of their lithology, facies or environment, lithologic units typically have a recognizable

**Fig. 6.1** Resistivity map of the Ireton Formation (Devonian), central Alberta. High values indicate an increase in carbonate content adjacent to reef bodies (McCrossan 1961)

profile on petrophysical logs, which enables them to be traced and correlated from well to well. The ease and reliability of this mapping technique depends on how rapidly lithologic units change in thickness and facies laterally. Where there is considerable lateral variability, as in channelized deposits such as fluvial systems and submarine fans, it may be difficult or impossible to correlate reliably between drill holes. Much depends on hole spacing. Figure 6.2 is an example of a lithologic succession that shows little change from hole to hole, over distances of several kilometres, which permits every major unit to be defined and correlated on the stratigraphic cross section. By contrast, in Fig. 6.3, with greater well spacing, several units thin, change lithology or thin out altogether across the section, which makes correlation more difficult. The use of log shape analysis to interpret vertical facies successions and characterize certain sedimentary environments is described in Sect. 3.5.10.

The application of sequence-stratigraphic techniques, including the identification of systems tracts, may

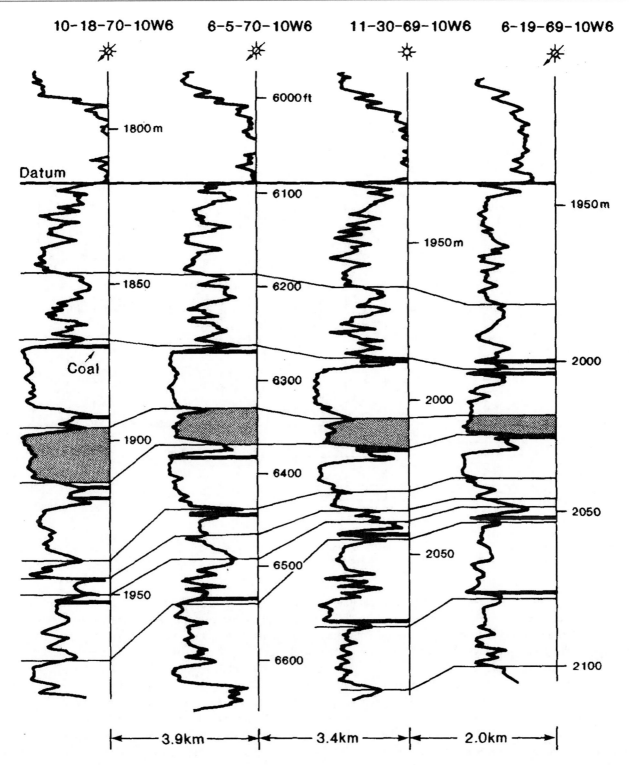

**Fig. 6.2** A suite of gamma-ray logs, showing how log-shape may be used to carry out subsurface lithostratigraphic correlation (Cant 1992)

considerably improve the practice of log correlation because of their predictive value. The various elements of many standard sequence models commonly have distinctive lithologies and geometries, and typically occur in the same vertical succession, which provides useful clues for interpretation and correlation (Chap. 5). For example, incised valleys that commonly cut down into highstand deposits may consist of distinctive coarse channel-fill deposits incised into layered, fine-grained shelf deposits; sequence boundaries in clastic successions are commonly

**Fig. 6.3** Use of gamma ray (*GR*), spontaneous potential (*SP*), and induction electric survey (*IES*) logs to correlate subsurface sections. This diagram shows that where lithostratigraphic units retain similar thickness and lithology they can readily be correlated using geophysical logs. Cretaceous, Alberta (Myhr and Meijer-Drees 1976)

marked by widespread coarse-grained deposits; maximum flooding surfaces may be recognized by widespread "hot" (radioactive) shales (usually detected by the use of high readings on the gamma-ray log) or by condensed deposits, and so on.

Log shape analysis may be a useful aid for interpreting depositional environment (Figs. 3.61–3.63). The analysis depends on the distinctiveness of the vertical profile through a sandstone body, and our ability to interpret this profile in terms of depositional processes and environment. The well-known bell-shaped gamma ray or spontaneous-potential log response yielded by a typical fluvial fining-upward cycle, is a classic example (Visher 1965). The study of an interpreted subsurface point-bar sandstone by Berg (1968) is an early example of the application of these ideas to a practical exploration problem.

Despite the dramatic increase in the volume and quality of reflection-seismic data, the correlation of petrophysical logs is still the best means to construct a sequence-stratigraphy for any stratigraphic volume that has an adequate network of exploration holes. For example, the huge data-base of well-logs stored by the Alberta government in Calgary has enabled A.G. Plint and his students and colleagues to extend parts of the high-resolution sequence framework of the Cretaceous succession over the entire province of Alberta and to extend it southward into Montana, where it can be tied into American work on the same part of the stratigraphic record. Shank and Plint (2013) constructed a network of petrophysical log-correlation cross-sections of the Cardium Allo-formation (Turonian-Coniacian) across the entire province, and Walaszczyk et al. (2014) tied this framework into the biostratigraphic and sequence framework of Montana, from whence it can be related to the cyclostratigraphic research underway across the Rocky Mountains states, using the so-called HiRES techniques first developed by Kauffman (1986, 1988). As discussed in Sect. 8.11, Locklair and

Sageman (2008) and Sageman et al. (2014) are developing a tightly constrained astrochronological time scale for equivalent rocks in the latter area.

### 6.2.2 Examples of Stratigraphic Reconstructions

Three modern examples of the use of petrophysical log correlation are discussed in this section. In the first, core and log data are used to provide a facies classification of the shallow-marine Bakken Formation in Saskatchewan where log correlation is made considerably easier by the recognition of key sequence surfaces and systems tracts. In the second example, a new model for the Marcellus Shale of New York shows how the correct choice of datum and the application of a new depositional model not only clarifies the stratigraphy but provides useful information for the understanding of the distribution of organic-rich shales, the source of the shale gas in this unit. In the last example, log shape

analysis is used to map the complex pattern of facies that constitute a Tertiary fluvial system in Texas.

The Bakken Formation is a marine siliciclastic unit of Upper Devonian and Lower Carboniferous age that extends across the Williston Basin (mainly North and South Dakota, Saskatchewan, southeastern Alberta and Montana,). In recent years, fracking operations in this unit in North Dakota have turned that state into one of the leading US producers of oil, gas and gas liquids. In this section we refer to a core and petrophysical log study of the unit carried out across southern Saskatchewan (Angulo and Buatois 2012). The formation has been subdivided into three members: lower and upper organic-rich black shale members and a middle member of dolomitic siltstone and sandstone, which constitutes the main hydrocarbon reservoir. Further subdivisions of the middle member are shown in Fig. 6.4. The three members may be traced on petrophysical logs throughout the Williston Basin, with the upper and lower members characterized by unusually high gamma-ray readings, reflecting

**Fig. 6.4** Core and log characteristics and stratigraphic subdivision of the Bakken Formation of southern Saskatchewan (Angulo and Buatois 2012). AAPG © 2012, reprinted by permission of the AAPG whose permission is required for further use

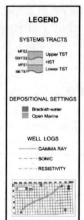

**Fig. 6.5** Stratigraphic cross-section through the Bakken Formation, Saskatchewan (Angulo and Buatois 2012). AAPG © 2012, reprinted by permission of the AAPG whose permission is required for further use

the concentration of radioactive elements in the organic shales. Eleven lithofacies types have been recognized by Angulo and Buatois (2012), falling into two broad facies associations, open-marine, and brackish-water marginal-marine. As shown in Fig. 6.4, a sequence analysis demonstrates that the lower member and the lower part of the middle member comprises a regressive (transgressive to highstand) systems tract, with a sequence boundary occurring at the base of a brackish-marine shoreface sandstone in the middle of the middle member, and the upper part of the Bakken Formation interpreted as a transgressive systems tract. Figure 6.5 shows how core and petrophysical-log data may be used to correlate these units across Saskatchewan. The distinctive deflections and shapes of the gamma ray, sonic and resistivity logs through each of these members are apparent from this diagram.

The second example described here is a re-analysis of the basin setting and configuration of the stratigraphic section that contains the important gas bearing unit, the Marcellus Shale (Smith and Leone 2014). Well logs were first calibrated against calculated values for total organic carbon and calcium carbonate obtained from core analyses (Fig. 6.6). Earlier basin models had suggested that the Marcellus Shale was a distal component of the Catskill clastic wedge that downlapped westward into an anoxic basin up to 250 m deep constituting the Appalachian foreland basin. However, the stratigraphic setting of nearby outcrops of similar organic-carbon-rich rocks suggests that they were deposited in shallow water over the basin forebulge in water depths probably less than 50 m. Instead of basing stratigraphic interpretations on the concept of a basal downlap surface, Smith and Leone (2014) used markers higher in the section as datum and constructed a different stratigraphic model (Fig. 6.7). In this model organic shales thin but become more organic rich westward as they interbed with progressively

thicker carbonate units. This configuration led to a different depositional model. Instead of a deep anoxic basin in which the organic shales were deposited, the authors suggested the occurrence of seasonal anoxia in a shallow-water basin, with recycling of nutrients delivered from the Catskill "delta" complex. The richest organic shales are those further to the west, the most distal from the clay influx from the delta. They onlap multiple unconformities westward onto a tectonically positive forebulge. In an eastward direction they are interbedded with turbidites derived from the Appalachian orogen. For other examples of petrophysical (and seismic) mapping of fine-grained rocks, the reader is referred to Bohacs et al. (2002).

The third example discussed briefly here demonstrates how petrophysical log correlation and the analysis of log shape was used to document the stratigraphy of a coastal clastic system and identify and map the array of lithofacies constituting the deltaic-barrier-lagoon system within which these deposits were formed (Zeng et al. 1996). Figure 6.8 is a dip-oriented cross-section through two progradational shelf-shoreface-coastal successions separated by a surface of transgression. Fourteen sandstone units have been identified and correlated across the study area. These show considerable lateral persistence, but display significant differences in log shape across the area. Facies analysis of each sandy depositional unit was derived from SP log character (and sometimes from resistivity log shape) and sandstone geometry. Some sidewall cores provided information on grain size and mineral identification. No conventional cores were available for this study. SP log patterns for 11 sandy units in the middle of the modeled section distinguished three major log facies: blocky, upward fining, and mixed (Fig. 6.9). These patterns were used to group sandy units developed in similar depositional environments. The distribution of these patterns in within three of the sandstone units is shown in the

**Fig. 6.6** Calibration of
petrophysical logs against
compositional data derived from
core analyses, Marcellus Shale,
New York (Smith and Leone
2014). AAPG © 2014, reprinted
by permission of the AAPG
whose permission is required for
further use

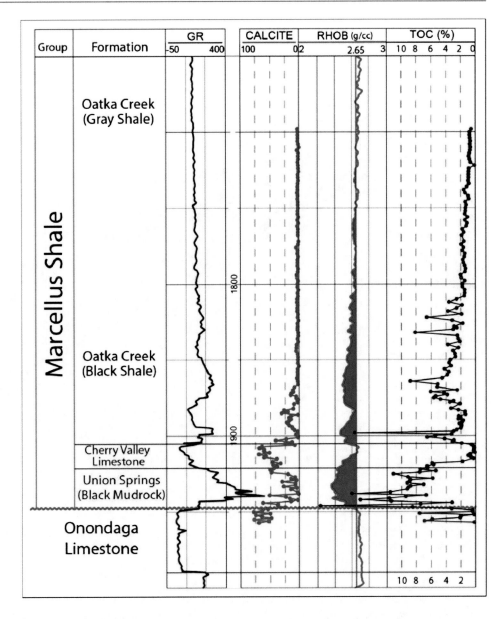

maps in Fig. 6.9, and sandstone thickness, measured from SP log response, is shown in Fig. 6.10. The thicknesses and relative arrangements of the various facies, with supplementary information on lithology from sidewall cores, were then used to make paleoenvironmental interpretations for each of the sandy units identified in the cross-section. Three of the resulting maps are shown in Fig. 6.11.

### 6.2.3   Problems and Solutions

There are several difficulties inherent in the use of petrophysical logs for stratigraphic work, which stem from the fact that drill holes necessarily represent a very limited sample of the rocks under study (the same problems arise with the use of outcrop sections for correlation except in areas of exceptional exposure, where correlations may be "walked out" on the ground). Even with the very close well spacing that characterize enhanced recovery projects (a few hundred metres) it is commonly the case that the lateral dimensions of lithofacies units may be less than the distance between wells, which means that correlations may be unreliable, and attempts to characterize and model lithofacies architecture for production purposes subject to wide margins of error (Martin 1993). An example is illustrated in Fig. 6.12, which shows three different approaches to the correlation of a suite of fluvial sand bodies in the subsurface where well spacing averages 1.54 km. This example is discussed by Miall (2014a, Sect. 2.3.2), as part of a general discussion of the difficulty of carrying out facies modeling in the subsurface. In other environments where lateral variability is not so marked, stratigraphic correlations may be

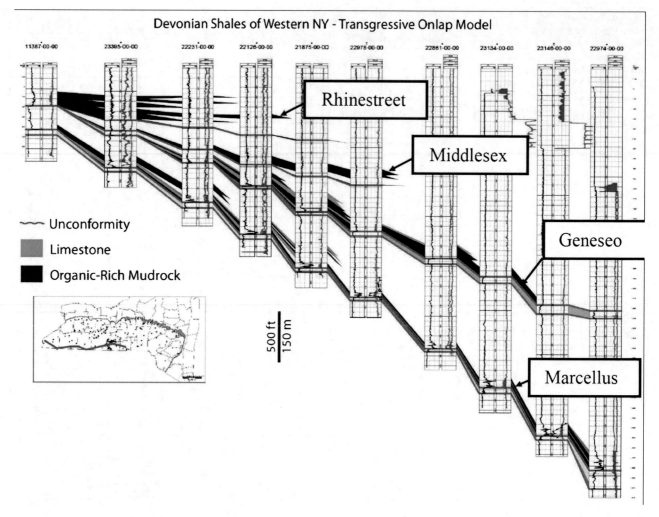

**Fig. 6.7** A cross-section through the Marcellus and other distal parts of the Catskill complex in western New York State showing the westward change in facies and thicknesses toward the onlap of the basin forebulge (Smith and Leone 2014). AAPG © 2014, reprinted by permission of the AAPG whose permission is required for further use

more reliable. The example of a complex facies mosaic illustrated in Figs. 6.8, 6.9, 6.10 and 6.11 shows where a degree of certainty may be assumed from the reconstructions showing that the extent of most of the facies units is considerably greater than well spacings, which range up to 5 km. In the case of some open-marine facies, such as transgressive sand bodies or turbidites, correlations of individual units over even greater distances may be possible.

Log correlation is usually guided by the identification of key surfaces and units that display characteristic log responses, and may be aided by biostratigraphic, chemostratigraphic, or other tools, some of which are discussed later in this chapter. The development of sequence stratigraphy has made an enormous difference to the work of subsurface stratigraphic correlation. Sequence models have a predictive power that can help to clarify what might otherwise be very confusing vertical and lateral lithofacies relationships. Key surfaces, such as subaerial erosion surfaces,

maximum flooding surfaces, or transgressive sands, may be traceable for many kilometres and may serve to vertically align well sections in arrangement that reveal the larger stratigraphic architecture. Much may depend on the choice of the datum on which to hang correlations.

In stratigraphic work the choice of datum may be key to the successful resolution of stratigraphic complexity. This is a surface selected because of its assumed original horizontality. Lining up sections by aligning a key surface at a horizontal datum removes structural disturbance from a section, and may move facies volumes and their interpreted depositional environments into relationships that make their interpretation much simpler. Bhattacharya (2011) provided an extensive discussion on this point, and provided many examples of how confusing facies and stratigraphic reconstructions can be clarified by the appropriate choice of datum and the application of the appropriate sequence model. Several examples follow. He noted that in marine deposits,

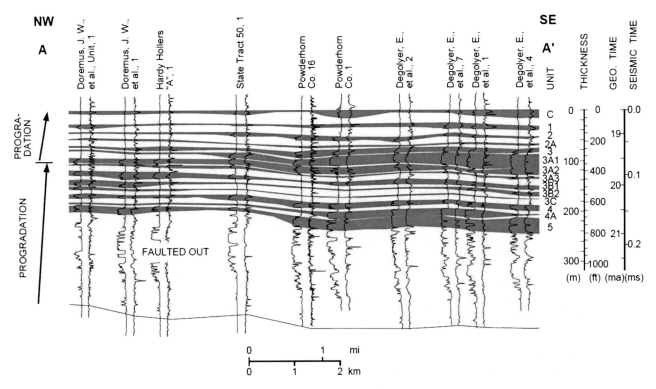

**Fig. 6.8** Dip-oriented stratigraphic cross-section through a Miocene coastal deltaic system in the Powderhorn field, Texas, showing the definition of a series of sand-dominated lithostratigraphic units, grouped into two progradational successions (Zeng et al. 1996, Fig. 4, p. 21). AAPG © 1996, reprinted by permission of the AAPG whose permission is required for further use

stratigraphic packages exhibiting significant depositional dip are extremely common. These are termed **clinoforms**, a term coined by Rich (1951). They range in scale from the dipping units at the front of deltas, typically as little as a few metres in amplitude where the delta progrades into shallow water, to the shelf clinoforms developed where a continental margin is building into an ocean, with vertical amplitudes up to several kilometres. Reconstructing clinoforms from outcrop or petrophysical data is extremely difficult because the depositional dip may be less than 1°. In fact, it was not until the advent of seismic stratigraphy that earth scientists fully appreciated the ubiquitous nature of the clinoform architecture in the stratigraphic record. Figure 6.13 illustrates the importance of correlating correctly. The reconstruction shown in Fig. 6.13a is based on traditional lithostratigraphic principles, using a prominent facies change as a marker and datum. However, Fig. 6.13b is the correct interpretation, in which units of gradually changing lithofacies with a significant depositional dip prograded from left to right.

What may be the first reconstruction of clinoform architecture to be based on petrophysical data is illustrated in Fig. 6.14. Earlier mapping of the reefal Woodbend Group (Upper Devonian) in central Alberta by McCrossan (1961) had suggested the presence of depositional dips in the Ireton Formation, a mixed shale-limestone unit which was recognized as constituting the talus deposits flanking prominent

reef carbonate units. However, it was the later work of Oliver and Cowper (1963, 1965) that specifically identified the three depositional environments of Rich (1951)—the undaform, clinoform and fondoform. Their reconstruction (Fig. 6.14) is based on the correlation of picks in resistivity logs. Thin, resistive beds are carbonate talus units deposited on the flanks of the reef. In this study, the top of the Ireton Formation was selected as the datum, but Bhattacharya (2011) argued that in some cases selecting an underlying surface as the datum is necessary for the correct reconstruction to be recognized. He made this point by presenting a redrawing of the seismic reflections within a modern delta (Fig. 6.15). The clinoform architecture of the delta is obvious from the original cross-section, in which every surface dips seaward, showing offlapping clinoform wedges and a dipping modern depositional surface (Fig. 6.15a). But if one were to imagine that each reflection was a pick on a petrophysical log and that the picks were correlated based on a flattening of the top surface, the architecture becomes severely distorted, with some beds actually dipping landward (Fig. 6.15b).

The evolving interpretation of the Wapiabi Formation in southern Alberta demonstrates the difficulty in making the appropriate reconstruction when several different types of criteria could be used to guide correlations. This example (Fig. 6.16) actually makes use of outcrop sections in an area where walking out correlations was not possible, but the

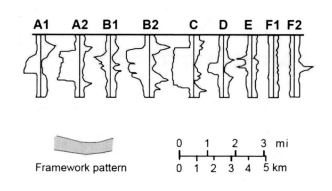

**Fig. 6.9** SP log patterns of three of the sandy depositional units identified in Fig. 6.4. Log shapes at each location are identified at lower right: *A1* fining upward, *A2* fining upward with shoulder, *B1* coarsening upward, *B2* coarsening upward and blocky, *C* blocky, *D* fingerlike, *E* serrate, *F1* straight with low resistivity, *F2* straight with high resistivity (Zeng et al. 1996, part of Fig. 5, p. 23). AAPG © 1996, reprinted by permission of the AAPG whose permission is required for further use

problems are identical to those encountered in the subsurface. In the first reconstruction (Fig. 6.16a) the top of the Chungo Member is used as datum. The Chungo Member constitutes a coarsening-upward and shallowing-upward succession in which diachronous and interfingering facies changes are indicated, but a clinoform architecture is not suggested or implied. In the second attempt (Fig. 6.16b) the top of the shoreface sandstones within the Chungo Member is used as datum, but this reconstruction provides no easily interpreted facies architectures. The third configuration (Fig. 6.16c) returns to the use of the top of the Chungo Member as the datum, and this time the clinoform configurations are specifically addressed in the suggested unit correlations. Given that the progradation of the Chungo shoreface must represent significant elapsed time, this implies a substantial time break above the fluvial Chungo preceding the Nomad transgression. Bhattacharya (2011, p. 130) reported that ancillary chronostratigraphic data supports this third reconstruction.

The reader is referred to Sect. 8.9 and to Bhattacharya's (2011) paper for further discussion of these issues, including the problems inherent in the mapping, correlation and chronostratigraphic interpretation of coastal fluvial and shallow-marine systems and incised valleys, and the use of supplementary correlation criteria, including coal seams and (particularly useful) bentonite beds. In Sect. 8.10 we discuss the use of Wheeler diagrams to illustrate correlations in terms of elapsed time—a technique that, in throwing light on chronostratigraphic relations, may raise important questions about sedimentation rates, timing, and the correlation of lithofacies units through areas of stratigraphic uncertainty.

## 6.3  Seismic Stratigraphy

Seismic stratigraphy is the study of stratigraphic units defined on the basis of their seismic characteristics (Cross and Lessenger 1988; Hart 2013).

An introduction to the use of seismic facies to map depositional systems and systems tracts is provided in Sect. 5.4. Problems of sequence definition are discussed in Sect. 7.7. In this section the practical methods of sequence

**Fig. 6.10**  Total sandstone thickness of sandy depositional units (Zeng et al. 1996, part of Fig. 6, p. 26). AAPG © 1996, reprinted by permission of the AAPG whose permission is required for further use

correlation and mapping are described and illustrated. This section does not deal with data processing, which is a very specialized field beyond the scope of the stratigraphic practitioner (see Veeken 2007, for an extensive introduction). The examples selected for discussion here make use of seismic data that has already received the necessary processing to highlight the relevant stratigraphic detail. If necessary, reference may be made back to the original articles from which the examples used here have been derived.

### 6.3.1  The Nature of the Seismic Record

A seismic reflection is generated at the interface between two materials with different acoustic properties. The latter is described as **acoustic impedance** and is given by the product of density and velocity. The larger the difference in impedance between two lithologies, the stronger the reflection. Where a soft lithology lies on a harder one, the reflection polarity is positive and is normally represented by a waveform deflection to the right, colored to render it more visible. Where the reverse situation applies, the polarity of the deflection is to the left (left blank). This is the origin of the familiar variable-area **wiggle traces**. Colors are now

commonly used to exaggerate amplitude and other differences, and these may emphasize major facies variations.

A seismic wave trace recorded at a geophone represents the modification of a source pulse by reflection from various layers below the surface. The nature of each wavelet depends on the nature of the impedance contrasts at each bed contact. An example of a synthetic seismogram illustrating the production of a seismic signal is shown in Fig. 6.17. Such seismograms are time sections; that is, their long axis is proportional to time, not depth. Complex calculations must be performed to convert time to depth. This is because seismic velocities generally increase downward as rock density increases and porosity diminishes because of mechanical compaction, pressure solution and cemetation. An empirical relationship, known as the **Faust equation**, has been demonstrated between seismic velocity and the depth and age of rock units (Sharma 1986):

$$V = 46.5(ZT)^{1/6}\,\mathrm{m/s},$$

where $V$ is seismic velocity, $Z$ is depth of burial, and $T$ is geological age in years.

It is important to remember that the seismic technique cannot resolve small-scale stratigraphic anomalies or very

**Fig. 6.11** Environmental interpretation of sandy depositional units. Symbols for various panel are as follows. Barrier bar/lagoon system (**e, f**): *BC* barrier core, *IF* inlet fill, *IN* inlet, *FTD* flood-tidal delta, *ETD* ebb-tidal delta, *LG* lagoon. Coastal stream plain (**a–c, g, i, j**): *FC* fluvial channel, *LV* levee, *CS* crevasse splay, *AC* abandoned channel, *FP* flood plain, *LK* lake, *BD* bayhead delta. Wave-dominated delta (**d, h, k**): *DC* distributary channel, *LV* levee, *CS* crevasse splay, *AC* abandoned channel, *DP* delta plain, *LK* lake, *SPL* strand plain, *SW* swale, *BR* beach ridge (Zeng et al. 1996, part of Fig. 7, p. 27). AAPG © 1996, reprinted by permission of the AAPG whose permission is required for further use

thin beds, and that the resolution decreases with depth, as seismic velocity increases. A general rule is that two reflecting horizons must be a minimum of about 1/4 of a wavelength apart if they are to be resolvable in a typical seismic section (Sheriff 1985). Thus, at shallow depths in loosely consolidated sandstones and mudstones, a typical seismic velocity would be on the order of 1800 m/s, and the typical frequency would be around 60 Hz. One quarter wavelength is therefore 7.5 m. Resolution is much poorer in older and deeper formations. Consider, for example, a Paleozoic carbonate section, with a seismic velocity of 4500 m/s and a frequency of 15 Hz. The thinnest unit resolvable in this situation would be 75 m. This lower limit is much larger (thicker) than many of the important facies details that can be seen in outcrops and high-resolution petrophysical logs, as shown dramatically in Fig. 7.1. High-resolution techniques are available for studying sediments at shallow depths and are becoming widely used in detailed studies of stratigraphic architecture in modern environments (Bouma et al. 1987);

but they are not yet available for the study of deep-basin structure and stratigraphy.

Published seismic sections show a series of vertical wave traces, each representing the recording made at a shot point. These are usually located about 25 to 50 m apart. Each trace is obtained by combining (**stacking**) the records of many adjacent geophones. The vertical scale of the section is **two-way travel time**, not depth. For each time horizon, depth varies according to velocity, as noted previously, so that, unless corrected, the structure revealed on a section may be spurious. For example, a mass of high-velocity rock, such as a dense carbonate reef or channel fill, may overlie what looks like a gentle anticline. This is a case of **velocity pull-up**, where the more rapid acoustic transmission through the dense material returns reflections from the underlying layers sooner than occurs on either side. Computer processing can produce a depth-corrected section if accurate velocity information is available, typically by making use of sonic petrophysical logs. This is discussed briefly below.

(a) Travis Peak Fm., Zone 1, interpretation by Tye (1991)

50 m                                                  5 km

Wells are shown regularly spaced. Actual spacing ranges from 0.8 to 2.2 km, averaging 1.54 km

(b) Travis Peak Fm., Zone 1, interpretation by Bridge and Tye (2000)

50 m                                                  5 km

7 m deep x 436 m wide

7 m deep x 1741 m wide

(c) Travis Peak Fm., Zone 1, model based on high sand and minimal relief on channel bounding surfaces

50 m                                                  5 km

**Fig. 6.12** Three interpretations of the braided-fluvial deposits of the Travis Peak Formation, Zone 1 (Early Cretaceous, East Texas). **a** Initial interpretation, by Tye (1991), based on detailed core and isopach mapping study. Arbitrary equal well-spacing is used in this and the subsequent diagrams; **b** A reinterpretation by Bridge and Tye (2000), based on assumptions of narrower channel belts. Rectangular boxes at the base of this panel indicate range of channel-belt sizes predicted from estimated bankfull depth, using the equations of Bridge and Mackey (1993). Their own model, shown here, does not make use of this range of values; **c** An alternative model developed by the present writer (Miall 2006), based on two basic guidelines for interpreting petrophysical logs: **a** channels normally have flat bases, and **b** the main sand bodies are indicated only by blocky-shaped, low-value gamma ray signatures (from Miall 2014a, Fig. 2.16)

It is important to remember that it is contacts that produce reflections, not rock bodies as such. For example, tight, cemented sandstone and dense, nonporous carbonate have higher impedances than most mudrocks, so that a contact of one of these with an overlying mudstone will yield a strong positive reflection. Conversely, a gas-filled sandstone has a very low impedance. Where it is overlain by a mudstone there will be a strong negative deflection. This appears as a

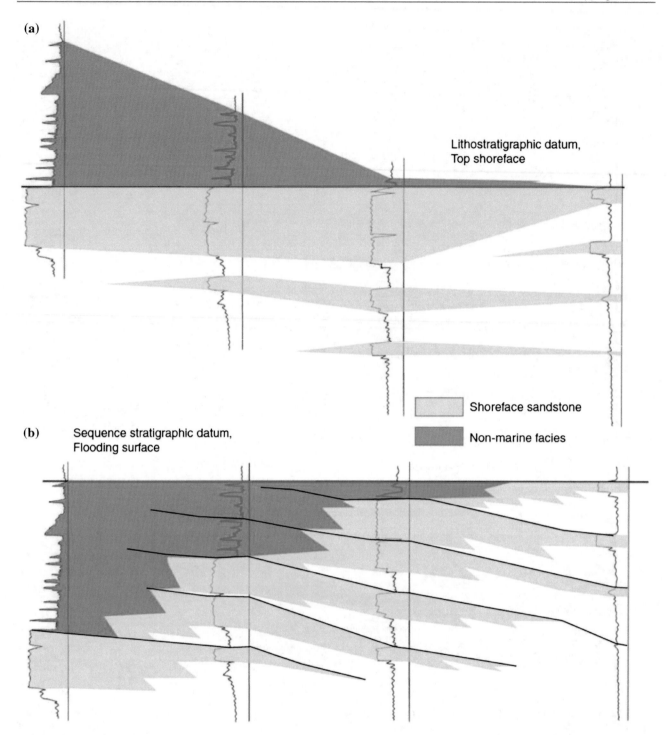

**Fig. 6.13** An example of how the choice of datum for the correlation of petrophysical logs can obscure or clarify depositional relationships. **a** In this reconstruction the assumption is made that the top of the marine sandstone constitutes a time line. **b** The presence of significant lateral facies change and the presence of a depositional dip characterizes this reconstruction. The succession constitutes a clinoform set within which facies change laterally from nonmarine to shoreface to offshore (Bhattacharya 2011, Fig. 5, p. 127)

blank area in the seismic record, and is the basis of the **bright-spot technique** for locating gas reservoirs. However, there are many conditions where interbedded units may have contrasting lithology but very similar impedances, so that they yield little or no seismic reflections. Examples would include a contact between a tight sandstone and a porous carbonate and a very porous oil- or water-saturated sandstone and a mudstone.

A very important assumption, first emphasized by Vail et al. (1977), and reemphasized by Cross and Lessenger

**Fig. 6.14** This is one of the first reconstructions of clinoform stratigraphy based on petrophysical log correlation (Oliver and Cowper 1965, Fig. 4, p. 1414). AAPG © 1965, reprinted by permission of the AAPG whose permission is required for further use

**Fig. 6.15 a** Seismic reflection through a modern shelf-edge delta (Rhone delta, Mediterranean Sea), showing a clear clinoform architecture. **b** Flattening the top surface of the delta, a common practice in the subsurface correlation of deposits if this type, results in a severe distortion of the internal geometry of the beds (Bhattacharya 2011, Fig. 10, p. 134)

**Fig. 6.16** Three different stratigraphic reconstructions of the Upper Cretaceous Wapiabi Formation in Alberta. These outcrop sections are tens of kilometres apart. **a** From Rosenthal et al. (1984; **b** From Rosenthal and Walker (1987); **c** From Leckie et al. (1994). See text for discussion and explanation

**Fig. 6.17** Synthetic reflection seismogram generated from a simple geological section. At each interface the input wavelet is modified by the reflection effects (Sharma 1986)

(1988), and Hart (2013) is that in most cases seismic reflections from conformable bedding surfaces are chronostratigraphic (isochronous) in nature. Facies contacts are diachronous, but they are too gradual to yield reflections. Facies changes may be deduced from subtle lateral changes in reflection characteristics within a single bed. The assumption that continuous seismic reflections represent "time lines" is of profound importance in the study of basin structure and stratigraphy and the analysis of stratigraphic sequences (Chap. 5).

Imagine an angular unconformity with lateral facies changes in the overlying unit and a gently dipping, truncated sequence of different units below. The seismic reflection from the unconformity could vary from strongly positive to strongly negative, with some areas completely invisible. This would be a difficult reflection to follow accurately on seismic records.

Seismic records are also complicated by **multiples** and **diffractions**. Multiples are reflections from a shallow layer that are reflected downward again from a near surface level such as the base of the weathering zone. This internal reflection is then redirected toward the surface from the original point and appears in the seismic record as a parallel layer several tenths of a second lower down in the section. Multiples may be strong enough to obscure deep reflections, but they are now relatively easy to remove by processing. Diffractions are reflections from steeply dipping surfaces, such as faults, channel margins, and erosional relief on an

unconformity. Their appearance in the record often occurs because seismic waves are not linear, laser-like energy beams but spherical wave fronts. The pulse from each shot point therefore sees reflection events over a circular region (the **Fresnel zone**), the diameter of which can be shown by calculation to range from about 100 to 1200 m, depending on depth and wavelength. A reflection on the periphery of the wave front will be seen later than one at the same depth directly below the shot point and, if at a high angle from the horizontal, may produce complex internal reflections. The resulting diffraction patterns may be very pronounced. They are a useful indicator of a steep reflecting event but may confuse other reflections and can be removed by a processing routine called **migration**, which repositions the reflections in their correct spatial position. The circular reflection area also accounts for the fact that diffractions can be produced by reflectors off the line of section, for example, if the seismic line skims past a reef body or a buried fault line scarp.

The seismic record can be considerably improved, and interpretation facilitated by the use of two special techniques, the construction of **synthetic seismograms**, and the use of **vertical seismic profiling** (VSP) (Hardage 1985). Synthetic seismograms are generated by the conversion of sonic and density data from petrophysical logs into a series of reflection coefficients (see Sect. 2.4 for a discussion of petrophysical logs). Vertical seismic profiling is the recording and analysis of seismic signals received from a geophone lowered downhole. Signals generated at the surface are recorded as they descend and as they return to the surface. The geophone is moved in increments, the length of each being set at less than half the shortest wavelength of the signals being analyzed. A new recording is made at each position. A comparison between synthetic, VSP, and field recorded data is shown in Fig. 6.18. Both synthetic and VSP data are extremely valuable for the purposes of calibrating the seismic record. For example, they enable a detailed record of seismic velocities to be generated that can be used to produce depth-corrected cross sections. However, the VSP technique is far superior for various reasons. It can, for example, be used to extend the calculations below the total depth (TD) of the well, which synthetic seismograms cannot (Fig. 6.18). The synthetic method relies on well log data, which only reflect conditions immediately adjacent to the hole. In the case of poor hole conditions or seismic reflectors having a very small horizontal extent (less than one Fresnel zone), the synthetic method may not provide a record typical of the area. For example, in Fig. 6.18 there is a discontinuous calcareous sandstone (denoted by the letters SD) at a depth of about 3070 ft. It is visible as a strong reflector at about

2.15 s in the synthetic seismogram, but not on the VSP trace, indicating that the bed does not extend very far from the borehole. Other possible occurrences of the bed are shown by the letters SD on the field record. VSP data are now routinely used to improve subsurface imaging around critical boreholes, to carry out precise time-depth conversions, and to remove multiples.

One of the most important features of **seismic stratigraphy**, as first described by Vail et al. (1977), is the concept **of seismic facies,** defined to mean an areally restricted group of seismic reflections whose appearance and characteristics are distinguishable from those of adjacent groups (Sangree and Widmier 1977). This is introduced in Sect. 5.2.2, and discussed in some detail below (Sect. 6.3.3).

A revolutionary advance in seismic methods is the use of **three-dimensional seismic data.** According to Davies et al. (2004b) the first commercial 3-D survey was run in the North Sea in 1975, but it was not until the mid-1980 s that the power of the new tool became more widely known (Brown 1985, 1986). Developments in computer processing and visualization were necessary before this could happen.

**Fig. 6.18** Comparison of VSP data with a synthetic seismogram and that produced by normal surface field recording. The surface section is split in two at the location of a well down which the synthetic record and the VSP traces were generated. At *left* is shown the VSP traces and above is the sonic log of the well. Total well depth is at 2.41 s. Notes: *1* Calcareous sandstone (see text); *2* Curvature of VSP reflections implies structural dip; *3* Termination of reflection implies presence of a fault near the borehole; *4* VSP data from below TD (Hardage 1985)

A tight grid of seismic lines is run across an area at line spacings of a few tens of meters (25 m is typical). By correlation of key reflectors from line to line this permits the plotting of horizontal seismic sections at selected depths. Correlation of key reflectors can be carried out by computers using automatic tracking techniques, and this provides a facility for plotting seismic maps of selected stratigraphic horizons. Where the structure is simple or can be removed by suitable processing routines the resulting plots display areal amplitude variations for a single stratigraphic horizon, which may be highlighted by the use of false colours. These variations depend on lithology, so that the sections may reveal detailed subsurface facies variations that can readily be interpreted in terms of depositional environment and paleogeography. Channels, sand bars, and other depositional features may be clearly delineated. For example, Zeng et al. (1996) explored the potential for individual depositional elements to be revealed by 3-D seismic by developing synthetic models based on petrophysical data (see Figs. 6.8, 6.9, 6.10 and 6.11 and discussion in Sect. 6.2.2).

One of the most important aspects of 3-D data is the three-dimensional visualization it provides of depositional systems. This is a hugely important development. Traditionally, 2-D cross-sections have been used to illustrate proximal-distal variability in stratigraphy and interpreted depositional systems. Typically, thicknesses and facies undergo significant changes from basin-margin to basin-centre, and dip-oriented 2-D cross-sections have been the preferred way to illustrate this. However, sediment-dispersal is a three-dimensional process—long-shore drift, geostrophic currents, diverging distributaries on deltas and submarine fans, are all examples of where significant sediment transport may take place out of the plane of such 2-D sections. For example, Moscardelli et al. (2012) discuss this important point with respect to a suite of shelf-margin deltas in Trinidad (a paper to which we refer below). In addition, orthogonal, linear subsidence or rotational subsidence on an axis parallel to the basin margin may not be the norm for any given basin, an important attribute that is difficult to capture with 2-D sections. Interpretations of the timing of important stages in sequence development, and quantitative modeling of a basin, such as source-to-sink calculations, may be in significant error if these three-dimensional characteristics of the basin cannot be fully taken into account.

Processing and interpretation of 3-D data require the use of sophisticated computer workstations and interactive graphics. Color display and plotting are essential. Savit and Changsheng (1982) provided an early example of the application of these ideas; the latest edition of the classic memoir by Brown (2011) brings them up to date. Hart (2013) offered a very useful review of modern seismic-stratigraphic methods and potential. Many examples of the interpretation of depositional features, such as

channels and bars, from horizontal seismic sections have now been published (Posamentier 2000). Henry Posamentier has led the introduction of an entire new subdiscipline, termed **seismic geomorphology**, which uses horizontal seismic sections to explore and interpret entire depositional systems (Posamentier 2000; Posamentier and Kolla 2003; Davies et al. 2004a, 2007; Sect. 6.3.4). Chronostratigraphic interpretations of 3-D seismic volumes may be constructed using techniques of "computational seismic chronostratigraphy" (Stark et al. 2013), as discussed in Sect. 8.10.2.

### 6.3.2 Constructing Regional Stratigraphies

The use of conventional 2-D seismic-reflection data to provide reconnaissance information in frontier basins has been standard practice since shortly after the end of the First World War. Since 1968, the year the *Glomar Challenger*, the first ship of the Deep Sea Drilling Project set sail, marine seismic has been a standard, essential component of the global effort to map the geology of the continental margins and deep oceans. Seismic data are calibrated against well and core records. In the case of onshore and offshore frontier exploration, the seismic data are correlated to the first suites of exploration drill holes. Seismic stratigraphy revolutionized the practice of petroleum exploration and development when it was introduced in the late 1970s, a development credited mainly to Peter Vail and his colleagues at Exxon Corporation, (Vail et al. 1977) and to Bert Bally at Shell (see, in particular, Bally 1987).

What follows are five examples of what has become standard subsurface methodology. Reconnaissance 2-D seismic is used first, in the exploration of new frontier basins, and the first exploration drill holes are tied into the seismic data in order to provide calibrations of lithologies, biostratigraphic ages, and depth-to-sonic velocity relationships. Prominent seismic reflections are labeled and codified (commonly, important reflectors are shown in distinctive colours) and are traced throughout the study area using the loop-correlation method, whereby horizons are followed as far as possible around an area using intersecting survey lines, ultimately bringing the correlation back to its starting to point to ensure that the correlations have not inadvertently shifted into a different stratigraphic level. Further ideas about the workflow to be used in basin analysis are provided in Sect. 1.5.

Figure 6.19 illustrates a seismic line from the West Antarctic Rift System, a series of marine rift basins located between the Antarctic Peninsula and the Ross Sea. Earlier seismic and drilling activity had provided the basis for the stratigraphic subdivisions shown on this section, which provides a clear picture of the subsurface stratigraphy and structure of the area. Seismic facies and structural relations in this and other sections may be interpreted in terms of the

**Fig. 6.19** Seismic stratigraphy of the Antarctic margin. This section shows rift architecture close to the western edge of the Victoria Land Basin (Western Shelf). *Inset* shows close-up view of the principal seismic facies of the Early Rift (*navy blue*—Ra to *yellow*—Rb reflectors) and Main Rift (*yellow*—Rb to *purple-grey*—Rc) successions. Thermal relaxation phase: *Purple* reflectors and Rf–Rg reflectors.

In the Early Rift section, mainly concordant, moderately continuous, high amplitude events with some low-angle clinoforms pass downdip into a chaotic seismic fabric adjacent to the half-graben-bounding fault. In the Main Rift section, mainly concordant events of similar character to the underlying section dominate, and are continuous across the top of earlier extensional topography (Fielding et al. 2008, Fig. 2, p. 13)

phases of development of the rifts basins: early rift, main phase, and thermal-relaxation phase. The illustrated section follows what is now standard practice, to use different colours for the key reflectors. Initially this adheres to the empirical approach necessary for building a stratigraphic reconstruction directly from the data and free of regional assumptions or other biases. Eventually these colour-coded reflections are used as the basis for stratigraphic subdivision. Many of them will subsequently be defined as sequence boundaries or other important surfaces.

As described by Fielding et al. (2008, p. 13), based on an interpretation of the section shown here as Fig. 6.19:

> In the area of Roberts Ridge, the Early Rift interval forms an eastward-thickening wedge that is abruptly truncated against the first major basement-offsetting fault. Within the Early Rift section, the principal seismic facies are 1. irregular to chaotic, discontinuous, moderate to high amplitude reflections in the updip area, passing downdip into 2. mainly concordant, parallel to somewhat irregular, semi-continuous reflections of moderate to high amplitude and moderate frequency, with local development of clinoform sets, in turn passing into 3. chaotic to reflection-free facies in the immediate hanging wall region of the half-graben-bounding fault.

Wherever possible, the calibration between petrophysical logs or cores and seismic data should be clearly shown. Core and log data help to clarify the interpretation of seismic

facies, and both together may be used in the definition of the sequence stratigraphy and regional tectonic reconstructions. Figure 6.20 is an example from a very complex area, the Sigsbee Knolls area, on the continental slope off Texas, in the northwest Gulf of Mexico. Vast amounts of clastic debris have been shed into the Gulf since the Jurassic, with many major depositional systems having prograded southward, extending the continental margin by tens of kilometres during that time interval. The Brazos-Trinity slope system, which developed during the Pleistocene, is one of the last of these depositional phases. It represents the distal end of a nonmarine-shallow-marine and shelf-edge delta system, the source for which lay within the Brazos-Trinity watersheds, that extend northwestward across Texas and into New Mexico (Pirmez et al. 2012). The Sigsbee Knolls is an area along the continental slope where evaporites of Jurassic age rise in multiple complex diapiric bodies to form topographic highs on the sea floor, between which clastic sediment is ponded in multiple intra-slope basins. The sediment influx is controlled by the shifting deltaic depositional systems to the north, by the modifications to the basins by salt tectonics, and by sea-level changes. Steep, local intra-basin slopes created by salt activity are thought to be capable of triggering local sediment gravity flows. Considerable exploration work has been carried out in this broad area because

**Fig. 6.20** Seismic section through the Brazos-Trinity depositional system (Pleistocene) in an intraslope basin off the Texas margin continental shelf. High-resolution 2D seismic data are correlated to relevant well logs, showing key surfaces, and positions of cores. Logs are plotted in depth scaled using an average velocity; note the good relationship between the key mapped horizons and major lithologic changes. Core columns contain, from *left* to *right*: (with respective core number), core recovery (*black* recovered, *white/blue* no core, *red* core interval with gas expansion), and lithologic column (with legend key to colors) (Pirmez et al. 2012, Fig. 4, p. 16)

the deep-water sands deposited by these depositional systems have proved to be very hydrocarbon rich. Local stratigraphic variability is therefore considerable, and explorationists must rely on, and reconcile, multiple data sets in order to generate reliable regional reconstructions.

The reflectors seen in Fig. 6.20 may be traced throughout the area. According to Pirmez et al. 2012, p. 115):

> Each key horizon is mappable throughout the area and has distinguishing geometric or seismic facies characteristics. Horizon 10 separates laterally continuous, moderate- to high-amplitude reflectors below, from a laterally continuous low-amplitude to transparent unit above. Horizon 20 is defined on the basis of onlapping reflections above it, observed particularly in Basin IV [Fig. 6.20] and Basin …. Horizon 20 can be traced to Basin II, where it shows conformable reflections above and below. Horizon 30 is a laterally continuous reflector defined by onlapping reflections above it in all basins; locally it is truncated by an acoustically transparent unit [Fig. 6.20]. Horizons 40, 50, and 60 are laterally continuous horizons that mark a vertical change in seismic facies. Horizon 40 is an erosional surface locally near the exit of Basin II … Horizon 61 [Fig. 6.20] is an erosional surface with onlapping reflections above. It is a very weak reflection and is mappable only in Basin IV. Horizon 70 is well defined in Basin IV, where it forms a baselap surface at the base of a fan wedge sourced from the eastern channel [Fig. 6.20].

This example of basin analysis is discussed further in Sect. 8.10.1, where some of the details of the advanced chronostratigraphic dating and correlation techniques are briefly summarized.

The third example discussed briefly here is a regional study of the Cretaceous Chalk in the Norwegian portion of the North Sea's Central Graben (from Gennaro et al. 2013). As the authors noted (p. 236), "In the last decade, the classical paradigm of chalk as a monotonous and flat-lying deposit has been significantly revised and increasing evidence from seismic data has revealed the role that bottom currents have on chalk deposition. These currents sculpted the chalk sea-floor relief creating important topographic features such as channels, drifts and valleys." Sediment gravity flows and contourites composed of chalk have been mapped, and there is increasing evidence for syndepositional tectonic activity. The stratigraphic subdivisions and petrophysical and seismic character of the Chalk are shown in Fig. 6.21. These are based on seismic facies characteristics and the mapping of reflection terminations. Examples of the regional stratigraphic architecture are illustrated in Fig. 6.22, and an enlarged portion of this section, showing facies characteristics in more detail, is shown in Fig. 6.23.

Detailed mapping of the thicknesses and facies variations of the sequences identified in this study have been interpreted in terms of three tectonic phases, pre-inversion, syn-inversion and post-inversion. Although the details of this study are not relevant for our present purposes, it is of interest to note how the seismic data illustrates the onlap of the top Narve and top Thud reflectors onto anticlinal highs in Fig. 6.22. These features of the seismic architecture are

**Fig. 6.21** Allostratigraphic and lithostratigraphic subdivision of the Chalk Group in the southern Norwegian sector of the North Sea and the corresponding seismostratigraphic units used in this study. The graph on the tectonic activity indicates the relative magnitude of the tectonic inversion. The tectonic activity is considered to be continuous, but with culminations. *Green* well curve, gamma ray log increasing from *left* to *right*; *black* well curve, sonic log increasing from *right* to *left* (Gennaro et al. 2013, Fig. 2, p. 238)

interpreted as the result of syndepositional uplift during the inversion phase, following which erosion of the resulting highs provided a source for allochthonous chalk during the post-inversion phase.

One of the more important improvements in the seismic-reflection technique has been the development of the ability to image the complex structures associated with di-apiric salt, including the edges and underlying boundaries of salt bodies and the stratigraphy beneath. A description of the processing techniques that have permitted this are beyond the scope of this book, but include the use of what is termed "pre-stack migration". Traditionally, seismic traces are stacked to form a single trace, and then undergo migration to restore distortion caused by varying seismic velocities. With the increased processing speed and power of modern computer methods, it is now possible to migrate each trace before combining, or "stacking" them, which preserves considerably more detail. An example is shown in Fig. 6.24. It is this breakthough in processing that has permitted deep-water exploration of pre-salt structure and stratigraphy in such basins as the Gulf of Mexico and the deep offshore Atlantic Ocean off Brazil, Angola, etc.

The final example in this section touches on one of the most frequently encountered seismic facies, that of the complex clinoform set. The example illustrated (Fig. 6.25) is from the Barents shelf margin, located between northern Norway and Svalbard. An active phase of subsidence commenced in the mid-Eocene as rifting accompanied the opening of the Norwegian-Greenland Sea. High relief, a narrow shelf and an abundant sediment supply led to the

**Fig. 6.22** Seismic sections of the southern Norwegian Central Graben with interpreted intra-Chalk seismic sequences LC1 to DN1. Regional structural elements and tectonic lineaments are also indicated. The rectangular areas labeled Fig. 5 is reproduced here as Fig. 6.23 (Gennaro et al. 2013, part of Fig. 4, p. 242)

**Fig. 6.23** Reflection configuration and terminations within the seismic units of the Chalk Group. **a** Uninterpreted and **b** interpreted 3D seismic profile (Gennaro et al. 2013, Fig. 5, p. 244)

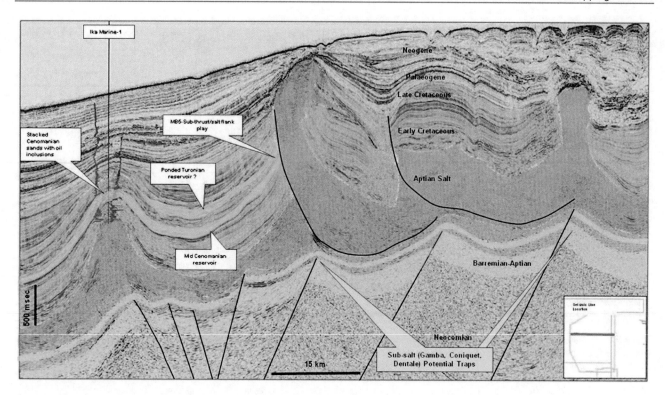

**Fig. 6.24** An interpreted seismic line from offshore Gabon, showing the extremely complex structure and stratigraphy associated with diapiric salt (from Presalt.com; a seismic line from joint exploration by Ophir and Petrobras)

development of thick prograding clinoform sets within which slumping and subsidence remobilized large volumes of sand to form injectites. A single well with two short cores has been drilled within the study area, although not along the line of the illustrated section. The major stratigraphic subdivisions that may be recognized in this section consist of prograding clinoform units on the continental slope, consisting of resedimented shelf deltaic and other marine sediment. These units pass down into what are labeled as "depositional units" that have been interpreted as submarine fan complexes. Analysis of the architecture of thee clinoforms provides considerable information about the history of subsidence and sea-level, a subject touched on in the next section.

### 6.3.3  Seismic Facies

The early work on seismic stratigraphy, in the 1970s, provided many useful broad generalizations about stratigraphic architecture that focused on large-scale geometric features, particularly the nature of reflection terminations (Sect. 5.2.2). These were illustrated in what is by now the classic diagram shown in Fig. 5.8, and constitute the larger scale elements of what is termed **seismic facies**, a term defined to mean "areally restricted group[s] of seismic reflections whose appearance and characteristics are distinguishable from those of adjacent groups" (Sangree and Widmier 1977). Reflection amplitude

and continuity are additional elements of seismic facies that have become increasingly important as processing and visualization power have improved. The concept of seismic facies is most usefully applied where the primary data consists of 2-D cross-sections, for which stratigraphic and sedimentological interpretations may not be immediately obvious. But the ground-truthing of seismic facies in terms of lithofacies may be difficult in the absence of well data (particularly cores) in key locations, because few seismic facies have unique interpretations (clinoform facies are the simplest to interpret directly). This problem is much less acute where 3-D data are available, because of the essential attribute of the 3-D volume that entire depositional systems may be visualized.

The other essential aspect of seismic-stratigraphic data, particularly 3-D data, is the scale of features that can now be imaged in their entirety. Complete depositional systems may be sliced and viewed in any vertical orientation and at different horizontal levels, over horizontal dimensions of tens of kilometres. This is orders of magnitude greater than even the best outcrop sections, which means that we can now "see" things that have never been seen before. This is particularly the case with deep-marine depositional systems, such as submarine fans and contourite drifts, for which outcrop data has mostly been very scrappy. The first depositional models for these features (Mutti and Ricci-Lucchi 1972; Walker 1978) were based on limited outcrop data and,

**Fig. 6.25** Clinoform sets of Eocene-Oligocene age on the margin of the Barents Shelf. **a** In this section, three deep-water depositional units (1–3) pass up dip into prograding units, four of which (units *A–D*) have been mapped. **b** Uninterpreted and interpreted sections showing Prograding units *A–D* in the northern part of the study area (the section outlined by the *blue rectangle* in the upper-right portion of diagram A) (Safronova et al. 2014, Fig. 3, p. 520). AAPG © 2014, reprinted by permission of the AAPG whose permission is required for further use

as has become apparent, provide very little real guidance for the interpretation of the large submarine channel and fan systems that form important reservoirs in many deep-water base-of-slope frontier basins, in part because of the enormous variability we now know exists in the size, shape and composition of submarine fans. This point can readily be appreciated by comparing these early, simple models, with the realities of a wide array of modern fans surveyed using side-scan sonar technology, summary details of which were assembled by Bouma et al. (1985).

Davies et al. (2004b) discussed the practicalities and pitfalls of the use of 3-D seismic data. The knowledge and experience of both geophysics and geology are required, in order to fully understand what can be done with processing, and the limitations inherent in the seismic method (geophysics) and the realities of stratigraphy and of sedimentological interpretation (geology). A common error is the attempt at direct interpretation of clastic lithofacies from reflection strength and continuity. There are many criteria that govern this feature of seismic facies (fluid content and diagenetic modification being obvious ones), and all of these need to be taken into account.

Facies models and sequence models provide guidance for the interpretation of the real world (Chaps. 4 and 5). It would be an ideal world if seismic facies could be categorized in the same way. Many researchers have provided categorizations of the seismic facies in their project areas, but as several authors have noted (e.g., Futalan et al. 2012), calibration between seismic attributes and lithofacies is to a considerable extent basin specific, and so generalizations must be accorded a considerable latitude for inherent variability. Two seismic-facies classification tables are shown here (Figs. 6.26 and 6.27).

The first illustration (Fig. 6.26) shows a facies classification for a mixed carbonate-clastic continental-margin assemblage in the Philippines. The lithologic attribute of most of these facies vary from clastic to carbonate or mixed, as indicated by well calibration, but the geometrical attributes are distinctive. For example, facies D represents turbidites, facies E clearly represents clinoforms, and facies F images channels.

Sayago et al. (2012, p. 1844) stated "It is known that 40 % or more of recoverable hydrocarbons in carbonate deposits are trapped at stratigraphic unconformities, and in most cases, they are of karst origin." Their study is of a largely carbonate succession of Pennsylvanian-Permian age on the Barents Shelf in the Norwegian Sea. The analysis commenced with a multi-attribute classification study of the seismic volume. Eighteen seismic attributes covering the whole range of seismic information were computed (e.g., time, amplitude, frequency, and attenuation). Comparisons and crossplots between attributes were used to define the seismic facies. This classification was then applied to samples of the seismic survey volume. An initial "unsupervised" classification

| Seismic Facies | | Reflection Attributes |
|---|---|---|
| | | *(a – external geometry, b – internal configuration, c – continuity, d) amplitude strength)* |
| **(a)** | | a) sheet to wedge<br>b) parallel to wavy<br>c) high continuity<br>d) moderate to high |
| **(b)** | | a) sheet to wedge<br>b) parallel to wavy<br>c) semi-continuous to high continuity<br>d) low to moderate |
| **(c)** | | a) sheet to mound<br>b) wavy to hummocky<br>c) disrupted to discontinuous<br>d) moderate to high |
| **(d)** | | a) sheet to wedge<br>b) parallel to subparallel<br>c) semi-continuous to disrupted<br>d) low to moderate |
| **(e)** | | a) lens to wedge<br>b) subparallel to convergent to oblique<br>c) semi-continuous to high continuity<br>d) low to moderate |
| **(f)** | | a) lens to channel-shaped<br>b) wavy to chaotic<br>c) discontinuous<br>d) low to moderate |

**Fig. 6.26** A seismic facies classification for a mixed carbonate-clastic assemblage on the continental margin of the Philippines (Futalan et al. 2012, Table 2). AAPG © 2012, reprinted by permission of the AAPG whose permission is required for further use

represents an automated analysis of the seismic data; a "supervised" classification includes some operator adjustments to better correlate the seismic data to information from core. The "training data" used in this classification is shown in Fig. 6.27. An example of a 2-D seismic cross-section analysed in this way is shown in Fig. 6.28a. Four seismic attributes have been calculated for the classification (instantaneous bandwidth, gradient magnitude, envelope, and dominant frequency), as shown in diagram B. Diagram C shows an "unsupervised classification" based entirely on numerical manipulations of the digital seismic data. Seismic facies 1 (SF1) is present not only in the flat-topped carbonate interval, but also in the vicinity of the main seismic reflectors occupying the outer slopes of the structural high, that is, top basement, top Gipsdalen, and top Bjarmeland (white arrows). Strongly chaotic patterns and complex alternation of seismic facies 2 (SF2), seismic facies 3 (SF3), seismic facies 4 (SF4),

| Seismic Facies | Reflection geometry | Amplitude characteristic | Spatial distribution | Example (Vertical bars represent 100 ms) |
|---|---|---|---|---|
| SF1 | Parallel continuous | High amplitude | Occurs mainly at the crest of the structural high and at the top of basement | |
| SF2 | Parallel continuous | Medium to low amplitude | Occurs mostly towards the flanks of the Loppa High | |
| SF3 | Parallel discontinuous | Medium to low amplitude | In overlying Triassic clastics and some areas of the carbonate intervals | |
| SF4 | Chaotic | Low amplitude | Present at the core of the buildups and in the basement | |
| SF5 | Semiparallel dipping discontinuous | Medium amplitude | Occurs mostly in the slopes of buildups | |

**Fig. 6.27** A seismic facies classification for a carbonate succession, including paleokarst breccias. See text for explanation (Sayago et al. 2012, Table 4, p. 1860). AAPG © 2012, reprinted by permission of the AAPG whose permission is required for further use

and seismic facies 5 (SF5) are observed. "Supervised classification" results are shown in Diagram D. Seismic facies 1 is mostly concentrated in the topmost area (crest) of the Loppa High where it correlates with the parallel high-amplitude training points (the volumes selected by the user to fine-tune the classification). It is interpreted as consisting of paleokarst breccia units and its distribution in the supervised classification is associated with a major regional unconformity. Seismic facies 2 is restricted to the slopes of the structural high, high-lighting the main seismic boundaries.

An instructive clastic example of seismic-facies and sequence interpretation is that of Moscardelli et al. (2012) who examined the development of shelf-edge deltas on the Atlantic shelf margin off Trinidad, in an area where the continental-margin tectonic environment changes from transpressive to extensional within a space of 100 km. Figure 6.29 illustrates a composite northwest-southeast-oriented 3-D seismic line along strike, showing the shelf-edge region in eastern offshore Trinidad. Diagram (A) shows the seismic section; diagram (B) is a line drawing showing the seismic facies (SF) distribution and sequence-stratigraphic

interpretation of this section. The seismic character and distribution of the main sequence-stratigraphic units significantly vary from south to north; these variations are thought to be associated with underlying structural controls and changes in sedimentation rates across the margin. Contrasts in tectonic setting between the northern and southern parts of the project areas described in this study are marked. Steep thrust faults are important in the north; extensional growth faults are common in the south. These faults have guided the development of submarine canyons and therefore the downslope distribution of detritus. Tidal and along-shore currents are important in sediment dispersal, and gravity sliding and slumping have had a significant effect on the ultimate architecture of depositional units. This is clearly an example where the traditional 2-D approach to stratigraphic and sedimentological interpretation would not be successful. A seismic-facies classification developed for this area is shown in Fig. 6.30.

In this study and that of Safronova et al. (2014), analysis of the internal composition and architecture of clinoform units provides essential clues for the analysis of depositional and structural history. For example, in Fig. 6.25, the trajectory of the shelf-slope break is highlighted in the bottom diagram, providing insights into the changes in sediment supply and changes in relative sea-level through the accumulation of this unit.

Zeng et al. (2013) pointed out that seismic resolution may be a problem in the identification and mapping of clinoform sets, particularly those developed by deltaic progradation, which typically develop in shallow water and are much smaller than shelf-margin clinoforms.

> It seems that the type of seismic clinoform configuration may also be related to data frequency. An oblique clinoform seismic configuration in higher frequency data … tends to become a shingled configuration in the lower frequency data … As a result, shingled facies observed in seismic data are not necessarily truly representative of geologic clinoform architecture. The merging of seismic responses of the thinner, low-angle downdip portion of clinoforms with that from underlying flat host rocks in low-frequency data appears to distort the seismic facies. (Zeng et al. 2013, p. SA45)

A process termed spectral balancing may be used to enhance the high-frequency end of the data spectrum, and this can reveal a subtle clinoform signature (Fig. 6.31).

The last example in this section is derived from the work of Burgess et al. (2013) who provided criteria for the identification of isolated carbonate buildups in the stratigraphic record. As they pointed out, these may be difficult to distinguish from volcanoes, erosional remnants and tilted fault blocks. In tropical and subtropical areas, such buildups may be common features of the continental shelf, but in many areas, such as continental margins off southeast Asia, characterized by rapid tectonism and a high sediment supply, local conditions may change over geologically rapid time scales, resulting in highly complex stratigraphic relationships. Aids to the identification of carbonate buildups (which can provide excellent petroleum reservoirs) have therefore been found useful. Figure 6.32 provides some examples of the distinctive seismic facies characteristics of carbonate buildups developed by Burgess et al. (2013) from a large seismic data bank assembled from offshore southeastern Asia. Key criteria include geometry (the vertical nature of the buildup, commonly with an overlying structural drape), seismic terminations (e.g., onlap of buildup margin by draping clastic units), and contrasts in reflection amplitude and continuity between the buildup and flanking strata.

### 6.3.4 Seismic Geomorphology

The development of seismic geomorphology represents a huge step forward in methods for petroleum exploration and development. High quality 3-D data are required, with

**Fig. 6.28** Shelf-margin seismic section showing how the seismic facies classification is derived. **a** The uninterpreted seismic section; **b** Four seismic attributes calculated for the classification; **c** Unsupervised classification of the original seismic data; **d** supervised classification (Sayago et al. 2012, Fig. 9, p. 1858). AAPG © 2012, reprinted by permission of the AAPG whose permission is required for further use

**Fig. 6.29** Strike-oriented seismic section (**a**) and its interpretation (**b**), along the Atlantic shelf-margin of Trinidad. Terms in quotation marks indicate that the indicated units and surfaces do not present the typical seismic characteristics defined by the sequence-stratigraphic approach; however, these intervals are time equivalent to units to the south that fit the traditional sequence-stratigraphic definition for key surfaces and system tracts. *TWTT* two-way traveltime; *LST* lowstand systems tract; *HST* highstand systems tract; *TST* transgressive systems tract; *mfs* maximum flooding surface; *lw* lowstand wedge; *bff* basin-floor fan (Moscardelli et al. 2012, Fig. 3, p. 1489). AAPG © 2012, reprinted by permission of the AAPG whose permission is required for further use

processing by skilled geophysicists, but a deep understanding of sequence stratigraphy and sedimentology is essential for in-depth interpretations of these data. Where such a combination of data and skillsets is available it may render all earlier sedimentological approaches to subsurface analysis obsolete. The simplistic facies models used through the 1960s to 1990s have been replaced by actual visualizations of entire depositional systems. However, the understanding of depositional processes that evolved during the facies-models era form an essential background to seismic analysis. For example, the earlier ideas about inner- mid- and outer-fan that were introduced in the 1970s (Mutti and Ricci-Lucchi 1972; Walker 1978) were still found to be useful when Bouma et al. (1985) compiled numerous studies of modern fans using side-scan sonar, and some of the defining elements of these three subdivisions do appear on seismic images of some submarine fans, a category of depositional system that we now know, through such studies, to be vastly more complex and varied than was imagined a few decades ago. For example, Fig. 6.33 illustrates three cross-sections though a fan complex on the Borneo continental margin, in which the three-fold subdivision that had

been proposed for fans appears to be applicable—an inner fan characterized by a prominent channel-levee complex (commonly a single main channel fed from a canyon point source), a mid-fan with a much more subdued distributary channel-level system, and an outer fan with a smooth, unchannelized, convex-up cross-section representing the accumulation of sheet-like turbidites.

The most spectacular 3-D interpretations of depositional systems are those of channelized environments, particularly fluvial and submarine fan settings, with their complexes of levees, crevasse splays, and overbank areas (e.g., Posamentier and Kolla 2003; see Figs. 6.34, 6.35, 6.36 and 6.37 of this book). For fluvial systems there is, of course, a virtually limitless suite of modern analogues for comparison with the ancient examples. Comparisons of horizontal seismic sections with selected images from Google Earth or some comparable source of aerial imagery make for very powerful and effective interpretations. Many such studies have now been published, many including quantitative analyses of channel and meander characteristics.

For submarine fans, imagery of modern fans is limited to what can be obtained from side-scan sonar data, and

| SF | Seismic Facies Description | |
|---|---|---|
| 1 | Steep Clinoform Package — Topset Incisions — Outer Shelf Delta | Clinoform heights 10 s and 200 m<br>Topset: Low-angle, high-amplitude, and relative continuous reflectors<br>Topsets are incised by channelized-like features.<br>Foreset: Steep slopes, medium- to low-amplitude reflectors<br>Rollover points do not surpass outer shelf region and overall rollover trajectory is mostly progradational. |
| 2 | Sigmoid Clinoform Package — Wedging and Onlapping — Small-Scale Faults — Sliding and Slumping — Shelf Edge Delta | Clinoform heights, >200 m<br><br>Topset: Low-angle, high-amplitude and continuous reflectors; wedging and onlapping against basal surface<br><br>Foreset: Steep slopes, high- and low-amplitude continuous reflectors that transition into low-amplitude, discontinuous, and sometimes chaotic reflectors<br><br>Bottomset: Low-amplitude to discontinuous reflectors<br>Rollover points are strongly progradational, sometimes displaced by sliding. |
| 3 | Onlapping Wedge — Onlapping against Slope — Slope Fan Deposits | Thickness of individual packages containing SF3 facies are in the range of 75 to 150 m.<br><br>High-amplitude continuous reflectors that are arranged on an onlapping wedge against the slope.<br><br>This facies is not volumetrically significant in the study area. |
| 4 | Continuous and Parallel Reflectors — Prodelta/Upper Slope Deposits | Thickness of individual packages containing SF4 facies are approximately 100 m.<br><br>This facies is characterized by high-amplitude, continuous, and parallel seismic reflectors.<br><br>Lateral or vertical variations within seismic reflectivity and continuity are limited. |
| 5 | Chaotic Seismic Facies — Mass Transport Deposits | Thickness of individual packages containing SF5 facies can reach up to 250 m<br><br>This facies is characterized by low-amplitude, semitransparent, chaotic reflections that can transition into low- and high-amplitude semicontinuous reflections in some areas.<br><br>This facies is volumetrically abundant on the lower slope region. |
| 6 | Gull-Wing Seismic Facies — Levee Channel Deposits | Thickness of individual packages containing SF6 facies can reach up to 250 m<br><br>This facies is characterized by high- and low-amplitude reflectors with variable continuity. Packages within SF6 facies oftentimes dissect each other vertically and laterally.<br><br>This facies is volumetrically abundant on the lower slope region. |

**Fig. 6.30** A seismic-facies classification of the shelf-margin depositional systems, Atlantic margin of Trinidad (Moscardelli et al. 2012, Fig. 6, p. 1493). AAPG © 2012, reprinted by permission of the AAPG whose permission is required for further use

ancient-to-modern comparisons are less abundant. Descriptions and illustrations of some of the world's best outcrops of ancient deep-marine systems were collected in a large-format volume by Nilsen et al. (2007).

A few seismic-geomorphic studies of carbonate environments have been published, but few that make use of horizontal sections to image a carbonate system. It would appear that such sections are of limited usefulness, compared to traditional 2-D seismic.

The other major use of seismic geomorphology has been to image erosion surfaces, including unconformities and such features as incised valleys and paleokarst surfaces (Figs. 6.38, 6.39, 6.40 and 6.41). A range of examples of modern applications is contained in the volumes edited by Davies et al. (2004a, 2007).

Turning to fluvial examples, Ethridge and Schumm (2007) provided a discussion of modern concepts in fluvial geomorphology, focusing on what could be deduced from imagery of a random segment of a fluvial system. The use of modern aerial photographs or satellite imagery makes such

interpretations relatively straightforward. Figure 6.34 is one of the first illustrations of a horizontal seismic section through a fluvial system to be published. It shows a meandering system in the shallow subsurface of the Gulf of Thailand. False colour imagery is used to highlight the amplitude differences between the meandering channels and the overbank areas. Miall (2002) used successive time slices from a Pleistocene system to explore the architecture of an incised-valley system and its badland-style tributaries.

A recent example of the use of horizontal seismic sections to explore an ancient fluvial system was provided by Hubbard et al. (2011) from the Alberta Oil Sands (Fig. 6.35). The strata in this study area are between 300 and 400 m below the surface, and seismic resolution is less than 5 m. A modern analogue of the meandering fluvial style was suggested, in the form of an aerial photograph of the Sittang Estuary, Myanmar. The addition of a map of gamma-ray readings from wells that penetrate this section provides an essential cross-calibration between seismic and gamma-ray data that can then be used for the interpretation of lithofacies

**(a)**

**(b)**

Amplitude

QAe1699

- +

**Fig. 6.31** Reducing ambiguity in interpreting clinoform prograding sequences by spectral balancing. **a** Original stacked and migrated seismic section in Abo Kingdom carbonate field of west Texas, with a flat (*dashed line*) event and some toplapped events (*arrows*) underneath. **b** The same section after spectral balancing processing. The flat event in the original data has been broken up into clinoforms (*dashed lines*) having slopes similar to those of surrounding events. The toplaps disappear (Zeng et al. 2013, Fig. 18, p. SA48). AAPG © 2013, reprinted by permission of the AAPG whose permission is required for further use

and reservoir properties in other areas. Hubbard et al. (2011) provided a detailed analysis of the lithofacies and architectural elements of this section, in terms of point bars, counterpoint bars and other features.

The use of perspective views of depositional systems can considerably enhance the visual effectiveness of 3-D seismic imagery. Figure 6.36 displays a turbidite channel in the Gulf of Mexico. The relevant portion of a vertical 2-D section that intersects the horizontal section provides the necessary

information on the vertical seismic characteristics of the succession. False colour variations, which display seismic amplitude variations, may be correlated locally to lithologic variability across the area, from channel-fill to overbank.

Another example of a submarine fan is shown in Fig. 6.37. This is a small fan, subdivided into upper-, mid- and lower-fan segments, according to Walker's (1978) model. The stratal slice image shown in Fig. 6.37b is not a structurally horizontal seismic section, but follows the synclinal dip of the strata, as shown by the dashed line in the vertical section, above. This image was derived from the top of a seismically chaotic section, which is interpreted by Zeng et al. (1998) as representing the lithologically heterogeneous assemblage characteristic of the fan setting.

The remaining examples in this section deal with the imaging of important surfaces in the subsurface. Unconformities may contain significant information about contemporary erosional patterns, and their topography may have useful implications for petroleum trapping configurations. A particularly interesting unconformity is the surface that marks the top of the Upper Miocene Messinian stage in the Mediterranean basin. Since the first deep-sea drilling project in the Mediterranean Sea in 1970 geologists have known about a prominent, widespread unconformity that spans most of the basin margins, and a thick evaporite section occupying the basin centre. During the Miocene the Mediterranean basin became isolated from the Atlantic Ocean, and evaporated to dryness several times (a drop in sea level of up to 2 km), leaving a deeply eroded basin margin cut by fluvial canyons (Ryan and Cita 1978). Figure 6.38 illustrates a seismic section across the basin margin off the Ebro delta in northeastern Spain. The erosional relief on what is interpreted as the Messinian unconformity is clear on this section, as is the truncation of the units below and the progradation of the Pliocene section that draped and buried the unconformity. A 3-D seismic survey of this area yielded the perspective map view shown in Fig. 6.39. The dendritic drainage pattern revealed by this map is very similar to that of "badland" topography in desert areas, such as that in southern Spain and parts of the southwest United States. About 400 m of erosional relief is visible in the seismic data from this area (Martinez et al. 2004).

As noted in the previous section, paleokarst deposits may be very important as petroleum traps, and 3-D seismic data has proved very successful in identifying and mapping these deposits (Sayago et al. 2012; Zeng et al. 2011; Ahlborn et al. 2014). Figure 6.40 is a 2-D seismic section through a succession of Lower Paleozoic carbonate deposits cut by several unconformities that are underlain by what are interpreted as paleokarst deposits. As each surface was developing by surface exposure and erosion, fluctuating water tables led to the development of subterranean drainage systems, with sinkholes, tunnels and caves, many of which underwent

**Fig. 6.32** Seismic facies of isolated carbonate buildups. **a** An antecedent topographic high beneath an isolated carbonate buildup. **b** Significant localized thickening within an isolated carbonate buildup. **c** Onlap of overburden onto the margins of an isolated carbonate buildup (*left*), contrasted with a situation where depositional relief on the margins of the isolated carbonate buildup was lower because of contemporaneous infill of the adjacent basin (*right*). In this case, carbonate material from the platform top was transported away from the platform margin to produce depositional wings that interfinger with the basin-fill strata. **d** Platform margin trajectories with phases of progradation, aggradation, and retrogradation, which can be indicative of an isolated carbonate buildup. **e** The thin-thick-thin pattern commonly developed on isolated carbonate buildups that are shedding material from the platform top through a bypass zone to be redeposited in deeper water adjacent to the platform. *TC* top carbonate; *BC* base carbonate. (diagrams assembled from several illustrations in Burgess et al. 2013). AAPG © 2013, reprinted by permission of the AAPG whose permission is required for further use

significant collapse, resulting in disruption of the stratigraphic layering. Surface erosion, in some cases, ultimately led to the development of fluvial gullies and channels, with isolated karts towers left behind. An analogy with the karst topography of parts of the Guilin Province in southern China was suggested by Zeng et al. (2011, p. 2076). Extraction of stratal slice data from seismic volumes through these types of succession may provide detailed maps of the surface karst topography. Figure 6.41 provides an oblique view of such a surface, and contains graphic imagery of the surface erosional and collapse features of a paleokarst surface of Ordovician age.

## 6.4    Directional Drilling and Geosteering

Historically, most drill holes have been drilled vertically or as near vertically as possible, but for many reasons, directional drilling may have advantages, both geologically and for practical reasons. For example, access to potential reservoir beneath environmentally sensitive lands, or under water, may be accessed by directional drilling, and the technology now allows multiple holes to be drilled in every direction from a single well pad, which reduces the surface impact of the operation. Geologically, the advantage of directional drilling is that it creates the possibility to access specific subsurface targets by the use of real-time downhole steering. In the 1980s the development of the technology permitted wells to be completed with horizontal segments that could be steered to penetrate a reservoir along its length, generating substantially greater exposure of a hydrocarbon-bearing unit to the producing well. Horizontal segments of up to 3 km are now possible. Drilling motors are mounted at the end of the drill pipe, driven by the drilling mud, and subsurface navigation is achieved by the use of gyroscopes, with steering achieved on the basis of navigational information sent electronically from the drill head in real time. Horizontal drilling has become widely used since the 1990s

**Fig. 6.33** Three strike-sections across the submarine fan formed on the basin floor, Borneo (Saller et al. 2004). AAPG © 2004, reprinted by permission of the AAPG whose permission is required for further use

in the production of shale gas and tight oil, and is also an integral part of the technique of in situ production of oil from the Alberta Oil Sands.

Geological mapping techniques may be used as an aid in subsurface navigation. This is termed **geosteering**, and may be carried out using detailed mapping information obtained before drilling commences, or may be done in real time, for example by using petrophysical log data recorded as the well is being drilled or by using well cuttings that may be monitored for lithological content, microfossils or palynomorphs that are known to characterize the target unit and the

overlying and underlying formations. The objective is to ensure that the well continues to penetrate the desired stratigraphic level as drilling proceeds, although allowance must be made for the delay caused by the travel time of the cuttings from the drill face. Three examples are briefly discussed here. In each case the operator stressed that considerable care needs to be taken to construct a very detailed and accurate three-dimensional model of the subsurface target volume before drilling commences, because the real-time data needs to be matched against the model as drilling proceeds.

**Fig. 6.34** A meandering fluvial channel in shallow sediments in the Gulf of Thailand. From the first (1986) edition of Brown's AAPG memoir (Brown 2011 is the latest edition). AAPG © 2011. Reprinted by permission of the AAPG whose permission is required for further use

Taylor (2010) described the use of real-time petrophysical monitoring during the drilling of a horizontal hole. The project began with the establishment of the petrophysical characteristics of the rocks to be drilled. From this information a model is developed of the petrophysical values to be predicted along the projected path of the hole, based on the three-dimensional reconstruction of the stratigraphic architecture (Fig. 6.42). Note how, in this diagram, the planned well path first passes nearly vertically through the stratigraphy, and is then designed to drill out straight legs at slightly different angles in order to follow the mapped structure, with the predicted petrophysical response along the course of the hole shown in the two graphs at the top. Figure 6.43 shows the results being obtained as drilling is in progress. Deviations from the modelled petrophysical responses may be fed back into the structural model of the target volume in order to permit corrections to be made, if necessary, to the path followed by the drill. Some earlier examples of the use of real-time "logging while drilling" were described by Bristow (2002).

Southcott (2014) described briefly how a 3-D seismic data volume was subjected to rigorous calibration and time-depth conversion with the use of relevant well

petrophysical and core data, and then used to plot the desired path of a projected lateral well.

In Sect. 6.6.3, brief reference is made to a study of "forensic chemostratigraphy" that was used by Hildred et al. (2014) to characterize the units that had been penetrated along a horizontal well through the Bakken petroleum system.

## 6.5 Older Methods: Isopleth Contouring

In the absence of seismic data, the explorationist and the basin analyst must maximize the data available from well or outcrop data. This was the situation after the Second World War, before the seismic technique had evolved to the point that stratigraphic information could be extracted from the data. As noted in Sect. 1.2.3, the need for a "big picture" of a sedimentary basin was the drive that underlay the work of W.C. Krumbein and his colleagues at Northwestern University (from which modern ideas about sequence stratigraphy emerged: see Sects. 1.2.9 and 1.2.10). Krumbein (1948) devised a suite of quantitative lithologic indices and ratios that could be calculated from the lithologic succession at each sample point. Contouring of the resulting data created "isopleths", which formed maps that could then be interpreted in terms of paleogeography. Some of these approaches have become part of the standard methodology of basin analysis. Net sand maps or sand-shale ratio maps are good examples. For example, Fig. 4.28 shows how the distribution of sand thicknesses may be used to define patterns that may be interpreted in terms of the sedimentary controls acting on deltas. Figure 4.47 shows how the ratio of clastic to carbonate lithologies defines the transition from a coastal clastic belt to a broad carbonate platform. The gamma-radiation map of Fig. 6.35c shows how petrophysical data may be used for a similar purpose, in this case the definition of net sand thickness, a parameter that could (in this case) be interpreted in terms of the patterns of channels and bars within a meandering fluvial system. Figure 6.44 illustrates the use of the net sand mapping technique to develop highly detailed facies interpretations, where adequate drill data are available. Figure 6.45 illustrates a modern example of the use of this technique. Seismic data were use to restore Middle Miocene topography in this basin in Trinidad. Net sand values reveal two major northeast to southwest sand trends and their relationship to growing structures. "Zones of absence of Middle Miocene Herrera sands coincided with paleo-highs on the restored Middle Miocene topography" (Moonan 2011).

Krumbein (1948) developed a clastic ratio parameter, calculated from the sum thickness of clastic lithologies divided by the sum thickness of carbonate and evaporite lithologies in a vertical section. Figure 6.46 is an example of the use of this ratio. He stated, of this map (p. 1913):

**Fig. 6.35** Seismic exploration of the Alberta Oil sands. Mannville Group, south of Fort McMurray, Alberta. **a** Time slice taken at a horizon, 8 ms (approx. 8 m) below a flooding surface at the top of the oil-bearing unit; **b** Interpretation in terms of scroll bars, point bars and channels; **c** Gamma-ray values at the selected horizon. From Hubbard et al. (2011, Fig. 2, p. 1126). AAPG © 2011. Reprinted by permission of the AAPG whose permission is required for further use

… the upper map shows generalized isopachs, clastic ratio lines, and sand-shale ratio lines. The three sets of lines render the map somewhat cumbersome, and in practice the clastic ratio may be used alone with the isopachs, if the section shows a balance between clastics and non-clastics. … The lower map … is a combination of the clastic and sand-shale ratios in the upper map, and it shows the statistical lithologic differentiations of the map area. The map is based on combinations of the clastic ratio and the sand-shale ratio in terms of certain limiting values.

The regional pattern revealed by this map is of a broad clastic belt in the east (Montana-Colorado) bordering the Canadian Shield, passing gradually westward into a carbonate-dominated area near the continental margin, running north-south through Idaho. As an aid to constructing regional paleogeography this is a useful map, but clearly many data points would be required to plot individual depositional systems on a more local, $10^3$–$10^4$ m scale.

Extensive use of these various isopleth maps was made in the early regional stratigraphic and sequence work of Sloss et al. (1949), but these techniques have largely fallen into disuse in recent decades, because of the more powerful methods of facies mapping that emerged in the 1960s and 1970s, and the seismic-stratigraphic and sequence-

**Fig. 6.36** A map of amplitude variations draped on a gently dipping stratal surface, illustrating a meandering turbidite channel in shallow sediments in the Gulf of Mexico (Posamentier et al. 2007, Fig. 2, p. 3)

stratigraphic techniques that evolved more recently. An exception would be the net-sand map, which is still used to illustrate sand reservoirs in stratigraphic traps (e.g., Fig. 6.45).

## 6.6  Mapping on the Basis of Detrital Composition

Traditional methods of correlating petrophysical logs are discussed in Sect. 6.2. A wide range of additional techniques has evolved over the years for the examination of subsurface petrophysical, core and sample data. Compilations of such studies have been published by Hailwood and Kidd (1993) and Ratcliffe and Zaitlin (2010). These methods have several purposes: not just to aid in correlation, but to provide additional information, such as provenance studies, which can help in reconstructions of paleogeography, regional stratigraphic trends, and tectonic history. Some of these techniques provide information on numerical ages, and are discussed in Chaps. 7 and 8. Others are useful for local or regional correlation or provenance studies but do not necessarily provide such age information. A list of some of the latter follows:

Characterization by clastic petrofacies, including
Heavy mineral stratigraphy
Geochemical fingerprinting of sandstones and mudstones (chemostratigraphy)
Magnetostratigraphic susceptibility
Detrital zircon provenance studies

### 6.6.1  Clastic Petrofacies

The grain-size, grain shape, and composition of the clastic detritus present in a stratigraphic unit depends initially on the nature of the source, or **provenance**, of the detritus, but the detritus may undergo many modification processes during its derivation, transport, deposition, and burial, so that the characteristics of the original source become obscured. By careful study, the effects of some of these modification processes can be identified, providing much valuable information to a basin analysis.

Detrital composition depends on these main factors:

1. geological composition of the source area;
2. climate and relief of the source area and the area through which the detritus is transported;
3. detrital dispersal and mixing patterns brought about by the transport processes;
4. chemical and mechanical abrasion, winnowing, and breakdown that occur during transport and at the depositional site; and
5. diagenetic changes during burial.

Despite the many changes that the detritus undergoes during transport and deposition, the nature of the original source area can usually be identified. Certain compositional ranges are characteristic of particular types of tectonic source environment. This is the subject of **petrofacies analysis** and is not discussed here (see Miall 1999, Sect. 9.4). Without

**Fig. 6.37** A submarine fan shown on a stratal slice taken from the Pleistocene, deep water offshore Louisiana. **a** A vertical profile with a dominant frequency of 40 Hz; **b** A stratal slice from the 1400 to 1800 m (1750–1950 ms) interval, along the dashed line in (**a**) (Zeng et al. 1998, Fig. 6, p. 521)

**Fig. 6.38** Dip-oriented seismic cross-section across the Mediterranean continental margin off Northeastern Spain, showing the deeply incised Messinian unconformity (MUS). Megasequence A (MA) consists of a shallow-upward deep-marine to deltaic succession of Miocene age. Megasquence B (MB) comprises a Prograding slope wedge of Pliocene age. The labelled rectangles refer to enlarged portions of this section that are not shown here (Martinez et al. 2004, Fig. 4, p. 93)

**Fig. 6.39** A perspective view of the Messinian unconformity on the Mediterranean continental margin of northeastern Spain. The 2D section shown in the previous figure crosses this area from *centre-left* to *lower right* (Martinez et al. 2004, Fig. 8, p. 96)

going into the details of plate-tectonic setting, information regarding the geology of the local source area may be of considerable value in a basin analysis, because of the information it provides regarding transport directions, drainage areas, location of sources of economically valuable minerals (such as those occurring in placer deposits), uplift and **unroofing history** of the source area, and so on. This is the main subject for discussion in this section.

Initially, of course, petrographic composition of primary detritus is similar to that of the rocks exposed in the

**Fig. 6.40** Vertical seismic section through an Ordovician-Silurian carbonate succession in the Tarim Basin, western China, flattened on the Lower Silurian (Tg4p reflection) in order to show the original topography at the top of the Ordovician unconformity (Tg5). At this and older paleokarst surfaces, collapsed caves and sinkhole are shown by the short coloured vertical lines, which mark disrupted or discontinuous reflections (Zeng et al. 2011, Fig. 10, p. 2074). AAPG © 2011, reprinted by permission of the AAPG whose permission is required for further use

**Fig. 6.41** A perspective view of the Ordovician unconformity extracted from 3-D seismic data, showing the morphology of the eroded, paleokarst surface (Zeng et al. 2011, Fig. 14, p. 2077). AAPG © 2011, reprinted by permission of the AAPG whose permission is required for further use

sediment source area, but changes begin to occur even as the rocks weather and become part of the subsoil. After only a short transport distance, the petrography of the detritus may not necessarily bear a simple relationship to that of the source area because of the destruction of unstable grains in transit, particularly in humid environments where chemical weathering may be rapid (e.g., Franzinelli and Potter 1983; Basu 1985). Nevertheless, studies of petrographic variation within the basin itself may provide much useful information. Downcutting by rivers reveals progressively older rocks, so that the derived sediments may display an **inverted stratigraphy**, older sediments containing detritus from the young cover rocks and the upper strata containing material derived from the local basement (Graham et al. 1986). Documentation of this unroofing sequence may help clarify the timing of orogeny, volcanism, plutonism, rates of uplift, etc. (Cerveny et al. 1988). Areal variations in petrographic composition depend on transport directions which, in turn, reflect tectonic control of the paleoslope. As clastic sediments are dispersed by currents or sediment-gravity flows, they are subjected to size sorting, abrasion, breakage, and rounding. They may retain a distinctive detrital composition, although easily weathered minerals, such as ferromagnesian grains, will gradually be destroyed. As a result of these

**Fig. 6.42** At *right* density and resistivity logs through a target interval; at *lower left*, a two-dimensional cross-section of the modeled structure and the planned path of the directed well bore; at *top left* the expected log response along the well (Taylor 2010). AAPG © 2010, reprinted by permission of the AAPG whose permission is required for further use

processes, a particular lithosome may become fingerprinted with a distinctive size range and composition, and down-current changes in the various petrographic parameters may be used as paleocurrent indicators. Interpretation of the source area for marine rocks may be a more complex procedure. For example, submarine fans may contain deltaic detritus plus shelf carbonate or siliciclastic material swept down to the continental rise via submarine canyons.

Petrographic data are simplest to use where the sediments have been deposited by unidirectional currents, as in fluvial, deltaic, and submarine-fan environments. Size and sorting parameters do not convey much, if any, directional information when the sediments have been formed by multidirectional currents, as in most shallow-marine and shelf environments. Few petrographic studies of this type have been carried out in submarine fans, in part because most such deposits contain abundant paleocurrent indicators such as sole markings (Sect. 2.2.2.7), and in part because most exposed fans are in complex structural belts where areal trends are difficult to reconstruct. Most of the remaining discussion in this section relates to fluvial and fluviodeltaic

deposits, for which petrographic studies may have uses in surface or subsurface exploration for petroleum, coal, uranium, or other economic deposits.

Grain-size changes caused by abrasion and breakage during downstream transport are rapid in gravel detritus, but the rate of change with distance of transport drops off asymptotically, becoming slow in coarse sand and negligible in fine sand to silt. Much experimental work has been carried out in an attempt to quantify the rate of change, so that grain-size in the field could be interpreted directly in terms of distance of transport. However, each detrital lithology responds differently to current transport, and rivers vary markedly in competency at any one point, so that research of this type is unlikely to meet with any success. Nevertheless, mapping of relative grain-size changes in conglomeratic and coarse sandy sediments has considerable use as a paleocurrent indicator.

What measurements should be made? For grain-size studies in conglomeratic rocks, the simplest and most useful mapping parameter is maximum clast size, which is usually given as the average of the intermediate clast diameter of the

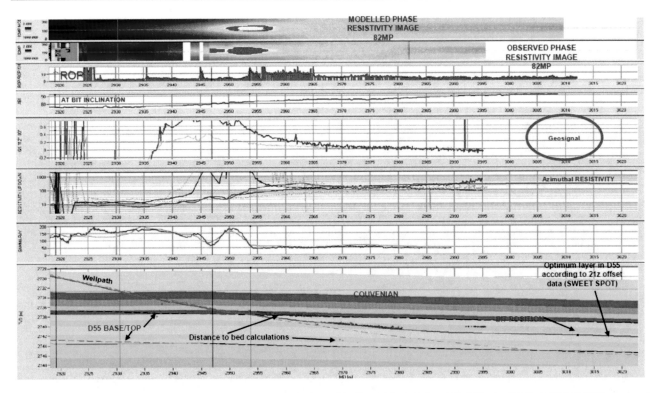

**Fig. 6.43** Data being obtained in real time as the directional well planned in Fig. 6.42 was being drilled (Taylor 2010). AAPG © 2010, reprinted by permission of the AAPG whose permission is required for further use

10 largest clasts present at a sample point. For sand-sized grains mean or maximum grain-size may be used. Both can be measured in an outcrop or in a thin section by visual estimation against a grain-size chart (Fig. 2.5). This permits measurements to the nearest 0.5 $\phi$, which is quite accurate enough for our purposes (sieve analyses and other laboratory studies are time consuming, and usually add little).

Detrital composition of conglomerates can be studied by counting and identifying 100 or more clasts at each sample point. Sandstone composition should be studied in thin-section using a moveable microscope stage and a point counter. It is recommended that at least 300 grains be identified and counted per sample point. The stage should be set to advance at each count by an amount equal to or slightly greater than the mean grain diameter. The number of points counted can be adjusted according to the compositional variability of the sample. Pure quartz arenites do not need such intensive study, but immature, lithic arenites with many varieties of rock fragments may need more detailed examination. Care should be taken to recognize and distinguish **extrabasinal** and **intrabasinal** detritus, particularly in the case of carbonate and volcanic fragments. Zuffa (1985b, 1987) discussed criteria for distinguishing these classes of particles. Mineral types have different densities and so they vary in their hydrodynamic behavior. They also have different fragmentation patterns. The detrital composition of a clastic rock may therefore vary markedly with grain-size (Fig. 6.47)

and so, for mapping purposes, it is advisable to control measurements according to a standard grain-size range, for example 1 $\phi$. class. These methods will usually result in a measurement repeatability of two or three percentage points or better, and this, again, is accurate enough for our purposes. Analytical methods are discussed in detail by Dickinson (1970), Pettijohn et al. (1973), and Zuffa (1985a, b, 1987). It is important to standardize laboratory procedures in order to facilitate comparisons between data sets. Zuffa (1985b, 1987) and Ingersoll et al. (1984) recommended use of the so-called **Gazzi-Dickinson point-counting method**, which counts coarse grains (>0.0625 mm) in terms of their component mineral grains rather than as rock fragments. The method also requires that the observer attempt to reconstruct the original composition of altered grains. This leads to a better clustering of data points and increased accuracy in provenance determinations. However, there is not universal agreement that the Gazzi-Dickinson method is suitable for all purposes. For example, Suttner and others (1985), in two separate discussions of the paper by Ingersoll et al. (1984), pointed out that the method of classifying coarse grains results in a loss of certain types of data, such as information regarding lithic grains. Careful study of these discussions and the replies by Ingersoll and his colleagues is recommended for anyone contemplating detailed petrographic work. Howard (1993) discussed the statistics of clast counting, sources of error, and field procedures.

**Fig. 6.44** Definition of a crevasse splay and a channel sand body by close-spaced drilling in Cenozoic fluvial deposits of the Gulf Coast (Galloway 1981)

**Fig. 6.45** Net sand map, derived from well data in an oil field, southwest Trinidad (Moonan 2011). AAPG © 2011, reprinted by permission of the AAPG whose permission is required for further use

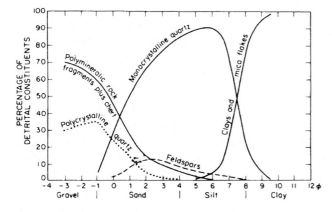

**Fig. 6.46** A classic ratio map for Cambrian strata in the western United States (Krumbein 1948, Fig. 3, p. 1914). AAPG © 1948, reprinted by permission of the AAPG whose permission is required for further use

**Fig. 6.47** Relationship between grain size and composition of the detrital fraction in siliciclastic sediments. (From Blatt, H., Middleton, G.V., and Murray, R.C., Origin of sedimentary rocks, 2nd ed., 1980, p. 321. Reprinted by permission of Prentice-Hall Inc., Englewood Cliffs, New Jersey.)

Much use has been made of heavy minerals (density greater than about 2.8) in provenance studies, even though most sands typically contain less than 1 % by weight. They are extracted from loose or disaggregated sand by separation in dense liquids, such as bromoform, and examined in grain mounts. Distribution is affected by hydraulic sorting during transport and by intrastratal solution following deposition (Morton 1985a). Considerable skill is needed to identify grain types correctly, and the technique is, in general, more difficult to use than thin-section analysis of the light fraction. We do not discuss this research theme further (see Morton and Hallsworth 1994, 1999).

Grain-size and compositional measurements may be averaged over a vertical section at each mapping location or restricted to individual correlatable stratigraphic units. Plotting of both areal and vertical changes may provide useful paleogeographic information, and it is up to the basin analyst to exploit the data in the most effective way. Petrographic data may be analyzed manually, graphically, or statistically to define petrographic assemblages or provinces, typifying a particular source area geology (Suttner 1974). Alternatively, particular mineral species may be treated separately.

Some workers have concentrated on identifying and mapping varieties of quartz, feldspar, zircon, or other grain types, requiring a great deal of highly specialized research, but possibly providing provenance information unobtainable in any other way (Allen 1972; Heim 1974; Schnitzer 1977; Helmold 1985). Folk's (1968) work on quartz, chert, chalcedony and their varieties, is particularly noteworthy. Sophisticated analytical techniques are increasingly being used to define mineral varieties based on trace-element composition (Matter and Ramseyer 1985; Morton 1985a, b). For example, Morton (1985b) employed an electron microprobe to discriminate three populations of detrital garnets in a stratigraphic unit and related them to probable sources. Heller and Frost (1988) explored a huge new area: the study of stable and radiogenic isotope distributions in detrital grains and their use in fingerprinting source terranes.

Resistant grains, such as quartz and chert, plus the heavy minerals zircon, tourmaline, rutile, and garnet, are commonly recycled through several episodes of erosional derivation, sedimentation, lithification, uplift, and re-erosion. Usually, they show abundant evidence of this in the form of a high degree of rounding and the presence of authigenic overgrowths. It is usually difficult or impossible to determine the source of such grains because, in most cases, they could have come from several different stratigraphic units in the source area. Variations in color, crystal habit, or inclusions may help, but require more detailed research (Zuffa 1985b, 1987). Zircon studies have recently assumed a new importance as provenance indicators, as discussed in Sect. 6.6.2.

Extensive treatments of sandstone petrology and its application in sedimentological studies were given by

Pettijohn et al. (1973) and Blatt et al. (1980). Zuffa (1985a) edited a useful compilation of research and review papers, several of which are quoted above. In the remainder of this section, some examples of field studies are discussed to illustrate the kind of contribution petrography can make to regional basin analysis.

Two examples of the analysis of clastic detritus are provided here in order to illustrate the main uses of the method. Miall (1979) analyzed the petrography of the Isachen Formation, a Lower Cretaceous braided river deposit in Banks Basin, Arctic Canada. The bulk of the formation consists of texturally and mineralogically mature medium to coarse sandstone, showing abundant evidence of a recycled origin. Samples were obtained from outcrops divided into two broad groups, from the north and south parts of the island. Paleocurrent evidence indicates northward flowing trunk rivers with tributaries entering from the east (the west side of the basin is covered by younger strata but may also have been a source area). Initial attempts to recognize petrographic provinces in the sandstone were unsuccessful. Ternary plots show that the samples from the north and south end of the basin contain essentially the same mineral composition. Cluster analysis revealed a weak trend for feldspar to be more common in the north, and sedimentary rock fragments are slightly more abundant in the south. This was then emphasized by re-plotting the thin-section data in terms of three minor components, which shows that chert is also slightly more abundant in the north. It was found that a few rare components, such as detrital quartz-feldspar intergrowths, quartz with mica, tourmaline, or zircon inclusions, and zoned quartz, are all present almost exclusively in the north end of the basin. Analyses of these data in terms of potential sediment sources resulted in the map shown in Fig. 6.48. The bulk of the quartz sand and the sedimentary rock fragments were derived from mature Proterozoic sediments, chert was fed into the north end of the basin from various Lower Paleozoic sources, and the feldspar plus rare quartz and feldspar components appear to have been derived from a small Archean granodiorite pluton. A Middle to Upper Devonian sandstone unit that outcrops over wide areas adjacent to the north end of the basin appears to have contributed little detritus. It is too fine grained and has the wrong texture and mineralogy.

Rahmani and Lerbekmo (1975) carried out a detailed heavy-mineral analysis of the Cretaceous-Paleocene molasse sandstones derived from the Canadian Rocky Mountains. Factor analysis was used to determine mineral associations, and these resulted in the definition of a series of petrographic provinces (Fig. 6.49). It was found, unexpectedly, that these provinces were distributed in belts subparallel to structural strike. It was concluded that the dominant fluvial systems flowed longitudinally down the depositional basin from source areas to the west and northwest, not transversely, as would initially have been expected. The mineral provinces

could be distinguished because of variations in source-area geology.

In southern Alberta the Lower Cretaceous stratigraphy is characterized by a suite of thin marine-nonmarine sequences featuring multiple cycles of incision, which have created a very complex stratigraphy in which it may be quite difficult to correctly correlate individual units. Petrographic and chemostratigraphic techniques have been employed to supplement conventional petrophysical log correlation work. We report here on a petrographic study by Zaitlin et al. (2002). Figure 6.50 was compiled from thin section studies of core samples. Zaitlin et al. (2002, p. 38) described this compilation as follows:

> Summary lithostratigraphic comparison of the Basal Quartz as proposed in this study. A ternary diagram with quartz, chert, and clay-rich grains at the apices is effective in partitioning the petrographic data into distinctive populations. A second set of ternary diagrams, with intergranular, intragranular, and micro-porosity pore types at the apices is used to illustrate porosity fabric and reservoir quality. Key diagnostic grains and alteration features are used to exhibit the two cycles of mineralogical, textural, and reservoir quality cyclicity.

The stratigraphy is defined as two cycles, as shown in Fig. 6.50, but in practice the cycles consist of a complex of sheet sandstones with mutually erosive incised valleys. The petrographic work helps to differentiate these stratigraphic fragments. An example of the resulting reconstructions is shown in Fig. 6.51.

### 6.6.2  Provenance Studies Using Detrital Zircons

The heavy mineral zircon has been found to be a particularly valuable subject for provenance studies. It is highly resistant and survives multiple recycling events, which means that it can be used to identify the ultimate source of the detritus in which it is found. Zircon grains are now readily dated with the Uranium/Lead method, and the advent of sensitive high-resolution ion microprobe (SHRIMP) and laser-ablation–inductively coupled plasma–mass spectrometry (LA-ICP-MS), means that large populations of detrital zircon grains can be analyzed quickly (Thomas 2011). With care, it is possible to use zircon assemblages to map sediment sources as an aid to paleogeographic reconstruction. It is, of course, necessary to take into account the issue of recycling, which can be done by examining all aspects of a sandstone's detrital petrography. For example, in the case of the Banks Island study discussed briefly above (Fig. 6.48) it is highly relevant that the sand grains of the unit under discussion (the Isachsen Formation) are much coarser and more mature than those of the Devonian Melville Island Group, which is the immediately underlying unit around much of the basin margin, indicating that this unit could not

**Fig. 6.48** The depositional basin (*shaded area*) of the Lower Cretaceous Isachsen Formation of the Banks Island area, Arctic Canada showing interpreted source rocks of the Isachsen sandstones (Miall 1979)

have provide the bulk of the sand in the Isachsen Formation. Given these types of caveats, zircon studies in North America have generated some surprising results, as discussed below.

Zircon studies on the North American continent began with a detailed study of the distribution and ages of zircons in potential sources terranes in the Cordilleran province along the western margin of the continent (Gehrels et al. 1995). Given the long-continued history of terrane accumulation that has built the continent, this information provides a quantitative basis for assigning zircons to ultimate sources based on their ages. In most sandstones, analysis demonstrates that there are several zircon populations exhibiting distinct age ranges, and it is usually a simple matter to relate these age populations to the age distributions of the ultimate sources. For example, Fig. 6.52 illustrates the

distribution of zircon ages in five samples from the Oil Sands of northern Alberta. This shows the importance of the Grenville as a sediment source, a feature of most zircon distributions from North American sources, and which indicates dramatic changes in sediment sources and the paleogeography of the major river systems through the Phanerozoic. Dickinson and Gehrels (2003) were amongst the first to realize the importance of distant sediment sources, such as the Grenville and Appalachian orogens, in the sourcing of much of the sand now constituting the Late Paleozoic and Mesozoic clastic successions of the Colorado Plateau area of the southwestern United States.

Raines et al. (2013) investigated the zircon geochronology of Upper Jurassic and lowermost Cretaceous strata in the Alberta foreland basin. Conventional petrographic and paleocurrent studies had long been interpreted in terms of

**Fig. 6.49** Heavy mineral provinces, as determined by factor analysis, in three formations within the upper molasse wedge of Alberta. *Arrows* show interpreted dispersal directions. **a** Belly River Formation (Upper Cretaceous); **b** Edmonton Formation (Cretaceous-Tertiary); **c** Paskapoo Formation (Tertiary) (Rahmani and Lerbekmo 1975)

sources to the south and west, but considerable additional detail was obtained by these new methods. The zircon data indicate sediment in the early foreland basin was delivered via two principal sedimentary systems: a south-to- north axial river system, and transverse fluvial systems that emanated from the adjacent Cordillera (Fig. 6.53). Of particular interest are derived zircons from contemporary (Jurassic) strata preserved in the Michigan basin, which confirm the importance of an Appalachian source for some of the detritus. Speculating from this single data point, it is suggested that a network of tributaries fed sediment into the axial system flowing through the Alberta Basin from a wide area to the southeast (Fig. 6.53).

Combining these studies and others, of sands in the southern United States, Blum and Pecha (2014) generated two speculative maps that showed the likely complete re-organization of the interior drainage of the North American continent that evolved with the elevation of the Cordilleran orogen in the mid-Mesozoic (Fig. 6.54). Prior to this time, large-scale drainage systems appeared to have delivered abundant detritus from the Appalachian orogen to the west of the continent (Dickinson and Gehrels 2003). Large-scale river systems developed oriented in the reverse direction once significant sediment sources arose in the paths of these rivers, with significant transport eastward. Drainage into Hudson Bay, as outlined in the Paleocene map, was earlier suggested by McMillan (1973) and Duk-Rodkin and Hughes (1994).

### 6.6.3 Chemostratigraphy

The development of high-precision analytical methods in the last few decades has facilitated the development of a range of new techniques for supplementing lithostratigraphic correlations in regional studies, and several detailed studies have been reported.

Pearce et al. (1999) provided a general discussion of chemostratigraphic methods, based on the few published sources available at that time, and also drawing on unpublished, proprietary studies carried out in petroleum industry laboratories. They made a number of useful general points. For example, studies of sandstones need to take into account grain size effects. Absolute elemental concentrations decrease in coarser-grain size fractions owing to dilution by quartz. Also, it is difficult to develop homogenous samples in coarser sandstone fractions, and so finer sands are preferred. Diagenesis may modify compositions, with Ca, Na, K, Pb, Rb, Sr and certain REEs liable to reduce in concentration during the dissolution of feldspars. Therefore chemostratigraphic work makes preferential use of the more

**Fig. 6.50** Characterization of the petrography of units within the Lower Cretaceous "Basal Quartz" in southern Alberta. These data are used to identify and correlate these units in areas of complex stratigraphy (Zaitlin et al. 2002, Fig. 5, p. 38)

immobile elements Al, Ti, Zr, Nb and Cr. Wright et al. (2010) pointed out that sandstone and mudstones have completely different whole-rock geochemestries. Sandstones, which typically contain a significant proportion of quartz, are silica dominated, whereas mudrocks are rich in the cations that constitute the lattice in clay minerals.

Wray and Gale (1993) focused on marl bands in the Cretaceous Chalk of the Anglo-Paris Basin, and demonstrated that they could be characterized (fingerprinted) based on their trace-element compositions, including Ba, Cu, Li, Na, Sc, Ti, V, Y, Zn, and Zr. They used this characterization to trace individual bands from the deeper parts of the basin into "poorly correlated, condensed successions" deposited over structurally positive areas. These correlations yielded insights into syndepositional fault activity and differential subsidence across the project area.

Wright et al. (2010) presented a detailed chemostratigraphic study of the Mannville and Colorado Group (Aptian-Albian) strata of southern Alberta, covering much the same stratigraphic internal in much the same area as the petrographic study by Zaitlin et al. (2002) reported in the previous section. In both cases, the objective was to develop practical tools for correlation in areas of complex stratigraphy. During the Early Cretaceous this area was undergoing repeated cycles of small-scale relative changes in sea level, but because this part of the Western Canada Sedimentary Basin was located on the distal fringe of the foreland basin,

accommodation changes were very slow. The result is an assemblage of thin, high-frequency sequences with multiple, mutually-erosive incised valleys, leading to a highly complex stratigraphic configuration (Fig. 6.55; see also Fig. 6.51). An example of the discrimination between depositional units that can be achieved using whole-rock geochemistry is shown by the binary plots in Fig. 6.56. Figure 6.57 consists of what are termed "synthetic chemical logs" based on core samples. An immediate practical use of these data is shown in Fig. 6.58. Petrophysical logs cannot distinguish different sandstone units, and in core it is not possible to discriminate between a local fluvial scour surface and a major sequence boundary. However, as demonstrated here, the chemostratigraphic signature of the successive incised-valley-fills is sufficiently different to permit discrimination, which is an essential starting point for constructing regional stratigraphic syntheses.

Wright et al. (2010) discussed the origins of these geochemical differences between the sandstones. Some are clearly related to different provenances, an interpretation supported by surface paleocurrent data, but some appear to have originated as changes in source area geology within a single source area, presumably related to erosional unroofing or to upstream changes in fluvial dispersal directions. All of these potential causes may become of importance if efforts are being made to reconstruct the paleogeographic evolution of the area.

**Fig. 6.51** An example of the complex Lower Cretaceous stratigraphy of southern Alberta. Individual lithosomes have been identified using the petrographic data shown in Fig. 6.50 (Zaitlin et al. 2002, Fig. 16, p. 50)

Experimental use of "forensic chemostratigraphy" has been used to map the path of a horizontal well. Hildred et al. (2014) used chemostratigraphic analysis of cuttings from vertical drill holes, to characterize four units, including a producing dolomite in the Bakken petroleum system of southwestern Alberta, and the formations that bracket it. They then took a series of cutting that had been generated along a horizontal well through this system, and were able to show how the hole had passed up and down through this stratigraphy, which contains several faulted offsets.

## 6.7  Paleocurrent Analysis

### 6.7.1  Introduction

The art of paleocurrent analysis was invented by the quintessential nineteenth century English amateur, Henry Clifton Sorby, who published his first paper on the subject in 1852. He understood the formation of "ripple drifted" and "drift bedded" structures (climbing ripples and large-scale crossbedding, respectively) and, in seven years, claimed to have made about 20,000 observations, a total that has probably never been equaled. Most of Sorby's work remained unpublished. He was far ahead of his time, as paleocurrent work did not become a routine analytical procedure until the 1950s. His work remained unappreciated and largely unknown until exhumed by Pettijohn (1962) and reinterpreted by Allen (1963b) (see also Miall 1996, Chap. 2).

The technique could not become a routine component of basin analysis until proper attention was paid to the description and classification of sedimentary structures (McKee and Weir 1953; Pettijohn 1962; Allen 1963a, b), and a start was made on the investigation of bedform hydraulics. The major breakthrough here was the development of the flow-regime concept by Simons and Richardson (1961) and Simons et al. (1965). These two topics are discussed at some length in Sects. 1.2.5 and 3.5.4.

**Fig. 6.52** Distribution of zircon ages in five sandstone samples from the Oil Sands of northern Alberta. Zircons of Grenville age (1.0–1.3 Ga) are particularly common, suggesting sources from the Grenville Province, which forms the edge of the Canadian Shield along the entire eastern margin of the continent (Blum and Pecha 2014)

Paleocurrent analysis is primarily an outcrop study. Oriented core can be used, but is rarely available. Subsurface information may also be obtained with the use of borehole imaging tools. The dipmeter, which makes use of four resistivity logs run at 90° around a borehole, may be used to reconstruct the attitude of surfaces intersecting the hole, including bedding planes, and fractures. The orientation of crossbedding is, theoretically, obtainable from such data, and service companies have promoted this use for the tool since the 1970s (see also Miall 1999, Chap. 5), but thin-bed resolution may be a problem, and the tool has not often been used. Modern digital imaging tools, including Schlumberger's Formation MicroScanner (FMS) and other, more recently developed tools, provide a full scan of the hole, and may be more practical (Pöppelreiter et al. 2010). In petroleum exploration wells, unoriented core may be oriented using a dipmeter, but correlation between the two is commonly difficult (Davison and Haszeldine 1984; Nelson et al. 1987). Figure 6.59 illustrates a modern borehole image of planar-tabular crossbedding and the dips calculated from it, indicating consistent westerly to northwesterly flow directions. Borehole imaging has not been widely used, mainly because of cost considerations, but as Fox et al. (2015) argued, this may be a false economy. For coal, metals, and other types of mining, where there may be large surface or underground outcrops, paleocurrent analysis may, in contrast, be extremely useful (e.g., Minter 1978; Long 2006).

Paleocurrent analysis can provide information on five main aspects of basin development:

1. the direction of local or regional paleoslope, which reflects tectonic subsidence patterns,
2. the direction of sediment supply,
3. the geometry and trend of lithologic units,
4. the depositional environment,
5. the architecture of bars and bedforms at the outcrop scale.

Examples of these applications are discussed in the ensuing paragraphs.

### 6.7.2  Types of Paleocurrent Indicators

Sedimentary structures and fabrics used in facies and paleocurrent studies are described and illustrated in Sect. 2.2.6. The following notes explain briefly how they yield current directions.

1. For ripple marks and crossbedding (Figs. 2.8 and 2.9), the inclination of foreset directions is generally downcurrent, because of the grain avalanching mechanism. Needless to say, it is not always this simple. Smith (1972) demonstrated that planar-crossbed sets in rivers commonly advance obliquely to the flow direction. Trough crossbeds do not yield accurate flow directions unless the analyst can observe the orientation of the trough axis (see below). Hunter (1981) discussed the flow of air around large eolian dunes and suggested that these, too, commonly advance obliquely to the wind direction. The use of statistical procedures and appropriate facies data usually circumvents these problems (this is also discussed below).
2. Channels and scours (Fig. 2.13) occur in many environments and may indicate the orientation of major erosive currents, such as those generating river or tidal channels and delta or submarine fan distributaries. However, the larger channels, which are those most likely to be of regional significance, are usually too large to be preserved in outcrops. They may be readily apparent on seismic records, particularly 3-D, as discussed in this chapter (which obviates the need for paleocurrent studies!).
3. Parting lineation or primary current lineation, the product of plane-bed flow conditions (Fig. 2.9d), is only visible on bedding-plane exposures. It usually yields directional readings of low variance because it forms during high-energy flow in river, delta, or tidal channels, when bars or other obstructions to flow are under water and flow-sinuosity is low. The structure indicates orientation but not direction of flow, because of the ambiguity between two equally possible readings at 180° to each

**Early Foredeep Deposits**
1: Elbow Falls, AB (H*)
2: Cadomin townsite, AB (H*)
3: Grande Cache, AB (H*)
4: Williston Lake, BC (*)
5: Southern Montana (H)
6: Northern Montana (I)

**Potential Northern Sandstone Sources**
7: Devonian (E)
8: Cambrian (E)
9: Pennsylvanian/Permian (E)
10: Ordovician (E)
11: Triassic (F)
12: Ordovician (E)
13: Pennsylvanian/Permian (H)
14: Cambrian (E)
15: Neoproterozoic (E)
16: Triassic (F)
17: Pennsylvanian/Permian (E)

**Potential Southern Sandstone Sources**
18: Ordovician (K)
19: Ordovician (L)
20: Jurassic Eolianite, CP3 (M)
21: Permian (G)
22: Pennsylvanian (G)
23: Devonian (G)
24: Cambrian (G)
25: Neoproterozoic (G)
26: Jurassic Eolianite, CP16 (M)
27: Jurassic Eolianite, Wingate (M)
28: Jurassic Eolianite, Entrada (M)
29: Jurassic Eolianite, Navajo (M)
30: Jurassic Eolianite, CP20 (M)
31: Jurassic Eolianite, CP12 (M)
32: Jurassic Eolianite, CP20 (M)
33: Jurassic Eolianite, CP24 (M)
34: Jurassic Eolianite, CP54 (M)
35: Upper Jurassic Fluvial x8 (N)
36: Jurassic, Michigan - Fluvial (J)

**Sources of Data**
* This study
A: Stott, 1998
B: Miles et al., 2012
C: Hamblin and Walker, 1979
D: Turner and Peterson, 2004
E: Gehrels and Ross, 1998
F: Ross et al., 1997
G: Lawton et al., 2010
H: Leier and Gehrels, 2011
I: Fuentes et al., 2009
J: Dickinson et al., 2010
K: Smith and Gehrels, 1994
L: Gehrels and Dickinson, 1995
M: Dickinson and Gehrels, 2009
N: Dickinson and Gehrels, 2010

**Fig. 6.53** Proposed paleogeography showing the sediment sources and interpreted (schematic) rivers systems for the Western Interior Basin during the Late Jurassic and earliest Cretaceous (Raines et al. 2013, Fig. 11, p. 752)

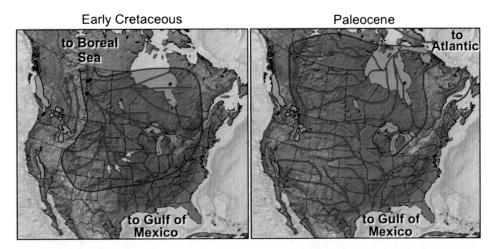

**Fig. 6.54** Early Cretaceous to Paleocene continental-scale drainage reorganization. The major Early Cretaceous (Aptian ca. 125–113 Ma) drainage system was sourced in the Appalachian Mountains, and routed through the Assiniboia paleovalley to the Boreal Sea (*green*). Tributaries joined from the southwestern United States and eastern Canada. Only a small part of North America south of the Appalachian-Ouachita cordillera (*orange*) was routed to the Gulf of Mexico. A similar map for the early Albian (ca. 113–108 Ma) would include tributaries from arc-dominated terrains in the northwestern United States and southwestern Canada. *Red arrows* indicate paleocurrents for basal Cretaceous fluvial deposits, whereas *yellow arrows* indicate orientations of bedrock paleovalleys that define the sub-Cretaceous unconformity. By the Paleocene, fluvial systems of the western United States were routed directly to the Gulf of Mexico, or to the Mississippi embayment, where they joined fluvial systems from the Appalachian Mountains: much of western Canada may have drained to the Atlantic. Locations of fluvial axes are schematic. *Pinkish areas* indicate locations of Cordilleran arc batholiths (after DeCelles 2004), (Blum and Pecha 2014, Fig. 4, p. 609)

**Fig. 6.55** The stratigraphy of the Upper Mannville-Colorado Group (Albian) of south-central Alberta, showing the complexity of the numerous incised valley systems that developed in this low-accommodation setting (Wright et al. 2010, Fig. 2, p. 96)

other. Usually, this can be resolved with reference to other structures nearby or to regional facies trends.

4. Clast transport by traction or in sediment-gravity flows commonly produces a measurable gravel fabric. Imbrication in traction current deposits occurs where platy clasts are stacked up in a shingled pattern, with their flattest surface dipping upstream and resting on the next clast downstream (Fig. 2.12). Because gravel is only moved under high-energy flow conditions, it tends to show low directional variance, like parting lineation.

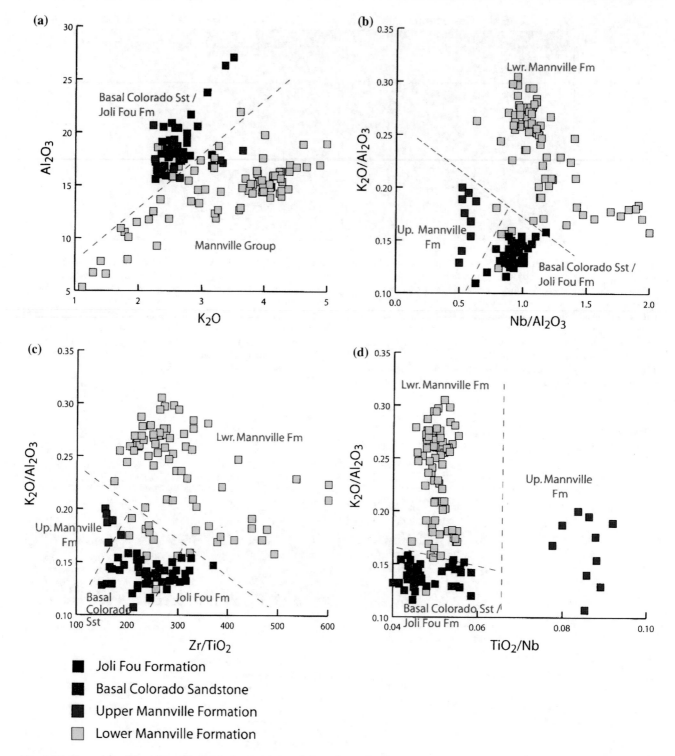

**Fig. 6.56** Examples of the binary diagrams showing how whole-rock geochemistry may be used to characterize stratigraphic units in the Mannville-Colorado Group (Aptian-Albian) of south-central Alberta (Wright et al. 2010, Fig. 3, p. 97)

Rust (1975) found that in gravel rivers variance decreased with increasing clast size, suggesting reduced flow sinuosity with increased stream power. Interpretation of clast fabrics in subaqueous, poorly-sorted conglomerates (diamictites) can help distinguish depositional

mechanisms, such as debris flow, ice rain-out, and glacial lodgment, as well as provide paleocurrent information (Eyles et al. 1987).

5. Sole markings (Fig. 2.15) are typically associated with the deposits of turbidity currents and fluidized or liquefied

**Fig. 6.57** "Synthetic chemical logs" developed from the sampling of cores through the Upper Mannville Formation and Colorado Group of south-central Alberta (Wright et al. 2010, Fig. 6, p. 100)

**Fig. 6.58** Whole-rock geochemical data are used here to differentiate incised-valley-fill systems within a single well-bore through the Lower Mannville Formation, south-central Alberta (Wright et al. 2010, Fig. 9, p. 105)

flow, where vortex erosion may occur at the base of a flow, and "tools" can be swept down a bedding surface. They also occur less commonly in other clastic environments. Their greatest use, however, is in the investigation of submarine-canyon and fan deposits. They are best seen on the undersides of bedding surfaces, where sandstone has formed a cast of the erosional feature in the underlying bed. Tool markings yield information on orientation but not direction, like parting lineation; flute marks are longitudinally asymmetric, with their deepest end lying upstream (Fig. 2.16).

6. Oriented plants, bones, shells, etc., do not respond systematically to the aligning effects of currents unless they are elongated. There may be ambiguity as to whether they are oriented transverse or perpendicular to current patterns, and there are other difficulties, for example, the tendency of fossils such as high-spired gastropods to roll in an arc. Fossils are usually only useful for local, specialized paleocurrent studies.

7. Slump structures generated on depositional slopes contain overfolds that may be aligned parallel to strike. They are therefore a potential paleoslope indicator, particularly on the relatively steep slopes of prodeltas and actively prograding continental margins. Friction at the margin of the slump may cause it to rotate out of strike alignment, so that a statistical approach to measurement is desirable. Potter and Pettijohn (1977, Chap. 6), Rupke (1978), and Woodcock (1979) reviewed their use as paleoslope indicators.

Several other paleocurrent indicators are reviewed by Potter and Pettijohn (1977), including some, such as sand-grain orientation and magnetic anisotropy, requiring detailed laboratory analysis of oriented samples. They are mainly of academic interest, where they may help solve a particular local problem, although laboratory techniques have improved in recent years to the point where studies of magnetic fabric could become routine (Hrouda and Stránik 1985; Eyles et al. 1987). The seven types of paleocurrent indicators listed above can be quickly and routinely measured in the field and have been widely used in basin studies.

### 6.7.3 Data Collection and Processing

Paleocurrent data should be carefully documented in the field. In the past, paleocurrent trends have sometimes been reported on the basis of a geologist's mental estimates of the range of indicated directions, but even if this is correct, it may result in a significant loss of useful information. For every paleocurrent observation recorded in field notes, the following information should be included:

1. location and (if relevant) precise position in a stratigraphic section;
2. structure type;

**Fig. 6.59** A borehole image log, showing planar-tabular crossbedding and the dip readings obtained from image interpretation. The red tadpoles indicate stratification and the green tadpoles are from crossbed foresets. Two separate representations of the same scan are shown, which were developed using different false-colour processing and visualization of the resistivity data. This image is of bitumen-saturated sandstone in the McMurray Formation of northern Alberta (Fox et al. 2015, reproduced with permission)

3. indicated current direction;
4. scale of structure (thickness of crossbed, depth and width of channel, mean or maximum clast size in imbricate gravels); and
5. local structural dip.

Current directions should be measured to the nearest 5° with a magnetic or sun-compass and corrected to true north wherever necessary. In the case of parting lineation and tool markings, the correct orientation of two possibilities at 180° can usually be identified by referring to other types of current structures nearby. Measurement accuracy greater than $\pm5°$ is difficult to achieve and is, in any case, rarely important. Difficulties in field measurement commonly arise because of incomplete exposure of crossbed sets. For example, two-dimensional exposures in flat outcrop faces cannot provide precise orientation information. DeCelles et al. (1983) and Bradley (1983) discussed the problem of incomplete or partial exposure of trough crossbed sets and ripple cross-lamination, and offered some solutions involving field measurement techniques and statistical treatment of the data.

Indicated current directions may need to be corrected for structural dip and fold plunge, otherwise significant directional distortions may result. This should be carried out as soon as possible, preferably after every day's work, in order that possible errors can be detected while there is still time to rectify them in the field. For linear structures such as parting lineation or sole markings, structural dips as high as 30° can be safely ignored, as they result in errors of less than 4°. However, the foreset dip orientation of crossbedding is significantly affected by structural dip, and should be corrected wherever the structural dip exceeds 10°. This subject is discussed further by Potter and Pettijohn (1977, Chap. 10), who illustrated a correction technique using a stereonet. Ramsey (1961) provided graphical solutions. Parks (1970) presented computer routines for the necessary trigonometric calculations. Wells (1988) provided a practical guide to field procedures, and recommended the use of a pocket calculator to perform structural corrections and data synthesis in the field.

Sooner or later the question will arise as to how many readings should be made? There is no single or simple answer to this problem because it depends on how many measurable current indicators are available for observation and what the objectives of the study are. Olson and Potter (1954) discussed the use of grid-sampling procedures and random selection of structures to measure, followed by calculation of reliability estimators to determine how many readings were necessary in order to be sure of determining correct directional trends. In this study, they were concerned only with determining regional paleoslope. We now understand a great deal about air and water flow patterns in different depositional environments, and a case could be made for measuring and recording every visible sedimentary structure. Such detailed data can be immensely useful in amplifying environmental interpretations and clarifying local problems, as discussed Sect. 6.7.5. Some selection may have to be made in areas of particularly good exposure. A practical compromise is to record every available structure along measured stratigraphic sections and to fill in the gaps between sections with spot (gridded or random) samples. This procedure permits the elaboration of both local and regional paleocurrent trends. When constructing lateral profiles of large outcrops, it is essential to document these with abundant paleocurrent determinations. Such data provide the third dimension to a two-dimensional outcrop and yield essential information regarding the shape and orientation of architectural elements (Sect. 3.5.11).

If the trend itself is important, 25 readings per sample station is commonly regarded as the minimum necessary for statistically significant small samples. However, the same or fewer readings plotted in map or section form can yield a great deal of environmental detail, whether or not their mean direction turns out to be statistically significant. Several hundred or a few thousand readings may be necessary for a thorough analysis of a complete basin.

A variety of statistical data-reduction and data-display techniques is available for paleocurrent work. The commonest approach is to group data into subsets according to stratigraphic or areal distribution criteria, display them visually in current-rose diagrams, and calculate their mean and standard deviation (or variance). The method of grouping the data into subsets has an important bearing on the interpretations to be made from them, as discussed in the next section. A current rose diagram is simply a histogram converted to a circular distribution. The compass is divided into 20, 30, 40 or 45° segments, and the rose is drawn with the segment radius proportional to number of readings or percent of total readings. A visually more correct procedure is to draw the radius proportional to the square root of the percent number of readings, so that the segment area is proportional to percent. Examples are given in Fig. 6.60. Wells (1988) warned that the choice of origin for rose diagrams and the method of subdivision of the segments can affect the visual appearance of the resulting diagram and the orientation of the modes. The calculation of means, which is done from raw data, is not affected, but Wells' warning indicates the need for caution in the visual interpretation of rose diagrams.

Potter and Pettijohn (1977) and Curray (1956) discussed arithmetic and vector methods for calculating mean, variance, vector strength, and statistical tests for randomness. Miall (1976) and Cant and Walker (1976) demonstrated the display and interpretation of data collected from vertical stratigraphic sections.

Statistical procedures, such as moving averages and trend analysis, are available for smoothing local detail and determining regional trends. An arbitrary grid may be drawn on the map; the mean current direction for the data in each group of four squares is then calculated and shown by an arrow at the center of this area. Each data point is thus used four times (except at the edges of the map). Examples of this method applied to paleocurrent data have been published by Potter and Pettijohn (1977).

The basin analyst should be wary of becoming too deeply enmeshed in the refinements of statistical methods. The use of probability tests is a useful curb on one's wilder flights of interpretive fancy, but there is a vast literature on the statistics of the circular distribution that seems to detract attention from very simple questions: Do the data make geological sense? Can they be correlated with trends derived from other methods, such as lithofacies mapping or petrographic data? Moving average and trend analyses are useful techniques for reducing masses of data to visually appealing maps, but they inevitably result in a loss of much interesting detail. The regional trends that emerge from such smoothing techniques are, in any case, usually readily deduced from modern stratigraphic data, such as grain-size trends, and the patterns of progradation and retrogradation within sequences. The more interesting approach is to relate individual readings to the architectural setting in which the structure resides, and to incorporate the data into an analysis of the bedform and lithosome hierarchy, as discussed in the next section.

### 6.7.4  The Bedform Hierarchy

A geologist collects paleocurrent measurements over individual bedding planes, through a local sequence in a quarry, along a lengthy dipping section in a river cut, or from several such sections throughout a map area. These readings differ from each other by varying amounts at different levels of this sampling hierarchy, but what do these differences mean? Statistically this is a classic example of a problem in analysis of variance (Olsen and Potter 1954; Kelling 1969). But what do the differences mean geologically? The larger the sampling area or the thicker the sampled section the greater will be the number of depositional events that contributed to the data set, and normally it will mean an increase in directional variance. This generalization can be systematized by the

**Fig. 6.60** Typical regional paleocurrent map, showing current rose diagrams and vector mean arrows for each of 10 outcrop areas, with station numbers. The map represents 165 field readings. Devonian fluvial sandstones, Somerset Island, Arctic Canada Most readings are consistent with an interpretation of deposition taking place on ephemeral sandy alluvial fans radiating from multiple point sources along the front the of the Boothia Uplift. Station 11C(G) represents giant crossbedding, interpreted as eolian dune stratification (Miall and Gibling 1978)

concept of the **bedform hierarchy,** as first noted by Allen (1966, 1967). Flow fields and the sedimentary structures arising from them are of up to six orders of magnitude, ranging from individual ripples to entire depositional systems. Allen (1966) and Miall (1974) applied this idea to river systems (Figs. 3.67 and 6.61). In Table 3.3 an attempt is made to expand it to other environments. Jackson (1975) showed that these ranks depend on three orders of time scale and physical scale, which we can now interpret as representing a range of Sedimentation Rate Scales (SRS) (See Sect. 8.4). Jackson (1975) proposed that they be grouped into three dynamic types:

1. **Microforms** are structures controlled by turbulent eddies in the fluid boundary layer. The time scale of variation is on the order of hours to seconds (deposits of SRS 1–3).
2. **Mesoforms** are structures that form in response to what Jackson (1975) termed **dynamic events**, such as hurricanes, seasonal floods, or eolian sand storms, when disproportionately large volumes of sediment are moved in geologically instantaneous time periods. The system may remain virtually unchanged between dynamic events (SRS 4–5).
3. **Macroforms** represent the long-term accumulation of sediment in response to major tectonic, geomorphic, and climatic controls, such as major bars and channels (SRS 6–7).

It is important for mapping purposes to attempt to relate the sampling scale to this hierarchy, so that the sources of directional variance can be interpreted intelligently. For example, a small quarry might be located entirely within the deposit of a single point bar or tidal delta, whereas a township (6 x 6 miles or 9.6 x 9.6 km) could encompass an entire delta lobe or barrier-inlet system. When interpreting microforms and mesoforms (the usual source of paleocurrent data) over large areas, it is important to remember that directional variance within each rank is summed to that of the higher ranks as far as is locally appropriate. Miall (1974) demonstrated the implications of this for fluvial-deltaic systems, such as those shown in Fig. 6.61, by compiling data for the various ranks summed to that of the single river or meander belt. A further discussion of the hierarchical nature of sedimentary facies is presented in Sect. 3.5.11.

### 6.7.5  Environment and Paleoslope Interpretations

A recommended first level of analysis is to plot current-rose diagrams and mean vectors for each outcrop or local outcrop group, if necessary separating the readings according to major facies variations. An example is illustrated in Fig. 6.60. If contemporaneous basin shape and orientation can be deduced from these data or from other mapping techniques, the paleocurrent data can be used interactively with lithofacies and biofacies criteria to interpret the depositional environment and to outline major depositional systems, such as deltas or submarine fan complexes. The geologist should examine the relationships between mean current directions in different lithofacies and in different outcrops. The number of modes in the rose diagrams (modality) and their orientation with respect to assumed

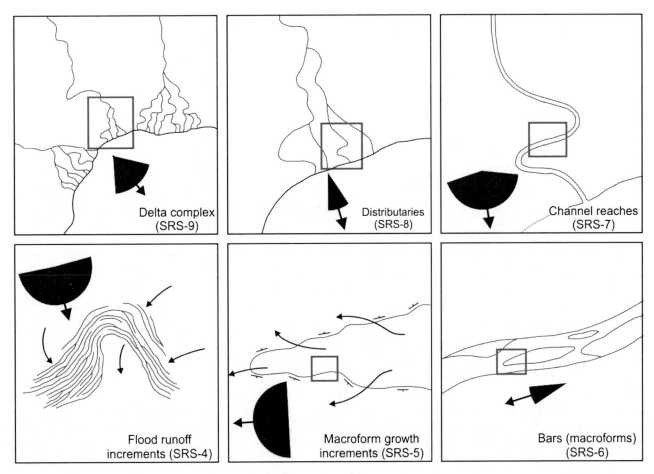

**Fig. 6.61** Paleocurrent data plotted according to geographic scale, encompassing increasingly large areas moving anticlockwise from the *bottom left*. Each diagram corresponds to the area shown in the *red square* in the next diagram in an anticlockwise direction (adapted from Miall 1974)

shoreline or lithofacies contours are also important information. Another useful approach is to plot individual readings or small groups of readings collected in measured sections at the correct position in graphic section logs (e.g., Fig. 6.65). Some examples of how to use these data are discussed below. Potter and Pettijohn (1977) provided extensive examples and an annotated bibliography. Advanced methods for use in the analysis of fluvial systems are described by Miall (1996, Chap. 9).

Useful introductions to paleocurrent models have been published by Pettijohn (1962) and Selley (1968), but our increased understanding of clastic (siliciclastic, plus clastic-carbonate and evaporite) depositional systems requires a fresh look at the problem, because some of the earlier models are now seen to be simplistic. Many ideas are to be found in Reading (1996), although this book does not deal explicitly with paleocurrent models. Harms et al. (1975, 1982) provided several useful case studies.

Figure 6.62 provides examples of paleocurrent data plotted regionally according to stratigraphic position. They indicate broad regional changes in dispersal directions that are interpreted in terms of changes in regional dispersal in response to tectonic tilting of the margin of the foreland basin.

Local paleocurrent distributions are commonly categorized as unimodal, bimodal, trimodal and polymodal. Each reflects a particular style of current dispersion. For example, the rose diagrams in Fig. 6.60 are unimodal to weakly bimodal, and vector mean directors are all oriented in easterly directions, roughly normal to the basin margin, with the exception of station 11C(G). The data were derived mainly from trough and planar crossbedding and parting lineation, and the environment of deposition is interpreted as braided fluvial (Miall and Gibling 1978). In some areas, for example, stations 6A, B, and C, there is a suggestion of a fanning out of the current systems, indicating possible deposition on large, sandy alluvial fans. The anomalous data set of station 11C(G) was derived from giant crossbed sets ranging from 1 to 6 m in thickness. They are interpreted as eolian dunes.

In high-sinuosity rivers, point bars dip at high angles to channel trends, whereas minor structures (mesoforms and

**Fig. 6.62** An example of regional paleocurrent distributions. Each map shows outcrop summaries, and the variability between the maps indicates significant changes in regional paleocurrent dispersal, reflecting tectonic tilting (Willis 2000; Miall and Arush 2001a)

microforms) migrating down the point bars are oriented subparallel to channel direction and therefore parallel to the strike of point-bar (epsilon) crossbedding (Fig. 6.63). This paleocurrent pattern is a useful diagnostic criterion for recognizing point bars and distinguishing them from other large, low-angle crossbed types, such as Gilbertian deltas, or the downstream-accretion surfaces that are particularly common in braided rivers (e.g., Puigdefábregas 1973; Mossop and Flach 1983; Miall 1996). An example of how meticulous field measurements may be used to make detailed reconstructions of fluvial architecture is shown in Fig. 6.64. Such measurements are typically made as part of the analysis of large two- or three-dimensional outcrops (e.g., Fig. 2.3).

Although fluvial deposits typically yield unimodal paleocurrent patterns on an outcrop scale, on a larger scale they may show much more complex patterns, for example centripetal patterns indicating internal drainage (e.g., Friend and Moody-Stuart 1972) or bimodal patterns with the two modes at 90° and occurring in different lithofacies assemblages. Rust (1981) reported two examples of the latter case. In each case, one mode, occurring in coarse conglomerates, was interpreted as the product of alluvial fans prograding transversely out from the basin margin. The other mode, in interbedded sandstones, represented a trunk-river system draining longitudinally.

Coastal regions where rivers debouch into an area affected by waves and tides can give rise to very complex paleocurrent patterns (Fig. 4.14). Bimodal, trimodal, or polymodal distributions may result, for example, in the tide-swept offshore bar described by Klein (1970). However, time-velocity

asymmetry of tidal currents can result in local segregation of currents, so that they are locally unimodally ebb- or flood-dominated (Fig. 3.27). This could cause confusion at the outcrop level, because the tidal deltas showing such paleocurrent patterns may consist of lithofacies that are very similar to some fluvial deposits (although recent work has demonstrated the existence of a range of very subtle facies criteria that can be used to recognize a tidal signature: see Sect. 3.5.4). In wave-dominated environments, current reversals generate distinctive internal ripple-lamination patterns and herringbone crossbedding, as illustrated in Figs. 2.9 c and 3.27. Paleocurrent analyses of these bimodal crossbeds may yield much useful information on the direction of wave attack and hence on shoreline orientation. The term paleoslope means little in this environment because there are a diversity of local slopes and because waves and tides are only marginally influenced by the presence and orientation of bottom slopes.

Several examples of sandy shoreline deposits and their paleocurrent patterns are given by Harms et al. (1975, 1982). One of the most instructive is the section through the Cretaceous Gallup Sandstone, New Mexico, illustrated in Fig. 6.65. Individual trough and ripple orientations are shown at the left (north toward top). The shoreline is known from regional mapping to be oriented northwest-southeast. The top 2–3 m of section consists of low-angle, cross-stratified, beach-accretion sets dipping at a few degrees toward the northeast (offshore). This structure is typical of intertidal wave-swash zones (Fig. 4.14) Facies 2 (about 3–14 m in depth) consists of fine- to medium-grained sandstone with

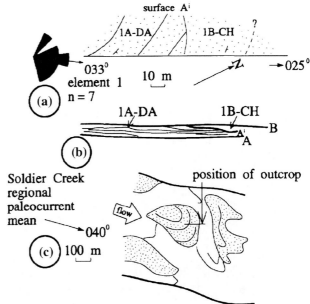

**Fig. 6.64** An example of the type of architectural interpretation that may be made from the analysis of a vertical outcrop face, using paleocurrent data. A sketch of the relevant portion of the outcrop is shown in diagram B. Diagram A, above, shows a reconstruction of a plan view of the top bounding surface (surface B), with the suggested orientations of accretion surface, based on the measurements of di and strike made at the outcrop face (dip-and-strike symbols). Diagram C shows the suggested interpretation of the outcrop, incorporating all orientation measurements of crossbedding (summarized in the rose diagram at top left). This is one of several worked examples provided by Miall (1994b)

**Fig. 6.63** Hypothetical examples of fluvial styles, indicating the range of orientations of dipping accretion surfaces and the variance of cross-bed orientations. **a** High-sinuosity river, such as a typical meandering river. Note that although there is a regional overlap of cross-bed and accretion-surface orientations, locally the two dips are oriented nearly perpendicular to each other, and the rose diagrams reflect this divergence. A counterpoint bar is shown. Such bars can be distinguished from the more common point bars by the fact that the accretion surfaces show curvature in plan view that is concave in a downdip direction, contrasting with the convex curvature of point-bar surfaces. Commonly such bars are characterized by finer grain sizes than other macroforms. **b** Low-sinuosity river—a typical sandy-braided river. Such rivers may show relatively high local channel sinuosity, in which case macroforms accreting downstream may also be oriented at a high angle to the regional trend. In the example shown here, mean directions are skewed to the southeast, diverging from the overall south-southeast channel orientation because of a major channel and bar complex (near the center of the reach) oriented in an eastward direction (Miall 1994b)

trough and some planar cross-stratification. Foreset-dip and trough-axis orientations are mainly toward the northwest, southwest, and south to southeast, with the latter predominating, suggesting onshore-directed, wave-generated currents with local reversals and longshore currents directed toward the southeast. The same longshore-current pattern is apparent in the subtidal lower-shoreface deposits, which are dominated by storm-generated hummocky cross-stratification. Wave ripples are present in the lowest facies, again showing predominantly onshore-directed currents.

Submarine-fan and other deep-marine deposits show three main paleocurrent patterns:

1. Individual submarine fans prograde out from the continental slope and therefore show radial paleocurrent patterns with vector mean directions oriented perpendicular to the regional basin strike (e.g., Stow et al. 1996; see Fig. 6.66). This pattern is typical of many continental margins, particularly divergent margins.

**Fig. 6.65** Lithofacies profile, crossbed data, and interpreted depositional environments of the Gallup Sandstone (Cretaceous), New Mexico (Harms et al. 1975)

2. In narrow oceans and many arc-related basins, deep-water sedimentation takes place in a trough oriented parallel to tectonic strike. Sediment-gravity flows, particularly low-viscosity turbidity currents, emerge from submarine fans and turn 90°, to flow longitudinally downslope, possibly for hundreds of kilometers. Current directions may be reversed by tilting of the basin. Many examples of this pattern have been published (e.g., Trettin 1970; Hesse 1974). Pickering and Hiscott (1985) used detailed grain-orientation measurements as a basis for reinterpreting anomalous paleocurrent measurements in a turbidite sequence. It was shown that supposed backset cross-lamination, attributed to upcurrent antidune migration, was in fact the product of turbidity currents that reversed direction as a result of deflection and reflection from the basin margins and internal topographic highs. Here is a case where subtle use was made of paleocurrent data to improve the sophistication of a basin-analysis case study.

3. Contour currents or boundary undercurrents flow parallel to continental margins and generate paleocurrent patterns oriented parallel to basin margins. Stow and Lovell (1979) and Stow and Holbrook (1984) discussed the use of facies criteria and paleocurrent data in distinguishing these deposits.

**Fig. 6.66 a** Isopach thickness map of Fan 3 constrained from integration of the core and outcrop data. Note the high rate of thinning to the south compared to the rest of the isopach map. Circles with crosses are outcrop logs, squares are borehole sites. **b** Paleocurrent arrows illustrate the general paleocurrent readings from Fan 3 across the entire study area from outcrop and Formation MicroImager measurements from boreholes. Overall the trend is to the north; however, note the strong eastward trend in the south of the study area where stratigraphic thinning rates are highest. The east-directed readings are interpreted as turbidity-current directions prior to deflection northwards off an ENE–WSW trending slope, possibly the basin-margin slope (Hodgson et al. 2006, Fig. 9, p. 28)

# References

Ahlborn, M., Stemmerik, L., and Kalstø, T.-K., 2014, 3D seismic analysis of karstified interbedded carbonates and evaporites, Lower Permian Gipsdalen Group, Loppa High, southwestern Barents Sea: Marine and Petroleum Geology, v. 56, p. 16-33.

Allen, J. R. L., 1963a, The classification of cross-stratified units, with notes on their origin: Sedimentology, v. 2, p. 93-114.

Allen, J. R. L., 1963b, Henry Clifton Sorby and the sedimentary structures of sands and sandstones in relation to flow conditions: Geologie en Mijnbouw, v. 42, p. 223-228.

Allen, J. R. L., 1966, On bed forms and paleocurrents: Sedimentology, v. 6, p. 153-190.

Allen, J. R. L., 1967, Notes on some fundamentals of paleocurrent analysis, with reference to preservation potential and sources of variance: Sedimentology, v. 9, p. 75-88.

Allen, P., 1972, Wealden detrital tourmaline: implications for north-western Europe: Journal of the Geological Society of London, v. 128, p. 273-294.

Angulo, S., and Buatois, L. A., 2012, Integrating depositional models, ichnology, and sequence stratigraphy in reservoir characterization: The middle member of the Devonian-Carboniferous Bakken Formation of subsurface southeastern Saskatchewan revisited: American Association of Petroleum Geologists Bulletin, v. 96, p. 1017-1043.

Bally, A. W., ed., 1987, Atlas of seismic stratigraphy: American Association of Petroleum Geologists Studies in Geology 27, in 3 vols.

Basu, A., 1985, Influence of climate and relief on composition of sands released at source areas, in Zuffa, G. G., ed., Provenance of arenites: D. Reidel Publishing Company, Dordrecht, p. 1-18.

Berg, R. R., 1968, point-bar origin of Fall River Sandstone reservoirs, northeastern Wyoming: American Association of Petroleum Geologists Bulletin, v. 52, p. 2116-2122.

Bhattacharya, J. P., 2011, Practical problems in the application of the sequence stratigraphic method and key surfaces: integrating observations from ancient fluvial-deltaic wedges with Quaternary and modelling studies: Sedimentology, v. 58, p. 120-169.

Blatt, H., Middleton, G. V., and Murray, R., 1980, Origin of sedimentary rocks; 2nd edition: Prentice-Hall Inc., Englewood Cliffs, New Jersey, 782 p.

Blum, M., and Pecha, M., 2014, Mid-Cretaceous to Paleocene North American drainage reorganization from detrital zircons: Geology, v. 42, p. 607-610.

Bohacs, K. M., Neal, J. E., and Grabowski, G. J., Jr., 2002, Sequence stratigraphy in fine-grained rocks: Beyond the correlative conformity: 22nd Annual Gulf Coast Section SEPM Foundation Bob F. Perkins Research Conference, p. 321-347.

Bouma, A. H., Normark, W. R., and Barnes, N. E., eds., 1985, Submarine fans and related turbidite systems: Springer-Verlag Inc., Berlin and New York, 351 p.

Bouma, A. H., Stelting, C. E., and Feeley, M. H., 1987, High-resolution seismic reflection profiles, in Bally, A. W., ed., Atlas of seismic stratigraphy: American Association of Petroleum Geologists Studies in Geology 27, v. 1, p. 72-94.

Bradley, D. C., 1983, Paleocurrent directions from two-dimensional exposures of cross laminae in the Devonian flysch of Maine: Journal of Geology, v. 95, p. 271-279.

Bridge, J. S., and Mackey, S. D., 1993, A theoretical study of fluvial sandstone body dimensions, in Flint, S. S., and Bryant, I. D., eds., The geological modelling of hydrocarbon reservoirs and outcrop analogues: International Association of Sedimentologists Special Publication 15, p. 213-236.

Bridge, J. S., and Tye, R. S., 2000, interpreting the dimensions of ancient fluvial channel bars, channels, and channel belts from wire-line logs and cores: American Association of Petroleum Geologists Bulletin, v. 84, p. 1205-1228.

Bristow, J. F., 2002, Real-time formation evaluation for optimal decision making while drilling: examples from the southern North Sea, in Lovell, M., and Parkinson, N., eds., Geological applications of well logs: American Association of Petroleum Geologists Methods in Exploration 13, p. 1-13.

Brown, A. R., 1985, The role of horizontal seismic sections in stratigraphic interpretation, in Berg, O. R., and Woolverton, D. G., eds., Seismic stratigraphy II: American Association of Petroleum Geologists Memoir 39, p. 37-47.

Brown, A. R., 1986, Interpretation of three-dimensional seismic data: American Association of Petroleum Geologists Memoir 42, 194 p.

Brown, A. R., 2011, Interpretation of three-dimensional seismic data, seventh edition, American Association of Petroleum Geologists Memoir 42, 646 p.

Burgess, P. M., Winefield, P., Minzoni, M., and Elders, C., 2013, Methods for identification of isolated carbonate buildups from seismic reflection data: American Association of Petroleum Geologists, v. 97, p. 1071-1098.

Cant, D. J., 1992, Subsurface facies analysis, in Walker, R. G., and James, N. P., eds., Facies models: response to sea level change: Geological Association of Canada, St. John's, Newfoundland, p. 27-45.

Cant, D. J., and Walker, R. G., 1976, Development of a braided-fluvial facies model for the Devonian Battery Point Sandstone, Quebec; Canadian Journal of Earth Sciences, v. 13, p. 102-119.

Cerveny, P. F., Naeser, N. D., Zeitler, P. K., Naeser, C. W., and Johnson, N. M., 1988, History of uplift and relief of the Himalaya during the past 18 million years: evidence from fission-track ages of detrital zircons from sandstones of the Siwalik Group, in Klein-spehn, K. L., and Paola, C., eds., New perspectives in basin analysis: Springer-Verlag Inc., Berlin and New York, p. 43-61.

Cross, T. A., and Lessenger, M. A., 1988, Seismic stratigraphy: Annual Review of Earth and Planetary Sciences, v. 16, p. 319-354.

Curray, J. R., 1956, The analysis of two-dimensional orientation data: Journal of Geology, v. 64, p. 117-131.

Davies, R. J., Cartwright, J. A., Stewart, S. A., Lappin, M., and Underhill, J. R., eds., 2004a, 3D Seismic technology: Application to the exploration of sedimentary Basins: Geological Society, London, Memoir 29, 355 p.

Davies, R. J., Stewart, S. A., Cartwright, J. A., Lappin, M., Johnston, R., Fraser, S. I., and Brown, A. R., 2004b, 3D seismic technology: are we realising its full potential? In Davies, R. J., Cartwright, J. A., Stewart, S. A., Lappin, M., and Underhill, J. R., eds., 3D Seismic technology: Application to the exploration of sedimentary Basins: Geological Society, London, Memoir 29, p. 1-9.

Davies, R. J., Posamentier, H. W., Wood, L. J., and Cartwright, J. A., eds., 2007, Seismic geomorphology: applications to hydrocarbon exploration and production: Geological Society, London, Special Publication 277, 274 p.

Davison, I., and Haszeldine, R. S., 1984, Orienting conventional cores for geological purposes: a review of methods: Journal of Petroleum Geology, v. 7, p. 461-466.

DeCelles, P. G., 2004, Late Jurassic to Eocene evolution of the Cordilleran thrust belt and foreland basin system, western USA: American Journal of Science, v. 304, p. 105–168.

DeCelles, P., Langford, R. P., and Schwartz, R. K., 1983, Two new methods of paleocurrent determination from trough cross-stratification: Journal of Sedimentary Petrology, v. 53, p. 629-642.

Dickinson, W. R., 1970, Interpreting detrital modes of graywacke and arkose: Journal of Sedimentary Petrology, v. 40, p. 695-707.

Dickinson, W. R., and Gehrels, G. E., 2003, U-Pb ages of detrital zircons from Permian and Jurassic eolian sandstones from the

Colorado Plateau, USA: paleogeographic implications: Sedimentary Geology, v. 163, p. 29-66.

Duk-Rodkin, A., Hughes, O.L., 1994. Tertiary–Quaternary drainage of the pre-glacial Mackenzie Basin. Quaternary International v. 22/ 23, p. 221 – 241.

Ethridge, F. G., and Schumm, S. A., 2007, Fluvial seismic geomorphology: a view from the surface, in Davies, R. J., Posamentier, H. W., Wood, L. J., and Cartwright, J. A., eds., Seismic geomorphology: applications to hydrocarbon exploration and production: Geological Society, London, Special Publication 277, p. 205-222.

Eyles, N., Day, T. E., and Gavican, A., 1987, Depositional controls on the magnetic characteristics of lodgement tills and other glacial diamict facies: Canadian Journal of Earth Sciences, v. 24, p. 2436-2458.

Fielding, C. R., Whittaker, J., Henrys, S. A., Wilson, T. J., and Naish, T. R., 2008, Seismic facies and stratigraphy of the Cenozoic succession in McMurdo Sound, Antarctica: implications for tectonic, climatic and glacial history, Palaeogeography, Palaeoclimatology, Palaeoecology, v. 260, p. 8-29.

Folk, R. L., 1968. Petrology of Sedimentary Rocks. Hemphill's, Austin, 170 p.

Fox, A., and Vickerman, K., 2015, The value of borehole image logs: Reservoir, Canadian Society of Petroleum Geologists, January, p. 30-34.

Franzinelli, E., and Potter, P. E., 1983, Petrology, chemistry and texture of modern river sands, Amazon river system: Journal of Geology, v. 91, p. 23-40.

Friend, P.F., and Moody-Stuart, M., 1972, Sedimentation of the Wood Bay Formation (Devonian) of Spitsbergen: regional analysis of a late orogenic basin: Norsk Polarinstitutt Skrifter Nr. 157.

Futalan, K., Mitchell, A., Amos, K., and Backe, G., 2012, Seismic facies analysis and structural intepretation of the Sandakan sub-basin, Sulu Sea, Philippines: American Association of Petroleum Geologists, Search and Discovery Article 30254.

Galloway, W. E., 1981, Depositional architecture of Cenozoic Gulf Coastal Plain fluvial systems, in Ethridge, F. G., and Flores, R. M., eds., Recent and ancient nonmarine depositional environments: models for exploration: Society of Economic Paleontologists and Mineralogists Special Publication 31, p. 127-156.

Gehrels, G. E., Dickinson, W. R., Ross, G. M., Stewart, J. H., Howell, D. G., 1995. Detrital zircon reference for Cambrian to Triassic miogeoclinal strata of western North America. Geology v. 23, p. 831-834.

Gennaro, M., Wonham, J. P., Gawthorpe, R. and Saelen, G., 2013, Seismic stratigraphy of the Chalk Group in the Norwegian Central Graben, North Sea: Marine and Petroleum Geology, v. 45, p. 236-266.

Graham, S. A., Tolson, R. B., DeCelles, P. G., Ingersoll, R. V., Bargar, E., Caldwell, M., Cavazza, W., Edwards, D. P., Follo, M. F., Handschy, J. F., Lemke, L., Moxon, I., Rice, R., Smith, G. A., and White, J., 1986, Provenance modeling as a technique for analyzing source terrane evolution and controls on foreland sedimentation, in Allen, P. A., and Homewood, P., eds., Foreland basins: International Association of Sedimentologists Special Publication 8, p. 425-436.

Hailwood, E. A., and Kidd, R. B., eds., 1993, High resolution stratigraphy: Geological Society, London, Special Publication 70, 357 p.

Hardage, B. A., 1985, Vertical seismic profiling—measurement that transfers geology to geophysics, in Berg, O. R., and Woolverton, D. G., eds., Seismic stratigraphy II: American Association of Petroleum Geologists Memoir 39, p. 13-34.

Harms, J. C., Southard, J. B., Spearing, D. R., and Walker, R. G., 1975, Depositional environments as interpreted from primary sedimentary structures and stratification sequences: Society of Economic Paleontologists and Mineralogists Short Course 2, 161 p.

Harms, J. C., Southard, J. B., and Walker, R. G., 1982, Structures and sequences in clastic rocks: Society of Economic Paleontologists and Mineralogists Short Course 9, Calgary.

Hart, B. S., 2013, Whither seismic stratigraphy: Interpretation: A journal of subsurface characterization: Society of Exploration Geophysicists and American Association of Petroleum Geologists, v. 1, p. SA3-SA20.

Heim, D., 1974, Uber die feldspate in Germanischen Buntsandstein, ihre Korngrossenabhangigkeit, verbreitung undpaleogeographische bedeutung: Geol. Rundschau, v. 63, p. 943-970.

Heller, P. L., and Frost, C. D., 1988, Isotopic provenance of clastic deposits: applications of geochemistry to sedimentary provenance studies, in Kleinspehn, K. L., and Paola, C., eds., New perspectives in basin analysis: Springer-Verlag, New York, p. 27-42.

Helmold, K. P., 1985, Provenance of feldspathic sandstones—the effect of diagenesis on provenance interpretations: a review: in Zuffa, G. G., ed., Provenance of arenites: D. Reidel Publishing Company, Dordrecht, p. 139-163.

Hesse, R., 1974, Long-distance continuity of turbidites: possible evidence for an Early Cretaceous trench-abyssal plain in the East Alps: Geological Society of America Bulletin, v. 85, p. 859-870.

Hildred, G. V., Martinez-Kulikowski, N., and Zaitlin, B. A., 2014, forensic chemostratigraphy: A tool to determine lateral wellbore placement: Search and Discovery Article 41359.

Hodgson, D. M., Flint, S. S., Hodgetts, D., Drinkwater, N. J., Johannessen, E. P., and Luthi, S. M., 2006, Stratigraphic evolution of fine-grained submarine fan systems, Tanqua depocenter, Karoo Basin, South Africa: Journal of Sedimentary Research, v. 76, p. 20-40.

Howard, J. L., 1993, The statistics of counting clasts in rudites: a review, with examples from the upper Paleogene of southern California, USA: Sedimentology, v. 40, p. 157-174.

Hrouda, F., and Stránik, Z., 1985: The magnetic fabric of the Zdanice thrust sheet of the Flysch belt of the West Carpathians: sedimentological and tectonic implications: Sedimentary Geology, v. 45, p. 125-145.

Hubbard, S. M., Smith, D. G., Nielsen, H., Leckie, D. A., Fustic, M., Spencer, R. J., and Bloom, L., 2011, Seismic geomorphology and sedimentology of a tidally influenced river deposit, Lower Cretaceous Athabasca oil sands, Alberta, Canada: American Association of Petroleum Geologists Bulletin, v. 95, p. 1123-1145.

Hunter, R. E., 1981, Stratification styles in eolian sandstones: some Pennsylvanian to Jurassic examples from the Western Interior U.S. A., in Ethridge, F. G., and Flores, R., eds., Recent and ancient nonmarine depositional environments: models for exploration: Society of Economic Paleontologists and Mineralogists Special Publication 31, p. 315-329.

Ingersoll, R. V., Bullard, T. F., Ford, R. L., Grimm, J. P., Pickle, J. D., and Sares, S. W., 1984, The effect of grain size on detrital modes: a test of the Gazzi-Dickinson point-counting method: Journal of Sedimentary Petrology, v. 54, p. 103-116.

Jackson, R. G., II, 1975, Hierarchical attributes and a unifying model of bed forms composed of cohesionless material and produced by shearing flow: Geological Society of America Bulletin, v. 86, p. 1523-1533.

Kauffman, E. G., 1986, High-resolution event stratigraphy: regional and global bio-events, in Walliser, O. H. ed., Global bioevents: Lecture Notes on Earth History: Springer-Verlag, Berlin, p. 279–335.

Kauffman, E. G., 1988, Concepts and methods of high-resolution event stratigraphy: Annual Reviews of Earth and Planetary Sciences, v. 16, p. 605-654.

Kelling, G., 1969: The environmental significance of cross-stratification parameters in an Upper Carboniferous fluvial basin: Journal of Sedimentary Petrology, v. 39, p. 857-875.

Klein, G. deV., 1970, Depositional and dispersal dynamics of intertidal sand bars: Journal of Sedimentary Petrology, v. 40, p. 1095-1127.

Krumbein, W. C., 1948, Lithofacies maps and regional sedimentary-stratigraphic analysis: American Association of Petroleum Geologists, v. 32, p. 1909-1923.

Locklair, R. E., and Sageman, B. B., 2008, Cyclostratigraphy of the Upper Cretaceous Niobrara Formation, Western Interior, U.S.A.: A Coniacian-Santonian timescale: Earth and Planetary Science Letters, v. 269, p. 539-552.

Long, D. G. F., 2006, Architecture of pre-vegetation sandy-braided perennial and ephemeral river deposits in the Paleoproterozoic Athabasca Group, northern Saskatchewan, Canada as indicators of Precambrian fluvial style: Sedimentary Geology, v. 190, p. 71-95.

Leckie, D. A., Bhattacharya, J., Gilboy, C. F. and Norris, B., 1994, Colorado\Alberta Group strata of the Western Canada Sedimentary Basin, in: Mossop, G. D., and Shetsen, I., eds., Geological Atlas of the Western Canada Sedimentary Basin: Canadian Society of Petroleum Geologists and Alberta Research Council, Calgary. p. 335–352.

Martin, J. H., 1993, A review of braided fluvial hydrocarbon reservoirs: the petroleum engineer's perspective, in: Best, J. L., and Bristow, C. S., eds., Braided rivers: Geological Society, London, Special Publication 75, p. 333-367.

Martinez, J. F., Cartwright, J. A., Burgess, P. M., and Bravo, J. V., 2004, 3D seismic interpretation of the Messinian unconformity in the Valencia Basin, Spain, in Davies, R. J., Cartwright, J. A., Stewart, S. A., Lappin, M., and Underhill, J. R., eds., 3D Seismic technology: Application to the exploration of sedimentary Basins: Geological Society, London, Memoir 29, p. 91-100.

Matter, A., and Ramseyer, K., 1985, Cathodoluminescence microscopy as a tool for provenance studies of sandstones, in Zuffa, G. G., ed., Provenance of arenites: D. Reidel Publishing Company, Dordrecht, p. 191-211.

McCrossan, R. G., 1961, Resistivity mapping and petrophysical study of Upper Devonian inter-reef calcareous shales of central Alberta, Canada: American Association of Petroleum Geologists Bulletin, v. 45, p. 441-470.

McKee, E. D., and Weir, G. W., 1953, Terminology for stratification in sediments: Geological Society of America Bulletin, v. 64, p. 381-389.

McMillan, N.J., 1973, Shelves of Labrador Sea and Baffin Bay, Canada, in: McCrossan, R. G., ed., The Future Petroleum Provinces of Canada; Their Geology and Potential, Canadian Society of Petroleum Memoir 1, p. 473-517.

Miall, A. D., 1974, Paleocurrent analysis of alluvial sediments: a discussion of directional variance and vector magnitude: Journal of Sedimentary Petrology, v. 44, p. 1174-1185.

Miall, A. D., 1976, Palaeocurrent and palaeohydrologic analysis of some vertical profiles through a Cretaceous braided stream deposit: Sedimentology, v. 23, p. 459-483.

Miall, A. D., 1979, Mesozoic and Tertiary geology of Banks Island, Arctic Canada: the history of an unstable craton margin: Geological Survey of Canada Memoir 387.

Miall, A. D., 1994b, Reconstructing fluvial macroform architecture from two-dimensional outcrops: examples from the Castlegate Sandstone, Book Cliffs, Utah: Journal of Sedimentary Research, v. B64, p. 146-158.

Miall, A. D., 1996, The geology of fluvial deposits: sedimentary facies, basin analysis and petroleum geology: Springer-Verlag Inc., Heidelberg, 582 p.

Miall, A. D., 1999, Principles of sedimentary basin analysis, Third edition: New York, N. Y.: Springer-Verlag Inc., 616 p.

Miall, A. D., 2002, Architecture and sequence stratigraphy of Pleistocene fluvial systems in the Malay Basin, based on seismic time-slice analysis: American Association of Petroleum Geologists Bulletin, v. 86, p. 1201-1216.

Miall, A. D., 2006, Reconstructing the architecture and sequence stratigraphy of the preserved fluvial record as a tool for reservoir development: a reality check; American Association of Petroleum Geologists Bulletin v. 90, p. 989-1002.

Miall, A. D., 2014a, Fluvial depositional systems: Springer-Verlag, Berlin 316 p.

Miall, A. D., and Arush, M., 2001a, The Castlegate Sandstone of the Book Cliffs, Utah: sequence stratigraphy, paleogeography, and tectonic controls: Journal of Sedimentary Research, v. 71, p. 536-547.

Miall, A. D., and Gibling, M. R., 1978, The Siluro-Devonian clastic wedge of Somerset Island, Arctic Canada, and some regional paleogeographic implications: Sedimentary Geology, v. 21, p. 85-127.

Minter, W. E. L., 1978, A sedimentological synthesis of placer gold, uranium and pyrite concentrations in Proterozoic Witwatersrand sediments, in Miall, A. D., ed., Fluvial Sedimentology: Canadian Society of Petroleum Geologists Memoir 5, p. 801-829.

Moonan, X. R., 2011, 4-D understanding of the evolution of the Penal/Barrackpore Anticline, southern sub-basin, Trinidad: American Association of Petroleum Geologists, Search and Discovery Article 90135.

Morton, A. C., 1985a, Heavy minerals in provenance studies, in Zuffa, G. G., ed., Provenance of arenites: D. Reidel Publishing Company, Dordrecht, p. 249-277.

Morton, A. C., 1985b, A new approach to provenance studies: electron microprobe analysis of detrital garnets from Middle Jurassic sandstones of the northern North Sea: Sedimentology, v. 32, p., 553-566.

Morton, A. C., and Hallsworth, C. R., 1994, identifying provenance-specific features of detrital heavy mineral assemblages in sandstones: Sedimentary Geology, v. 90, p. 241-256.

Morton, A. C., and Hallsworth, C. R., 1999, Processes controlling the composition of heavy minderal assemblages in sandstones: Sedimentary Geology, v. 124, p. 3-29.

Moscardelli, L., Wood, L. J., and Dunlap, D. B., 2012, Shelf-edge deltas along structurally complex margins" A case study from eastern offshore Trinidad: American Association of Petroleum Geologists, v. 96, p. 1483-1522.

Mossop, G. D., and Flach, P. D., 1983, Deep channel sedimentation in the Lower Cretaceous McMurray Formation, Athabasca Oil Sands, Alberta: Sedimentology, v. 30, p. 493-509.

Mutti, E., and Ricci-Lucchi, F., 1972, Le turbiditi dell'Appennino settentrionale: introduzione all' analisi di facies: Soc. Geol. Ital. Mem. 11, p. 161-199.

Myhr, D. W., and Meijer-Drees, N. C., 1976, Geology of the southeastern Alberta Milk River gas pool; in Lerand, M. M., ed., The sedimentology of selected clastic oil and gas reservoirs in Alberta: Canadian Society of Petroleum Geologists, p. 96-117.

Nelson, R. A., Lenox, L. C., and Ward, B. J., Jr., 1987, Oriented core: its use, error, and uncertainty: American Association of Petroleum Geologists Bulletin, v. 71, p. 357-367.

Nilsen, T., Shew, R., Steffens, G., and Studlick, J., eds., 2007, Atlas of deep-water outcrops: American Association of Petroleum Geologists, Studies in Geology 56, 504 p.

Oliver, T. A., and Cowper, N. W., 1963, Depositional environments of the Ireton Formation, central Alberta: Bulletin of Canadian petroleum Geology, v. 11, p. 183-202.

Oliver, T. A., and Cowper, N. W., 1965, Depositional environments of the Ireton Formation, central Alberta: Bulletin of the American Association of Petroleum Geologists, v. 49, p. 1410-1425.

Olson, J. S., and Potter, P. E., 1954, Variance components of crossbedding direction in some basal Pennsylvanian sandstones of the eastern Interior Basin: statistical methods: Journal of Geology, v. 62, p. 26-49.

Parks, J. M., 1970, Computerized trigonometric solution for rotation of structurally tilted sedimentary directional features: Geological Society of America Bulletin, v. 81, p. 537-540.

Pearce, T. J., Besly, N. M., Wray, D. S, and Wright, D. K., 1999, Chemostratigraphy: a method to improve interwell correlation in barren sequences — a case study using onshore Duckmantian/Stephanian sequences (West Midlands, U.K.): Sedimentary Geology, v. 124, p. 197-220.

Pettijohn, F. J., 1962, Paleocurrents and paleogeography: American Association of Petroleum Geologists Bulletin, v. 46, p. 1468-1493.

Pettijohn, F. J., Potter, P. E., and Siever, R., 1973, Sand and sandstone: Springer-Verlag, New York, 618 p.

Pickering, K. T., and Hiscott, R. N., 1985, Contained (reflected) turbidity currents from the Middle Ordovician Cloridorme Formation, Quebec, Canada: an alternative to the antidune hypothesis: Sedimentology, v. 32, p. 373-394.

Pirmez, C., Prather, B. E., Mallarino, G., O'Hayer, W. W., Droxler, A. W, and Winker, C. D., 2012, Chronostratigraphy of the Brazos-Trinity depositional system, western Gulf of Mexico: implications for deepwater depositional models, in Applications of the Principles of seismic geomorphology to continental slope and base-of-slope systems: case studies from seafloor and near-seafloor analogues: Society for Sedimentary Geology Special Publication 99, p. 111-143.

Pöppelreiter, M., Garcia-Carballido, C., and Kraaijveld, M. A., eds. 2010, Dipmeter and borehole image log technology, American Association of Petroleum Geologists, Memoir 92.

Posamentier, H. W., 2000, Seismic stratigraphy into the next millennium: a focus on 3-D seismic data: American Association of Petroleum Geologists, Annual Conference, New Orleans, p. 16-19.

Posamentier, H. W., and Kolla, V., 2003, Seismic geomorphology and stratigraphy of depositional elements in deep-water settings: Journal of Sedimentary Research, v. 73, p. 367-388.

Posamentier, H. W., Davies, R. J., Cartwright, J. A., and Wood, L., 2007, Seismic geomorphology – an overview, in Davies, R. J., Posamentier, H. W., Wood, L. J., and Cartwright, J. A., eds., Seismic geomorphology: applications to hydrocarbon exploration and production: Geological Society, London, Special Publication 277, p. 1-14.

Potter, P. E., and Pettijohn, F. J., 1977, Paleocurrents and basin analysis, 2nd edition: Academic Press, San Diego, California, 296 p.

Puigdefábregas, C., 1973, Miocene point-bar deposits in the Ebro Basin, northern Spain: Sedimentology, v. 20, p. 133-144.

Rahmani, R. A., and Lerbekmo, J. F., 1975, Heavy mineral analysis of Upper Cretaceous and Paleocene sandstones in Alberta and adjacent areas of Saskatchewan, in Caldwell, W. G. E. ed., The Cretaceous system in the Western Interior of North America: Geological Association of Canada Special Paper 13, p. 607-632.

Raines, M. K., Hubbard, S. M., Kukulski, R. B., Leier, A. L., and Gehrels, G. E., 2013, Sediment dispersal in an evolving foreland: detrital zircon geochronology from Upper Jurassic and lowermost Cretaceous strata, Alberta Basin, Canada, Geological Society of America Bulletin, v. 125, p. 741-755

Ramsey, J. G., 1961, The effects of folding upon the orientation of sedimentation structures: Journal of Geology, v. 69, p. 84-100.

Ratcliffe, K. T., and Zaitlin, B. A., eds., 2010, Application of modern stratigraphic techniques: theory and case histories: Society for Sedimentary Geology (SEPM) Special Publication 94, 241 p.

Reading, H. G., ed., 1996, Sedimentary environments: processes, facies and stratigraphy, third edition: Blackwell Science, Oxford, 688 p.

Rich, J. L., 1951, Three critical environments of deposition and criteria for recognition of rocks deposited in each of them: Geological Society of America Bulletin, v. 62, p. 1-20.

Rosenthal, L. R. P. and Walker, R. G., 1987, Lateral and vertical facies sequences in the Upper Cretaceous Chungo Member, Wapiabi Formation, southern Alberta: Canadian Journal of Earth Sciences, v. 24, p. 771–783.

Rosenthal, L. R. P., Leckie, D. A. and Nadon, G., 1984, Depositional cycles and facies relationships within the Upper Cretaceous Wapiabi and Belly River formations of west-central Alberta: Canadian Society of Petroleum Geologists, Field Trip Guide Book, 54.

Rupke, N. A., 1978, Deep clastic seas, in Reading, H. G., ed., Sedimentary environments and facies: Blackwell, Oxford, p. 372-415.

Rust, B. R., 1975, Fabric and structure in glaciofluvial gravels, in Jopling, A. V., and McDonald, B. C., eds., Glaciofluvial and glaciolacustrine sedimentation: Society of Economic Paleontologists and Mineralogists Special Publication 23, p. 238-248.

Rust, B. R., 1981, Alluvial deposits and tectonic style: Devonian and Carboniferous successions in eastern Gaspe, in Miall, A. D., ed., Sedimentation and tectonics in alluvial basins: Geological Association of Canada Special Paper 23, p. 49-76.

Ryan, W. B. F., and Cita, M. B., 1978, The nature and distribution of Messinian erosional surfaces — indicators of a several-kilometres-deep Mediterranean in the Miocene: Marine geology, v. 27, p. 193-230.

Safronova, P. A., Henriksen, S, Andreassen, K., Laberg, J. S., and Vorren, T. O., 2014, Evolution of shelf-margin clinoforms and deep-water fans during the middle Eocene in the Sørvestsnaget Basin, southwest Barents Sea: American Association of Petroleum Geologists Bulletin, v. 98, p. 515-544.

Sageman, B. B., Singer, B. S., Meyers, S. R., Siewert, S. E., Walaszczyk, I., Condon, D. J., Jicha, B. R., Obradovich, J. D., and Sawyer, D. A., 2014, Integrating $^{40}Ar/^{39}Ar$e,U-Pb and astronomical clocks in the Cretaceous Niobrara Formation, Western Interior Basin, USA: Geological Society of America Bulletin, v. 126, p. 956-973.

Saller, A. H., Noah, J. T., Ruzuar, A. P., and Schneider, R., 2004, Linked lowstand delta to basin-floor fan deposition, offshore Indonesia: an analog for deep-water reservoir systems: American Association of Petroleum Geologists Bulletin, v. 88, p. 21-46.

Sangree, J. B., and Widmier, J. M., 1977, Seismic stratigraphy and global changes of sea level, part 9: Seismic interpretation of clastic depositional facies, in Payton, C. E., ed., Seismic stratigraphy — applications to hydrocarbon exploration: American Association of Petroleum Geologists Memoir 26, p. 165-184.

Savit, C. H., and Changsheng, W. V., 1982, Geophysical characterization of lithology—application to subtle traps; in Halbouty, M. T., ed., The deliberate search for the subtle trap: American Association of Petroleum Geologists Memoir 32, p. 11-30.

Sayago, J., Di Lucia, M., Mutti, M.,Cotti, A., Sitta, A., Broberg, K., Przybylo, A., Buonaguro, R. and Zimina, O., 2012, Characterization of a deeply buried paleojarst terrain in the Loppa High using core data and multiattribute seismic facies classification: American Association of Petroleum Geologists, v. 96, p. 1843-1866.

Schnitzer, W. A., 1977, Die quarzkornfarben-methoden und ihre Bedeutung fur die stratigraphische und palaogeographische erforschung psammitischer sedimente: Erlanger Geol. Abh., Heft 103, 28 p.

Selley, R. C., 1968, A classification of paleocurrent models: Journal of Geology, v. 76, p. 99-110.

Shank, J. A., and Plint, A. G., 2013, Allostratigraphy of the Upper Cretaceous Cardium Formation in subsurface and outcrop in southern Alberta, and correlation to equivalent strata in northwestern Montana: Bulletin of Canadian Petroleum Geology, v. 61, p. 1-40.

Sharma, P. V., 1986, Geophysical methods in geology: 2$^{nd}$. Edition: Elsevier Scientific Publications, Amsterdam, 442 p.

Sheriff, R. E., 1985, Aspects of seismic resolution, in Berg, O. R., and Woolverton, D. G., eds., Seismic stratigraphy II: American Association of Petroleum Geologists Memoir 39, p. 1-10.

Simons, D. B., and Richardson, E. V., 1961, Forms of bed roughness in alluvial channels: American Society of Civil Engineers Proceedings, v. 87, No. HY3, p. 87-105.

Simons, D. B., Richardson, E. V., and Nordin, C. F., 1965, Sedimentary structures generated by flow in alluvial channels, in Middleton, G. V., ed., Primary sedimentary structures and their hydrodynamic interpretation: Society of Economic Paleontologists and Mineralogists Special Publication 12, p. 34-52.

Sloss, L. L., Krumbein, W. C., and Dapples, E. C., 1949, Integrated facies analysis; in Longwell, C. R., ed., Sedimentary facies in geologic history: Geological Society of America Memoir 39, p. 91-124.

Smith, N. D., 1972, Some sedimentological aspects of planar cross-stratification in a sandy braided river: Journal of Sedimentary Petrology, v. 42, p. 624-634.

Smith, T., and Leone, J., 2014, Shallow onlap model for Ordovician and Devonian organic-rich shales, New York State: American Association of Petroleum Geologists Search and Discovery Article #50911.

Southcott, A., 2014, 3D seismic proves its value in Bakken geosteering: Search and Discovery article 41435.

Stark, T. J., Zeng, H., and Jackson, A., 2013, An introduction to this special section: Chronostratigraphy: The Leading Edge, v. 32, p. 132-138.

Stow, D. A. V., and Holbrook, J. A., 1984, North Atlantic contourites: an overview, in Stow, D. A. V., and Piper, D. J. W., eds., Fine-grained sediments: deep-water processes and facies: Geological Society of London Special Publication 15, p. 245-256.

Stow, D. A. V., and Lovell, J. P. B., 1979, Contourites: their recognition in modern and ancient sediments: Earth Science Reviews, v. 14, p. 251-291.

Stow, D. A. V., Reading, H. G., and Collinson, J. D., 1996, Deep seas, in Reading, H. G., Sedimentary environments: processes, facies and stratigraphy, 3$^{rd}$ edition: Blackwell Science, Oxford, p. 395-453.

Suttner, L. J., 1974, Sedimentary petrographic provinces: an evaluation, in Ross, C. A., ed., Paleogeographic provinces and provinciality: Society of Economic Paleontologists and Mineralogists Special Publication 21, p. 75-84.

Suttner, L. J., Basu, A., Decker, J., Helmhold, K. P., Ingersoll, R. V., et al., 1985: The effect of grain size on detrital modes: a test of the Gazzi-Dickinson point-counting method: discussions and replies: Journal of Sedimentary Petrology, v. 55, p. 616-621.

Taylor, M. S. G., 2010, Visualization and the use of real time data while geosteering – onshore Algeria: Search and Discovery Article 40592.

Thomas, W. A., 2011, detrital-zircon geochronology and sedimentary provenance: Lithosphere, Geological Society of America, v. 3, p. 304-308.

Trettin, H. P., 1970, Ordovician-Silurian flysch sedimentation in the axial trough of the Franklinian Geosyncline, northeastern Ellesmere Island, Arctic Canada, in Lajoie, J., ed., Flysch sedimentology in North America: Geological Association of Canada Special Paper 7, p. 13-35.

Tye, R. S., 1991, Fluvial-sandstone reservoirs of the Travis Peak Formation, East Texas Basin, in Miall, A. D., and Tyler, N., eds., The three-dimensional facies architecture of terrigenous clastic sediments, and its implications for hydrocarbon discovery and recovery: Society of Economic Paleontologists and Mineralogists Concepts and Models Series, v. 3, p. 172-188.

Vail, P. R., Mitchum, R. M., Jr., Todd, R. G., Widmier, J. M., Thompson, S., III, Sangree, J. B., Bubb, J. N., and Hatlelid, W. G., 1977, Seismic stratigraphy and global changes of sea-level, in Payton, C. E., ed., Seismic stratigraphy - applications to hydrocarbon exploration: American Association of Petroleum Geologists Memoir 26, p. 49-212.

Veeken, 2007, Seismic stratigraphy, basin analysis and reservoir characterization: Elsevier, Amsterdam, Seismic Exploration, v. 37, 509 p.

Visher, G. S., 1965, Use of vertical profile in environmental reconstruction; American Association of Petroleum Geologists Bulletin: v. 49, p. 41-61.

Walaszczyk, I., Shank, J. A., Plint, A. G., and Cobban, W. A., 2014, Interregional correlation of disconformities in Upper Cretaceous strata, Western Interior Seaway: Biostratigraphic and sequence-stratigraphic evidence for eustatic change: Geological Society of America Bulletin, v. 126, p. 307-316.

Walker, R. G., 1978, Deep-water sandstone facies and ancient submarine fans: models for exploration for stratigraphic traps: American Association of Petroleum Geologists Bulletin, v. 62, p. 932-966.

Wells, N. A., 1988, Working with paleocurrents: Journal of Geological Education, v. 35, p. 39-43.

Willis, A. J., 2000, Tectonic control of nested sequence architecture in the Sego Sandstone, Neslen Formation, and Upper Castlegate Sandstone (Upper Cretaceous), Sevier Foreland Basin, Utah, U.S. A.: Sedimentary Geology, v. 136, p. 277–318.

Woodcock, N. H., 1979, The use of slump structures as palaeoslope orientation estimators: Sedimentology, v. 26, p. 83-99.

Wray, D. S., and Gale, A. S., 1993, Geochemical correlation of marl bands in Turonian chalks of the Anglo-Paris basin, in Hailwood, E. A., and Kidd, R. B., eds., High resolution stratigraphy: Geological Society, London, Special Publication 70, p. 211-226.

Wright, A. M., Ratcliffe, K. T., Zaitlin, B. A., and Wray, D. S., 2010, The application of chemostratigraphic techniques to distinguish compound incised valleys in low-accommodation incised-valley systems in a foreland-basin setting: an example from the Lower Cretaceous Mannville Group and Basal Colorado Sandstone (Colorado Group), Western Canadian Sedimentary Basin, in Ratcliffe, K. T., and Zaitlin, B. A., eds., Application of modern stratigraphic techniques, Society for Sedimentary geology, Special Publication 94, p. 93-107.

Zaitlin, B. A., Warren, M. J., Potocki, D., Rosenthal, L., and Boyd, R., 2002, depositional styles in a low accommodation foreland basin setting: an example from the Basal Quartz (Lower Cretaceous), southern Alberta: Bulletin of Canadian Petroleum Geology, v. 50, p. 31-72.

Zeng, H., Backus, M., Barrow, K. T., and Tyler, N., 1996, Facies mapping from three-dimensional seismic data: potential and guidelines from a Tertiary sandstone-shale sequence model, Powderhorn field, Calhoun County, Texas: American Association of Petroleum Geologists Bulletin, v. 80, p. 16-46.

Zeng, H., Henry, S. C., and Riola, J. P., 1998, Stratal slicing, Part II: Real 3-D seismic data: Geophysics, v. 63, p. 514-522.

Zeng, H., Loucks, R., Janson, X., Wang, G., Xia, Y., Yuan, B., and Xu, L., 2011, Three-dimensional seismic geomorphology and analysis of the Ordovician paleokarst drainage system in the central Tabei Uplift, northern Tarim Basin, western China: American Association of Petroleum Geologists, v. 95, p. 2061-2083.

Zeng, H., Zhu, X., and Zhu, R., 2013, new insights into seismic stratigraphy of shallow-water progradational sequences: subseismic clinoforms: Interpretation: A journal of subsurface characterization:

Society of Exploration Geophysicists and American Association of Petroleum Geologists, v. 1, p. SA35-SA51.

Zuffa, G. G., ed., 1985a, Provenance of arenites: D. Reidel Publishing Company, Dordrecht, 408 p.

Zuffa, G. G., 1985b, Optical analyses of arenites: influence of methodology on compositional results; in Zuffa, G. G., ed., Provenance of arenites: D. Reidel Publishing Company, Dordrecht, p. 165-189.

Zuffa, G. G., 1987, Unravelling hinterland and offshore paleogeography from deep-water arenites, in Leggett, J. K., and Zuffa, G. G., eds., Marine clastic sedimentology, concepts and case studies: Graham and Trotman, London, p. 39-61.

## Contents

## 7.1 Introduction

**Stratigraphy** is defined as the study of layered rocks. In the context of sedimentary geology in general, and of this book in particular, Stratigraphy is the discipline that pulls everything together. In Chaps. 2–5 of this book we deal with increasingly large and complex sedimentological concepts, and in Chap. 6 we discuss mapping methods, which are essentially methods for extending our interpretations beyond our immediate data points by interpolation and extrapolation. Here in this chapter we add the elements of chronostratigraphic dating and correlation, and demonstrate the interdisciplinary nature of modern stratigraphic methods.

Interpretations that focus on local to regional facies studies or sequence stratigraphy rely on the principles of Stratigraphy to create and confirm local and regional correlations. Rock units and the depositional environments in which they formed may be related to each other in a way that enables additional correlations to be made to, for example, regional and global climatic and tectonic events. Stratigraphic methods are required to construct such correlations, and a dependence on these methods increases with the scale of a project, from the local to the regional to the continental to the global.

Formal stratigraphic practices, including the definition of formations and of stages, had their origins in nineteenth-century field geology, beginning with William Smith (Conkin and Conkin 1984; Miall 2004), and have evolved into a set of carefully defined procedures for naming and correlating the various kinds of stratigraphic unit (Salvador 1994; NACSN 2005). They have also, for many years, been applied successfully to subsurface well data. These methods are mainly based on detailed lithostratigraphic and biostratigraphic information, the analysis of which is discussed later in this Chapter, along with other important aids to correlation, including radiometric dating and magnetic reversal stratigraphy. Lithostratigraphic classification of the sedimentary record remains the basic descriptive process for stratigraphic documentation, but the ultimate goal is now to develop

© Springer International Publishing Switzerland 2016
A.D. Miall, *Stratigraphy: A Modern Synthesis*, DOI 10.1007/978-3-319-24304-7_7

stratigraphic frameworks based on sequence stratigraphy (Catuneanu 2006; Catuneanu et al. 2011; see Sect. 7.7).

A broader, more regional approach to correlation generally is taken by those dealing with reflection seismic data (Veeken 2007). Commercial seismic work in frontier regions and the deep reflection profiles produced by groups such as the Consortium for Continental Reflection Profiling (COCORP) in the United States, Canada's Lithoprobe Project, and many other international projects, provide sweeping regional cross sections, within which correlations at the detailed level may be far from clear. This work may be supplemented by detailed three-dimensional reconstructions created using 3-D seismic methods applied to prospective rock volumes (Brown 2011).

With outcrop and well data, the problem may be to establish the regional framework from a mass of local detail, whereas with seismic data it is the detail that may be hard to see (depending on depth, on which seismic resolution depends). The ideal combination is, of course, a basin with a network of key exploration holes tied to regional seismic lines, plus local 3-D seismic volumes. Most advanced petroleum exploration projects now achieve this level of detail.

These differences in data type and scale have led to two different approaches to stratigraphic correlation. In industry exploration work in frontier regions, particularly the great offshore basins, a rather informal, pragmatic approach may be taken to such topics as biostratigraphic zonation and the naming of formations. The broad picture can be derived from seismic cross-sections, and the details gradually resolve themselves as more well data become available. Application of various basin mapping methods (Chap. 6) and use of the genetic, depositional systems approaches and sequence stratigraphy concepts (Chap. 5) are of particular value here. Some examples of detailed mapping and correlation and the kinds of research questions these projects raise are discussed in Chap. 8.

In the absence of seismic data, it is necessary to construct the forest from the individual trees (the second approach). The stratigraphic framework in most well explored (mature) basins was built up this way using lithostratigraphic methods, and in the past the work has usually been accompanied by considerable controversy, as local specialists have argued about the relationships between the successions in different parts of a basin or between outcrop and subsurface units. Sequence concepts, because of their predictive power, can now make this task easier, although the work is usually not without its problems.

Whether a basin analysis exercise starts from seismic sections or from outcrop work, it is desirable, eventually, to document the fine detail of the stratigraphy by establishing a sequence-stratigraphic framework and, ideally, to tie this to the global time scale (formal sequence-stratigraphic methods have yet to be finalized; see Sect. 7.7). Every local

biostratigraphic, radiometric, and magnetostratigraphic study can potentially contribute to the long-term effort to perfect a **global chronostratigraphic (time) scale**. This last step is beyond the needs of most exploration companies and is an area of research commonly taken over by state geological surveys and individuals in academic institutions, although the data base and expertise built up in industry may form an essential component. This is one area in which the government core and data repositories can prove their usefulness.

Several commissions and subcommissions and numerous working groups of the *International Union of Geological Sciences* (IUGS), mostly under the auspices of the *International Commission on Stratigraphy* (ICS), have been carrying this work for many years, as have many national groups. The *North American Commission on Stratigraphic Nomenclature* has been particularly influential. Some of the results are reported in this chapter, and the interested reader may wish to examine the IUGS journal *Episodes*, which reports on the activities of these various groups and announces important publications. Other important publications include *Newsletters on Stratigraphy* (published by E. Schweizerbart, founded in 1965), and the journal *Stratigraphy* (published by Micropaleontology Press, founded by William Berggren and John A. Van Couvering in 2004). They both publish original research articles and reviews, and may also include background information on ongoing work of ICS and the *International Subcommission on Stratigraphic Classification* (ISSC). The North American Commission published an updated, comprehensive guide to stratigraphic procedure in 2005. Recent synthesis publications describing the Geological Time Scale have brought this topic to a high level of sophistication (Gradstein et al. 2004a, 2012). We discuss this later in this chapter (Sect. 7.8). The website www. stratigraphy.org is the official website of the ICS.

This chapter is intended to provide an introduction to practical working methods. Chronostratigraphic (including biostratigraphic) research must form an integral part of any ongoing basin analysis project unless it is strictly local in scope. The work is usually performed by specialists. Correlation methods based on lithostratigraphy and sequence stratigraphy are also compared and contrasted here, although the procedures for erecting formal, named units (included here for consistency and completeness) can be left to advanced stages of the analysis (and do not yet include procedures relating to sequence stratigraphy). Such naming is best carried out by individuals with sedimentological training, so that the depositional systems and sequence concepts described in Chaps. 3–5 can be incorporated into the work. Research trends, emerging problems, and some new developments, are discussed in Chap. 8.

As discussed in detail in Chap. 8, the sedimentary record is "more gap than record". For a hundred years, since Barrell (1917) published his ground-breaking work on the rates of sedimentary processes, the fragmentary nature of the record has largely been ignored by practicing stratigraphers and sedimentologists. Typically, in many stratigraphic sections, only about 10 % of elapsed time is represented by preserved rock at a $10^6$-year time scale. All the developments in dating and correlation methods, and the emergence of the powerful new descriptive-interpretive methods of sequence stratigraphy have all been accomplished despite this fact. Stratigraphy continues to "work" as a practical method of basin mapping and resource exploration. How can this be? It is because there is a limited suite of natural processes that deposit and preserve sediment, and these predominate in the development of the stratigraphic record, including both the accumulated sedimentary successions and the hiatuses that separate them. For example, modern work on sequence stratigraphy has identified a limited range of allogenic processes, all characterized by a specific range of time scales, that are now known to generate sequences (Miall 1995). The order-of-magnitude range of time scales over which these processes operate was the basis for the original hierarchical "order" classification of sequences (Vail et al. 1977). The range of natural processes that build the sedimentary record, from the burst-and-sweep turbulence of a traction current to the accumulation of a basin-fill over millions of years, constitutes a crudely fractal distribution of rates and time scales; but these processes are genetically unrelated. This is important because it means that the distribution of relevant time scales is not mathematically precise, and that, therefore, quantification of processes, for example on the basis of fractal theory, may not be particularly useful, or illuminating or, indeed, relevant. Also, as noted elsewhere, interpretations of sedimentary processes that imply continuity, such as sediment-transport mass-balance calculations and the reconstruction of shoreline trajectories through transgressions and regressions, need to take into account the interruptions in the record represented by hiatuses, the frequency and duration of which have consistently been ignored. We address these issues in more detail in Chap. 8.

## 7.2 Types of Stratigraphic Unit

Rocks may be described in terms of any of their physical, chemical, organic, or other properties, including lithology, fossil content, geochemistry, petrology, mineralogy, electrical resistivity, seismic velocity, density (gravity), magnetic polarity, or age. Theoretically, any of these properties may be used for description and correlation, and most are so used for various purposes. In practice lithology is the most important criterion; fossil content is also crucial for rocks of Phanerozoic age. Magnetic polarity is finding increasing application as a correlation tool, particularly for the younger

Mesozoic and Cenozoic, and radiometric ages are used to assign numerical ("absolute") ages to biostratigraphic, magnetic and other chronostratigraphic units. A standard oxygen isotope scale is available for the late Cenozoic. Other geophysical properties are used in the early reconnaissance stages of exploration of a sedimentary basin. Not all these properties will necessarily give rise to the same correlations of a given rock body; for example, it is commonly difficult to relate geophysical properties precisely to lithology. Therefore, no single type of stratigraphic unit can be used to define all the variability present in nature.

Reflection-seismic data have been widely used in the exploration of sedimentary basins since the 1970s, making use of the concepts of seismic stratigraphy (Sect. 6.3), and with particular application to the interpretive field of sequence stratigraphy (Chap. 5). Sheriff (1976) pointed out that for reflecting events to be distinguished on seismic data they must each represent a clear velocity contrast and must be at least the equivalent of a quarter wavelength apart. At shallow depths, velocities are in the range of 1.5–2.5 km/s and reflections are relatively high frequency, about 5–100 Hz, so that a quarter wavelength is on the order of 5–12 m. At greater depths, typical reflection wavelengths increase considerably. Therefore stratigraphic resolution is fairly coarse (Fig. 7.1). Early work in seismic sequence stratigraphy had portrayed sequence boundaries as distinctive reflection surfaces correlatable over wide distances (Mitchum et al. 1977), but improved acquisition and processing techniques have demonstrated that the broad, through-going correlations on which this early work were based may be suspect. Cartwright et al. (1993) demonstrated that many of the critical features of sequence architecture, including the sequence boundaries, break down upon detailed examination into complex reflection patterns representing local facies variability, and major through-going surfaces may not be present or easy to trace. Seismic data are essential for studying large-scale stratigraphic features such as depositional systems and regional (or global) sequences, but may be of less use in the development of the refined stratigraphic subdivisions that are the subject of this chapter. The reader interested in the application of seismic methods to basin analysis may wish to consult standard textbooks in this area, such as Catuneanu (2006) or Veeken (2007).

Two basic categories of stratigraphic information are essential for the complete documentation of the stratigraphic record: (1) descriptive lithic units; and (2) geochronologic information, dealing with correlation and age of the strata (Harland 1993).

The most important types of stratigraphic units are:

*Lithostratigraphic units*: these are strictly empirical, based on observable lithologic features including composition and

**Fig. 7.1** Scale of a typical seismic wave form as compared to an outcrop (*left*), and compared to a wireline log (*right*). The frequencies characteristically used in petroleum exploration seismic work (10–60 Hz) have long wavelengths. Seismic resolution is therefore limited to large-scale stratigraphic features (based on an idea from A.E. Pallister and A.E. Wren)

tens of metres

grain-size and possibly also including certain basic sedimentological information, such as types of sedimentary structures and cyclic successions.

*Biostratigraphic units*: these are based on fossil content. Life forms evolve with time, permitting subdivision into **biozones** on the basis of changes in the fauna or flora. The first and last appearance of particular species or variants may also serve as useful time markers.

*Unconformity-bounded units*: these are units bounded above and below by unconformities. They may consist of any kinds of rocks, igneous, metamorphic or sedimentary. Unconformity-bounded sedimentary successions may be formalized using the empirical, descriptive classification procedures of **allostratigraphy** (see NACSN 2005), but increasingly geologists now employ the interpretive procedures and models of **sequence stratigraphy** as the main basis for subdivision and mapping of the basin fill.

*Chronostratigraphic units*: these comprise an interpretive stratigraphy, in contrast to lithostratigraphic and biostratigraphic units, which are strictly descriptive. Chronostratigraphy concerns itself with correlation and the age of the strata, which may be determined by a variety of means, of which the most important are fossil content, radiometric dating, magnetic polarity (for the post Middle Jurassic) and the oxygen isotope record (for the Cenozoic).

Both lithostratigraphic and biostratigraphic units may be local in extent. Lithologic character depends on the depositional environment, sediment supply, climate, rate of subsidence, etc., all of which can vary over short distances. Lithostratigraphic units are diachronous to a greater or lesser degree, that is, they represent a different time range in different places, reflecting gradual shifts in environment, for example, during transgression or regression. The limits of a stratigraphic unit are either its erosional truncation at the surface or beneath an unconformity or a facies change into a contemporaneous unit of a different lithology. A special type

of lithostratigraphic unit is that formed by short-lived stratigraphic events, which are those that have widespread depositional effects within very short time spans (Ager 1981). Examples of such events are volcanic ash falls, the deposits of violent storms and tsunamis, and certain regional or global sea-level changes. Such event deposits may prove very useful as local correlation tools, as discussed below (Sect. 7.8.4).

Biostratigraphic units are based on fauna or flora, the distribution of which is ecologically controlled. Also, contemporaneous faunas located in ecological niches that are similar but geographically isolated may show subtly different evolutionary patterns, making comparisons or correlations between the areas difficult (this is the subject of **biogeography**). All life forms evolve with time, so that faunas and floras show both spatial and temporal limits on their distribution.

The unconformities that demarcate unconformity-bounded units are caused by subaerial erosion during times of low stands of sea level, by submarine erosion, or by sudden environmental change. These events typically are widespread. Sea-level change may be caused by tectonic elevation of the basement or by eustatic changes in sea level. In either case, the unconformity surfaces define units of considerable lateral extent. Rapid changes in depositional environment may generate what Schlager (1989) termed **drowning unconformities**. Some unconformities may be of global significance, although this is typically difficult to demonstrate (Miall 2010, Chap. 14) Unconformities provide an excellent basis for the regional subdivision of a basin fill, and their interpretation may throw considerable light on regional tectonic evolution. **Sequence stratigraphy** has become the method of choice for subsurface mapping by the petroleum industry precisely because of its practical utility in focusing on and documenting these unconformity-bounded successions (Chap. 5).

Chronostratigraphy attempts to resolve the difficulties in regional and global correlation by establishing a global, time-based reference frame. A standard **Geological Time**

Scale (GTS) has gradually evolved since the discovery of radiometric dating early in the twentieth century. However, the accuracy of chronostratigraphic correlation is only as good as that of the time-diagnostic criteria on which it is based. Imprecision and error remain (Sects. 7.8, 8.10).

The evolution of these four types of units has had a long and complex history (Hancock 1977; Conkin and Conkin 1984; Miall 2004) and there has been controversy about definitions. Hedberg (1976), Hancock (1977), and Harland (1978, 1993) discussed some of the early practical and philosophical problems. Nineteenth-century geological practice did not distinguish lithology from age, causing severe correlation problems wherever a facies change or a diachronous boundary occurred. More recently there has been controversy over whether the rocks (lithologic units) or interpreted age range should form the primary basis of chronostratigraphy (e.g., Zalasiewicz et al. 2004). The discussions are likely to seem somewhat academic and theoretical to the average basin analyst and will not be discussed at length here.

## 7.3 The Six Steps Involved in Dating and Correlation

Six main "steps" are involved in the dating and correlation of stratigraphic events (Miall 1994). Figure 7.2 summarizes these steps and provides generalized estimates of the magnitude of the uncertainty associated with each aspect of the correlation and dating of the stratigraphic record. Some of these errors may be cumulative, as discussed in the subsequent sections. The assignment of ages and of correlations with global frameworks is an iterative process that, in some areas, has been underway for many years. There is much feed-back and cross-checking from one step to another. What follows should be viewed, therefore, as an attempt to break down the practical business of dating and correlation into more readily understandable pieces, all of which may be employed at one time or another in the unravelling of regional and global stratigraphies. The main steps are as follows:

**Fig. 7.2** Steps in the correlation and dating of stratigraphic events. e = typical range of error associated with each step. **a** In the case of the sequence framework, location of sequence boundaries may not be a simple matter, but depends on interpretation of the rock record using sequence principles. **b** Assignment of the boundary event to the biozone framework. An incomplete record of preserved taxa (almost always the case) may lead to ambiguity in the placement of biozone boundaries. **c** The precision of biozone correlation depends on biozone duration. Shown here is a simplification of Cox's (1990) summary of the duration of zones in Jurassic sediments of the North Sea Basin. **d** The building of a global stage framework is fundamental to the development of a global time scale. However, global correlation is hampered by faunal provincialism. Shown here is a simplification of the faunal provinces of Cretaceous ammonites, shown on a mid-Cretaceous plate-tectonic reconstruction. Based on Kennedy and Cobban (1977) and Kauffman (1984). **e** The assignment of numerical ages to stage boundaries and other stratigraphic events contains inherent experimental error and also the error involved in the original correlation of the datable horizon(s) to the stratigraphic event in question. Diagrams of this type are a standard feature of any discussion of the global time scale (e.g., Haq et al. 1988; Harland et al. 1990). The establishment of a global biostratigraphically-based sequence framework involves the accumulation of uncertainty over steps **a–d**. Potential error may be reduced by the application of radiometric, magnetostratigraphic or chemostratigraphic techniques which, nonetheless, contain their own inherent uncertainties (step **e**)

1. Identification of the units or stratigraphic events to be correlated, and development of regional correlation frameworks, including the mapping of hiatuses, unconformities and other key surfaces. Local correlations may be based on lithostratigraphy, but sequence stratigraphic concepts and methods are now practically universal. Correlations may be guided or constrained by supplementary data, such as biostratigraphic zonation. Determining the position of events such as sequence boundaries may or may not be a straightforward procedure, and requiring the application of facies mapping and sequence mapping techniques (Chaps. 3, 4, and 5).

2. Determining the extent and chronostratigraphic significance of unconformities. Unconformities, including sequence boundaries, represent finite time spans which vary in duration from place to place. In any given location this time span could encompass the time span represented by several different sedimentary breaks at other locations. Resolving such problems may require that some of the other steps be completed, particularly step 3.

3. Determination of the biostratigraphic framework. One or more fossil groups is used to assign the selected event to a biozone framework, and zones are defined and correlated from section to section. Error and uncertainty may be introduced because of the incompleteness of the fossil record. Graphic correlation or other quantitative techniques may be employed (Sect. 7.5).

4. Assessment of relative biostratigraphic precision. The length of time represented by biozones depends on such factors as faunal diversity and rates of evolution. Durations of biozones vary considerably through geological time and between different fossil groups

5. Correlation of biozones with the global stage framework (Sect. 7.8.3). Much of the existing stage framework was initially, with notable exceptions, built from the study of macrofossils in European type sections, although microfossils have become increasingly important for subsurface work and for global studies (McGowran 2005). Correlation with this framework raises questions of environmental limitations on biozone extent, our ability to inter-relate zonal schemes built from different fossil groups, and problems of global faunal and floral provinciality and diachroneity.

6. Assignment of numerical ("absolute") ages (Sect. 7.8.2). The use of radiometric and magnetostratigraphic dating methods, plus the increasing use of chemostratigraphy (oxygen and strontium isotope concentrations) permits the assignment of numerical ages in years to the biostratigraphic framework. Such techniques also constitute methods of correlation in their own right, especially where fossils are sparse. The Geological Time Scale (GTS) has become an instrument of considerable geological importance and practical utility in recent years, contributing to the emergence of what Miall (2013) termed "Sophisticated Stratigraphy" (Sect. 7.8; see Fig. 7.33).

## 7.4 Lithostratigraphy

Until the 1980s it was standard practice to describe and map stratigraphic successions on the basis of lithostratigraphic principles (Fig. 7.3). In the field, particularly in arid regions where the rocks are well exposed, it is still the historically established formations that are the basis for field location and identification. Such is the case, for example, in the Grand Canyon and Canyonlands areas of the United States, and the Front Ranges of the Rocky Mountains in Alberta. It is necessary, therefore, to be able to read older publications

Fig. 63.—Section across the London Basin, showing the probable Ridge of old Rocks. (After W. Whitaker, Geol. Survey.)
*a* Lower Bagshot Sand (of Hampstead)   *b* London clay   *c* Woolwich and Reading Beds (including the Oldhaven Beds, which occur on the south only).
*d* Thanet Sand (crops out on the south only).   *e* Chalk with flints.   *f* Chalk without flints.
*g* Upper Greensand (crops up on the south only).   *h* Gault.   *i* Lower Greensand.
*k* Wealden Beds (on the south only).
*l* Oolitic Clays (shown only on the north, but proved to occur on the south, beyond range of the section, by the Sub-Wealden boring, near Battle, in Sussex).
*x* Old Rocks (Palæozoic).

**Fig. 7.3** A cross-section of the London Basin, England, showing the development of descriptive terminology for stratigraphic units. From "Geology of the counties of England and Wales", by Harrison (1882) and maps and understand what type of information they convey.

Amongst the problems with lithostratigraphy as a method of description is that the defined units carry no meaning regarding the origins or age of the units. Formations are commonly diachronous, and many stratigraphic names were established many decades ago, long before the advent of modern facies and sequence analysis. Older literature may therefore be replete with the names of local, poorly defined units, with a given body of rocks defined and named differently in different parts of a basin. Procedures are available (e.g., see NACSN 2005) for the revision and redefinition of units as new information becomes available from surface mapping or subsurface exploration.

### 7.4.1 Types of Lithostratigraphic Units and Their Definition

A hierarchy of units has been developed based on the **formation**, which is the primary lithostratigraphic unit (NACSN 2005).

Group
Formation
Member
Tongue or lentil
Bed.

*The formation.* An important convention has long since been established that all sedimentary rocks should be subdivided (when sufficient data have been collected) into formations. No other types of lithostratigraphic subdivision need be used, although convenience of description may require them.

What is a formation? There are no fixed definitions that deal with the scale or variability of what should constitute a formation, although the procedures for establishing limits (contacts) and names are well established (e.g., see NACSN 2005). Figure 7.4 provides a good example of the way in which stratigraphic successions are subdivided on the basis of lithology. The lithologies, colors and weathering characteristics of the rocks suggest a fourfold subdivision of the exposure. Comparisons with other exposures nearby and the presence of distinctive fossils permits three of the subdivisions to be assigned to previously existing formations, while the fourth (oldest) unit is different from the local succession, and has yet to be given a name. This outcrop is large enough that the angular unconformity between two of the units (the Nansen and the Barrow formations) can clearly be seen.

The degree of lithologic variability required to distinguish a separate formation tends to reflect the level of information available to the stratigrapher. Formations may be only a few meters or several thousands of meters in thickness; they may be traceable for only a few kilometers or for thousands of kilometers. Formations in frontier basins usually are completely different in physical magnitude from those in populated, well-explored basins, such as much of western Europe and the United States. In exploration in frontier basins proceeds, some of the larger formations first defined on a reconnaissance basis may have subsequently been subdivided into smaller units and the ranking of the names

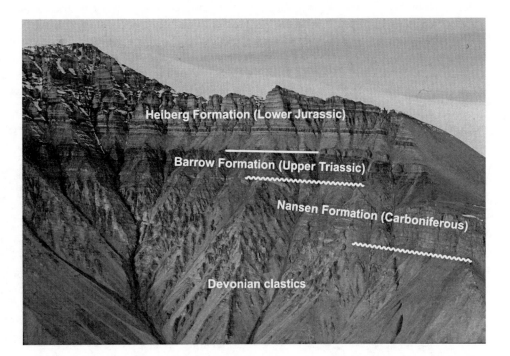

**Fig. 7.4** An example of a lithostratigraphic subdivision of a rock succession. Stratigraphic units exposed in the mountains of northern Ellesmere Island, Arctic Canada (*photo* A.F. Embry)

changed. NACSN (2005) provides the procedures for making these kinds of revisions.

The most important criteria for establishing a formation are its usefulness in subdividing stratigraphic cross sections and its "mappability". For reconnaissance mapping, a thin unit that cannot accurately be depicted at a scale of, for example, 1:250,000, may be of little use, although the definition and mapping of thin but widespread marker units may be of considerable utility. For more detailed work, mappability at a scale of 1:50,000 or even 1:10,000 may be a more useful criterion. Problems of consistency may arise when detailed work is conducted around a mine site within what is otherwise a poorly explored frontier basin.

Formations should not contain major unconformities, although minor disconformities may be acceptable (indeed, as we now recognize, they are all but unavoidable). The contacts of the formation should be established at obvious lithologic changes. These may be sharp or gradational. An unconformity is a logical choice for a formation contact. Where lithologies change gradually, either vertically or laterally, it may be difficult to choose a logical place to draw the contact. For example, a mudstone may pass up into a sandstone through a transitional succession with sandstone beds becoming thicker and more abundant upward. The mudstone-sandstone formation contact could be drawn at the oldest thick, coarse sandstone (with thickness and coarseness carefully spelled out), at the level where sandstone and mudstone each constitute 50 % of the section, or at the youngest extensive mudstone bed. The choice is arbitrary, and it is immaterial which method is selected as long as the same method is used as consistently as possible throughout the extent of the formation.

Other problems of definition arise where there are lateral lithologic changes, requiring definition of a new formation. A simple diachronous contact is not a problem, but where the two units intertongue with one another, it may be virtually impossible to draw a simple formation contact. One solution is to give each tongue the same name as the parent

formation. A section passing through the transition region may then show the two formations succeeding each other several times. The only problem this causes is if formation contact and thickness data are stored in a data bank and used in automated contouring programs. Without additional input from the operator, a computer program might not be able to handle this type of data. Other alternatives are to define the whole transitional rock volume in terms of one of the parent units, to separate the transitional lithologies as a separate lithostratigraphic entity, or to give separate tongues their own bed, tongue, or member names. Published stratigraphic codes (e.g., Salvador 1994; NACSN 2005) provide procedures and practical solutions but do not specify any rigid rules for the resolution of such problems. The main criteria should be practicality, convenience, and consistency.

The sometimes arbitrary nomenclatural issues raised by lithostratigraphic methods may be clarified or avoided by the use of modern sequence methods, but it should not be forgotten that sequence stratigraphy is an interpretive approach to the rocks, and the inductive, empirical nature of lithostratigraphy will likely remain as an essential underpinning of basinal stratigraphic frameworks for some time to come.

A range of other terms is used to group or subdivide stratigraphic successions on the basis of lithostratigraphy (Fig. 7.5). Vertical and lateral contacts between units may be defined on the basis of clear lithologic change, but are commonly somewhat arbitrary, and as noted above, such subdivisions contain no useful information about the depositional relationships of the strata. We address these issues later, in the discussion on sequence stratigraphy (Sect. 7.6), where it is demonstrated how this genetic approach to stratigraphy can lead to much more meaningful reconstructions and interpretations.

*The group.* All other stratigraphic units are based on the formation. A group consists of two or more formations related lithologically. In the past, named groups have been established for thick and varied successions without first

**Fig. 7.5** A correlation table for a hypothetical basin fill, drawn to illustrate the various ways in which lithostratigraphic terminology may be adapted to best capture existing stratigraphic variability defining the constituent formations. This is not recommended practice. By contrast, formations defined during reconnaissance exploration may be subdivided into constituent formations and the original name retained and elevated to group status if detailed mapping subsequently provides appropriate data. Groups should not contain major unconformities.

The component formations of a group may not everywhere be the same. Lateral facies changes requiring definition of different formations can occur within a single group. In contrast, a component formation may extend laterally from one group to another. Groups are normally defined for regions of complex stratigraphy. Toward the basin margin or basin center, the component formations may lose their individuality, in which case the group may there be "demoted" to a formation, while still retaining the same name.

The terms supergroup and subgroup are occasionally used to provide an additional hierarchy of subdivision. Usually, there are historical reasons for this; some of the higher ranking names may have started out as member, formation or group names, with reclassification and promotion being required as additional work demonstrated the need for further subdivision.

*The member.* This is the next ranking unit below formation. Not all formations need be divided into members, and formal names need be used for only a few, one, or none of the constituent members, depending on convenience or the level of information available.

There are no standards for the thickness or extent of members, and commonly it is difficult to decide whether to define a given lithostratigraphic unit as a member or a formation. However, recommended practice is that all parts of a succession be subdivided into formations, and so this is the best level at which to start. A member cannot be defined without its parent formation.

For mapping and other purposes, it commonly is convenient to establish informal units such as the lower sandstone member, which do not require formal names.

*Tongue or lentil.* These are similar to members. Because of their geometric connotations, the terms are useful for parts of formations where they interfinger with each other. Formal names may be established for one, several, or all such units, depending on convenience and practicality.

*Bed.* This is the smallest formal, named unit in the hierarchy of lithostratigraphic units. Normally, only a few parts of a stratigraphic succession will be subdivided into named beds. Coal seams in mine areas, prominent volcanic tuff horizons, and other marker beds are typical examples.

Certain stratabound ore-bearing beds, such as placer units may also be named.

## 7.4.2 The Names of Lithostratigraphic Units

When establishing a named unit, it is standard practice to give it a geographical name, chosen to suggest the location or areal extent of the unit. This may be a river, lake, bay, headland, hill, mountain, town, village, etc. Permanent names are preferable. Subsurface work in frontier basins, particularly in offshore areas, may rapidly use up all the available names, in which case the name of the well chosen as the type section may be used. Failing this, names may have to be invented.

In most cases, the geographical name will be followed by the rank designation, for example, Wilcox Group, Pocono Formation. For beds, this is commonly not done, particularly in the case of coal seams, for which a complex mine terminology may have evolved. Many older stratigraphic units use a lithologic term instead of a rank term, for example, Gault Clay, Dakota Sandstone, Austin Chalk (e.g., see Fig. 7.3), but this is not recommended because the rank of the unit is not clear from the name alone (NACSN 2005).

Workers should beware of using a geographic name that has already been employed in a different context or renaming units without justification. Geological survey organizations commonly retain a file of current and obsolete stratigraphic names that the worker may wish to consult. Formal naming of units requires that the name be published in a recognized publication, such as a national or international journal. Information required to establish a name includes a designated type section or **stratotype**, with a detailed description of the succession and information about the distribution of the unit and its relationship to overlying, underlying or age-equivalent units in adjacent locations. Further details on the establishment of stratotypes is provided in Sect. 7.8: Chronostratigraphy.

Lithostratigraphic units may be changed in rank as the level of knowledge improves. For example, the Cornwallis Group in the Canadian Arctic Islands started as the Cornwallis Formation and was raised to group rank when it was realized that it contained three mappable units of formation rank. Conversely, the Eureka Sound Group was named for a thick and varied clastic succession, but it was never subdivided into named constituent formations by the original author and was reduced to formation status. The unit has now been subdivided and has been formally redefined as a group once again.

When a unit is raised in rank, the original name should not be used for any of the subdivisions, but is best retained for the higher ranking unit or abandoned altogether.

## 7.5  Biostratigraphy

Biostratigraphy is the study of the relative arrangement of strata based on their fossil content. Descriptive or empirical biostratigraphy is used in erecting zones for local or regional stratigraphic correlation and forms the basis for a global system of chronostratigraphic subdivision.

Fossil content varies through a stratigraphic succession for two main reasons: evolutionary changes and ecological differences, such as changes in climate or depositional environment. Biostratigraphy should be based only on evolutionary changes, but it is always difficult to distinguish these from changes that take place in a biostratigraphic assemblage as a result of ecological modifications, and this problem is a cause of continuing controversy for many fossil groups.

Biostratigraphy obviously can only be studied and a classification erected where fossils are present. This rules out all of the Precambrian, except for the concluding subdivision of the Proterozoic—the Ediacaran (Knoll et al. 2006). Even in the Phanerozoic, there are many rock units for which the fossil record is very sparse, and biostratigraphic subdivision is correspondingly crude. This is particularly the case in nonmarine strata or those (particularly carbonates) in which fossil remains have been destroyed by diagenesis.

Biostratigraphy is a study for specialists. Refined work requires intimate knowledge of the phylogeny of a large number of fossil groups and their regional or global distribution. To accumulate this knowledge may take half a lifetime, and the subject is an excellent example of a science in which the practitioner seems to spend inordinate amounts of time "learning more and more about less and less". Some of the leading authorities in a particular fossil group may be able to discuss the cutting edge of their research with only half a dozen other colleagues around the world. This gives them considerable value if one happens to find their kind of fossil, but it may somewhat restrict their scientific scope. Geologists engaged in basin analysis of the Phanerozoic are very rarely such specialists. Biostratigraphers are therefore employed by many organizations to provide these specialized service skills, or they function independently as consultants. They may be engaged much of the time in pursuing paleontological research, but are able to provide biostratigraphic diagnoses for selected fossil types over a specified age range.

Professional biostratigraphic work may take a great deal of field and laboratory time. Sections that a sedimentologist may dismiss as sparsely fossiliferous may yield hundreds or even thousands of specimens to the careful collector (see Fig. 3.50). Laboratory extraction of microfossils or palynomorphs may yield similar numbers. It is this kind of work that is necessary for modern, refined biostratigraphic studies. Much of the submitted material, particularly that from frontier exploration wells, may itself provide the basis for new biostratigraphic zoning schemes.

Basin analysts should understand what they are getting when they submit their own material to a specialist for identification. Commonly, they are interested in two items of information: (1) the age of the enclosing rock, information that can be used for correlation purposes, and (2) information regarding the ecological environment of the fossils, which can aid in the interpretation of depositional environments. Age is a chronostratigraphic interpretation based on taxonomic descriptions, but commonly there are problems of fossil identification or interpretation, particularly where the fossil record is sparse or the material is from a new, poorly studied area. The purpose of this section, therefore, is to discuss some of the problems of the biostratigraphic record, and to describe the methods biostratigraphers use in plying their trade.

### 7.5.1  The Nature of the Biostratigraphic Record

*Biofacies and Biogeography*: The geographical distribution of taxa reflects the restriction of ranges due to ecological variations and the geographical isolation of populations. Two topics are included under this heading, the facies control of faunas and floras and the problem of faunal and floral provincialism.

Some taxa are adapted to a benthonic (bottom-dwelling) mode of life and others to a nektonic (swimming) or planktonic (floating) habit. Ideally nektonic, or planktonic forms are greatly to be preferred for biostratigraphic purposes because of the likelihood of their being more widely distributed and therefore more broadly useful. Benthonic forms tend to be more facies dependent because of their need for certain water conditions or sediment types for feeding and dwelling behavior. However, in practice, benthonic forms are widely used by professional biostratigraphers. Even such static forms as corals, burrowing molluscs, or anchored brachiopods have been found to be invaluable for zoning the deposits of the continental shelves. Many benthonic taxa have a planktonic larval stage that ensures wide distribution via marine circulation. Conversely, many planktonic forms, such as the graptolites, are too fragile to survive in agitated, shallow water environments and are therefore just as facies-bound as their benthonic contemporaries. In practice, virtually every taxonomic group has some biostratigraphic utility, although considerable problems may arise in attempts to determine the relationships between the

**Fig. 7.6** The interbedding of
three biofacies resulting from
shifting of environments.
Typically such interbedding is
caused by sea-level change
(McGowran 2005, Fig. 1.4, p. 8)

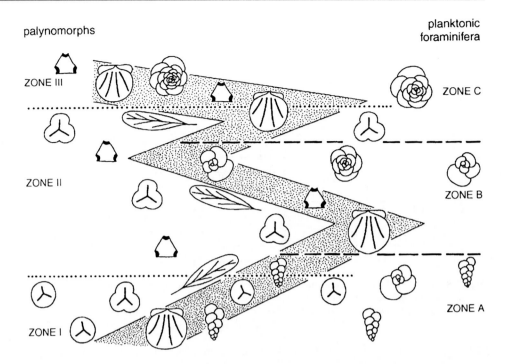

various facies-bound faunas, unless environmental fluctuations cause lithofacies of different types, with their accompanying faunas or floras, to become interbedded (Fig. 7.6).
Where well exposed, such mixed successions are of great
value in establishing a global chronostratigraphic framework. Figure 7.6 illustrates such a scenario schematically,
where marine foraminiferal zones can be correlated to nonmarine paynomorphs because of the interfingering of these
facies zones. The zig-zag "shazam" interfingering configuration used to illustrate facies relationships in this diagram is
an overly simplistic representation of the progradation and
retrogradation that occurs as a result of changes in relative
sea-level, but helps to make the point that the vertical range
of a given assemblage may be **diachronous**, changing in age
laterally as a result of shifts through time of facies belts.

Classic examples of facies-bound faunas widely used by
biostratigraphers are the **shelly** and **graptolitic** faunas of the
lower Paleozoic. The shelly fauna actually includes two
more or less distinct sub-faunas, one in the inner, shallower
shelf dominated by brachiopods and the other on the outer
shelf, characterized by trilobites. The graptolitic fauna is
confined mainly to low-energy deposits of the continental
slope, rise, and abyssal plain (Berry 1977). In Newfoundland, on what was the ancient eastern continental margin of
North America during the Cambrian and Ordovician, these
two facies interfinger. Carbonate turbidites and debris flows,
derived from collapse of the continental margin, are
interbedded with graptolitic shales at the base of the continental slope (Fig. 7.7). At Green Point, on the west coast,
this assemblage straddles the Cambrian-Ordovician boundary, and the location has been established as the stratotype

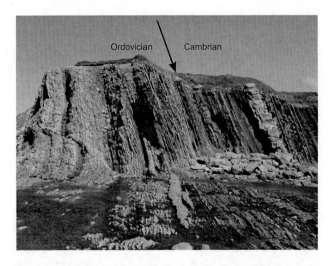

**Fig. 7.7** Interbedding of a shelly fauna in limestone sediment-gravity
flows and a graptolitic fauna in interbedded shales,
Cambrian-Ordovician continental-margin deposits, Green Point, Newfoundland. The beds here are overturned. This location has been
designated as a stratotype for the Cambrian-Ordovician boundary

for this important chronostratigraphic boundary (the GSSP
for the base of the Tremadocian Series: Cooper et al. 2001).

A good example of facies control of what might appear at
first sight to be a recurrent, biostratigraphically controlled
fauna is provided by the brachiopod communities of the
Upper Ordovician to Middle Silurian of the Welsh Borderlands. Ziegler et al. (1968) showed that at many localities
there is a sequence of assemblages containing, in upward
stratigraphic order, *Lingula*, *Eocoelia*, *Pentamerus*, *Stricklandia*, and *Clorinda*, followed by a graptolitic fauna.

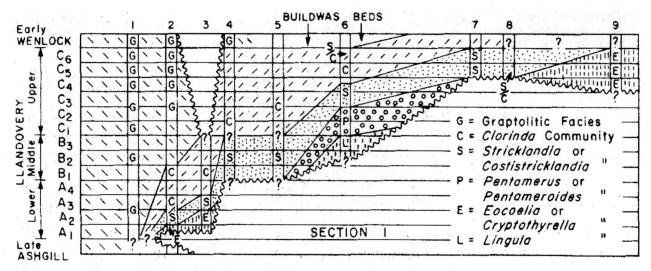

**Fig. 7.8** An example of facies-bound faunas. The succession of brachiopod communities in each of these sections is the same (*L, E, P, S, C*), but use of graptolite biostratigraphy (biozones *A1–C7*) shows that they are markedly diachronous and therefore facies controlled. Silurian, Welsh Borderland (Ziegler et al. 1968; McKerrow 1971)

Careful correlation of these sections using graptolites showed that the brachiopod sequence is markedly diachronous. This is shown in Fig. 7.8, in which the graptolite zones are shown by horizontal correlation lines labeled A1–C7. One might ask, why are the graptolites trusted more than the brachiopods for the purposes of chronostratigraphic correlation? The answer is that brachiopods are benthonic organisms known to be prone to facies control, whereas graptolites are tried and tested biostratigraphic indicators. The correlations shown in Fig. 7.8 are supported by additional work on two of the brachiopod genera, *Eocoelia* and *Stricklandia*. When examined in detail, evolutionary trends can be detected within the populations of these genera as they are traced from west to east across the line of section shown in Fig. 7.8 (and other sections not shown). The overall interpretation of these faunal data is that the sections reveal a gradual eastward marine transgression and deepening of the water, such that successive brachiopod communities represent ecological adjustments to increased depths (*Lingula* inhabited brackish waters, and in part the subsequent succession represents changing shell thicknesses in response to wave and tidal energy). There is no obvious relationship between brachiopod assemblage and sediment type in this case.

*Biogeography and evolution*: William Smith recognized the significance of the succession of faunas in the stratigraphic record long before Darwin's theory of evolution was established. The point is that it is not necessary to understand evolutionary relationships of fossil groups in order to make use of them as tools in the inductive establishment of stratigraphic order and correlation. However, it can certainly help to explain the nature of taxonomic change and the distribution of distinct groups through time and space.

Molecular biology shows that the underlying process of evolution is genetic drift, the gradual accumulation of random incremental change in gene variants. Natural selection favors some mutations over others, which generates steady change. Isolation of populations will also tend to increase genetic divergence, eventually to the point that populations will be unable to interbreed and then constitute distinct species (see review of current ideas by Kelley et al. 2013).

Three styles of evolution were described by Eldridge and Gould (1977), and are illustrated in Fig. 7.9. The first, termed **phyletic gradualism** or **transformational evolution** (McGowran 2005, p. 382), refers to long-term evolutionary change, typically in response to geographical, climatic, or other environmental pressures (Fig. 7.9a). Certain varieties of a species may be favored by these changes, so that there is a gradual adjustment in the stock until a distinctive new species appears.

Kauffman (1977) described two examples of phyletic gradualism in Cretaceous pelecypods. Figure 7.10 illustrates a series of histograms of height-width ratios of the *Inoceramus pictus* lineage, derived from populations collected at about 40-cm intervals (lower graph), and the number of growth ridges in the first 25 mm of shells of *Mytiloides labiatus* (upper graph). The data permit subdivision of the population into species (S), subspecies (SS), and morphological zones (MZ), as indicated in the adjacent columns.

Another example of gradual evolution is provided by the foraminifera *Globigerina* and *Orbulina* (Fig. 7.11). The species listed in the top are an evolutionary series, three of which are illustrated (1: *Globigerina quadrilobatus*; 2: *G. bisphericus*; 3: *Orbulina universa*). The gradual variation between these types has permitted the erection of seven zones, as indicated by the horizontal lines at the left.

**Fig. 7.9** Three models of
evolution: **a** phyletic gradualism;
**b** punctuated equilibrium; and
**c** reticulate speciation. *A*, *B*, etc.,
refer to successive varieties or
species; *1*, *2*, etc., refer to
chronology of events. In **b**, the
same area at *left* of diagram is
successively occupied by three
species *A*, *B*, and *C*, which evolve
elsewhere and migrate in
(Sylvester-Bradley 1977)

Two types of environmental adaptation can occur, that
which is accompanied by permanent genetic change and that
which can, to some extent (never precisely), reverse itself to
recreate the same variety or race of a species more than once,
whenever the same environmental conditions are repeated
(homeomorphy). Clearly, the first type is the only one of use
to biostratigraphers, but the literature is replete with
ambiguous biostratigraphic determinations that may be fal-
sely based on diachronous environmental change. For
example, this has been a serious problem with the ammonites (Kennedy and Cobban 1977), one of the best of biostratigraphic indicators.

Taxa that evolve by phyletic gradualism have the most
potential for refined biostratigraphic zonation, but they
require specialist study to recognize the very subtle changes
between the varieties. This type of work is beyond the
abilities of the generalist basin analyst.

The second style of evolution was named **punctuated
equilibrium** by Eldridge and Gould (1972, 1977). The
concept was adapted from an earlier term, **allopatric speciation**, and is based on the premise that in a successful,
widely distributed taxon the population is genetically con-
servative. Evolution is thought to occur only where extreme
variants are selected by environmental pressures on the
fringes of the species range. Rather than a gradual adaptation
to an ecological niche or a broadening of a species range by

extending slowly into subtly different niches, as in phyletic
gradualism, the hypothesis of punctuated equilibrium pro-
poses the spasmodic occurrence of bursts of relatively rapid
evolutionary change. Extreme variants of a species can only
evolve into a new species if they become isolated by changes
in environment, climate, or geography, as through the rifting
and drifting apart of continental plates. Also, the catastrophic
extinction of organisms by bolide impacts or other catas-
trophes empties out many ecological niches and permits
rapid adaptive radiation and the explosive development of
many new taxa in the period immediately following these
extinctions (Fig. 7.9b).

Sylvester-Bradley (1977) proposed a third style of evolution, which he termed **reticulate speciation** (Fig. 7.9c).
This combines, on a small-scale, the mechanisms of both the
other two evolutionary styles. Sylvester-Bradley offered the
modern common vole as an example of a taxon that has
evolved in this way. The vole is distributed virtually globally
and comprises numerous races reflecting adaptation to local
variations in climate, vegetation, altitude, isolation on
islands, etc. These varieties have evolved in response to
rapid global changes following the Pleistocene ice age, and
they demonstrate the rapidity with which geographical and
ecological changes may bring about evolution. The apparent
stability of the species of several taxonomic groups for
several million years or more at times during the geological

**Fig. 7.10** Evolutionary trends in two pelecypod species, Cretaceous of Western Interior. Symbol X in lithologic column indicates bentonite beds used for radiometric age dating. Suggested systematics and zonal subdivisions are shown in the numbered columns: *MZ* morphological biozone; *SS* subspecies; *S* species (Kauffmann 1977)

past contrasts with the rapid adaptability of the modern vole. To what extent reticulate speciation will be recognized for fossil groups remains to be determined. However, to recognize this style of evolution would seem to require an immense bank of detailed descriptive data, and therefore it is very much a subject of study for specialists.

A review of modern concepts in evolution as applied to the fossil record is provided by Kelley et al. (2013).

In the geological record, many distinct populations have been recognized, based on geographical distributions, and are defined as **faunal provinces**. These are much discussed by biostratigraphers. To the non-specialist such concepts as the Malvinokaffric Province or the Tethyan or Boreal Realms are sometimes difficult to understand. The definition of what constitutes a given faunal province requires a great deal of specialist knowledge, but even the specialists had difficulty before the advent of plate tectonics in comprehending why many of these provinces existed. Until the 1970s there was much discussion of the appearance and disappearance of strange, narrow "land bridges" to explain the merging and divergence of provincial variations. But in fact, faunal provincialism provides some of the most

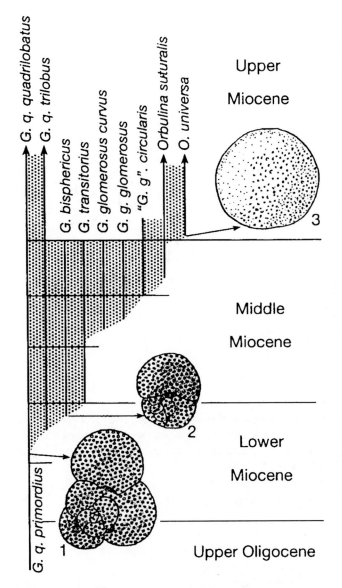

**Fig. 7.11** Evolution of the foraminifera *Globigerina*, culminating in the different genus *Orbulina*, in southern Australia. The stippling pattern indicates continuous variation between the various morphotypes, and their ranges. Seven zones are defined by the *horizontal lines at left* (from McGowran 2005, Fig. 4.1, p. 95)

convincing geological lines of evidence for plate tectonics (Tarling 1982).

One of the most famous of these is the example of the trilobite faunas flanking Iapetus, the proto-Atlantic Ocean that developed between Laurentia (ancient North America) and Baltica and Africa in the earliest Paleozoic. Specialists noted significant differences in Cambrian trilobites between those found in England and those in Scotland, and between those from western and eastern occurrences in Newfoundland, whereas Scottish trilobites are similar to those in western Newfoundland, and English trilobites are similar to those in eastern Newfoundland (Fig. 7.12). Given current

geographic distributions of the fossils these similarities and differences make no sense. However, Wilson (1966) cited these distributions as one of several lines of evidence in his proposal—now universally accepted—that the line demarking the two distinct trilobite faunal provinces constitutes an ancient continental suture that formed when a former ocean closed as a result of subduction and continental collision. That ocean, now called Iapetus, occupied approximately the position of the present Atlantic Ocean, but many significant continental fragments changed margins when the present ocean developed; that is, the continental rift occurred along a somewhat different line. Landing et al. (2013) reviewed in detail the faunal provincialism that developed in the Cambrian following the breakup of Rodinia.

The ammonites provide an excellent example of the various biogeographic styles that can occur in organisms, some offering considerable advantages to the biostratigrapher, others a severe hindrance. Many ammonites underwent a planktonic larval stage that may have lasted from hours to weeks. Where this occurred, it would have been of some importance to the distribution of the species. Not all ammonites showed this. Distribution patterns and varying degrees of facies independence show that some ammonites were benthonic in adult life habitat, some were nektonic, and some may have been planktonic. Their facies distribution and provincial tendencies thus varied considerably. Some ammonites may have drifted long distances after death. The modern Nautilus, the only living relative of the ammonites, has a buoyant shell after death, and observations in modern oceans suggests that the shell may drift for hundreds, if not thousands of kilometers. Geologically this could be of great importance, but Kennedy and Cobban (1977) suggested that many ammonites in fact became rapidly waterlogged after death and did not float appreciable distances.

Kennedy and Cobban (1977) summarized much of the data for Cretaceous ammonite distribution and concluded that there were five types of faunal provinces. Some genera have a virtually worldwide, or **pandemic**, distribution. Pandemic taxa would seem to offer the best possibilities for global correlation. They are relatively facies independent, but it turns out that many are long-ranging forms and thus of limited biostratigraphic usefulness.

Some ammonites have **latitudinally restricted** distributions, reflecting their preference for waters of a certain temperature or salinity and their tolerance of seasonal fluctuations. Some examples are shown in Fig. 7.13. They define two provinces, the northern, colder water Boreal province (open symbols) and the more tropical Tethyan province (closed symbols). In many parts of Europe and North America, faunal fluctuations through stratigraphic successions between Tethyan and Boreal (and other) faunas have been cited as evidence of the existence of connecting seaways and transgressions across otherwise barren areas.

**Fig. 7.12** The Cambrian trilobite faunas of the Iapetus margins. Based on Wilson (1966)

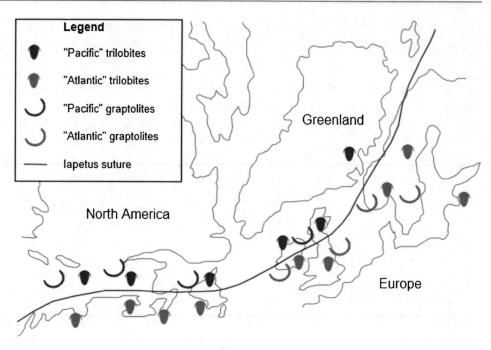

**Fig. 7.13** The biogeography of selected Cretaceous ammonites, plotted on a Cretaceous plate-tectonic reconstruction of the continents. Climatic tolerances underlie the different geographic spread of Boreal versus Tethyan forms (an example of latitudinally restricted distributions), while the gradually widening Atlantic Ocean caused the gradual isolation and separate evolutionary development of ammonite faunas on different sides of the ocean (longitudinally restricted distributions). These faunas occasionally mixed in the Western Interior Seaway of North America because of shifts in climate and changes in sea level that caused local faunal migrations along a north-south axis (Kennedy and Cobban 1977)

**Longitudinal restrictions** on distribution, such as the presence of land masses or large ocean basins, are a cause of further provincialism. These are added to latitudinal restrictions in the generation of the third type of faunal province: **endemic** distributions. Note that in Fig. 7.13 the Tethyan genera show no longitudinal restriction, whereas the four Boreal genera are typical endemic taxa, restricted to either Eurasia or North America (these are examples chosen to illustrate a point and should not be taken to define a universal difference between the Boreal and Tethyan provinces). Endemic ammonites have been shown to have evolved rapidly and thus are of prime biostratigraphic importance, although their provincialism has hindered intercontinental correlation.

**Disjunct** distributions are those of scattered but nevertheless widely distributed taxa. The distributions are not thought to represent inadequate data or severe facies control, but probably reflect very low population densities.

As noted, some ammonite taxa may drift in oceanic currents after death. In extreme cases, where an endemic form is involved, such drifted or **necrotic** distributions may prove invaluable for long-distance correlation.

In general, the provinciality of taxa increases the difficulty by which they may be related to the global time scale, because provincialism reduces the variety of forms that may be used to establish relative ages.

### 7.5.2  Biochronology: Zones and Datums

By the mid-nineteenth century, the work of the early stratigraphers, following Smith, had clearly established the value of fossil assemblages for the establishment of stratigraphic order and for the purposes of comparison between stratigraphic sections. The similarity of the succession of faunas or floras between sections in different basins, even different continents—termed **homotaxis**—was well established. The main elements of the Phanerozoic geological time scale had been defined, including all the names of the periods (Berry 1968), and the concepts of the biozone and the stage were already developed (Hancock 1977). However, there remained the issue of age. In the absence of a clear understanding of how faunas and floras changed with time, and without the tools to establish numerical age, it remained a legitimate question whether homotaxis could be equated with **synchrony**. Darwin's *The Origin of Species* was published in 1859, yet in 1862 T.H. Huxley stated "for anything that geology or palaeontology are able to show to the contrary, a Devonian fauna or flora in the British Isles may have been contemporaneous with Silurian life in North America, and with a Carboniferous fauna and flora in Africa."

Eventually this philosophical dilemma was resolved by developments in the understanding of the processes of evolution, coupled with the establishment of ever more detailed systems of zonation and correlation, which left little room for doubt regarding the reality of the principle of relative age and time correlation based on fossil content. Some of the steps in the evolution of thought are described by Hancock (1977), Miall (2004) and McGowran (2005, pp. 54–65).

It is now universally accepted that formally established biozones represent specific intervals of time, subject to two important caveats, (1) the changes in taxonomy that lead to the definition of discrete datum planes of change, or zones of similarity in the fossil record, are not globally instantaneous. The appearance of a successful evolutionary step requires a discrete period of time for it to spread throughout its full range. This time period may be short, in geological terms, but it is not instantaneous, and may be measureable; we return to this problem in Sect. 7.5.3. (2) the geographic range of a biostratigraphic datum or zone is limited by ecological factors. A biozone may not, therefore, have exactly the same age range everywhere.

There is a variety of methods available for making the most efficient use of fossil occurrences. Particularly distinctive and/or abundant forms may serve to represent a specific span of time. Such forms are called **index fossils**. The first or last appearance of particular, distinctive species, are commonly employed as biomarkers. These horizons are termed the **first-appearance datum (FAD)** and the **last-appearance datum (LAD)**. Suites of fossils may be used to define **biozones**. This may be done in several different ways (Fig. 7.14). Figure 7.15 illustrates one of the more common methods of defining a biozone, which takes advantage of the fact that the ranges of different species typically overlap. In this diagram a **concurrent-range zone** is defined as that interval of the rocks within which all three of the fossils A, B and C are present, and falls between the FAD of species B and the LAD of species A.

It is important to be aware that the time range indicated by a biozone is not necessarily the same everywhere. Figure 7.16 illustrates a hypothetical example where the diachronous spread of a fossil taxon and subsequent variations in its range owing to climatic factors has led to significant variability in the local range of the biozone. Diachroneity is discussed further in the next section. There are sophisticated methods for managing these issues, and, as discussed in Sect. 7.8, biostratigraphy is still the major foundation of the Geological Time Scale for the Phanerozoic.

In addition to ecological factors, there are considerations of preservation and potential sampling bias. Diagenetic destruction of fossils is common, and sampling bias might

**Fig. 7.14** Types of biozone, as defined in the International Stratigraphic Guide (Salvador 1994. Diagram from Pearson 1998, Fig. 5.2, p. 126)

**Fig. 7.15** The use of first- and last-appearance datums to define a concurrent-range biozone. The area colored in *grey* is the only part of the section where all three species, *A*, *B* and *C*, are expected to be present

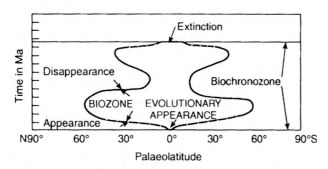

**Fig. 7.16** The diachroneity of a biozone. A taxon first appears near the equator, and takes a discrete amount of time (up to 2 million years) to migrate into higher latitudes north and south of the equator. The range does not then remain constant because of ecological factors, in this case, climate change, which leads to contraction and then renewed expansion of the range, until the taxon become extinct. The biochronozone (or biochron) of this taxon (the time span it represents) is global in extent but, in practice, the time span is that which is indicated by the actual presence of the taxon, which varies from place to place and, in this illustration, is absent altogether in high latitudes (McGowran 2005, Fig. 2.11, p. 38; after Loutit et al. 1988)

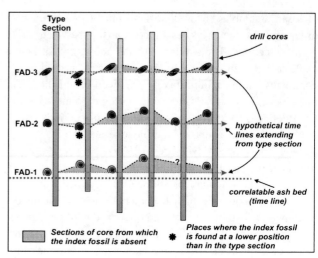

**Fig. 7.17** First- and last-appearance datums define ideal time lines in the rocks. However, in this example, many first occurrences are higher (younger) than the hypothetical time lines would predict, and a few are lower (older). The reasons for this are discussed in the text

examination of an outcrop for fossil content, a professional biostratigrapher is likely to carry out more thorough investigations. For example, soft sands and clays may be run through a sieve or water-washed on site to isolate macrofossils, or large lithologic samples may be collected at routine stratigraphic intervals for water or acid treatment to extract microfossils or palynomorphs. In the subsurface, drilling disturbance may constitute a major problem. The drilling process penetrates layers from the top down, so the stratigraphically last appearance of a fossil (the LAD) will be the first encounter with a given taxon, and the level of this horizon is one that can be trusted. However, the tendency for holes to cave can lead to fossils (and rock cuttings) being fed into the mud stream after the drill bit has passed on down their point of origin. For this reason, the FAD of a fossil taxon in the subsurface needs to be treated with caution.

As Sadler et al. (2014, p. 4) stated: "Signals and noise mingle among these contradictions. Records of real ecological patch dynamics, biogeographical habitat shifts and evolutionary turnover are confounded by incomplete preservation and collection."

Figure 7.18 is an example of palynological zonation of two wells through a Cretaceous succession in Delaware (from Doyle 1977). The wells are represented by their gamma ray logs with sample collection depths given in feet. The ranges of the principal angiosperm pollen types are shown by vertical bars, shown dashed where identification is uncertain. Concurrent-range zones are delimited by dashed lines perpendicular to the depth scale and are numbered I to IV next to the Series, Stage and Formation designations. It was found that the zones could be most easily defined on the basis of first (oldest) appearance of a taxon, partly because

simply reflect bad luck in choice of sampling site, or a bias induced by poor collection practices (Fig. 7.17). For example, whereas a field geologist undertaking a reconnaissance mapping exercise might be satisfied with a cursory

**Fig. 7.18** Use of palynological concurrent-range biozones to correlate two subsurface wells (as shown by gamma ray logs). Correlation brackets (*double-headed arrows*) terminate just above and below samples in the other well that *bracket* the age of the indicated sample (Doyle 1977)

extinct species tended to be reworked, and partly because taxa were found to die out slowly at the upper limit of their range. Work of this type required the counting and documentation of several hundreds of individual pollen grains in each sample. Several or many complete sections through the succession of interest may be required before the data are adequate for the definition of the biozones. Range charts, such as that illustrated here, must be prepared for each and carefully compared.

Another way to define biozones focuses on the gradual change in the anatomy of a particular evolutionary lineage. This is called a **lineage zone**. McGowran (2005) used an

example from late Cenozoic mammalian evolution to illustrate this concept (Fig. 7.19). The evolution of dental morphology provided much of the information on which this zonal scheme was based.

### 7.5.3  Diachroneity of the Biostratigraphic Record

Another common item of conventional wisdom is that evolutionary changes in faunal assemblages are dispersed so rapidly that, on geological time scales, they can essentially

**Fig. 7.19** Lineage zones defined by rodent evolution (McGowran 2005, Fig. 4.31, p. 150; based on Fejar and Heinrich 1989)

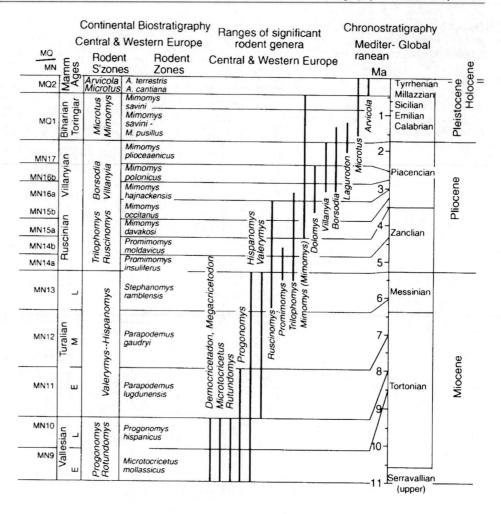

be regarded as instantaneous. This argument is used, in particular, to justify the interpretation of FADs as time-stratigraphic events (setting aside the problems of preservation discussed above). However, this is not always the case. Some examples of detailed work have demonstrated considerable diachroneity in important pelagic fossil groups. Landing et al. (2013) provided an overview of the problem, focusing in particular on the Cambrian. Cramer et al. (2015) pointed out that many "events" that are assumed to be instantaneous on a geological time scale, may, in fact, be diachronous on a finer time scale. With our increasing ability to provide chronostratigraphic control on events in deep time to a ±0.1 % level of accuracy (Sect. 8.10) quantifying diachroneity may become increasingly important.

MacLeod and Keller (1991) explored the completeness of the stratigraphic sections that span the Cretaceous-Tertiary boundary, as a basis for an examination of the various hypotheses that have been proposed to explain the dramatic global extinction occurring at that time. They used graphic correlation methods, and were able to demonstrate that many foraminiferal FADs and LADs are diachronous. Maximum diachroneity at this time is indicated by the species

*Subbotina pseudobulloides*, the FAD of which may vary by up to 250 ka between Texas and North Africa. However, it is not clear how much of this apparent diachroneity is due to preservational factors.

An even more startling example of diachroneity is that reported by Jenkins and Gamson (1993). The FAD of the Neogene foraminifera *Globorotalia truncatulinoides* differs by 600 ka between the southeast Pacific Ocean and the North Atlantic Ocean, based on analysis of much DSDP material. This is interpreted as indicating the time taken for the organism to migrate northward from the South Pacific following its first evolutionary appearance there. As Jenkins and Gamson (1993) concluded

> The implications are that some of the well documented evolutionary lineages in the Cenozoic may show similar patterns of evolution being limited to discrete ocean water masses followed by later migration into other oceans … If this is true, then some of these so-called 'datum planes' are diachronous.

This conclusion is of considerable importance, because the result is derived from excellent data, and can, therefore, be regarded as highly reliable, and relates to one of the most universally preferred fossil groups for Mesozoic-Cenozoic biostratigraphic purposes, the foraminifera. It would appear to suggest a limit of up to about one half million years on the precision that can be expected of any biostratigraphic event. Figure 7.16 illustrates the general problem.

Cody et al. (2008) provided another example of diachroneity. They reported on the distribution of diatoms in 32 Neogene cores from the Southern Ocean, which allowed an estimate to be made of the differences between the levels of the observed local First Occurrence and Last Occurrence events and the projected levels of the global FADs and LADs for the same species. Around 50 % of local event levels do not accurately record the timing of the global event: a few are off by 4 Ma or more from the total global FAD and LAD, due mostly to a small set of individually incomplete local ranges.

The cases reported here may or may not be a fair representation of the magnitude of diachroneity in general, although this is recognized as a general problem (Smith et al. 2015). After a great deal of study, experienced biostratigraphers commonly determine that some species are more reliable or consistent in their occurrence than others. Such forms may be termed **index fossils**, and receive a prominence reflecting their usefulness in stratigraphic studies. Studies may indicate that some groups are more reliable than others as biostratigraphic indicators. For example, Ziegler et al. (1968) demonstrated that brachiopod successions in the Welsh Paleozoic record were facies controlled and markedly diachronous, based on the use of the zonal scheme provided by graptolites as the primary indicator of relative time (Fig. 7.8). Armentrout (1981) used diatom zones to demonstrate that molluscan stages are time transgressive in the Cenozoic rocks of the northwest United States. Wignall (1991) demonstrated the diachroneity of Jurassic ostracod zones. Landing et al. (2013, p. 136) offered this general caution.

> Use of the local FADs of a fossil for correlation between sections without rigorous supplementary information will lead to errors in correlation or poorly defined chronostratigraphic units because significant time intervals likely will separate the local FADs. Each FAD must mirror biological phenomena ranging from evolutionary origination, to dispersal, successful local colonisation, and appearance of facies that allow a species' fossilisation

### 7.5.4 Quantitative Methods in Biochronology

*The graphic correlation technique.* This method was first described in a landmark book by Shaw (1964). Useful explanations of the technique were given by Miller (1977), and Edwards (1984, 1985). Mann and Lane (1995) edited a research collection devoted to the application of this topic to practical problems in basin analysis. Gradstein et al. (2004b)

discussed the use of the method in the construction of the Geological Time Scale. The example used herein has been borrowed from Miller (1977).

As with conventional biostratigraphy, the graphic method relies on the careful field or laboratory recording of occurrence data. However, only two items of data are noted for each taxon, the first (oldest) and last (youngest) occurrence (the FAD and LAD). These define a local range for each taxon. The objective is to define the local ranges for many taxa in at least three complete sections through the succession of interest. The more sections that are used, the more nearly these ranges will correspond to the total (true) ranges of the taxa. To compare the sections, a simple graphical method is used.

One particularly complete and well-sampled section is chosen as a **standard reference section**. Eventually, data from several other good sections are amalgamated with it to produce a **composite standard reference section**. A particularly thorough paleontologic study should be carried out on the standard reference section, as this enables later sections, for example, those produced by exploration drilling, to be correlated with it rapidly and accurately.

The graphic technique, which will now be described, is used both to amalgamate data for the production of the composite standard and for correlating the standard with new sections. Figure 7.20 shows a two-dimensional graph in which the thicknesses of two sections X and Y have been marked off on the corresponding axes. The first occurrence of each taxon is marked by a circle on each section and the

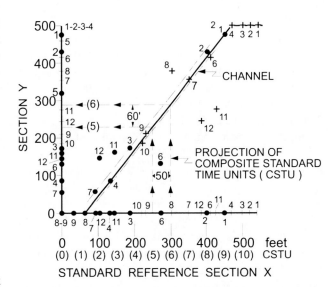

**Fig. 7.20** This and the next three figures illustrate Shaw's (1964) graphic correlation method as discussed by Miller (1977). This plot shows the distribution of first occurrences (*open circles*) and last occurrences (*crosses*) in two sections and positioning of the line of correlation. The channel is the zone on either side of the line of correlation encompassing observation error

**Fig. 7.21** The method used to compile a composite standard reference section. Data from new sections may be used to extend the range of occurrence of taxa that do not show their full range in the standard section (lowest occurrence of species 6, highest occurrence of species 8) and may also be used to transfer data on to the standard section, such as the range of species 5, which does not occur in the latter (Miller 1977)

last occurrence by a cross. If the fossil taxon occurs in both sections, points can be drawn within the graph corresponding to first and last occurrences by tracing lines perpendicular to the X and Y axes until they intersect. For example, the plot for the top of fossil 7 is the coincidence of points X = 350 and Y = 355.

If all the taxa occur over their total range in both sections and if sedimentation rates are constant (but not necessarily the same) in both sections, the points on the graph fall on a straight line, called the line of correlation. In most cases, however, there will be a scatter of points. The X section is chosen as the standard reference section, and ranges will presumably be more complete there. The line of correlation is then drawn so that it falls below most of the first occurrence points and above most of the last occurrence points. First occurrence points to the left of the line indicate late first appearance of the taxon in section Y. Those to the right of the line indicate late first appearance in section X. If X is the composite standard, it can be corrected by using the occurrence in section Y to determine where the taxon should have first appeared in the standard. The procedure is shown in Fig. 7.21. Arrows from the first occurrence of fossil 6 show that in section X the corrected first appearance should be at 165 ft. The same arguments apply to the points for last appearances. Corrections of the kind carried out for fossil 6 (and also the last occurrence of fossil 8) in Fig. 7.21 enable refinements to be made to the reference section. Combining

several sections in this manner is the method by which the composite standard is produced. Data points can also be introduced for fossils that do not occur in the reference section. In Fig. 7.21, arrows from the first and last occurrences of fossil 5 in section Y show that it should have occurred between 320 and 437 ft. in section X.

If the average, long-term rate of sedimentation changes in one or other of the sections, the line of correlation will bend. If there is a hiatus (or a fault) in the new, untested sections (sections Y), the line will show a horizontal terrace. Obviously, the standard reference section should be chosen so as to avoid these problems as far as possible. Harper and Crowley (1985) pointed out that sedimentation rates are in fact never constant and that stratigraphic sections are full of gaps of varying lengths (we discuss this problem in this chapter). For this reason, they questioned the value of the graphic correlation method. However, Edwards (1985) responded that when due regard is paid to the scale of intraformational stratigraphic gaps, versus the (usually) much coarser scale of biostratigraphic correlation, the presence of gaps is not of critical importance. Longer gaps, of the scale that can be detected in biostratigraphic data (e.g., missing biozones) will give rise to obvious bends in the line of correlation, as noted previously.

The advantage of the graphic method is that once a reliable composite standard reference section has been drawn up it enables chronostratigraphic correlation to be determined between any point within it and the correct point on any comparison section. Correlation points may simply be read off the line of correlation. The range of error arising from such correlation depends on the accuracy with which the line of correlation can be drawn. Hay and Southam (1978) recommended using linear regression techniques to determine the correlation line, but this approach assigns equal weight to all data points instead of using one standard section as a basis for a continuing process of improvement. But as Edwards (1984) noted, all data points do not necessarily have equal value; the judgment and experience of the biostratigrapher are essential in evaluating the input data. For this reason, statistical treatment of the data is inappropriate.

Figures 7.22 and 7.23 illustrate an example of the use of the graphic method in correlating an Upper Cretaceous succession in the Green River Basin, Wyoming, using palynological data (from Miller 1977). The composite standard reference section has been converted from thickness into **composite standard time units**, by dividing it up arbitrarily into units of equal thickness. As long as the rate of sedimentation in the reference section is constant, these time units will be of constant duration, although we cannot determine by this method alone what their duration is in years. Figure 7.22 shows the method for determining the position of selected time lines on each test section, and in Fig. 7.23 the time units are used as the basis for drawing

**Fig. 7.22** An example of the use of the graphic method, showing plot of data from one well against the composite standard. *Break* and *shaded area* on line of correlation are interpreted as an unconformity. Reference section has been divided into arbitrary thickness units (composite standard time units) (Miller 1977)

**Fig. 7.23** Correlation of four wells in an area of marked lateral facies change using the composite standard time units from Fig. 7.22 (Miller 1977)

correlation lines between four such sections. Note the unconformity in each illustration and the variation in sedimentation rates in Fig. 7.23.

The value of the graphic method for correlating sections with highly variable lithofacies and no marker beds is obvious, and it is perhaps surprising that the method is not more widely used. An important difference between this method and conventional zoning schemes is that zoning methods provide little more than an ordinal level of correlation (biozones, as expressed in the rock record, have a finite thickness which commonly cannot be further subdivided), whereas the graphic method provides interval data (the ability to make graduated subdivisions of relative time). Given appropriate ties to the global time frame the composite standard time units can be correlated to absolute ages in years, and used to make precise interpolations of the age of any given horizon (such as a sequence boundary) between fossil occurrences and tie points (Gradstein et al. 2004b). The precision of these estimates is limited solely by the accuracy and precision obtainable during the correlation to the global standard. MacLeod and Keller (1991) provided excellent examples of this procedure, and their results suggest an obtainable precision of less than ±100 ka.

*Constrained optimization (CONOP)*: Gradstein et al. (2004b) pointed out several disadvantages of the basic graphic correlation method. It relies on only a few sections, placing particular importance on a single section that becomes the basis for the composite standard. A superior, automated correlation method, called constrained optimization, has been used in the construction of several parts of the Phanerozoic time scale. This method automates graphic correlation so that multiple sections are compared and correlated simultaneously. In this way, gaps and changes in sedimentation rate in the initial standard section do not influence the outcome. The method is described by Kemple et al. (1995), Sadler (1999a), and critically evaluated by Smith et al. (2015).

Several of the time scales for the Phanerozoic periods described in detail in Gradstein et al. (2004a) were developed using these techniques. For example, the Ordovician scale in Gradstein et al. (2004a) made use of 669 graptolite taxa in 119 sections. In the recent study of Ordovician-Silurian graptolite biostratigraphy by Sadler et al. (2009) they noted (p. 887) that "The graptolite zones vary widely in duration from as short as 0.1 million years to nearly 5.0 million years. The mean duration of zones or zonal groupings calibrated here is 1.44 million years in the Ordovician and 0.91 million years in the Silurian." Current developments in the biostratigraphic basis of chronostratigraphy are discussed further in Sect. 8.10.1.

## 7.6    Unconformity-Bounded Units

Unconformities have commonly served as convenient boundaries for various types of stratigraphic units, particularly those based on lithostratigraphy. In many areas, regional unconformities have long been used to define natural subdivisions of the stratigraphic record (Blackwelder 1909). As originally defined by Sloss et al. (1949), sequences were defined as operational units separated by "marked discontinuities in the stratal record of the craton which may be traced and correlated for great distances on the objective bases of lithologic and faunal 'breaks' ..." The use of unconformities as boundaries is now avoided in the definition of chronostratigraphic units, for reasons explained in Sect. 7.7. However, it is increasingly, being recognized that the stratigraphic record is subdivided into unconformity-bounded units of regional and possibly even global extent, caused by widespread changes in sea level or by regional tectonic or climatic events (Emery and Myers 1996; Miall 2010). The methods of sequence stratigraphy have become virtually universal (Catuneanu 2006).

Unconformity-bounded units include other types of unit within them, such as biostratigraphic and lithostratigraphic units. They are not the same as these and are not necessarily equivalent to chronostratigraphic units, because the ages of the bounding unconformities may change from place to place. However, unconformity-bounded units have a certain chronostratigraphic significance because, with certain unusual exceptions, all the rocks below an unconformity are older than all of those above, and time lines do not cross unconformity surfaces. Some exceptions to these rules include (1) where a disconformity surface is caused by submarine erosion by a deep oceanic current that changes position with time, as a result of changing configuration of the ocean basin (Christie-Blick et al. 1990), but this is not a problem that is likely to be encountered very frequently; (2) Ravinement surfaces are time-transgressive (Nummedal and Swift 1987); (3) Large-scale incised valleys, caused by base-level fall at a coastline, may be significantly time-transgressive (Strong and Paola 2008).

Proposals for the formal definition of unconformity-bounded units were given by the International Subcommission on Stratigraphic Classification (1987; see also Salvador 1994) and are also contained in the North American Commission on Stratigraphic Nomenclature (1983, 2005). The term **sequence**, as originally used by Sloss et al. (1949) was not recommended by these authorities, because this term had come to be used in a number of slightly but significantly different ways. In common geological parlance, the term *sequence* has commonly been used as a synonym for *succession*, a practice that needs to be discouraged because of the value now associated with the sequence method. The ISSC (1987) recommended the formal term **synthem**, a proposal followed by Salvador (1994), whereas the North American Stratigraphic Commission proposed the definition of **allostratigraphic units**, including alloformation, allogroup, and allomember (NACSN 1983). The term synthem has not been accepted by the stratigraphic community. **Allostratigraphy**, as a formal system of definition and naming, has had a modest success. The intent was that major unconformity-bounded units would be termed alloformations, with minor units nested within them, such as parasequences, labeled allomembers (NACSN 2005). The intent of this approach was to be purely descriptive, with no genetic connotations built into the terminology or methods.

Several groups of workers have made use of allostratigraphic methodology. For example, Autin (1992) subdivided the terraces and associated sediments in a Holocene fluvial floodplain succession into alloformations. R.G. Walker and his coworkers employed allostratigraphic terminology in their study of the sequence stratigraphy of part of the Alberta Basin, Canada. Their first definition of unconformity-bounded units is described in Plint et al. (1986), where the defining concepts were referred to as **event stratigraphy**, following the developments of ideas in this area by Einsele and Seilacher (1982). With the increasing realization that sequences and their bounding surfaces may be markedly diachronous, we no longer refer to sequences and sequence boundaries as "events". Walker (1990) discussed some of the practical problems in making use of sequence and allostratigraphic concepts (see Sect. 1.2.13). Explicit use of allostratigraphic terms appears in later papers by this group (e.g., Plint 1990). A text on facies analysis that built extensively on the work of this group recommends the use of allostratigraphic methods and terminology as a general approach to the study of stratigraphic sequences (Walker 1992). Martinsen et al. (1993) compared lithostratigraphic, allostratigraphic and sequence concepts as applied to a stratigraphic succession in Wyoming. As they were able to demonstrate, each method has its local advantages and disadvantages. However, it is now safe to say that, in 2013, where the data justify it, allostratigraphy has largely been bypassed in favour of the much more powerful, genetic concepts and interpretive methods associated with sequence stratigraphy.

## 7.7 The Development of Formal Definitions for Sequence Stratigraphy

Allostratigraphy represented one of the first, hesitant, attempt to incorporate unconformity-bounded units into the formal framework of stratigraphy (see Sect. 1.2.9 for earlier ideas on this topic). Acceptance of its successor, sequence stratigraphy, by the "official" international community—the *International Commission on Stratigraphy*—has nearly been achieved. Controversies about how to define sequences have hindered the development of formal procedures.

One of the commonest complaints about sequence stratigraphy was that it is "model-driven." Catuneanu (2006, p. 6–9) summarized the various approaches that have been taken to defining sequences, and argued the case that the differences between the various models is not important, so long as sequences are described properly with reference to a selected standard model, with correct and appropriate recognition of systems tracts and bounding surfaces. His comparison diagram is reproduced here as Fig. 7.24, and the suite of important surfaces that are used in sequence and systems-tract definition is shown in Fig. 7.25. The major difference between the sequence models is where different workers have chosen to place the sequence boundary. It should be noted that in each of the sequence definitions shown in Fig. 7.26, a similar set of systems tracts is shown in much the same relationship to each other. Exceptions include the T-R sequence, which makes use of a simplified definition of systems tracts, and such differences as that between the "late highstand" of depositional sequence III and the "falling stage" of depositional sequence IV.

There has been extensive discussion between the original proponents of the modern sequence models (P.R. Vail, H. Posamentier and their colleagues at Exxon) and others, regarding sequence definitions, centered on such characteristics as the facies shifts that take place within sequences and their significance with regard to the base-level cycle. This discussion has led to a number of different ways of defining sequences (Figs. 7.24–7.26). In a masterly synthesis of the controversies, Catuneanu (2006) showed how to resolve these differences, and in three major collaborative publications (Catuneanu et al. 2009, 2010, 2011) discussed proposals for formal definitions that could be accepted by all workers (Fig. 7.26). As he demonstrated, "all approaches are correct under the specific circumstances for which they were proposed" (Catuneanu et al. 2011, pp. 232–233). Additional details, including descriptions and illustrations of the key bounding surfaces that develop on continental shelves, are described by Zecchin and Catuneanu (2013).

**Fig. 7.24** The evolution of sequence definitions. *T-R* transgressive-regressive. From Catuneanu et al. (2011)

| Base level | Events | Surfaces |
|---|---|---|
| | Onset of forced regression | ← Basal surface of forced regression ** |
| | End of transgression | ← Maximum flooding surface |
| | | Transgressive ravinement surfaces |
| | End of regression | ← Maximum regressive surface |
| | End of forced regression | ← Correlative conformity * |
| | | Subaerial unconformity and Regressive surface of marine erosion |
| | Onset of forced regression | ← Basal surface of forced regression ** |

* *sensu* Hunt and Tucker (1992)
** = correlative conformity *sensu* Posamentier et al. (1988)

**Fig. 7.25** Stratigraphic surfaces used in the definition of sequences and systems tracts, and their timing, relative to the cycle of base-level change (Catuneanu 2006, Fig. 4.7)

| Events and stages \ Sequence model | Depositional Sequence I | Depositional Sequence II | Depositional Sequence III | Depositional Sequence IV | Genetic Sequence | T-R Sequence |
|---|---|---|---|---|---|---|
| HNR | | HST | early HST | HST | HST | RST |
| *end of T* | | | | | MFS | |
| T | Sequence | TST | TST | TST | TST | TST |
| *end of R* | | | | | | MRS |
| LNR | | late LST (wedge) | LST | LST | late LST (wedge) | |
| *end of RSL fall* | | | | CC** | | |
| FR | | early LST (fan) | late HST | FSST | early LST (fan) | RST |
| *onset of RSL fall* | CC* | CC* | | | | |
| HNR | | HST | early HST | HST | HST | |

———— sequence boundary
——— systems tract boundary
········· within-sequence surface
- - - - - within-systems tract surface

end of RSL fall

end of transgression

time

onset of RSL fall

end of regression

**Fig. 7.26** Nomenclature of systems tracts, and timing of sequence boundaries for the various sequence stratigraphic approaches. Abbreviations: *RSL* relative sea level; *T* transgression; *R* regression; *FR* forced regression; *LNR* lowstand normal regression; *HNR* highstand normal regression; *LST* lowstand systems tract; *TST* transgressive systems tract; *HST* highstand systems tract; *FSST* falling-stage systems tract; *RST* regressive systems tract; *T-R* transgressive-regressive; *CC\** correlative conformity in the sense of Posamentier and Allen (1999); *CC\*\** correlative conformity in the sense of Hunt and Tucker (1992); *MFS* maximum flooding surface; *MRS* maximum regressive surface (from Catuneanu et al. 2011)

In the original Exxon model (Vail et al. 1977) the sequence boundary (commonly abbreviated as SB on diagrams) was drawn at the subaerial unconformity surface, following the precedent set by Sloss (1963), an approach which readily permits the sequence framework to be incorporated into an allostratigraphic terminology, at least for coastal deposits, where the subaerial erosion surface is readily mapped. Offshore may be a different story. The first sequence model (Vail et al. 1977) did not recognize the falling-stage systems tract. The highstand of one sequence was followed directly by the lowstand of the next sequence, with the sequence boundary falling between the two systems tracts. With the recognition of the process of forced regression (Plint 1988), the forced regressive deposits could be assigned either to a "late highstand" or to an "early lowstand." Based on assumptions about the changing rate of sea level fall, the sequence boundary—the coastal equivalent of the subaerial erosion surface—was initially placed at the **basal surface of forced regression** (assumed commencement of forced regression). This placement of the sequence boundary is the basis for what Catuneanu (2006) refers to as "depositional sequence II" (Fig. 7.26). The problem with this definition is addressed below. In addition, the early Exxon work defined several different types of sequence-bounding unconformity. Vail and Todd (1981) recognized three types, but later work (e.g., Van Wagoner et al. 1987) simplified this into two, termed type-1 and type-2 unconformities, based on assumptions about the rate of change of sea level and how this was reflected in the sequence architecture. In the rock record, they would be differentiated on the basis of the extent of subaerial erosion and the amount of seaward shift of facies belts. However, Catuneanu (2006, p. 167) pointed out the long-standing confusions associated with these definitions and recommended that they be abandoned. These "types" are not discussed further in this book. Schlager (2005, p. 121) recommended the separate recognition of a third type of sequence boundary, one which forms "when sea level rises faster than the system can aggrade, such that a transgressive systems tract directly overlies the preceding highstand tract often with a significant marine hiatus. … Marine erosion frequently accentuates this sequence boundary, particularly on drowned carbonate platforms."

Hunt and Tucker (1992) were amongst the first (since Barrell!) to point out that during sea-level fall, subaerial erosion continues until the time of sea-level lowstand, with the continuing transfer of sediment through clastic delivery systems to the shelf, slope and basin, and with continuing downcutting of the **subaerial erosion surface** throughout this phase. The age of the subaerial erosion surface, therefore, spans the time up to the end of the phase of sea-level fall, a time substantially later than the time of initiation of forced regression. The use of the basal surface of forced regression as a sequence boundary, as in "depositional sequence II" is, therefore, not an ideal surface at which to define the sequence boundary, although, as Catuneanu (2006) reports, it is commonly a prominent surface on seismic-reflection lines. In fact, as Embry (1995) pointed out (see Fig. 7.29), there is no through-going surface associated with forced regression that can be used to extend the

subaerial erosion surface offshore for the purpose of defining a sequence boundary. He argued that "from my experience I have found that the most suitable stratigraphic surface for the conformable expression of a sequence boundary is the transgressive surface" (Embry 1995, p. 4). This meets his criterion—one with which all stratigraphers would agree—that "one of the main purposes of sequence definition [is] a coherent genetic unit without significant internal breaks" (Embry 1995, p. 2). His preferred definition of sequences, the T-R sequence, places the sequence boundary at the TR surface, at the end of the phase of regression and the time of initial transgression (Figs. 7.28 and 7.29). There is, of course, a delay in time between the end of downcutting of the subaerial erosion surface during the falling stage, and the flooding of the same surface during transgression. The results of the two processes may coincide in the rocks, which is why this surface may provide a good stratigraphic marker; but it is important to remember that the surface is not a time marker, but represents a time gap, with the gap decreasing in duration basinward.

Highlighting the timing of development of the subaerial erosion surface by Hunt and Tucker (1992) also served to highlight the inconsistency of assigning the main succession of submarine fan deposits on the basin floor to the lowstand systems tract, as shown in the Vail et al. (1977) and Posamentier et al. (1988) models, in which these deposits are shown resting on the sequence boundary. Notwithstanding the discussion of the Hunt and Tucker (1992) paper by Kolla et al. (1995), who defended the original Exxon models, this is an inconsistency that required a redefinition of the standard sequence boundary. It is now recognized that in the deep offshore, within submarine-fan deposits formed from sediment delivered to the basin floor during a falling stage (the period of most active sediment delivery to the continental margin; see Fig. 5.13), there may be no sharp definition of the end of the falling stage nor of the turn-around and subsequent beginning of the next cycle of sea-level rise and, therefore, no distinct surface at which to draw the sequence boundary. The boundary here is a correlative conformity, and may be very difficult to define in practice.

An alternative sequence model, termed the **genetic stratigraphic sequence** (Figs. 7.26 and 7.28), was defined by Galloway (1989), building on the work of Frazier (1974). Although Galloway stressed supposed philosophical differences between his model and the Exxon model, in practice, the difference between them is simply one of where to define the sequence boundaries. The Exxon model places emphasis on subaerial unconformities, but Galloway (1989) pointed out that under some circumstances unconformities may be poorly defined or absent and, in any case, are not always easy to recognize and map. Galloway's (1989) preference is to draw the sequence boundaries at the maximum flooding surface, which corresponds to the highstand downlap surfaces. He claims that these surfaces are more prominent in the stratigraphic record, and therefore more readily mappable.

Galloway's proposal has not met with general acceptance. For example, Walker (1992) disputed one of Galloway's main contentions, that "because shelf deposits are derived from reworked transgressed or contemporary retrogradational deposits, their distribution commonly reflects the paleogeography of the precursor depositional episode." Galloway (1989) went on to state that "these deposits are best included in and mapped as a facies element of the underlying genetic stratigraphic sequence." However, as Walker (1992) pointed out, most sedimentological parameters, including depth of water, waves, tides, basin geometry, salinity, rates of sediment supply, and grain size, change when an unconformity or a maximum flooding surface is crossed. From the point of view of genetic linkage, therefore, the only sedimentologically related packages lie (1) between a subaerial unconformity and a maximum flooding surface, (2) between a maximum flooding surface and the next younger unconformity, or (3) between a subaerial erosion surface and the overlying unconformity (an incised-valley-fill) (Walker 1992, p. 11).

However, some workers have found Galloway's use of the maximum flooding surface much more convenient for sequence mapping, for practical reasons. For example, it may yield a prominent gamma-ray spike in wireline logs (Underhill and Partington 1993), or it may correspond to widespread and distinctive goniatite bands (Martinsen 1993), or it may provide a more readily traceable marker, in contrast to the surface at the base of the lowstand systems tract, which may have irregular topography and may be hard to distinguish from other channel-scour surfaces (Gibling and Bird 1994). In nonmarine sections it may be hard to find the paleosol on interfluves that correlates with the sequence-bounding channel-scour surface (Martinesen 1993). In some studies (e.g., Plint et al. 1986; Bhattacharya 1993) it has been found that ravinement erosion during transgression has removed the transgressive systems tract, so that the marine flooding surface coincides with the sequence boundary.

Catuneanu et al. (2011, p. 183) pointed out that however sequences are defined, there may be unconformities contained within them. For example, Galloway's (1989) genetic stratigraphic sequences are explicitly defined with the sequence boundary at the maximum flooding surface, not the subaerial erosion surface. Therefore, sequences can no longer be defined as "unconformity-bounded." Given the need to encompass all types of sequence model, Catuneanu et al. (2009, p. 19) proposed to redefine a sequence as "a succession of strata deposited during a full cycle of change in accommodation or sediment supply".

As described in Chaps. 3 and 4, many stratigraphic successions contain small-scale cycles nested within a sequence. Van Wagoner et al. (1988, 1990) erected the term **parasequence** to encompass the shoaling-upward successions, capped by flooding surfaces, that are common in coastal clastic successions. The term was proposed originally as part of a hierarchy of terms, the bed, bedset, parasequence, parasequence set, and sequence. Prograding delta lobes, regressing clastic shorelines, and peritidal carbonate cycles are examples of parasequences that are particularly common in the geological record. Catuneanu et al. (2009, p. 19) noted that the term has also been used for cyclic deposits in some fluvial and deep-marine deposits, where the concept of "flooding surface" is irrelevant. A particular source of confusion comes from the incorporation of the word "sequence" within the term parasequence. The nomenclature problem was not improved by the usage employed by Mitchum and Van Wagoner (1991), who equated parasequences with "4th-order paracycles." Sequences are allogenic products of regional controls, whereas parasequences may be a product of autogenic processes, such as delta lobe switching (Fig. 7.27). This has been demonstrated to be the case in the example of the Dunvegan delta illustrated in Fig. 5.11 (Bhattacharya 1991). The shingles and their bounding flooding surfaces are therefore local in distribution, and their development has little, if anything, to do with the allogenic mechanisms that generate sequences. However, to apply to these successions a term that contains the word "sequence" in it is inevitably to introduce the implication that they are allogenic in origin and constitute regionally correlatable units. The correct interpretation clearly depends on good mapping to determine the extent and correlatability of each shingle, and it would seem advisable not to use a term in a descriptive sense that carries genetic implications. Yet Mitchum and Van Wagoner (1991) illustrated parasequences/paracycles with a dip-oriented cross-section 90 km long, which claims to show correlations between several separate depositional systems. This, and the implication that parasequences are the same thing as minor sequences, added to the confusion. Furthermore, some high-frequency sequences formed by Milankovitch processes are comparable in thickness and outcrop appearance. It would be this author's preference that the term parasequence be abandoned entirely, but Catuneanu et al. (2009, 2011) (papers to which this author contributed!) recommended that the term may continue to be used if it is restricted to its original definition.

The selection of which sequence definition or model to use in any given stratigraphic setting is a matter of choice. The original Sloss/Vail **depositional sequence** model (Sloss 1963; Vail et al. 1977) and the **T-R model** of Embry and Johannessen (1992) use the subaerial erosion surface as the

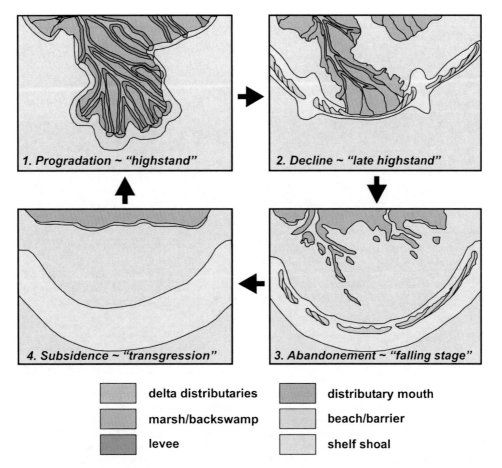

**Fig. 7.27** The development and abandonment of delta lobes in a river dominated, Mississippi-type delta (e.g., see Fig. 4.21), based on detailed analysis of the Mississippi delta system by Boyd and Penland (1988). In *stage 1*, progradation develops an upward-shoaling deltaic succession. Abandonment, followed by subsidence (resulting from compaction) cause the upper layers of the succession to be reworked (*stage 2*), resulting in the development of an extensive barrier island system (*stage 3*). Finally, the deposit undergoes transgression and is covered by marine shale (*stage 4*). Repetition of this succession of events when a new delta lobe progrades back over the older deposit results in shoaling upward successions bounded by transgressive flooding surfaces, that is, parasequences. In this case, however, they are clearly of autogenic origin. Systems-tract designations for each of the four stages are indicated in *parenthesis*

**Fig. 7.28** Selection of sequence boundaries according to the 'depositional', 'genetic stratigraphic' and 'transgressive–regressive' sequence models. The choice of sequence boundary is less important than the correct identification of all sequence stratigraphic surfaces in a succession (Fig. 7.25). Abbreviations: *SU* subaerial unconformity; *CC1* correlative conformity sensu Posamentier and Allen (1999); *CC2* correlative conformity sensu Hunt and Tucker (1992); *MFS* maximum flooding surface; *MRS* maximum regressive surface. The subaerial unconformity is a stage-significant surface, whereas all other surfaces shown in this diagram are event-significant (Catuneanu et al. 2009, Fig. 24, p. 17)

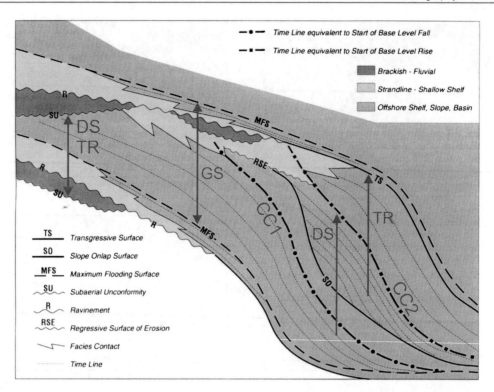

**Fig. 7.29** Schematic cross-section through an ideal continental-margin sequence, showing the relationships between the major surfaces. *Red arrows* and *letters* denote three alternative ways to define sequences. In coastal and nearshore deposits the sequence boundary for depositional sequences (*DS*) and TR sequences (*TR*) is defined by the subaerial erosion surface (*SU*). Correlating this surface offshore may be difficult in the absence of high-quality seismic-reflection data. Genetic stratigraphic sequences (*GS*) are defined by the maximum flooding surfaces. Depositional sequences III and IV use the correlative conformity corresponding to the end of base-level fall as the sequence boundary (*CC\*\** of Fig. 7.26, *CC2* of Fig. 7.28 and this figure). This may be indicated by a change from erosional to aggradational deposition in the forced-regressive deposits. The time of initiation of base-level fall corresponds to correlative conformity *CC\** of Fig. 7.26 and *CC1* of Fig. 7.28 and this figure, and defines the contact between early and late highstand in depositional sequence III. The *top* of a DS is indicated here by the *arrow* labeled DS that terminates at the surface marking the beginning of base-level rise. TR sequences are defined in the offshore by the end of regression. In this diagram the wedge of regressive deposits is truncated by the transgressive surface, which here marks the *top* of a TR sequence. In depositional sequence terminology, this latest wedge of regressive deposits is classified as the lowstand systems tract (adapted from Embry 1995, Fig. 1)

sequence boundary (Figs. 7.28 and 7.29). The sequence then represents a full cycle of increasing and decreasing accommodation preserved between two erosion surfaces. The Galloway (1989) **genetic stratigraphic sequence** model uses the maximum flooding surface (MFS) as the sequence boundary (Figs. 7.28 and 7.29). In many stratigraphic settings, recognizing and correlating a maximum flooding surface is much more easily accomplished than mapping the subaerial erosion surface. The MFS is commonly represented by a marine shale or a condensed section which, because it is deposited at a time of maximum transgression, is typically widespread and forms a distinctive marker bed between packages of coarser clastic or carbonate/evaporite facies. Subaerial erosion surfaces may be characterized by significant erosional relief, which may make them difficult to trace within suites of wireline logs. Furthermore, where this surface occurs within successions of nonmarine strata it can be very difficult to distinguish the sequence boundary from local fluvial channel scour surfaces.

However, these differences in mappability do not need to be reflected in the choice of sequence model.

The TR sequence model uses the subaerial erosion surface as the sequence boundary for the nonmarine and nearshore portions of a sequence, but differs from the depositional sequence where sequences are traced into the offshore.

In a nonmarine to coastal to marine section, the maximum seaward extent of the subaerial erosion surface depends on the amplitude of relative sea-level change. Traced far enough seaward there will be a point beyond which water depths are such that exposure and erosion do not take place during base-level fall. Correlating sequence boundaries and systems tracts into the offshore may be difficult. The erosion surface generated by forced regression (RSE in Fig. 7.29) defines the base of the falling-stage systems tract but, as demonstrated by Plint (1988), this surface is not at a stratigraphically consistent level, but may recur at stratigraphically higher positions in the offshore direction as base-level fall continues. The offshore limit of the RSE records the end of the phase of base-level fall, and this point is used to define the sequence boundary in **depositional sequences III** and **IV** (Fig. 7.29). Regressive sedimentation is likely to continue, until the rate of rising accommodation during the subsequent cycle of base-level rise equals the rate of sediment supply. At this time, regression ends, and this point is used by Embry (1995) and Embry and Johannessen (1992) to define the sequence boundary for *TR cycles*. This coincides with the transgressive surface marking the beginning of the next cycle of base-level rise (TS in Fig. 7.29). In many shelf settings, the TS is marked by lag deposits or a condensed section and is readily recognizable in cores and on wireline logs. It passes landward into the ravinement surface (R in Fig. 7.29), a diachronous surface formed by wave erosion during transgression.

In earlier definitions of the depositional sequence (**depositional sequences I** and **II** in Fig. 7.26), the seaward correlative of the subaerial erosion surface was equated with the beginning of base-level fall and this surface, and its seaward extension as a correlative conformity (CC1 in Fig. 7.28; this surface is also shown in Fig. 7.29), was used to demarcate the sequence boundary. The inconsistencies in this approach were pointed out by Hunt and Tucker (1992). Subaerial erosion continues throughout the falling stage of the base-level cycle, and so the erosion surface that is subsequently preserved by being transgressed and buried during the succeeding base-level rise will correspond in age to the end of the falling stage, not the beginning. The first regressive surface marking the beginning of the falling stage may be followed by others, further seaward and stratigraphically younger, as the falling-stage proceeds. Another

of the inconsistencies of the early sequence definitions was the assignment of offshore submarine fan deposits primarily to the "lowstand wedge" formed (according to this earlier interpretation) during the beginning of the next cycle of base-level rise. Given that much of the sediment supply for submarine fans comes from subaerial erosion during falling base level (Fig. 5.13), this does not seem plausible.

Embry and Johannessen (1992) claimed that the TR sequence definition is the only one that adheres to objective criteria, meaning the designation of the sequence boundary at a mappable surface. This is correct in that, as shown in Fig. 7.29, the surfaces marking the beginning and end of a cycle of base-level change are not necessarily associated with any recognizable facies change in the rocks, meaning that their recognition and mapping in surface outcrops or in subsurface cores, logs or seismic data, requires interpolation and extrapolation. The surface marking the end of regression is a lithologically mappable surface, but it will be directly mappable only so far seaward as regressive deposits extend. Beyond that point, correlation may be possible by the tracing of reflections in seismic data, but recognition of the surface of maximum regression in cores or logs through deep marine deposits is not likely to be possible, so this particular feature of the TR definition is not significantly better than that of the depositional sequence.

A generalized sequence model is shown in Fig. 7.30. In this model the base-of-slope submarine fan and other deep-water deposits are assigned to the falling stage systems tract, contrary to the first sequence definitions (depositional sequences I and II). The sequence boundary (surface 4) includes an incised valley, which is filled by the lowstand systems tract and during transgression (surfaces 5–7).

**Fig. 7.30** A generalized sequence model

**Fig. 7.31** Seismic line in the Gulf of Mexico showing different genetic types of deposits (forced regressive, normal regressive, transgressive) and stratigraphic surfaces that may serve as sequence boundaries according to different sequence stratigraphic models (modified from Posamentier and Kolla 2003). Abbreviations: *FR* forced regressive; *LNR* lowstand normal regressive; *T* transgressive; *SU* subaerial unconformity; *CC\** correlative conformity sensu Posamentier and Allen 1999 (=basal surface of forced regression); *CC\*\** correlative conformity sensu Hunt and Tucker 1992; *MRS* maximum regressive surface; *MFS* maximum flooding surface. The *line* displays the typical stacking patterns and stratal terminations associated with forced regression (offlap, downlap, toplap, truncation), normal regression (downlap, topset), and transgression (onlap) (Catuneanu et al. 2009, Fig. 7, p. 6)

Commonly, incised valleys are filled towards the end of the transgressive phase, with deposition extending beyond the margins of the valley. Elsewhere on the continental margin transgressive deposits are characteristically thin, and may consist of condensed deposits or a coarse lag. Hardgrounds and extensive bioturbation are common.

A significant flaw of the TR model is that the surface of maximum regression may not correlate with the subaerial erosion surface. As Embry and Johannessen (1992) noted, regression may continue in time beyond the end of the cycle of base-level fall, which corresponds to the age of the subaerial erosion surface. The deposits that form after this time are assigned to the lowstand systems tract, and may comprise a substantial thickness of regressive deposits, accumulated on the shelf, slope and deep basin during the beginning of the next cycle of transgression. Two seismic lines are shown here which contain this feature (Figs. 7.31 and 7.32). The lowstand normal regressive deposits (LNR) in Fig. 7.31 are the deposits which form after the end of base-level fall—indicated by the correlative conformity (CC\*\*) extending basinward from the subaerial erosion surface—and are capped by the maximum regressive surface (MRS). A particularly thick succession of shelf and slope deposits are assigned to the LNR in Fig. 7.32. This section illustrates the migration through time of the shelf margin and the shoreline. In this example a high sediment supply was maintained throughout transgression, resulting in a thick transgressive shelf section. This is commonly not the case, with transgression commonly recorded as a lag deposit or a condensed section.

| --- ▶ | Shelf-edge trajectory | | Sequence stratigraphic surfaces: |
|---|---|---|---|
| | Shoreline trajectory: | ▬▬▬ | subaerial unconformity |
| --- ▶ | lowstand normal regression | ---- | correlative conformity (*sensu* Hunt and Tucker, 1992) |
| --- ▶ | transgression | -- -- | maximum regressive surface |
| --- ▶ | highstand normal regression | -·-·- | maximum flooding surface |

**Fig. 7.32** Dip regional seismic profile from the Pelotas Basin, southern Brazil (modified from Abreu 1998), showing large-scale (high-rank) lowstand normal regressive (*LNR*), transgressive (*T*) and highstand normal regressive (*HNR*) systems tracts. Lower rank sequences are nested within these higher rank systems tracts. The transgressive systems tract thickens landward, which reflects the direction of shift of the depocenter. Individual backstepping parasequences are difficult to observe within the transgressive systems tract due to the limitation imposed by vertical seismic resolution. The shoreline trajectory and the shelf-edge trajectory may coincide during lowstand normal regression, but are separate during transgression and highstand normal regression. The change in depositional trends from dominantly progradational to dominantly aggradational is typical for lowstand normal regressions. Conversely, the change in depositional trends from dominantly aggradational to dominantly progradational is typical for highstand normal regressions. *Horizontal scale* approximately 50 km. *Vertical scale* 2 s two-way travel time (Catuneanu et al. 2009, Fig. 19, p. 14)

## 7.8   Chronostratigraphy and Geochronometry

### 7.8.1   The Emergence of Modern Methods

**Geochronometry** is the study of the continuum of geologic time. Geologic events and rock units may be fixed within this time frame by a variety of methods, of which radiometric dating is the most direct.

**Chronostratigraphy** is the study of the standard stratigraphic scale, comprising the familiar eras, periods, and ages (e.g., Paleozoic, Triassic, Campanian) (Table 7.1).

For the Phanerozoic, biostratigraphy is the main basis of the chronostratigraphic method, but radiometric methods and magnetic reversal stratigraphy are essential for providing numerical ("absolute") ages, and chemostratigraphy is becoming increasingly important. Radiometric dating is the only chronostratigraphic tool available for the Precambrian, and is essential for providing a calibration scale of biostratigraphic subdivisions in the Phanerozoic (Harland et al. 1964, 1990; Berggren et al. 1995). Magnetic reversals are now of great importance for studying Upper Cretaceous and Cenozoic strata, but are difficult to use in older rocks owing to the difficulty of obtaining complete reversal sequences and the problems of post-depositional modification (Kennett 1980).

Early speculations about the age of the Earth, and of rates of geological processes, began to be situated within a modern quantitative framework with the development of radiometric dating by Ernest Rutherford in 1905. The English geologist Arthur Holmes made the development of a radiometrically calibrated geologic time scale a central part of his life's work, and it could be argued that the first modern era in the quantification of geologic time ended with his death in 1965. Two major events occurred at about this time. The second edition of his great book "*Principles of Physical Geology*" appeared (Holmes 1965), containing, among other things, a lengthy treatment of radiometric dating (plus what were then his very advanced ideas about continental drift), and the Geological Society of London published the first comprehensive compilation of data in

support of a modern geological time scale (Harland and Francis 1964; Harland et al. 1971).

It is important to distinguish between the two quite different, but interrelated concepts of time, geochronometry, the measurement of time, in standard units such as the year and the second, and chronostratigraphy, the compilation of standard rock units. A distinction between "time" and "rocks" is essential, given the incompleteness of the stratigraphic record, and the need to continually revise, expand and update the means by which we relate the chronostratigraphic to the geochronometric scale. Harland (1978) and Harland et al. (1990, Chaps. 1–3) provided detailed explanations of the theory and terminology surrounding these terms. Because of the very fragmentary nature of the sedimentary record, there is still no chronostratigraphic standard for the Precambrian, which is subdivided mainly on a geochronometric basis, except for the Ediacaran System (635–542 Ma), the youngest part of the Precambrian, which is the only exception (Knoll et al. 2006).

The subdivision of the stratigraphic record into ever smaller units based on detailed studies of their fossil content had reached a remarkably sophisticated level by the early twentieth century, based on the specialized study of some unique units in which the fossils are abundant and contain readily measureable indicators of rapid evolution. The ammonites of the Jurassic in southern Britain figure prominently in this history, as described by Callomon (1995, 2001). It can be demonstrated that local subdivisions of relative stratal time representing time spans in the range of $10^5$ years are possible based on such work (this is discussed further in Chap. 8). But the question of how to incorporate this information systematically in the construction of a chronostratigraphic scale that could be used worldwide continued to be controversial and problematic until the 1970s (this early history is summarized by Miall 2004, pp. 6–11). There are three obvious reasons: the record is everywhere locally incomplete, fossils are facies bound, and with very rare exceptions it is not possible to assign specific numerical ages to fossil horizons without interpolations and extrapolations that might incorporate the other two problems. Some boundaries, including many established early in the history of the science, had been very poorly defined. For example, the Silurian-Devonian boundary, first recognised in Britain, occurs at a major angular unconformity, or within a marine to nonmarine transition, within which biostratigraphic correlation was problematic, raising (but only much later, when time began to be quantified) the question of how to define and categorize the time undocumented at the break.

Solutions to these problems began to emerge in the 1960s, and might be said to constitute the beginning of the second phase of the modern era of modern stratigraphy. The key development was the evolution of the concept of the **Global Stratigraphic Sections and Points (GSSP)**.

**Table 7.1** The conventional hierarchy of formal chronostratigraphic and geochronologic terms

| Chronostratigraphic | Geochronologic |
| --- | --- |
| Eonothem | Eon |
| Erathem | Era |
| System | Period |
| Series | Epoch |
| Stage | Age |
| Substage | Subage |

The ideas appear to be primarily British in origin (e.g., Ager 1964; Bassett 1985; Cowie 1986; Holland 1986). They encompass two important concepts: the idea of designating key marker boundaries at specific type locations within continuous sections, and the idea that multiple criteria—biostratigraphy and radiometric dating, and more recently magnetostratigraphy and chemostratigraphy, wherever applicable, should be used to nail down the precise age of the boundary "by all available means," to quote Torrens (2002, p. 256). McLaren (1970, p. 802) explained the desirability of defining boundaries within continuous sections in this way:

> There is another approach to boundaries, however, which maintains that they should be defined wherever possible in an area where "nothing happened." The International Subcommission on Stratigraphic Classification, of which Hollis Hedberg is Chairman, has recommended in its Circular No. 25 of July, 1969, that "Boundary-stratotypes should always be chosen within sequences of continuous sedimentation. The boundary of a chronostratigraphic unit should never be placed at an unconformity. Abrupt and drastic changes in lithology or fossil content should be looked at with suspicion as possibly indicating gaps in the sequence which would impair the value of the boundary as a chronostratigraphic marker and should be used only if there is adequate evidence of essential continuity of deposition. The marker for a boundary-stratotype may often best be placed *within* a certain bed to minimize the possibility that it may fall at a time gap." This marker is becoming known as "the Golden Spike."

By "nothing happened" is meant a stratigraphic successions that is apparently continuous. The choice of boundary is then purely arbitrary, and depends simply on our ability to select a horizon that can be the most efficiently and most completely documented and defined (just as there is nothing about time itself that distinguishes between, say, February and March, but to define a boundary between them is useful for purposes of communication and record). This is the epitome of an empirical approach to stratigraphy. Choosing to place a boundary where "nothing happened" is to deliberately avoid having to deal with some "event" that would require interpretation (see Miall 2004 for a discussion of the importance of this point). This recommendation was accepted in the first International Stratigraphic Guide (Hedberg 1976, pp. 84–85). The concept also includes the proviso that only the base of a unit is so defined at the chosen location, not the top of the underlying unit, lest future work determine that at the type section the boundary is marked by a hiatus—hence the term **topless stage**. The "missing time" so identified is then assigned to the underlying unit, permitting continuous revision without the need for new boundary definitions.

The importance of using "all available means" to identify, calibrate and date GSSPs cannot be over-emphasized. As noted in Sect. 7.5.3 the use of single criteria, such as FADs and LADs can lead to errors associated with biostratigraphic diachroneity. Quantitative methods, particularly graphic correlation and its advanced version, constrained optimization, make use of multiple criteria. As Smith et al. (2015) noted, despite the best efforts of stratigraphers, it has emerged that some GSSPs have been defined at sedimentary breaks, and others still over-rely on few chronostratigraphic criteria, such as the range of a single key taxon. Some 74 of currently defined GSSPs actually only make use of biostratigraphic criteria. Smith et al. (2015) also noted that most GSSPs have been defined at outcrop sections of shallow-marine rocks (where the probability of multiple hiatuses and much missing section is greatest: see Chap. 8), with little use being made of subsurface drill-core sections, and none have been established in deep-marine sediments, where the preserved record is likely to contain fewer hiatuses.

The concept of the Global Stratigraphic Section and Point (GSSP), informally called the **Golden Spike** concept, was rapidly accepted, and has led to an explosion of specialized work under the auspices of the *International Commission on Stratigraphy* (ICS), a division of the *International Union of Geological Sciences* (IUGS), to correlate key sections worldwide in order to develop internationally recognized markers for epochs and stages that could then become part of the standard chronostratigraphic scale (e.g., Remane 2000a). This work is regularly reported in the journal IUGS *Episodes* and, nowadays, on the website www.stratigraphy.org maintained by the ICS. A discussion of selected GSSPs is presented in Sect. 7.8.3. About 100 Phanerozoic boundaries have been established as requiring definition in the form of a GSSP, and according to Smith et al. (2015) 65 have been ratified by the *International Union of Geological Sciences* (as of the summer of 2013) with several awaiting ratification.

It is instructive to review the development of the modern geologic time scale through from 1964 to the present. Amongst the first to appear were scales for the Jurassic and Cretaceous developed by van Hinte (1976a, b). Useful discussions of stratigraphic concepts were compiled by Cohee et al. (1978). Numerous attempts to synthesize existing data for the Phanerozoic have been made, notably by British (Harland et al. 1982, 1990), French (Odin 1982) and American (Berggren et al. 1995) groups. Each group drew on its own data base and made different interpolations and extrapolations, with the result that there are significant differences in the assigned ages of many of the important chronostratigraphic boundaries.

A quantum leap forward was achieved by the ICS with the publication in 2004 of its updated Geologic Time Scale (GTS2004: Gradstein et al. 2004a). Gradstein et al. (2012) subsequently published their own updated version (although this is not an official product of the *International Stratigraphic Commission*). The 2004 version incorporates numerous new data points, documented with the use of quantitative biostratigraphy, much-improved radiometric dating methods, chemostratigraphy and (for the Neogene)

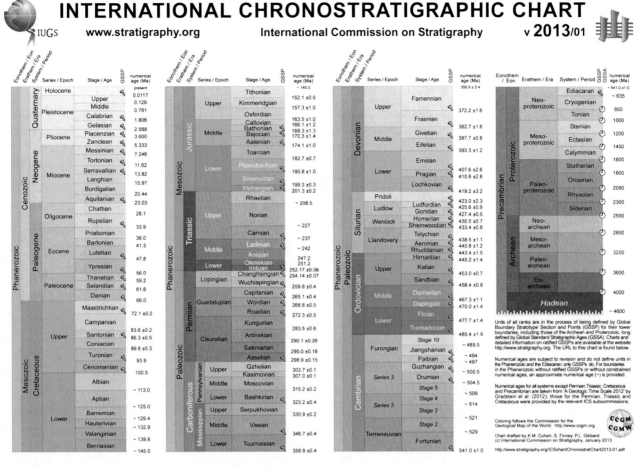

**Fig. 7.33** The Geological Time Scale prepared by the International Commission on Stratigraphy. Internationally agreed GSSPs are marked with a golden-spike icon. This *figure* shows the version current as of January 2013 (www.stratigraphy.org)

cyclostratigraphy (Fig. 7.33). The new scale presents us with unprecedented opportunities for the comparison and calibration of detailed local and regional studies of rates and processes. Paleogene, Mesozoic and most Paleozoic ages are given to the nearest 100,000 years, although for parts of the scale, potential errors of >1 million years. remain. This scale, like all before it, incorporates numerous revisions of assigned ages. Almost all major chronostratigraphic boundaries in the Mesozoic and Paleozoic have been revised by several million years relative to earlier scales, such as that of Berggren et al. (1995), reflecting new data or changing interpretations of earlier data. There is no sign, yet, that the time scale has finally stabilized, although the incremental changes from one scale to the next do appear to be getting smaller. We return to this point later.

The Time Scale now undergoes continuous revision, under the auspices of the ICS. For example, a 2012 version was published in time for the 34th international Geological Congress at Brisbane (Cohen et al. 2012) (Fig. 7.33). An example of the detail involved in the construction of the scale is shown in Fig. 7.34.

## 7.8.2 Determining the Numerical ("Absolute") Age of a Stratigraphic Horizon

Direct dating of sedimentary rocks by radiometric dating can only be carried out on a few potassium-bearing minerals, such as glauconite, because potassium is the only constituent of authigenic sedimentary minerals that contains naturally occurring radiogenic isotopes. Potassium-argon and argon-argon methods are the most common ones employed. However, the ages determined by the use of this method may relate to diagenetic age rather than depositional age, and there are problems associated with the loss of the daughter product, argon.

More commonly, radiometric ages are determined for interbedded volcanic horizons, especially ash beds or bentonites, and the stratigraphic age of the rocks of interest are then determined by interpolation, as shown in Fig. 7.35. In order to make use of this method, two major assumptions have to be invoked, firstly, that sedimentation was continuous and, secondly, that sedimentation was at a constant rate. In most cases, neither of these assumptions can be assumed to

**Fig. 7.34** Part of the GTS2004 scale for the early Cretaceous, showing the magnetic polarity chrons and selected biozone schemes. *Arrows* in the microfossil columns indicate FADs (*pointing upward*) and LADs (*pointing downward*) (Gradstein et al. 2004a, Fig. 19.1, p. 358)

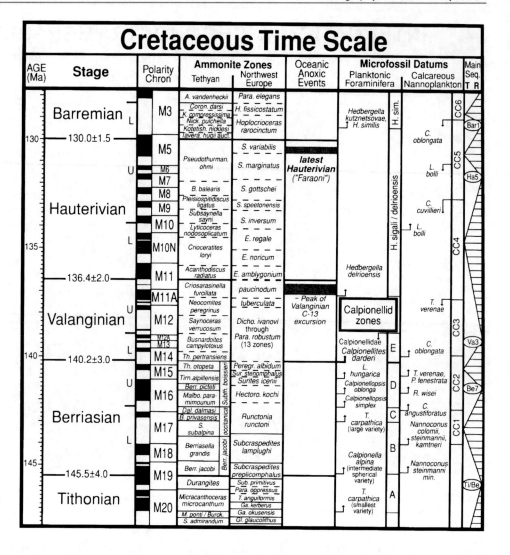

be fulfilled. This does not mean that the technique is useless, only that it must be employed with care. Ideally, many radiometric determinations should be made on successive volcanic units, and then interpolations and extrapolations can be fine-tuned by methods of averaging and by making local corrections. We come back to this issue in Chap. 8, where we examine further some deeper questions regarding the interpretation of time as preserved in the rock record.

*Modern radiometric dating methods.* Not surprisingly, the methods of radiometric dating and the data base have undergone orders-of-magnitude transformations over the last half-century (Mattinson 2013). Holmes (1960) provided a list of 63 ages, mostly obtained by the K/Ar method. Of these, nineteen were used to construct a linear time scale, adjusting the space allotted to each system according to cumulative stratal thickness, it being assumed at the time that this constituted an indication of the duration of the period. By the time of the 1964 Geological Society of London scale in 1964, the data base consisted of some 337 measurements. The Harland et al. (1990, Table 4.2) compilation lists some

750 entries. Figure 7.36 illustrates the increased accuracy and precision of the dating of the Permian-Triassic boundary over the 22-year period from 1991 to 2011. The current ICS date for this boundary is $252.17 \pm 0.06$ Ma.

Amongst the improvements in methods has been the establishment of the chronogram method by Harland et al. (1982, pp. 3–4) for plotting and visualization of measurement error, a standardization of decay constants, much enhanced analytical methods, and a better appreciation of the systematics of the various isotopic decay paths. The K/Ar and Rb/Sr methods have been found to be less reliable than those obtained by the $^{40}Ar/^{39}Ar$ and U–Pb methods (Villeneuve 2004; Mattinson 2013). The use of ages derived from glauconites, popular in the 1980s (Odin 1982), has decreased, because of the increasing realization of the unreliability of the method. It has been demonstrated repeatedly that daughter products, Sr or Ar, may be lost from the mineral grains, yielding ages that are too young.

Radiometric age determinations are characterized by a normal experimental error. Current practice may achieve a

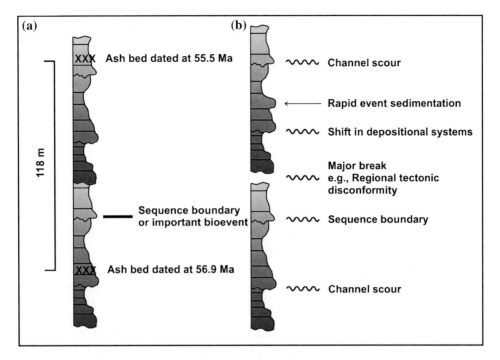

**Fig. 7.35** The determination of "absolute" ages of stratigraphic events by using the method of "bracketing" the event by a pair of radiometric ages; The principle of the method (*left*) and the reality in practice (*right*). **a** On the *left*, 118 m of beds are shown accumulating in 56.9 −55.5 million years = 1.4 million years. The average sedimentation rate is therefore 118/1.4 m/million years = 84 m/million years. The sequence boundary of interest is 30.3 m above the lower ash bed. Assuming a constant sedimentation rate and no hiatuses, the 30.3 m of beds accumulated in 30.3/84 = 0.36 million years. Therefore the age of this boundary is 56.9−0.36 = 56.54 Ma. However, as shown in **b**, at *right*, real sections are full of sedimentary *breaks*, which represent missing time, and variations in sedimentation rate, all of which render the concept of the average sedimentation rate suspect

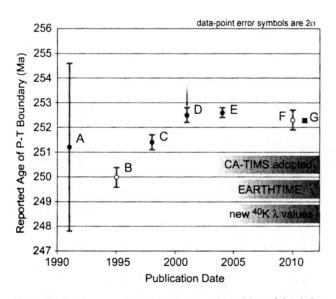

**Fig. 7.36** The improvements in accuracy and precision of the dating of the Permian-Triassic boundary from 1991 to 2011. *Solid circles* are dates based on U–Pb zircon analysis; *open circles* are dates based on Ar–Ar analysis of sanidine (Mattinson 2013, Fig. 11, p. 315)

precision of ±0.1 % or better (Mattinson 2013, p. 310; Smith et al. 2015; Cramer et al. 2015), e.g., ±100,000 years at 100 Ma. However, the attainment of such accuracy and precision in the stratigraphic record depends on the availability of appropriate datable material at the rights places in the rock record, and the accuracy of the existing time scale varies from stage to stage because of this. Refinements in decay constants and inter-laboratory calibration become ever more important. As Sageman et al. (2014) demonstrated by their discussion of U–Pb and $^{40}Ar/^{39}Ar$ data sets from an Upper Cretaceous interval in the Western Interior Basin, dating accuracy and precision is now such that differences between the two methods may relate to internal differences in the processes that set the final isotopic ratios—differences in Ar closure temperature between different minerals and the length of the cooling phases of igneous bodies from which such minerals as zircon are ultimately derived. There may be differences of several hundred thousand years between the results from different methods, that then need to be reconciled by calibration against other data sets, including biochronology and magnetostratigraphy.

Even given a very precisely dated GSSP for a key chronostratigraphic boundary, it is then quite another matter to exploit this information to date and calibrate new sections elsewhere. This exercise requires a biostratigraphic, radiometric or chemostratigraphic record capable of yielding correlations of equivalent accuracy and precision.

*Magnetostratigraphy*: The first major development, the erection of a satisfactory geomagnetic polarity scale for the last 4.5 Ma, was published by Cox (1969), based primarily on the sampling of successions of lava flows. The standard scale now in use was developed primarily from the study of cores through undisturbed deep-sea sediments. The first use of such cores predated the *Deep Sea Drilling Project* (DSDP), and the cores were very short (see review by Kennett 1980). Harrison and Funnell (1964) were the first to combine biostratigraphy and chronostratigraphy in an attempt to correlate a reversal event. Later, Opdyke et al. (1966) studied some longer cores, up to 12 m in length and were able to correlate reversal events with the land-based lava sequence. Radiolarian zone boundaries closely parallel the reversal correlations (Fig. 7.37). The DSDP started in 1968 and began to have an important effect on chronostratigraphy. However, difficulties were encountered in establishing magnetic stratigraphy directly from the cores because of drilling disturbance and bioturbation. The practice developed of calibrating polarity and biostratigraphic data by correlating deep marine with exposed on-land sections and dating the latter radiometrically.

Since the 1960s, magnetostratigraphy has become almost equal in importance to biostratigraphy and radiometric dating in establishing a time scale for the last 160 Ma of Earth's history. The establishment of a reliable reversal sequence requires close sampling of an assumed, undisturbed, continuous section, and relies on the ability to recognize a kind of "bar-code" pattern of normal and reversed intervals. However, undisturbed sections for the pre-Late Jurassic are rare—there is no undisturbed sea floor of greater age—and it therefore much more difficult to standardize the scale for these older rocks.

The technique is most useful for time spans characterized by frequent reversals. During the Cenozoic, reversals occurred two or three times each million years, providing a very distinctive pattern that has enabled the establishment of 29 reversal intervals, named *chrons* for that era (Ogg and Smith 2004). The availability of age information on a $10^4$ to $10^5$-year time scale has provided powerful new tools for exploring the rates of sedimentary and stratigraphic processes in the geological record. Reversal sequences for older parts of the Mesozoic and for the Paleozoic have been assembled from partial sections (Langereis et al. 2010), but are not yet reliable enough to become part of the standard geological time scale.

Magnetostratigraphy may also be used to correlate local sections with each other without regard to the global scale. However, because reversal events are not unique, it is not possible to correlate them by matching sequences from

**Fig. 7.37** Correlation of magnetic stratigraphy and radiolarian zones in seven cores from the Antarctic Ocean (Opdyke et al. 1966)

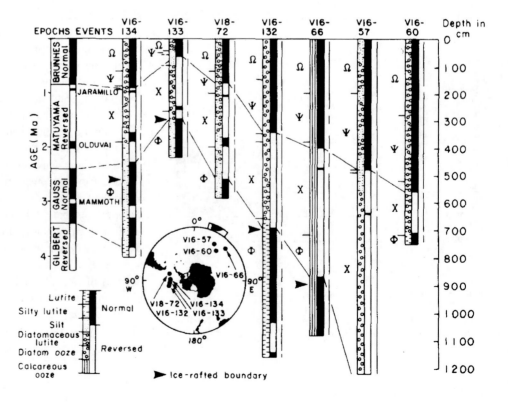

different stratigraphic sections unless they contain particularly distinctive long or short polarity intervals. Some supplementary criteria may be required to assist in matching, such as marker beds or biostratigraphic zonation. Picard (1964) and Irving (1966) were among the first Western workers to use paleomagnetic correlation for sediments on the continents. The technique has become widely used for nonmarine sediments because of the scarcity of other means of precise correlation.

A single example of the use of magnetostratigraphy in a practical field problem is described here briefly. For many years, a team has been exploring the nonmarine Siwalik Group (Oligocene-Quaternary) of Pakistan, in part because of its rich vertebrate fauna and in part because of the information the sediments yield about the tectonics of the Himalayas, from which the sediments were derived. Magnetostratigraphy, coupled with radiometric dating of several ash beds, has provided a useful means of local correlation between sections that show marked lateral facies changes (Keller et al. 1977; Barndt et al. 1978; Johnson et al. 1979). It was also possible to propose a correlation with the global scale. Figure 7.38 illustrates the correlation of three closely spaced sections in the Pabbi Hills area, near Jhelum. From three to five oriented rock specimens were collected from

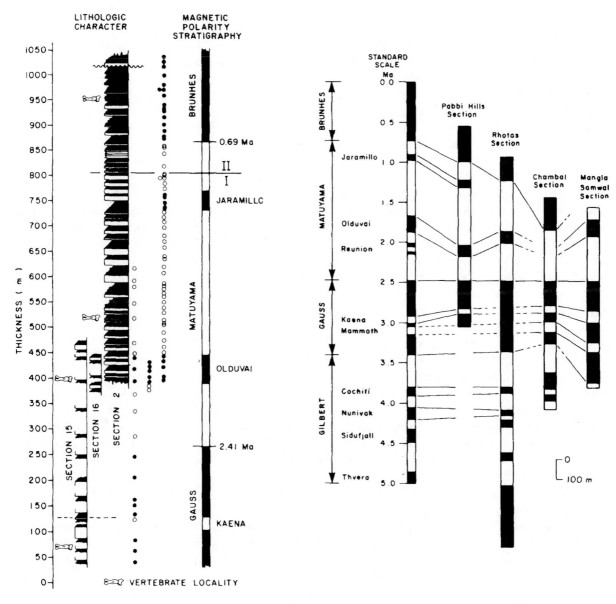

**Fig. 7.38** The practical use of magnetostratigraphy to correlate sections of nonmarine strata in Pakistan. Three partial sections are compiled at *left* to provide a single composite section for the Pabbi Hills region. Key vertebrate locations are shown, and field samples showing normal and reversed magnetic polarities are indicated by the *vertically* arranged suites of *black* and *white circles*. Correlation of the Pabbi Hills section with the standard scale and with other nearby stratigraphic sections is shown at *right*. Note the presence of missing reversals in some of the sections, indicating either incomplete sampling, or reversals that are actually missing because of local erosion (redrawn from Johnson et al. 1979)

each of 113 sites within the sections. These were subjected to laboratory tests to determine the stability of the field and the absence of magnetic overprinting, and the pole positions obtained corrected for structural dip. The reversal zones were correlated with the standard scale of Opdyke (1972) using the following argument. The oldest remains of *Equus* (horse) were found at the 400 m level in the composite section. The oldest occurrence of *Equus* in North America is dated 3.5 Ma and in Asia 2.5 Ma. It is considered unlikely that the Pabbi Hills fossils are older than 3.5 Ma. Note that the fossil locality is in a short normal polarity sequence within a long reversed interval. Only two such dominantly reversed intervals are present in the magnetostratigraphic scale, the Matuyama and the Gilbert zones. The Gilbert zone extended from 5.1 to 3.3 Ma, and the Matuyama from 2.41 to 0.70 Ma. (Revised ages of Opdyke 1972; modified from Cox 1969). The evidence of *Equus* suggests that this is the Matuyama zone. The two short normal events then correlate with the Olduvai and Jaramillo subzones. The Gilsa event (subzone) of Fig. 7.38 is not universally recognized and does not appear on Opdyke's (1972) chart.

Four composite sections have been correlated using the presence of two tuff horizons (Visser and Johnson 1978) and the polarity zones, as shown in Fig. 7.38.

The importance of this work is that it permits precise local and global correlation of vertebrate localities, permits accurate calculations of sedimentation rates, and provides accurate control for studying sedimentological characteristics, basin architecture, and tectonic events. Some of these aspects have been explored in later papers by this research group and other workers in the area. For example, Johnson et al. (1985) were able to determine the various time scales represented by fluvial cycles and to explore the implications for rates of channel wandering and the nature of tectonic and climatic controls on sedimentation. Behrensmeyer (1987) developed detailed two-dimensional reconstructions of the stratigraphic architecture as a basis for an examination of the taphonomy of the vertebrate remains.

*Chemostratigraphy*: Oxygen isotope stratigraphy has made an enormous contribution to the development of stratigraphy. The method depends on measurements of the $^{16}O/^{18}O$ ratio. Because $^{16}O$ is the lighter of the two isotopes, water molecules containing this light oxygen are preferentially evaporated from seawater. During times of ice-free global climate they are recycled to the oceans and the isotopic ratio remains in a stable balance corresponding to the natural proportions of the two isotopes in the hydrosphere. However, glacial ice is composed of condensed $^{16}O$-enriched water, so that continental ice buildups are preferentially enriched in $^{16}O$, with the result that the $\delta^{18}O$ content of the oceans is increased. Sediments preserve the isotopic ratios of the oxygen that existed in the hydrosphere at the time the sediments were formed. Measurements are made on the carbonate comprising foraminiferal tests. Emiliani (1955) was the first to demonstrate that the oxygen isotope record is cyclic, and was amongst the first to argue that the fluctuations should reflect the high-frequency oscillations between glacial and interglacial stages that have characterized the Cenozoic record.

We now know that the $^{16}O/^{18}O$ ratio is a highly sensitive indicator of global ocean temperatures and ice cover, and can therefore be used as an analog recorder of ice volumes (Hays et al. 1976; Shackleton and Opdyke 1976; Matthews 1984a, 1988). Matthews (1984a) suggested a calibration value of $\delta^{18}O$ variation of about 0.011 ‰ per meter of sea-level change.

A modern oxygen isotope scale developed from the early systematic work of the SPECMAP (SPEctral MAPping) project (Imbrie et al. 1984; Imbrie 1985). Following, in particular, the pioneering work of Hays et al. (1976) the SPECMAP project established a detailed record for the late Pleistocene (the last 780 ka) which calibrated the scale against an insolation energy index calculated from the integration of the three major orbital cycles, obliquity, precession and eccentricity. Since that time it has become standard procedure to record the oxygen isotope signal in deep sea cores, and numerous studies have contributed to the refinement of the scale.

The current scale used in GTS2004 includes a systematic series of marine cycles back to 5.5 Ma (Gradstein et al. 2004a, Fig. 21.5) and a more detailed presentation of the 106 Marine Isotope Stages (MIS) for the last 2.65 million years of Earth history (Fig. 7.39; Gradstein et al. 2004a, Fig. 22.1) —an average duration of 25,000 years for each stage. As discussed in Sects. 7.8.7 and 8.11, the calibration of this scale against the record of orbital variations through the last few million years of Earth history is now providing the basis for a new, astrochronological time scale.

Strontium isotope stratigraphy is a relatively new topic. It was not mentioned at all by Harland et al. (1990). In the late 1970s it began to be recognised that the ratio $^{87}Sr/^{86}Sr$ varied systematically through time, as recorded in marine sediments (Burke et al. 1982). Veizer (1989) provided a detailed review of the origins of strontium in seawater and the processes that affected the preservation of the signal, including the effects of diagenesis. He argued that the rate of mixing of ocean waters and the long-residence time of strontium in seawater could potentially yield a reliable signal. McArthur (1994, 1998; McArthur and Howarth 2004) has taken the lead in the development and application of the tool to age determination and standardization of the method and its integration into the process of refining the time scale.

The $^{87}Sr/^{86}Sr$ ratio varies with age in the carbonate of various marine shells, mainly because of fluctuations in the rates and types of continental weathering, and the measurement of this value enables a sample to be situated on a

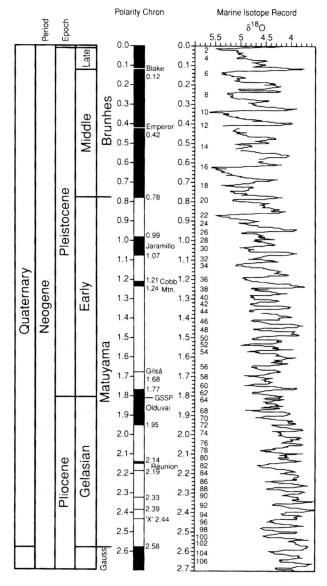

**Fig. 7.39** The GTS2004 time scale for the last 2.7 Ma, showing the magnetic polarity chrons and the marine oxygen isotope record (Gradstein et al. 2004a, Fig. 22.1, p. 444)

### 7.8.3 Stages and Boundaries

The evolution of the **stage** concept was a confused one. D'Orbigny and Oppel in the mid-nineteenth century were the first to use the term with essentially its present meaning (Hancock 1977; McGowran 2005).

Stages are essentially convenient groupings of biozones. Stage boundaries may be drawn at the top or base of a particularly well or widely developed biozone or at a prominent faunal change. Many of our modern stage names were rather loosely defined, perhaps on only a handful of taxa when originally established, but have subsequently been refined by more and more detailed study using many different life forms. Biozones are now usually established using only the members of a single phylum (except, perhaps, Oppel zones), whereas many stages have now been defined in many different ways. For example, Devonian stages are based mainly on brachiopods, corals, trilobites, fish, conodonts, and palynomorphs; Cretaceous stages are based mainly on ammonites, pelecypods, brachiopods, foraminifera, nannofossils, and palynomorphs.

What has happened over the years is that stage terms became so useful that geologists began to define the same stage in different ways. They added descriptions of suites of different kinds of fossils, which helped to define parallel successions of biozones and to transcend biogeographic problems, and then when radiometric and magnetostratigraphic data became available, this information was added in as well. Gradually the rock-term stage became a more broadly-based term referring globally to a particular interval of time. The modern use of stage concepts is well illustrated in current treatments of the Geologic Time Scale, e.g., in Gradstein et al. (2004a, 2012). The stage is now regarded as "the basic working unit of chronostratigraphy" (Hedberg 1976; Salvador 1994; Smith et al. 2015).

In any given sedimentary basin, the recognition of a stage and its component chronozones depends on the nature of the fossil and other evidence that can be compiled for analysis. Surface studies may be able to benefit from the very detailed studies of macrofossils that have been carried out over the last 150 years. For example, paleontologists have found that ammonites evolved so rapidly during the Mesozoic that they have been able to erect scores of biozones. Modern dating methods have shown that for some parts of the Mesozoic these zones may represent time intervals as small as 250,000 years, which permits an astonish precision in dating where the record is complete enough (Callomon 1995). An extreme example is shown in Fig. 7.40 (we return to a discussion of this particular example in Chap. 8). For parts of the Paleozoic, trilobites and graptolites offer similar biostratigraphic precision.

graph showing the variation in composition with time. The graph for the Phanerozoic, as currently in use (McArthur and Howarth 2004, Fig. 7.1) is remarkably similar to that compiled by Veizer (1989), which indicates that the technique rapidly reached maturity.

The values of the ratio are not unique with respect to time, because it has risen and fallen in a crudely cyclic manner within the range 0.7070–0.7090 during the last 450 million years, so that some intermediate values correspond to several different ages on the curve. It is therefore necessary to know, within a few millions of years, which part of the graph to use for reading off the age against the calculated ratio.

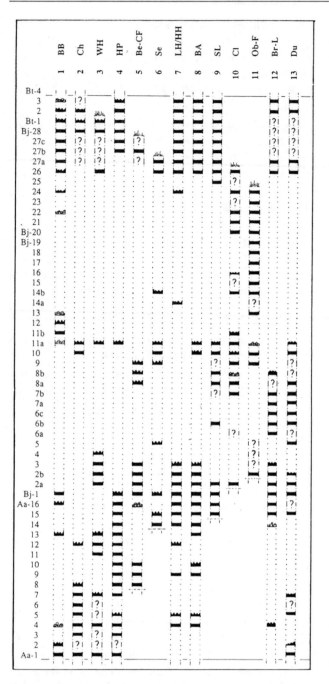

**Fig. 7.40** The 56 ammonite faunal horizons recognized in 13 sections of the Inferior Oolite of Dorset and Somerset, England. The sections average 5 m in thickness, are spaced out over a total distance of about 80 km, and span about 5 Ma of the Middle Jurassic; therefore each faunal horizon represents an average of about 90,000 years. *Aa* Aalenian, *Bj* Bajocian, *Bt* Bathonian stages (Callomon 1995, Fig. 5, p. 143)

Subsurface studies must rely on microfossil or palynological evidence (McGowran 2005). Foraminifera evolved rapidly during the Cenozoic, but for most of geologic time microfossils, though commonly occurring in great abundance, seemed to have evolved more slowly, and so individual biozones necessarily represent longer intervals of time.

Most fossil groups tends to be facies bound, and the key to successful stage correlation is to locate sections where more than one useful fossil group is present or where repeated facies variations cause ecologically incompatible taxa to occur close together by interdigitation. This is largely a matter of chance. Much depends on lucky exposure of the right section, and the literature is replete with obscure geographic localities that have attained a specialized kind of fame because of the excellence of the biostratigraphic work carried out there. Examples are such small Welsh towns as Llanvirn and Llandeilo (Ordovician), the Eifel district, Belgium (Devonian), and the Barrandian area of the Czech Republic (Silurian-Devonian boundary). Geologists worldwide knew about Maastricht long before the average European politician had heard about this minor city. Most stages are named after such places. The end result of this extended effort is that many stages can now be recognized globally, with varying degrees of confidence. Plate-tectonic movements cause faunal provincialism to vary in many taxonomic groups simultaneously, and so the perfection of this correlation varies from place to place and time to time. For example, Berry (1977) reported that correlation of Early and Middle Ordovician graptolitic faunas between Europe and North America has been fraught with controversy because of faunal provincialism. At that time, the proto-Atlantic Ocean (Iapetus) was at its widest. We earlier referred to the comparable problem of trilobite correlation across Iapetus (Fig. 7.12).

As an example of modern work in establishing stages by means of detailed faunal work on several animal groups, a brief discussion is presented here of the Pridolian and Lochkovian stages, as defined in the Barrandian area of the Czech Republic. Chlupác (1972) published a detailed description of the faunas and revised the earlier biostratigraphic subdivisions of these rocks. His faunal list for the two stages includes over 300 species, of which graptolites, conodonts, and trilobites constitute the most important biostratigraphic indicators. Other groups providing subsidiary control include eurypterids, phyllocarids, ostracodes, echinoids, cephalopods, gastropods, pelecypods, and brachiopods. Figure 7.41 illustrates one of several short but critical sections measured through the Pridolian-Lochkovian boundary near Klonk. Limestone beds are numbered 1–53. This section constitutes the very first internationally agreed GSSP, having been ratified by the International Union of Geological Sciences in 1972. It constitutes the stratotype for the Silurian-Devonian boundary. A vital characteristic of the Barrandian area is that here the section straddling the Silurian-Devonian boundary is basinal in origin, and therefore preserves a more continuous marine section relative to the marginal-marine, nonmarine or unconformable contact in the British type areas of the Silurian and Devonian.

The Pridolian-Lochkovian succession in this area consists mainly of thinly interbedded, grayish-black, calcareous

**Fig. 7.41** The principle of the Golden Spike. The boundary between the Silurian and the Devonian was defined here in 1968 at a bed a certain distance below the first occurrence (FAD) of the graptolite *Monograptus uniformis*. This definition also serves to define the boundary between the Pridolian and the Lochkovian stages (Adapted from Chlupác 1972). This became the first internationally recognized GSSP

**Chronostratigraphic boundaries should be established within continuous successions, i.e., where "nothing happens."**

**This is the principal of the GOLDEN SPIKE**

*chronostratigraphic boundary*

**The Silurian-Devonian boundary at Klonk, Czech Republic**

mudstone and grayish-black to dark gray, fine-grained (micritic), skeletal, platy, weathering limestone, with subordinate beds of pale limestone and coarser, detrital limestone (calcarenite to calcirudite). Graptolites are abundant in the mudstones; trilobites and conodonts occur sparsely in the limestones, particularly in the paler, less muddy units, and become more abundant north of Klonk, where the rocks undergo a facies change into a predominantly pure carbonate succession. This interbedding of different facies with their contrasting faunas is one of the most important features of these central European sections from the point of view of biostratigraphic stratotype definition.

Two graptolite biozones have been recognized in the Pridolian and form the main basis for the definition of the unit. They are the *Monograptus ultimus* zone below and the *M. transgrediens* zone above. The latter does not reach to the top of the Pridolian, but is followed by a graptolite interregnum containing only sparse, nondiagnostic forms (Fig. 7.41). The base of the Lochkovian is defined by the sudden widespread appearance and abundance of *Monograptus uniformis* in bed 20. Other species of *Monograptus* and *Linograptus* appear in the upper part of the lower Lochkovian, while in the upper Lochkovian, *M. hercynicus* is typical. Some of these species occur in the pure carbonate facies, permitting close correlation with the shelly fauna. It is interesting to note that in spite of the effort biostratigraphers have made to formalize their biozone types Chlupác's work is typical of many in that no attempt is made to state what kind of biozone is in use. The older of these graptolite biozones appear to be single-taxon range biozones, with concurrent-range biozones for the upper part of the lower Lochkovian and the upper Lochkovian.

The trilobite *Warburgella (Podolites) rugulosa rugosa* is of primary importance in delineating the lower boundary of the Lochkovian. In the Klonk section, it appears in limestone bed 21, immediately above the first appearance of *M. uniformis* in the upper part of bed 20 (Fig. 7.41).

The conodont *Icriodus woschmidti* defines a range biozone corresponding approximately with the lower part of the Lochkovian, although in the Barrandian area, it ranges down through the graptolite interregnum into the top of the *M. transgrediens* biozone. Conodonts are not common in the somewhat argillaceous facies at Klonk.

This discussion could be extended considerably into a consideration of other faunal groups and some of the subsidiary species that define concurrent-range zones. It is to be hoped, however, that by this time the reader can discern the main threads of a procedure that has now been followed innumerable times by many different workers.

Barnes et al. (1976) discussed the correlation between graptolites, conodonts, trilobites, and brachiopods for the Ordovician rocks of Canada. This is an interesting case in that the benthonic fauna was used to establish a North American stage nomenclature in the craton (the Richmond, Maysville, etc.), whereas graptolites occur mainly in deeper water deposits and were correlated with the classic British stages (Ashgill, Caradoc, etc.), in spite of the difficulties of faunal provincialism across the Iapetus Ocean referred to earlier. To be of any regional use, these biostratigraphic schemes had to be integrated with each other, and this depended on finding locations, as at Klonk, where the different biofacies are interbedded. In order to cover the entire Ordovician System, it was necessary to study partial sections in the Canadian Rocky Mountains, Nevada, Texas, Newfoundland, the St. Lawrence Lowlands, and parts of the Canadian Arctic Islands. One of the key sections was the Lower to Middle Ordovician Cow Head Group of Newfoundland. At Green Point, where the beds are overturned, the stratotype for the Cambrian-Ordovician boundary has been established (Fig. 7.7; Cooper et al. 2001). Here, trilobites and conodonts occur in limestone boulders slumped from the shelf into the

deep water basin, where graptolite-bearing shales buried them. Integration of the biozone schemes had to allow for the fact that the boulders were probably slightly younger than the enclosing shales. Elsewhere, transgressions and regressions caused intermingling of faunas from different facies and different faunal provinces.

Another example of GSSP is that for the base of the Turonian at Pueblo, Colorado, the description of which was provided by Kennedy et al. (2005). The succession here consists of alternating limestones and shales (Fig. 7.42). The limestones are fossiliferous biomicrites. Much of the

**Fig. 7.42** The type section of the base of the Turonian. The lithological succession is that of the Bridge Creek Member of the Greenhorn Limestone on the north side of the Pueblo Reservoir State Recreation area, Colorado. The GSSP for the base of the Turonian is placed at the base of bed 86 (Kennedy et al. 2005, Fig. 8, p. 101)

**Fig. 7.43** The GSSP for the base of the Turonian. See Fig. 7.42 for stratigraphic location (photo by Brad Sageman)

succession is bioturbated. The alternation of the two major lithologies is attributed to orbitally-forced climatic cyclicity. The base of the Turonian has been placed at the base of bed 86, which corresponds to the first occurrence of the ammonite *Watinoceras devonense* (Fig. 7.43). There are many secondary biostratigraphic indicators of the boundary, including bivalves, dinoflagellates, and other ammonites. Figure 7.42 shows the ranges of the inoceramids in this section, and the inoceramid and ammonites zones. The boundary is also indicated by a carbon-isotope excursion which can be correlated world-wide, and corresponds to an Oceanic Anoxic Event. The numerical age of the boundary has been derived from radiometric dating of several bentonite beds that be traced throughout much of the Western Interior Basin. The bentonites at the Pueblo location are too weathered to be dated, but Kennedy et al. (2005) provide dates determined from six other samples collected from correlative units in other parts of the basin.

### 7.8.4 Event Stratigraphy

The term **event stratigraphy** has been attributed to Ager (1973), although, as noted by Torrens (2002, p. 258), geologists have been aware of the importance of sudden events for some time. Typical geological "events" include sudden sedimentary events, such as storms and sediment gravity flows; volcanic events, generating widespread ash beds; earthquakes; biologic events such as first- and last-appearances of taxa, and mass extinction events; and chemostratigraphic events, such as "carbon excursions". Comprehesive treatments of this topic have been provided by Einsele and Seilacher (1982) and Kauffman (1988).

Where the products of specific events can be recognized reliably they provide invaluable markers for local and regional correlation. Bentonite layers, because they can be characterized by their petrology and radiometric age, have been used for this purpose for many years. Some events appear to be truly unique, such as the K-T boundary event as the product of a meteorite impact, as first proposed by Alvarez et al. (1980); a hypothesis now almost universally accepted. However, there are many other types of event that are repeatable (e.g., major sediment-gravity flows; liquefaction events attributable to earthquakes), which means that their reliability as markers may be limited, unless correlation can be substantiated by secondary means.

Kauffman (1988) documented in detail the methods of what he termed **High-Resolution Event Stratigraphy**, in which he made use of bentonites, biomarkers, and other events to construct regional stratigraphic correlations using the methods of graphic correlation, with an estimated accuracy of ±100 ka. In Sect. 7.8.6 we discuss a debate regarding the use of event stratigraphy in the construction of the standard suite of GSSPs. We return to this topic in Sect. 8.11 as part of a brief introduction to modern research to develop the astrochronological time scale.

### 7.8.5 Absolute Ages: Their Accuracy and Precision

Time scales like the one in Fig. 7.33 look finished. It would appear that if fossils can be assigned to one or other of the stages shown in this diagram, it should be possible to determine the age of a section almost anywhere within the Phanerozoic to within a fraction of a million years. Often this is, in fact the case, but the precision of the scale is deceptive. It represents geologists' best guesses, given all the available information at the time of compilation. It does not show the amount of interpolation and extrapolation that has gone into the construction of the scale. To get an idea about this look at Fig. 7.44. This is a table that compares the assigned ages of the Mesozoic stages between eleven different published compilations. Each of these syntheses represents research of the highest international standard.

Why are there so many variations in assigned age from one scale to the next? For example, the age of the Jurassic-Cretaceous boundary has moved up and down by 15.3 Ma years since 1982. How can this be?

The answer is that many of the assigned ages depend on extrapolation or interpolation from a limited number of well-known fixed points. It is rare for an important biomarker, such as a key FAD, to occur stratigraphically next to a datable ash bed or a magnetic reversal event. Many event ages are determined using the kinds of calculations shown in Fig. 7.35. A key event is bracketed by two or more dated

**Fig. 7.44** Comparison of time
scales for the Mesozoic
(Gradstein et al. 2004a, Fig. 1.6b)

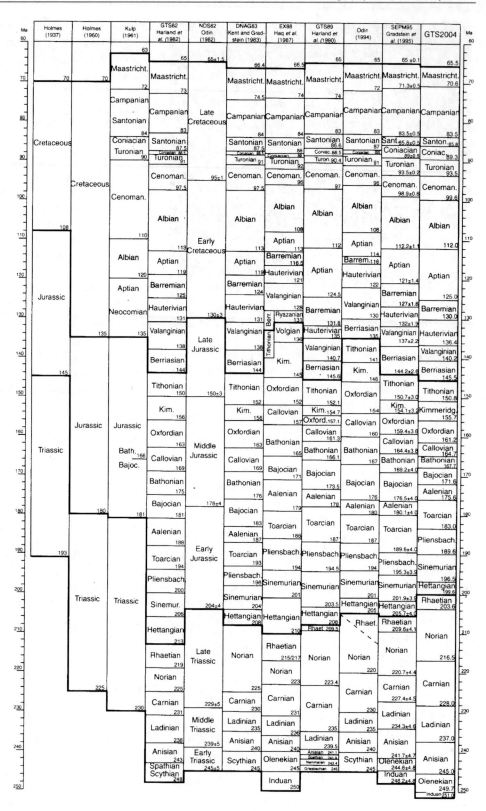

horizons, and the age worked out by assuming continuous sedimentation at a constant rate. Both these assumptions are commonly incorrect. Making the same kind of calculations at several locations where the same kinds of assumptions have to be made may result in a range of ages for the same event. The geologist then makes a best guess, picking the one that seems to be the most reliable, or averaging them in some way. Opinions may differ about which particular field locality or which dating exercise offers the most reliable tie point for the scale.

In the absence of sufficient datable field locations, all kinds of assumptions have had to be made. For example, assumptions that the species of a given group of animals evolve at a constant rate, so that their zones can all be assumed to be of the same duration. The first work to date the reversal events in the magnetostratigraphic scale was carried out on the magnetized oceanic crust, and made use of an assumption that sea-floor-spreading rates were constant, so that the width of the reversal zone could be transformed into a value for its duration. This assumption has since been proved to be completely wrong—sea-floor spreading rates vary almost continuously.

Figure 7.45 illustrates the problem, using some real data from a well off the Florida coast. Which of the two interpretations is correct? Both will yield approximate ages for samples lying between the dated points, but the differences, though small, could be important. Where the slope between

two points in this diagram is low, this means that the thickness of strata between the points is untypically small, given the indicated age span. A simple explanation would be that there is a local disconformity. An unusually steep slope could mean the occurrence of a rare event, such as the passage of a turbidity current, which locally thickens the stratal record. Careful re-examination of the original rock record could help to resolve this problem.

Further developments in chronostratigraphic accuracy and precision are discussed in Sects. 8.10 and 8.11.

### 7.8.6 The Current State of the Global Stratigraphic Sections and Points (GSSP) Concept, and Standardization of the Chronostratigraphic Scale

Despite the apparent inductive simplicity of the approach described here to the refinement of the time scale, the completion of the necessary suite of GSSPs has been slow, in part because of the inability of some working groups to arrive at agreement (Vai 2001). In addition, two contrasting approaches to the definition of chronostratigraphic units and unit boundaries have now evolved, each emphasizing different characteristics of the rock record and the accumulated data that describes it (Smith et al. 2015). Castradori (2002)

**Fig. 7.45** The problem of calibration. The *data points* in both graphs represent the same set of real data from a well in the Gulf Coast, off Florida (from Roof et al. 1991). *Each point* is a biomarker that has been dated against the available time scale. The question then arises, does this data represent continuous sedimentation at a constant rate? The *straight line* of correlation in the *left-hand figure* suggests that this is the case, making allowance for small errors in the dating of the fossils.

But it is possible to look at the data a different way, as seen in the *right-hand figure*. Sedimentation rate may vary, and there may be several minor unconformities. Careful examination of the section is necessary to determine the nature of facies changes, the presence of breaks in sedimentation, and so on, in order to test the validity of the interpretations suggested in the *right-hand diagram* provided an excellent summary of what has become a lively controversy within the *International Commission on Stratigraphy*. The first approach, which Castradori described as the **historical and conceptual approach**, emphasizes the historical continuity of the erection and definition of units and their boundaries, the data base for which has continued to grow since the nineteenth century by a process of inductive accretion. Aubry et al. (1999, 2000) expanded upon and defended this approach. As noted by Smith et al. (2015), in some cases, precedence and historical continuity have had to be set aside in favour of choosing new stratotypes that provide a greater data base for correlation purposes.

The alternative method, which Castradori terms the **hyper-pragmatic approach** (a very misleading label, in this writer's opinion), focuses on the search for and recognition of significant "events" as providing the most suitable basis for rock-time markers, from which correlation and unit definition can then proceed. The followers of this methodology (see response by Remane 2000b, to the discussion by Aubry et al. 2000) suggest that in some instances historical definitions of units and their boundaries should be modified or set aside in favour of globally recognizable event markers, such as a prominent biomarker, a magnetic reversal event, an isotopic excursion, or, eventually, events based on cyclostratigraphy. This approach explicitly sets aside McLaren's (1970) recommendation (cited above) that boundaries be defined in places where "nothing happened," although it is in accord with suggestions in the first stratigraphic guide that "natural breaks" in the stratigraphy could be used or boundaries defined "at or near markers favorable for long-distance time-correlation" (Hedberg 1976, p. 71, 84). The virtue of this method is that where appropriately applied it may make field recognition of the boundary easier. The potential disadvantage is that is places prime emphasis on a single criterion for definition, and relies on assumptions about the superior time-significance of the selected boundary event. The deductive flavour of hypothesis is therefore added to the methodology. In this sense the method is not strictly empirical (as discussed below, assumptions about global synchroneity of stratigraphic events may in some cases be misguided. See Miall and Miall 2001).

Smith et al. (2015) noted that even where the traditional approach has been used to define a GSSP, there has tended to be an overreliance on single biostratigraphic criteria for the definition, which may limit their usefulness and flexibility. To the key defining criterion of (for example) the first- or last-appearance datum of a chosen taxon could be added the stratigraphic distance above or below the FAD or LAD of other taxa. They also noted the importance of supplementary criteria.

The hyper-pragmatic approach builds assumptions into what has otherwise been an inductive methodology free of all but the most basic of hypotheses about the time-significance of the rock record. The strength of the historical and conceptual approach is that it emphasizes multiple criteria, and makes use of long-established practices for reconciling different data bases, and for carrying correlations into areas where any given criterion may not be recognizable. For this reason, this writer is not in favour of the proposal by Zalasiewicz et al. (2004) to eliminate the distinction between time-rock units (chronostratigraphy) and the measurement of geologic time (geochronology). Their proposal hinges on the supposed supremacy of the global stratotype boundary points. History has repeatedly demonstrated the difficulties that have arisen from the reliance on single criteria for stratigraphic definitions, and the incompleteness of the rock record, which is why "time" and the "rocks" are so rarely synonymous in practice (see also Aubry 2007, on this point; and Heckert and Lucas 2004, for other comments on the Zalasiewicz et al. proposal). Some of the current controversies surrounding the placement of GSSPs in the Cenozoic are discussed by Berggren (2007) and Walsh (2004). The latter paper also contains a lengthy discussion regarding the controversies surrounding the definitions and useages of the key terms, including stage, boundary stratotypes, GSSP, etc., most of which is beyond the concerns of the practicing stratigrapher.

A different debate has arisen since the power of astrochronological calibration of the time scale became evident, a topic we take up in the next section, and again in Sect. 8.11. Astrochronology is based on the assertion that, in certain, carefully selected sections, a complete record of orbital forcing is preserved by cyclic variations in facies and thicknesses in the sedimentary record, and that by counting the cycles and correlating them to numerical ages derived from radiometric dating or some other means, a precise time scale with an accuracy and precision in the $10^4$-year range may be established.

As noted above, the traditional (at least since the 1970s) method of defining the time scale has been by the erection of GSSPs and topless stages, the purpose of the latter feature being to automatically allow for the presence of hidden hiatuses in the succession at the stratotypes. However, the relevance of this issue has been called into question when the method of selecting and dating a GSSP involves the necessary assumption of stratigraphic completeness, either in the actual stratotype or in the composite section upon which the stratotype is based. Such is the case with astrochronology. As noted by Hilgen et al. (2006, p. 117):

> ... all late Neogene GSSPs are by now defined in land-based deep marine sections. All these sections have an integrated high-resolution stratigraphy, uniting detailed cyclo-, magneto- and biostratigraphies and have been astronomically tuned. Moreover, they cover the entire interval of the stage in a demonstrable continuous succession. As such, the sections perfectly embody the concept of a stage and may serve as unit stratotype for that stage in addition to accommodating its GSSP.

These authors argued for the extension of the concept to the remainder of the Cenozoic and also to the Mesozoic, as the astrochronological data base becomes more complete (Hilgen et al. 2015; see Sect. 8.11). They also pointed out that if the GTS is based on complete sections then the distinction between "rock" and "time" becomes unnecessary. However, I remain convinced that we need to be careful about abandoning the long-held cautions that support this dual terminology. The preservation of sections that are complete at the $10^4$ to $10^5$-year time scale is likely to be very unusual, requiring especially undisturbed basinal conditions, and the identification of the sections that are complete enough to be used in the establishment of an astrochronological time scale should be considered a rare event (see next section). The issue of the incompleteness of the stratigraphic record is discussed further in Chap. 8.

## 7.8.7 Cyclostratigraphy and Astrochronology

**Cyclostratigraphy**: the subdiscipline of stratigraphy that deals with the identification, characterization, correlation, and interpretation of cyclic variations in the stratigraphic record.

**Astrochronology**: The dating of sedimentary units by calibration of the cyclostratigraphic record with astronomically tuned time scales.

**Tuning**: Adjusting the frequencies, including harmonics, of a complex record preserved in a natural succession to best-fit a predicted astronomical signal.

Croll (1864) and Gilbert (1895) were the first to realize that variations in the Earth's orbital behavior may affect the amount and distribution of solar radiation received at the Earth's surface, by latitude and by season, and could be the cause of major climate variations. Several classic studies were undertaken to search for orbital frequencies in the rock record, and theoretical work on the distribution of insolation was carried out by the Serbian mathematician Milankovitch (1930, 1941), who showed how orbital oscillations could affect the distribution of solar radiation over the earth's surface. However, it was not for some years that the necessary data from the sedimentary record was obtained to

support his model. Emiliani (1955) was the first to discover periodicities in the Pleistocene marine isotopic record, and the work by Hays et al. (1976) is regarded by many (e.g., de Boer and Smith 1994) as the definitive study that marked the beginning of a more widespread acceptance of orbital forcing, the so-called **Milankovitch processes**, as a major cause of stratigraphic cyclicity on a $10^4$ to $10^5$-year frequency— what is now termed the **Milankovitch band**. The model is now firmly established, particularly since accurate chronostratigraphic dating of marine sediments has led to the documentation of the record of faunal variations and temperature changes in numerous upper Cenozoic sections (Gradstein et al. 2004a; Hilgen et al. 2015; see summary in Miall 2010, Sects. 7.2, 11.3). These show remarkably close agreement with the predictions made from astronomical observations. Many high-frequency sequence records are now interpreted in terms of the orbital-forcing model (summaries and reviews in Miall 2010, Chap. 11; Hilgen et al. 2015).

There are several separate components of orbital variation (Fig. 7.46). The present orbital behavior of the earth includes the following cyclic changes (Schwarzacher 1993):

1. Variations in orbital **eccentricity** (the shape of the Earth's orbit around the sun). Several "wobbles", which have periods of 2035.4, 412.8, 128.2, 99.5, 94.9, and 54 ka. The major periods are those at around 405 and 100 ka.
2. Changes of up to 3° in the **obliquity** of the ecliptic, with a major period of 41 ka, and minor periods of 53.6 and 39.7 ka.
3. **Precession** of the equinoxes. The Earth's orbit rotates like a spinning top, with a major period of 23.7 ka. This affects the timing of the perihelion (the position of closest approach of the earth to the sun on an elliptical orbit), which changes with a period of 19 ka.

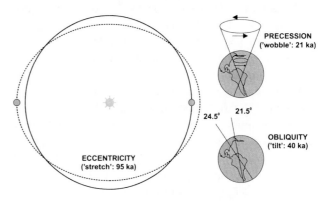

**Fig. 7.46** Perturbations in the orbital behaviour of the earth, showing the causes of Milankovitch cyclicity. Adapted from Imbrie and Imbrie (1979)

Imbrie (1985, p. 423) explained the effects of these variables as follows:

> Variations in obliquity alter the income side of the radiation budget in two fundamental ways: they modulate the intensity of the seasonal cycle, and they alter the annually integrated pole-to-equator insolation gradient on which the intensity of the atmospheric and oceanic circulations largely depend. (Low values of obliquity correspond to lower seasonality and steeper insolation gradients.) Variations in precession, on the other hand, alter the structure of the seasonal cycle by moving the perihelion point along the orbit. The effect of this motion is to change the earth-sun distance at every season, and thereby change the intensity of incoming radiation at every season.

Each of these components is capable of causing significant climatic fluctuations given an adequate degree of global sensitivity to climate forcing. For example, when obliquity is low (rotation axis nearly normal to the ecliptic), more energy is delivered to the equator and less to the poles, giving rise to a steeper latitudinal temperature gradient and lower seasonality. Variations in precession alter the structure of the seasonal cycle, by moving the perihelion point along the orbit. This changes the earth-sun distance at every season, thus changing the intensity of insolation at each season. "For a given latitude and season typical departures from modern values are on the order of ∼5 %" (Imbrie 1985, p. 423). Because the forcing effects have different periods they go in and out of phase. One of the major contributions of Milankovitch was to demonstrate these phase relationships on the basis of laborious time-series calculations. These can now, of course, be readily carried out by computer. The success of modern stratigraphic work has been to demonstrate the existence of curves of change in temperature, redox state, carbonate content, organic productivity, and other variables in the Cenozoic record that can be correlated directly with the curves of Fig. 7.47. For this purpose sophisticated time-series spectral analysis is performed on various measured parameters, such as oxygen-isotope content or cycle thickness. This approach has led to the development of a special type of quantitative analysis termed **cyclostratigraphy** (House 1985).

The Earth became highly sensitized to orbital variations during the cool climates of the Late Cenozoic, and it is now generally accepted, following the work of Hays et al. (1976) that the fluctuations in glaciation that characterized the Neogene were driven by orbital forcing, a process that gave us the oxygen isotope time scale (Emiliani 1955; Imbrie and Imbrie 1979; Imbrie 1985; Imbrie et al. 1984).

To develop a time scale from the orbital record in sedimentary successions requires several important assumptions, given that, unlike, for example, bioevents or datable ash beds, cyclostratigraphy consists of a succession of identical, or near identical, cyclic fluctuations in some primary depositional characteristic, such as carbon or calcium carbonate

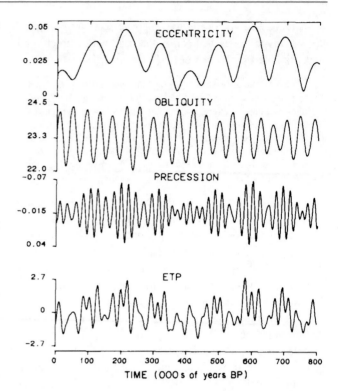

**Fig. 7.47** The three major orbital-forcing parameters, showing their combined effect in the eccentricity-tilt-precession (*ETP*) curve at bottom. Absolute eccentricity values are shown. Obliquity is measured in degrees. Precession is shown by a precession index. The ETP scale is in standard deviation units (Imbrie 1985)

content, redox state, bed thickness or, more generally, facies. The assumptions include the following:

1. The section is continuous, or
2. (alternate): Discontinuities in the section can be recognized and accounted for in the subsequent analysis;
3. Sedimentation rate was constant, or events (such as turbidites) beds can all be recognized and discounted.
4. Orbital frequencies may be reconstructed for the distant past based on astronomical calculations of planetary motions.
5. Thickness can be converted to time using a simple sedimentation-rate transformation;
6. The variabilities in stratigraphic preservation (facies changes, hiatuses) can be effectively managed by pattern-matching techniques.
7. Orbital frequencies can be reconstructed from the rock record of the distant geological past, based on independent age-bracketing of the section.

In general, these assumptions are more likely to be met in deep-marine and lacustrine settings, where it may be expected that the allogenic forcing of sedimentary processes by orbital mechanisms might be expected to overwhelm local autogenic influences. However, one of the more convincing studies of a high-frequency sequence stratigraphic record being used as the basis for a cyclostratigraphic time scale is that of the shallow-marine Plio-Pleistocene cycles of the Wanganui Basin in North Island, New Zealand (Fig. 7.48). Also, Hilgen et al. (2015) cite several studies of fluvial systems in which it would appear that orbital forcing is the primary determinant of stratigraphic cyclicity.

The first major modern study of cyclostratigraphy (Hays et al. 1976) demonstrated that the last 800 ka of Earth time was dominated by a 100-ka cyclicity, as indicated by the cyclic climatic fluctuations from glacial to interglacial stages. These are recorded in fine detail by the oxygen-isotope record, particularly as this is measured in

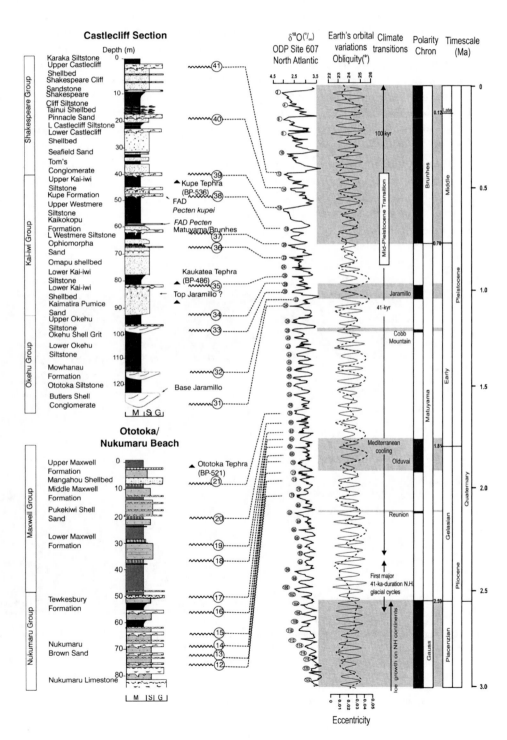

**Fig. 7.48** Composite stratigraphic columns for the Nukumaru and Castlecliff coastal sections, Wanganui Basin, New Zealand, showing lithostratigraphy, sequence stratigraphy, and correlations with the oxygen isotope timescale (Naish et al. 2005). This particularly complete Pliocene to Pleistocene section serves as a regional standard for geological time (Naish et al. 2005)

ODP cores (Fig. 7.39). On the longer term, extending back into the Paleogene and earlier, the 405-ka eccentricity cycle seems to be the most stable and the most likely to provide the basis for astrochronology.

There have been several focused attempts to examine the use of the orbital "pacemaker" as the basis for a high-precision time scale (Herbert et al. 1995; House and Gale 1995; Shackleton et al. 1999), and several important regional studies that began to make a substantial contribution to the growth of this field of research. The stratigraphic and sedimentologic basis for this research was summarized by Miall (2010, Chap. 11). A range of indicators may be used to examine for orbital cyclicity, beyond the physical "cyclic" appearance of the rocks themselves. These include bed thickness, oxygen isotope ratios, weight-percent calcium carbonate, grayscale pixel data (from core scans), magnetic susceptibility, and resistivity data from a microimaging scanner.

Pioneering studies to establish an astrochronological time scale have been led by Fritz Hilgen. A reliable cyclostratigraphic (astrochronologic) time scale was first established for the youngest Cenozoic strata, back to about 5 Ma (Hilgen 1991; Berggren et al. 1995; Hilgen et al., 2006; see Figs. 7.49 and 7.50). Over the succeeding decade, astronomically

calibrated sections were used to extend the astrochronological time scale back to 14.84 Ma, the base of the Serravallian stage, in the mid-Miocene (www.stratigraphy.org), and research is proceeding to extend the time scale not only to the base of the Cenozoic, but through at least the Mesozoic (Hinnov and Ogg 2007; Hilgen et al. 2006, 2015). Figure 7.51 illustrates the correlation of Upper Miocene sections in the Mediterranean basin to the orbital scale. Westphal et al. (2008) offered a sharply critical review of the field data base on which part of the astrochronological time scale is based, pointing to problems of diagenesis and differential compaction of contrasting lithologies, that render direct one-for-one correlations between the critical field sections problematic. A careful comparison of the correlations between two critical field sections shows that the correlations of the astronomical cycles are not always supported by the correlations of bioevents. In some cases, there are different numbers of cycles between the occurrences of key bioevents. Multiple cross-checks are required to evaluate and, if possible, correct for such discrepancies.

For the older part of the geological record (particularly the Mesozoic and Paleozoic), several studies have now established convincing "floating" scales for specific stratigraphic intervals; that is, scales that exhibit reliable orbital

**Fig. 7.49**  The Punta di Maiata section on Sicily. Punta di Maiata is the middle partial section of the Rossello Composite and part of the Zanclean unit stratotypes, which defines the base of the Pliocene (Van Couvering et al. 2000; Hilgen et al. 2006). Larger-scale eccentricity-related cycles are clearly visible in the weathering profile of the cape. Small-scale quadripartite cycles are precession-related;

precession-obliquity interference patterns are present in particular in the older 400-kyr carbonate maximum indicated in blue. All cycles have been tuned in detail and the section has an excellent magnetostratigraphy, calcareous plankton biostratigraphy and stable isotope stratigraphy (Hilgen et al. 2006)

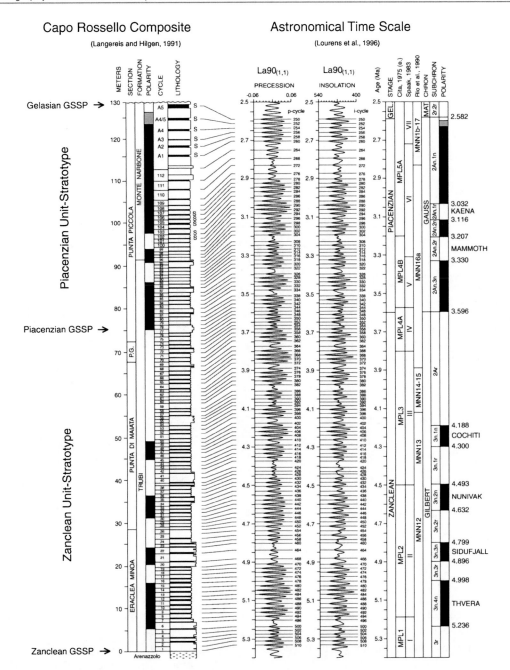

**Fig. 7.50** The Rossello Composite Section (RCS, Sicily, Italy) the unit stratotypes for series spanning the base of the Pliocene, incorporating the orbital tuning of the basic precession-controlled sedimentary cycles and the resulting astronomical time scale with accurate and precise astronomical ages for sedimentary cycles, calcareous plankton events and magnetic reversal boundaries. The Zanclean and Piacenzian GSSPs are formally defined in the RCS while the level that time-stratigraphically correlates with the Gelasian GSSP is found in the topmost part of the section. The well tuned RCS lies at the base of the Early–Middle Pliocene part of the astrochronological time scale and the Global Standard Chronostratigraphic Scale and as such could serve as unit stratotype for both the Zanclean and Piacenzian Stage (Hilgen et al. 2006)

**Fig. 7.51** Astronomical tuning of sapropels and associated grey marls in land-based deep marine sections in the Mediterranean for the interval between 10 and 7 Ma (upper Miocene). Colours in the lithological columns indicate sapropels (*black*), associated grey marls (*grey*) and homogeneous marls (*yellow*). Colours in the magnetostratigraphic columns indicate normal polarities (*black*), reversed polarities (*white*) and uncertain polarities (*grey*). Sapropels and associated grey marls have been numbered per section and lumped into large-scale groups (*roman numerals*) and small scale groups. The initial age model is based on magnetobiostratigraphy. Phase relations between sapropel cycles and the orbital parameters/insolation used for the tuning are based on the comparison of the sapropel chronology for the last 0.5 million years with astronomical target curves (Hilgen et al. 2015, Fig. 2)

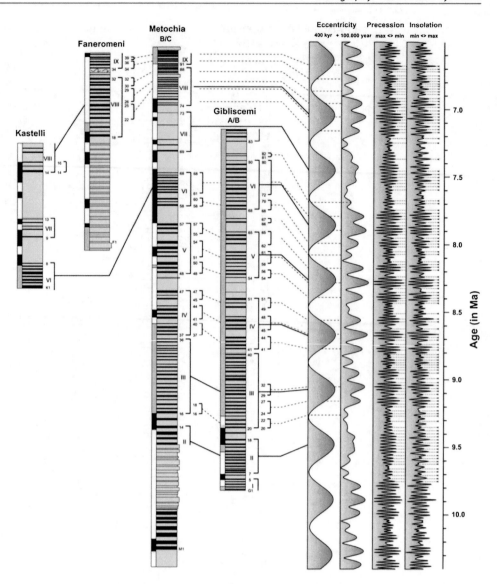

frequencies, once tuned, but that cannot be precisely correlated to the numerical time scale because of residual imprecisions in numerical dating methods (Hilgen et al. 2015).

A discussion of some current research in this area is provided in Sect. 8.11.

# References

Abreu, V., 1998. Evolution of the conjugate volcanic passive margins: Pelotas Basin (Brazil) and offshore Namibia (Africa): Implication for global sea-level changes. Unpublished Ph.D. thesis, Rice University, Houston.

Ager, D. V., 1964, The British Mesozoic Committee: Nature, v. 203, p. 1059.

Ager, D. V., 1973, The nature of the stratigraphical record: New York, John Wiley, 114 p.

Ager, D. V., 1981, The nature of the stratigraphical record (second edition): John Wiley, New York, 122 p.

Alvarez, L. W., Alvarez, W., Asaro, F., and Michel, H. V. 1980, Extraterrestrial cause for the Cretaceous-Tertiary extinction: Science, v. 208, p. 1095-1108.

Armentrout, J. M., 1981, Correlation and ages of Cenozoic chronostratigraphic units in Oregon and Washington: Geological Society of America Special Paper 184, p. 137-148.

Aubry, M.-P., 2007, chronostratigraphic terminology: Building on Principles: Stratigraphy, v. 4, p. 117-125.

Aubry, M.-P., Van Couvering, J., Berggren, W. A., and Steininger, F., 1999, Problems in chronostratigraphy: stages, series, unit and boundary stratotype section and point and tarnished golden spikes: Earth-Science Reviews, v. 46, p. 99-148.

Aubry, M.-P., Van Couvering, J., Berggren, W. A., and Steininger, F., 2000, Should the gold spike glitter: Episodes, v. 23, p. 203-210.

Autin, W. J., 1992, Use of alloformations for definition of Holocene meander belts in the middle Amite River, southeastern Louisiana: Geological Society of America Bulletin, v. 104, p. 233-241.

Barndt, J., Johnson, N. M., Johnson, G. D., Opdyke, N. D., Lindsay, E. H., Pilbeam, D., And Tahirkheli, R. A. H., 1978, The magnetic polarity stratigraphy and age of the Siwalik Group near Dhok Pathan Village, Potwar Plateau, Pakistan: Earth Planetary Science Letters, v. 41, p. 355-364.

Barnes, C. R., Jackson, D. E., and Norford, B. S., 1976, Correlation between Canadian Ordovician zonations based on graptolites, conodonts and benthic macrofossils from key successions, in Bassett, M. G., ed., The Ordovician System: proceedings of a Palaeontological Association symposium, Birmingham, September 1974: University of Wales and National Museum of Wales, Cardiff, p. 209-225.

Barrell, Joseph, 1917, Rhythms and the measurement of geologic time: Geological Society of America Bulletin, v. 28, p. 745-904.

Bassett, M. G., 1985, Towards a "common language" in stratigraphy: Episodes, v. 8, p. 87-92.

Behrensmeyer, A. K., 1987, Miocene fluvial facies and vertebrate taphonomy in northern Pakistan;. in Ethridge, F. G., Flores, R. M., and Harvey, M. D., eds., Recent developments in fluvial sedimentology: Society of Economic Paleontologists and Mineralogists Special Publication 39, p. 169-176.

Berggren, W. A., 2007, Status of the hierarchical subdivision of higher order marine Cenozoic chronostratigraphic units: Stratigraphy, v. 4, p. 99-108.

Berggren, W. A., Kent, D. V., Aubry, M.-P., and Hardenbol, J., eds., 1995, Geochronology, time scales and global stratigraphic correlation: Society for Sedimentary Geology Special Publication 54, 386 p.

Berry, W. B. N., 1968, Growth of prehistoric time scale, based on organic evolution: W.H. Freeman and Co., San Francisco, 158 p.

Berry, W. B. N., 1977, Graptolite biostratigraphy: a wedding of classical principles and current concepts; in Kauffman, E. G., and Hazel, J. E., eds., Concepts and methods of biostratigraphy: Dowden, Hutchinson and Ross Inc., Stroudsburg, Pennsylvania, p. 321-338.

Bhattacharya, J., 1991, Regional to sub-regional facies architecture of river-dominated deltas, Upper Cretaceous Dunvegan Formation, Alberta subsurface, in Miall, A. D., and Tyler, N., eds., The three-dimensional facies architecture of terrigenous clastic sediments and its implications for hydrocarbon discovery and recovery, Society of Economic Paleontologists and Mineralogists, Concepts in Sedimentology and Paleontology, v. 3, p. 189-206.

Bhattacharya, J. P., 1993, The expression and interpretation of marine flooding surfaces and erosional surfaces in core; examples from the Upper Cretaceous Dunvegan Formation, Alberta foreland basin, Canada, in Posamentier, H. W., Summerhayes, C. P., Haq, B. U., and Allen, G. P., eds., Sequence stratigraphy and facies associations: International Association of Sedimentologists Special Publication 18, p. 125-160.

Blackwelder, E., 1909, The valuation of unconformities: Journal of Geology, v. 17, p. 289-299.

Boyd, R., and Penland, S., 1988, A geomorphic model for Mississippi delta evolution: Gulf Coast Association of Geological Societies, Transactions, v. 38, p. 443-452.

Brown, A. R., 2011, Interpretation of three-dimensional seismic data, seventh edition, American Association of Petroleum Geologists Memoir 42, 646 p.

Burke, W. H., Denson, R. E., Hetherington, E. A., Koepnick, R. B., Nelson, H. F., and Otto, J. B., 1982, Variations of seawater $^{87}Sr/^{86}Sr$ through Phanerozoic time: Geology, v. 10, p. 516-519.

Callomon, J. H., 1995, Time from fossils: S. S. Buckman and Jurassic high-resolution geochronology, in Le Bas, M. J., ed., Milestones in Geology: Geological Society of London Memoir 16, p. 127-150.

Callomon, J. H., 2001, Fossils as geological clocks, in C. L. E. Lewis and S. J. Knell, eds., The age of the Earth: from 4004 BC to AD 2002: Geological Society of London Special Publication 190, p. 237-252.

Cartwright, J. A., Haddock, R. C., and Pinheiro, L. M., 1993, The lateral extent of sequence boundaries, in Williams, G. D., and Dobb,

A., eds., Tectonics and seismic sequence stratigraphy: Geological Society, London, Special Publication 71, p. 15-34.

Castradori, D., 2002, A complete standard chronostratigraphic scale: how to turn a dream into reality? Episodes, v. 25, p. 107-110.

Catuneanu, O., 2006, Principles of sequence stratigraphy: Elsevier, Amsterdam, 375 p.

Catuneanu, O., Abreu, V., Bhattacharya, J. P., Blum, M. D., Dalrymple, R. W., Eriksson, P. G., Fielding, C. R., Fisher, W. L., Galloway, W. E., Gibling, M. R., Giles, K. A., Holbrook, J. M., Jordan, R., Kendall, C. G. St. C., Macurda, B., Martinsen, O. J., Miall, A. D., Neal, J. E., Nummedal, D., Pomar, L., Posamentier, H. W., Pratt, B. R,. Sarg, J. F., Shanley, K. W., Steel, R. J., Strasser, A., Tucker, M. E., and Winker, C., 2009, Toward the Standardization of Sequence Stratigraphy: Earth Science Reviews, v. 92, p. 1-33.

Catuneanu, O., Bhattacharya, J. P., Blum, M. D., Dalrymple, R. W., Eriksson, P. G., Fielding, C. R., Fisher, W. L., Galloway, W. E., Gianolla, P., Gibling, M. R., Giles, K. A., Holbrook, J. M., Jordan, R., Kendall, C. G. St. C., Macurda, B., Martinsen, O. J., Miall, A. D., Nummedal, D., Posamentier, H. W., Pratt, B. R,. Shanley, K. W., Steel, R. J., Strasser, A., and Tucker, M. E., 2010, Sequence stratigraphy: common ground after three decades of development: First Break, v. 28, p. 21-34.

Catuneanu, O., Galloway, W.E., Kendall, C.G.St.C., Miall, A.D., Posamentier, H.W., Strasser A., and Tucker M.E., 2011, Sequence Stratigraphy: Methodology and Nomenclature: Report to ISSC: Newsletters on Stratigraphy, v. 4 (3), p. 173-245.

Chlupác, I., 1972, The Silurian-Devonian boundary in the Barrandian: Bulletin of Canadian Petroleum Geology, v. 20, p. 104-174.

Christie-Blick, N., Mountain, G. S., and Miller, K. G., 1990, Seismic stratigraphy: record of sea-level change, in Revelle, R., ed., Sea-level change: National Research Council, Studies in Geophysics, Washington, National Academy Press, p. 116-140.

Cody, R. M., Levy, R. H., Harwood, D. M., and Sadler, P. M., 2008, Thinking outside the zone: high-resolution quantitative biochronology for the Antarctic Neogene: Palaeogeography, Palaeoclimatology, Palaeoecology, v. 260, p. 92 - 121.

Cohee, G. V., Glaessner, M. F., and Hedberg, H. D., eds., 1978, Contributions to the geologic time scale: American Association of Petroleum Geologists Studies in Geology No. 6.

Cohen, K.M., Finney, S., and Gibbard, P.L., 2012, International Chronostratigraphic Chart: International Commission on Stratigraphy, www.stratigraphy.org.

Conkin, B. M., and Conkin, J. E., eds., 1984, Stratigraphy: foundations and concepts: Benchmark Papers in Geology, New York: Van Nostrand Reinhold, 363 p.

Cooper, R. A., Nowlan, G. S., and Williams, S. H., 2001, Global Stratotype Section and Point for base of the Ordovician System: Episodes, v. 24, p. 19-28.

Cowie, J. W., 1986, Guidelines for boundary stratotypes: Episodes, v. 9, p. 78-82.

Cox, A., 1969, Geomagnetic reversals: Science, v. 163, p. 237-245.

Cox, B. M., 1990, A review of Jurassic chronostratigraphy and age indicators for the UK, in Hardman, R. F. P., and Brooks, J., eds., Tectonic events responsible for Britain's oil and gas reserves: Geological Society, London, Special Publication 55, p. 169-190.

Cramer, B. D., Vandenbroucke, T. R. A., and Ludvigson, G. A., 2015, High-resolution event stratigraphy (HiRES) and the quantification of stratigraphic uncertainty: Silurian examples of the quest for precision in stratigraphy: Earth Science reviews, v. 141, p. 136-153.

Croll, J., 1864, On the physical cause of the change of climate during geological epochs: Philosophical Magazine, v. 28, p. 435-436.

de Boer, P. L., and Smith, D. G., 1994, Orbital forcing and cyclic sequences, in de Boer, P. L., and Smith, D. G., eds., Orbital forcing and cyclic sequences: International Association of Sedimentologists Special Publication 19, p. 1-14.

Doyle, J. A., 1977: Spores and pollen: the Potomac Group (Cretaceous) Angiosperm sequence; in Kauffman, E. G., and Hazel, J. E., eds., Concepts and methods of biostratigraphy: Dowden, Hutchinson and Ross Inc., Stroudsburg, Pennsylvania, p. 339-364.

Edwards, L. E., 1984, Insights on why graphic correlation (Shaw's method) works: Journal of Geology, v. 92, p. 583-597.

Edwards, L. E., 1985, Insights on why graphic correlation (Shaw's method) works: A reply [to discussion]: Journal of Geology, v. 93, p. 507-509.

Einsele, G., and Seilacher, A., eds., 1982, Cyclic and event stratification: Springer-Verlag Inc., Berlin, 536 p.

Eldredge, N., and Gould, S. J., 1972, Punctuated equilibrium: an alternative to phyletic gradualism, in Schopf, T. J. M., ed., Models in paleobiology: San Francisco, Freeman, Cooper and Company, p. 82-115.

Eldredge, N., and Gould, S. J., 1977, Evolutionary models and biostratigraphic strategies; in Kauffman, E. G., and Hazel, J. E., eds., Concepts and methods of biostratigraphy: Dowden, Hutchinson and Ross Inc., Stroudsburg, Pennsylvania, p. 25-40.

Embry, A. F., 1995, Sequence boundaries and sequence hierarchies: problems and proposals, in Steel, R. J., Felt, V. L., Johannessen, E. P., and Mathieu, C., eds., Sequence stratigraphy on the Northwest European margin: Norsk Petroleumsforening Special Publication 5, Elsevier, Amsterdam, p. 1-11

Embry, A. F., and Johannessen, E. P., 1992, T-R sequence stratigraphy, facies analysis and reservoir distribution in the uppermost Triassic-Lower Jurassic succession, western Sverdrup Basin, Arctic Canada, in Vorren, T. O., Bergsager, E., Dahl-Stamnes, O. A., Holter, E., Johansen, B., Lie, E., and Lund, T. B., eds., Arctic geology and petroleum potential: Norwegian Petroleum Society Special Publication 2, p. 121-146.

Emery, D., and Myers, K. J., 1996, Sequence stratigraphy: Blackwell, Oxford, 297 p.

Emiliani, C., 1955, Pleistocene temperatures: Journal of Geology, v. 63, p. 538-578.

Fejfar, O., and Heinrich, W. D., 1989, Muroid rodent biochronology of the Neogene and Quaternary, in Lindsay, E. H., Fahlbusch, V., and Mein, P., eds., European mammal chronology: NATO Advanced Research Workshop, p. 91-118.

Frazier, D. E., 1974, Depositional episodes: their relationship to the Quaternary stratigraphic framework in the northwestern portion of the Gulf Basin: Bureau of Economic Geology, University of Texas, Geological Circular 74-1, 26 p.

House. M. R., and Gale, A. S., eds., 1995, Orbital forcing timescales and cyclostratigraphy: Geological Society, London, Special Publication 85, 210 p.

Galloway, W. E., 1989, Genetic stratigraphic sequences in basin analysis I: Architecture and genesis of flooding-surface bounded depositional units: American Association of Petroleum Geologists Bulletin, v. 73, p. 125-142.

Gibling, M. R., and Bird, D. J., 1994, Late Carboniferous cyclothems and alluvial paleovalleys in the Sydney Basin, Nova Scotia: Geological Society of America Bulletin, v. 106, p. 105-117.

Gilbert, G. K., 1895, Sedimentary measurement of geologic time: Journal of Geology, v. 3, p. 121-127.

Gradstein, F. M., Ogg, J. G., and Smith, A. G., eds., 2004a, A geologic time scale: Cambridge University Press, Cambridge, 610 p.

Gradstein, F. M., Cooper, R. A., and Sadler, P. M., 2004b, Biostratigraphy: time scales from graphic and quantitative methods, in Gradstein, F. M., Ogg, J. G., and Smith, A. G., eds., A geologic time scale: Cambridge University Press, Cambridge, p. 49-54.

Gradstein, F. M., Ogg, J. G., Schmitz, M. D., and Ogg, G. M., 2012, The Geologic time scale 2012: Elsevier, Amsterdam, 2 vols., 1176 p.

Hancock, J. M., 1977, The historic development of biostratigraphic correlation, in Kauffman, E. G. and Hazel, J. E., eds., Concepts and

methods of biostratigraphy: Dowden, Hutchinson and Ross Inc., Stroudsburg, Pennsylvania, p. 3-22.

Haq, B. U., Hardenbol, J., and Vail, P. R., 1988, Mesozoic and Cenozoic chronostratigraphy and cycles of sea-level change, in Wilgus, C. K., Hastings, B. S., Kendall, C. G. St. C., Posamentier, H. W., Ross, C. A., and Van Wagoner, J. C., eds., Sea-level Changes: an integrated approach: Society of Economic Paleontologists and Mineralogists Special Publication 42, p. 71-108.

Harland, W. B., 1978, Geochronologic scales, in Cohee, G. V., Glaessner, M. F. and Hedberg, H. D., eds., Contributions to the Geologic time scale: American Association of Petroleum Geologists Studies in Geology 6, p. 9-32.

Harland, W. B., 1993, Stratigraphic regulation and guidance: a critique of current tendencies in stratigraphic codes and guides: Discussion: Geological Society of America Bulletin, v. 105, p. 1135-1136.

Harland, W. B., and Francis, H., eds., 1964, The Phanerozoic time scale (A symposium dedicated to Professor Arthur Holmes): Quarterly Journal of the Geological Society of London, v. 120s, 458 p.

Harland, W. B., and Francis, H., eds., 1971, The Phanerozoic time scale — A Supplement. Geological Society of London Special Publication 5, 356 p.

Harland, W. B., Smith, A. G., and Wilcock, B., eds., 1964, The Phanerozoic time scale (A symposium dedicated to Professor Arthur Holmes): Geological Society of London, Supplement to Quarterly Journal, v. 120 s, 458 p.

Harland, W. B., Cox, A. V., Llewellyn, P. G., Pickton, C. A. G., Smith, A. G., and Walters, R., 1982, A geologic time scale: Cambridge Earth Science Series, Cambridge University Press, Cambridge, 131 p.

Harland, W. B., Armstrong, R. L., Cox, A. V., Craig, L. E., Smith, A. G., and Smith, D. G., 1990, A geologic time scale, 1989: Cambridge Earth Science Series, Cambridge University Press, Cambridge, 263 p.

Harper, C. W., Jr., and Crowley, K. D., 1985, Insights on why graphic correlation (Shaw's method) works: A discussion: Journal of Geology, v. 93, p. 503-506.

Harrison, W. J., 1882, Geology of the counties of England and Wales, London, Kelly and Company, 346 p.

Harrison, C. G. A., and Funnell, B. M., 1964, Relationship of palaeomagnetic reversals and micropalaeontology in two Late Cenozoic cores from the Pacific Ocean: Nature, v. 204, p. 566.

Hay, W. W., and Southam, J. R., 1978, Quantifying biostratigraphic correlation: Annual Review of Earth and Planetary Sciences, v. 6, p. 353-375.

Hays, J. D., Imbrie, J., and Shackleton, N. J., 1976, Variations in the earth's orbit: pacemaker of the ice ages: Science, v. 194, p. 1121-1132.

Heckert, A. B., and Lucas, S. G., 2004, Simplifying the stratigraphy of time: Comments and Reply: Geology, v. 32, p. e58.

Hedberg, H. D., ed., 1976, International Stratigraphic Guide: Wiley, New York, 200 p.

Herbert, T. D., Premoli Silva, P, Erba, E., and Fischer, A. G., 1995. Orbital chronology of Cretaceous-Paleocene marine sediments, in Berggren, W. A., Kent, D. V., Aubry, M.-P., and Hardenbol, J., eds., Geochronology, time scales and global stratigraphic correlation: Society for Sedimentary Geology Special Publication 54, p. 81-93.

Hilgen, F. J., 1991. Extension of the astronomically calibrated (polarity) time scale to the Miocene/Pliocene boundary: Earth and Planetary Sciences Letters, v. 107, p. 349-368.

Hilgen, F. J., Brinkhuis, H., and Zachariasse, W. J., 2006, Unit stratotypes for global stages. The Neogene perspective: Earth Science Reviews v. 74, p. 113-125.

Hilgen, F. J., Hinnov, L. A., Aziz, H. A., Abels, H. A., Batenburg, S., Bosmans, J. H. C., de Boer, B., Hüsings, S. K., Kuiper, K. F., and Lourens, L. J., 2015, Stratigraphic continuity and fragmentary sedimentation: the success of cyclostratigraphy as part of integrated stratigraphy in Smith, D. G., Bailey, R., J., Burgess, P., and Fraser, A., eds., Strata and time: Geological Society, London, Special Publication 404, p. 157-197.

Hinnov, L. A., and Ogg, J. G., 2007, Cyclostratigraphy and the astronomical time scale: Stratigraphy, v. 4, p. 239-251.

Holland, C. H., 1986, Does the golden spike still glitter? Journal of the Geological Society, London, v. 143, p. 3-21.

Holmes, A., 1960, A revised geological time-scale: Transactions of the Edinburgh Geological Society, v. 17, p. 183-216.

Holmes, A., 1965: Principles of Physical Geology, second edition, Nelson, London, 1288 p.

House, M. R., 1985, A new approach to an absolute timescale from measurements of orbital cycles and sedimentary microrhythms: Nature, v. 315, p. 721-725.

Hunt, D., and Tucker, M. E., 1992, Stranded parasequences and the forced regressive wedge systems tract: deposition during base-level fall: Sedimentary Geology, v. 81, p. 1-9.

Imbrie, J., 1985, A theoretical framework for the Pleistocene ice age: Journal of the Geological Society, London, v. 142, p. 417-432.

Imbrie, J., and Imbrie, K. P., 1979, Ice ages: solving the mystery: Enslow, Hillside, New Jersey, 224 p.

Imbrie, J., Hays, J. D., Martinson, D. G., McIntyre, A., Mix, A. C., Morley, J. J., Pisias, N. G., Prell, W. L., Shackleton, N. J., 1984. The orbital theory of Pleistocene climate: support from a revised chronology of the marine $\delta^{18}O$ record, in: Berger, A. L., Imbrie, J., Hays, J., Kukla, G., and Saltzman, B., eds., Milankovitch and Climate. D. Reidel, Norwell, Mass, p. 269-305.

International Subcommission on Stratigraphic Classification, 1987, Unconformity-bounded stratigraphic units: Geological Society of America Bulletin, v. 98, p. 232-237.

Irving, E., 1966: Paleomagnetism of some Carboniferous rocks from New South Wales and its relation to geological events: Journal of Geophysical Research, v. 71, p. 6025-6051.

Jenkins, D. G., and Gamson, P., 1993, The late Cenozoic *Globorotalia truncatulinoides* datum plane in the Atlantic, Pacific and Indian Oceans, in Hailwood, E. A., and Kidd, R. B., eds., High resolution stratigraphy: Geological Society, London, Special Publication 70, p. 127-130.

Johnson, G. D., Johnson, N. M., Opdyke, N. D., and Tahirkheli, R. A. K., 1979, Magnetic reversal stratigraphy and sedimentary tectonic history of the Upper Siwalik Group, eastern Salt Range and southwestern Kashmir, in Farah, A., and DeJong, K. A., eds., Geodynamics of Pakistan: Geological Survey of Pakistan, Quetta, Pakistan, p. 149-165.

Johnson, N. M., Stix, J., Tauxe, L., Cerveny, P. F., and Tahirkheli, R. A. K., 1985, Paleomagnetic chronology, fluvial processes, and tectonic implications of the Siwalik deposits near Chinji Village, Pakistan: Journal of Geology, v. 93, p. 27-40.

Kauffman, E. G., 1984, Paleobiogeography and evolutionary response dynamic in the Cretaceous Western Interior Seaway of North America, in Westerman, G. E., ed., Jurassic-Cretaceous biochronology and paleogeography of North America: Geological Association of Canada Special Paper 27, p. 273-306.

Kauffman, E. G., 1988, Concepts and methods of high-resolution event stratigraphy: Annual Reviews of Earth and Planetary Sciences, v. 16, p. 605-654.

Kauffmann, E. G., 1977, Evolutionary rates and biostratigraphy; in Kauffman, E. G., and Hazel, J. E., eds., Concepts and methods of biostratigraphy: Dowden, Hutchinson and Ross Inc., Stroudsburg, Pennsylvania, p. 109-142.

Keller, H. M., Tahirkheli, R. A. K., Mirza, M. A., Johnson, G. D., and Johnson, N. M., 1977, Magnetic polarity stratigraphy of the Upper Siwalik deposits, Pabbi Hills, Pakistan: Earth and Planetary Science Letters, v. 36, p. 187-201.

Kelley, P. H., Fastovsky, D. E., Wilson, M. A., Laws, R. A., and Raymond, A., 2013, From paleontology to paleobiology: A half-century of progress in understanding life history, in Bickford, M. E., ed., The web of geological sciences: Advances, impacts and interactions: Geological Society of America Special Paper 500, p. 191-232.

Kemple, W. G., Sadler, P. M., and Strauss, D. J., 1995, Extending graphic correlation to many dimensions: stratigraphic correlation as constrained optimization, in Mann, K. O., and Lane, H. R., eds., Graphic correlation: Society for Sedimentary Geology, Special Publication 53, p. 65-82.

Kennedy, W. J., and Cobban, W. A., 1977, The role of ammonites in biostratigraphy; in Kauffman, E. G. and Hazel, J. E., eds., Concepts and methods of biostratigraphy: Dowden, Hutchinson and Ross Inc., Stroudsburg, Pennsylvania, p. 309-320.

Kennedy, W. J., Walaszczyk, I., and Cobban, W. A., 2005, The Global Boundary Stratotype Section and Point for the base of the Turonian Stage of the Cretaceous, Pueblo, Colorado, U.S.A.: Episodes, v. 28, p. 93-104.

Kennett, J. P., ed., 1980, Magnetic stratigraphy of sediments: Dowden, Hutchinson and Ross Inc., Stroudsburg, Pennsylvania, Benchmark Papers in Geology 54, 438 p.

Knoll, A., Walter, M. R., Narbonne, G. M., and Christie-Blick, N., 2006, The Ediacaran Period: a new addition to the geologic time scale: Lethaia, v. 39, p. 13-30.

Kolla, V., Posamentier, H. W., and Eichenseer, H., 1995, Stranded parasequences and the forced regressive wedge systems tract: deposition during base-level fall—discussion: Sedimentary Geology, v. 95, p. 139-145.

Landing, E., Geyer, G., Brasier, M. D., and Bowring, 2013, S. A., Cambrian evolutionary radiation: context, correlation and chronostratigraphy—Overcoming deficiencies of the first appearance datum (FAD) concept: Earth Science Reviews, v. 123, p. 133-172.

Langereis, C. G., Krijgsman, W., Muttoni, G., and Menning, M., 2010, Magnetostratigraphy—concepts, definitions, and applications: Newsletters on Stratigraphy, v. 43, p. 207-233.

Loutit, T. S., Hardenbol, J., Vail, P. R., and Baum, G. R., 1988, Condensed sections: the key to age dating and correlation of continental margin sequences, in Wilgus, C. K., Hastings, B. S., Kendall, C. G. St. C., Posamentier, H. W., Ross, C. A., and Van Wagoner, J. C., eds., Sea-level Changes: an integrated approach: Society of Economic Paleontologists and Mineralogists Special Publication 42, p. 183-213.

MacLeod, N., and Keller, G., 1991, How complete are Cretaceous/Tertiary boundary sections? A chronostratigraphic estimate based on graphic correlation: Geological Society of America Bulletin, v. 103, p. 1439-1457.

Mann, K. O., and Lane, H. R., eds., 1995, Graphic correlation: Society for Sedimentary Geology, Special Publication 53, 263 p.

Martinsen, O. J., 1993, Namurian (Late Carboniferous) depositional systems of the Craven-Askrigg area, northern England: implications for sequence-stratigraphic models, in Posamentier, H. W., Summerhayes, C. P., Haq, B. U., and Allen, G. P., eds., Sequence stratigraphy and facies associations: International Association of Sedimentologists Special Publication 18, p. 247-281.

Martinsen, O. J., 1993, Namurian (Late Carboniferous) depositional systems of the Craven-Askrigg area, northern England: implications for sequence-stratigraphic models, in Posamentier, H. W., Summerhayes, C. P., Haq, B. U., and Allen, G. P., eds., Sequence stratigraphy and facies associations: International Association of Sedimentologists Special Publication 18, p. 247-281.

Martinsen, O. J., Martinsen, R. S., and Steidtmann, J. R., 1993, Mesaverde Group (Upper Cretaceous), southeastern Wyoming: allostratigraphy versus sequence stratigraphy in a tectonically active area: American Association of Petroleum Geologists Bulletin, v. 77, p. 1351-1373.

Matthews, R. K., 1984a, Oxygen-isotope record of ice-volume history: 100 million years of glacio-isostatic sea-level fluctuation, in Schlee, J. S., ed., Interregional unconformities and hydrocarbon accumulation: American Association of Petroleum Geologists Memoir 36, p. 97-107.

Matthews, R. K., 1988, Sea level history: Science, v. 241, p. 597-599.

Mattinson, J. M., 2013, The geochronology revolution, in Bickford, M. E., The web of geological sciences: advances, impacts and interactions: Geological Society of America Special paper 500, p. 303-320.

McArthur, J. M., 1994, Recent trends in strontium isotope stratigraphy, Terra Nova, v. 6, p. 331-358.

McArthur, J. M., 1998, Strontium isotope stratigraphy, in Doyle, P. and Bennett, M. R., eds., Unlocking the stratigraphical record: John Wiley and Sons, Chichester, p. 221-241.

McArthur, J. M., and Howarth, R. J., 2004, Strontium isotope stratigraphy, in Gradstein, F. M., Ogg, J. G., and Smith, A. G., eds., A geologic time scale: Cambridge University Press, Cambridge, p. 96-105.

McGowran, B., 2005, Biostratigraphy: Microfossils and Geological Time: Cambridge University Press, Cambridge, 459 p.

McKerrow, W. S., 1971, Palaeontological prospects—the use of fossils in stratigraphy: Journal of the Geological Society, London, v. 127, p. 455-464.

McLaren, D. J., 1970, Presidential address: time, life and boundaries: Journal of Paleontology, v. 44, p. 801-813.

Miall, A. D., 1994, Sequence stratigraphy and chronostratigraphy: problems of definition and precision in correlation, and their implications for global eustasy: Geoscience Canada, v. 21, p. 1-26.

Miall, A. D., 1995, Whither stratigraphy? Sedimentary Geology, v. 100, p. 5-20.

Miall, A. D., 2004, Empiricism and model building in stratigraphy: the historical roots of present-day practices. Stratigraphy: American Museum of Natural History, v. 1, p. 3-25.

Miall, A. D., 2010, The geology of stratigraphic sequences, second edition: Springer-Verlag, Berlin, 522 p.

Miall, A. D., 2013, Sophisticated stratigraphy, in Bickford, M. E., ed., The web of geological sciences: Advances, impacts and interactions: Geological Society of America Special Paper 500, p. 169-190.

Miall, A. D., and Miall, C. E., 2001, Sequence stratigraphy as a scientific enterprise: the evolution and persistence of conflicting paradigms: Earth Science Reviews, v. 54, #4, p. 321-348.

Milankovitch, M., 1930, Mathematische klimalehre und astronomische theorie der klimaschwankungen, in Koppen, W., and Geiger, R., eds., Handbuch der klimatologie, I (A); Gebruder Borntraeger, Berlin.

Milankovitch, M., 1941, Kanon der Erdbestrahlung und seine Anwendung auf das Eiszeitenproblem: Akad. Royale Serbe, 133, 633 p.

Miller, F. X., 1977, The graphic correlation method in biostratigraphy, in Kauffman, E. G., and Hazel, J. E., eds., Concepts and methods in biostratigraphy: Dowden, Hutchinson and Ross, Inc., Stroudsburg, Pennsylvania, p. 165-186.

Mitchum, R. M., Jr., and Van Wagoner, J. C., 1991, High-frequency sequences and their stacking patterns: sequence-stratigraphic evidence of high-frequency eustatic cycles: Sedimentary Geology, v. 70, 131-160.

Mitchum, R. M., Jr., Vail, P. R., and Sangree, J. B., 1977, Seismic stratigraphy and global changes of sea level, Part six: Stratigraphic interpretation of seismic reflection patterns in depositional

sequences, in Payton, C. E., ed., Seismic stratigraphy—applications to hydrocarbon exploration; American Association of Petroleum Geologists Memoir 26, p. 117-133.

Naish, T. R, Field, B. D., Zhu, H., Melhuish, A., Carter, R. M., Abbott, S. T., Edwards, S., Alloway, B. V., Wilson, G. S., Niessen, F., Barker, A., Browne, G. H., and Maslen, G., 2005, integrated outcrop, drill core, borehole and seismic stratigraphic architecture of a cyclothemic, shallow-marine depositional system, Wanganui Basin, New Zealand: Journal of the Royal Society of New Zealand, v. 35, p, 91-122.

North American Commission on Stratigraphic Nomenclature, 1983, North American Stratigraphic Code: American Association of Petroleum Geologists Bulletin, v. 67, p. 841-875.

North American Commission on Stratigraphic Nomenclature (NACSN), 2005, North American Stratigraphic Code: American Association of Petroleum Geologists Bulletin, v. 89, p. 1547-1591.

Nummedal, D., and Swift, D. J. P., 1987, Transgressive stratigraphy at sequence-bounding unconformities: some principles derived from Holocene and Cretaceous examples, in Nummedal, D., Pilkey, O. H., and Howard, J. D., eds., Sea-level fluctuation and coastal evolution; Society of Economic Paleontologists and Mineralogists Special Publication 41, p. 241-260.

Odin, G. S., 1982, Numerical dating in Stratigraphy, v. 1 and 2, Chichester: Wiley-Interscience.

Ogg, J. G., and Smith, A. G., 2004, The geomagnetic polarity time scale,, in Gradstein, F. M., Ogg, J. G., and Smith, A. G., eds., A geologic time scale: Cambridge University Press, Cambridge, p. 63-86.

Opdyke, N. D., 1972, Paleomagnetism of deep-sea cores; Review of Geophysics and Space Physics, v. 10, p. 213.

Opdyke, N. D., Glass, B., Hays, J. D., and Foster, J., 1966, Paleomagnetic study of Antarctic deep-sea cores: Science, v. 154, p. 349-357.

Pearson, P. N., 1998, Evolutionary concepts in biostratigraphy, in Doyle, P. and Bennett, M. R., eds., Unlocking the stratigraphical record: John Wiley and Sons, Chichester, p. 123–144.

Picard, N. D., 1964, Paleomagnetic correlation of units within the Chugwater (Triassic) Formation, west-central Wyoming: American Association of Petroleum Geologists Bulletin, v. 48, p. 269-291.

Plint, A. G., 1988, Sharp-based shoreface sequences and "offshore bars" in the Cardium Formation of Alberta: their relationship to relative changes in sea level, in Wilgus, C. K., Hastings, B. S., Kendall, C. G. St. C., Posamentier, H. W., Ross, C. A., and Van Wagoner, J. C., eds., Sea-level Changes: an integrated approach: Society of Economic Paleontologists and Mineralogists Special Publication 42, p. 357-370.

Plint, A. G., 1990, An allostratigraphic correlation of the Muskiki and Marshybank Formations (Coniacian-Santonian) in the foothills and subsurface of the Alberta Basin: Bulletin of Canadian Petroleum Geology, v. 38, p. 288-306.

Plint, A. G., Walker, R. G., and Bergman, K. M., 1986, Cardium Formation 6. Stratigraphic framework of the Cardium in subsurface: Bulletin of Canadian Petroleum Geology, v. 34, p. 213-225.

Posamentier, H. W., and Allen, G. P., 1999, Siliciclastic sequence stratigraphy—concepts and applications: Society for Sedimentary Geology (SEPM), Concepts in sedimentology and paleontology 7, 210 p.

Posamentier, H. W., and Kolla, V., 2003, Seismic geomorphology and stratigraphy of depositional elements in deep-water settings: Journal of Sedimentary Research, v. 73, p. 367-388.

Posamentier, H. W., Jervey, M. T., and Vail, P. R., 1988, Eustatic controls on clastic deposition I—Conceptual framework, in Wilgus, C. K., Hastings, B. S., Kendall, C. G. St. C., Posamentier, H. W., Ross, C. A., and Van Wagoner, J. C., eds., Sea level Changes - an integrated approach: Society of Economic Paleontologists and Mineralogists Special Publication 42, p. 109-124.

Remane, J., 2000a, Explanatory note and international stratigraphic chart: UNESCO, Division of Earth Sciences, Paris.

Remane, J., 2000b, Should the golden spike glitter?—Comments to the paper of M.-P. Aubry et al.: Episodes, v. 23, p. 211-213.

Roof, S. R., Mullins, H. T., Gartner, S., Huang, T. C., Joyce, E., Prutzman, J., and Tjalmsa, L., 1991, Climatic forcing of cyclic carbonate sedimentation during the last 5.4 million years along the west Florida continental margin: Journal of Sedimentary Research, v. 61, p. 1070-1088.

Sadler, P. M., 1999a, Constrained optimization approaches to stratigraphic correlation and seriation problems. A user's guide and reference manuals to the CONOP program family: University of California, Riverside, 142 p.

Sadler, P. M., Cooper, R. A., and Melchin, M., 2009: High-resolution, early Paleozoic (Ordovician-Silurian) time scales: Geological Society of America Bulletin, v. 121, p. 887-906.

Sadler, P. M., Cooper, R. A., and Crampton, J. S., 2014, High-resolution geobiologic time-lines: progress and potential, fifty years after the advent of graphic correlation: The Sedimentary Record, v. 12, #3, p. 4-9.

Sageman, B. B., Singer, B. S., Meyers, S. R., Siewert, S. E., Walaszczyk, I., Condon, D. J., Jicha, B. R., Obradovich, J. D., and Sawyer, D. A., 2014, Integrating $^{40}$Ar/$^{39}$Are, U-Pb and astronomical clocks in the Cretaceous Niobrara Formation, Western Interior Basin, USA: Geological Society of America Bulletin, v. 126, p. 956-973.

Salvador, A., ed., 1994, International Stratigraphic Guide, Second edition: International Union of Geological Sciences, Trondheim, Norway, and Geological Society of America, Boulder, Colorado, 214 p.

Schlager, W., 1989, Drowning unconformities on carbonate platforms, in Crevello, P. D., Wilson, J. L., Sarg, J. F., and Read, J. F., eds., Controls on carbonate platforms and basin development: Society of Economic Paleontologists and Mineralogists Special Publication 44, p. 15-25.

Schlager, W., 2005, Carbonate sedimentology and sequence stratigraphy: SEPM Concepts in Sedimentology and Paleontology #8, 200p.

Schwarzacher, W., 1993, Cyclostratigraphy and the Milankovitch theory: Elsevier, Amsterdam, Developments in Sedimentology 52, 225 p.

Shackleton, N. J., and Opdyke, N. D., 1976, Oxygen isotope and paleomagnetic stratigraphy of equatorial Pacific core V28-239, Late Pliocene to latest Pliocene: Geological Society of America Memoir 145, p. 449-464.

Shackleton, N. J., McCave, I. N., and Weedon, G. P., eds., 1999, Astronomical (Milankovitch) calibration of the geological time-scale: Philosophical Transactions of the Royal Society, London, Series A, v. 357, p 1731-2007.

Shaw, A. B., 1964, Time in stratigraphy: McGraw Hill, New York, 365 p.

Sheriff, R. E., 1976, Inferring stratigraphy from seismic data; American Association of Petroleum Geologists Bulletin: v. 60, p. 528-542.

Sloss, L. L., 1963, Sequences in the cratonic interior of North America: Geological Society of America Bulletin, v. 74, p. 93-113.

Sloss, L. L., Krumbein, W. C., and Dapples, E. C., 1949, Integrated facies analysis; in Longwell, C. R., ed., Sedimentary facies in geologic history: Geological Society of America Memoir 39, p. 91-124.

Smith, A. G., Barry, T., Bown, P., Cope, J., Gale, A., Gibbard, P., Gregory, J., Hounslow, M., Kemp, D., Knox, R., Marshall, J., Oates, M., Rawson, P., Powell, J., and Waters, C., 2015, GSSPs, global stratigraphy and correlation: in Smith, D. G., Bailey, R., J., Burgess, P., and Fraser, A., eds., Strata and time: Geological Society, London, Special Publication 404, p. 37-67.

Strong, N., and Paola, C., 2008, Valleys that never were: time surfaces versus stratigraphic surfaces: Journal of Sedimentary Research, v. 78, p. 579-593.

Sylvester-Bradley, P.C., 1977: Biostratigraphical tests of evolutionary theory; in Kauffman, E. G., and Hazel, J. E., eds., Concepts and methods of biostratigraphy: Dowden, Hutchinson and Ross Inc., Stroudsburg, Pennsylvania, p. 41-64.

Tarling, D. H., 1982, Land bridges and plate tectonics: Geobios, v. 15, Supplement 1, p. 361-374.

Torrens, H. S., 2002, Some personal thoughts on stratigraphic precision in the twentieth century, in Oldroyd, D. R., ed., The Earth inside and out: some major contributions to geology in the twentieth century: Geological Society, London, Special Publication 192, p. 251-272.

Underhill, J R., and Partington, M. A., 1993, Jurassic thermal doming and deflation in the North Sea: implications of the sequence stratigraphy evidence, in Parker, J. R., ed., Petroleum geology of northwest Europe: Proceedings of the 4$^{th}$ Conference, Bath, Geological Society, London, v. 1, p. 337-346.

Vai, G. B., 2001, GSSP, IUGS and IGC: an endless story toward a common language in the Earth Sciences: Episodes: v. 24, p. 29-31.

Vail, P. R., and Todd, R. G., 1981, Northern North Sea Jurassic unconformities, chronostratigraphy and sea-level changes from seismic stratigraphy, in Illing, L. V., and Hobson, G. D., eds., Petroleum Geology of the continental shelf of northwest Europe: Institute of Petroleum, London, p. 216-235.

Vail, P. R., Mitchum, R. M., Jr., Todd, R. G., Widmier, J. M., Thompson, S., III, Sangree, J. B., Bubb, J. N., and Hatlelid, W. G., 1977, Seismic stratigraphy and global changes of sea-level, in Payton, C. E., ed., Seismic stratigraphy - applications to hydrocarbon exploration: American Association of Petroleum Geologists Memoir 26, p. 49-212.

Van Couvering, J. A., Castradori, D., Cita, M. B., Hilgen, F. J., and Rio, D., 2000, The base of the Zanclean Stage and of the Pliocene Series: Episodes, v. 23, p. 179-187.

Van Hinte, J. E., 1976a, A Jurassic time scale: American Association of Petroleum Geologists Bulletin, v. 60, p. 489-497.

Van Hinte, J. E., 1976b, A Cretaceous time scale: American Association of Petroleum Geologists Bulletin, v. 60, p. 498-516.

Van Wagoner, J. C., Mitchum, R. M., Jr., Posamentier, H. W., and Vail, P. R., 1987, Seismic stratigraphy interpretation using sequence stratigraphy, Part 2: key definitions of sequence stratigraphy, in Bally, A. W., ed., Atlas of seismic stratigraphy: American Association of Petroleum Geologists Studies in Geology 27, v. 1, p. 11-14.

Van Wagoner, J. C., Posamentier, H. W., Mitchum, R. M., Jr., Vail, P. R., Sarg, J. F., Loutit, T. S., and Hardenbol, J., 1988, An overview of the fundamentals of sequence stratigraphy and key definitions, in Wilgus, C. K., Hastings, B. S., Kendall, C. G. St. C., Posamentier, H. W., Ross, C. A., and Van Wagoner, J. C., eds., Sea level Changes - an integrated approach: Society of Economic Paleontologists and Mineralogists Special Publication 42, p. 39-45.

Van Wagoner, J. C., Mitchum, R. M., Campion, K. M. and Rahmanian, V. D. 1990, Siliciclastic sequence stratigraphy in well logs, cores, and outcrops: American Association of Petroleum Geologists Methods in Exploration Series 7, 55 p.

Veeken, 2007, Seismic stratigraphy, basin analysis and reservoir characterization: Elsevier, Amsterdam, Seismic Exploration, v. 37, 509 p.

Veizer, J. 1989, Strontium isotopes in seawater through time: Annual Review of Earth and Planetary Science Letters, v. 17, p. 141-167.

Villeneuve, M., 2004, Radiogenic isotope geochronology, in Gradstein, F. M., Ogg, J. G., and Smith, A. G., eds., A geologic time scale: Cambridge University Press, Cambridge, p. 87-95.

Visser, C. F., and Johnson, G. D., 1978, Tectonic control of Late Pliocene molasse sedimentation in a portion of the Jhelum re-entrant, Pakistan: Geologische Rundschau, v. 67, p. 15-37.

Walker, R. G., 1990, Facies modeling and sequence stratigraphy: Journal of Sedimentary Petrology, v. 60, p. 777-786.

Walker, R. G., 1992, Facies, facies models and modern stratigraphic concepts, in Walker, R. G. and James, N. P., eds., Facies models: response to sea-level change: Geological Association of Canada, p. 1-14.

Walsh, S. L., 2004, Solutions in chronostratigraphy: the Paleocene/Eocene boundary debate, and Aubry vs. Hedberg on chronostratigraphic principles: Earth Science Reviews, v. 64, p. 119-155.

Westphal, H., Munnecke, A., and Brandano, M., 2008, Effects of diagenesis on the astrochronological approach of defining stratigraphic boundaries in calcareous rhythmites: The Tortonian GSSP: Lethaia, v. 41, p. 461-476.

Wignall, P. B., 1991, Ostracod and foraminifera micropaleontology and its bearing on biostratigraphy: a case study from the Kimmeridgian (Late Jurassic) of north west Europe: Palaios, v. 5, p. 219-226.

Wilson, J. T., 1966, Did the Atlantic close and then re-open? Nature, v. 211, p. 676-681.

Zalasiewicz, J., Smith, A., Brenchley, P., Evans, J., Knox, R., Riley, N., Gale, A., Gregory, F. J., Rushton, A., Gibbard, P., Hesselbo, S., Marshall, J., Oates, M., Rawson, P., and Trewin, N., 2004, Simplifying the stratigraphy of time: Geology, v. 32, p. 1-4.

Zecchin, M., and Catuneanu, O., 2013, High-resolution sequence stratigraphy of clastic shelves I: units and bounding surfaces: Marine and Petroleum Geology, v. 39, p. 1-25.

Ziegler, A. M., Cocks, L. R. M., and McKerrow, W .S., 1968, The Llandovery transgression of the Welsh borderland: Paleontology, v. 11, p. 736-782.

# The Future of Time

**8**

## Contents

## 8.1 Introduction

The purpose of this chapter is to assess the current status of the measurement of geologic time as stored in the stratigraphic record, and to present some arguments regarding present pitfalls and future potential regarding the measurement and interpretation of geological time.

Modern methods of relative and "absolute" age dating of the stratigraphic record, and the current status of the Geological Time Scale, are discussed in the previous chapter. What these modern methods reveal is that the stratigraphic record is far more fragmentary than most geologists are accustomed to thinking. Many key concepts in sedimentary geology carry an implication of continuity in the sedimentary record, including the practices of stratigraphic classification and correlation, Walther's law, cyclic sedimentation, facies models, and sequence stratigraphy. However, as argued in this chapter, it is becoming increasingly clear that we need to carefully re-evaluate these assumptions of continuity. A key criterion to keep in mind is that of time scale. We tend to assume that ancient sedimentary records representing very long time intervals on the human time scale ($>10^4$ a) may be reliably compared with observations made over the much shorter time scales accessible to human observation ($\leq 10^2$ a). This is the basis of the Hutton-Lyellian aphorism "the present is the key to the past" (and the reverse). But a question that persists is that concerning the relevance and significance of transient processes and ephemeral modern deposits to the interpretation of the rock record, given questions about the highly variable preservability of different sedimentary facies.

© Springer International Publishing Switzerland 2016
A.D. Miall, *Stratigraphy: A Modern Synthesis*, DOI 10.1007/978-3-319-24304-7_8

## 8.2   Where We Are Now and How We Got Here

Barrell (1917), in a review and discussion that was about seven decades ahead of its time, was probably the first to fully recognize the incompleteness of the stratigraphic record. Barrell (1917, Fig. 8.7) constructed a diagram showing the "Sedimentary Record made by Harmonic Oscillation in Baselevel" (Fig. 5.2). This is remarkably similar to diagrams that have appeared in some of the Exxon sequence model publications since the 1980s (e.g., Van Wagoner et al. 1990, Fig. 8.59; see Fig. 5.5 of this book), and represents a thoroughly modern deductive model of the way in which "time" is stored in the rock record. Curve A-A simulates the record of long-term subsidence and the corresponding rise of the sea. Curve B-B simulates an oscillation of sea levels brought about by other causes. Barrell discussed diastrophic and climatic causes, including glacial causes, and applied these ideas to the rhythmic stratigraphic record of the "upper Paleozoic formation of the Appalachian geosyncline" in a discussion that would appear to have provided the foundation for the interpretations of "cyclothems" that appeared in the 1930s. Barrell showed that when the long-term and short-term curves of sea-level change are combined, the oscillations of base level provide only limited time periods when sea-level is rising and sediments can accumulate. "Only one-sixth of time is recorded" by sediments (Barrell 1917, p. 797).

Wheeler (1958, 1959) developed the concept of the chronostratigraphic cross-section, in which the vertical dimension in a stratigraphic cross-section is drawn with a time scale instead of a thickness scale (Fig. 8.1). In this way, time gaps (unconformities) become readily apparent, and the nature of time correlation may be accurately indicated. Such graphs have come to be termed "Wheeler diagrams." They are commonly constructed for use as stratigraphic tables in regional reports, but typically the time scale is arbitrary and variable, which means that information about missing time is distorted and usually not considered beyond qualitative statements about the significance of regional unconformities.

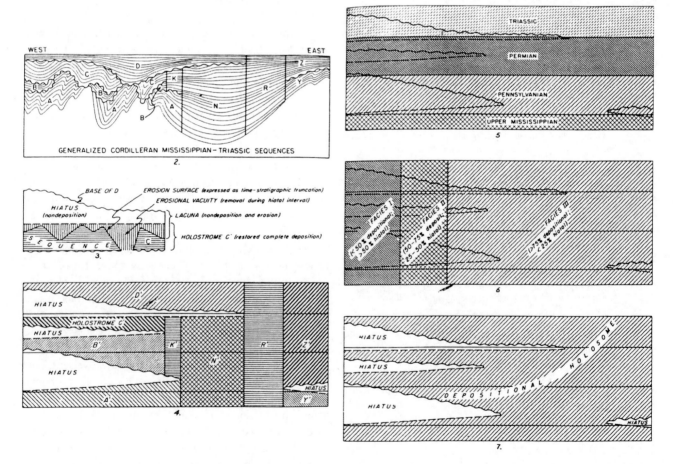

**Fig. 8.1** Wheeler's development of chronostratigraphic diagrams. *Diagram 2* shows a complex stratigraphic cross-section containing numerous unconformities. *Diagram 3* explains some of the terminology used, and the remaining diagrams are "Wheeler plots" of the section in *Diagram 2*, ornamented in various ways to highlight different geological features. (Wheeler 1958). AAPG © 1958, reprinted by permission of the AAPG whose permission is required for further use

In fact, this quote from Grabau (1960, p. 1097) would seem to imply that stratigraphic continuity is the norm:

> It is of first importance to the chronographer of earth history that he should find a continuous record, in order that he may have a measure by which to judge the partial records of any given region and to discover the breaks and imperfections in the local records thus presented. The question then arises: under what conditions may we expect to obtain a continuous record and how are we to guard against the introduction of errors?

We return to the use of Wheeler plots later in this chapter. Ager's (1981) famously remarked that the sedimentary record is "more gap than record." In a later book he expanded on the theme of gaps. Following a description of the major unconformities in the record at the Grand Canyon, he said, (Ager 1993, p. 14):

> We talk about such obvious breaks, but there are also gaps on a much smaller scale, which may add up to vastly more unrecorded time. Every bedding plane is, in effect, an unconformity. It may seem paradoxical, but to me the gaps probably cover most of earth history, not the dirt that happened to accumulate in the moments between. It was during the breaks that most events probably occurred.

Much has now been learned about the significance of unconformities. Sequence boundaries are unconformities, but assigning an age to an unconformity surface is not necessarily a simple matter, as illustrated in the useful theoretical discussion by Aubry (1991). An unconformity represents a finite time span at any one location; it may have a complex genesis, representing amalgamation of more than one event. It may also be markedly diachronous, because the transgressions and regressions that occur during the genesis of a stratigraphic sequence could span the entire duration of the sequence. Ravinement surfaces, which commonly form sequence boundaries, are the product of diachronous erosion during transgression (Nummedal and Swift 1987). Kidwell (1988) demonstrated that transgression results in an offset in sequence-boundary unconformities by as much as one half of a cycle between basin centre and basin margin. Christie-Blick et al. (1990, 2007) pointed out many geological a where unconformities may be diachronous. On a small scale, it has been demonstrated by flume experiments and observations of the rock record that laminated mudstones may preserve less than 10 % of elapsed time as a result of rapid deposition and frequent reworking (Trabucho-Alexandre 2015).

The subaerial erosion surfaces that underlie incised valley-fills at continental margins, and which commonly provide important sequence boundaries, are good examples of unconformities that evolve over time and are not everywhere the same age. Strong and Paola (2008) pointed out an important distinction between a *topographic surface* and a *stratigraphic surface*. The topographic surface that corresponds to the subaerial erosion surface undergoes continual change until it is finally buried and preserved. It is the stratigraphic surface that we map in the rock record, but this is a surface that never actually existed in its entirety as a topographic surface in its final preserved form, because it undergoes continuous modification by erosion or sedimentation until final burial. The subaerial erosion surface may also violate one of the fundamental principles of an unconformity, which is that all the deposits below the unconformity surface are older than all the beds above the surface. Deepening and widening of an alluvial valley may continue during the final stages of the evolution of a subaerial surface, even while a turn-around in the base-level cycle has begun to transgress and bury the surface during the beginning of a transgression. Channel or overbank deposits that are preserved as terrace remnants, resting on the basal erosion surface, could therefore predate the coastal deposits formed during the final stages of base-level fall, and would therefore be older than the sequence boundary at the coast, although resting on it.

Miall (2010, Chap. 14) provided an extensive discussion of the dating and correlation of sequence boundaries.

One of the most significant achievements of modern stratigraphic work is the increasing accuracy and precision of the Geological Time Scale and our ability to provide ever more refined dating of the geological record (Sect. 7.8.6). A particularly instructive example of this advance (although dependent on detailed paleontological work that goes back to the early nineteenth century!) is the biostratigraphic subdivision of parts of the European Jurassic section. In Chap. 7 the work of Callomon (1995) is highlighted (see Fig. 7.40 of this book). The shallow-marine Inferior Oolite, which in the study area of southern Britain is only some 5 m thick, may be subdivided into as many as 56 faunal zones, although the interesting point that is relevant to our discussion here is that none of the thirteen sections examined for this project contains a complete record of the zones, and the preserved record is different in every section. This formation is replete with local diastems that record local areas of negative accommodation (Fig. 8.2). Buckman (1910, p. 90) concluded, after many years of meticulous analysis of this unit that the Inferior Oolite of Dorset, an extremely condensed shelf sequence, '…might be defined as a series of gaps united by thin bands of deposits …. the deposits are so local, the deposits of one place correspond to the gaps of another…'.

Overall, it is impossible to distinguish any ordered pattern to the record of sedimentation and erosion in these sections. How typical is this of shallow-marine sedimentation in general? Does the availability of an unusually-detailed ammonite biostratigraphy enable us to develop a much more detailed record of local change than would otherwise be available? And should this section therefore be regarded as a model for the interpretation of other shallow-marine

**Fig. 8.2** Three of the stratigraphic sections through the Inferior Oolite studied by Callomon (1995) showing (at *left*) the series of faunal zones recorded in the rocks and (at *right*) *red arrows* indicating the likely position of diastems where one or more zones are missing

carbonate sections? Would it be correct to conclude that many other shallow marine (and nonmarine?) successions should similarly be regarded as containing numerous local diastems? If so, what does this tell us about short- to long-term sedimentary processes?

As the study of modern sedimentary environments and facies interpretations evolved during the 1960s (Sect. 1.2.6 and 1.2.7), it began to be realized that there is a wide variation in the energy, magnitude and time scales of sedimentary processes. Some facies, such as turbidites and storm deposits, are clearly formed rapidly over geologically insignificant periods of time. Although he was not the first to discuss this, Ager (1973) is credited with the coining of the term **event sedimentation** to encompass this pattern of rapid deposition. One of his stratigraphic studies was carried out in order to raise the question of the time significance of the preserved record. He re-evaluated the sedimentary history of the Sutton Stone, a beach conglomerate that forms the base of the Blue Lias succession in Glamorgan, South Wales, where it ranges from 10 to 13.5 m in thickness. The

conglomerate rests on a transgressive surface, and had long been interpreted as the product of slow sedimentation during a protracted transgression. According to Wobber (1965) it contains the fossils representing at least four ammonite zones of Hettangian and Lower Sinemurian age which, according to the data provided by Gradstein et al. (2004a, Fig. 18.1) span a total of about 4 million years. However, Ager (1986) suggested that the deposit was formed in a single major tropical storm or hurricane. He stated (p. 35) "I do not think it took the three or four million years or so of three or four or five ammonite chronozones." His one-line conclusion: "It all happened one Tuesday afternoon."

The "Tuesday afternoon" remark is, of course deliberately provocative (one might ask, why not Wednesday?), but Ager's point was to emphasize the rapidity by which certain geological processes may accomplish spectacular results. Storms do, indeed, accomplish most of their erosional and depositional work within a space of a few hours or days, at most. A four million-year period includes 208,000,000 Tuesdays. Are we to understand that only one of these days left a sedimentary record? The question is fascinating, and points to a nagging issue regarding sedimentation rates and preservation that has gradually emerged with the accumulation of an ever-increasing volume of data concerning modern and ancient sedimentary processes. The central point is that when an ancient deposit is compared to the sediments accumulating in an equivalent environment at the present day, the rates of modern sedimentary processes and stratigraphic accumulation, if applied to the rock record, would allow for the accumulation of the ancient deposit in a fraction of the time that chronostratigraphic data indicate is available. It is not at all uncommon to find that the accumulation of a given thickness of sediment in any given environment could take place in as little as one tenth of the available time. What happened during the rest of the time that passed, according to the chronostratigraphy of the unit? Callomon's demonstration of numerous diastems in the Inferior Oolite might be pointing the way to a more general conclusion.

It does not materially undermine Ager's (1986) broader arguments about event sedimentation to note that later workers have disputed his claim that the Sutton Stone was deposited in a single event. Evidence of encrusting organisms at more than one level within the conglomerate, and the observations of varying conglomerate facies through the unit suggested an alternative model of cliff collapse on an exposed rocky shoreline, perhaps subsequently influenced by several or many storms (Fletcher et al. 1986; Johnson and McKerrow 1995).

Sadler (1981) compiled thousands of records concerning sedimentation in deposits ranging in age from ancient to modern, and demonstrated that the relationship between sedimentation rate and elapsed time (the time period over which the sedimentation rate is measured) is linear and inverse on a log-log scale (Fig. 8.3). His data base consisted of 25,000 records of accumulation rates. Measured sedimentation rates vary by eleven orders of magnitude, from $10^{-4}$ to $10^{7}$ m/ka. This huge range of values would seem to suggest the presence of an increasing number and duration of intervals of non-deposition or erosion that become factored into the measurements as the length of the measured stratigraphic record increases. A comparable range of sedimentation rates was revealed in his later compilation of records focusing on shallow-marine carbonate deposits (Kemp and Sadler 2014).

Although Sadler's (1981) synthesis has been in the public domain for more than thirty years, few stratigraphers have attempted to wrestle with its significance for the geological interpretations that we make based on the Hutton-Lyell Principle of Uniformitarianism: "The present is the key to the past." Although Gould (1965) argued, on logical grounds, that this Principle is no longer necessary, given the assumption of the invariance of natural laws, it nonetheless exerts a powerful influence on the methods of modern

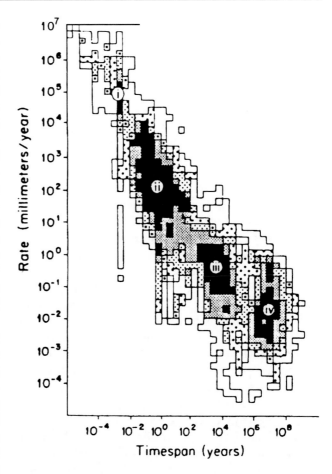

**Fig. 8.3**  The relationship between sedimentation rate and elapsed time in the stratigraphic record (Sadler 1981)

sedimentology, which are very firmly based on the practice of seeking modern analogues for an ancient deposit of interest. As argued in this chapter, it is becoming clear that there are some important provisos and limitations that must now be inserted into this practice.

## 8.3   A Natural Hierarchy of Sedimentary Processes

Bailey and Smith (2010, pp. 57–58) pointed out the ephemeral nature of most sedimentary processes:

> There would seem to be a very small chance of the preservation in 'stratigraphic snapshots' of, say, one particular ripple-marked shoreface out of the thousands or millions, created and destroyed diurnally through geologic time. Such instances suggest that such stratigraphic records are better viewed as the outcome of temporary cessation of the erosion and redistribution of sediment: 'frozen accidents' of accumulation.

It is now widely recognized that not only the durations of the gaps, but also the distribution of layer thicknesses and sedimentation rates in stratigraphic successions have fractal-like properties (Plotnick 1986; Sadler 1999; Schlager 2004; Bailey and Smith 2005; Smith et al. 2015). However, the practical development of this concept is hampered by the current methods of stratigraphic documentation. Bailey and Smith (2010, p. 58) noted that current classifications of stratigraphic units, based as they are on "a human scale observer" (e.g., lithostratigraphy, sequence stratigraphy) constitute hierarchies that are somewhat arbitrary, and make statistical analysis of bedding and its contained gaps difficult. In addition, it is extremely difficult to operationalize a lithologically, petrologically and statistically reliable practical field definition of what constitutes a "layer", at all scales from the lamina to the basin fill.

To circumvent this problem Bailey and Smith (2010) and Bailey and Schumer (2012) developed a method of analyzing the stratigraphic record termed the "Layer Thickness Inventory". The analysis is carried out on continuous, digitized records, such as wireline logs. The gamma-ray log is particularly suitable for this purpose, because it is readily interpreted in lithologic terms. A computer routine works its way through the digital data records one at a time, searching the data string above and below for records in which the GR reading is higher and lower than that at each sample point, and records the calculated thicknesses. All lithologically defined layers are thereby recorded, ignoring hiatuses. The procedure records layers within layers, which therefore overlap, and in this way "it recognizes that the various sedimentary influences on lithology operate in overlapping time frames, rather than as the succession of discrete process-response effects suggested by conventional hierarchical stratigraphic subdivisions." (Bailey and Smith 2010, p. 59).

Their analysis of a range of geological examples demonstrated that log-log plots of bed thickness against number of records always generate linear distributions, regardless of the scale of the stratigraphic section and the nature of the lithology, suggesting "that there is a universal relationship between layer thickness and frequency of occurrence in the record." (Bailey and Smith 2010, p. 62).

Bailey and Smith (2010) raised the question of the degree to which the evidently fractal record is representative of past surface processes, and made the following points:

- The notion of continuous deposition, on which the historicity of the record depends, has no theoretical or evidential basis. In relation to the accumulation of particulate solids it is, in fact, an impossibility. At best, it is a scale-dependent descriptive convenience.
- If there is no continuity in accumulation, the sequential preservation of laterally contiguous facies, according to Walther's Law, becomes questionable.
- Stratigraphic hierarchies are constructs, commonly tailored to human-scale analysis of the fractal record. They

are a practical, convenient, but incomplete, representation of this record.
- Currently-observable sedimentary processes and facies underpin uniformitarian stratigraphic interpretations. Yet there is no way of determining whether a present day deposit will be preserved millions of years hence. Equally, if there is "more gap" there is the question of the degree to which the preserved record is representative of the continuous operations of past sedimentary systems. Specifically, are the snapshot "frozen accidents of preservation" representative?
- As Sadler (1999) has shown, local calculations of accumulation rate are time-scale dependent.

Is the stratigraphic record fundamentally unrepresentative of the geological past? These conclusions would appear to invalidate virtually the whole of the last two centuries of stratigraphic progress!

However, all is not lost!

Miall (1991) suggested that the sedimentary time scale constitutes a natural hierarchy corresponding to the natural hierarchy of temporal processes (diurnal, lunar, seasonal, geomorphic threshold, tectonic, etc.) and the main purpose of this chapter is to develop this idea further, making extensive use of modern quantitative data dealing with sedimentation rates and accumulation rates. Most of the discussion that follows relates to clastic sedimentation in shallow-marine and nonmarine environments. I return to the above discussion points in a later section of this chapter, where it is argued that most may be managed within the context of the appropriate time frame.

In an attempt to understand the log-log sedimentation rate: duration plot of Sadler (1981), Miall (1991, 2010, Chap. 13) undertook an analysis of the relationship between sedimentation rate and sedimentary process in the published record of shallow-marine stratigraphy. It emerged that there does, in fact, appear to be a natural hierarchy of process and preservation based on the natural time scales of sedimentary processes. The model which emerged from this looks remarkably like a fractal analysis, although it was generated entirely in the absence of any guidance from fractal theory (Fig. 8.4). Two cycles with frequencies in the million-year range are plotted on a chronostratigraphic scale (column MC), and successively broken down into components that reflect an increasingly fine scale of chronostratigraphic subdivision. The second column shows hundred-thousand-year cycles (HC), followed by depositional systems (DS) and individual lithosomes (L), such as channels, deltas, beaches, etc. At this scale chronostratigraphic subdivision is at the limit of line thickness, and is therefore generalized, but does not represent the limit of subdivision that should be indicated, based on the control of deposition

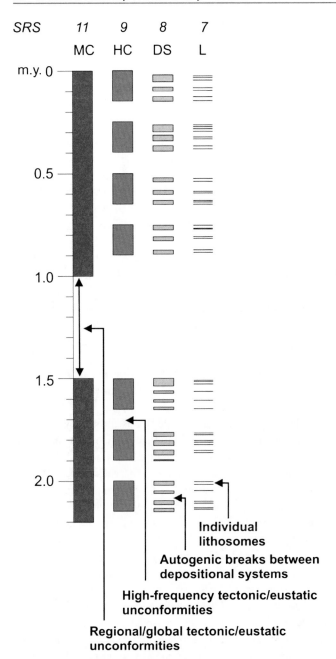

*SRS*    *11*    *9*    *8*    *7*

**Individual lithosomes**

**Autogenic breaks between depositional systems**

**High-frequency tectonic/eustatic unconformities**

**Regional/global tectonic/eustatic unconformities**

**Fig. 8.4** A demonstration of the predominance of missing time in the sedimentary record

by events of shorter duration and recurrence interval (e.g., infrequent hurricanes, seasonal dynamic events, etc.).

The record shows that when measurements of sedimentation rate are calculated at the appropriate scale, they are internally consistent and can be related to natural processes occurring within that time scale. Miall (1991) identified ten informal groupings of sediment packages, based on sedimentation rate. This classification, updated by Miall (2010,

Table 8.1; Miall 2015), is reproduced here as column 1 in Table 3.3. It constitutes a natural range of *Sedimentation Rate Scales* (SRS), now expanded to twelve groupings. Column 2 indicates the time scale of measurement, and column 3 provides the instantaneous sedimentation rate for deposits formed over that time scale. Stratigraphic and sedimentologic studies ranging from the micro scale to the regional, and based on time scales ranging from the short-term (e.g., studies of processes in laboratory models or modern settings) to the long-term (e.g., the evolution of major sedimentary basins), are best carried out at the appropriate *SRS*, much as photography uses lenses of different focal length, from macro to telephoto to wide-angle, to focus in on features at the desired scale.

The rest of Table 3.3 is discussed in the section "Accommodation and Preservation".

Many detailed chronostratigraphic compilations have shown that marine stratigraphic successions commonly consist of intervals of "continuous" section representing up to a few million years of sedimentation, separated by disconformities spanning a few hundred thousand years to more than one million years (e.g., MacLeod and Keller 1991, Fig. 15; Aubry 1991, Fig. 8.8). The first column of Fig. 8.4, labeled MC (for cycles in the million-year range), illustrates an example of such a succession. Detailed studies of such cycles demonstrate that only a fraction of elapsed time is represented by sediment. For example, Crampton et al. (2006) demonstrated that an average of 24 % of time is recorded in a suite of Upper Cretaceous sections in New Zealand, when measured at a $10^6$-year time scale, whereas in a suite of drill cores through the Lower Cretaceous to Miocene stratigraphic record of New Jersey, the plots of Browning et al. (2008) show that about 82 % of elapsed time is represented by sediments, although some sections are more complete than others. Each million-year cycle may be composed of a suite of high-frequency cycles, such as those in the hundred-thousand-year range, labeled HC in Fig. 8.4.

Chronostratigraphic analyses of many cyclic successions demonstrate that the hiatuses between the cycles represent as much or more missing time than is recorded by actual sediment (e.g., Ramsbottom 1979; Heckel 1986; Kamp and Turner 1990). Sedimentation rates calculated for such sequences (Table 3.3) confirm this, and the second column of Fig. 8.4 indicates a possible chronostratigraphic breakdown of the third-order cycles into component Milankovitch-band cycles (labelled HC), which may similarly represent incomplete preservation. For example, a detailed chronostratigraphic correlation of the coastal Wanganui Basin sequences in New Zealand ("5th-order Milankovitch cycles of *SRS 8* in Table 3.3) shows that at the $10^5$-year time scale only 47 % of elapsed time is represented by sediments (Kamp and Turner 1990). Each Milankovitch cycle consists of superimposed depositional systems (column DS) such as delta or barrier-strandplain complexes, and each of these, in turn, is made up of individual lithosomes (column L), including fluvial and tidal channels, beaches, delta lobes, etc.

Devine's (1991) lithostratigraphic and chronostratigraphic model of a typical marginal-marine sequence demonstrates the importance of missing time at the sequence boundary (his subaerial hiatus). Shorter breaks in his model, such as the estuarine scours, correspond to breaks between depositional systems (the DS column in Fig. 8.4), but it is suggested that more are present in such a succession than Devine (1991) has indicated. Additional discontinuities at the lithosome level (L in Fig. 8.4) correspond to the types of breaks in the record introduced by switches in depositional systems, channel avulsions, storms and hurricanes, etc.

According to the hierarchical breakdown of Table 3.3, the four columns in Fig. 8.4 correspond to sediment *SRSs 11, 9, 8, and 7*, in order from left to right. In each case, moving (from left to right) to a smaller scale of depositional unit focuses attention on a finer scale of depositional subdivision, including contained discontinuities. The evidence clearly confirms Ager's (1973, 1981) assertion that the sedimentary record consists of "more gap than record".

Fluvial deposits have long been known to consist of a hierarchy of depositional units accumulated over a wide range of time scales and sedimentation rates (Miall 1978a). An architectural approach to field studies was introduced by Allen (1983), developed by Miall (1988a, b, 1996) and has now been fully documented in a number of detailed field studies (e.g., Holbrook 2001). However, owing to the difficulty of dating nonmarine deposits, assigning ages and calculating sedimentation rates for these deposits is usually not possible.

## 8.4   Sedimentation Rates

Examples of measured and calculated sedimentation rate from modern and ancient environments provide the basis for the ranges of values indicated in column 3 in Table 3.3 and in Fig. 8.5.

*SRS 2:* small-scale ripples typically migrate a distance equivalent to their own wavelength in 20–60 min (Southard et al. 1980). A 5-cm-high ripple that forms in 30 min is equivalent to an instantaneous sedimentation rate of 876,000 m/ka. Clearly, this number is meaningless, but it will serve to emphasize the extremes of sedimentation rate, to compare with more geologically typical rates discussed later.

*SRS 3:* Tidal sand waves have similarly very high instantaneous rates. In the Bay of Fundy Dalrymple (1984) demonstrated that in one tidal cycle sand waves migrate a distance about equivalent to their average height, which is 0.8 m. Bay of Fundy tides are semi-diurnal, and so this migration is equivalent to a sedimentation rate of 584,000 m/ka.

*SRS 5:* The deposits formed by seasonal or more irregular runoff events have extremely variable instantaneous sedimentation rates. The flood deposit in Bijou Creek, Colorado, described by McKee et al. (1967) was formed by the most violent flood in 30 years. It formed 1–4 m of sediment in about 12 h, an instantaneous sedimentation rate of 730,000–2,920,000 m/ka. Assuming no erosion, and a repeat of such floods every 30 years this translates into a rate of 33–133 m/ka averaged over a few hundred years. In fact, scour depths during the flood ranged from 1.5 to 3 m, and true net preservation of any one flood deposit over periods of hundreds or thousands of years may be negligible. Long-term rates measured over hundreds to thousands of years are likely to be an order of magnitude less, in the range of $10^{-1}$ m/ka. Leclair (2011) demonstrated that the large-scale dunes that may characterize seasonal to longer-term floods do not necessarily have a higher preservation potential than the deposits formed during non-peak flood periods.

In the Rio Grande valley in Texas, Dean et al. (2011) used tree-ring studies to establish the recent sedimentation history. Over a 13-year period they determined that overbank floods occurred with a recurrence interval of 1.5–7 years, at accretion rates of 16–35 cm/year, an instantaneous sedimentation rate of $10^2$ m/ka.

*SRS 6:* The point bars in the Wabash River are on average about 5 m thick. The active area of each bar is about 200 m wide. If the bars migrate at a maximum rate of 2 km in 50 years (Jackson 1976) they would take 5 years to migrate one point-bar width, which is equivalent to an instantaneous sedimentation rate of 1,000 m/ka.

Comparable rates may be calculated from the migration of distributary mouth bars. The Southwest Pass of the

**Fig. 8.5** Rates and durations of sedimentary processes. Numerals refer to the Sedimentation Rate Scale (see also Table 3.3)

Mississippi delta migrated a distance of 9 km in 100 years (Gould 1970, Fig. 20). The mouth-bar deposits, from the mouth of the channel to the toe of the distal bar, are about 4 km wide (in a dip direction) and about 70 m thick. This lateral migration is equivalent to an instantaneous sedimentation rate of 730 m/ka. Oomkens (1970) quoted sedimentation rates of 35 cm/year (350 m/ka) for the delta front of the modem Rhone River. A short-term rate of 4.4 cm/month (528 m/ka) was determined close to the mouth of the Yangtze River, China, by McKee et al. (1983), who studied the decay of short-lived radionuclides in the uppermost 15 cm of recent deposits, representing about 100 days of accumulation (*SRS 5*). The uppermost 200 cm of section, representing about 100 years of accumulation (*SRS 6*), yielded a rate an order of magnitude lower, 5.4 cm/a (54 m/ka).

Data from modem Dutch tidal deposits summarized by Yang and Nio (1989) showed that ebb-tide deltas accumulate at rates of 100 to 450 m/ka for periods of about 20 years, before abandonment occurs. Van den Bergh et al. (2007) used $^{210}$Pb methods, which provide age information on the 100-a scale (*SRS 6*), to assess sedimentation rates on the prodelta of the Red River off Vietnam. They range from high values of 330 to 940 m/ka on the proximal prodelta slope, to less than 10 m/ka on the distal margin. The distal sedimentation rate is at the lower limit for *SRS 6* sedimentation rates when measured at the appropriate 100-year scale.

Rates calculated for the upper Bengal submarine fan, the world's largest and most active depositional system are consistent with the *SRS 5-6* time scales. Using $^{14}$C age determinations for the last 10,000 years of sedimentation, Weber et al. (1997, p. 317) calculated that "on the shelf, sedimentation rates are currently extremely high in the foreset region of the recent delta (as much as 8 cm/a)" where water depth is 30–70 m, "and especially in the head of the canyon" where sedimentation rates reach as much as 1 m/a ($10^3$ m/ka). A sedimentation rate of 1 m/a places this system in the range of *SRS 5* and *6*, for which such sedimentation rates occur within time spans of $10^0$–$10^3$ years. In this case the preservation machine at work is the removal of sediment by slope failure and slumping, with huge volumes of sediment moving downslope as sediment-gravity flows, Weber et al. (1997, p. 317) noted that "This sediment load would fill the entire canyon in less than 1000 years. Therefore, we conclude that, because of the steep gradients at the head of the canyon, frequent slumping and formation of turbidity currents occur even during the current sea-level highstand." The frequency of turbidite occurrence ranges between 500 and 10,000 years (Stow et al. 1983, p. 58), which is well within the $10^0$–$10^3$-year time range.

*SRS 7:* Most studies of post-glacial sedimentation are carried out on a "long-term" ($10^3$–$10^4$-year) time scale that is assigned to *SRS 7*. Rates of post-glacial sea-level rise (and accommodation generation) reached as high as 18 m/ka (e.g., East China Sea: Wellner and Bartek 2003), although rates of 1–6 m/ka were more characteristic.

In the coastal river valleys of Texas, Blum (1993) demonstrated that late Pleistocene-Holocene cycles of degradation and aggradation depended on climatically controlled variations in discharge and sediment supply, not on sea-level change. Humid periods corresponded to episodes of aggradation. During two such periods, lasting 6 ka, as much as 10 m of valley-fill sediment accumulated in the upper Colorado drainage, indicating sedimentation rates of 1.7 m/ka. Long-term ($10^4$-a) floodplain aggradation rates reported by Bridge and Leeder (1979) ranged between 0.035 and 0.2 cm/a (0.35–2 m/ka).

Turning to the deltaic environment, if we assume that a distributary will only build out across a given area of the delta front once during the migration of one major delta lobe, we can calculate the sedimentation rate of the mouth bar averaged over the life of the lobe. In the post-glacial Mississippi delta, major lobes are formed and abandoned in about 1000 years (Kolb and Van Lopik 1966; Frazier 1967), giving an average sedimentation rate for that period of 70 m/ka. Lobe sedimentation took place during the post-glacial period of sea-level rise.

The recurrence interval of delta lobes themselves depends on subsidence rates, sediment supply and the configuration of the continental shelf. In the case of the Mississippi complex the river is attempting to switch discharge (and delta construction) to the Atchafalaya River, where one of the earliest lobes developed about 6–8 ka BP. On this scale deposition of a 45-m-thick mouth-bar deposit represents a sedimentation rate of 5.6–7.5 m/ka, although this calculation does not take into account the bay-fill and other facies interbedded with the mouth bar deposit. This compares with typical values for Holocene sedimentation rates of 6–12 m/ka that are commonly quoted for the Mississippi delta complex (e.g., Weimer 1970), and the Rhone delta, which has a thickness of only 50 m, accumulated since about 8.2 ka, indicating an average sedimentation rate of 6.1 m/ka (Oomkens 1970).

Dating of peat layers and other units in the Rhine-Meuse delta indicated that aggradation of ribbon-like (anastomosed) channel belts kept pace with the average sea-level rise of 1.5 mm/a, at sedimentation rates in the order of $10^0$ m/ka (Blum and Törnqvist 2000). Highly detailed studies of the development of the Rhine-Meuse system using multiple $^{14}$C dates have provided a wealth of detail regarding the aggradation history (Stouthamer et al. 2011). Regional flood basin accumulation rates for the upper delta are in the range of 0.3–1 mm/a (0.3–1 m/ka). Some channel belts indicate local rates as high as 2.8 mm/a (2.8 m/ka). The ratio of local to regional aggradation rate ranges between 0.4 and 4.0, but all these values, measured over time periods of $10^3$–$10^4$ years, are within the *SRS 7* range.

A few sedimentation rates can be calculated for tidal-inlet and barrier deposits. The Galveston Island barrier is 12 m thick and 3.5 ka old at its base (Bernard et al. 1962, Fig. 8.60), indicating an average sedimentation rate of 3.4 m/ka. A tidal inlet at Fire Island, New York has migrated 8 km in 115 years (Kumar and Sanders 1974). The depositional slope from spit crest to channel floor is about 500 m wide, suggesting that at any one point the entire tidal-inlet fill could form by lateral accretion in about 7 years. The sequence is 12 m thick, indicating an instantaneous sedimentation rate of 1,714 m/ka. This migration rate is unusually rapid. Tidal inlets at Sapelo Island, Georgia appear to have migrated only about 2.5 km since the post-glacial sea-level rise (Hoyt and Henry 1967, Fig. 8.6c), indicating a sedimentation rate of 4.5 m/ka. Sommerfield (2006) calculated accumulation rates and stratigraphic completeness for modern oceanic continental margins. His research was based on measures of mass per unit area with time, and translates into sedimentation rates in the *SRS 7–8* range.

Coal seams are estimated to represent 4,000–12,000 years of peat accumulation, at accumulation rates of 1–3 m/ka (Nemec 1988; Phillips and Bustin 1996). Allowing for a gradual 3:1 compaction during accumulation, Nemec (1988, p. 163) calculated an accommodation generation rate in a mire of 0.4–1.1 m/ka, which, as he noted, is one to two orders of magnitude greater than the long-term subsidence typical of the cratonic basins he studied (Illinois, South Wales, SW Poland).

Rates of sedimentation on modern alluvial fans and fluvial floodplains have been measured using $^{14}C$ dates on plant material, and tephrochronology. Available data were summarized by Miall (1978b) and shown to encompass a wide range, from 0.08 to 50 m/ka. However, these measurements have not been correlated to specific scales of architectural units, such as the depositional groups defined here.

Plio-Pleistocene slope and basin deposits in the Gulf of Mexico are characterized by sedimentation rates ranging from 0.16 to 6.45 m/ka ($10^1$–$10^0$ m/ka). The lower values are from areas characterized by slow subsidence rates and a high proportion of hemipelagic sedimentation; the high values were derived from areas with a high proportion of sediment-gravity-flow deposits and accumulation in salt-withdrawal basins. These values were calculated by Fiduk and Behrens (1993) for tectonostratigraphic sequences representing between 0.3 and 1.25 Ma. These are unusually rapid rates of accumulation, the rates corresponding to *SRS 7* or *8*, over time scales of *SRS 9*, confirming the unique setting of the Gulf Coast basins or, alternatively, suggesting that the geological preservation machine (see below) may not yet have completed its work.

Scarponi et al. (2013) used amino-acid racemization data to demonstrate varying accumulation rates at the *SRS 7-8* scale, within the topmost 100-ka Holocene coastal sequence of the Po river coastal plain, Italy. Lowest rates (0.2–

0.7 m/ka) were recorded within the nonmarine lowstand systems tract, increasing to 1.4–2.5 m/ka in the transgressive systems tract and to as much as 10 m/ka in the highstand. This variability was interpreted as a response to the systematic change in the probability of preservation of depositional events as accommodation increased with rising base-level, versus changing sediment supply in the shallow-marine realm as the depositional systems shifted from retrogradational to progradational. Higher values of net accumulation rate were recorded in the transgressive and highstand systems tracts, and lower values (with concomitantly more diastems) in the condensed section formed around the time of the maximum flooding surface. The highest values, up to 19.6 m/ka were recorded in homogeneous sandy sediment that may incorporate fewer diastems, in accord with Sadler's (1981) analysis that high accumulation rates scale inversely with time span.

The compilation of sedimentation rates in shallow-marine carbonate sediments by Kemp and Sadler (2014) indicated that when normalized to a 5-ka time scale—the "the typical duration of Holocene and Pleistocene sections lacking major exposure surfaces" (Kemp and Sadler 2014, p. 1290) maximum accumulation rates range between 10 and 60 m/ka, depending on latitude, with rates dropping off in the higher latitudes.

*SRS 8 and 9*: At time scales of $10^4$–$10^5$ years, measured sedimentation rates are in the range of 0.1–1.0 m/ka (*SRS 8*). At time scales of $10^5$–$10^6$ years, rates are 0.01–0.1 m/ka (*SRS 9*).

Stratigraphic studies based on magnetostratigraphic dating and correlation, and studies of Late Cenozoic deposits using such techniques as high-resolution reflection-seismic data provide the appropriate focus. Cyclic successions developed by orbital forcing typically fall into this category. Examples of such deposits include the Quaternary shelf-margin sequences of Suter et al. (1987), the classic 41-ka cycles of the Wanganui Basin, New Zealand (Pillans et al. 2005), and the minor and major cyclothems of Heckel (1986).

The sequences described from the Gulf Coast by Suter et al. (1987) averaged 25,000 years in duration and range in thickness from about 25 to 160 m, indicating average accumulation rates of 1–6.4 m/ka. Heckel (1986) documented the chronology of 55 cycles of Westphalian-Stephanian age in the U.S. Midcontinent. Estimates of the length of this time span range from 8 to 12 Ma The thickness of the succession varies from 260 m in Iowa to 550 m in Kansas. These values indicate an average accumulation rate of between 0.02 and 0.07 m/ka (*SRS 12*). Many of the cycles contain substantial fluvial-deltaic sandstone units and, according to Ramsbottom (1979), who studied similar cyclothems in Europe, rates of lateral deltaic growth must have been about as rapid as that of the modern

Mississippi; yet the average sedimentation (vertical aggradation) rate is two orders of magnitude less than that of the Holocene Mississippi delta complex and its Pleistocene shelf-margin precursors on the Louisiana Gulf Coast. Part of the explanation for this marked contrast is that the Carboniferous cyclothems that were the subject of Heckel's study are located in a cratonic region, where subsidence rates would be expected to be substantially lower than on the continental margin of the Gulf Coast. As demonstrated by Runkel et al. (2007, 2008) cratonic sequences may develop by very low-angle lateral accretion.

Oxygen isotope and magnetostratigraphic data for the Wanganui Basin sections confirm the predominance of the 41-ka orbital cycle. In several composite sections it can be demonstrated that about 1 km of section accumulated in 1.2 Ma between the Olduvai and Brunhes paleomagnetic stages, indicating an average sedimentation rate of 0.8 m/ka (Pillans et al. 2005, Fig. 11).

A well-known ancient example of interpreted orbitally-forced cyclic sedimentation comprises the Newark-type lacustrine cycles of eastern North America. Olsen (1990), reconstructed characteristic orbital cyclic frequencies based on available chronostratigraphic information, and this yielded average sedimentation rates of 0.27 m/ka at the $10^4$–$10^5$-year time scale.

A magnetostratigraphic study of the Siwalik fluvial deposits of Pakistan provides some control data for an ancient fluvial system (Johnson et al. 1985, 1988). In the Miocene Chinji formation, 400–500 m thick, and deposited over a time span of approximately 3.5 Ma, fluvial cycles representing channel belts up to several kilometres wide (Johnson et al. 1988, Fig. 9.3) range from 12 to 50 m in thickness. At an average sedimentation rate of 0.12 m/ka, these represent cycle return periods of between $10^4$ and $10^5$ years. Detailed studies reported in their second paper indicate considerable variation in local sedimentation rate, with evidence that specific magnetostratigraphic reversals are missing, indicating gaps in sedimentation on a $10^4$-year scale. The formation as a whole corresponds to a *SRS-9* assemblage, but it is likely that the fluvial cycles, containing missing intervals, represent channel belts of *SRS-8*.

Jones et al. (2004) explored sedimentation rates and sediment transport rates in a foreland basin in Spain. The long-term sedimentation rate for their complete section averaged 0.075 m/ka over 12 Ma (*SRS-9*). Magnetostratigraphic dating of short intervals within this section indicated sedimentation rates varying between 0.03 and 0.2 m/ka (*SRS 8-9*), with much of the local variability being attributed to syndepositional folding affecting accommodation rates.

Fluvial cycles representing similar long-term avulsion processes were described by Hofmann et al. (2011) from the Cretaceous Piceance Basin of Colorado. The cycles average 120 m in thickness and are estimated to represent about 400 ka, accumulating at an average sedimentation rate of 0.305 m/ka (*SRS 8*). Clusters of channels develop an alluvial belt, the topographic elevation of which eventually leads to avulsive switching to lower areas on the floodplain, a process termed *compensational stacking*. Magnetostratigraphic studies of the Eocene Escanilla Formation, a braided-stream deposit, in the Spanish Pyrenees yield similar *SRS* values, at 0.17–0.57 m/ka over time periods of $10^5$ years (Bentham et al. 1993).

Cores through interpreted precessional cycles of Cretaceous age on the floor of the Atlantic Ocean off tropical west Africa (66 22-ka cycles totalling 37 m of core: Beckmann et al. 2005) yield sedimentation rates of 0.025 m/ka. This is an order of magnitude slower than the *SRS-8* range characteristic of high-frequency orbital cycles.

The long-term sedimentation rates of *SRSs 8* to *12* depend largely on long-term rates of generation of sedimentary-accommodation space. This depends both on basin subsidence, which is controlled by tectonic setting, and by changes in base level, such as eustasy. Miall (1978b) showed that most nonmarine basins, in various tectonic settings, have sedimentation rates averaged over millions of years of 0.03–1.5 m/ka.

*SRS 10:* Basins in convergent margins are provided their own category in Table 3.3 and Fig. 8.5, because of the exceptionally high rates of subsidence and sedimentation that have been recorded in this tectonic setting. Miall (2010, Table 8.2, pp. 280–281) summarized data from settings such as the Banda Arc, the Himalayan foreland basin, the Cretaceous forearc basin of Baja California and a forearc basin in Japan where sedimentation rates of $10^{-1}$–$10^0$ m/ka have been measured over durations in the order of $10^6$ years. Magnetostratigraphic calibration of several sections in Andean foreland basin strata of Argentina indicated sedimentation rates ranging between 0.22 and 1.71 m/ka over intervals ranging between approximately 0.5 and 5 Ma (Echavarria et al. 2003). Hiatuses lasting up to 2 Ma, bring the average sedimentation rates, measured over total sections representing between 7 and 13.5 Ma, down to between 0.183 and 0.571 m/ka.

Growth strata that develop adjacent to active structures, such as basin-margin thrust faults are typically deposited at *SRS-10* rates. Burbank et al. (1996, Fig. 8.8) provided an example where accumulation rates averaged 0.117 m/ka over 1.7 Ma Data provided by Medwedeff (1989) indicate a growth rate of 0.305 m/ka over 8 Ma At the margins of the Tarim Basin, in western China, Sun et al. (2010) used magnetostratigraphic data to determine the rates of accumulation of "growth strata" in proximity to a growing anticline. Sedimentation rates increased from 0.325 m/ka prior to the syndepositional movement of the growth structure, to 0.403 m/ka during the period of active tectonism, over a total time span of about 10 Ma.

Rapid subsidence is indicated in basins developed along the San Andreas Fault system. Dorsey et al. (2011) calculated subsidence rates of between 0.4 and 2.1 mm/a ($10^{-1}$–$10^{0}$ m/ka) measured over periods of a few million years.

Not all convergent-margin basins are characterized by high sedimentation rates. In deep-water basins, where sediment supply is low, sedimentation rates may be much less. Finney et al. (1996) compiled data indicating that Paleozoic graptolitic shales in the Taconic foreland basin of the southeast USA accumulate at average rates of 0.01–0.03 m/ka measured over time periods of a few million years (comparable to *SRS 9*).

*SRS 11:* This is the rate characteristic of long-term geological processes. Aschoff and Steel (2011) calculated sedimentation rates for the Upper Cretaceous clastic wedge of the Book Cliffs (Utah-Colorado) in order to explore relationships between sedimentation and tectonism in the Sevier foreland basin. The range of rates is 0.047–0.14 m/ka, calculated over stratigraphic times spans of between 2.1 and 6.5 Ma These are within the range for *SRS 11*, but are low relative to those recorded in some Andean basins (*SRS-10*, above). The Catskill Delta of New York-Pennsylvania accumulated at comparable rates. Data provided by Ettensohn (2008) indicate a maximum rate for the proximal part of the "delta" (in reality a major clastic wedge deposited in a range of nonmarine to shallow-marine environments) of 0.096 m/ka (maximum thickness of 3 km accumulated over about 9 Ma between the Givetian and the Famennian).

Aschoff and Steel (2011) speculated about the possible influence of basement uplift within the Sevier foreland basin, which would tend to cancel out some of the subsidence due to flexural loading. This seems particularly likely for the middle portion of the clastic wedge, that characterized by the Castlegate sandstone, the sheet-like nature of which has, for some time, been attributed to a slow rate of regional subsidence (Yoshida et al. 1996). The incipient activation of Laramide structures within the basin, as suggested by Aschoff and Steel (2011), would be consistent with these characteristics of the clastic wedge. However, no such special influence on rates of accommodation has been suggested for the Appalachian basin.

*SRS 12:* Long-term sedimentation rates in cratonic environments include the lowest that have been recorded. Runkel et al. (2008) demonstrated that the deposits flanking the Transcontinental Arch in Wisconsin-Minnesota accumulated at an average rate of 0.4 m/ka, measured over a 15-Ma Upper Cambrian depositional record. Detailed biostratigraphic studies indicate that that the section is relatively complete, and accumulated as a series of thin, offlapping shingles. The thickness of each shingle, as indicated by the biostratigraphic zonation, varies from 50 to as little as 5 m, indicating an order of magnitude variation in sedimentation rate within this average rate. The Lower Mannville

formation of southern Alberta consists of a series of high-frequency sequences with multiple phases of incised valley cut-and-fill. Long-term sedimentation rates range between 0.0013 and 0.02 m/ka (Zaitlin et al. 2002).

## 8.5   The Fractal-Like Character of Sedimentary Accumulation

The log-linear relationship between sedimentation rates and time span was addressed by Plotnick (1986), Middleton et al. (1995), and Sadler (1999), who demonstrated that it could be interpreted using the fractal "Cantor bar" model of Mandelbrot (1983). Using the process described by Mandlebrot (1983) as "curdling", Plotnick (1986, p. 885) developed a Cantor bar for a hypothetical stratigraphic section by successively emplacing hiatuses within portions of the section, at ever increasing levels of detail (the result is illustrated in Fig. 8.6):

> Assume a sedimentary pile 1000 m thick, deposited over a total interval of 1,000,000 years. The measured sedimentation rate for the entire pile is, therefore, 1 m/1000 years. Now assume that a recognizable hiatus exists exactly in the middle of the section, corresponding to a third of the total time (i.e., 333,333 years). The subpiles above and below the hiatus each contain 500 m of sediment, each deposited over 333,333 years, so that the measured sedimentation rate for each subpile is 1.5 m/1000 years. We now repeat the process, introducing hiatuses of 111,111 years in each of the two subpiles. This produces 4 subpiles, each 250 m thick, each with a duration of 111,111 years. The measured sedimentation rate is now 2.25 m/1000 years. The process can be reiterated endlessly, producing subpiles representing progressively shorter periods of time with higher sedimentation rates [Fig. 8.6 of this book]. Nevertheless, because the total sediment thickness is conserved at each step, the sedimentation rate of the entire pile remains 1 m/1000 years.

The selection of one third as the length of the hiatus, or gap (g in Fig. 8.6) is arbitrary. Other gap lengths generate Cantor bars that differ only in detail. Figure 8.6 is remarkably similar to Fig. 8.4, which was constructed by Miall

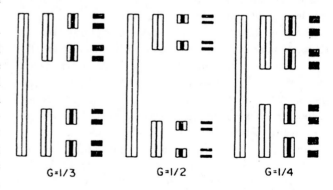

G=1/3                    G=1/2                    G=1/4

**Fig. 8.6** Cantor bars generated using three different gap sizes (G) (Plotnick 1986, Fig. 1)

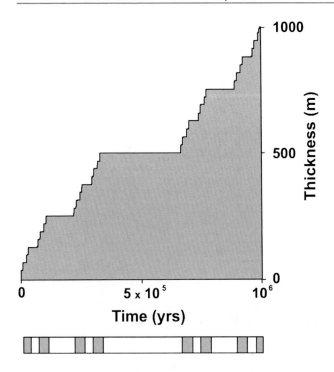

**Fig. 8.7** Stratigraphic thickness accumulation viewed as a Cantor function—what has been termed a Devils's staircase—constructed with G = 1/2. The corresponding *Cantor bar* is shown at a lower resolution *below* the graph (adapted from Plotnick 1986, Fig. 8.7)

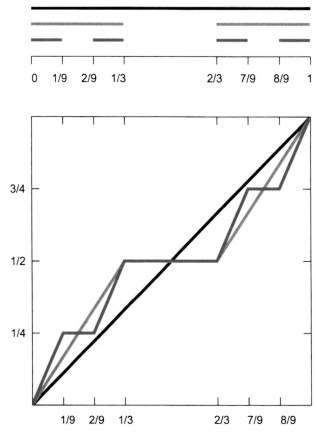

**Fig. 8.8** A fractal plot of elapsed time (X-axis) versus sediment thickness (Y-axis) showing how the choice of gap length determines the distribution of hiatuses and sedimentation rate (from Schlager 2005)

(1997) based on the hierarchies of sedimentation rates compiled by Miall (1991), but with no knowledge of fractals.

The dependence of sedimentary accumulation on the availability of accommodation was understood by Barrell (1917; Fig. 5.2 of this book) and is the basis of modern sequence stratigraphy (Van Wagoner et al. 1990). The fractal model provides an elegant basis for integrating this knowledge with the data on varying sedimentation rates and varying scales of hiatuses discussed in the paragraphs above. Mandelbrot (1983) and Plotnick (1986) provided a version of an accumulation graph, called a Devil's staircase (Fig. 8.7). This shows how sediments accumulate as a series of clusters of varying lengths. Vertical increments of the graph correspond to intervals of sedimentation; horizontal plateaus represent periods of non-accumulation (or sedimentation removed by erosion). Sequence stratigraphy is essentially a study of the repetitive cycle of accumulation followed by the next gap, at various scales. The larger, more obvious gaps (the longer plateaus in Fig. 8.7) define for us the major sequences, over a range of time scales. The prominence of particular ranges of "accumulation + gap" length in the first data sets compiled by Vail et al. (1977) was what led to the establishment of the sequence hierarchy of first-, second-order, and so on. That this has now been shown to be an incomplete representation of nature (Miall

1997, 2010; Schlager 2005) does not alter the fact that there is a limited range of processes that control accumulation, and these have fairly well defined rates which, nevertheless, overlap in time to some extent.

As Plotnick (1986, Table 1) and Sadler (1999) demonstrated, the incompleteness of the stratigraphic record depends on the scale at which that record is examined. Sections spanning several million years may only represent as little as 10 % of elapsed time at the 1000-year measurement scale, although this is below the resolution normally obtainable in geological data. Sequences, as we know them, each essentially consist of clusters of the shorter "accumulation + gap" intervals separated by the longer gaps—those more readily recognizable from geological data.

Schlager (2005) illustrated the relationship between sedimentation date and time span in a different way (Fig. 8.8). This diagram highlights the central point that the length and frequency of the gaps determines the calculated sedimentation rate. Almost all actual geological data sets yield correlation lines similar to that indicated by the black line of correlation. This line suggests a relatively slow rate of

**Fig. 8.9** Drill core from the modern Guadalquivir foreland basin in SW Spain. This is the Lebrija borehole, drilled from the surface of the modern, active floodplain, down through post-glacial estuarine deposits and into the older stratigraphic record. Sedimentation rates calculated from paleomagnetic calibration of the section, are indicated by the values, in cm/ka, along the diagonal line of correlation. Adapted from Salvaney et al. (2011), with permission

## 8.6 Apparent Anomalies of High Sedimentation Rate Versus Slow Rate of Accommodation Generation

Bailey (2011) highlighted the preservation of fossil tree trunks in some coal-bearing strata as examples of the apparent dilemma posed by what appear to be exceptional modes of stratigraphic preservation. Tree trunks would be expected to decay rapidly, probably within decades (But were decay rates this rapid in the Carboniferous? Was there the same range of microorganisms that we observe at the present?), so the preservation of tree trunks up to 12 m high, in good condition, appears to argue for an unusually rapid rate of burial. Bailey (2011) suggested a rate of ~100 m/ka, well in excess of the 0.005–0.1 m/ka rates of accommodation generation indicated by the setting of the fossil trees

sedimentation with few, widely dispersed hiatuses. Such an interpretation inevitably follows from (1) the limited ability of geological methods to provide numerous tightly constrained age dates, and (2) the cryptic nature of most sedimentary hiatuses. Reality might be much closer to one or other of the red or orange correlation lines, which incorporate closer spacings of hiatuses and higher short-term sedimentation rates.

within orbital cycles accumulating within tectonically active basins (the *SRS-9* time scale). The trees would seem to qualify as "frozen accidents", to use Bailey and Smith's (2010) term. But how unusual is this? And does it require a special explanation, as Bailey (2011) suggested? He argued for episodic, rapid (indeed, instantaneous) seismogenic subsidence to create the necessary accommodation.

Another example of an apparent stratigraphic puzzle is shown in Fig. 8.9. This is a carefully calibrated and dated stratigraphic record from the Guadalquivir foreland basin in SW Spain. High-precision dating for this Pliocene to Recent succession has been provided by the magnetostratigraphic record. For the purpose of this discussion, the interesting point is the increase in sedimentation rate following a hiatus at 1.6 Ma (calculated sedimentation rates are indicated next to the line of correlation). Prior to the hiatus, sedimentation rates were in the order of $10^{-2}$ m/ka (*SRS 9*). After the break they rose to 0.5 m/ka ($10^{-1}$ m/ka: *SRS 8*) and then to 3.5 cm/ka ($10^{0}$ m/ka: *SRS 7*). Why? What happened?

Surface sediments in this basin are not yet as compacted as in the ancient record, but this is unlikely to account for more than a few percent of the total thickness of the post-1.6 Ma section. I suggest that what we are seeing is typical sediment accumulations that have yet to be completely processed by the geological preservation machine.

The post-11.5 ka section represents sedimentation in accommodation generated by the post-glacial sea-level rise.

The calculated sedimentation rate is entirely in accord with sedimentation rates calculated over a 1000-year time scale (*SRS 7*). Future events could include a fall in sea level over a $10^4$-year period, if the orbital cycles that have characterized Earth history for the last 2.5 Ma continue. That could potentially remove most or all of the top 95 m of the sedimentary record, as rivers grade themselves to the lower base level, thereby completing the work of the geological preservation machine at the *SRS 8* or *9* time scale. It is suggested that the pre-1.6 Ma section consists of short intervals of stratigraphy which accumulated at rates comparable to that calculated for the top of the section, separated from each other by numerous unrecognized hiatuses, a pattern comparable to the red line of correlation in Fig. 8.8. Sadler (1999) had explained a similar pattern of apparently accelerating accumulation in younger sediments, based on his study of sedimentation rates and time scales.

Whereas most published stratigraphic data sets contain the generalization exemplified by the black line of correlation in Fig. 8.8, the reality for the top of the Guadalquivir foreland basin section (Fig. 8.9), as measured at appropriate times scales, would be closer to the red line of correlation in this diagram, where short intervals of time characterized by high rates of sedimentation are separated from each other by hiatuses. For example, this interpretation can explain the intervals of rapid sedimentation (*SRS 5* or *6*) required to preserve the tree trunks described by Bailey (2011). The rapid rate can then be seen as part of a predictable spectrum of sedimentation rates, when measured at the appropriate time scale. As noted above, coal seams represent *SRS-7* deposits. Valley fills and deltas lobes, where many coal seams (and fossil trees) accumulate, are *SRS-7* and *8* deposits.

In a specific attempt to "disentangle time" in the preserved rock record, de Natris (2012) and de Natris and Helland-Hansen (2012) used rates of sedimentation derived from modern shallow-marine environments to calculate elapsed time in the Tarbert Formation (Mid-Late Jurassic) of the northern North Sea. They applied these rates to facies successions interpreted to have been deposited in these environments. A summation of rate versus thickness for each facies explained only 7 % of the 2.8 Ma elapsed time span of the Tarbert Formation as measured at geological time scales (*SRS 9* and higher). However, these calculations were carried out using rates in the *SRS 6-7* range, and therefore did not account for the longer-term events recorded in the system.

Scott and Stephens (2015) addressed the issue of missing time through detailed calculations of sedimentation rates of Carboniferous coal-bearing successions in Britain. They cited sedimentation rates of $10^0$–$10^{-2}$ m/ka, based on studies of estuarine and deltaic environments. These are equivalent to SRS 7, 8 or 9. Their results indicate a representation of

elapsed time ranging between 4.9 and 33 %, depending on the interval selected and such factors as corrections for compaction of the peat to coal.

## 8.7   Accommodation and Preservation

The general question arises from the cases discussed in the preceding section: how could sediments accumulate at rates an order of magnitude or more greater than the local rate of accommodation generation (these processes are summarized in columns 4, 5 and 6 of Table 3.3). Blum and Törnqvist (2000, p. 20) noted:

> It … seems that accommodation, as it is commonly used, somewhat imprecisely mixes processes that operate over a range of rates and temporal scales; it is difficult to reconcile the time-scales over which sediments are deposited in the first place, whether or not those deposits will be preserved in the stratigraphic record, and the manner in which ancient alluvial successions are interpreted in terms of changes in accommodation or an accommodation/sediment supply ratio.

Six important points help to explain the types of apparent anomalies described above:

- While accommodation is typically quantified in terms of vertical space relative to sea level (base level) and the rate at which it is created or removed, many important sedimentary processes are dominated by lateral sedimentary accretion. Sediments accumulate on mid-channel bars and on meander bends by lateral (cross-channel) and downstream accretion. Deltas and continental margins accumulate by oceanward progradation.

- Fluvial, tidal, and other channels, and valleys, ranging up in scale to major incised valley systems, are locations where accommodation is not controlled by base level but are best understood with reference to the buffer concept of Holbrook et al. (2006). Accommodation generation on geomorphic time scales is therefore not dependent on tectonic subsidence rates and may be substantially higher (at the appropriate *SRS*).

- In two other major settings, accommodation is not restricted by base level: Deposition landward of the shoreline and in inland nonmarine basins is constrained by depositional slopes that are dependent on upstream controls, such as rates of tectonic uplift, river discharge and sediment load (Holbrook et al. 2006). Also, deep marine sediments are not in any way constrained by rates of accommodation generation, but are largely dependent on sediment supply and slope.

- Allogenic and autogenic sedimentary processes may generate predictable, ordered stratigraphic patterns at all time scales. The order and predictability may include erosional processes as well as processes of accumulation. This has always been the basis for Walther's Law and, more recently, sequence stratigraphy. Therefore, contrary to the random or chaotic processes of accumulation implied by Bailey and Smith (2010) stratigraphic order, including cyclicity, may be preserved in the rock record and may be understood and interpreted within the focus of the appropriate *SRS*.

- Although sediment preservation is extremely discontinuous and spasmodic at any one location, the shifting locus of accumulation (aggrading channels, delta lobes, prograding clinoforms, etc.) means that substantially more elapsed time is represented by preserved sediment in three dimensions than the percentages relating to vertical accumulation noted earlier in this section. At intermediate time scales (*SRS* 5-8) (and in the experiments of Sheets et al. 2002), sedimentation is continuous for lengthy periods of time, but distributed across an entire depositional system. Our tools for reconstructing these processes in the ancient record are quite limited.

- Actual sedimentation rates in most geological settings are always likely to be much higher—typically orders of magnitude higher—than those calculated from the rock record, based on observable geological data, such as extrapolations from datable ash beds or biohorizons, or rates based on regional rates of accommodation generation. There is no conflict between the rapid sedimentation that can commonly be observed in modern settings, and the rates that prevailed in the past. In that limited sense, traditional uniformitarianism ("the present is the key to the past") is correct, but with the additional proviso that analyses of the past must take into account the ubiquitous hiatuses, many quite cryptic, that occur at all time scales.

There is not necessarily a simple relationship between forcing processes and a stratigraphic result. Many surface processes are characterized by a **geomorphic threshold**, whereby a steady or unsteady process may operate without significant effect until a particular critical level is reached, whereupon sudden and dramatic change may take place (Schumm 1973). Theoretical and experimental studies of nonmarine and coastal environments have confirmed the non-linear relationship between such forcing functions as tectonism and climate change and the resulting effects on erosion patterns and sediment delivery (e.g., Allen and Densmore 2000; Kim and Paola 2007; Allen 2008). In fact, Jerolmack and Paola (2010) refer to the "shredding of environmental signals by sediment transport". Allen (2008, p. 20) suggested that "Large alluvial systems with extensive floodplains should therefore strongly buffer any variations in sediment supply with frequencies of less than $10^5$–$10^6$ years. This has strong implications for the detection of high-frequency

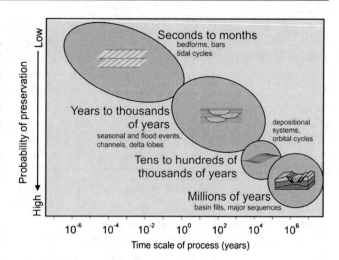

**Fig. 8.10** The geological preservation machine (adapted from Fig. 8.5). Geological processes are grouped for the purposes of discussion into four major groups, based on sedimentation rates and durations. The preservation of any unit depends initially on processes operating at a given time scale, but longer-term preservation depends on the set of processes operating at the next longer-term time scale (to the right and down)

driving mechanisms in the stratigraphy of sedimentary basins." Jones et al. (2004) demonstrated that in a foreland basin in Spain, the effects of basin-margin thrust faulting and erosional sediment unroofing would take in the order of 1 Ma to be recorded in petrographic changes 20 km into the basin.

Jerolmack and Sadler (2007) examined the relationship between the transience of autogenic processes and the persistence of the "nested hierarchy of beds and bedding planes" constituting "the patchwork record of former landscape surfaces" that are ultimately preserved over the longer-term. The purpose of their paper was to develop a quantitative stochastic diffusion model to simulate the multitude of overlapping processes.

Sadler supplemented his data compilation on vertical aggradation rates (Sadler 1981) with a similar compilation on lateral accumulation rates (the lateral growth of ripples, bars, delta lobes, continental margin clinoforms, etc.) (Sadler and Jerolmack 2012, 2015). They demonstrated that, in cross-sectional area perpendicular to strike, growth rates of fluvial, coastal and shelf clastics are fairly constant at around 1 m²/a, at all time scales, as measured over time spans from months to hundreds of millions of years. This continuity of sediment flux is, of course, not evident from the geologic record but, as noted above, non-deposition or erosion in one location is expected to be contemporaneous with deposition elsewhere,

In the next sections I discuss how specific sedimentary processes operating over a range of rates can lead to the generation of "frozen accidents" over a range of time scales (see also Fig. 8.10).

### 8.7.1 Preservation at a Scale of Seconds to Months

Ripples and dunes form and migrate continuously under running water. Typically, trains of bedforms migrate down channels and across bar flanks, with one bedform replacing another, resulting in no net sedimentation. Temporary accumulations may form by lateral accretion where bedform trains build into areas of increasing water depth, such as scour pools or the flanks of bars. Under conditions of high bedload, ripple sets may become superimposed, to form climbing sets. Sedimentation on tidal flats can be affected by the lunar cycle from neap- to spring-tide conditions, which has been observed in some cases to be recorded as rhythmicity in lamina thickness (e.g., the Dutch tidal flats: Visser 1980). Mud in marine systems of all types is constantly undergoing deposition, erosion and transportation as tides, storms and other processes affect the sea floor (Traboucho-Alexandre 2015). How do such ephemeral deposits become preserved? Some must be so preserved, because we see them in ancient deposits in settings that clearly indicate the types of environmental processes just described (e.g., Archer et al. 1991).

Channels in flowing systems are ephemeral over a wide range of physical and time scales. Bank erosion, bar and meander migration are a result of the ever-changing structure of turbulence in the system. Minor changes in discharge or the direction of flow in one channel may trigger a cascading set of changes downstream. Sediment movement and deposition are therefore dynamic and ever changing. However, these autogenic processes will lead to the abandonment of channel reaches and bars, which serve as areas of temporary sediment storage, even while sedimentation may be relatively continuous when the system is considered as a whole. When we walk across a tidal flat or a fluvial point bar, it is these temporary deposits that we see. Returning to the same location days or years later, the deposits may have the same appearance, but there is a high degree of probability that the specific deposits we took note of have been replaced by others. What happens next is again a matter of chance. Long-term preservation depends on the events at the next time scale.

### 8.7.2 Preservation at a Scale of Years to Thousands of Years

The time scale of years to thousands of years is what Sheets et al. (2002) termed the stratigraphic "mesoscale." Within this time frame, "the depositional pattern shifts from reflecting the short-term flow pattern to reflecting long-term basinal accommodation. Individual events are averaged to produce large-scale stratal patterns" (op. cit., p. 288). This process is what Duller et al. (2012) termed the transition from "noisiness" to "drift."

Walther's Law is based on the concept of shifting depositional environments that are represented by deposits stacked in a vertical succession. Many of the environments to which this law has been applied generate the characteristic vertical profiles (e.g., fluvial fining-upward cycles, deltaic mouth bars and crevasse splays) over time spans of $10^1–10^3$ years, and may, therefore be interpreted within the framework of SRS 5-7. Examples are summarized below. Longer-term processes are described in the next section.

Most changes in fluvial systems take place during episodes of maximum discharge, which may be regular spring floods, or rarer flood events. Tidal systems are most affected during spring tides and during storms. These can result in large-scale changes in channel and bar location and orientation. In braided fluvial systems, entire minor channel systems may be abandoned. In meandering systems, meander chute and neck cutoff can leave earlier deposits behind. The deposits that fill scours, such as those which form at channel confluences, have a particularly high preservation potential at this time scale. Holbrook et al. (2006) defined what they termed the buffer zone, the zone of instantaneous preservation space for fluvial systems. For graded rivers, this is the space between the deepest level of scour and the highest level to which levees and floodplains can aggrade during normal year-to-year flow conditions.

At a somewhat longer time scale, from hundreds to a few thousand years, nodal avulsion of river channels is an important process for deposit abandonment (Schumm 1977), which considerably increases the preservability of the deposits so affected.

Deposits that form by flow expansion, including alluvial fans, crevasse splays and delta lobes are abandoned by switching of the distributary system as a result of slope advantages. This is a well-known process based on the example of the Mississippi delta, from the scale of the interdistributary bay-fill delta (Coleman and Gagliano 1964) up to the scale of the major delta lobes (Kolb and Van Lopik 1966; Frazier 1967). Once abandoned, the larger delta lobes undergo natural compaction and subsidence, which gradually takes them below the level of active scour and increases their chance of long-term preservation.

In tidal systems, channels, bars and tidal deltas evolve in the same way. Long-shore drift can displace inlet mouths, with resultant abandonment of inlet fill and bar-flank deposits. Scarponi et al. (2013), as noted earlier, demonstrated how accumulation rates and the probability of preservation of individual depositional events changed through a 100-ka base-level cycle.

In the experimental braid-delta constructed by Sheets et al. (2002), localized episodes of rapid aggradation occurred by avulsive channel switching across the experimental tank, eventually evening out "regional" deposition to the point that average aggradation equaled subsidence. Sheets et al. (2002, p. 300) scaled this up to a scenario whereby at a long-term aggradation rate of 1 m/ka the "depositional transition from flow control to subsidence control would occur on a time scale of the order of 15,000–30,000 years." This corresponds exactly to the conditions prevailing within the SRSs 7 and 8 scale of depositional units. Sheets et al. (2002) demonstrated that in their model, after the deposition of a sediment layer equivalent to between five and ten channel-depths, which required an equivalent number of avulsion events to occur, the resultant layer had evolved a relatively consistent thickness and that the regional variation in this thickness could be related to the pattern of subsidence. What this means from the perspective of this chapter is that sedimentation rates for individual channels in a fluvial or deltaic system differ from the sedimentation rate for the entire depositional system by one half to about one order of magnitude (SRS 5-6 versus SRS 7-8).

Fans and deltas illustrate the important point that (as expected following the arguments of Sadler and Jerolmack 2012, 2015) a continuous sediment flux still yields a discontinuous record because of the patterns of channel, channel-belt and lobe switching that take place as a result of natural avulsion processes. Sediment bypass in one area is contemporaneous with sediment accumulation elsewhere.

At these time scales, several processes are repetitive, and can result in deposits that preserve an element of internal repetition or cyclicity. Channel aggradation and bar accretion will record the range of flow conditions from higher-velocity flow at the base of the channel, the bar toe or the base of the bar flank, to low-flow at the bar crest or the channel bank. Seasonal flooding and flash floods can impose a crude cyclicity of minor erosion followed by sedimentation that decreases in grain size as flow energy dissipates. All these processes tend to generate upward-fining successions which have a chance of being preserved at the decadal to millennial scale. Progradational deposits develop an upward-coarsening profile as deeper water environments are gradually filled with sediment.

All the processes described here include erosional episodes operating at the same time scale, and constituting integral components of the geological preservation machine.

### 8.7.3  Preservation at the Scale of Tens of Thousands to Hundreds of Thousands of Years

At this time scale, sedimentary processes (the generation and removal of accommodation, sediment accumulation and erosion) may be dominated by high-frequency tectonic processes, or by Milankovitch processes (including sea-level

change), or by both. Walther's Law may be applicable to processes operating at SRS 7-9. The section illustrated in Fig. 8.9 illustrates the geological preservation machine in action at this scale, the topmost 100 m of the section representing preservation at the SRS 7-8 scale, but with long-term geological processes, that would most likely remove much of this section, still to come (Fig. 8.10).

As the discussion in this chapter has demonstrated, in most settings, local sedimentation rates are more rapid than the rate of accommodation generation, and complete preservation of any succession at any time scale is unlikely. For example, the detailed Jurassic ammonite studies of Callomon (1995) reveal an extremely variable and fragmentary record of shallow-marine preservation over the $10^5$-year time scale in the Jurassic deposits of Dorset (see discussion of Fig. 8.2).

Clastic wedges or *tectonic cyclothems* (Blair and Bilodeau 1988) are cycles developed under tectonic control over time scales of $10^4$–$10^7$ years. Accommodation is generated by differential movement at basin margins. Episodic thrust loading within a foreland-basin setting may generate regional basement adjustments at this time scale, and has been suggested as one of the generating mechanisms for tectonic cyclothems (Peper et al. 1992). The angle or direction of tilt of depositional slopes may be changed by changes in intraplate stress, by extensional subsidence or faulting, or by changes in the supracrustal load in the case of contractional settings (e.g., foreland basins). Heller et al. (1993) modeled localized crustal flexure (uplift and subsidence) at rates of up to 0.16 m/ka over time periods of $10^5$ years caused by tectonic reactivation of lines of crustal weakness in response to far-field intraplate stresses. Zecchin et al. (2010) compared the architecture of Milankovitch cycles generated under conditions of high-frequency sea-level change and tectonism, with accommodation changing at rates of 1–10 m/ka.

Entire depositional systems may be affected by these processes, leading to formation, and then abandonment and preservation of previously formed deposits. Kim and Paola (2007) modeled a coastal fluvial-deltaic system and demonstrated that autogenic cycles of delta and channel switching may, under the influence of fault movement, develop cyclothem-like cycles over time periods of $10^5$ years (SRS 8-9). Allen (2008) suggested that the response time of fluvial systems to tectonic perturbations of an alluvial landscape would be in the order of $10^{5-6}$ years. Tectonic cyclothems are discussed further in the next section.

Autogenic switching of alluvial channel belts at a $10^3$–$10^5$-year time scale has been described by Hajek et al. (2010) and Hofmann et al. (2011). Clusters of channels generate an alluvial ridge leading to instability and avulsion into neighbouring low areas on the alluvial valley. A greater degree of compaction of the adjacent floodplain units relative to the channel deposits is a factor in creating the additional accommodation. This process, termed *compensational stacking*, generates cycles of about 120 m in thickness, and clearly requires the switching mechanism to be superimposed on a long-term process of tectonic subsidence.

In nonmarine successions, major surfaces of nondeposition may be difficult to distinguish from autogenic scour surfaces. Miall and Arush (2001b) labeled such unconformities "cryptic sequence boundaries." They may be recognized by careful petrographic work, and also by the application of special petrophysical methods that reflect the subtle diagenetic signatures of such surfaces (Filomena and Stollhofen 2011).

Sequence models for nonmarine systems have long incorporated the concept of variable stacking patterns of channelized sand bodies. Following the modeling experiments of Bridge and Leeder (1979), the standard interpretation has been that the architecture of channels and channel belts depends largely on the balance between the rate of avulsion and the rate of accommodation (Wright and Marriott 1993; Shanley and McCabe 1994). The models assert that rapid accommodation generation may increase the likelihood of a given channel deposit being buried as an isolated sandbody by floodplain deposits before channel migration and scour removes it from the record. Slow accommodation generation favours the accumulation of channel bodies that erode laterally into each other, as a fluvial system slowly migrates across a floodplain, developing laterally-amalgamated sand bodies. Such interpretations must, however, be consistent in terms of time scale. Alluvial architecture—the preserved complex of amalgamated macroforms, is determined by sedimentary processes that occur within the *SRS-6* to *SRS 8* range, that is, on time scales of $10^2$–$10^5$ years and sedimentation rates of $10^{-1}$ to $10^2$ m/ka. These are the rates used by Bridge and Leeder (1979) based on information they compiled from modern rivers. However, this means that the formation of the elements of alluvial stratigraphy occurs within time scales that are several orders of magnitude more rapid than the rate at which accommodation is typically generated by regional geological subsidence (*SRS 11*). Therefore it seems likely that in the ancient record there would be a genetic relationship between fluvial architecture and rates of accommodation only in the case of high-frequency sequences, those formed within the *SRS-8* range, e.g., orbital cycles. Nonmarine sequences formed over longer time periods (many have been documented on a $10^6$-year time scale) require different interpretations, in which changes in alluvial architecture are related to migration of facies belts or to changes in fluvial style in response to tectonic or climatic forcing.

Orbital cycles are superbly exposed in pelagic sedimentary records in southern Italy, and have been studied as a basis for erecting a cyclostratigraphic time scale (Hilgen 1991; Hilgen et al. 2007, 2015). In this setting, accommodation is not an issue. Sea level, water chemistry, sediment supply, and every other important aspect of the environment, including even the regularity of processes generating hiatuses, is controlled by orbital forcing, and a clear orbital signal and a representative sedimentary record can be expected to be preserved. Calibration of the cyclicity against a retrodicted astronomical record confirms this. Lacustrine settings may also provide the necessary environment of total control by orbitally-forced parameters for development and preservation of a cyclostratigraphic signature (e.g., Aziz et al. 2003), and the classic cyclothems of the US Midcontinent, controlled as they were by substantial climatic and glacioeustiatic sea-level change, also provide excellent examples of preservation at this time scale (other examples include the Triassic cycles of the eastern US: Olsen 1990; Green River Formation of Wyoming: Fischer and Roberts 1991). Some ancient carbonate-platform margins also offer convincing cases. In other settings, however, questions of preservability of the cyclostratigraphic record arise, particularly because of the likelihood of overprinting by autogenic processes (Miall and Miall 2004; Bailey 2009; Miall 2010). As Kemp (2012) demonstrated by numerical modeling, in the case of marine cycles, if sea-level changes occurring at a higher than Milankovitch frequency are embedded in the preserved record, the evidence of orbital control may be masked. Even in deep-marine sections, the sedimentary record may not be complete (Trabucho-Alexandre 2012), and so caution must be applied. Each putative example needs to be examined on its merits. Strasser et al. (2006, p. 81) said:

> It is clear that astronomical climate forcing is most accurately recorded in depositional settings where the preservation potential is highest (deep marine basins, rapidly subsiding shelves, long-lasting deep lakes).

Recent developments in the field of cyclostratigraphy and astrochronology are discussed in Sect. 8.11.

### 8.7.4 Preservation at the Scale of Millions of Years

On extensional continental margins, subsidence rates ranges from 0.2 m/ka at the initiation of rifting, decreasing to less than 0.05 m/ka during the flexural subsidence phase, at time scales of $10^6$–$10^7$ years. Foreland basins subside at rates of 0.2–0.5 m/ka, and cratonic basins at 0.01–0.04 m/ka (rates from Allen and Allen 2005, pp. 364–365). Intraplate stress changes can generate regional accommodation changes at 0.01 to 0.1 m/ka at time scales of $10^6$ a. (Cloetingh 1988, p. 216). Detailed studies of offsets on growth faults in the

Niger delta, based on measuring the displacement of maximum flooding surfaces, indicated rates of accommodation generation on the downthrown side of 0.01–0.12 m/ka, over time scales of $10^5$–$10^6$ years (Pochat et al. 2009). These rates are all within the range of *SRS 9-11*.

Where accommodation generation is rapid, sedimentation rates comparable to the "long-term geomorphic" rates of *SRS 7* ($10^0$ m/ka) have been recorded over intervals of several millions of years. Such settings include forearc and foreland basins and some basins associated with strike-slip faults (Miall 2010, pp. 280–281). These are categorized as convergent-margin basins (including transpressive settings) in Fig. 8.5 and assigned to *SRS-10*. As discussed above, variations on these long term rates may provide clues concerning subsidence and uplift mechanisms.

Sedimentation and long-term preservation by lateral accretion may be the key to understanding the thin sedimentary succession present in cratonic-interior settings (*SRS 12*). It required the highly detailed chronostratigraphic reconstructions of Runkel et al. (2007, 2008) to demonstrate that the Upper Cambrian-Lower Ordovician succession on the flank of the Trans-Continental Arch in Wisconsin and Minnesota developed by gradual offlapping of successive sedimentary shingles on a very gently-dipping ramp.

On continental margins, given the importance of lateral progradation, total accumulation is not limited by vertical accommodation generation, and sedimentation is supply-dominated. The following generalizations are adapted from Stow et al. (1983, p. 58): Typical long term accumulation rates on carbonate or clastic shelves are from 10 to 40 m/ka ($10^1$ m/ka). Pelagic ooze sedimentation will normally not exceed 0.03 m/ka, although under upwelling areas it may reach 0.1 m/ka ($10^{-1}$ m/ka). Resedimentation of material to deeper water results in accumulation rates on modern deep-sea fans from 0.1 to 2 m/ka, and up to 10 m/ka ($10^{-1}$–$10^1$ m/ka) in small tectonically active basins.

## 8.8 Implications of Missing Time for Modern Stratigraphic Methods

### 8.8.1 Sequence Stratigraphy

It has been reasoned that sequences are scale independent. Catuneanu (2006, p. 10) argued this point, citing the observations by Posamentier et al. (1992) on a small delta only 1 m across that was observed building into a pool of water from a gully, where every sequence process associated with base-level change and systems tract development could be observed on a tiny scale. Paola et al. (2009) also argued this point, suggesting that their scaled laboratory experimental systems run over periods of seconds to hours in a tank a few

meters across could legitimately be interpreted to explain geological-scale processes in major sedimentary basins.

Fragmentary the stratigraphic record might be, but as the construction of the Devil's staircase (Fig. 8.7) indicates, the fractal nature of the record means that it consists of intervals of succession fragments separated by larger gaps that developed at higher time scales. These larger gaps can legitimately be considered as the sequence boundaries. Several decades of analysis have now indicated that there is a limited number of sequence types, which develop because of the occurrence of particular allogenic processes that are characterized by particular time scales (Miall 1995a, 2010). These natural time scales, because of their predominance, tend to lead to enhanced preservability, and it is for this reason that sequence stratigraphy "works." The original concept of a sequence hierarchy—the five or six "orders" of Vail et al. (1977) has been shown to be unworkable (Schlager 2004), but a crude hierarchy does exist, based on the nature of the processes that develop sequences, which range over time scales from $10^4$ to $10^8$ years (Table 3.3; Miall 1995a, 2010, Table 4.1).

### 8.8.2 Implications for Stratigraphic Continuity, the Concept of Correlation and the Principal of the GSSP

The discipline of stratigraphy is dependent on the principle that the sedimentary record is amenable to dating and correlation. The thrust of this chapter has been to demonstrate the fragmentary nature of the stratigraphic record, whereas the existence of the property of correlatability—which has been amply demonstrated by two hundred years of stratigraphic practice, implies continuity. There is no contradiction here. Consider the evolution of time at *SRS 1*, the present moment. At this time more than half of the Earth's surface is under water and accumulating water-laid deposits in fluvial and marine environments, or otherwise in a condition which is favourable to instantaneous sediment accumulation, e.g., the migration of an eolian dune, or alluvial fan deposits banking up against a fault. Therefore there are countless locations where the deposits that could constitute a future GSSP for "Now" are accumulating. Many of these will survive the rigours of *SRS 2* to *10*, *11* or *12* (depending on tectonic location) to provide the framework for the correlation of "Now" at some distant point in the future. This work of the geological preservation machine is all that stratigraphers have ever assumed and depended upon. However, as Smith et al. (2015) have pointed out, most of the officially ratified GSSPs that currently form the basis of the Geological Time Scale are defined in shallow-marine strata, where there is a high probability of multiple hiatuses and significant missing time.

## 8.8.3  Discussion

It has long been known that the sedimentary record is fragmentary. However, this has not stopped stratigraphers from making calculations about sedimentation rates and the ages of key beds, based on assumptions of continuous sedimentation and extrapolation from known horizons. For example, the study of cyclostratigraphy requires the conversion of the "depth domain" to the "time domain" using age calibration points (Strasser et al. 2006, p. 82). This type of analysis should always be treated with considerable caution, and the arguments presented here only serve to emphasize this point. This could be argued before on an ad hoc basis, but not until the advent of the fractal concept has it been possible to systematize these observations and place them into a framework that suggests a continuity of process over all time scales. The fractal framework constitutes a useful method of statistical description of the geological preservation machine, but because the time frames of geological processes are not genetically related, there is no reason to expect that the framework will constitute anything other than a mathematical approximation. Accordingly, it may be more appropriate to describe the relationships as "fractal-like."

The significant differences highlighted in this chapter between (1) the preservation of the products of modern sedimentary processes, (2) those preserved in the recent (post-glacial) record, and (3) those preserved in the more ancient record, indicate the need for a modified use in geological work of the concepts of *uniformitarianism*, hence the title of the paper from which this section has been adapted (Miall 2015). The same applies to the comparable term "*actualism*", which is "the principle that the same processes and natural laws applied in the past as those active today" (Donaldson et al. 2002). With the exception of some unique conditions in the Precambrian relating to marine and atmospheric chemistry (Eriksson et al. 1998), the *processes* of sediment creation have been comparable throughout geological time. It is the issue of *preservation* that has required a reevaluation. Interpretations of the geological record that use modern, active, post-glacial depositional systems as analogues (e.g., deltas, valley-fills, prograding continental margins) need to take into account that these deposits can only illustrate the working of the geological preservation machine up to the time scale of *SRS 7* or *8* Fig. 8.10). Application of uniformitarianist concepts to the longer-term geological time scale (*SRS 9-12*) needs to be carried out with these cautions in mind. In this sense, the concept of uniformitarianism, as applied in practice, is incomplete.

These concepts likely hold a key to an improved way of studying and interpreting the sedimentary record, requiring us to go back and look at that record again, ironically, to document what is not there in greater detail: the record of missing time. Sedimentary units at all scales need to be evaluated in terms of the sedimentation rates they indicate over the full range of scales, at the appropriate *SRS*, in order to unravel the complexity of preservation and removal. De Natris (2012) and de Natris and Helland-Hansen (2012) have made a start at this form of analysis. Amongst other consequences, this research will re-emphasize the value of the "mesoscale" experiments described by Paola et al. (2009). For example, he and his colleagues have demonstrated that incised valleys formed by shoreline incision and fill during cycles of base-level change do not represent single chronostratigraphic surfaces (as commonly assumed for the purpose of sequence definition) but represent amalgamated fragments of surfaces and deposits that evolve both during falling and rising stages of the base-level cycle (Strong and Paola 2008). The experiments of Sheets et al. (2002) provided a quantitative basis for the transition from an autogenic to an allogenic time scale in the gradual filling of an alluvial-deltaic basin by avulsive switching of channel belts. All of these findings will help to clarify the concepts of time and the behaviour of the geological preservation machine.

Quantitative stratigraphic studies (e.g., those based in time-series analysis) are becoming increasingly popular. However, lest the new fractal concepts tempt geologists to focus in future on quantitative studies based on fractal theory, the warning of the field sedimentologist needs to be heard. Quantitative analyses too frequently ignore the field reality of the rocks under study (e.g., see Sect. 8.11). Without careful analyses of facies details, a careful search for grain size and lithologic changes, and a focus on the nature of facies contacts (sharp versus transitional), researchers can mistake mathematical or statistical rigour for geological reality.

The examples used here to illustrate the *Sedimentation Rate Scales* merely brush the surface of a potentially instructive form of deductive investigation in which tectonic and geomorphic setting, sedimentary processes and preservation mechanisms can be evaluated against each other both qualitatively and quantitatively, leading to more complete quantitative understanding of the geological preservation machine, and a more grounded approach than earlier treatments of stratigraphic completeness. For example, wide variations in sedimentation rate in foreland basins, and in continental margin sedimentation have been touched on here, and may help to refine future geological interpretations. Once tectonic setting is taken into account, the variability in the data in Sadler's (1981, 1999) linear log-log plot become comprehensible. When accommodation generation is particularly rapid, as in many convergent-margin settings, and where accommodation is essentially limitless, as on continental margins, and sedimentation is supply-dominated,

**Fig. 8.11** Location of the Book Cliffs study area, showing the locations of detailed sections and other illustrations

**Fig. 8.12** Regional cross section of the Mesaverde Group. Adapted from Seymour and Fielding (2013), after the detailed mapping of Young (1955)

long-term sedimentation rates may be one to two orders of magnitude greater than Sadler's compilation would suggest. Sediment flux may be relatively constant over a wide range of time scales (Sadler and Jerolmack 2012, 2015), but processes of sediment distribution, sedimentation and preservation "shred" the resulting record, as explained in this chapter. We return to this debate in Sect. 8.10.2, which is a discussion of the construction and interpretation of chronostratigraphic charts (Wheeler diagrams) for stratigraphic successions.

It is now clear that the stratigraphic record is more than just incomplete. To extend Ager's famous thought: there are gaps within the gaps, and the record is permeated with them, at every scale. The frozen accidents that the gaps enclose can

still tell us a great deal, but only if we get the time scale right.

---

## 8.9  An Example of the Evaluation of Missing Time: The Mesaverde Group of the Book Cliffs, Utah

The Mesaverde Group is well exposed in the Book Cliffs of east-central Utah and western Colorado. It constitutes a classic foreland-basin clastic wedge, which developed in response to uplift and erosion along the Sevier Orogen during the Late Cretaceous. A resistant nonmarine sandstone, the Castlegate Sandstone, constitutes a prominent cliff-forming unit that caps the cliffs over a distance of about 200 km (Figs. 8.11 and 8.12).

The stratigraphy of the group was mapped in detail by Spieker and Reeside (1925) and by Young (1955, 1957). In the 1990s these rocks served as one of a set of worked examples used to explain the principles of sequence stratigraphy (Van Wagoner et al. 1990; Van Wagoner and Bertram 1995), and largely because of this work the area has been a favorite destination for field trips by academic and corporate groups (e.g., Cole and Friberg 1989; Van Wagoner et al. 1991).

In more recent years the Mesaverde Group has become the focus for examinations of many concepts about stratigraphy and sedimentation. The most detailed treatment, in which the author explored many quantitative models concerning sedimentation in an ancient foreland basin, was provided by Hampson (2010). However, the issue of fragmentary preservation of the record has not been discussed. This is not an inconsequential issue because, as the work of Hampson (2010) exemplifies, the stratigraphic and sedimentologic detail that are now available for the Mesaverde Group are amongst the most extensive available for clastic wedges of this type, and the unit should, therefore, provide an ideal test bed for the exploration of advanced sedimentological concepts. Amongst the few workers who have addressed the issue of missing time in the Mesaverde Group, and speculated about its implications for correlations and sequence stratigraphy, are Howell and Flint (2003), who reviewed the chronostratigraphy of the succession and the presence of gaps at the parasequence scale, and Bhattacharya (2011), who discussed alternative sequence models for the Castlegate Sandstone. The discussion presented here was first published in Miall (2014b).

### 8.9.1   Chronostratigraphy of the Mesaverde Group

The Mesaverde Group of the Book Cliffs encompasses two formations, the Blackhawk and the Castlegate. The Blackhawk Formation consists of undifferentiated fluvial, shoreface, and deltaic deposits in the western Book Cliffs (Hampson et al. 2013), and passes eastward into a series of shoreface to shallow-marine tongues that were assigned to six members by Young (1955) (Fig. 8.12). The shoreface sandstones constituting the thicker, proximal portions of these members form a series of cliffs (Fig. 8.13), the appearance of which, from a distance, is what suggested to early explorers the pages of a book lying on its side; hence the name Book Cliffs.

As summarized by Hampson (2010), the six members of the Blackhawk Formation have now been subdivided into a total of 23 submembers. The first sequence-stratigraphic studies of these rocks interpreted them as the product of eustatic sea-level change (Van Wagoner et al. 1990), but these units have subsequently been much discussed as examples of a type of high-frequency sequence stratigraphy controlled by regional tectonism and its effect on accommodation (e.g., Krystinik and de Jarnett 1995; Hampson 2010; Aschoff and Steel 2011).

The Castlegate Sandstone, named after the Castle Gate, a prominent landmark in the Price River Canyon north of Helper, was interpreted as a third-order sequence by Olsen et al. (1995), but comprises two sequences, according to Miall and Arush (2001a), although, as suggested by Bhattacharya (2011), and as discussed later, other interpretations must be considered.

The age range of the Mesaverde Group and its constituent units has been interpreted primarily from the ammonite fauna contained in the marine portions of the Blackhawk and in the Mancos Shale, with which it is interbedded to the east. Howell and Flint (2003), Hampson (2010), Aschoff and Steel (2011), and Seymour and Fielding (2013) provided overviews of the biostratigraphic and other work that has been carried out on these rocks since mapping began in the 1920s. Three of the ammonite zones can be correlated to the global time scale, providing numerical-age tie points. The following data are taken from Aschoff and Steel (2011).

The base of the Blackhawk Formation is Lower Campanian in age, and is dated at 81.86 Ma. The top is placed at 79 Ma, and the top of the Castlegate is Middle Campanian, at 77 Ma. The time span of the Mesaverde Group is therefore 4.86 million years. Near Price the section is up to 700 m thick, which indicates an average sedimentation rate of 0.14 m/ky. This rate ($10^{-1}$ m/ky) is at the upper end (more rapid) of rates characteristic of long-term geological subsidence measured over periods of $10^5$–$10^7$ year, including

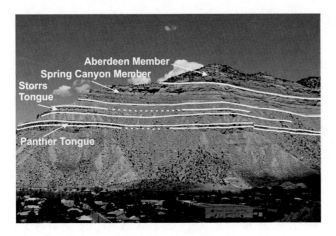

**Fig. 8.13** View from Helper of a portion of the Mesaverde Group

thermal subsidence and flexural loading of the crust (*SRS 10-11*), and is comparable to rates associated with low-frequency orbital cycles (*SRS 9*) (Miall 2013). If all six members of the Blackhawk plus the two constituent sequences of the Castlegate Formation are assumed to represent equal time spans (for which there is no evidence), this equates to an average duration of 607 ky per unit, a value that corresponds to no known geological frequency. Furthermore, each of the members has now been subdivided into submembers, totalling 23 for the Blackhawk Formation (the individual shoreface tongues in Fig. 8.12), averaging 125 ky in duration. At this time there is no method by which the ages and time spans of the constituent units of the Mesaverde group may be individually dated with greater precision.

A comparison of Book Cliffs stratigraphy with that of the Henry Mountains area, some 200 km to the south of Prince (Seymour and Fielding 2013) indicates that there is no detailed correlation between the Blackhawk members and comparable units in that section. This, and the lack of spatial regularity of the Blackhawk members, argues against orbital cyclicity as a major mechanism for the generation of the members and submembers of the Blackhawk Formation.

Following what is essentially traditional practice, correlation diagrams for the Mesaverde Group, including chronostratigraphic charts, show the Blackhawk Formation and its constituent members as the product of continuous sedimentation (e.g., Hampson 2010, Figs. 2, 7; Aschoff and Steel 2011, Fig. 3; Seymour and Fielding 2013, Figs. 2, 7). That continuous sedimentation is unlikely is indicated by the large lateral shifts of proximal and distal facies and the facies contrasts across the sequence boundaries, which indicate significant changes in accommodation and/or sediment supply.

What now follows is speculative, representing an attempt to apply the *Sedimentation Rate Scale* concept to the Blackhawk formation. Following Sadler (1981, 1999), it may be suspected that the representation of elapsed time by the preserved record of this formation would be very low (cf. de Natris 2012). The log-normal distribution in time of sedimentation rates that was demonstrated by Sadler (1981, 1999) is likely paralleled by a log-normal distribution in the duration of sedimentary gaps. Miall (2015; Figs. 8.5 and 8.6 of this book) suggested that these distributions are fractal-like in character. A significant proportion of the elapsed time might, therefore, be represented by intraformational hiatuses, member and submember boundaries and other surfaces of erosion or nondeposition. There is little evidence of major erosion at the Blackhawk member or submember boundaries, except for

**Fig. 8.14** An interpretation of the chronostratigraphy of the Mesaverde Group. Age information is from Hampson (2010) and Seymour and Fielding (2013)

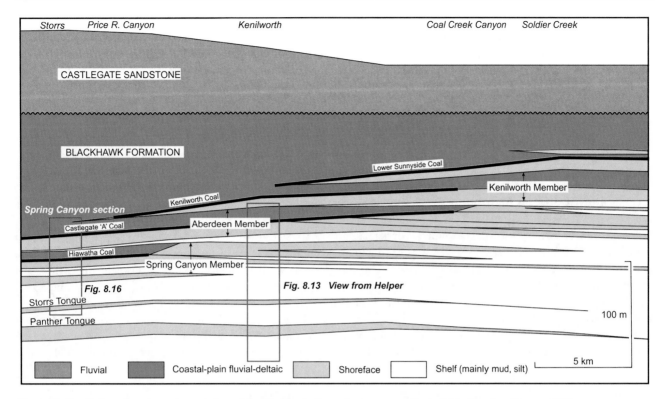

**Fig. 8.15**  Detailed stratigraphy and correlation of the Mesaverde Group in the western Book Cliffs. Based on Young (1955)

the rare presence of incised valleys that indicate erosion during the falling stage of the base-level curve at the end of each progradational episode. East of Green River, erosional relief at the base of the Desert Member reaches 30 m (Van Wagoner 1995, p. 151), and incised valleys at the top of the Grassy Member are up to 20 m deep (O'Byrne and Flint 1995, p. 242). It seems very likely that at least the member and submember boundaries represent significant breaks in sedimentation, with durations in the high $10^4$-year to $10^5$-year range. This is consistent with the pattern that emerged from a forward-modeling exercise by Howell et al. (1999). As a preliminary model, the six member boundaries are set here arbitrarily so that the preserved sedimentary record represents a total of approximately half of the elapsed time, with an average sedimentation rate for the preserved deposits of 0.29 m/ky. Sediment bypass, with sedimentation taking place by lateral progradation of shelf clinoforms, may explain much of the time "missing" through any given vertical section.

This rate is comparable to that of deposits measured at the *SRS 8* time scale of $10^4$–$10^5$ years, including the late Cenozoic glacioeustastic cycles of the Wanganui Basin in New Zealand and the Triassic Newark cycles of the eastern United States (details for all these comparative rates are, unless otherwise stated, provided in Miall 2015).

A speculative chronostratigraphic chart for the Mesaverde Group that includes the suggestion of significant gaps at the sequence boundaries is presented here as Fig. 8.14. This follows the approximate timing of the sequence boundaries shown by Hampson (2010, his Fig. 7), but otherwise differs substantially from that chart in the following principal ways: (1) in the updip coastal-plain region represented by the Blackhawk Formation, most of the elapsed time is completely unrepresented. Some of this unit consists of fluvial deposits formed above marine base level (Hampson et al. 2013) and here, accommodation can be explained by the buffer concept of Holbrook et al. (2006), as discussed later. These deposits are shown schematically in Fig. 8.14. (2) The arrangement of sequence boundaries in Hampson's (2010, Fig. 7) chart suggests that the completeness of the Blackhawk and the contemporary shallow-marine record (that of the Blackhawk members) decreases basinward, whereas the opposite is more likely to be the case. Sedimentation of the Mancos Shale towards the basin center is likely to be much more continuous and therefore more complete at the $10^4$–$10^6$ time scale than the proximal sediments of the coastal region, but may contain substantial disconformities, as shown in Fig. 8.14. The pattern shown here in Fig. 8.14 is more like that developed by Krystinik and DeJarnett (1995, Fig. 3), in which it was suggested that the proximal region to the west of the basin was a region of uplift and erosion. Units of condensed (slow) sedimentation may be expected to develop offshore at times of regional transgression, and there is also evidence that sediment gravity flows occurred on distal clinoform slopes during the falling stage of some of the sequence cycles (Pattison 2005; as shown by Hampson (2010, his Fig. 7), but not included in Fig. 8.14 of this book). (3) Following from the previous point, the time span represented by the member (sequence) boundaries increases landward, and the undifferentiated proximal, deltaic Blackhawk formation represents, in total, much less than one half of the available elapsed time, with substantial erosional breaks embedded in this coastal succession.

## 8.9.2  Chronostratigraphy of the Spring Canyon and Aberdeen Members

The next step in this analysis is to examine sedimentation rates and preservation at shorter time scales. *SRS 7* is the time scale for long-term geomorphic processes, those occurring over time periods of $10^3$–$10^4$ years. Such processes include the development of major delta lobes and alluvial channel belts, regressive shoreface complexes, major coal seams, etc.

Sedimentation rates that have been calculated for such deposits include the following examples of post-glacial successions (details in Sect. 8.4): Fluvial channel aggradation in coastal-plain rivers in Texas, up to 1.7 m/ky; modern floodplain rates compiled by Bridge and Leeder (1979) as the basis for their simulation experiments, 0.35–2 m/ky; the Holocene Mississippi Delta, 6–12 m/ky; Rhone delta, 6.1 m/ky; Rhine-Meuse system, 1.5 m/ky; Galveston barrier island, 3.4 m/ky; Sapelo Island tidal inlet, 4.5 m/ky. The rate of post-glacial sea-level change is comparable, at 1–18 m/ky. All of these rates are in the range of $10^0$–$10^1$ m/ky, which is up to an order of magnitude greater than the *SRS 8* range, which applies to high-frequency orbital cycles, and up to two orders of magnitude greater than long-term geological rates (*SRS 11*).

To study sedimentation at this scale, a more detailed examination is presented here of the Spring Canyon section, located west of Helper (location shown in Fig. 8.11). The stratigraphic framework of this section is shown in Fig. 8.15, and the section is illustrated in Fig. 8.16. The section is shown in Fig. 8.16 in the traditional form, as a continuous succession of facies units, just as it appears in actuality in the field. However, as first pointed out by Barrell (1917), this

**Fig. 8.16** The section at Spring Canyon. Based on Cole and Friberg (1989) and Kamola and Van Wagoner (1995). The section is subdivided into 22 lithofacies units, as numbered at *left*

**Fig. 8.17** The Spring Canyon Member dismembered. Bed unit numbers are as in Fig. 8.16. At *left*, the section is plotted to correspond to a *SRS 8* time scale, suggesting an approximate 273 ky timespan for the accumulation of the succession. At *right*, the same section is evaluated in terms of an *SRS 7* time scale, for which sedimentation rates are an order of magnitude more rapid. The section is subdivided into a set of progradational coastal-plain and shoreface successions

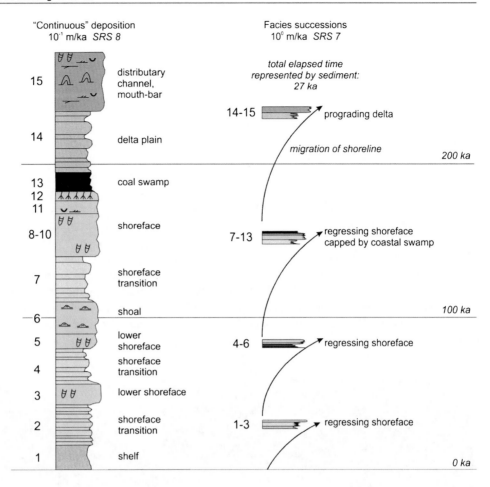

form of plot obscures the numerous cryptic hiatuses that permeate all stratigraphic sections.

In Fig. 8.17 units 1–15 of the same section are plotted against an *SRS 8* time scale. At the *SRS 8* sedimentation rate of 0.29 m/ky employed in the previous section, the 79.2 m of Spring Canyon section (units 2–15) shown in this figure would represent 273 ky of elapsed time. The same fifteen units are shown at the right re-plotted using *SRS 7* rates. At an *SRS 7* sedimentation rate range of 1.5–6 m/ky, this 79.2 m of section would accumulate over a time span of between 52.8 and 13.2 ky. A mid-range of 27 ky used here (arbitrarily) as an average for illustration, is one tenth the time assumed for the sequences timed at the longer-term *SRS 8* rate in the previous section (and comparable to the 7 % of elapsed time represented by sediment that was calculated by de Natris 2012). How is preservation and non-preservation distributed through the estimated 273 ky represented by the section?

As suggested in Fig. 8.17, and following all previous interpretations, the section can be interpreted in terms of four progradational successions, or parasequences. Using the same lithofacies unit numbers as in the original section, these are displayed at the right in this figure. Three

regressive shoreface successions, the last capped by a coal swamp (unit 13 is the Hiawatha coal), were followed by a progradational delta.

Figure 8.17 provides an illustration of the problem that sedimentologists have yet to address fully, that of the lengthy periods of empty time, time for which there is no sedimentary record within apparently continuous successions. For example, Hampson's (2010) detailed examination of the clastic wedge includes an analysis of the trajectory of shorelines in time and space through the evolution of the Mesaverde Group. The outcome of the analysis (Hampson 2010, his Fig. 14) is displayed in the form of a height-versus-distance diagram that provides quantitative estimates of forestepping (progradation) and backstepping (flooding) and is interpreted in terms of changing bathymetry and sediment supply in the basin as the succession gradually prograded eastward across the foreland basin. The analysis is keyed to the evolution of the Blackhawk and Castlegate members, but timing is not discussed beyond the labelling of each eastward movement in the shoreline with the name of the appropriate prograding submember. However, the plotting of the trajectory as a continuous zig-zag line implies continuity, and does not include the possibility that the shoreline may have regressed entirely across the basin (leading to lengthy periods of exposure and erosion), even though this possibility is hinted at by the way his chronostratigraphic diagram is constructed (Hampson 2010, his Fig. 7), nor does the trajectory diagram provide any indication of the point argued in this section, that there were lengthy periods of time for which we have no preserved record from which to construct such a diagram. This problem is addressed further in the next section.

### 8.9.3  The Representation of Time in a Coastal Clastic Succession

The Book Cliffs should offer an excellent opportunity to examine the problem of how time is represented in the sedimentary record, given the quality and completeness of the exposure and the amount of detail now amassed by recent field studies. However, having said this, they also exemplify the major problem with such analyses. The

absence of any techniques for chronostratigraphic dating with a greater precision than a $10^6$-year time scale means that all such analyses are speculative. Some of the key portions of the Spring Canyon section are illustrated in Fig. 8.18, and are discussed next.

The grouping of lithofacies units into progradational successions, and their interpretation in terms of the *SRS 7* time scale in Fig. 8.17, is based on the analysis of comparable successions, which accumulated over a similar time scale in the post-glacial record. As discussed elsewhere (Miall 2015) it is possible to extend such an analysis to a shorter time scale and correspondingly higher sedimentation rates. For example, each of the major units identified in the progradational shoreface successions illustrated in Fig. 8.18c, d consist, in turn, of thinner component units a few meters thick, or less, that can be interpreted by comparison with modern shoreface successions that accumulate from individual storm events or by fair-weather aggradation over periods of a few years to a few hundreds of years (*SRS 5, 6*). Episodes of non-deposition or bottom storm scour

**Fig. 8.18  a** The top of a distributary-channel and mouth-bar complex that forms the top of the Spring Canyon member. Unit 15 is 12 m thick. **b** The top of the Aberdeen member, consisting of a shoreface to foreshore complex. Person for scale. **c** The Storrs Tongue, a shoreface succession 9 m thick exposed at the base of the Spring Canyon section.

It is capped by a ravinement surface. **d** A shoreface-transition to lower-shoreface succession that forms the base of the Aberdeen Member. Person for scale. Unit numbers for **a, b,** and **d** are as in Fig. 8.16

appear as distinct bedding surfaces in these outcrops, but their time significance, which could be considerable ($10^0$–$10^3$ years), is cryptic.

At the *SRS 7* scale, the most significant time-related events represented in the Spring Canyon outcrops are the ravinement surfaces that cap the shoreface successions (Fig. 8.18a–c), and include the surfaces that cap each of the Blackhawk constituent members. Ravinement occurs during a rise in relative sea level. Nummedal and Swift (1987), in their original definition of this process, described various examples of ravinement developed during post-glacial sea-level rise, a process that during the Holocene, continued for thousands of years. We cannot know the duration of this process during the Cretaceous, but we can speculate. The driving process for episodicity in stratigraphic accommodation seems likely to include flexural subsidence and changes in intraplate stress (Miall 2010; Aschoff and Steel 2011). These are processes that operate over rates and time scales in the *SRS 8-9* range, that is, time periods of $10^4$–$10^6$ years. It is conceivable, therefore, and consistent with the interpretations of Howell and Flint (2003), that lateral clinoform progradation and the subsequent flooding and ravinement that preceded each of the progradational shoreface successions could have taken many tens of thousands of years, a significant proportion of the time available for each of these cycles.

Would it not be expected that there would be more evidence of such significant episodes in the development of the clastic wedge? Not necessarily. Ravinement is an erosional process and may cut down through significant thicknesses of shelf and shoreface deposits. Complete removal of contemporary deposits emplaces the post-ravinement shelf deposits over the eroded top of the preceding cycle, and in that case the ravinement surface corresponds to the sequence boundary (Nummedal and Swift 1987). There are many examples of such configurations now described in the literature. One of the first was the description of the E/T surfaces in the Cardium Sandstone of Alberta by Plint et al. (1986). When first proposed, the sequence interpretation was highly controversial, and one of the reasons for this was the implication that basin-wide sequence boundaries extending for hundreds of kilometers could be accompanied by little to no evidence of the elapsed time, and of the depositional and erosional events that must have transpired during their formation and subsequent burial. Draping the transgressive surfaces in some places are beach conglomerates, which the model indicates were transported there initially by fluvial processes from the basin margin hundreds of kilometers away during the falling stages of the base-level cycles. But the evidence for such rivers, in the form of remnant valleys and fluvial deposits draping the surfaces, is almost entirely absent.

Likewise in the Blackhawk Formation the ravinement surfaces could represent periods of up to $10^5$ years, with all evidence of basin-margin exposure and the basinward transport of continental detritus long gone, except for the rare presence of incised valleys at the tops of the members, as noted above. Hampson (2010, p. 107) refers to paleocurrent data in the shoreface sandstone indicating "persistent longshore currents" directed towards the south and southwest. These could have been the currents that removed the evidence of long-lasting episodes of exposure and of ravinement erosion. Some of this detritus may have contributed to the contemporaneous Muley Canyon and Masuk formations in the Henry Mountains area to the south (Seymour and Fielding 2013), although longshore drift is not described by these authors.

### 8.9.4 Sequence Stratigraphy of the Nonmarine Facies of the Blackhawk Formation and Castlegate Sandstone

The first sequence-stratigraphic analysis of the Castlegate Sandstone was carried out by Olsen et al. (1995). They divided the formation near the type section (Castle Gate, at the mouth of the Price River Canyon, north of Helper; location shown in Fig. 8.19) into a lower, sandstone-dominated member, deposited in a braided-stream environment, following the fluvial architectural analyses of Miall (1993, 1994), and an upper member containing a significant proportion of interbedded mudstones, units of inclined heterolithic strata (terminology of Thomas et al. 1987), and evidence of marine, tidal influence in the form of flaser bedding and *Skolithos* burrows. They interpreted the formation thus subdivided as a "third-order" sequence. To explain the subdivision into the two members Olsen et al. (1995) turned to the fluvial models of Wright and Marriott (1993) and Shanley and McCabe (1994). The sandstone-rich lower member and the more heterogeneous upper member were interpreted, respectively, as low- and high-accommodation systems tracts. Miall (2014a, Sect. 6.2) has argued that such interpretations are untenable, given the issue of dramatically different sedimentation rates for the models and for the Castlegate Sandstone.

In a further analysis McLaurin and Steel (2000) subdivided the upper member into five higher-order (fourth-order) sequences and mapped a transition within these sequences between the fluvial deposits in the west into barrier, deltaic, and estuarine deposits near Green River, and ultimately into the offshore mudstones of the Buck Tongue. However, Willis (2000), who recognized a high-frequency sequence stratigraphy in the Sego Sandstone east of Green River, was unable to trace these sequences westward into the predominantly fluvial upper Castlegate Sandstone.

In an alternative analysis based on detailed mapping north of Green River, Yoshida et al. (1996), Willis (2000), and

Yoshida (2000) argued that the Buck Tongue is truncated by the upper member of the Castlegate, at an angular unconformity that cuts gradually down section along the Book Cliffs to the northwest. According to this interpretation (shown in Fig. 8.19) the beds overlying the unconformity above the Buck Tongue in the east (Sego Sandstone) are stratigraphically equivalent to the upper part of the lower Castlegate Sandstone at the type section. As noted by Miall and Arush (2001a), based on this interpretation, the truncation of the Buck Tongue implies that updip from the pinch-out of this unit, approximately 1 million year of section is missing in proximal parts of the Book Cliffs, including at the type section of the Castlegate Sandstone.

The evidence to enable a choice to be made for any of these interpretations depends on the ability to trace ("walk out") key surfaces between sections. Even in the case of the Book Cliffs, where exposure is much better than average, it is not possible reliably to trace key surfaces based on facies and outcrop characteristics for long distances within what is a very heterogeneous succession. Accordingly, Miall and Arush (2001a) sought to develop other means to analyze the stratigraphy and determined, on the basis of petrographic evidence, that the best evidence for missing time at the type section consists of changes in detrital composition and evidence for early cementation at surface "D" in the type section (Fig. 8.20). According to this interpretation, the lower Castlegate at this location comprises parts of two sequences (sequences 1 and 2 in Fig. 8.19), and the upper part of this unit at the type section passes laterally (downdip) through a

facies transition into the more heterogeneous beds of the upper Castlegate and the Sego Sandstone to the east and southeast (Willis 2000).

Yet another interpretation of Castlegate sequence stratigraphy was offered by Bhattacharya (2011, his Fig. 17), in which he speculated about the relationship between the lower Castlegate Sandstone and the underlying Desert Member in the area east of Green River. The original interpretation of Van Wagoner et al. (1990) and Van Wagoner (1995) was that the Desert Member is entirely older than the Castlegate. However, Bhattacharya (2011), referring to a discussion by Van Wagoner (1995) about the whereabouts of the coastal marine equivalents of the Castlegate fluvial sandstones (i.e., where are the mouths of the Castlegate rivers?), suggested that the Castlegate may in fact comprise a suite of high-frequency sequences, each with its own attached "Desert" shoreface (Bhattacharya, 2011, his Fig. 18). Again, this is a debate that could be answered only by detailed local correlations for which the evidence is not available, and a definitive answer is beyond the scope of this discussion to provide. However, we can speculate. It is entirely possible that the Castlegate Sandstone consists of a set of sequence fragments, number unknown.

Turning to the Blackhawk Formation, Hampson et al. (2013) documented the changes in fluvial style through the Blackhawk Formation in the proximal part of the Book Cliffs, west of Price, where the formation is not differentiated into members (see Fig. 8.12). Detailed reconstructions of alluvial architecture revealed a range of fluvial styles

**Fig. 8.19** Stratigraphy of the Castlegate Sandstone (from Miall and Arush 2001a; based on Willis 2000, and Yoshida 2000)

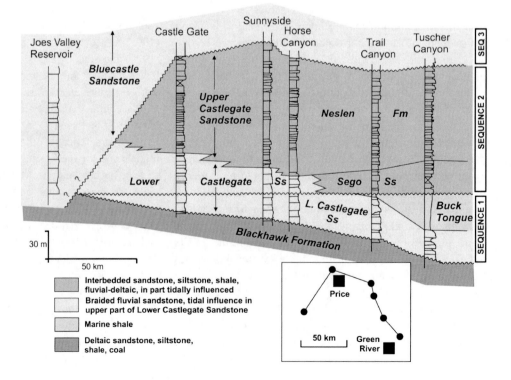

throughout the Blackhawk Formation but no systematic changes with stratigraphic position. They also made reference to earlier studies (Adams and Bhattacharya 2005; McLaurin and Steel 2007) which demonstrated no systematic variation in fluvial style through the overlying sequence boundary into the Castlegate Formation. The conclusion of Hampson et al. (2013, p. 166) is that the "internal architectures of the sandbodies do not result from systematic, short-term changes in accommodation such as those that characterize incised-valley fills formed by relative sea-level change in coastal-plain settings" but they "appear to reflect local changes in the balance of sediment flux and transport capacity due to upstream controls, such as high-frequency climatic variations, and autogenic responses." As noted above, Miall (2014a, Sect. 6.2) has argued that the Wright and Marriott (1993) and Shanley and McCabe (1994) fluvial sequence models are not applicable to interpretations of fluvial sequences that represent long-term geological accommodation rates (*SRS 9-12*). The interpretations by Hampson et al. (2013, pp. 166–167) are consistent with the alternative explanations offered by Miall (2014a).

Preservation of a nonmarine deposit such as the Castlegate Sandstone and the nonmarine Blackhawk depends on a different set of controls than those that operate in marine environments. At time scales up to *SRS 7* or *8*, deposition and preservation are limited by the alluvial "buffer", in the sense described by Holbrook et al. (2006). The buffer is defined by two surfaces, one above and one below a river's graded profile. The lower surface corresponds to the maximum depth of scour, the upper surface to the maximum height to which the floodplain could accumulate fine-grained overbank sediment. Aggradation within this buffer zone will typically generate a fining-upward succession. The position of the buffer in space over time depends on long-term changes to the alluvial basin, including subsidence of the basin, and changes to the elevation of the source area, which, in turn, depend on the balance between tectonic uplift and

erosion. Long-term preservation at *SRS 9-12* time scales requires that accommodation be generated rapidly enough that part or all of the buffer zone descends below the lower scour limit. This can happen either by basinal subsidence or by source-area uplift. Preservation of a complete buffer-zone succession will depend on such a succession being abandoned by changes in position of the channel belt for a long-enough time that it is not partly or entirely removed by erosion at a later time. Fragmentary preservation, typically of the lower portion of a buffer succession, will generally be the norm, and this will tend to favor the coarser, channel-fill parts of the succession unless increases in accommodation are unusually rapid. Variations between a succession dominated by the coarse (channel-fill) member and those in which significant thickness of floodplain fines are preserved may simply represent chance changes in erosion versus preservation of the buffer zone as alluvial systems migrate or avulse across an alluvial plain, or they may represent back-and-forth migration of facies belts in response to changes in paleoslope or sediment supply, as in the experiments performed by Strong et al. (2005). It has been a common misconception to interpret such changes in vertical profile only in terms of changing fluvial style (Miall 1980).

### 8.9.5   The Representation of Time in a Fluvial Succession

The largely unconsidered element in the preceding discussion is that of time. The 170 m of Castlegate Sandstone at the type section (Fig. 8.20) represents about 2 million years, at an average *SRS 9* sedimentation rate of 0.085 m/ky. However, shorter-term sedimentation rates reveal the extent of missing time in this section (Fig. 8.22). The 170 m of section at the type section, at a *SRS 7-8* sedimentation rate of between $10^{0}$ and $10^{-1}$ m/ky, represents an elapsed time of between a minimum of 17 ky and maximum of

**Fig. 8.20** The type section of the Castlegate Sandstone, with key bounding surfaces. Surface D of Miall and Arush (2001a). is tentatively identified as a major intraformational unconformity and sequence boundary. Width of field of view is about 250 m

1.7 million years. At an average rate of 0.29 m/ky, the same *SRS 8* rate used for the Spring Canyon Section in Fig. 8.17, the preserved sediment would represent an elapsed time of 586 ky. Many of the numerous well-developed near planar bounding surfaces visible in all Castlegate sections may account for significant time spans. Surface "D" provided some evidence of significant missing time, and was labelled a "cryptic sequence boundary" by Miall and Arush (2001a, b). As noted earlier, surface "D" at the type section could represent about half of the total 2-million years time span, or it is possible that this missing time is distributed through more than one of the prominent surfaces mapped at and near the type section (Figs. 8.20 and 8.21). Fragmentary preservation with more time missing than preserved as sediment seems the more likely scenario (and the same also seems

more probable for the undifferentiated Blackhawk Formation). In Fig. 8.14 the Castlegate Sandstone is shown as a set of three superimposed units separated by significant time breaks, following the suggestion of Bhattacharya (2011), but other interpretations are possible.

Figure 8.22 represents a speculative attempt to account for the distribution of elapsed time between preserved deposits and bounding surfaces in the Castlegate Sandstone at the type section. At left is the sequence model of Olsen et al. (1995). The interpretation of the two-part succession in terms of low- and high-accommodation depositional environments is not consistent with the data now available on the relationship between channel migration and avulsion rates and rates of accommodation (Miall 2014a, Sect. 6.2). Fluvial sequence models, such as those of Wright and Marriott

**Fig. 8.21** The Lower Castlegate Sandstone south of the type section. The south face of the Castle Gate is seen at *left*. Lettering of the surfaces is as in Fig. 8.20. Width of field of view is about 300 m

**Fig. 8.22** Different interpretations of the Castlegate Sandstone at the type section. See text for explanation

(1993) and Shanley and McCabe (1994) are based on observations at *SRS-7* rates, that is, a time scale of $10^3$–$10^4$ years, and sedimentation rates of $10^0$–$10^1$ m/ky. They used modern studies and simulations that assume an accommodation rate up to three orders of magnitude more rapid than is typically represented in the preserved ancient record.

During this study it was not possible to replicate the two-part subdivision of the Castlegate Sandstone proposed by Olsen et al. (1995), at least not at the type section. This exposure is shown in Fig. 8.20, and the bounding surfaces there are repeated in Figs. 8.21 and 8.22 (the lettering is shown for convenience, using the original labeled surfaces A, D, and H from Miall and Arush 2001a). The type section consists of a succession of braided sandstone sheets bounded by surfaces of at least fifth-order rank (Miall 1993), in the terminology of Miall (1996). At least one of these, surface D of Miall and Arush (2001a), is interpreted as a sequence boundary (a sixth-order surface) but we have no evidence about the greater or lesser significance of the other surfaces in this outcrop. More than one could be "cryptic" sequence boundaries, in the terminology of Miall and Arush (2001a, b).

At the right hand side of Fig. 8.22, two other scenarios for the Castlegate Sandstone are shown. One shows a version of Bhattacharya's (2011) speculation about three Castlegate sequences. The sequences would likely represent a response to allogenic forcing, such as flexural loading and/or changes in intraplate stress (Deramond et al. 1993; Naylor and Sinclair 2007). The sequence boundary between the two lower sequences is correlated to surface D at the type section. The upper sequence boundary cannot be located in the type section. None of the surfaces between D and H exhibit any features, such as cut-and-fill relief, extensive lag deposits, or evidence of early cementation, that would indicate their significance. This could be a characteristic of a "cryptic" sequence boundary, of the type suggested by Miall and Arush (2001b). The observation of tidal sedimentary structures in the upper part of the Castlegate Sandstone at locations east of the type section (Olsen et al. 1995; Yoshida et al. 1996) may be explained by long-term (SRS 9-11) changes in paleoslope that permitted periodic tidal influx into the Castlegate rivers. The three sequences are envisaged as sequences formed at *SRS 8* rates, deposited at average sedimentation rates of 0.29 m/ky and each representing 195 ky of elapsed time. As seen in Fig. 8.22, this leaves a substantial amount of "Castlegate" time unrepresented, with only 29 % of the 2 million years of time allotted to this formation represented by sediments, at the *SRS 8* time scale.

Another interpretation of the Castlegate Sandstone is that it consists simply of a set of unrelated braided sandstone sheets, some formed successively over a limited time range, some separated by longer intervals such as the unconformity represented by surface D. These would represent long-term geomorphic processes, and should be evaluated at *SRS 7*.

This is how they are presented at the right side of Fig. 8.22. Nine braided sandstone sheets, averaging 19 m thick (bounded by the ten surfaces A to H at the type section), deposited at an average *SRS 7* rate of 3 m/ky would require in total only 57 ky to accumulate, which is less than 3 % of the 2 million years age range of the Castlegate Sandstone. Each sheet would represent an average of about 6 ky. How to account for the remaining elapsed time? Intervals of non-deposition or erosion between each sheet would average 216 ky. The sandstone sheets are probably accidental remnants of long-lived braid-plain deposits across which temporary sediment storage and remobilization were the norm, with preservation taking place only because of abandonment following avulsion events. The lengthy intervals between each sheet have not left any identifiable signature, such as mature paleosoils, or evidence of early cementation (except for surface D) or of deep erosion. As Willis (2000, his Fig. 17; Fig. 6.62 of this book) demonstrated, paleocurrent patterns shifted significantly during deposition of the Castlegate Sandstone, from southeastward during the deposition of sequence 1 (sequence definitions as in Fig. 8.19), to S to SSE in sequence 2, to E to NE in sequence 3. Such shifts presumably reflect subtle tectonic tilts in regional paleoslope at SRS 9-11 rates, and could have facilitated switches in local flow directions, particularly where aggradation of a channel belt created slope advantages for alternative flow directions, the same type of process that leads to the fan shape of alluvial fans and the distributary pattern of deltas. Some of the younger sheets contain evidence of tidal influence, indicating a slow but steady rise of sea level during the deposition of this formation.

It is possible that both the *SRS 8* and *SRS 7* scenarios of Fig. 8.22 are correct, with the latter nested within the former. In other words, the Castlegate Sandstone consists of sequences that are themselves composed of unrelated braided sandstone sheets.

### 8.9.6   Conclusions

Sedimentation rates derived from studies of modern non-marine and shallow-marine environments deposited over periods of decades (*SRS 5*) are in the range $10^2$–$10^3$ m/ky. This is up to four orders of magnitude more rapid than the rates derived from geological studies, such as that of the Mesaverde Group. When rates comparable to the post-glacial record ($10^4$–$10^5$-year records; *SRS 7, 8*) are applied to the preserved facies successions that comprise this group, less than 10 % of the 4.86 million years elapsed time represented by the group can be accounted for. The remainder is "missing" at bedding planes and other bounding surfaces of all types, a pattern predicted by Sadler (1999). Nearshore sediment bypass with progradation of shelf clinoforms could account for a significant proportion of the elapsed time. In the shallow-marine environment, widespread non-deposition or erosion are thought to have preceded ravinement and the development of flooding surfaces from which shoaling-upward successions (parasequences) then pro-graded. In the Mesaverde Group there are 23 such major surfaces that define the boundaries of the six members and submembers. Each of these could account for a time span of up to ~ 100,000 years. The major sequence boundary at the base of the Castlegate Sandstone, which can be traced for more than 150 km eastward into Colorado, could represent as much as 1 million year, and the surface that truncates the Buck Tongue could also represent a comparable interval of missing time. Many other bounding surfaces simply repre-sent long-term sediment bypass, with no net accumulation. The Castlegate Sandstone may consist of a set of unrelated fragments of braided sandstone sheets separated by signifi-cant erosional or non-depositional breaks.

Analyses of long-term processes, including mass-balance transport models, and interpretations of shoreline trajectories through time, need to take into account the fact that far more time is missing than is represented. Continuity is the exception in nonmarine and shallow-marine stratigraphy. Bounding surfaces are the "dark matter" of sedimentology. We know they are there, but means are not yet available for a complete analysis of their range of characteristics and time significance.

In the next section we close in somewhat on this problem, by examining some examples of detailed stratigraphic studies where much better chronostratigraphic control is available, providing some insights on the level of interpre-tive detail that is now available using modern, current, chronostratigraphic methods.

## 8.10   The Future of Conventional Chronostratigraphy

### 8.10.1   Current Examples of Outstanding Work

The modern geological time scale (GTS2004: Gradstein et al. 2004a, b, 2012, and the updated scale maintained at www.stratigraphy.org; see Fig. 7.33 of this book) takes into account the types of potential error discussed in Sect. 7.8.5 by collating data from multiple sources. The construction of Composite Standard Reference Sections using graphic cor-relation methods (Sect. 7.5.4) is part of this process. Cur-rently finalized global stratotypes for systems, series and stages are identified in Gradstein et al. (2004a, b, 2012) and on the website, with references to published documentation, most of which consists of reports in the journal *Episodes* by representatives from boundary working groups. Realistic error estimates are provided for Phanerozoic stages, and

range from very small values ($10^4$-year range) for most of the Cenozoic, the time scale for which is increasingly linked to an astrochronological record, to as much as ±4 mil-lion years for several stages between the Middle Jurassic and Early Cretaceous. Figure 8.23 illustrates the expected error in age estimation through the Phanerozoic, based on infor-mation available in 2007. The refinement of the scale expected to accrue from the integration of astrochronology into the data base (Fig. 8.23e) may be regarded as optimistic, but the key workers in this field make a good case for such a development, as discussed in Sect. 8.11.

Miall (2010, Chap. 14) provided a discussion of the global scale as applied to the issue of the global correlation of se-quences, particularly as this relates to the possible global extent of sequences based on their presumed origin as a result of eustatic changes of sea level. An example of this type of work is illustrated in Figs. 8.24 and 8.25. The stratigraphy of the New Jersey continental margins has been extensively studied by K. Miller and his colleagues, in part to provide a chronostrati-graphic baseline for global sequence studies. Figure 8.24 shows the sequence subdivision and generalized lithologies in relevant boreholes along a transect across a portion of the continental margin which consists of a series of prograding Oligocene cli-noform sets. Figure 8.25 is an example of the highly detailed chronostratigraphic documentation of the Oligocene-Miocene stratigraphy that can now be developed, based on integrated biostratigraphy, oxygen isotope stratigraphy, strontium isotope stratigraphy and magnetostratigraphy. The details of this work are presented several major regional studies (Miller et al. 1998, 2008; Browning et al. 2008; Kominz et al. 2008) and are dis-cussed in detail by Miall (2010, Sect. 14.6.1). The following is a brief summary.

Figure 8.26 reproduces a correlation diagram developed by Betzler et al. (2000) comparing the Miocene-Pliocene $10^6$-year sequence record on the Bahamas Bank and the Queens-land plateau, to which I have added the sequence boundaries from the same interval of the New Jersey continental margin succession (from Kominz et al. 2008). The Queensland-Bahamas study is based on correlations of cal-careous nannoplankton and planktonic foraminifera. The density of the biostratigraphic data ranges from one to three biohorizons per million years, calibrated to the Berggren et al. (1995) time scale. Gradstein et al. (2004a, b, Fig. 1.7) indi-cated that stage-boundary ages have been adjusted by between about 0.1 and 0.8 million years. from the Berggren et al. (1995) scale, not enough to significantly affect the level of correlation with the more recently dated New Jersey section.

Figure 8.26 indicates a high degree of correlation between Queensland and the Bahamas. Betzler et al. (2000, p. 727) stated that throughout the Miocene-Pliocene "The isochroneity of sea level lowstands in two tectonically unrelated carbonate platforms is strong evidence for eustatic

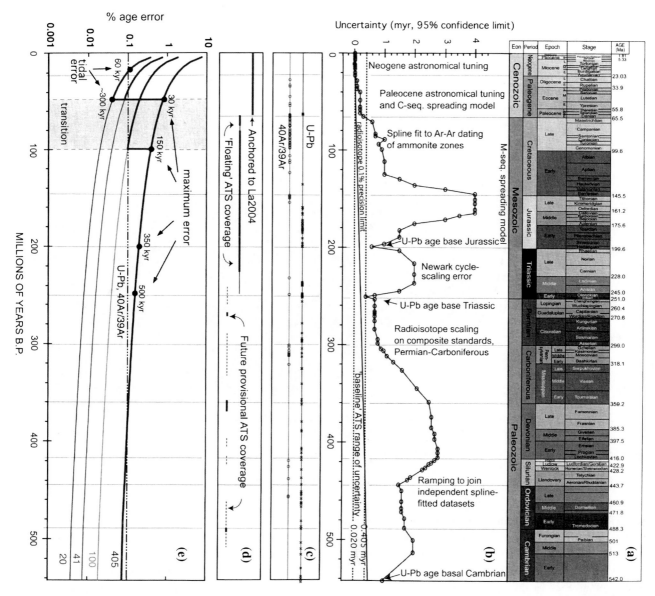

**Fig. 8.23** **a** The standard divisions of the Phanerozoic International Geologic Time Scale (Gradstein et al. 2004a). **b** Estimated uncertainty (95 % confidence level) in the ages of stage boundaries. **c** Distribution of radiometric ages used in the construction of the time scale. **d** Documented and potential astrochronological time series. **e** Estimated error to be expected when the astrochronological time scale through the Phanerozoic is completed and integrated into the International Geologic Time Scale (Hinnov and Ogg 2007)

sea level changes as the controlling factor on large to medium-scale (1–5 Ma) stratigraphic packaging." Betzler et al. (2000) went on to note that the number of sequence boundaries in their data does not correspond to the number of boundaries in the Hardenbol et al. (1998) chart—an issue that, as discussed by Miall and Miall (2001) constitutes one of the criteria for sequence correlation that has consistently misled and confused stratigraphers. We do not consider the lack of correlation with the Hardenbol chart further, for reasons discussed elsewhere (Miall 2010). Betzler et al. (2000) suggested that the accuracy of their sequence boundary correlation might be improved by more intensive

biostratigraphic analysis. However, for our purposes, it is much more significant to evaluate the Queensland and Bahaman correlations with the sequence boundary record from New Jersey (accurate sequence boundary ages for the succession younger than 10 Ma are not available from this area). This is a comparison between two independent studies and therefore fulfills one of the most important criteria for a meaningful test of the global-eustasy paradigm. In this case, the evidence indicates a strong degree of correlation between New Jersey, The Bahamas, and Queensland. Despite the small differences in the time scales used in these two different interpretations, the comparisons between the data sets

**Fig. 8.24** Details of the Oligocene sequences O1–O6 forming a set of seaward-dipping clinoforms along a NW-SE transect across the New Jersey continental margin. Gross lithologies for each borehole are indicated. Upper-case letters to the left of each borehole indicate foraminiferal biofacies, identified by factor analysis (Kominz and Pekar 2001)

are striking. The adjustments that would be required to recalibrate the Betzler et al. (2000) study to GTS2004 would not significantly change the pattern visible in Fig. 8.26. The case for eustatic control of Miocene stratigraphy is considerably strengthened by this comparison.

Also shown in Fig. 8.26, for reference purposes, are the sequence boundaries documented in the "great Neogene sedimentary wedge" (wording of McGowran 2005, p. 190) reconstructed by Vail et al. (1991), based in part on an analysis of the Antarctic margin (Bartek et al. 1991). Their model of the Neogene wedge claims to illustrate a global pattern of sea-level change and accompanying stratigraphic architecture that can be recognized worldwide. However, there is very little correspondence between these sequence boundaries and the New Jersey-Queensland-Bahamas events documented in Fig. 8.26, which calls into question the chronostratigraphic basis on which the wedge model was drawn. McGowran (2005, p. 191) indicated that the number and ages of the sequence boundaries has been modified by Hardenbol et al. (1998), but the boundaries and ages indicated in his redrawn version of the Vail et al. (1991, Fig. 12)

diagram are the same as in that diagram, which were, in turn, based on the Haq et al. (1987, 1988) global cycle chart.

This discussion of the New Jersey margin and what the chronostratigraphy tells us provides an excellent example of the capabilities of modern chronostratigraphic methods. The demonstration of the correlatability of Neogene sequences between tectonically unrelated areas that is shown in Fig. 8.26 supports the interpretation of glacioeustatic control for the sequence stratigraphy, a not unexpected result given the likely extensive development of the Antarctic ice cover by the Neogene (Miller et al. 2005).

The second example discussed here is the detailed study of late Cenozoic sedimentation of the Brazos-Trinity depositional system in the Gulf Coast margin by Pirmez et al. (2012) and Prather et al. (2012). This study was discussed in Sect. 6.3.2 as an example of the construction of a regional stratigraphy using 2-D seismic data integrated with well and core data. Here I focus on the chronostratigraphic methods used in the study. Figure 8.27 provides the basis for chronostratigraphic dating and correlation used in this study. Biostratigraphic zonation of planktonic

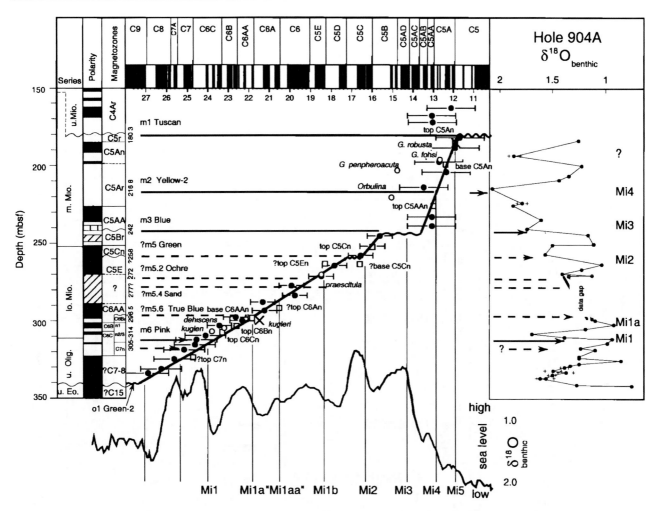

**Fig. 8.25** Age-depth diagram for Hole 904, New Jersey continental slope. The core record, with magnetozones, is indicated at *left*. Sequence boundaries, recognized from seismic and facies studies, are shown by the *horizontal lines* extending from the core log across to the *right*, to the sloping line of correlation. Sequences are named, from the top down, "m1 Tuscan", etc. The standard magnetostratigraphic time scale is shown across the *top* of the figure and the $\delta^{18}O$ record is shown

along the *bottom*. Dated sample points are indicated by symbols along the line of correlation: strontium isotope values, with error bars *black circles*), planktonic foraminifer lowest occurrences (*open circles*) and highest occurrences (*crosses*), and magnetostratigraphic reversal boundaries (*squares with chron number*). $\delta^{18}O$ from hole 904A are indicated in the *box* at right. Unconformities are indicated by *wavy lines* in the line of correlation (Miller et al. 1998)

foraminifera and nannofossils obtained from core was supplemented by oxygen isotope measurements. Additional age and correlation information was provided by volcanic ash geochemistry (correlated to Caribbean volcanic events). For deposits younger than 40 ka, radiocarbon dating was used on planktonic foraminifera, reworked plant debris and shell material. Magnetostratigraphy proved not be useable because of the presence of sand and silt layers, which did not provide magnetic information, and because of highly variable sedimentation rates, making the "bar-code" matching of the magnetic signal to the standard record difficult to impossible. The absolute ages of first and last appearance datums provided key datable and mappable horizons, as shown in Fig. 8.28. Their ages are based on standard studies referenced in Pirmez et al. (2012, p. 119).

The average sedimentation rate for the entire succession examined in this project averages 2.1 m/ka (a maximum of 300 m of section deposited in about 140 ka). This corresponds in time range and sedimentation rate to *SRS-7* (Table 3.3). However, the detailed chronostratigraphy reveals that there are significant sedimentation breaks and condensed intervals, and intervals within which sedimentation rates are substantially higher (Fig. 8.29). Hemipelagic units reveal sedimentation rates between 0.1 and 9 m/ka, which is a similar order of magnitude to the long-term rate but in at least one instance more than four times the long-term rate, which emphasizes the importance of knowing the time scale over which the measurement is made. Sand-rich ponded apron deposits yield sedimentation rates an order of magnitude greater, ranging between 20 and 64 m/ka, which corresponds to the higher-rate end of *SRS-7*.

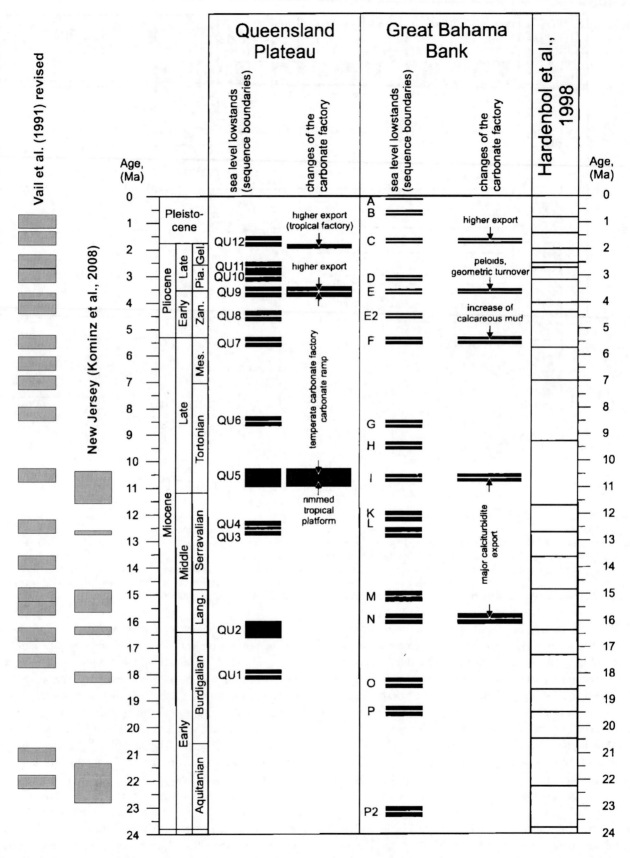

**Fig. 8.26** Comparison of Neogene sequence boundaries in the platform carbonate margins of Queensland and the Bahamas Bank. The sequence boundaries from the Hardenbol et al. (1998) scale are shown at *right*. From Betzler et al. (2000). Two additional sets of sequence boundaries have been added, at *left*, the sequence boundaries from the Miocene portion of the New Jersey record (from Kominz et al. 2008), and the Neogene record of the Antarctic (from Vail et al. 1991, with boundary ages revised by Hardenbol et al. 1998)

**Fig. 8.27** Chronostratigraphic chart used for the Brazos-Trinity depositional system. *Mwp* meltwater pulse. *LO/FO* last/first occurrence locally observed in the area. *FAD/LAD* first/last appearance datum. Ash layers are identified by an *asterisk* (Pirmez et al. 2012, Fig. 7, p. 119)

Many additional details may be gleaned from detailed study of these data, including the partitioning of the sediment flux between different basins within the project area. This, in turn, throws light on the evolution of the dispersal system as it crossed the shelf and fed several separate basins arranged along a downstream trend through the Sigsbee Knolls.

The first example discussed in this section, the Neogene stratigraphy of the New Jersey continental margin, indicates

that modern conventional chronostratigraphic techniques may be able to provide ages and regional to global correlations for the late Paleogene (Oligocene) to within about ±0.5 million years. In the case of the second example, much younger rocks were under discussion, and accuracy appears to be in the $10^5$-year range, with at least an order of magnitude greater accuracy attainable for the last 40 ka of stratigraphic time through the use of radiocarbon dating.

**Fig. 8.28** Correlation of wells in the Brazos-Trinity depositional system, showing identified unit boundaries at *left* (Pirmez et al. 2012, Fig. 10, p. 122)

Each of these examples were approached by their authors as studies of typical stratigraphic problems, making use of all available chronostratigraphic data. This is different from the objective of perfecting the geological time scale, when field locations are selected specifically to maximize the chronostratigraphic information available, rather than to answer some sedimentological or tectonic question. This is an important distinction because, as noted earlier (see Fig. 8.23), accuracy and precision of the Geological Time Scale in the $10^5$-year range is expected to be attainable for the Mesozoic when cyclostratigraphic studies are completed. Even if this proves to be correct, this does not mean that every Mesozoic section will then be capable of dating and correlation at that level of accuracy, because accuracy depends on the chronostratigraphic tools available in each case for each section.

### 8.10.2   The Use of Wheeler Diagrams

The construction of Wheeler diagrams focuses attention on the issue of elapsed and missing time (Fig. 8.1). Given the modern improvements in chronostratigraphy and the availability of modern sequence-stratigraphic methods, the use of this device can help to throw considerable light on sedimentary processes, and provide diagrammatic constructs that

force the stratigrapher to think through the implications of their reconstructions.

A Wheeler diagram for the Brazos-Trinity deposits discussed in the previous section (Fig. 8.30) helps to emphasize the substantial variability in sedimentation rates that characterized this region. Most of the elapsed time between 140 and 25 ka is represented by hemipelagic or condensed intervals. Sand-rich intervals (seismic units 30–70) were deposited rapidly during a relatively very short interval (25–15 ka), corresponding to the Last Glacial Maximum, when sea-levels were low, and immediately before a meltwater pulse from the Mississippi system corresponding to the commencement of rapid post-glacial sea-level rise.

The use of sequence methods for mapping and correlation, while providing much more logical architectural reconstructions of complex sedimentary units (Sect. 6.2.3), may raise questions that can take interpretations to an entirely new level (Bhattacharya 2011). For example, Fig. 8.31 is a detailed sequence correlation of a delta in the Ferron Sandstone of Utah. At first sight the incised valley-fill complexes (units at several levels coloured red in the figure) that mark the base of sequences 1 and 2 would appear to define a relatively simple, almost layer-cake stratigraphy. This is essentially the interpretation shown in Fig. 8.32a, in which all the incised valleys are assigned to Sequence 2. However, there is no definitive proof that this is the correct

**Fig. 8.29** Age-depth curve for two wells in the Brazos-Trinity depositional system. Numbers 10–79, at *right*, refer to seismic units (see Fig. 6.20) (Pirmez et al. 2012, Fig. 12, p. 125)

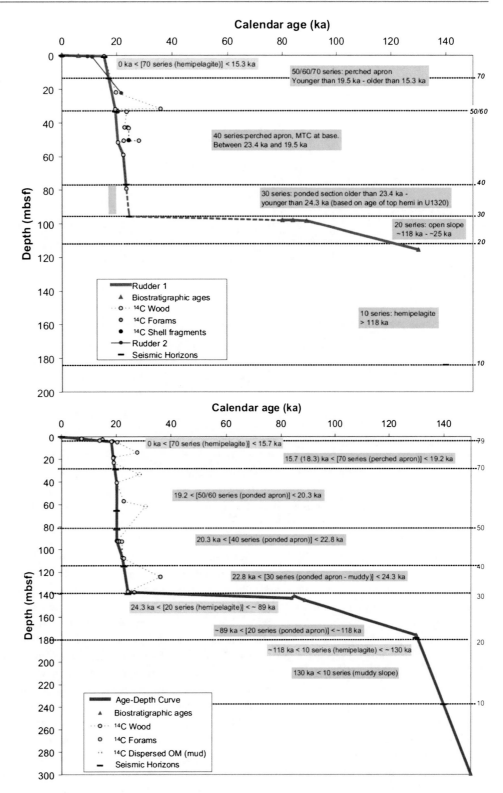

interpretation. The regional erosion surface that marks the base of sequence 2 across the left-hand end of the section could be a composite of erosion surfaces cut at different times after the end of deposition of parasequence 12. This is the basis for the interpretation shown in Fig. 8.32b, in which

it is suggested that the incised valley-fills that comprise the first valley-fill complex were formed at different times following the deposition of para sequence 9, each one correlated in time to a different shoreface tongue. The alternative interpretations may be followed through with questions

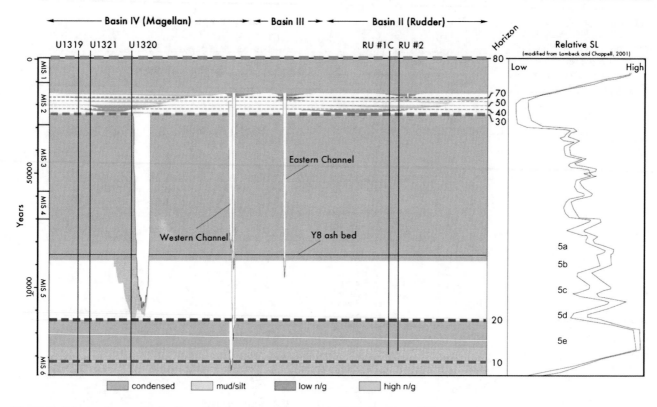

**Fig. 8.30**  Wheeler diagram for the Brazos-Trinity system (Pirmez et al. 2012, Fig. 14, p. 128)

about sedimentation rates and missing time, of the type raised in Sect. 8.9. What differences in fluvial style, if any, may be suspected from the different interpretations shown in Fig. 8.32? Was there a single rapid episode of changing accommodation or sediment supply that gave rise to a single, amalgamated incised valley complex (Fig. 8.32a)? Or was this a more drawn-out process, reflecting longer-term processes, or a more distant signal that became distorted ("shredded"?) down the transport direction, which is what could be implied by the interpretation indicated in Fig. 8.32b?

Comparable questions may asked regarding the age and correlation of the sandstone sheets comprising the Castlegate Sandstone (Sect. 8.9.5). The original interpretation of this unit as a single sheet extending basinward for more than 150 km (e.g., Van Wagoner et al. 1990) was called into question by the suggestion of Bhattacharya (2011, Fig. 18) that it comprises several sequences, each correlated to a shoreface tongue representing part of the Desert Member.

Developments in the acquisition and processing of 3-D seismic data have provided the possibility for the automated generation of Wheeler diagrams for subsurface sections, including four-dimensional visualizations that illustrate paleogeographic evolution in a temporal framework. Stark et al. (2013) described the evolution of a technique for extracting time information from seismic data called

"computational seismic chronostratigraphy." Qayyum et al. (2015) discussed the method and provided worked examples. Other worked examples are available in the special issue of "The Leading Edge" edited and introduced by Stark et al. (2013).

The procedure is as follows: automatic tracking software is used to generate multiple mapped horizons within the seismic volume. Seismic terminations, corresponding to such real structural features as unconformities or downlap surfaces, can be recognised and mapped by the software by the convergence of reflections to within a pre-set spacing value. A Wheeler diagram may then be constructed by flattening each surface to the horizontal. In this way thickness, the Z dimension (the y-axis of a 2-D plot), is converted to relative time. Structural and thickness information are lost in this process, and the resulting diagram emphasizes non-depositional and erosional hiatuses by the presence of blank space. The horizontal extent of each surface reflects either (1) its erosional extent, for example the technique provides a clear time-space map of subaerial unconformities, or (2) the original depositional extent, such as that of a clinoform unit, extending from its coastal origins at the mouth of a delta, down the continental margin to its lapout at the foot of the continental slope (this is, of course, within the limit of seismic resolution. In fact, distal clinoform sets may extend for considerable distances into a deep basin as thin

**Fig. 8.31** Cross-section through the Ferron Notom delta (Turonian), Utah, hung on bentonites. Para sequences are numbered 1–17, and are grouped into six sequences. Two alternative Wheeler diagrams for this cross-section are shown in Fig. 8.32 (Bhattacharya 2011, Fig. 11, p. 135)

pelagic units). The vertical dimension of the plot depends on the number of autotracked horizons, which means that it varies in scale depending on the heterogeneity of the section and is also dependent on seismic resolution. The time scale to be derived from the vertical axis is therefore relative, and variable. Nonetheless, the resulting ability to visualize missing section, reflecting non-depositional or erosional hiatuses may provide a valuable addition to the tools for sequence interpretation and the development of paleogeographic models.

An example of the use of this technique is illustrated in Figs. 8.33 and 8.34. The gradual stepping out of the axis of deposition as the clinoforms prograde across the continental margin is graphically seen in the Wheeler plot, and the substantial age span of the hiatus at right representing a subaerial unconformity is particularly striking. This hiatus spans essentially the entire interval from one basal surface of forced regression to the next such surface above.

One problem with this method of developing Wheeler plots is the distortion of time that results from variable sedimentation rates and would also result from long-continued deposition of homogeneous sediment, resulting in reflection-free records. For example, in Fig. 8.33 note the two pale-coloured reflections labelled "condensed section". The labeling arises from the seismic interpretation of the section, which suggests a thin unit that blankets sequence 1. The term "condensed section" typically indicates condensation of time, such as is commonly seen in the thin units that form offshore during rapid transgression

(Loutit et al. 1988). However, in Fig. 8.33 there is no suggestion of the condensation of time because time, here, is in fact a distorted representation of seismic heterogeneity. Compare this with the lengthy intervals of condensation shown in Fig. 8.30, which are based on careful chronostratigraphic calibration of the seismic and well records. The Wheeler plot shown here was drawn by the seismic interpretation algorithms, but it is possible to imagine a further step, in which information on sedimentation rates and chronostratigraphic tie points is used to adjust the vertical scale (y-axis) of the plot to more accurately represent elapsed time.

### 8.10.3 Improving Accuracy and Precision

Several developments are underway that promise to improve chronostratigraphic accuracy and precision, both for the purposes of local to regional correlation (for which, relative dating would suffice) and also for the purpose of improving the Geological Time Scale. Developments in cyclostratigraphy and astrochronology are discussed in Sect. 8.11. In this section we discuss briefly the work to maximize the information available from the biostratigraphic record.

The rapid increase in computer power, in addition to enormously increasing the utility of 3-D seismic data, is also providing the opportunity for many different approaches to the consolidation, sorting, selection and correlation of all the different types of chronostratigraphic information that are

**Fig. 8.32** Two alternative
Wheeler diagrams for the
cross-section shown in Fig. 8.31.
The differences between them
focus primarily on the
chronostratigraphic interpretation
of the incised valley-fills
(Bhattacharya 2011, Fig. 14,
p. 138)

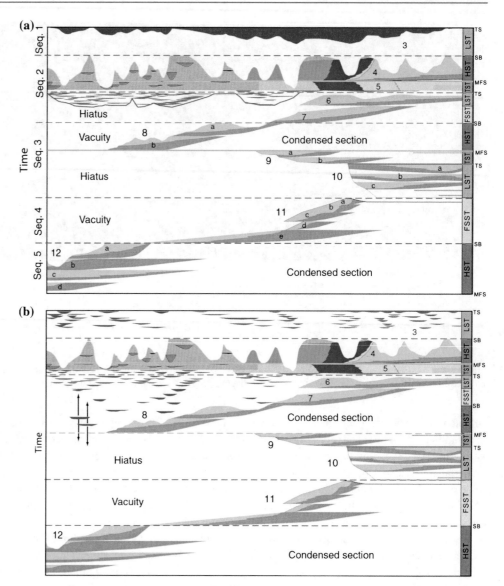

increasingly available from the rock record. The concept of the **composite standard reference section** (Sect. 7.5.4) can be vastly expanded by adding such observational data as isotopic excursions, key seismic reflections, paleomagnetic reversals, chemostratigraphic determinations, dated volcanic ash beds and bentonites to the data on taxon ranges. Taxon ranges, as observed in any given section, may be incomplete, for reasons explained in Sect. 7.5, and where subsurface information is used, borehole caving may introduce inaccuracies into the record, which is why the use of the composite standard reference section method is so powerful, because it constitutes a method that includes the opportunity for continuous correction and refinement. There are now numerous algorithms available for sorting and standardizing the various types of data now being assembled in digital data bases. Sadler et al. (2014) provided a useful summary.

Figure 8.35A provides an example of the problems that may arise with the plotting of raw chronostratigraphic data points. Because of ecological selection, selective preservation, or incomplete sampling, the indicated correlations of taxa between sample points may include crossing correlations. First-appearance datums may appear too high in some sections because of these problems, whereas last-appearance datums might be too low because of caving in the subsurface, or because of reworking. By contrast, unless borehole caving is a problem, such data points as dated ash beds can be regarded as fixed points that may be used as "nailed" datums (Sadler et al.'s 2014 term; see Fig. 8.36), around which bioevents may be adjusted. Corrections to local taxon ranges may involve expansion to fit the composite standard ("jacked" apart in Fig. 8.36) or reductions, where corrections from other data, including fixed or "nailed" datums are

**Fig. 8.33 a** Seismic transect through two wells with overlain systems-tract interpretation of a Pliocene deltaic unit in the North Sea. **b** Automated Wheeler diagram of the studied interval, with sequence and systems-tract interpretation at left. The Y-axis of the diagram represents relative geological time. The colour-coded lines are autotracked events, and the colours represent relative thickness within each systems tract. Note that *Sequence 1* shows a low rate of sedimentation in the basinward direction. In seismic Wheeler diagrams, such condensed sections show up as hiatuses. *TL* transgressive lag; *BSFR* basal surface of forced regression; *MFS* maximum flooding surface; *MRS* maximum regressive surface; *SU* subaerial unconformity; *CC* correlative conformity (Qayyum et al. 2015, Fig. 5)

present ("clamped: Fig. 8.36"). Figure 8.35c shows the result of a fully automated correlation exercise that has adjusted and corrected the raw chronostratigraphic data to provide a best-fit solution. It may be compared with Fig. 8.35b, which represents a lithostratigraphic correlation guided by older, conventional biostratigraphy.

An insight into the accuracy and precision that is now available for the chronostratigraphic calibration of the distant geologic past was provided by Sadler et al. (2009), who reported on the use of a global graptolite data base for the Ordovician-Silurian interval comprising 17,861 locally observed range-end events for 1983 taxa, and 131 local observations of 57 other events (23 dated events and 34

marker beds) in 446 sections worldwide. This paper describes the details of how data points are ordered, culled, corrected, and calibrated to generate a time scale. The resolving power of the resulting scale is mostly in the $10^4$- to $10^5$-year time range (Figs. 8.37 and 8.38), which is at least an order of magnitude greater precision than that obtainable by conventional biostratigraphically-based methods for this time interval (Fig. 8.23). To repeat a caution made elsewhere in this book, the availability of such a precise scale does not automatically mean that any graptolite-bearing Ordovician-Silurian section may be dated with a comparable precision. The application of the scale, in practice, depends on the locally available suite of datable materials.

**Fig. 8.34** 4-D Wheeler diagram for the Pliocene interval shown in Fig. 8.31. The three-dimensional space (two horizontal dimensions plus relative time in the vertical dimension) is filled with information from the 4th dimension, systems tracts thickness in this case. The bottom slice is a colour-blended spectral decomposition attribute slice for a particular reflection event (in this case, a MFS, maximum flooding surface). Along the surface, several geomorphological features are identifiable—NE- to SW-direction flowing deep water channels (*blue arrows*) and NW-SE oriented elongated features that are interpreted as sand ridges comparable to those existing in the North Sea at the present day (Qayyum et al. 2015, Fig. 6)

## 8.11   High-Resolution Event Stratigraphy, Cyclostratigraphy and Astrochronology

Cyclostratigraphy and astrochronology are amongst the most active areas of stratigraphic research at the time of writing. While there are many success stories, including the development of reliable astrochronological time scales, with associated GSSPs, for much of the Cenozoic, and many individual studies of "floating" scales for parts of the older Phanerozoic (Hilgen et al. 2015), most of the successes relate to studies of deep-marine and lacustrine successions, where it could be expected that orbital control may act essentially undisturbed over millions of years within environments where autogenic influences are weak and other allogenic controls are longer-term and readily recognized. The cautions expressed by Miall and Miall (2004) should continue to provide guidance for researchers in this field. Amongst these are the following issues, which relate to all current research into high-resolution stratigraphic definitions and correlation, whether concerned with cyclostratigraphy or not:

*The continuity of the stratigraphic record:* The discussion in this chapter has focused on the issue of sedimentation rates and the ubiquity of "missing" time in the stratigraphic record. Hilgen et al. (2015) asserted that when an integrated approach is used to build a high-resolution time scale it can be demonstrated that many successions are complete at the Milankovitch time scale ($10^4$–$10^5$ years), although it is conceded that not all cycles, especially those of short period, may be recorded. It is conceivable that in some settings orbital forcing may be the dominant allogenic mechanism at orbital time scales, overwhelming or at least modifying autogenic processes, such as the control of fluvial style through a climatic control on fluvial runoff and sediment supply. In such cases, even the distribution of hiatuses in the record might illustrate orbital control. A case study of the Coniacian-Santonian (Upper Cretaceous) interval in the Western Interior of the United States (Locklair and Sageman 2008; Sageman et al. 2014) provides an excellent example of the use of multiple chronostratigraphic criteria in the establishment of more accurately dated stage boundaries and biozones, and the documentation of a 400-ka cyclicity in the Upper Cretaceous of the Western Interior Basin. They

**Fig. 8.35** Correlation of
foraminifera, nannofossils,
dinoflagellates, spores and pollen
from 8 wells in New Zealand's
Taranaki Basin. **a** Raw correlation
of range tops of the 87 most
reliable of 351 taxa. Range bases
were potentially compromised by
borehole caving.
**b** Lithostratigraphy and
traditional biostratigraphic
correlation. **c** Outcome of
automated correlation to a
fully-resolved, composite, scaled
time-line by minimal adjustment
of observed ranges and insertion
of missing taxa (Sadler et al.
2014, Fig. 2, p. 6)

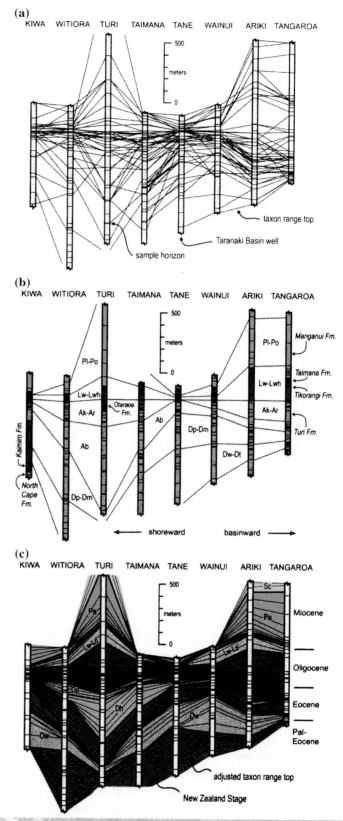

*Figure 2: Correlation of foraminifera, nannofossils, dinoflagellates, spores and*

Figure 3: A tiny fraction of end-Ordovician (Hirnantian) information: 9 taxon ranges and 3 stable isotope excursions constrained by 92 local

**Fig. 8.36** A sample of end-Ordovician (Hirnantian) information: 9 taxon ranges and 3 stable isotope excursions constrained by 92 local event observations from 7 Estonian cores (see Sadler et al. 2014 for references). To align all cores with one sequence of events, some observed event pairs may be jacked farther apart (e.g. taxon range ends); others may be clamped closer together (e.g. uncertainty intervals on parts of stable isotope excursions); and some must be left in-place—the nailed horizons of steepest onset (*thick dashed line*) of the Hirnantian Isotopic Carbon Excursion (HICE) (Sadler et al. 2014, Fig. 3, p. 8)

**Fig. 8.37** Variation of potential resolving power as a function of age in an Ordovician-Silurian time scale generated by the computer-based optimization process described by Sadler et al. (2009). Differences in interpolated age were determined for all 3904 pairs of adjacent events. These highly variable data are summarized by 10-point (*thin gray line*) and 200-point moving averages (*thick gray line*), plotted at the center point of the moving window (Sadler et al. 2009, Fig. 8, p. 900)

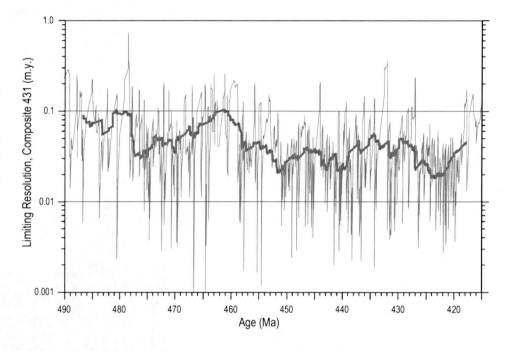

commented that "The use of accurate radioisotope data sets to determine how time is distributed in the strata of interest becomes ever more challenging as the level of temporal resolution increases" (Sageman et al. 2014, p. 966).

*Tuning-induced Milankovitch spectra:* "The potential for Milankovitch frequencies to be introduced into records through astronomical tuning is well known and is a serious drawback" (Hilgen et al. 2015). Another issue is that autogenic and other allogenic processes can generate a cyclicity that may mimic an orbital cycle thickness spectrum (Algeo and Wilkinson 1988). The use of various statistical tests and a careful attention to the real cyclicity in actual stratigraphic data should alleviate this problem.

*Potential error in the assignment of numerical ages to orbital successions:* Until relatively recently, the ability of radiogenic dating systems to provide sufficiently accurate age determinations was a major stumbling block in the research on orbital control. As noted in Sect. 7.8.2, it is now

**Fig. 8.38** An
Ordovician-Silurian time scale
developed by the automated
processing of a
chronostratigraphic data base
consisting primarily of global
graptolite occurrences. *Grey
intervals in the "uncertainty"
column* show the quantified
potential error associated with
each zone boundary (Sadler et al.
2009, Fig. 9, p. 901)

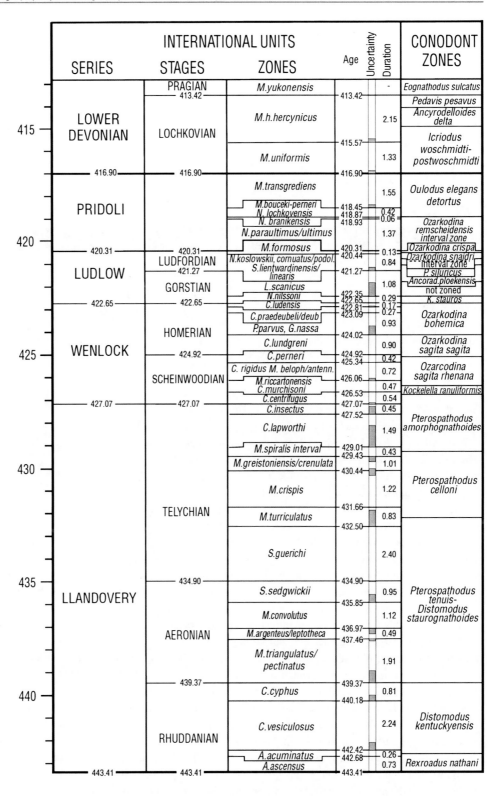

possible, under some circumstances, to develop numerical ages with uncertainties of ±0.1 % (e.g., 100,000 years at 100 Ma). This is at the scale of the long eccentricity cycle (405 ka) for Mesozoic rocks, which means that the extension of an astrochronological scale back into the Mesozoic may be possible. Systematic uncertainties in numerical dating,

such as possible errors in a decay constant, are not a concern in the calculation of orbital frequencies and relative timing for floating sections, so long as the same systematics are used for all relevant age determinations. However, such potential error will still impact the development of a global time scale so long as these imprecisions remain of an order

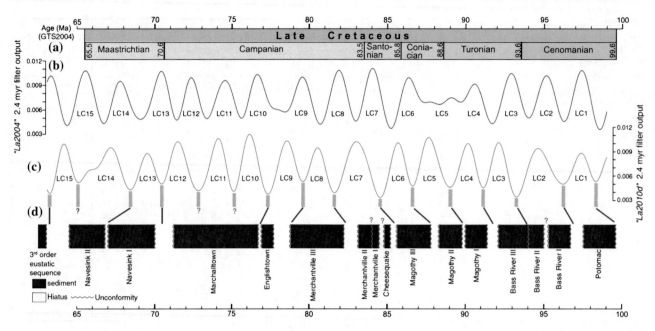

**Fig. 8.39** Comparison of the Late Cretaceous (∼64 to ∼99 Ma) third-order eustatic sequences of New Jersey (Kominz et al. 2008; Browning et al. 2008) with ∼2.4 million years orbital eccentricity cycles (Boulila et al. 2011, Fig. 8, p. 104)

of magnitude comparable to that of orbital frequencies. Current work to develop a High-Resolution Event Stratigraphy (HiRES) is discussed below.

*The stability of astronomical frequencies through geologic time.* The retrodiction of astronomical frequencies back through geologic time is itself an outcome of the careful, integrated stratigraphic studies emphasized in modern research (Hilgen et al. 2015). The most recent astronomical solutions of planetary behaviour suggest that eccentricity cycles have remained stable for the last 50–55 million years, with the 405 ka cycle unchanged for at least 250 million years. It is the 405-ka cycle that guides tuning for the early Cenozoic, and the floating scales of the Mesozoic and Paleozoic.

Many papers have been published in recent years offering cyclostratigraphic interpretations of local stratigraphic successions. It is common practice to show the sections in the time domain, which makes correlation to a calculated astrochronological scale straightforward. However, caution is to be recommended, because this practice may hide irregularities in the succession caused, for example, by autogenic processes. Time series analysis of sections in the thickness domain cannot be used to explore orbital control where there are significant autogenic effects on lithofacies and unit thicknesses. A recent cyclostratigraphic interpretation of the so-called "third-order" sequences of the New Jersey continental margin is a case in point (Boulila et al. 2011; e.g., Fig. 8.39 of this book). A time-depth plot for the five wells drilled through the Oligocene-Miocene succession on the continental slope (ODP Leg 150) reveals very irregular

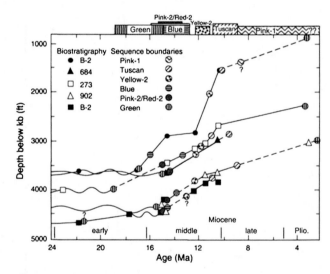

**Fig. 8.40** Subsidence plots for drill holes on the New Jersey continental margin (Miller and Mountain 1994)

sedimentation rates (Miller and Mountain 1994; Fig. 8.40 of this book). Facies analyses of these deposits (Browning et al. 2008) indicates that they were deposited under a range of coastal, deltaic and shelf environments where autogenic redistribution of clastic sediment by wave, tide and other processes is ubiquitous (Fig. 8.41). Sequence analysis based strictly on empirical seismic, facies and biostratigraphic interpretations, indicates that, both in the thickness and time domains, sequence boundaries are quite irregularly spaced and there is no obvious cyclicity in the lithofacies successions (Kominz et al. 2008). Yet Boulila et al. (2011, Figs. 4, 6)

**Fig. 8.41** A depositional model for the Niger delta, used as an illustration of the depositional environments and autogenic processes that, it is interpreted, were active on the New Jersey continental shelf during the deposition of the Early Cretaceous and Paleogene (from Browning et al. 2008, Fig. 7, p. 235)

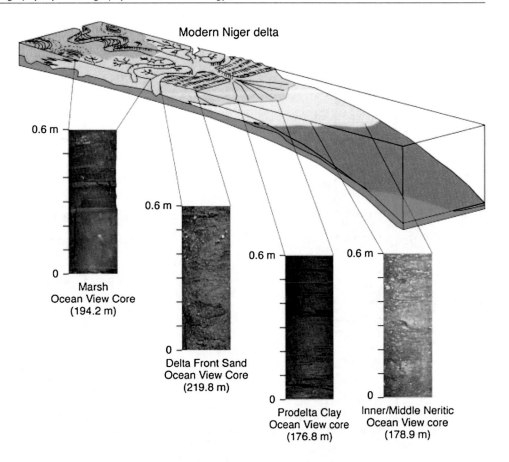

Modern Niger delta

0.6 m

0

Marsh
Ocean View Core
(194.2 m)

0.6 m

0

Delta Front Sand
Ocean View Core
(219.8 m)

0.6 m

0

Prodelta Clay
Ocean View core
(176.8 m)

0.6 m

0

Inner/Middle Neritic
Ocean View core
(178.9 m)

show correlations to a calculated obliquity sequence characterized by 1.2-Ma cycles, or by 2.4-Ma cyclicity (Fig. 8.39). The sequence record and the obliquity record for the Oligocene-Miocene record show the same "number" of sequence boundaries for the 5–34 Ma time period, but there is not the same exact "match" for the Late Cretaceous cycles shown in Fig. 8.39, and otherwise there is nothing about the stratigraphy or the sedimentology of the succession that would suggest cyclostratigraphic control.

A more convincing example of cyclostratigraphic analysis is that of the Green River Formation, by Aswasereelert et al. (2013). Long identified as an example of an orbitally-forced stratigraphic succession (Bradley 1929), this recent study used the ages of six tephra layers interbedded with the approximately 350 m of section to generate a thickness-to-time transformation. Ages of the tephras were obtained with maximum errors of ±0.21 million years, and provided a model of modest, smooth changes in sedimentation rate. Detailed stratigraphic analysis confirmed the importance of the 100-ka astronomical cycle in sedimentological forcing that generated the distinctive layering in this well-known formation.

Where the evidence of autogenic activity, or of tectonic influence acting within a similar time scale to that of cyclostratigraphy ($10^4$–$10^5$-a) is strong, it is clearly incumbent on proponents of cyclostratigraphic control (especially for hanging sections representing the distant geological past)

to do more than provide statistical "proof" of their reality, such as from amplitude spectra of time series studies. Stratigraphic sections should be shown in the depth domain, and a detailed facies analysis performed. Statistical analysis cannot take account of changes in facies or sedimentation rate, the presence of cryptic hiatuses, etc., without the use of special methods. The facies successions should be clearly cyclic, with regularity of facies and of unit thicknesses. Bailey (2009) discussed these and other problems, pointing out the difficulty of designing sufficiently rigorous statistical and other tests for the detection of orbital control. Meyers (2012) discussed the issue of "red noise" in stratigraphic sections, proposing statistical tests to evaluate the suitability of given data sets for cyclostratigraphic analysis. As he pointed out, it is possible for "noise to look like signal." Even random variation can generate "cycles" (e.g., Hiscott 1981).

The current "best practices" for chronostratigraphic analysis make use of the techniques of High-Resolution Event Stratigraphy, or HiRES. "The cornerstone of HiRES is the integration of every available piece of stratigraphic information into a single set of data that can be cross-correlated and internally calibrated" (Cramer et al. 2015, p. 138). Originally conceived by Kauffman (1986, 1988), the guiding principal is that there are many features of a sedimentary section, beyond biostratigraphy, that may be used to develop tools for correlation and dating.

# High-Resolution Event Stratigraphy (HiRES)

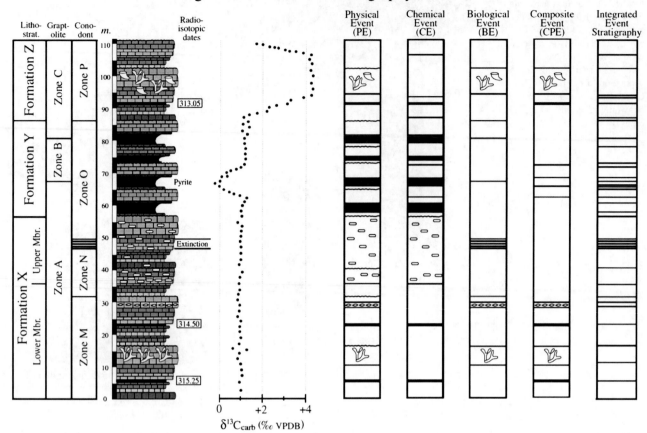

**Fig. 8.42** Demonstration of High-Resolution Event Stratigraphy (HiRES) concepts and methods by using a hypothetical stratigraphic section and data (modified from Kauffman 1988). Lithostratigraphic nomenclature, biostratigraphy, lithostratigraphy, biotic events, radioisotopic age determinations, and stable carbon isotope stratigraphy shown at the *left*. The *right five columns* illustrate the principles of HiRES in which all stratigraphic information is included and a series of 'events' is delimited within the section. All of the chronostratigraphically useful horizons are combined into the integrated event stratigraphy at the far right. In principle, the lithostratigraphic names and biozones at the far left provide a total of 12 discreet horizons for correlation while the integrated event stratigraphy at the far right provides many more potential discreet horizons for correlation and an improved chronostratigraphic resolution (Cramer et al. 2015, Fig. 2, p. 139)

Stratigraphic events, such as storm beds, tephras, and flooding events, may have limited regional extents, but when used in combination across multiple stratigraphic sections and integrated with chemostratigraphic, magnetostratigraphic and other indicators, they may permit highly detailed stratigraphic syntheses to be developed. Figure 8.42 illustrates the principals involved in the development of an "integrated event stratigraphy." The application of this methodology is not necessarily straightforward. Individual events may be diachronous to a greater or lesser degree, particularly first and last occurrence taxon data, reflecting environmental control and migration patterns. For example, for HiRES work in the Paleozoic record, it remains a question whether conodonts or graptolites provide the most chronostratigraphically reliable information (Cramer, 2015).

Other events, such as storm beds may be very confined in their distribution. Chemostratigraphic signatures vary in their diachroneity. Cramer et al. (2015, pp. 148–149) suggested that given the long residence time of strontium and the magnitude of the Sr reservoir in the marine environment, the $^{87}Sr/^{86}Sr$ composition of ocean waters should be quite stable over long time scales, whereas the shorter residence time of $^{13}C$ makes $\delta^{13}C$ a higher-precision chronostratigraphic tool, but one with an imprecision in the $10^{3-4}$-year range reflecting a mixing time of a few thousand years.

Cramer et al. (2015, p. 144) suggested that the process of establishing and refining a HiRES chronostratigraphic record could be clarified by the use of three terms:

**Geochronometry**: refers to the numerical dating of stratigraphic events by the use of radioisotpes.

**Fig. 8.43** Lithostratigraphic and biostratigraphic correlation of Coniacian-Campanian sections from Colorado to Montana, showing the 40-ka cycle record derived by time-series analysis (Sageman et al. 2014, Fig. 2, p. 959)

**Geochronology**: is the synthetic art of reconciling all available geochronologic and chronostratigraphic data to provide temporal calibration of the stratigraphic record.

**Chronostratigraphy** refers to the documentation, testing and calibration of a regional to global framework of age-dated stratigraphy.

The use of these three terms should help to illuminate what exactly is being done in each specific research endeavour. Thus, for example, a discussion of the potential error in correlation, in stratigraphic interval (expressed in metres) of a GSSP is an exercise in chronostratigraphy. The numerical calibration of a specific section is an exercise in geochronology. The laboratory processing of a sample and assignment of an age and range of error is geochronometry. In this last case, consideration of the specifics of precision and accuracy and the difference between them may become significant.

Locklair and Sageman (2008) and Sageman et al. (2014) reported on a study to investigate the cyclostratigraphy of the Niobrara Formation (Santonian-Coniacian) and laterally equivalent units in the Western Interior Basin of the United States. The chalk-marl cycles of the Niobrara were digitized using microscanner resistivity data from drill core. The 85-m core was divided into eight segments, corresponding to the lithologically-defined members of the formation. Ages were assigned using the available time scale for this unit, which is characterized by one of the best established chronostratigraphic records in the Phanerozoic time scale. The depth-to-time conversion and the analysis for orbital cyclostratigraphy were carried out using a succession of statistical methods. An essential step is a technique termed evolutive harmonic analysis (Locklair and Sageman 2008, Fig. 5), which is designed to take into account gradual changes in average sedimentation rate reflecting allogenic forcing at a time scale longer than that of orbital forcing. Calculated rates vary between 0.0075 and 0.0235 m/ka with an average of about 0.07 m/ka, which corresponds to SRS-9 and is a typical accumulation rate for orbital cycles (Miall 2015). The cyclostratigraphic analysis, carried out over successive segments of the core, suggested that sedimentation of this section at this locality was continuous at the $10^5$-year scale, a conclusion supported by biostratigraphic analysis and bentonite correlations. Tuning to the temporally stable 400-year eccentricity period revealed obliquity and precessional frequencies in the record. The later study (Sageman et al. 2014) extended the cyclic analysis to other cores and to a series of outcrop sections in New Mexico, Utah and Montana (Fig. 8.43), in which the cyclostratigraphic analysis was combined with new radioisotope data with the objective of refining laboratory methods, recalibrating decay constants and developing more precise correlations between methods in order to increase the accuracy and precision of the Coniacian-Santonian time scale. The possible presence of hiatuses in the sampled sections was

recognized to be a potential problem. Breaks with a length equivalent to a whole cycle or multiple of whole cycles may not be detected by spectral analysis (Meyers and Sageman 2004). The duration of several gaps that were suggested by biostratigraphic zonation were within the potential error of radioisotopic calibrations ($\sim 10^5$ years), which means that they could not be ruled out in the cyclostratigraphic analysis and require further study by means of regional correlations and additional data. This means that correlation to specific cycles in the 400-ka record could be in error by one or more cycles, but the results, nevertheless, represent a substantial increase in the accuracy and precision of this part of the Cretaceous time scale, and constitute a substantial contribution in the project to extend an astrochronological time scale back through the Mesozoic.

## 8.12 Conclusions

Two hundred years ago, William Smith gave us the first complete geological map (Smith 1815), and started us on the road to an understanding of Earth's geologic history. Lyell (1830–1833) gave us the necessary tools for interpretation, based on the principle of uniformitarianism, and Holmes, beginning a little more than one hundred years ago, began the development of the modern geological time scale (culminating in his first book: Holmes 1913). Barrell (1917) was the first to attempt to synthesize these critical developments, but most of his ideas were forgotten or ignored for decades. The development of the formal principles of stratigraphy, the evolution of sedimentology as a mature discipline, the stimulus provided by seismic stratigraphy, all have been necessary developments in the evolution of the modern synthesis that constitutes the science described in this book.

So where have we arrived at today, and what may we predict as possible future developments, and outcomes from the application of this science?

Analysis of depositional and erosional processes over the full range of time scales provides insights into long-term geological preservation. At time scales of days to months ($10^{-1}$–$10^{-6}$ years), preservation of individual lithofacies units is essentially random, reflecting autogenic processes, such as diurnal changes in current speed and tidal activity. Packages of strata that survive long enough are subject to the next cycle of preservational processes, such as autogenic channel switching, the "100-year flood", or storm activity at the $10^0$–$10^3$-year time scale. At the $10^4$–$10^5$-year time scale, so-called "high-frequency" geological processes come into play, including orbital forcing of climate and sea level, and local tectonic episodicity. Ultimately, all remaining stratigraphic accumulations are subject to the long-term ($10^6$–$10^7$-year) geological (largely tectonic) controls on basin accommodation (Miall 2014b).

An important insight emerging from this analysis has critical implications for the application of uniformitarianist principles. Geological processes interpreted from successions accumulated over the post-glacial period, such as those of the Mississippi and Rhine-Meuse deltas (descriptions and analyses of which comprise substantial contributions to sedimentological literature on fluvial and deltaic systems) can only be used as analogs for interpretations of geological processes up to the $10^4$-year time scale. The geological record contains many examples of coastal fluvial-deltaic successions spanning millions of years, but a $10^6$–$10^7$-year record cannot be interpreted simply by "scaling-up" an analysis carried out on a $10^4$–$10^5$-year time scale. Firstly, coastal succession such as those on present-day continental margins could be largely eliminated by subaerial erosion during the next glacial cycle of lowered sea-level, as the geological preservation machine begins to operate over the next longer-time scale. Secondly, the time-scales implied for sedimentary processes would be wrong. For example, sequence models for fluvial systems, which relate channel stacking behaviour to rates of sedimentary accommodation, are largely based on measurements of rates of sedimentation and channel switching in modern rivers and in post-glacial alluvial valley-fills, at time scales no greater than $10^3$ years. Applications of these models to the ancient record deal mostly with so-called "third-order" sequences (durations in the $10^6$-year range) for which calculated accumulation rates are one to three orders of magnitude slower than those on which the sequence models are based (Miall 2014a). Colombera et al. (2015) confirmed, by a detailed study of twenty ancient fluvial systems, that there is no relationship between channel stacking pattern and aggradation rate. It is suggested that the observed changes in channel-stacking patterns that have been observed in the rock record are the product of longer-term procesess, such as tectonically-controlled changes in paleoslope or sediment supply (Miall 2014a).

Modern stratigraphic techniques are at the very core of current methods of petroleum exploration and development. Figure 8.44 is an excellent example of a continuing major thrust in the petroleum industry, the development of ever more sophisticated techniques of seismic data acquisition, processing and visualization. The image shown here is from an advertisement appearing in "*AAPG Explorer*", the monthly news magazine of the American Association of Petroleum Geologists. It shows an example of the type of seismic imagery now being developed in the highly competitive field of seismic exploration. It is from an area of the world that has received considerable attention from the industry in recent years because geophysicists have developed techniques for "seeing" through the thick layers of evaporites that occur near the bottom of the sedimentary layer on the floor of the Gulf of Mexico. The so-called

**Fig. 8.44** An image generate from three-dimensional seismic data showing the upper surface of a salt horizon in the Gulf Coast, intersected by vertical and horizontal 2-D sections (Petroleum Geo-Services, Houston)

"sub-salt play" has been the hottest area of exploration in North America for several years, and is finding increasing application in other comparable basinal settings worldwide, such as the deep Atlantic Ocean, offshore from Brazil and Angola. Other examples of advanced seismic techniques are illustrated and discussed in Chap. 6. The increasingly precise information now obtainable from seismic data, together with the modern techniques of directional drilling, and geosteering, backed up by sequence-stratigraphic mapping and detailed chronostratigraphic control, have substantially improved the odds of successful drilling in frontier areas. On a smaller scale, such detail is also an essential component of projects to efficiently exploit heavy oil using in situ methods. Issues such as shale distribution and bed length, and reservoir cap integrity, are key to the successful use of the steam-assisted gravity drainage (SAG-D) method for oil extraction in the Alberta Oil Sands.

Fresh water is a critical natural resource. About one quarter of the world's supply is stored in the ground (most of the rest is in major ice caps), and its health is very much a concern of sedimentary geologists. Groundwater movement depends on the porosity and permeability architecture of the rocks (mostly sediments) through which water moves. In some deep basins, groundwater bodies at depth contain enough heat to offer considerable potential for geothermal power generation. Hydrogeology involves many of the same stratigraphic approaches utilized by petroleum geologists. The tracking and treatment of pollutants involves adaptations of traditional geological methods of exploration and remote sensing, including down-hole petrophysics, high-resolution seismic exploration and ground-penetrating radar.

One of the lines of evidence that modern climate scientists point to as an indication of the anthropogenic influence on climate is the supposed rapidity with which global warming has occurred (essentially all of which has taken place since about 1850) compared with the supposedly much slower rates of change that can be observed in the geological record. In fact, earth scientists have not been able to either prove or disprove this assertion because the accuracy and precision of the geological time scale for deep time has not permitted refinements in the dating of geological events down to the $10^2$-year time scale. The development of the astrochronological time scale is getting is closer, with a time scale in a $10^{4-5}$-year accuracy range, but this gap is unlikely ever to be completely closed. However, refinements in the paleoclimatic reconstructions from the stratigraphic record, in terms of the scope, type, and rate of past climate change, should be able to provide much better comparisons, and analogues for testing climate models and understanding Earth's behaviour under different climatic conditions than have been available in the recent past.

High-resolution stratigraphic studies are increasingly pointing to a much more complex climatic history during the Cretaceous than the simple, long-term "greenhouse" that has long been assumed. Studies in both the American and Canadian portions of the Western Interior Basin, some of which are referred to elsewhere in this book, are providing ever more convincing evidence for what Miller et al. (2005) referred to as "cold snaps", in the isotopic record, in the recognition of cyclostratigraphic signatures, and in the record of erosion surfaces that can be traced for hundreds, if not thousands of kilometres, indicating likely glacioeustatic control on sedimentation.

In a more general sense, as the well-known sedimentologist Harvey Blatt said in 1982: "Sediments and sedimentary rocks are the only source of knowledge about conditions on the Earth's surface before the invention of written language a few thousand years ago." Historical geology depends largely on the study of the stratigraphic record. It is from the rocks that we know about the evolution of life, the plate-tectonic development of the earth's crust, and ancient climate changes. Continued detailed study, and the development of new techniques, can therefore be expected in the future.

# References

Abreu, V., 1998. Evolution of the conjugate volcanic passive margins: Pelotas Basin (Brazil) and offshore Namibia (Africa): Implication for global sea-level changes. Unpublished Ph.D. thesis, Rice University, Houston.

Adams, M.M., and Bhattacharya, J.P., 2005, No change in fluvial style across a sequence boundary, Cretaceous Blackhawk and Castlegate Formations of central Utah, USA: Journal of Sedimentary Research, v. 75, p. 1038–1051.

Ager, D. V., 1973, The nature of the stratigraphical record: New York, John Wiley, 114 p.

Ager, D. V., 1981, The nature of the stratigraphical record (second edition): John Wiley, New York, 122 p.

Ager, D., 1986, A reinterpretation of the basal 'Littoral Lias' of the Vale of Glamorgan, Proceedings of the Geologists Association, v. 97, p. 29-35.

Ager, D. V., 1993, The new catastrophism, Cambridge University Press, 231 p.

Aigner, T., 1985, Storm depositional systems: Springer-Verlag Inc., New York, Lecture Notes in Earth Sciences 3, 174 p.

Algeo, T. J., and Wilkinson, B. H., 1988, Periodicity of mesoscale Phanerozoic sedimentary cycles and the role of Milankovitch orbital modulation: Journal of Geology, v. 96, p. 313-322.

Allen, J. R. L., 1983, Studies in fluviatile sedimentation: bars, bar complexes and sandstone sheets (low-sinuosity braided streams) in the Brownstones (L. Devonian), Welsh Borders: Sedimentary Geology, v. 33, p. 237-293.

Allen, P. A., 2008, Time scales of tectonic landscapes and their sediment routing systems, in Landscape Evolution: Denudation, Climate and Tectonics Over Different Time and Space Scales, edited by K. Gallagher, S. J. Jones, and J. Wainwright, Spec. Publ. Geol. Soc., 296, 7–28,

Allen, P. A., and Allen, J. R., 2005, Basin analysis, principles and applications, second edition, Wiley-Blackwell, Hoboken, New Jersey, 500 p.

Allen, P. A., and Densmore, A. L., 2000, Sediment flux from an uplifting fault block: processes and controls in the stratigraphic development of extensional basins: Basin Research, v. 12, p. 367-380.

Archer, A. W., Kvale, E. P., and Johnson, H. R., 1991, Analysis of modern equatorial tidal periodicities as a test of information encoded in ancient tidal rhythmites, in Clastic tidal sedimentology: Canadian Society of Petroleum Geologists, Memoir 16, p. 189-196.

Aschoff, J. L., and Steel, R. J., 2011, Anomalous clastic wedge development during the Sevier-Laramide transition, North American Cordilleran foreland basin, USA: Geological Society of America Bulletin, v. 123, p. 1822-1835.

Aswasereelert, W., Mayers, S. R., Carroll, A. R., Peters, S. E., Smith, M. E., and Feigl, K. L., 2013, basin-scale cyclostratigraphy of the green River Formation, Wyoming: Geological Society of America Bulletin, v. 125, p. 216-228.

Aubry, M.-P., 1991, Sequence stratigraphy: eustasy or tectonic imprint: Journal of Geophysical Research, v. 96B, p. 6641-6679.

Aziz, A., Krijgsman, W., Hilgen, F. J., Wilson, D. S., and Calvo, J. P., 2003, An astronomical polarity time scale for the late middle Miocene based on cyclic continental sequences: Journal of Geophysical Research, v. 108B, doi:10.1029/2002JB001818.

Bailey, R. J., 2009, Cyclostratigraphic reasoning and orbital time calibration: Terra Nova, v. 21, #5, p. 340-351.

Bailey, R. J., 2011, Buried trees and basin tectonics: a Discussion: Stratigraphy, v. 8, p. 1-6.

Bailey, R. J., and Schumer, R., 2012, The statistical properties of stratigraphic layering and their possible significance: The Geological Society, London, William Smith Meeting 2012, Strata and Time: Probing the gaps in our understanding,

Bailey, R. J., and Smith, D. G., 2005, Quantitative evidence for the fractal nature of the stratigraphic record: results and implications: Proceedings of the Geologists' Association, v. 116, p. 129-138.

Bailey, R. J., and Smith, D. G., 2010, Scaling in stratigraphic data series: implications for practical stratigraphy: First Break, v. 28, p. 57-66.

Barrell, Joseph, 1917, Rhythms and the measurement of geologic time: Geological Society of America Bulletin, v. 28, p. 745-904.

Bartek, L. R., Vail, P. R., Anderson, J. B., Emmet, P. A., and Wu, S., 1991, Effect of Cenozoic ice sheet fluctuations in Antarctica on the stratigraphic signature of the Neogene: Journal of Geophysical Research, v. 96B, p. 6753-6778.

Beckmann, B., Flögel, S., Hofmann, P., Schulz, M., and Wagner, T., 2005, Orbital forcing of Cretaceous river discharge in tropical Africa and ocean response: Nature, v. 437, p. 241-244.

Bentham, P. A., Talling, P. J., and Burbank, D. W., 1993, Braided stream and flood-plain deposition in a rapidly aggrading basin: the Escanilla formation, Spanish Pyrenees, in Best, J. L., and Bristow, C. S., eds., Braided rivers: Geological Society, London, Special Publication 75, p. 177-194.

Berggren, W. A., Kent, D. V., Aubry, M.-P., and Hardenbol, J., eds., 1995, Geochronology, time scales and global stratigraphic correlation: Society for Sedimentary Geology Special Publication 54, 386 p.

Bernard, H. A., Leblanc, R. J., and Major, C. J., 1962, Recent and Pleistocene geology of southeast Texas, in Rainwater, E. H., and Zingula, R. P., eds., Geology of the Gulf Coast and central Texas: Geological Society of America, Guidebook for 1962 Annual Meeting., p. 175-224.

Betzler, C., Kroon, D., and Reijmer, J. G. J., 2000, Sychroneity of major late Neogene sea level fluctuations and paleoceanographically controlled changes as recorded by two carbonate platforms: Paleoceanography, v. 15, p. 722-730.

Bhattacharya, J. P., 2011, Practical problems in the application of the sequence stratigraphic method and key surfaces: integrating observations from ancient fluvial-deltaic wedges with Quaternary and modelling studies: Sedimentology, v. 58, p. 120-169.

Bird, J. M., and Dewey, J. F., 1970, Lithosphere plate-continental margin tectonics and the evolution of the Appalachian Orogen: Geological Society of America Bulletin, v. 81, p. 1031-1060.

Blair, T. C., and Bilodeau, W. L., 1988, Development of tectonic cyclothems in rift, pull-apart, and foreland basins: sedimentary response to episodic tectonism: Geology, v. 16, p. 517-520.

Blakey, R. C., and Middleton, L. T., 1983, Permian shoreline aeolian complexes in central Arizona; dune changes in response to cyclic sea level changes, in Brookfield, M. E., and Ahlbrandt, T. S., eds., Eolian sediments and processes: Elsevier, Amsterdam, p. 551-581.

Blum, M. D., 1993, Genesis and architecture of incised valley fill sequences: a late Quaternary example from the Colorado River, Gulf Coastal Plain of Texas, in Weimer, P., and Posamentier, H. W., eds., Siliciclastic sequence stratigraphy: Recent developments and applications: American Association of Petroleum Geologists Memoir 58, p. 259-283.

Blum, M. D., 1994, Genesis and architecture of incised valley fill sequences: a late Quaternary example from the Colorado River, Gulf Coastal Plain of Texas, in Weimer, P., and Posamentier, H. W., eds., Siliciclastic sequence stratigraphy: Recent developments and applications: American Association of Petroleum Geologists Memoir 58, p. 259-283.

Blum, M. D., and Törnqvist, T. E., 2000, Fluvial responses to climate and sea-level change: a review and look forward: Sedimentology, v. 47, p. 2-48.

Boulila, S., Galbrun, B., Miller, K. G., Pekar, S. F., Browning, J. V., Laskar, J., and Wright, J. D., 2011, On the origin of Cenozoic and Mesozoic "third-order" eustatic sequences: Earth Science Reviews, v. 109, p. 94-112.

Bradley, W. H., 1929, The Varves and Climate of the Green River Epoch: U.S. Geological Survey Professional Paper 158-E, 110 p.

Bridge, J. S. and Leeder, M. R., 1979, A simulation model of alluvial stratigraphy: Sedimentology, v. 26, p. 617-644.

Browning, J. V., Miller, K. G., Sugarman, P. J., Kominz, M. A., McLaughlin, P. P. Kulpecz, A. A., and Feigenson, M. D., 2008, 100 Myr record of sequences, sedimentary facies and sea level changes from Ocean Drilling Program onshore coreholes, US Mid-Atlantic coastal plain: Basin Research, v. 20, p. 227-248.

Buckman, S. S. 1910. Certain Jurassic ('Inferior Oolite') species of ammonites and brachiopoda. Quarterly Journal of the Geological Society, London, v. 66, p. 90-108.

Burbank, D. W., Meigs, A., and Brozovic, N., 1996, Interactions of growing folds and coeval depositional systems: Basin Research, v. 8, p. 199-223.

Callomon, J. H., 1995, Time from fossils: S. S. Buckman and Jurassic high-resolution geochronology, in Le Bas, M. J., ed., Milestones in Geology: Geological Society of London Memoir 16, p. 127-150.

Catuneanu, O., 2006, Principles of sequence stratigraphy: Elsevier, Amsterdam, 375 p.

Christie-Blick, N., Mountain, G. S., and Miller, K. G., 1990, Seismic stratigraphy: record of sea-level change, in Revelle, R., ed., Sea-level change: National Research Council, Studies in Geophysics, Washington, National Academy Press, p. 116-140.

Christie-Blick, N., Pekar, S. F., and Madof, A. S., 2007, Is there a role for sequence stratigraphy in chronostratigraphy? Stratigraphy, v. 4, p. 217-229.

Cole. R.D., and Friberg, J.F., 1989, Stratigraphy and sedimentation of the Book Cliffs, Utah, in Nummedal, D. and Wright, R., eds., Cretaceous Shelf Sandstone and Shelf Depositional Sequences, Western Interior Basin, Utah, Colorado, and New Mexico: 28th International Geological Congress, Field Trip Guidebook T119.

Cloetingh, S., 1988, Intraplate stress: a new element in basin analysis, in Kleinspehn, K., and Paola, C., eds., New Perspectives in basin analysis: Springer-Verlag, New York, p. 205-230.

Coleman, J. M., and Gagliano, S. W., 1964, Cyclic sedimentation in the Mississippi River deltaic plain: Transactions of the Gulf Coast Association of Geological Societies, v. 14, p. 67-80.

Colombera, L., Mountney, N. P., and McCaffrey, W. D., 2015, A meta-study of relationships between fluvial channel-body stacking pattern and aggradation rate: implications for sequence stratigraphy: Geology,

Cramer, B. D., Vandenbroucke, T. R. A., and Ludvigson, G. A., 2015, High-resolution event stratigraphy (HiRES) and the quantification of stratigraphic uncertainty: Silurian examples of the quest for precision in stratigraphy: Earth Science reviews, v. 141, p. 136-153.

Crampton, J. S., Schiøler, P., and Roncaglia, L., 2006, Detection of Late Cretaceous eustatic signatures using quantitative biostratigraphy: Geological Society of America Bulletin, v. 118, p. 975-990.

Dahlstrom, C. D. A., 1970, Structural geology in the eastern margin of the Canadian Rocky Mountains: Bulletin of Canadian Petroleum Geology, v. 18, p. 332-406.

Dalrymple, R. W., 1984, Morphology and internal structure of sandwaves in the Bay of Fundy: Sedimentology, v. 31, p. 365-382.

de Natris, M. F., 2012, Facies- and time-analysis of the upper part of the Brent Group (mid-upper Jurassic) in the Greater Oseberg area, northern North Sea: M.Sc., thesis, University of Bergen, 90 p.

de Natris, M., and Helland-Hansen, W., 2012, Where has all the time gone? Disentangling time in the Mid-Late Jurassic Tarbert Formation, northern North Sea: The Geological Society, London, William Smith Meeting 2012, Strata and Time: Probing the gaps in our understanding, Abstract, p. 40.

Dean, D. J., Scott, M. L., Shafroth, P. B., and Schmidt, J. C., 2011, Stratigraphic, sedimentologic, and dendrogeomorphic analyses of rapid floodplain formation along the Rio Grande in Big Bend National Park, Texas: Geological Society of America Bulletin, v. 123, p. 1908-1925.

Deramond, J., Souquet, P., Fondecave-Wallez, M.-J., and Specht, M., 1993, Relationships between thrust tectonics and sequence stratigraphy surfaces in foredeeps: model and examples from the Pyrenees (Cretaceous-Eocene, France, Spain), in Williams, G.D., and Dobb, A., eds., Tectonics and Seismic Sequence Stratigraphy: Geological Society of London, Special Publication 71, p. 193-219.

Devine, P. E., 1991, Transgressive origin of channeled estuarine deposits in the Point Lookout Sandstone, northwestern New Mexico: a model for Upper Cretaceous, cyclic regressive parasequences of the U. S. Western Interior: American Association of Petroleum Geologists Bulletin, v. 75, p. 1039-1063.

Dewey, J. F., and Bird, J. M., 1970, Plate tectonics and geosynclines, Tectonophysics, v. 10, p. 625-638.

Donaldson, J. A., Eriksson, P. G. and Altermann, W., 2002, Actualistic versus non-actualistic conditions in the Precambrian: a reappraisal of an enduring discussion, in: Altermann, W. and Corcoran, P. L., eds., Precambrian Sedimentary Environments: a Modern Approach to Ancient Depositional Systems. International Association of Sedimentologists Special Publication 33, p. 3-13.

Dorsey, R. J., Housen, B. A., Janecke, S. U., Fanning, C. M., and Spears, A. L. F., 2011, Stratigraphic record of basin development within the San Andreas fault system: Late Cenozoic Fish Creek-Vallecito basin, southern California: Geological Society of America Bulletin, v. 123, p. 771-793.

Duller, R. A., Kougioumtzoglou, I., Dunning, S., and Fedele, J., 2012, Vertical and lateral patterns of grain size in the stratigraphic record, The Geological Society, London, William Smith Meeting 2012, Strata and Time: Probing the gaps in our understanding, Abstract, p. 10.

Echavarria, L., Hernández, R., Allmendinger, R., and Reynolds, J., 2003, Subandean thrust and fold belt of northwestern Argentina: Geometry and timing of the Andean evolution: American Association of Petroleum Geologists Bulletin, v. 87, p. 965-985.

Eriksson, P. G., Condie, K. C., Tirsgaard, H., Mueller, W. U., Alterman, W., Miall, A. D., Aspler, L. B., Catuneanu, O., and Chiarenzelli, J. R., 1998, Precambrian clastic sedimentation systems, in Eriksson, P. G., ed., Precambrian clastic sedimentation systems: Sedimentary Geology, Special Issue, v. 120, p. 5-53.

Eriksson, P. G., Tirsgaard, H., and Mueller, W. U., eds., 1998, Precambrian clastic sedimentation systems: Sedimentary Geology, v. 120 (special issue), 346 p.

Ettensohn, F. R., 2008, The Appalachian foreland basin in eastern United States, in Miall, A. D., ed., The Sedimentary Basins of the United States and Canada: Sedimentary basins of the World, v. 5, K. J. Hsü, Series Editor, Elsevier Science, Amsterdam, p. 105-179.

Eyles, N., and Clark, B. M., 1986, Significance of hummocky and swaley cross-stratification in late Pleistocene lacustrine sediments of the Ontario basin, Canada: Geology, v. 14, p. 679-682.

Fiduk, J. C., and Behrens, E. W., 1993, A comparison of Plio-Pleistocene to recent sediment accumulation rates in the East breaks area, northwestern Gulf of Mexico, in Armentrout, J. M., Bloch, R., Olson, H. C., and Perkins, B. E., eds., Rates of Geologic Processes: Gulf Coast Section, Society of Economic Paleontologists and Mineralogists, Fourteenth Annual Research Conference, Houston, Texas, p. 41-55.

Filomena, C. M., Stollhofen, H., 2011, Ultrasonic logging across unconformities — outcrop and core logger sonic patterns of the Early Triassic Middle Buntsandstein Hardegsen unconformity, southern Germany, Sedimentary Geology, v. 236, p. 185-196.

Finney, S. C., Grubb, B. J., and Hatcher, R. D., Jr, 1996, Graphic correlation of Middle Ordovician graptolite shale, southern Appalachian: an approach for examining the subsidence and migration of a Taconic foreland basin: Geological Society of America Bulletin, v. 108, p. 355-371.

Fischer, A. G., and Roberts, L. T., 1991, Cyclicity in the Green River Formation (lacustrine Eocene) of Wyoming: Journal of Sedimentary Petrology, v. 61, p. 1146-1154.

Fletcher, C. J. N., Davies, J. R., Wilson, D., and Smith, M., 1986, The depositional environment of the basal 'Littoral Lias' in the Vale of Glamorgan – a discussion of the reinterpretation by Ager (1986): Proceedings of the Geologists' Association, v. 97, p. 383-384.

Frazier, D. E., 1967; Recent deltaic deposits of the Mississippi River— their development and chronology: Transactions of the Gulf Coast Association of Geological Societies, v. 17, p. 287-315.

Gould, S. J., 1965, Is uniformitarianism necessary?. American Journal of Science, v. 263, p. 223–228.

Gould, H. R., 1970, The Mississippi Delta complex, in Morgan, J. P., ed., Deltaic sedimentation: modern and ancient: Society of Economic Paleontologists and Mineralogists Special Publication 15, p. 3-30.

Grabau, A. W., 1960: Principles of stratigraphy: New York, Dover Publications, 1185 p.

Gradstein, F. M., Ogg, J. G., and Smith, A. G., eds., 2004a, A geologic time scale: Cambridge University Press, Cambridge, 610 p.

Gradstein, F. M., Cooper, R. A., and Sadler, P. M., 2004b, Biostratigraphy: time scales from graphic and quantitative methods, in Gradstein, F. M., Ogg, J. G., and Smith, A. G., eds., A geologic time scale: Cambridge University Press, Cambridge, p. 49-54.

Gradstein, F. M., Ogg, J. G., Schmitz, M. D., and Ogg, G. M., 2012, The Geologic time scale 2012: Elsevier, Amsterdam, 2 vols., 1176 p.

Hajek, E. A., Heller, P. L., and Sheets, B. A., 2010, Significance of channel-belt clustering in alluvial basins: Geology, v. 38, p. 535-538.

Hamblin, A. P., and Walker, R. G., 1979, Storm-dominated shallow marine deposits; the Fernie-Kootenay (Jurassic) transition, southern Rocky Mountains: Canadian Journal of Earth Sciences, v. 16, p. 1673-1690.

Hampson, G.J., 2010, Sediment dispersal and quantitative stratigraphic architecture across and ancient shelf: Sedimentology, v. 57, p. 96-141.

Hampson, G.J., Jewell, T.O., Irfan, N., Gani, M.R. and Bracken, B., 2013, Modest changes in fluvial style with varying accommodation in regressive alluvial-to-coastal-plain wedge: Upper Cretaceous Blackhawk Formation, Wasatch Plateau, central Utah, U.S.A.: Journal of Sedimentary Research, v. 83, p. 145-169.

Haq, B. U., Hardenbol, J., and Vail, P. R., 1987, Chronology of fluctuating sea levels since the Triassic (250 million years ago to present): Science, v. 235, p. 1156-1167.

Haq, B. U., Hardenbol, J., and Vail, P. R., 1988, Mesozoic and Cenozoic chronostratigraphy and cycles of sea-level change, in Wilgus, C. K., Hastings, B. S., Kendall, C. G. St. C., Posamentier, H. W., Ross, C. A., and Van Wagoner, J. C., eds., Sea-level Changes: an integrated approach: Society of Economic Paleontologists and Mineralogists Special Publication 42, p. 71-108.

Hardenbol, J., Thierry, J., Farley, M. B., Jacquin, T., Graciansky, P.-C., and Vailo, P. R., 1998. Mesozoic and Cenozoic sequence chronostratigraphic framework of European basins, in Graciansky, P.-C. de, Hardenbol, J., Jacquin, T., and Vail, P. R., eds., Mesozoic and Cenozoic sequence stratigraphy of European basins, Society for Sedimentary Geology (SEPM) Special Publication 60, p. 3-13.

Heckel, P. H., 1986, Sea-level curve for Pennsylvanian eustatic marine transgressive-regressive depositional cycles along midcontinent outcrop belt, North America: Geology, v. 14, p. 330-334.

Heller, P. L., Beekman, F., Angevine, C. L., and Cloetingh, S. A. P. L., 1993, Cause of tectonic reactivation and subtle uplifts in the Rocky Mountain region and its effect on the stratigraphic record: Geology, v. 21, p. 1003-1006.

Hilgen, F. J., 1991. Extension of the astronomically calibrated (polarity) time scale to the Miocene/Pliocene boundary: Earth and Planetary Sciences Letters, v. 107, p. 349-368.

Hilgen, F. J., Kuiper, K., Krijgsman, W., Snel, E., and van der Laan, E., 2007, Astronomical tuning as the basis for high resolution chronostratigraphy: the intricate history of the Messinian Salinity Crisis: Stratigraphy, v. 4, p. 231-238.

Hilgen, F. J., Hinnov, L. A., Aziz, H. A., Abels, H. A., Batenburg, S., Bosmans, J. H. C., de Boer, B., Hüsings, S. K., Kuiper, K. F., and Lourens, L. J., 2015, Stratigraphic continuity and fragmentary sedimentation: the success of cyclostratigraphy as part of integrated stratigraphy in Smith, D. G., Bailey, R., J., Burgess, P., and Fraser, A., eds., Strata and time: Geological Society, London, Special Publication 404, p. 157-197.

Hinnov, L. A., and Ogg, J. G., 2007, Cyclostratigraphy and the astronomical time scale: Stratigraphy, v. 4, p. 239-251.

Hiscott, R. N., 1981, Deep-sea fan deposits in the Macigno Formation (Middle-Upper Oligocene) of the Gordana valley, northern Apennines, Italy — Discussion: Journal of Sedimentary Petrology, v. 51, p. 1015-1033.

Hofmann, M. H., Wroblewski, A., and Boyd, R., 2011, Mechanisms controlling the clustering of fluvial channels and the compensational stacking of cluster belts: Journal of Sedimentary Research, v. 81, p. 670-685.

Holbrook, J. M., 2001, Origin, genetic interrelationships, and stratigraphy over the continuum of fluvial channel-form bounding surfaces: an illustration from middle Cretaceous strata, southeastern Colorado: Sedimentary Geology, v. 144, p. 179-222.

Holbrook, J., Scott, R. W., and Oboh-Ikuenobe, F. E., 2006, Base-level buffers and buttresses: a model for upstream versus downstream control on fluvial geometry and architecture within sequences: Journal of Sedimentary Research, v. 76, p. 162-174.

Holmes, A., 1913, The age of the Earth: Harper, London.

Howard, J. D., and Reineck, R.-E., 1981, Depositional facies of high-energy beach-to-offshore sequence: comparison with low energy sequence: American Association of Petroleum Geologists Bulletin, v. 65, p. 807-830.

Howell, J.A., and Flint, S.S., 2003, Siliciclastics case study: the Book Cliffs, in Coe, A., ed., The Sedimentary Record of Sea-level Change: Cambridge, U.K., Open University and Cambridge University Press, p. 135–208.

Howell, J., Hodgetts, D., Burgess, P., and Flint, S., 1999, The Blackhawk Formation of eastern Utah: Application of a high-quality outcrop dataset to calibrate modeling studies: American Association of Petroleum Geologists, International Conference and Exhibition, Birmingham, Search and Discovery Article #90923.

Hoyt, J. H., and Henry, V. J., Jr., 1967, Influence of island migration on barrier island sedimentation: Geological Society of America Bulletin, v. 78, p. 77-86.

Jackson, R. G. II, 1976, Depositional model of point bars in the lower Wabash River: Journal of Sedimentary Petrology, v. 46, p. 579-594.

Jerolmack, D. J., and Paola, C., 2010, Shredding of environmental signals by sediment transport: Geophysical Research Letters, v. 37, L10401, 5 p.

Jerolmack, D., and Sadler, P., 2007, Transience and persistence in the depositional record of continental margins: Journal of Geophysical Research, v. 112, F03S13, 14 p.

Johnson, M. E., and McKerrow, W. S., 1995, The Sutton Stone: an early Jurassic rocky shore deposit in South Wales: Palaeontology, v. 38, p. 529-541.

Johnson, N. M., Stix, J., Tauxe, L., Cerveny, P. F., and Tahirkheli, R. A. K., 1985, Paleomagnetic chronology, fluvial processes, and tectonic implications of the Siwalik deposits near Chinji Village, Pakistan: Journal of Geology, v. 93, p. 27-40.

Johnson, N. M., Sheikh, K. A., Dawson-Saunders, E., and McRae, L. E., 1988, The use of magnetic-reversal time lines in stratigraphic analysis: a case study in measuring variability in sedimentation rates, in Kleinspehn, K. L., and Paola, C., eds., New perspectives in basin analysis: Springer-Verlag Inc., Berlin and New York, p. 189-200.

Jones, M. A., Heller, P. L., Roca, E., Garcés, M., and Cabrera, L., 2004, Time lag of syntectonic sedimentation across an alluvial basin: theory and example from the Ebro basin, Spain, Basin Research, v. 16, p. 467-488.

Kamola, D.L., and Van Wagoner, J.C., 1995, Stratigraphy and facies architecture of parasequences with examples from the Spring Canyon Member, Blackhawk Formation, Utah, in Van Wagoner, J. C., and Bertram, G T., eds., Sequence Stratigraphy of Foreland Basin Deposits: American Association of Petroleum Geologists, Memoir 64, p. 27–54.

Kamp, P. J. J., and Turner, G. M., 1990, Pleistocene unconformity-bounded shelf sequences (Wanganui Basin, New Zealand) correlated with global isotope record: Sedimentary Geology, v. 68, p. 155-161.

Kauffman, E. G., 1986, High-resolution event stratigraphy: regional and global bio-events, in Walliser, O. H. ed., Global bioevents: Lecture Notes on Earth History: Springer-Verlag, Berlin, p. 279–335.

Kauffman, E. G., 1988, Concepts and methods of high-resolution event stratigraphy: Annual Reviews of Earth and Planetary Sciences, v. 16, p. 605-654.

Kemp, D. B., 2012, Stochastic and deterministic controls on stratigraphic completeness and fidelity: International Journal of Earth Sciences (Geol. Rundschau), v. 101, p. 2225-2238. DOI 10.1007/s00531-012-0784-1.

Kemp, D. B., and Sadler, P. M., 2014, Climatic and eustatic signals in a global compilation of shallow marine carbonate accumulation rates: Sedimentology, v. 61, p. 1286-1297.

Kidwell, S. M., 1988, Reciprocal sedimentation and noncorrelative hiatuses in marine-paralic siliciclastics: Miocene outcrop evidence: Geology, v. 16, p. 609-612.

Kim, W., and Paola, C., 2007, Long-period cyclic sedimentation with constant tectonic forcing in an experimental relay ramp: Geology, v. 35, p. 331-334.

Kocurek, G., and Dott, R. H., Jr., 1981, Distinctions and uses of stratification types in the interpretation of eolian sand: Journal of Sedimentary Petrology, v. 51, p. 579-595.

Kolb, C. R., and Van Lopik, J. R., 1966, Depositional environment of the Mississippi River deltaic plain - southeastern Louisiana, in Shirley, M. L., ed., Deltas in their geologic framework: Houston Geological Society, p. 17-61.

Kominz, M. A., and Pekar, S. F., 2001, Oligocene eustasy from two-dimensional sequence stratigraphic backstripping: Geological Society of America Bulletin, v. 113, p. 291-304.

Kominz, M. A., Browning, J. V., Miller, K. G., Sugarman, P. J., Mizintserva, S., and Scotese, C. R., 2008, late Cretaceous to Miocene sea-level estimates from the New Jersey and Delaware coastal plain coreholes: an error analysis: Basin Research, v. 20, p. 211-226.

Krystinik, L.F., DeJarnett, B.B., 1995, Lateral variability of sequence stratigraphic framework in the Campanian and Lower Maastrichtian of the Western Interior Seaway, in Van Wagoner, J.C., and Bertram, G.T., eds., Sequence Stratigraphy of Foreland Basin Deposits: American Association of Petroleum Geologists, Memoir 64, p. 11-25.

Kumar, N., and Sanders, J. E., 1974, Inlet sequence: a vertical succession of sedimentary structures and textures created by the lateral migration of tidal inlets: Sedimentology, v. 21, p. 491-532.

Leckie, D. A., Bhattacharya, J., Gilboy, C. F. and Norris, B., 1994, Colorado\Alberta Group strata of the Western Canada Sedimentary Basin, in: Mossop, G. D., and Shetsen, I., eds., Geological Atlas of the Western Canada Sedimentary Basin: Canadian Society of Petroleum Geologists and Alberta Research Council, Calgary. p. 335–352.

Leclair, S. F., 2011, Interpreting fluvial hydromorphology from the rock record: large-river peak flows leave no clear signature, in Davidson, S. K., Leleu, S., and North, C. P., eds., From river to rock record: Society for Sedimentary Geology (SEPM) Special Publication 97, p. 113-123.

Locklair, R. E., and Sageman, B. B., 2008, Cyclostratigraphy of the Upper Cretaceous Niobrara Formation, Western Interior, U.S.A.: A Coniacian-Santonian timescale: Earth and Planetary Science Letters, v. 269, p. 539-552.

Loutit, T. S., Hardenbol, J., Vail, P. R., and Baum, G. R., 1988, Condensed sections: the key to age dating and correlation of continental margin sequences, in Wilgus, C. K., Hastings, B. S., Kendall, C. G. St. C., Posamentier, H. W., Ross, C. A., and Van Wagoner, J. C., eds., Sea-level Changes: an integrated approach: Society of Economic Paleontologists and Mineralogists Special Publication 42, p. 183-213.

Lyell, C., 1830-1833, Principles of Geology, 3 vols.: John Murray, London (reprinted by Johnson Reprint Corp., New York, 1969).

MacLeod, N., and Keller, G., 1991, How complete are Cretaceous/Tertiary boundary sections? A chronostratigraphic estimate based on graphic correlation: Geological Society of America Bulletin, v. 103, p. 1439-1457.

Mandelbrot, B. B., 1983, The fractal geometry of nature: Freeman, New York, 468 p.

McGowran, B., 2005, Biostratigraphy: Microfossils and Geological Time: Cambridge University Press, Cambridge, 459 p.

McKee, E. D., ed., 1979, A study of global sand seas: U.S. Geological Survey Professional Paper 1052.

McKee, E. D, Crosby, E. J., and Berryhill, H. L., Jr., 1967, Flood deposits, Bijou Creek, Colorado, June 1965: Journal of Sedimentary Petrology, v. 37, p. 829-851.

McKee, B. A., Nittrouer, C. A., and Demaster, D. J., 1983, Concepts of sediment deposition and accumulation applied to the continental shelf near the mouth of the Yangtze River: Geology, v. 11, p. 631-633.

McLaurin, B.T., and Steel, R.J., 2000, Fourth-order nonmarine to marine sequences, middle Castlegate Formation, Book Cliffs, Utah: Geology, v. 28, p. 359-362.

McLaurin, B.T., and Steel, R.J., 2007, Architecture and origin of an amalgamated fluvial sheet sand, lower Castlegate Formation, Book Cliffs, Utah: Sedimentary Geology, v. 197, p. 291-311.

Medwedeff, D. A., 1989, Growth fault-bend folding at Southeast Lost Hills, San Joaquin valley, California: American Association of Petroleum Geologists Bulletin, v. 73, p. 54-67.

Meyers, S., 2012, Seeing red in cyclic stratigraphy, The Geological Society, London, William Smith Meeting 2012, Strata and Time: Probing the gaps in our understanding, Abstract, p. 18.

Meyers, S. R., and Sageman, B. B., 2004, detection, quantification, and significance of hiatuses in pelagic and hemipelagic strata: Earth and Planetary Sciences Letters, v. 224, p. 55-72.

Miall, A. D., 1978a, Fluvial sedimentology: an historical review, in Miall, A. D., ed., Fluvial Sedimentology: Canadian Society of Petroleum Geologists Memoir 5, p. 1-47.

Miall, A. D., 1978b, Tectonic setting and syndepositional deformation of molasse and other nonmarine-paralic sedimentary basins; Canadian Journal of Earth Sciences, v. 15, p. 1613-1632.

Miall, A. D., 1980, Cyclicity and the facies model concept in geology: Bulletin of Canadian Petroleum Geology, v. 28, p. 59-80.

Miall, A. D., 1988a, Reservoir heterogeneities in fluvial sandstones: lessons from outcrop studies: American Association of Petroleum Geologists Bulletin, v. 72, p. 682-697.

Miall, A. D., 1988b, Facies architecture in clastic sedimentary basins, in Kleinspehn, K., and Paola, C., eds., New perspectives in basin analysis: Springer-Verlag Inc., New York, p. 67-81.

Miall, A. D., 1991, Hierarchies of architectural units in terrigenous clastic rocks, and their relationship to sedimentation rate, in A. D. Miall and N. Tyler, eds., The three-dimensional facies architecture of terrigenous clastic sediments and its implications for hydrocarbon discovery and recovery: Society of Economic Paleontologists and Mineralogists, Concepts in Sedimentology and Paleontology, v. 3, p. 6-12.

Miall, A.D., 1993, The architecture of fluvial-deltaic sequences in the Upper Mesaverde Group (Upper Cretaceous), Book Cliffs, Utah, in Best, J. L., and Bristow, C. S., eds., Braided Rivers: Geological Society of London, Special Publication 75, p. 305-332.

Miall, A. D., 1995a, Whither stratigraphy? Sedimentary Geology, v. 100, p. 5-20.

Miall, A. D., 1996, The geology of fluvial deposits: sedimentary facies, basin analysis and petroleum geology: Springer-Verlag Inc., Heidelberg, 582 p.

Miall, A. D., 1997, The geology of stratigraphic sequences, First edition: Springer-Verlag, Berlin, 433 p.

Miall, A. D., 2010, The geology of stratigraphic sequences, second edition: Springer-Verlag, Berlin, 522 p.

Miall, A. D., 2013, Sophisticated stratigraphy, in Bickford, M. E., ed., The web of geological sciences: Advances, impacts and interactions: Geological Society of America Special Paper 500, p. 169-190.

Miall, A. D., 2014a, Fluvial depositional systems: Springer-Verlag, Berlin 316 p.

Miall, A. D., 2014b, The emptiness of the stratigraphic record: A preliminary evaluation of missing time in the Mesaverde Group, Book Cliffs, Utah: Journal of Sedimentary Research, v. 84, p. 457-469.

Miall, A. D., 2015, Updating uniformitarianism: stratigraphy as just a set of "frozen accidents", in Smith, D. G., Bailey, R., J., Burgess, P., and Fraser, A., eds., Strata and time: Geological Society, London, Special Publication 404, p. 11-36.

Miall, A. D., and Arush, M., 2001a, The Castlegate Sandstone of the Book Cliffs, Utah: sequence stratigraphy, paleogeography, and tectonic controls: Journal of Sedimentary Research, v. 71, p. 536-547.

Miall, A.D., and Arush, M., 2001b, Cryptic sequence boundaries in braided fluvial successions: Sedimentology, v. 48, p. 971-985.

Miall, A. D., and Miall, C. E., 2001, Sequence stratigraphy as a scientific enterprise: the evolution and persistence of conflicting paradigms: Earth Science Reviews, v. 54, #4, p. 321-348.

Miall, A. D., and Miall, C. E., 2004, Empiricism and Model-Building in stratigraphy: Around the Hermeneutic Circle in the Pursuit of Stratigraphic Correlation. Stratigraphy: American Museum of Natural History, v. 1, p. 27-46.

Middleton, G. V., Plotnick, R. E., and Rubin, D. M., 1995, Nonlinear dynamics and fractals; New numerical techniques for sedimentary data sets: Society for Sedimentary Geology, Tulsa, Oklahoma, Short Course No. 36, 174 p.

Miller, K. G., Browning, J. V., Aubry, M.-P., Wade, B., Katz, M. E., Kulpecz, A. A., and Wright, J. D., 2008, Eocene-Oligocene global climate and sea-level changes: St. Stephens Quarry, Alabama: Geological Society of America Bulletin, v. 120, p. 34-53.

Miller, K. G., and Mountain, G. S., 1994, Global sea-level change and the New Jersey margin, in Mountain, G. S., Miller, K. G., Blum, P., et al., eds., Proceedings of the Ocean Drilling Program, Initial Reports, v. 150, p. 11-20.

Miller, K. G., Mountain, G. S., Browning, J. V., Kominz, M., Sugarman, P. J., Christi-Blick, N., Katz, M. E., and Wright, J. D., 1998, Cenozoic global sea level, sequences, and the New Jersey transect: results from coastal plain and continental slope drilling: Reviews of Geophysics, v. 36, p. 569-601.

Miller, K. G., Wright, J. D., and Browning, J. V., 2005, Visions of ice sheets in a greenhouse world: Marine Geology, v. 217, p. 215-231.

Minter, W. E. L., 1978, A sedimentological synthesis of placer gold, uranium and pyrite concentrations in Proterozoic Witwatersrand sediments, in Miall, A. D., ed., Fluvial Sedimentology: Canadian Society of Petroleum Geologists Memoir 5, p. 801-829.

Naylor, M., and Sinclair, H.D., 2007, Punctuated thrust deformation in the context of doubly vergent thrust wedges: Implications for the localization of uplift and exhumation: Geology, v. 35, p. 559–562.

Nemec, W., 1988, Coal correlations and intrabasinal subsidence: a new analytical perspective, in Kleinspehn, K. L., and Paola, C., eds., New perspectives in basin analysis: Springer-Verlag Inc., Berlin and New York, p. 161-188.

Nummedal, D., and Swift, D. J. P., 1987, Transgressive stratigraphy at sequence-bounding unconformities: some principles derived from Holocene and Cretaceous examples, in Nummedal, D., Pilkey, O. H., and Howard, J. D., eds., Sea-level fluctuation and coastal evolution; Society of Economic Paleontologists and Mineralogists Special Publication 41, p. 241-260.

Nummedal, D., Pilkey, O. H., and Howard, J. D., eds., 1987, Sea-level fluctuation and coastal evolution: Society of Economic Paleontologists and Mineralogists Special Publication 41, 267 p.

O'Byrne, C.J., and Flint, S.S., 1995, Sequence, parasequence and intra-parasequence architecture of the Grassy Member, Blackhawk Formation, Book Cliffs, Utah, USA., in Van Wagoner, J.C., and Bertram, G.T., eds., Sequence Stratigraphy of Foreland Basin Deposits: American Association of Petroleum Geologists, Memoir 64, p. 225–255.

Olsen, P. E., 1990, Tectonic, climatic, and biotic modulation of lacustrine ecosystems—examples from Newark Supergroup of eastern North America, in Katz, B. J., ed., Lacustrine basin exploration: case studies and modern analogs: American Association of Petroleum Geologists Memoir 50, p. 209-224.

Olsen, T., Steel, R. J., Høgseth, K., Skar, T., and Røe, S.-L., 1995, Sequential architecture in a fluvial succession: sequence stratigraphy in the Upper Cretaceous Mesaverde Group, Price Canyon, Utah: Journal of Sedimentary Research, v. B65, p. 265-280.

Oomkens, E., 1970, Depositional sequences and sand distribution in the postglacial Rhône delta complex, in Morgan, J. P., ed., Deltaic sedimentation: modern and ancient: Society of Economic Paleontologists and Mineralogists Special Publication 15, p. 198-212.

Paola, C., Straub, K., Mohrig, D., and Reinhardt, L., 2009, The unreasonable effectiveness of stratigraphic and geomorphic experiments: Earth Science Reviews, v. 97, p. 1-43.

Pattison, S.A.J., 2005, Storm-influenced prodelta turbidite complex in the lower Kenilworth Member at Hatch Mesa, Book Cliffs, Utah, U. S.A.: implications for shallow marine facies models: Journal of Sedimentary Research, v. 75, p. 420–439.

Pearson, P. N., 1998, Evolutionary concepts in biostratigraphy, in Doyle, P. and Bennett, M. R., eds., Unlocking the stratigraphical record: John Wiley and Sons, Chichester, p. 123-144.

Pedersen, S. A. S., 2012, Glaciodynamic sequence stratigraphy, in Huuse, M., Redfern, J., LeHeron, D. P., Dixon, R. J., Moscariello, A., and Craig, J., eds., Glaciogenic reservoirs and hydrocarbon systems, Geological Society, London, Special Publication 368, p. 29-51.

Peper, T., Beekman, F., and Cloetingh, S., 1992, Consequences of thrusting and intraplate stress fluctuations for vertical motions in

foreland basins and peripheral areas: Geophysical Journal International, v. 111, p. 104-126.

Phillips, S., and Bustin, R. M., 1996, Sedimentology of the Changuinola peat deposit: organic and clastic sedimentary response to punctuated coastal subsidence: Geological Society of America Bulletin, v. 108, p. 794-814.

Pillans, B., Alloway, B., Naish, T., Westgate, J., Abbott, S., and Palmer, A., 2005, Silicic tephras in Pleistocene shallow-marine sediments of Wanganui Basin, New Zealand: Journal of the Royal Society of New Zealand, v. 35, p. 43-90.

Pirmez, C., Prather, B. E., Mallarino, G., O'Hayer, W. W., Droxler, A. W., and Winker, C. D., 2012, Chronostratigraphy of the Brazos-Trinity depositional system, western Gulf of Mexico: implications for deepwater depositional models, in Applications of the Principles of seismic geomorphology to continental slope and base-of-slope systems: case studies from seafloor and near-seafloor analogues: Society for Sedimentary Geology Special Publication 99, p. 111-143.

Plint, A. G., Walker, R. G., and Bergman, K. M., 1986, Cardium Formation 6. Stratigraphic framework of the Cardium in subsurface: Bulletin of Canadian Petroleum Geology, v. 34, p. 213-225.

Plotnick, R. E., 1986, A fractal model for the distribution of stratigraphic hiatuses: Journal of Geology, v. 94, p. 885-890.

Pochat. S., Castelltort, S., Choblet, G., and Driessche, J. V. Den, 2009, High-resolution record of tectonic and sedimentary processes in growth strata: Marine and Petroleum Geology, v. 26, p. 1350-1364.

Posamentier, H. W., Allan, G. P., and James, D. P., 1992, High-resolution sequence stratigraphy – the East Coulee Delta, Alberta: Journal of Sedimentary Petrology, v. 62, p. 310-317.

Prather, B. E., Pirmez, C., and Winker, C. D., 2012, Stratigraphy of linked intraslope basins: Brazos-Trinity System, western Gulf of Mexico, in Application of the Principles of Seismic Geomorphology to Continental-Slope and Base-of-Slope Systems: Case Studies from Seafloor and Near-Seafloor Analogues: SEPM Special Publication 99, p. 83–109.

Qayyum, F., de Groot, P., Hemstra. N., and Catuneanu, O., 2015, 4D Wheeler diagrams: concept and applications, in Smith, D. G., Bailey, R., J., Burgess, P., and Fraser, A., eds., Strata and time: Geological Society, London, Special Publication 404, p. 223-232.

Raines, M. K., Hubbard, S. M., Kukulski, R. B., Leier, A. L., and Gehrels, G. E., 2012, Sediment dispersal in an evolving foreland: detrital zircon geochronology from Upper Jurassic and lowermost Cretaceous strata, Alberta Basin, Canada, Geological Society of America Bulletin, v. 125, p. 741-755.

Ramsbottom, W. H. C., 1979, Rates of transgression and regression in the Carboniferous of NW Europe: Journal of the Geological Society, London, v. 136, p. 147-153.

Runkel, A. C., Miller, J. F., McKay, R. M., Palmer, A. R., and Taylor, J. F., 2007, High-resolution sequence stratigraphy of lower Paleozoic sheet sandstones in central North America: the role of special conditions of cratonic interiors in development of stratal architecture: Geological Society of America Bulletin, v. 119, p. 860-881.

Runkel, A. C., Miller, J. F., McKay, R. M., Palmer, A. R., and Taylor, J. F., 2008, The record of time in cratonic interior strata: does exceptionally slow subsidence necessarily result in exceptionally poor stratigraphic completeness? in Pratt, B. R., and Holmden, C., eds., Dynamics of epeiric seas: Geological Association of Canada Special Paper 48, p. 341-362.

Sadler, P. M., 1981, Sedimentation rates and the completeness of stratigraphic sections: Journal of Geology, v. 89, p. 569-584.

Sadler, P. M., and Jerolmack, D., 2012, Scaling laws for aggradation, denudation and progradation rates: the case for time-scale invariance at sedimentary sources and sinks: The Geological Society,

London, William Smith Meeting 2012, Strata and Time: Probing the gaps in our understanding, Abstract, p. 32.

Sadler, P. M., and Jerolmack, D., 2015, Scaling laws for aggradation, denudation and progradation rates: the case for time-scale invariance at sediment sources and sinks, in Smith, D. G., Bailey, R., J., Burgess, P., and Fraser, A., eds., Strata and time: Geological Society, London, Special Publication 404, p. 69-88.

Sadler, P. M., Cooper, R. A., and Melchin, M., 2009: High-resolution, early Paleozoic (Ordovician-Silurian) time scales: Geological Society of America Bulletin, v. 121, p. 887-906.

Sadler, P. M., Cooper, R. A., and Crampton, J. S., 2014, High-resolution geobiologic time-lines: progress and potential, fifty years after the advent of graphic correlation: The Sedimentary Record, v. 12, #3, p. 4-9.

Sageman, B. B., Singer, B. S., Meyers, S. R., Siewert, S. E., Walaszczyk, I., Condon, D. J., Jicha, B. R., Obradovich, J. D., and Sawyer, D. A., 2014, Integrating $^{40}$Ar/$^{39}$Are,U-Pb and astronomical clocks in the Cretaceous Niobrara Formation, Western Interior Basin, USA: Geological Society of America Bulletin, v. 126, p. 956-973.

Salvaney, J. M., Larrasoaña, J. G., Mediavilla, G., and Rebollo, A., 2011, Chronology and tecto-sedimentary evolution of the Late Pliocene to Quaternary deposits from the Lower Guadilquivir foreland basin, SW Spain: Sedimentary Geology, v. 241, p. 22-39.

Scarponi, D., Kaufman, D., Amarosi, A. and Kowalewski, M., 2013, Sequence stratigraphy and the resolution of the fossil record: Geology, v. 41, p. 239-242.

Schlager, W., 2004, Fractal nature of stratigraphic sequences, Geology, v. 32, p. 185-188.

Schlager, W., 2005, Carbonate sedimentology and sequence stratigraphy: SEPM Concepts in Sedimentology and Paleontology #8, 200p.

Schreiber, B. C., Friedman, G. M., Decima, A., and Schreiber, E., 1976, Depositional environments of Upper Miocene (Messinian) evaporite deposits of the Sicilian Basin: Sedimentology, v. 23, p. 729-760.

Schumm S. A., 1973, Geomorphic thresholds and complex response of drainage systems, in Morisawa,, M. ed., Fluvial geomorphology, Publications in Geomorphology, State University of New York, Binghamton, N .Y., p. 299-310.

Schumm, S. A., 1977, The fluvial system: John Wiley and Sons, New York, 338 p.

Scott, A. C., and Stephens, R. S., 2015, British Pennsylvanian (Carboniferous) coal-bearing sequences: where is the time? in Smith, D. G., Bailey, R., J., Burgess, P., and Fraser, A., eds., Strata and time: Geological Society, London, Special Publication 404, p. 283-302.

Seymour, D.L., and Fielding, C.R., 2013, High resolution correlation of the Upper Cretaceous stratigraphy between the Book Cliffs and the western Henry Mountains Syncline, Utah, U.S.A:, Journal of Sedimentary Research, v. 83, p. 475-494.

Shanley, K. W., and McCabe, P. J., 1994, Perspectives on the sequence stratigraphy of continental strata: American Association of Petroleum Geologists Bulletin, v. 78, p. 544-568.

Shaw, A. B., 1964, Time in stratigraphy: McGraw Hill, New York, 365 p.

Sheets, B. A., Hickson, T. A., and Paola, C., 2002, Assembling the stratigraphic record: depositional patterns and time-scales in an experimental alluvial basin: Basin Research, v. 14, p. 287-301.

Smith, W., 1815, A memoir to the map and delineation of the strata of England and Wales, with part of Scotland: London: John Carey, 51 p.

Smith, A. G., Barry, T., Bown, P., Cope, J., Gale, A., Gibbard, P., Gregory, J., Hounslow, M., Kemp, D., Knox, R., Marshall, J., Oates, M., Rawson, P., Powell, J., and Waters, C., 2015, GSSPs, global stratigraphy and correlation: in Smith, D. G., Bailey, R., J.,

Burgess, P., and Fraser, A., eds., Strata and time: Geological Society, London, Special Publication 404, p. 37-67.

Sommerfield, C. K., 2006, On sediment accumulation rates and stratigraphic completeness: lessons from Holocene ocean margins: Continental Shelf Research, v. 26, p. 2225-2240.

Southard, J. B., Bohuchwal, L. A., and Romea, R. D., 1980, Test of scale modelling of sediment transport in steady unidirectional flow: Earth Surface Processes, v. 5, p. 17-23.

Spieker, E.M., and Reeside, J.B., Jr., 1925, Cretaceous and Tertiary formations of the Wasatch Plateau, Utah: Geological Society of America, Bulletin, v. 36, p. 429-454.

Stark, T. J., Zeng, H., and Jackson, A., 2013, An introduction to this special section: Chronostratigraphy: The Leading Edge, v. 32, p. 132-138.

Stouthamer, E., Cohen, K. M., and Gouw, M. J. P., 2011, Avulsion and its implication for fluvial-deltaic architecture: insights from the Holocene Rhine-Meuse delta,, in Davidson, S. K., Leleu, S., and North, C. P., eds., From river to rock record: Society for Sedimentary Geology (SEPM) Special Publication 97, p. 215-231.

Stow, D. A. V., Howell, D. G., and Nelson, C. H., 1983, Sedimentary, tectonic, and sea-level controls on submarine fan and slope-apron turbidite systems: Geo-marine Letters, v. 3, p. 57-64.

Strasser, A., Hilgen, F. J., and Heckel, P. H., 2006, Cyclostratigraphy — Concepts, definitions, and applications: Newsletters in Stratigraphy, v. 42, p. 75-114.

Strong, N., and Paola, C., 2008, Valleys that never were: time surfaces versus stratigraphic surfaces: Journal of Sedimentary Research, v. 78, p. 579-593.

Strong, N., Sheets, B., Hickson, T., and Paola, C., 2005, A mass-balance framework for quantifying downstream changes in fluvial architecture, in Blum, M.D., Marriott, S.B., and Leclair, S.F., eds., Fluvial Sedimentology VII: International Association of Sedimentologists, Special Publication 35, p. 243-253.

Sun, J., Li, Y., Zhang, Z., and Fu, B., 2010, Magnetostratigraphic data on Neogene growth folding in the foreland basin of the southern Tianshan Mountains: Geology, v. 37, p. 1051-1054.

Suter, J. R., Berryhill, H. L., Jr., and Penland, S., 1987, Late Quaternary sea-level fluctuations and depositional sequences, southwest Louisiana continental shelf, in Nummedal, D., Pilkey, O. H., and Howard, J. D., eds., 1987, Sea-level fluctuation and coastal evolution: Society of Economic Paleontologists and Mineralogists Special Publication 41, p. 199-219.

Thomas, R. G., Smith, D. G., Wood, J. M., Visser, J., Calverley-Range, E. A., and Koster, E. H., 1987, Inclined heterolithic stratification - terminology, description, interpretation and significance: Sedimentary Geology, v. 53, p. 123-179.

Trabucho-Alexandre, J., 2012, Stratigraphic (in)completeness of marine fine-grained sedimentary successions: implications for palaeoenvironmental interpretations of proxy records, The Geological Society, London, William Smith Meeting 2012, Strata and Time: Probing the gaps in our understanding, Abstract, p. 52.

Trabucho-Alexandre, J. 2015. More gaps than shale: erosion of mud and its effect on preserved geochemical and palaeobiological signals. in: Smith, D. G., Bailey, R. J., Burgess, P. M., and Fraser, A .J., eds., Strata and Time: Probing the Gaps in Our Understanding: Geological Society, London, Special Publications, 404, p. 251-270.

Vail, P. R., Mitchum, R. M., Jr., Todd, R. G., Widmier, J. M., Thompson, S., III, Sangree, J. B., Bubb, J. N., and Hatlelid, W. G., 1977, Seismic stratigraphy and global changes of sea-level, in Payton, C. E., ed., Seismic stratigraphy - applications to hydrocarbon exploration: American Association of Petroleum Geologists Memoir 26, p. 49-212.

Vail, P. R., Audemard, F., Bowman, S. A., Eisner, P. N., and Perez-Crus, C., 1991, The stratigraphic signatures of tectonics, eustasy and sedimentology—an overview, in Einsele, G., Ricken, W., and Seilacher, A., eds., Cycles and events in stratigraphy: Springer-Verlag, Berlin, p. 617-659.

Van den Bergh, G. D., Boer, W., Schaapveld, M.A. S., Duc, D. M., and van Weering, T. C. E., 2007, Recent sedimentation and sediment accumulation rates of the ba Lat prodelta (Red River, Vietnam), Journal of Asian Earth Sciences, v. 29, p. 545-557.

Van Wagoner, J.C., 1995, Sequence stratigraphy and marine to nonmarine facies architecture of foreland basin strata, Book Cliffs, Utah, USA, in Van Wagoner, J.C., and Bertram, G.T., eds., Sequence Stratigraphy of Foreland Basin Deposits: American Association of Petroleum Geologists, Memoir 64, p. 137-223.

Van Wagoner, J.C., and Bertram, G.T., eds., 1995, Sequence Stratigraphy of Foreland Basin Deposits - Outcrop and Subsurface Examples from the Cretaceous of North America: American Association of Petroleum Geologists, Memoir 64, 489 p.

Van Wagoner, J. C., Mitchum, R. M., Campion, K. M. and Rahmanian, V. D. 1990, Siliciclastic sequence stratigraphy in well logs, cores, and outcrops: American Association of Petroleum Geologists Methods in Exploration Series 7, 55 p.

Van Wagoner, J.C., Jones, C.R., Tayler, D.R., Nummedal, D., Jennette, D.C., and Riley, G.W., 1991, Sequence stratigraphy: applications to shelf sandstone reservoirs: American Association of Petroleum Geologists, Field Conference, September 1991.

Visser, M. J., 1980, Neap-spring cycles reflected in Holocene subtidal large-scale bedform deposits: a preliminary note: Geology, v. 8, p. 543-546.

Walker, R. G. and James, N. P., eds., 1992, Facies models: response to sea-level change: Geological Association of Canada, Geotext 1, 409 p.

Weber, M.E., Wiedicke, M.H., Kudrass, H.R., Huebscher, C., and Erlenkeuser, H., 1997, Active growth of the Bengal Fan during sea-level rise and highstand: Geology, v. 25, p. 315–318.

Weimer, R. J., 1970, Rates of deltaic sedimentation and intrabasin deformation, Upper Cretaceous of Rocky Mountain region, in Morgan, J. P., ed., Deltaic sedimentation modern and ancient: Society of Economic Paleontologists and Mineralogists Special Publication 15, p. 270-292.

Wellner, R. W., and Bartek, L. R., 2003, The effect of sea level, climate, and shelf physiography on the development of incised-valley complexes: a modern example from the East China Sea: Journal of Sedimentary Research, v. 73, p. 926-940.

Wheeler, H. E., 1958, Time-stratigraphy: American Association of Petroleum Geologists Bulletin, v. 42, p. 1047-1063.

Wheeler, H. E., 1959, Stratigraphic units in time and space: American Journal of Science, v. 257, p. 692-706.

Willis, A. J., 2000, Tectonic control of nested sequence architecture in the Sego Sandstone, Neslen Formation, and Upper Castlegate Sandstone (Upper Cretaceous), Sevier Foreland Basin, Utah, U.S.A.: Sedimentary Geology, v. 136, p. 277–318.

Wobber, F. J., 1965, Sedimentology of the Lias (Lower Jurassic) of South Wales: Journal of Sedimentary Petrology, v. 35, p. 683-703.

Wright, V.P., and Marriott, S.B., 1993, The sequence stratigraphy of fluvial depositional systems: the role of floodplain sediment storage: Sedimentary Geology, v. 86, p. 203-210.

Yang Chang-shu, and Nio, S.-D., 1989, An ebb-tide delta depositional model - a comparison between the modern Eastern Scheldt tidal basin (southwest Netherlands) and the Lower Eocene Roda Sandstone in the southern Pyrenees (Spain): Sedimentary Geology, v. 64, p. 175-196.

Yoshida, S., 2000, Sequence stratigraphy and facies architecture of the upper Blackhawk Formation and the Lower Castlegate Sandstone (Upper Cretaceous), Book Cliffs, Utah, USA: Sedimentary Geology, v. 136, p. 239-276.

Yoshida, S., Willis, A., and Miall, A., 1996, Tectonic control of nested sequence architecture in the Castlegate Sandstone (Upper Cretaceous), Book Cliffs, Utah: Journal of Sedimentary Research, v. 66, p. 737-748.

Young, R.G., 1955, Sedimentary faces and intertonguing in the Upper Cretaceous of the Book Cliffs, Utah-Colorado: Geological Society of America, Bulletin, v. 66, p. 177-202.

Young, R.G., 1957, Late Cretaceous cyclic deposits, Book Cliffs, eastern Utah, American Association of Petroleum Geologists, Bulletin, v. 41, p. 1760-1774.

Zaitlin, B. A., Warren, M. J., Potocki, D., Rosenthal, L., and Boyd, R., 2002, depositional styles in a low accommodation foreland basin setting: an example from the Basal Quartz (Lower Cretaceous), southern Alberta: Bulletin of Canadian Petroleum Geology, v. 50, p. 31-72.

Zecchin, M., Caffau, M., Tosi, L., Civile, D., Brancolini, G., Rizzetto, F., and Roda, C., 2010, The impact of Late Quaternary glacio-eustasy and tectonics on sequence development: evidence from both uplifting and subsiding settings in Italy: Terra Nova, v. 22, p. 324-329.

# Author Index

# Subject Index

## A

Absolute ages, 343, 345
    accuracy and precision, 355
Abyss of time, 11
Accommodation, 11, 15, 135, 216, 217
    and time scale, 385
Accommodation, fluvial, 233, 234
Accommodation generation, 237
Accommodation potential, 220
Acid, 34, 39
Acoustic impedance, 255
Actualism, 3, 5, 391
Actualistic facies model, 89
Aerial photographs, 91
Age, 343
    assignment, 315
    definition, 12
    determination, 347
Aggradation, 4, 135, 219, 233, 276, 340
    cycles of, 379
Aggradation rate
    deltaic, 379
    floodplain, 379
Agulhas Bank, 183
Alberta Basin, 10
Alberta Oil Sands, 274, 279
Alignment, of fossils, 131
Alizarin red-S, 39
Allochems, 97
Allocyclic process, 3
Allogenic process, 3, 16, 135, 217. *See also* Allocyclic process
Allomember, 147
Allopatric speciation, 323
Allostratigraphy, 15, 216, 314, 334
Alluvial architecture and accommodation, 389
Alluvial fan profile, 136
Amino-acid geochronometry, 17
Amino-acid racemization, 380
Ammonite biogeography, 325
Amplitude, 43
    mapping of, 280
Anastomosed, 165
Anastomosed river, 165
Andros Island, 137
Anoxic basin, 250
Anoxic environment, 102, 103
Anoxic events, 240
Antarctic Peninsula, 263

Antidune, 105, 106, 108
Appalachian foreland basin, 250
Architectural element, 14, 95, 139, 142, 146, 148, 163
Architectural-element analysis, 88
Architectural scale, 142
Argon-argon method, 345
Arkose, 99
Ash, 181
Astrochronological calibration, 358
Astrochronological time scale, 249, 362, 405, 420
Astrochronology, 18, 358, 359, 413, 416
Astronomical forcing, 17. *See also* Orbital forcing
Astronomical frequencies, 420
Atchafalaya River, 178
Athabasca Oil Sands, 175
Atlantic shelf, 92
Austin Chalk, 230
Autocyclic process, 3
Autogenic process, 3, 16, 135, 424. *See also* Autocyclic process
Autogenic switching, 388
Autoincision, 16
Automated correlation, 415
Autosuspension, 122
Avulsion, 178, 234

## B

Backstepping, 135
Backstripping, 23
Bahama Platform, 89, 92, 101, 327
Bahamite, 8
Bakken Formation, 249, 250
Ball, 49
Bank, carbonate, sequence stratigraphy of, 237
Banks Basin, Arctic Canada, 288
Barchans, 113
Bar-code matching, 407
Bar-code pattern matching, 348
Barrier island, 173
Barrier island facies model, 7, 135
Barrier systems, 174
Basal surface of forced regression, 227, 337
Base-level, 11
Base-level controls, 233
Base level cycle, 216, 218, 230
Basin analysis, 2
Basin and Range Province, 10
Basin-fill complexes, 147

A.D. Miall, *Stratigraphy: A Modern Synthesis*, DOI 10.1007/978-3-319-24304-7